Digital Image Processing

Digital Image Processing

Theory, Practice, and AI Applications

Mahmood R. Azimi-Sadjadi

Professor Emeritus
Department of Electrical and Computer Engineering
Colorado State University
Fort Collins, CO, USA

WILEY

Published by John Wiley & Sons, Inc., Hoboken, New Jersey.

Library of Congress Cataloging-in-Publication Data:

Names: Azimi-Sadjadi, Mahmood R. author
Title: Digital image processing: theory and practice / Mahmood R.
  Azimi-Sadjadi.
Description: Hoboken, New Jersey: John Wiley & Sons, 2026. | Includes
  bibliographical references and index.
Identifiers: LCCN 2026022371 | ISBN 9781394240449 cloth | ISBN
  9781394240463 epdf | ISBN 9781394240456 epub | ISBN 9781394240470 obook
Subjects: LCSH: Image processing–Digital techniques | Digital images
Classification: LCC TA1637.A95 2026
LC record available at https://lccn.loc.gov/2026022371

Cover Design: Wiley
Cover Image: kanpisut/stock.adobe.com

On seeing the unseen: *The eye sees, the heart desires, and love is in between. The soul passes by, asking, "What did you see, and what do you desire?" by Mewlana Jalaluddin Rumi*

To the three people who were my guiding lights on this long journey: In memory of my brother-in-law, Reza Moezzi, whose spirit emboldened me to start; and to my sister, Manijeh (Mimi), and my wife, Sheida, for your unwavering love and encouragement to complete.

Contents

Preface *xvii*
About the Book *xvii*
Prerequisites and Course Organization *xviii*
Acknowledgments *xviii*
About the Author *xix*
Conventions and Notations *xxi*
About the Companion Website *xxiii*

1 Introduction *1*
1.1 DIP System Components *4*
1.2 Areas of DIP *5*
1.2.1 Image Representation and Modeling *5*
1.2.2 Image Enhancement *5*
1.2.3 Image Restoration *6*
1.2.4 Image Transforms *6*
1.2.5 Image Data Compression and Encoding *6*
1.2.6 Image Analysis and Computer Vision *7*
1.3 Final Remarks and Goals *7*
References *8*

2 Review of 1-D Signal Processing *9*
2.1 Discrete-Time Signals and Systems-Deterministic Case *9*
2.1.1 Discrete-Time Signals *9*
2.1.2 Discrete-Time Systems and Properties *11*
2.1.3 Representation of LSI Systems-Convolution Summation *11*
2.1.4 Representation of LSI Systems-Difference Equation *14*
2.1.5 Frequency Domain Analysis – Discrete-Time Fourier Transform *15*
2.1.5.1 Properties of DTFT *16*
2.1.6 Frequency Domain Analysis – Discrete Fourier Transform *17*
2.1.6.1 Properties of DFT *19*
2.1.6.2 Computing Linear Convolution Using DFT *20*
2.2 Discrete-Time Signals and Systems-Random Case *20*
2.2.1 Statistical Measures on Random Processes *21*
2.2.2 Wide Sense Stationary Random Processes *22*
2.2.2.1 Time-Domain Properties of WSS Processes *23*
2.2.3 Time Averages and Ergodicity *24*
2.2.4 Power Spectrum *24*
2.2.5 Cross Power Spectrum *25*

2.2.6 Response of LTI Systems to Random Inputs 25
2.2.7 Spectral Factorization 27
2.3 Problems 28
References 30

3 Theoretical Background 31
3.1 Discrete-Space 2-D Signals and Systems-Deterministic Case 31
3.1.1 Discrete-Space 2-D Signals 31
3.1.2 Discrete-Space 2-D Systems and Properties 33
3.1.3 Representation of 2-D LSI Systems-Convolution Summation 33
3.1.4 Representation of 2-D LSI Systems-Difference Equation 36
3.1.5 Frequency Domain Analysis – 2-D Continuous Fourier Transform 37
3.1.5.1 Remarks and Properties 38
3.1.6 Frequency Domain Analysis – 2-D Discrete Space Fourier Transform 40
3.1.7 Frequency Domain Analysis – 2-D DFT 42
3.1.7.1 Properties of 2-D DFT 44
3.1.8 Computing 2-D Linear Convolution Using 2-D DFT 45
3.2 Discrete-space 2-D Signals and Systems-Random Case 45
3.2.1 Statistical Measures on 2-D Random Processes 45
3.2.2 Power Spectra of 2-D Random Processes 46
3.2.3 Response of 2-D LSI Systems to Random Images 48
3.3 Computer Projects 49
3.4 Problems 50
References 52

4 Image Formation and Perception 53
4.1 Human Eye 53
4.2 Elements of Visual Perception 54
4.2.1 Brightness Sensitivity 54
4.2.2 Contrast Sensitivity and Spatial-Temporal Vision 56
4.2.3 Color Discrimination 58
4.3 Color Systems 61
4.3.1 The RGB System 61
4.3.2 The CMYK System 62
4.3.3 The HSI System 62
4.3.4 The YC_bC_r System 63
4.4 Problems 64
References 65

5 Image Sampling and Quantization 67
5.1 Image Sampling 67
5.1.1 2-D Sampling Theorem 67
5.1.2 Reconstruction 69
5.1.3 Undersampling and Aliasing Effects 69
5.1.4 Color Image Sampling 72
5.2 Image Quantization 72
5.2.1 Uniform Quantizer 73
5.2.2 Minimum Mean Squared or Lloyd-Max Quantizer 74
5.2.2.1 High-Rate Quantization 74
5.2.2.2 Special Case 77
5.2.3 Compandor Design 78
5.2.4 Color Image Quantization 80

5.3 Computer Projects *81*
5.4 Problems *82*
References *84*

6 Image Transforms *85*
6.1 2-D Discrete Fourier Transform *85*
6.1.1 Matrix Representation of 1-D DFT *85*
6.1.1.2 Properties of DFT Matrix *86*
6.1.2 Matrix Representation of 2-D DFT *87*
6.2 2-D Discrete Cosine Transform *87*
6.2.1 Matrix Representation of 1-D DCT *87*
6.2.1.1 Properties of DCT Matrix *88*
6.2.2 Matrix Representation of 2-D DCT *89*
6.3 2-D Walsh-Hadamard Transform *90*
6.3.1 Walsh Transform *90*
6.3.1.1 Properties of Walsh Transform *92*
6.3.2 Matrix Representation of 1-D WHT *92*
6.3.2.1 Properties of WHT Matrix *93*
6.3.3 Matrix Representation of 2-D WHT *94*
6.4 2-D Karhunen–Loeve Transform *96*
6.4.1 1-D KLT of Random Signals *96*
6.4.1.1 Properties of KLT Matrix *96*
6.4.1.2 KLT-Based Data Reduction Algorithm *97*
6.4.2 2-D KLT of Random Images *98*
6.4.3 An Efficient Algorithm for Eigen-Image and PCA of Images *98*
6.4.4 Face Representation Using Eigen-Faces *99*
6.4.5 PCA and Singular Value Decomposition (SVD) *100*
6.5 Computer Projects *101*
6.6 Problems *102*
References *104*

7 Wavelet Transform *107*
7.1 Short-Time Fourier Transform *107*
7.1.1 Time-Frequency Resolutions *108*
7.2 Continuous Wavelet Transform *109*
7.2.1 Properties of CWT *111*
7.2.1.1 Wigner–Ville Distribution *113*
7.3 Digitization of Shift-Scale Parameters and Wavelet Series *114*
7.3.1 Wavelet Frames *114*
7.4 Filter Banks and Discrete Wavelet Series *115*
7.4.1 Matrix Representation of DWS *116*
7.4.2 Orthonormal Filter Banks and DWT *118*
7.4.3 Iterated Filter Banks *120*
7.5 2-D DWT and Filter Banks *123*
7.6 Computer Projects *126*
7.7 Problems *127*
References *129*

8 Image Enhancement *131*
8.1 Point Operations *131*
8.1.1 Grayscale Mapping *131*
8.1.1.1 Special Cases *134*

8.1.2 Histogram Modeling *135*
8.1.2.1 Histogram Equalization *136*
8.1.2.2 Histogram Specification *137*
8.2 Spatial Operations *139*
8.2.1 Noise Reduction and 2-D Low-Pass Filtering *139*
8.2.1.1 Effects of Mask Size on Performance *141*
8.2.1.2 Edge-Preserving Spatial Filters *143*
8.2.2 Median Filtering *144*
8.2.2.1 Properties of Median Filter *144*
8.2.3 Edge Extraction and 2-D High-Pass Filtering (HPF) *146*
8.2.4 Correspondence and 2-D Matched Filtering *152*
8.2.5 Hybrid Operations *153*
8.3 Transform Domain Operations *154*
8.3.1 2-D Filtering in Transform Domain *154*
8.3.2 2-D Filtering Using Other Unitary Transforms *159*
8.3.3 2-D Matched Filtering in Transform Domain *159*
8.3.4 Root Filtering and Cepstral Domain Processing *160*
8.4 Computer Projects *162*
8.5 Problems *164*
References *169*

9 Image Modeling *171*
9.1 1-D Linear Models *171*
9.1.1 1-D Causal Representation *172*
9.1.1.1 Parameter Estimation for Causal Models *173*
9.1.1.2 Convergence Property *175*
9.1.1.3 Causal Minimum Variance Prediction *175*
9.1.2 1-D Noncausal Representation *176*
9.1.2.1 Parameter Estimation for Noncausal Models *176*
9.1.2.2 Convergence Property *177*
9.1.2.3 Noncausal Minimum Variance Prediction *177*
9.2 2-D Linear Image Models *178*
9.2.1 2-D Causal Representation *179*
9.2.1.1 Parameter Estimation of 2-D Causal AR Model *180*
9.2.1.2 Convergence Property *182*
9.2.2 2-D Semicausal Representation *182*
9.2.2.1 Parameter Estimation of 2-D Semicausal AR Model *183*
9.2.2.2 Convergence Property *184*
9.2.3 2-D Noncausal Representation *185*
9.2.3.1 Parameter Estimation of 2-D Noncausal AR Model *185*
9.2.3.2 Convergence Property *187*
9.2.4 Numerical Example *187*
9.3 Multivariate or Vector Linear Models *188*
9.3.1 Parameter Estimation of Vector AR Models *190*
9.4 Computer Projects *191*
9.5 Problems *191*
References *193*

10 Image Restoration *195*
10.1 Image Degradation Models *195*
10.2 Image Restoration Filters *197*
10.2.1 Deterministic Filters *198*
10.2.1.1 Direct Inverse Filter *198*

10.2.1.2 Regularized Inverse Filter *198*
10.2.1.3 The Least Squares (LS) Filter *199*
10.2.2 Stochastic Filters – 1-D Wiener Filter *199*
10.2.2.1 Properties and Remarks *202*
10.2.3 Stochastic Filters – 2-D Wiener Filter *203*
10.2.3.1 2-D Wiener Filter – Multiplicative and Additive Noise Model *204*
10.2.4 Stochastic Filters – Strip Kalman Filter *206*
10.2.4.1 Important Remarks *208*
10.2.5 Stochastic Filters – Reduced Update Kalman Filter (RUKF) *209*
10.2.5.1 RUKF Remarks *212*
10.2.5.2 RUKF for Speckle Noise Removal *214*
10.2.5.3 Special Case *215*
10.3 Computer Projects *216*
10.4 Problems *217*
References *221*

11 Image Compression and Encoding *223*
11.1 A Review of Information Theory *224*
11.1.1 Measure of Information and Entropy *224*
11.1.2 Entropy and Coding *225*
11.1.2.1 Shannon's Source Coding Theorem *225*
11.1.3 Rate Distortion Function *226*
11.2 Image Compression and Encoding Methods – Lossless *227*
11.2.1 Pulse-Code Modulation *228*
11.2.2 Huffman Encoding *228*
11.2.3 Arithmetic Encoding *230*
11.2.4 Run-Length Encoding *231*
11.3 Image Compression and Encoding Methods – Lossy *232*
11.3.1 Predictive Encoding *233*
11.3.1.1 Performance Analysis of DPCM *234*
11.3.2 Transform-Based Encoding *235*
11.3.2.1 Bit Allocation Strategy *236*
11.3.3 Wavelet-Based Encoding and New JPEG Standards *238*
11.3.4 Vector Quantization *241*
11.3.4.1 K-Means Clustering Algorithm *241*
11.4 Hybrid Methods *242*
11.5 Computer Projects *242*
11.6 Problems *243*
References *246*

12 Image Segmentation *249*
12.1 Similarity-Based Image Segmentation *250*
12.1.1 Histogram Thresholding *250*
12.1.2 Adaptive Threshold Selection *251*
12.1.2.1 Summary Remarks *255*
12.1.3 Region-Based Segmentation *256*
12.1.3.1 Summary Remarks *258*
12.1.4 Clustering Methods *258*
12.1.5 Markov Random Field Segmentation *260*
12.1.5.1 Iterated Conditional Modes (ICM) *262*
12.1.5.2 Summary Remarks *264*
12.2 Edge and Contour-Based Segmentation *264*
12.2.1 Active Contour Segmentation *264*

12.2.1.1 Active Contour or Snake Formulation *265*
12.2.2 Gradient Vector Flow (GVF) Snake *266*
12.3 Computer Projects *269*
12.4 Problems *271*
References *273*

13 Feature Extraction *275*
13.1 Objectives and Requirements *275*
13.1.1 Objectives *275*
13.1.2 Invariance and Equivariance Requirements *276*
13.2 Color Features *277*
13.3 Energy Features *279*
13.3.1 Invariance Properties *279*
13.4 Spectral Features *280*
13.4.1 Multi-Channel Spectral Features *280*
13.4.2 Single-Channel Spectral Features *282*
13.5 Textural Features *283*
13.5.1 Gray Level Co-Occurrence Matrix *283*
13.5.2 Gabor Filter *287*
13.6 Line and Contour Features *291*
13.6.1 Fourier Descriptor *291*
13.6.1.1 Invariance Properties *292*
13.6.2 Wavelet Descriptor *294*
13.6.3 Chain Code *296*
13.6.4 Hough Transform *297*
13.6.5 B-Spline *298*
13.7 Shape Features *301*
13.7.1 Moments Invariant *301*
13.7.2 Zernike Moments *302*
13.7.2.1 Invariance Properties *303*
13.8 Computer Projects *305*
13.9 Problems *307*
References *309*

14 Morphological Operations *311*
14.1 Basic Morphological Operations – Binary Images *312*
14.1.1 Dilation and Erosion *312*
14.1.1.1 Algebraic Properties *313*
14.1.2 Opening and Closing *314*
14.2 Morphological Transforms – Binary Images *316*
14.2.1 Hit-or-Miss *316*
14.2.2 Thinning *316*
14.2.3 Thickening *317*
14.2.4 Convex Hull *318*
14.2.5 Skeleton *319*
14.2.6 Pruning *321*
14.3 Morphological Operations – Grayscale Images *322*
14.3.1 Grayscale Dilation and Erosion *323*
14.3.2 Grayscale Opening and Closing *324*
14.3.3 Grayscale Skeleton and Shape Representation *325*
14.3.4 Extension of Morphological Operations to Functions *325*
14.3.5 Geodesic Transforms and Reconstruction Filters *326*

14.3.5.1 Binary Geodesic Dilation and Erosion 326
14.3.5.2 Binary Geodesic Reconstruction 327
14.3.5.3 Grayscale Geodesic Reconstruction 328
14.4 Applications of Morphological Operations 328
14.4.1 Morphological Edge Detection 329
14.4.2 Morphological Image Segmentation 330
14.5 Computer Projects 332
14.6 Problems 334
References 337

15 Image Reconstruction 339
15.1 Basics of Image Reconstruction 339
15.1.1 Radon Transform 340
15.1.2 Inverse Radon Transform 341
15.1.2.1 Projection-Slice Theorem 342
15.1.3 The Back-Projection Operator 343
15.2 Analytical Image Reconstruction Methods 344
15.2.1 Filtered Back-Projection 344
15.2.2 Discrete Implementation of Filtered Back-Projection 346
15.3 Iterative Image Reconstruction Methods 347
15.3.1 Algebraic Reconstruction Technique 347
15.3.1.1 Choice of Basis Functions 349
15.3.1.2 Stochastic Gradient Descent and ART 351
15.3.2 Statistical Image Reconstruction 351
15.3.2.1 Conjugate Gradient Method 352
15.4 Computer Projects 354
15.5 Problems 354
References 356

16 Traditional Image Classification 359
16.1 Deterministic Pattern Classification 359
16.1.1 Discriminant Functions and Linear Classifiers 360
16.1.2 Minimum Distance Classifier 362
16.2 Statistical Pattern Classification 363
16.2.1 Bayes Classifier 363
16.2.1.1 Special Cases 364
16.2.2 Error Analysis and Performance Metric 365
16.2.3 Naive Bayes Classifiers 367
16.2.3.1 Important Remark 371
16.3 Computer Projects 371
16.4 Problems 371
References 374

17 Modern Image Classification: Elements of Machine Learning 375
17.1 Biological Neural Networks 375
17.2 Artificial Neural Networks 376
17.2.1 Neuron Models 376
17.3 Learning Algorithms 379
17.3.1 Generalized Learning Rule 380
17.3.2 Performance Learning 381
17.3.2.1 ADALINE Learning 381
17.3.2.2 Gradient Descent Algorithm 383

17.3.2.3 Least Mean-Squared (LMS) Algorithm *384*
17.3.2.4 Perceptron Learning Rule *386*
17.3.2.5 Delta Learning Rule *387*
17.3.2.6 Logistic Regression and LMS Rule *388*
17.3.2.7 Multi-Class Logistic Regression *389*
17.3.3 Coincidence Learning *390*
17.3.4 Competitive Learning *392*
17.4 Computer Projects *395*
17.5 Problems *396*
References *398*

18 Modern Image Classification-Layered Machines *401*
18.1 Multilayer Perceptron Neural Networks *401*
18.1.1 Generalized Delta Rule and Error Back-Propagation *403*
18.1.1.1 MLP and Bayes Classification *406*
18.1.2 Practical Considerations and Issues *406*
18.2 Convolutional Neural Networks *408*
18.2.1 Learning Rule and Error Back-Propagation for CNN *410*
18.2.1.1 Loss Functions *411*
18.2.1.2 Error Back-Propagation in CNN *413*
18.2.2 Alternative Learning Algorithms for CNN *415*
18.3 Self-Organizing Feature Maps *417*
18.3.1 SOFM Unsupervised Training *418*
18.3.2 Practical Considerations and Issues *420*
18.3.3 Growing SOFM *421*
18.3.4 Learning Vector Quantization *422*
18.4 Computer Projects *424*
18.5 Problems *425*
References *428*

19 Dimensionality Reduction Networks and Autoencoders *431*
19.1 Linear Autoencoders *431*
19.1.1 PC Extraction Using RLS-Based Learning *433*
19.2 Nonlinear Dimensionality Reduction Using Manifold Learning *434*
19.2.1 Laplacian Eigenmaps *435*
19.3 Nonlinear Autoencoders *438*
19.3.1 Basic Autoencoders *438*
19.3.2 Sparse Autoencoders *439*
19.3.3 Denoising Autoencoders *441*
19.3.4 Contractive Autoencoders *442*
19.3.5 Relation of Autoencoders to Manifold Learning *443*
19.4 Computer Projects *444*
19.5 Problems *445*
References *446*

20 AI Applications in Digital Image Processing *449*
20.1 Disparity Estimation from Stereo Images *449*
20.1.1 Pyramid Stereo Matching Network *450*
20.2 Associative Memory for Pattern Retrieval *453*
20.2.1 Recurrent Neural Networks *454*
20.2.2 Gated Recurrent Units *457*

20.2.3 A Hybrid Architecture for Auto- and Hetero-Association Learning *458*
20.3 Computer Projects *461*
References *462*

Appendix-A: Review of 1-D *z*-Transform *465*
Appendix-B: 2-D *z*-Transform *471*
Index *473*

Preface

About the Book

Since the 1960s, the field of Digital Image Processing (DIP) has evolved rapidly, driven by remarkable advances in computing power, sensor technology, and high-resolution display systems. Beyond its foundational roles in medicine, defense, and industrial automation, DIP has become a ubiquitous technology that shapes many aspects of our daily lives. Today, the most transformative shift is the emergence of Artificial Intelligence (AI) and machine learning (ML) tools in many aspects of DIP. AI-driven DIP is redefining the boundaries of pattern recognition in medicine, remote sensing, autonomous systems, security, and bioinformatics. As computing hardware continues to miniaturize, sensor accuracy reaches new heights, and imaging technologies continue to evolve the influence of DIP is poised for exponential growth, becoming an even more seamless and indispensable part of the modern society.

Driven by the profound impact of AI on DIP systems, and informed by more than four decades of teaching and research at Colorado State University, I realized a critical need for a resource that bridges the gap between traditional and modern AI-based approaches. My goal was to develop a textbook that not only provides comprehensive and in-depth coverage of both classical and traditional methods, but also empowers a wide variety of readers, whether in academia, industry, or government labs, to become familiar with the many important practical aspects of this dynamic field. Accordingly, this book is structured to provide both foundational and advanced topics in DIP for a broad audience with diverse technical background and experience.

Although many excellent textbooks have been published in this area, several of which provide broad coverage of classical DIP techniques, this book offers a number of distinctive features and advantages that set it apart. These include:

- **Python-Centric and Collaborative:** Moving beyond proprietary software, this book leverages Python's powerful, free, and open-source libraries, including OpenCV, scikit-image, and PyTorch among many others. The use of the shared Google Colab platform enables students, researchers, and professionals to run, modify, and collaboratively develop code without the need for local software installation, fostering an interactive and highly accessible learning environment. Sample Python code for various DIP algorithms are provided in the relevant chapter folders, making the material accessible even to readers with limited Python programming experience. These examples illustrate practical applications across medical, remote sensing, and radar imaging areas. The students and readers can access the chapter folders through the book's GitHub repository at https://github.com/mgcy/Digital-Image-Processing.
- **Extensive Coverage of AI in DIP:** The final five chapters are devoted to important AI and ML techniques that have become integral to modern DIP. Topics include different supervised and unsupervised learning algorithms, generative and discriminative models, pattern classification, dimensionality reduction, latent feature learning, data compression, deep neural networks, autoencoders, and advanced topics such as associative memories, stereo imaging, and depth perception.
- **Comprehensive and In-Depth Coverage:** The book features 20 chapters that encompass fundamental principles and theory, classical techniques, and modern AI-driven methods of DIP. This extensive scope and breath of coverage provides instructors with the flexibility to tailor the material for undergraduate or graduate courses by selecting chapters that meet their specific needs. Topics are presented with theoretical depth and clear, concise explanations of how algorithms work and are implemented, utilizing illustrative examples to demonstrate the practical applications in

medical, remote sensing, and radar imaging areas. Almost all chapters conclude with extensive problem sets designed to challenge students and assess their mastery of the material.

- **Advanced Special-Purpose Topics:** The book is designed to serve a broad audience with diverse backgrounds, from students seeking foundational knowledge to researchers and practicing engineers. While the core chapters cover fundamental concepts and essential techniques and tools applicable to a broad range of image processing problems, three specialized chapters examine selected application areas in greater depth. These chapters are intended for readers seeking additional knowledge in specialized and more advanced areas of DIP.
- **Real-World Applications and Hands-On Learning:** Most chapters feature numerous Python-based computer projects that are specifically designed to demonstrate the practical implementation of the DIP algorithms across diverse fields, including medical imaging, satellite image analysis, radar image processing, and other imaging problems. To ensure accessibility and execution in a resource-rich environment, these computer projects leverage the computational power of the Google Colab platform. This platform conveniently allows users to utilize Graphical Processing Units (GPUs) for computationally intensive applications, particularly those involving deep learning and other large-scale AI-based models integral to modern DIP.

Prerequisites and Course Organization

To use this book as the primary text for a senior-level undergraduate or an early graduate level course in DIP, I recommend that students have completed appropriate prerequisite courses in *Signals and Systems* and *Probability and Random Variables,* which are typically taken during the junior year. For a graduate level DIP course, however, fundamental courses in *Digital Signal Processing (DSP)* and *Linear Algebra* are recommended. These prerequisites are generally met by most students enrolled in undergraduate or graduate programs in electrical and computer engineering, computer science, and related programs at universities in the United States and abroad.

Depending on the course objectives and number of credits, instructors may select topics from Chapters 1–14 and 16–18 to develop a course appropriate for either undergraduate or graduate students. Chapters 15, 19, and 20 cover specialized topics intended primarily for research students, practicing engineers, and professionals seeking advanced knowledge in selected areas of DIP and AI and gaining exposure to their real-world applications. Chapter 15 provides an in-depth and extensive coverage of the theoretical foundations and practical applications of image reconstruction. Particular emphasis is placed on important applications in computerized tomography (CT) for diagnostic medicine and non-destructive testing and evaluation (NDT/NDE) for industrial imaging systems. Chapter 19 explores the role of autoencoders in various image processing tasks including linear and nonlinear dimensionality reduction, latent feature learning, data visualization, and noise and perturbation removal. The final chapter presents two advanced topics. The first topic is disparity estimation in stereo imaging systems for data compression or depth perception. The second topic introduces a new a hybrid architecture leveraging an autoencoder and a recurrent neural network (RNN) to implement associative memories for grayscale and color image storage and retrieval applications.

Acknowledgments

This long journey could not have reached its conclusion without the diverse contributions of many individuals to whom I owe my deepest gratitude. First and foremost, I extend my deepest gratitude to my former colleague, Professor Louis L. Scharf, who meticulously reviewed and revised six chapters of the manuscript. His thoughtful comments, technical insight, and careful attention to detail significantly enhanced both the quality and clarity of the book. I am sincerely grateful for the considerable time and effort he devoted to this project. I would also like to thank my PhD students, Yifan Yang, Miles Brim, and Ungkana Srivitidkul, for their invaluable assistance with the Python implementations. I am particularly indebted to Yifan, whose contributions extended beyond coding to include generating a number of experimental results and organizing the Google Colab notebooks accompanying the text. Finally, my thanks go to the many graduate students who assisted with several experiments, including James Bryce, Pallavi Ramicetty, Alexander Max, Daniel Way, and Max Marshall. Their enthusiasm and dedication are sincerely appreciated.

About the Author

Dr. Azimi-Sadjadi received his MS and PhD degrees from the Imperial College of Science and Technology, University of London, England in 1978 and 1982, respectively, both in Electrical Engineering with specialization in Statistical Digital Signal and Image Processing. Since 1986, he has been with the Department of Electrical and Computer Engineering at Colorado State University (CSU), where he is currently a Professor and Director of the Digital Signal/Image Laboratory. His research interest include statistical signal and image processing, machine learning and adaptive systems, sensor array processing, and distributed sensor networks. He has authored or coauthored more than 250 journal and conference publications in these areas. He has also contributed several chapters in two books, namely *Digital Filtering in One and Two Dimensions* (Plenum Press, 1989) and *Machine Learning for Indoor Localization and Navigation* (Springer, 2023). Throughout his academic career, he has supervised more than 65 MS and PhD students and has conducted research in collaboration with numerous government agencies and industrial partners. His teaching and research over more than four decades have formed the foundation of this textbook.

Dr. Azimi-Sadjadi served as an Associate Editor of the *IEEE Transactions on Signal Processing* and the *IEEE Transactions on Neural Networks*. He is a Life Member of the IEEE.

Conventions and Notations

Throughout this book, the following notation and conventions are used unless stated otherwise.

Lowercase letters (e.g. a) and Greek symbols (e.g. α) denote **scalar** quantities.

Lowercase underline letters (e.g. $\underline{a}$) and Greek symbols (e.g. $\underline{\alpha}$) denote **vectors**.

Boldface letters (e.g. $\mathbf{x}$ and $\mathbf{X}$) or Greek symbols (e.g. $\boldsymbol{\Sigma}$) denote **matrices**.

The notation $\underline{f}(x)$ denotes a vector-valued function of scalar variable x.

The notation $\overline{f}(\underline{x})$ denotes a scalar-valued function of multivariate or vector $\underline{x}$.

The set of all real numbers is denoted by $\mathbb{R}$.

The notation $x(u, v)$ represents a continuous image with intensity x defined over the continuous spatial variables $u, v \in \mathbb{R}$.

The set of all integers is denoted by $\mathbb{Z}$.

The notation $x(m, n)$ represents a grayscale digital (sampled and quantized) image with pixel intensity $x \in \mathbb{Z}$ defined over the discrete spatial variables $m, n \in \mathbb{Z}$.

The notation $\underline{x}(m, n)$ represents a multichannel image, such as an RGB color image, multispectral image, or hyperspectral image, where $\underline{x}$ is the vector of channel intensities at the pixel location (m, n).

The notation $\mathbf{X} \in \mathbb{R}^{M \times N}$ represents a matrix of dimensions $M \times N$ with real-valued elements $x_{i,j}$.

The notation $\mathbf{X} \in \mathbb{Z}^{M \times N}$ represents a matrix of dimensions $M \times N$ with integer-valued elements $x_{i,j}$.

The Euclidean norm of vector $\underline{x} = [x_1, \cdots x_N]^T$ is denoted by $||\underline{x}|| = \sqrt{\sum_{i=1}^{N} x_i^2}$.

The notation $\det(\mathbf{A})$ represents the determinant of matrix $\mathbf{A}$.

The notation $\mathrm{tr}(\mathbf{A}) = \sum_{i=1}^{N} a_{i,i}$ represents the trace of matrix $\mathbf{A}$ with diagonal elements $a_{i,i}$.

The notation $\mathbf{A}^H$ denotes Hermitian or conjugate transpose operation.

The notation $||\mathbf{A}||_F$ denotes the Frobenius norm of matrix $\mathbf{A}$ and is defined as $||\mathbf{A}||_F = \left(\mathrm{tr}\left(\mathbf{A}\mathbf{A}^H\right)\right)^{\frac{1}{2}}$.

About the Companion Website

This book is accompanied by a companion website.

www.wiley.com/go/Azimi449

To access the website materials, click on the link under Related Resources on the book's page on wiley.com. You can also find this by searching www.wiley.com for the ISBN 9781394240449.

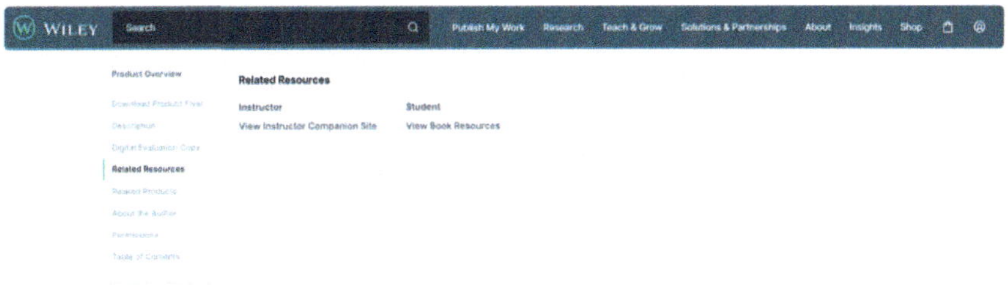

This website includes:

1. Solution Manual
2. Offset Figures

1

Introduction

The origins of image transmission in digitized form can be traced to the 1920s, when early systems first encoded pictures into discrete levels for telegraph-based communication. However, further development of digital image processing (DIP) systems was hampered by technological limitations such as lack of adequate computing and storage requirements. The field languished until the mid-1960s, when the demands of space exploration provided the necessary catalyst and resources. Following the success of Apollo missions, the Jet Propulsion Laboratory (JPL) found itself with abundance of low-quality lunar photos. The serious, foundational work in DIP began at JPL as scientists developed techniques for geometric correction, noise removal, and enhancement to produce the first detailed maps of the moon's surface. This critical application of DIP proved its value and launched the field into a new era of growth.

Originally conceived for analyzing and enhancing lunar images, DIP has grown into a vast field with numerous applications, driven by advances in dedicated algorithms and computer technology. Following its emergence in the 1960s, DIP saw rapid development in the 1970s, driven especially by the need to restore and enhance images from new medical imaging technologies such as Computerized Tomography (CT) and Magnetic Resonance Imaging (MRI). Since then, this expansion has impacted many technical areas, experiencing exponential growth. Prominent areas of growth for DIP include: medical imaging, data communication and compression, remote sensing, computer vision, and Artificial Intelligence (AI). Today, the focus of DIP is shifting toward real-time, intelligent systems that leverage machine learning (ML). Given the continuous evolution in computer engineering and AI, and the emergence of new applications, the potential for DIP appears limitless.

Among some specific applications of DIP are:

1. **Medical [1]:** Automatic detection and classification of abnormalities in medical images, such as X-rays, MRI, CT scans, and ultrasounds. For example, Figure 1.1 shows brain MRI images where the goal is to automatically detect, localize, and classify lesions or other irregularities (e.g. benign vs. malignant). Other applications include chromosome identification.
2. **Computer Vision [2]:** Applications of computer vision include part identification in assembly lines, robotics and autonomous systems, and tele-operation and tele-surgery. For example, a pick-and-place robot, illustrated in Figure 1.2, uses computer vision to sort objects based on their shape, size, or color, placing them in specific bins. For mobile robots and other autonomous systems, computer vision enables critical functions like obstacle avoidance, path planning, and navigation. This technology is also a core component of self-driving vehicles.
3. **Remote Sensing [3]:** Remote sensing applications include multispectral and hyperspectral imaging in meteorology, climatology, and military; tracking of earth resources; geographical mapping; predicting agricultural crops, urban growth, and weather patterns; and the early detection and mitigation of floods and fires. A digital image captured using one of the infrared (IR) channels of the Geostationary (GOES-16) satellite, which covers mostly the eastern part of the United States, is shown in Figure 1.3. The multispectral GOES imagery consists of many visible and IR channels that are used for applications including cloud type classification, precipitation estimation, and cloud height determination.
4. **Radar and Sonar [4]:** The role of DIP in these areas includes detection and recognition of various ground, airborne or underwater targets, as well as aircraft or missile guidance or maneuvering. Figure 1.4 shows a side-looking sonar image of an underwater environment. One of the main challenging issues in such situations is the detection and classification of targets of interest, e.g. underwater mines or other explosives, in heavy clutter due to the presence of corals, rocks, vegetation, and other anomalies. Moreover, the underwater environment is highly variable, hence making these tasks even more challenging and difficult for most detectors or classifiers [5].

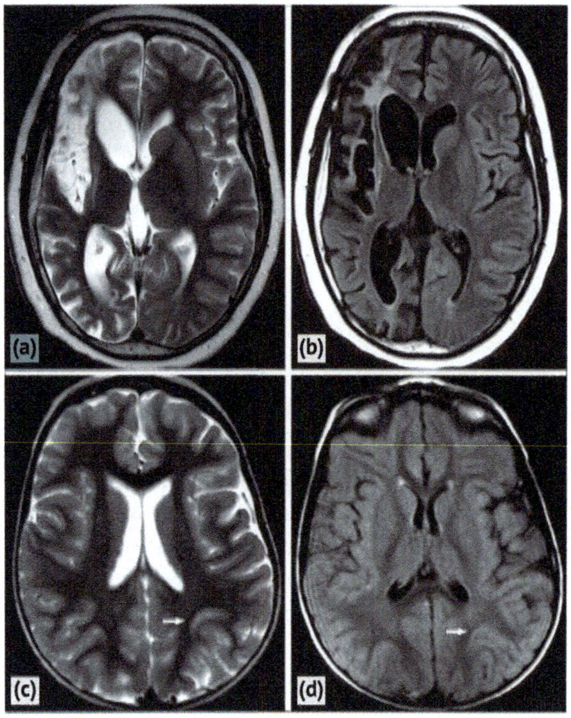

Figure 1.1 Examples of brain MRI digital images of two patients. Scans (a) and (c) capture water-rich tissue (T2) information, while (b) and (d) are the FLAIR (fluid-attenuated inversion recovery) scans for patients 1 and 2, respectively. *Source*: [11] / Springer Nature / https://www.researchgate.net/publication/236613159_Differential_expression_of_interferon-g_and_chemokine_genes_distinguishes_Rasmussen_encephalitis_from_cortical_dysplasia_and_provides_evidence_for_an_early_Th1_immune_response / CC BY 4.0 / last accessed on Jan 30, 2026. (The color rendition of this image is available to view at www.wiley.com/go/Azimi449.)

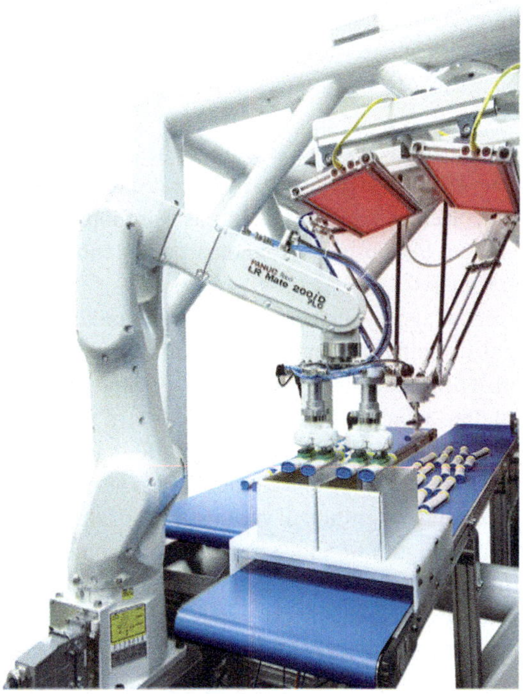

Figure 1.2 A pick and place robot with computer vision-Robotics Tomorrow.

Figure 1.3 GOES-16 IR image of land, water, and cloud areas. *Source*: National Oceanic and Atmospheric Administration (NOAA) / https://www.goes-r.gov/multimedia/dataAndImageryImagesGoes-16.html / Public domain / last accessed on Jan 30, 2026.

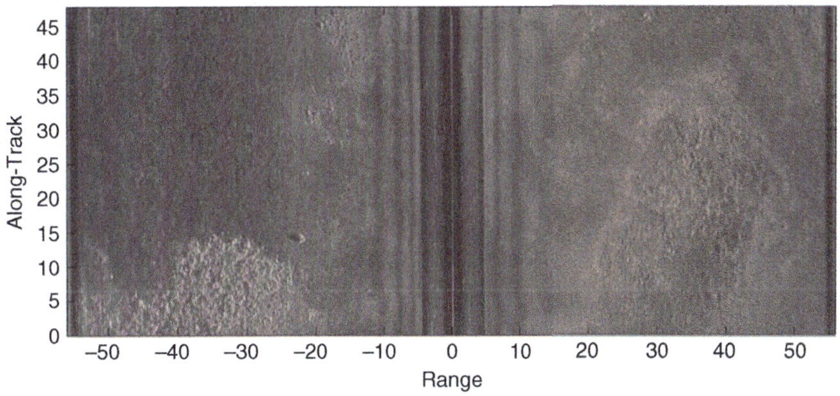

Figure 1.4 A synthetic aperture sonar (SAS) image of an underwater environment. *Source*: Courtesy of Office of Naval Research (ONR).

5. **Image Transmission [6, 7]:** Applications include: mobile devices, stereo or video image compression, teleconferencing and communications over computer networks, distributed sensor networks, space and satellite communication, etc. Figure 1.5 shows a pair of stereo images with both horizontal and vertical disparity due to the displacement of the left and right cameras. In such applications, to reduce the amount of data to be transmitted over a resource-limited channel, typically one image is used as a *reference* and encoded using the standard methods, e.g. JPEG-based, while disparity vectors are estimated and encoded for each block of the other image. These vectors are then used in conjunction with the reference image to reconstruct the second image block-by-block at the receiver [8].
6. **Office Automation [9]:** Document storage, retrieval and reproduction. Figure 1.6 depicts a schematic diagram of a typical information (e.g. document, images, audio signals, etc.) retrieval system. The user provides a query as text or an image, and the system breaks it down into tokens or attributes. It then searches the indexed documents to retrieve a rank-ordered set of relevant documents. In some cases, the user can also provide *relevance feedback* to the system to refine future searches [10]. The association between the query tokens/attributes and the retrieved documents is generally based on a form of *matching*, using traditional, statistical, Boolean-based, or trainable neural network-based models.
7. **Biometric Systems [12–14]:** Biometric recognition systems are generally categorized as either physiological or behavioral. Physiological biometrics uses inputs like facial, iris/retina, and fingerprint images, along with other information such as DNA. Behavioral biometrics, on the other hand, exploits unique cues like gait, signature, keystroke, mouse movements, and voice to authenticate users. Combining these two types of traits can produce a highly reliable biometric recognition system. These systems are widely used in banking, security, and forensic applications.

Figure 1.5 Stereo image pair horizontal and vertical disparity. *Source*: Kim Scarborough / https://commons.wikimedia.org/wiki/File:Kasuga_Lantern_(cross-eye_stereo_pair).jpeg / CC BY-SA 3.0 / last accessed on Jan 30, 2026.

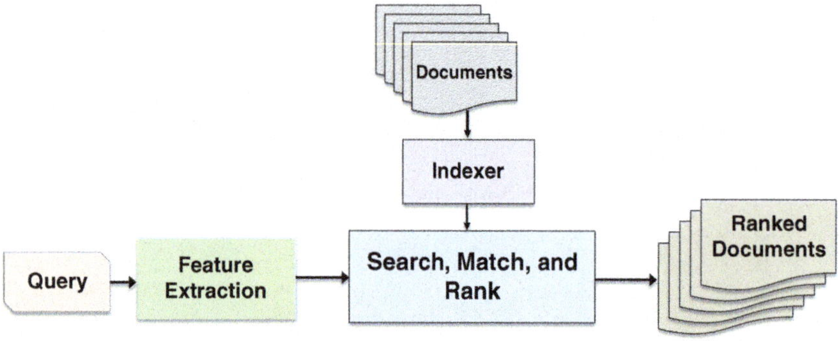

Figure 1.6 A typical document retrieval system.

1.1 DIP System Components

A digital image is a representation of a 2-D continuous function (image) acquired by an imaging sensor, e.g. optical or otherwise, converted into a discrete matrix of pixels through sampling and quantization that occur in the analog to digital converters (ADCs). In this process, sampling discretizes the spatial coordinates, dividing the continuous image into fixed size rectangular grids; while quantization converts the continuous range of intensity or color levels at each pixel into a finite set of discrete values. The diagram in Figure 1.7 depicts a typical DIP system and its components [15], where digital computers perform enhancement, analysis, and/or transmission operations on digital images. The results are buffered for display or storage purposes.

In general, the operations performed in digital computer in Figure 1.7 can be categorized into three types: *low-level*, *intermediate-level*, and *high-level* processing [15, 16]. Low-level processing corresponds to those operations for which both input and output are images, e.g. contrast enhancement, noise filtering, and edge extraction. Intermediate-level processing produces a set of features or attributes extracted from segmented regions in images, e.g. shape- or texture-dependent feature extraction methods. Finally, high-level processing typically takes the extracted features or attributes as input to generate decisions, assign labels, or produce symbolic representations of the original scene as output. An example of high-level processing is the use of a multimodal biometric system for identification, that is, one might be interested in using a combination of facial, iris, and gesture characteristics to recognize or identify persons of interest among many individuals.

Figure 1.8 illustrates a complete DIP system that includes all the processing levels mentioned above to provide aids to human observers or operators in analyzing the scene captured by the imaging system. As an example, let us assume that we want to design an automated diagnosis machine to help radiologists identify potential malignant calcifications in breast mammograms using digitized X-ray imaging. In this case, preprocessing is used to improve the contrast and accentuate the subtle edges in the images. Segmentation involves the detection and isolation of anything that resembles tumors or calcification based on their shapes, sizes, or other known characteristics. Once these potential candidate regions are separated, other more specific features, e.g. texture-based, will be extracted in order to facilitate classification of the detected regions into benign or malignant cases (classes). The structural analysis provides a description of the image, e.g. number, location, relationships of the tumors, or any other abnormalities, for review by the radiologists. The World Model or Knowledge Base (KB) monitors the entire process and provides input to each processing stage. The bidirectional arrows indicate that the

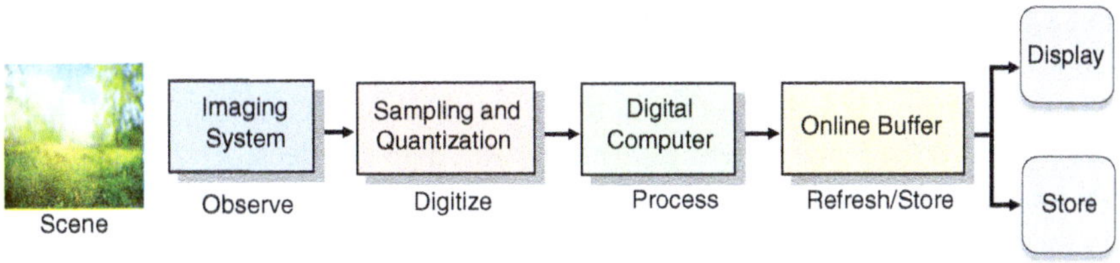

Figure 1.7 Components of a typical DIP system.

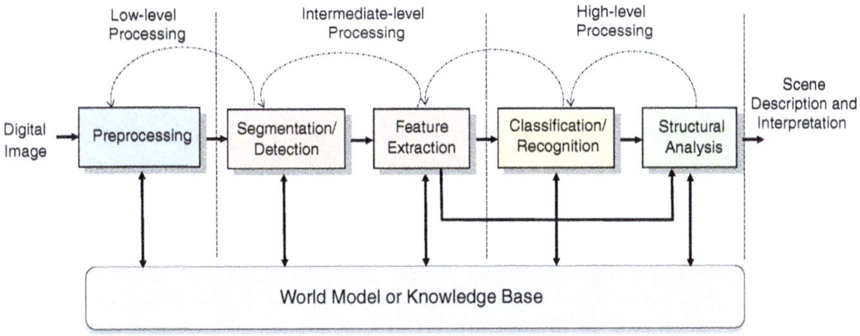

Figure 1.8 A DIP system consisting of low-level, intermediate-level, and high-level processing stages.

KB is updated as more images are processed, incorporating the new knowledge to improve detection and discrimination. During the design phase, feedback loops from each processing level to preceding ones are essential to refine and optimize the overall system efficiency.

1.2 Areas of DIP

1.2.1 Image Representation and Modeling

An image can be represented in either the spatial or transform domain. An important consideration here is the fidelity or intelligibility criterion for measuring quality of an image, e.g. contrast, spatial frequencies, edge sharpness, etc. Images represented in the spatial domain directly relate to the type and physical properties of the imaging sensors that acquired them; e.g. luminance of object in a scene for pictures taken by camera; absorption characteristics of body tissue for X-ray imaging; radar cross-section of a target for radar imaging; temperature profile of an object for IR imaging; and gravitational field in geophysical imaging. However, the downside is the large dimensionality and volume of the data to represent the image, especially for images with high spatial resolution.

Alternatively, images can be mapped to the transform domain using several possible 2-D transforms or any other image mapping method, leading to low-dimensional attributes such as the main spectral components, the principal components, the textural or any other salient and representative features. Unlike spatial domain representation, the attributes in the transformed domain tend to be more localized and compact. Thus, they are better suited for filtering, data compression, or image recognition.

In some cases, linear prediction models [16] are used to represent a set of images using a few model coefficients that are obtained based on second-order statistics of the images. The model coefficients can then be used to represent the images.

1.2.2 Image Enhancement

No imaging system can produce a perfect image. In image enhancement, the aim is to manipulate an image in order to improve its quality. To assess the quality of the processed image compared to the original image, an expert human observer is often required to evaluate and approve the processed images. However, human evaluation is highly dependent on the *subjective* judgment of the individual, which may be wise or fickle. Observers may have different preferences; for example, some might prioritize noise reduction at the expense of sacrificing sharp edges, while others prefer distinct edges even in presence of noise. Psychophysical factors play an important role in this decision-making and evaluation process. A comprehensive study, therefore, necessitates subjective testing by a group of human viewers. Moreover, the decision on the acceptability of a processed image generally depends on the nature of the subsequent chain of operations. For instance, images intended for object detection and classification generally benefit from sharp, well-defined edges.

Common image enhancement techniques include contrast adjustment, noise reduction, edge sharpening, magnification, and pseudo-coloring. The methods used in these areas are often heuristic and can be applied broadly to different images, regardless of their specific characteristics. This differs from image restoration, which uses specific models to remove the effects of known degradations, such as blur and additive/multiplicative noise.

1.2.3 Image Restoration

While image enhancement focuses on improving visual appearance, image restoration aims to recover or estimate the original image by reversing known degradation effects from both deterministic and stochastic sources. Deterministic degradation often results in blur caused by effects such as atmospheric turbulence in satellite imaging, relative motion between the camera and the object, and defocusing. Stochastic degradation is caused by noise that corrupts images additively or multiplicatively. Additive noise can stem from sensor imperfections, thermal effects, and quantization. Multiplicative noise, such as speckle [17], is common in coherent imaging systems like synthetic aperture radar (SAR), lasers, and ultrasound.

The main goal in image restoration is to utilize the knowledge of degradation processes, that is, the deterministic and stochastic representations of blur and noise, respectively, to devise an appropriate inverse process to estimate the original image. Traditional restoration techniques include deblurring, pseudo-inverse filtering, and 2-D Wiener and Kalman filters. Unlike image enhancement, which is often subjective, image restoration is objective, as it relies on specific, quantifiable information about both the image and the degradation processes to construct the restoration filter.

While some texts treat image reconstruction from projections as a separate field [16], it is indeed fundamentally related to image restoration. The goal of image reconstruction is to reconstruct 2-D or 3-D objects from several 1-D or 2-D projections. Examples are medical CT, MRI, astronomy, radar imaging, and nondestructive evaluation (NDE). Common methods in image reconstruction are the Radon transform, filtered back-projection, algebraic reconstruction, and statistical-based methods.

There are many reasons why image reconstruction and image restoration are considered related. Both are *inverse problems* that aim to recover an original, *unobserved* image from degraded measurements. The core challenge in both fields is to undo the effects of a physical process to arrive at a restored or reconstructed version of the original image. These inverse problems are often ill-posed or ill-conditioned in nature. This means that a unique, stable solution may not exist or is sensitive to noise in the measured data. To address this issue, both fields rely on *regularization* methods to constrain the solution and find a plausible outcome. These methods incorporate prior knowledge about the expected properties of the image to guide the reconstruction or restoration process. This shared foundation as inverse problems means that many concepts, tools, and challenges are relevant across both fields. Subsequent chapters will delve into the specific similarities and differences between the methodologies used in image restoration and image reconstruction.

1.2.4 Image Transforms

Image transform is a powerful tool for revealing hidden characteristics of an image and enabling more efficient processing to be carried out in the transform domain. That is, rather than processing an image in the spatial domain pixel-by-pixel, this process maps the image into a different representation, or transform domain, where its underlying information becomes more apparent. This change of perspective is achieved using various unitary transforms such as 2-D Discrete Fourier Transform (DFT), 2-D Discrete Cosine transform (DCT), 2-D Discrete Wavelet Transform (DWT), and 2-D Karhunen–Loeve (KL) transform [15, 16]. Image transformation performs both feature extraction and dimensionality reduction, which are crucial for applications such as data compression and pattern recognition.

1.2.5 Image Data Compression and Encoding

It is often required to significantly reduce the number of bits to represent digital images before transmission or storage. For example, the NOAA GOES-15 satellite produces a massive data stream every 15 minutes by capturing 5 spectral channels (1 visible, 4 IR) at 3K × 3K pixels each with a 10-bit gray-level resolution. The sheer scale of this data necessitates significant bit-rate reduction to manage storage capacity and network bandwidth.

The efficiency of image compression is directly proportional to the amount of statistical and psychovisual redundancy present in the source images. Higher redundancy enables a greater reduction in the number of bits required to represent the image without an appreciable loss of essential visual information. Standard pulse code modulation (PCM) represents a raw, uncompressed baseline, typically requiring 8 bits per pixel. However, efficient compression methods, such as predictive encoding (e.g. differential PCM) and transform-based techniques [16], can reduce this rate to under 1 bit per pixel while maintaining image fidelity.

For video images, motion-compensated encoding exploits redundancy due to still parts of the image and only detects and estimates motion parameters from image sequences and motion-compensated frame differences are transmitted [18]. A modern example is the new JPEG XS standard, which delivers visually lossless, low-latency, and highly efficient compression. This makes it ideal for a wide range of applications, including virtual and augmented reality, 5G devices, and autonomous vehicles. A related technique used in 3-D TV, wearable devices, augmented reality, and mobile platforms is stereo image compression, which estimates disparity information between two images, captured by two disparate cameras or other imaging systems [19].

1.2.6 Image Analysis and Computer Vision

The schematic diagram in Figure 1.8 illustrates a traditional image analysis system used for pattern recognition and scene perception. Such systems typically rely on a series of low-level and intermediate-level processing prior to final recognition and analysis. Segmentation is used to detect and isolate objects of interest, such as separating potential targets from the background in radar imaging, thereby facilitating feature extraction. A representative set of features is then extracted from each isolated object. The quantitative evaluation of these features allows for the classification and description of the objects. Ultimately, the goal is to create an automatic recognition system that can derive semantic descriptions from a visual scene. In essence, pattern recognition is the inverse of computer graphics; it transforms a picture into an abstract description, such as a set of labels, a string of symbols, or a scene graph.

While all stages of image analysis are essential, the feature extraction subsystem is arguably the most critical component in any traditional image analysis system. Its design and the choice of methodology profoundly influence the system's overall success and performance. Typical choices for feature extraction methods are those that rely on extracting: (i) edge and contour information; (ii) morphological attributes; (iii) 2-D or 3-D shape information; and (iv) textural features.

1.3 Final Remarks and Goals

The advent of AI has led to a dramatic shift in the way most image processing tasks are performed with much better precision and adaptability than using traditional methods [2, 20]. A key difference lies in the feature extraction process. Traditional image analysis systems are highly dependent on manually designed feature extractors, a task that requires significant human expertise and prior knowledge. In contrast, artificial neural network models, particularly deep neural networks, are trained to automatically extract features at multiple levels of abstraction directly from images, resulting in greater accuracy and robustness. Furthermore, traditional feature extraction methods often fail to capture the contextual and semantic information critical for complex tasks like scene analysis. Deep neural networks, however, can not only perform pattern recognition but also achieve a higher level of scene understanding and interpretation by discovering relationships between different parts of an image, capabilities that are a hallmark of biological visual systems. Finally, another key advantage is that unlike humans, whose decision-making can be compromised by the sheer volume of image data, physical impairment, or fatigue, AI-based systems maintain a consistent level of performance and analysis, making them highly reliable in such demanding scenarios.

However, a major challenge for these AI models is their reliance on large, well-labeled, and highly reliable training datasets. While extensive datasets are available for some applications, such as medical imaging, others face data scarcity. For instance, anomaly detection and identification in surveillance and reconnaissance systems often have limited instances of the phenomena they are trained to detect and classify. This is particularly problematic for supervised training, which necessitates human experts to provide the expensive and time-consuming process of labeling the data. Moreover, potential biases within training data, e.g. those arising from under-represented cases (instances of real targets), can critically impair the performance of AI-based image analysis systems.

This book introduces the reader to various areas of DIP, providing an understanding and appreciation for the scope of this dynamic field. We present fundamental concepts, methods, and theoretical background, enabling readers to understand, apply, and develop image processing algorithms for various applications. While most chapters cover traditional DIP methods, others explore more advanced, state-of-the-art algorithms using ML and AI tools. Through numerous practical examples from fields like medical imaging and remote sensing, we demonstrate the knowledge and tools needed to apply these methods to real-world problems. Specifically, our goals are to equip students, researchers, and readers with the mathematical tools and know-how to:

- Learn, apply, and implement various traditional and AI-based methods for specific DIP problems.
- Design and analyze various traditional and AI-based algorithms.
- Analyze DIP system performance using various metrics.
- Examine a wide range of applications of DIP algorithms.

References

1 G. Dougherty, "*Digital Image Processing for Medical Applications*," Academic Press, 2009.

2 R. Szeliski, "*Computer Vision: Algorithms and Applications*," Springer, 2022.

3 M. de Carvalho Alves and L. Sanches, "*Remote Sensing and Digital Image Processing with R-Lab Manual*," CRC Press, 1st Edition, 2023.

4 S. Stergiopoulos, "*Advanced Signal Processing Handbook: Theory and Implementation for Radar, Sonar, and Medical Imaging Real Time Systems*," CRC Press, 2001.

5 N. Klausner and M. R. Azimi-Sadjadi, "Performance prediction and estimation for underwater target detection using multi-channel sonar," *IEEE Journal on Oceanic Engineering*, vol. 45, pp. 534–546, 2020.

6 C. Solomon and T. Breckon, "*Fundamentals of Digital Image Processing: A Practical Approach with Examples in Matlab*," Wiley-Blackwell, 2011.

7 P. K. Devulapalli, S. Kumar, S. B. Maganti, and S. Rachapogula, "Image transmission in mobile wireless multimedia sensor networks using cat swarm optimization," *Multimedia Tools and Applications*, vol. 83, pp. 17557–17578, 2023.

8 S. H. Seo and M. R. Azimi-Sadjadi, "A 2-D filtering scheme for stereo image compression using sequential orthogonal subspace updating," *IEEE Transactions on Circuits and Systems for Video Technology*, vol. 11, pp. 52–64, 2001.

9 I. M. Hameed, S. H. Abdulhussain and B. M. Mahmmod, "Content-based image retrieval: A review of recent trends," *Cogent Engineering*, pp. 1–37, 2021.

10 M. R. Azimi-Sadjadi, J. Salazar, and S. Srinivasan, "An adaptable image retrieval system with relevance feedback using kernel machines and selective sampling," *IEEE Transactions on Image Processing*, vol. 18, pp. 1645–1659, 2009.

11 G. C. Owens, M. N. Huynh, J. W. Chang, D. L. McArthur, M. J. Hickey, H. V. Vinters, G. W. Mathern, and C. A. Kruse, "Differential expression of interferon-γ and chemokine genes distinguishes Rasmussen encephalitis from cortical dysplasiaand provides evidence for an early Th1 immune response," *Journal of Neuroinflammation,* vol. 10, p. 825, 2013.

12 A. Nait-Ali and R. Fournier (Editors), "*Signal and Image Processing for Biometrics*," Wiley, 1st Edition, 2012.

13 D. Zhang, X. Jing, and J. Yang, "*Biometric Image Discrimination Technologies*," IGI Global, 2006.

14 S. Y. Kung, M. W. Mak, and S. H. Lin, "*Biometric Authentication: A Machine Learning Approach*," Prentice-Hall, 2005.

15 R. Gonzalez and R. Woods, "*Digital Image Processing*," Pearson, 4th Edition, 2017.

16 A. K. Jain, "*Fundamentals of Digital Images Processing*," Prentice-Hall, 1989.

17 M. Tur, K. C. Chin, and J. W. Goodman, "When is speckle noise multiplicative?," *Applied Optics*, vol. 21, pp. 1157–1159, 1982.

18 Y. Q. Shi and H. Sun, "*Image and Video Compression for Multimedia Engineering Fundamentals, Algorithms, and Standards*," Chapman & Hall, 3rd Edition, 2019.

19 H-Y. Shum, S. B. Kang, and S-C. Chan, "Survey of image-based representations and compression techniques," *IEEE Transactions on Circuits and Systems for Video Technology*, vol. 13, pp. 1020–1037, 2003.

20 K. V. Arya, C. R. Rodriguez, S. Singh, and A. Singhal (Editors), "*Artificial Intelligence and Machine Learning Techniques in Image Processing and Computer Vision*," CRC Press, 2024.

2

Review of 1-D Signal Processing

Digital signal processing (DSP) has a rich history and its importance is evident in such diverse fields as biomedical, acoustic, sonar, radar, seismic, speech processing, communication systems, data compression, and even AI among many others. The fundamentals of DSP area stem from other well-established fields such as classical sampled-data control systems from the 1950s and also numerical analysis techniques developed in the 1960s. The evolution of a new point of view toward DSP was further accelerated in mid-1960 when an efficient algorithm was disclosed for computing the Discrete Fourier Transform (DFT), known as the fast Fourier transform (FFT) [1, 2]. FFT is inherently a discrete-time process, and it represents a shift away from the notion that signal processing on a digital computer is not merely an approximation to analog signal processing but rather its own area of significance.

Until a few decades ago, digital computers were used to simulate DSP algorithms with great flexibility. However, digital computers are not only bulky and costly for many commonly used DSP operations but also, unlike dedicated DSP hardware, lack the ability to process signals in real time. The advancement in field-programmable gate arrays (FPGAs) and dedicated DSP hardware system technologies in recent years has had a significant impact on the growth of this field and allowed for on-board implementation of many sophisticated signal processing operations in real time. DSP is still a rapidly growing field in particular due to the emergence of AI and machine learning applications, quantum computing [1], and many others.

The goal of this chapter is to review some fundamental concepts and techniques in both deterministic and statistical signal processing and to provide a foundation for the materials to be covered in the subsequent chapters. The reader is assumed to be familiar with the basic probability and random variables (r.v.s).

2.1 Discrete-Time Signals and Systems-Deterministic Case

2.1.1 Discrete-Time Signals

A signal is a function of an independent variable *time* and is designed to convey or represent information about the state or behavior of a physical system. Signals can be categorized into four groups which are: continuous-time, analog, discrete-time, and digital signals. A digital signal is a discrete-time signal whose amplitude is quantized, i.e. takes only a certain set of values. Figure 2.1 shows the process of sampling and quantization to convert a continuous time signal $x(t)$ to a digital signal $x(n)$ where $n \in \mathbb{Z}$ and $\mathbb{Z}$ represent a set of integers. The quantization level is determined by the type and bit length of the DSP system used.

Most of the methods developed for discrete-time signals can also be applied to digital signals with no modification. Therefore, in this chapter, we restrict our attention to these methods without considering the quantization effects. In Chapter 5, the effects of quantization for images will be studied.

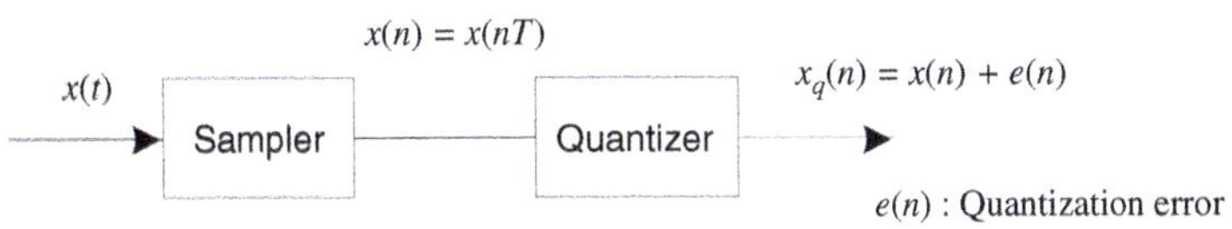

Figure 2.1 Sampling and quantization of 1-D signals.

A. Some Important 1-D Signals

The following signals play an important role in DSP.

1. *Unit Pulse or Kronecker Delta Function:* The unit pulse or Kronecker delta $\delta(n)$ plays the same role as the Dirac delta in the continuous-time signals and is defined as:

$$\delta(n) = \begin{cases} 1, & n = 0 \\ 0, & \text{otherwise} \end{cases}$$

Any signal $x(n)$ can be expressed as a linear combination of a sequence of delayed unit pulse signals $\delta(n-k)$ weighted by the amplitude of the signal $x(k)$ as,

$$x(n) = \sum_{k=-\infty}^{\infty} x(k)\delta(n-k), \tag{2.1}$$

where k is the time delay. This property will be used in Section 2.1.3 to arrive at convolution summation for linear systems.

2. *Unit Step Sequence:* The unit step sequence $u(n)$ is defined as:

$$u(n) = \begin{cases} 1, & n \geq 0 \\ 0, & n < 0 \end{cases}.$$

The relationships between unit step and unit pulse signals are $\delta(n) = u(n) - u(n-1)$ and $u(n) = \sum_{k=0}^{\infty} \delta(n-k)$ which are derived from their definitions.

3. *One-sided Real Exponential Sequence:* Let $0 < a < 1$ be a real constant. The one-sided (right) real exponential sequence is then defined as:

$$x(n) = a^n u(n).$$

This sequence is the discrete-time version of the continuous-time exponential function $e^{\alpha t}$, $t \geq 0$ or its sampled-data version with sampling interval T, i.e. $e^{\alpha t}\big|_{t=nT} = e^{\alpha Tn} = a^n$ where $a = e^{\alpha T}$.

4. *Complex Exponential Sequence:* The complex exponential sequence is defined as $x(n) = e^{j\Omega n}$, $\forall n \in \mathbb{Z}$ where $j = \sqrt{-1}$ and Ω is the frequency of the complex exponential. This function can also be viewed as the sampled version of the continuous-time function $e^{j\omega t}$, i.e. $e^{j\omega t}\big|_{t=nT} = e^{j\omega nT} = e^{j\Omega n}$ where $\Omega = \omega T$ represents the relationship between *analog* and *digital* frequencies.

B. 1-D Signal Types and Properties

1. *Periodic Versus Aperiodic Signals:* A discrete-time signal $x(n)$ is *periodic* with period N, if $x(n) = x(n + kN)$, $\forall k \in \mathbb{Z}$. The period N is the smallest positive integer for which equality holds. An example of a periodic signal is the complex exponential that is periodic in Ω with period 2π since $e^{j(\Omega+2\pi k)n} = e^{j\Omega n}$ due to $e^{j2\pi kn} = 1$. Consequently, it is adequate to define this signal for one period only, i.e. $0 \leq \Omega \leq 2\pi$ or $-\pi \leq \Omega \leq \pi$. In contrast, in the continuous-time case, the complex exponential signal $e^{j\omega t}$ is distinct for distinct values of ω.

Example 2.1 *Determine the condition under which the complex exponential signal $e^{j\Omega n}$ is periodic in n. What is the period, if any?*

Solution: For periodicity in n, we should have $e^{j\Omega(n+kN)} = e^{j\Omega n}$ which is satisfied when $\frac{\Omega}{2\pi} = \frac{m}{kN}$, $m \in \mathbb{Z}$. That is, $e^{j\Omega n}$ is periodic only when $\frac{\Omega}{2\pi}$ is a rational number. Assuming this is satisfied, the fundamental period, N_0, is the *smallest integer* which gives:

$$e^{jm\frac{2\pi}{kN}(n+kN_0)} = e^{jm\frac{2\pi}{kN}n}.$$

This requires $m\frac{2\pi N_0}{N} = 2\pi l$ or $m\frac{N_0}{N} = l$ for $l \in \mathbb{Z}$. If $\gcd(m, N)$ represents the greatest common divisor of m and N, then $m = m' \gcd(m, N)$ and $N = N' \gcd(m, N)$. Note that m' and N' have no common factors (prime numbers), therefore $m'\frac{N_0}{N'} = l \Rightarrow N_0 = N'$ or $N_0 = \frac{N}{\gcd(m,N)}$. If m and N have no common factors, then the fundamental period is $N_0 = N$ and the fundamental frequency is $\frac{2\pi}{N}$.

2. *Symmetric Versus Antisymmetric Signal:* A discrete-time signal $x(n)$ is called a symmetric or even signal if $x(-n) = x(n)$, whereas it is antisymmetric or odd if $x(-n) = -x(n)$. An arbitrary signal $x(n)$ can be expressed in terms of the sum of even and odd components, i.e.

$$x(n) = x_e(n) + x_o(n),$$

where $x_e(n) = \frac{1}{2}[x(n) + x(-n)]$ and $x_o(n) = \frac{1}{2}[x(n) - x(-n)]$ are the even and odd parts, respectively.

2.1.2 Discrete-Time Systems and Properties

A discrete-time system performs a transformation $T[.]$, which maps the input signal $x(n)$ to the output signal given by $y(n) = T[x(n)]$. The operator $T[.]$ may have many important properties that lead to different types of systems.

1. **Linearity:** The operator $T[.]$ is linear if and only if (iff)

$$T[ax_1(n) + bx_2(n)] = aT[x_1(n)] + bT[x_2(n)] = ay_1(n) + by_2(n) \tag{2.2}$$

where $y_1(n) = T[x_1(n)]$, $y_2(n) = T[x_2(n)]$ and a and b are arbitrary constants. Equation (2.2) implies that, if a system is linear, superposition of two or more inputs with arbitrary weighting constants leads to superposition of their corresponding outputs with the same weighting constants.

2. **Shift-Invariance:** Let $y(n) = T[x(n)]$ and $\hat{x}(n) = x(n - m)$ with $m \in \mathbb{Z}$ be the shifted version of $x(n)$ by the m samples. If $\hat{y}(n) = T[\hat{x}(n)] = y(n - m)$, then the system is said to be shift-invariant.
3. **Causality:** Causality is a requirement for all physically realizable 1-D systems, which states that the output $y(n)$ is dependent only on those samples of $x(m)$ for which $m \le n$, i.e. past and present but not future ones. This is always the case when the independent variable is *time*.

Example 2.2 *Determine if the system given by the input–output relation* $y(n) = T[x(n)] = \sum_{k=n_0}^{n} x(k)$ *is (a) linear, (b) shift-invariant, and (c) causal.*

Solution:

a. To check for linearity property, we let $x(n) = ax_1(n) + bx_2(n)$. Then, the output is:

$$T[ax_1(n) + bx_2(n)] = a\sum_{k=n_0}^{n} x_1(k) + b\sum_{k=n_0}^{n} x_2(k) = ay_1(n) + by_2(n),$$

That is, linear.

b. To determine shift invariance, we let $\hat{x}(n) = x(n - m)$ and apply this shifted signal to the system. Then

$$\hat{y}(n) = T[x(n - m)] = \sum_{k=n_0}^{n} \hat{x}(k) = \sum_{k=n_0}^{n} x(k - m)$$

Now, if we change $k - m = l$, then $\hat{y}(n) = \sum_{l=n_0-m}^{n-m} x(l)$. On the other hand, $y(n - m) = \sum_{k=n_0}^{n-m} x(k)$ thus $\hat{y}(n) \neq y(n - m)$ and the system is shift varying.

c. Since $y(n)$ depends on future values of $x(n)$ when $n < n_0$ the system is *not* causal.

2.1.3 Representation of LSI Systems-Convolution Summation

Consider a linear discrete-time system with an input–output relation $y(n) = T[x(n)]$. The response of a linear system to a unit pulse or an impulse input, $\delta(n - m)$, is called the unit pulse or the impulse response of the system. That is, the response at time instant n to a unit pulse input applied at time instant m is $h(n, m) = T[\delta(n - m)]$, where $T[.]$ is a linear operator. Using the property of $\delta(n)$ function in (2.1) and invoking linearity, we can write:

$$y(n) = T[\sum_{k=-\infty}^{\infty} x(k)\delta(n-k)] = \sum_{k=-\infty}^{\infty} x(k)T[\delta(n-k)]$$
$$= \sum_{k=-\infty}^{\infty} x(k)h(n,k).$$

If the system is linear and shift-invariant (LSI), then $h(n, m) = h(n - m)$ and the above equation becomes:

$$y(n) = \sum_{k=-\infty}^{\infty} x(k)h(n-k) = \sum_{k=-\infty}^{\infty} h(k)x(n-k)$$
$$= x(n) * h(n), \tag{2.3}$$

where $*$ stands for linear convolution operation.

Definition 2.1 A 1-D LSI system is *causal* if impulse response sequence satisfies $h(n) = 0, \forall n < 0$, *anti-causal* if $h(n) = 0, \forall n > 0$ and *non-causal* if it is neither causal nor anti-causal.

Using this definition for causal LSI systems, the *convolution summation* in (2.3) changes to:

$$y(n) = \sum_{k=0}^{\infty} h(k)x(n-k), \tag{2.4}$$

i.e. the output of an LSI system is obtained by convolving the input with the unit pulse response if all initial conditions of the system are zero. *Thus, the impulse response, $h(n)$, completely characterizes the behavior of an LSI system.*

A. Properties of Linear Convolution:

1. Convolution has the following important properties:
   Commutativity: Signals can be convolved in any order:

$$y(n) = \sum_{k=-\infty}^{\infty} h(k)x(n-k) = \sum_{k=-\infty}^{\infty} x(k)h(n-k)$$
$$= x(n) * h(n) = h(n) * x(n)$$

   Associativity: Convolution operations in series can be done in any order, i.e.

$$(x(n) * h(n)) * g(n) = x(n) * (h(n) * g(n))$$

   This property is important for finding the overall impulse response of a series combination of several LSI systems.
   Distributivity: Convolution is distributive over addition, i.e.

$$x(n) * [h(n) + g(n)] = x(n) * h(n) + x(n) * g(n)$$

   This property is important for finding the overall impulse response of a parallel combination of several LSI systems.
2. The result of the convolution of two finite-extent sequences of sizes M and N is another finite-extent sequence of size $M + N - 1$.
3. The result of linear convolution can be verified by checking the identity $\sum_n y(n) = (\sum_n h(n))(\sum_n x(n))$.

Convolution sum can be carried out graphically by keeping one signal, $x(k)$ (or $h(k)$), fixed, flipping the other signal, $h(k)$ (or $x(k)$), with respect to the vertical axis to produce $h(-k)$ (or $x(-k)$), and considering it as a moving frame shifted by n, i.e. $h(n - k)$ (or $x(n - k)$). This shifted signal is then overlayed on the other fixed signal, and the sum of products of the two signal samples within the region of their overlap gives the result $y(n)$ at time n.

Example 2.3 *Perform linear convolution $y(n) = h(n) * x(n)$ graphically where:*

$$h(n) = \begin{cases} \alpha^n, & 0 \le n \le N \\ 0, & \text{elsewhere} \end{cases} \quad \text{and} \quad x(n) = \begin{cases} \beta^{n-n_0}, & n_0 \le n \\ 0, & n < n_0 \end{cases}$$

Solution: Since $h(n)$ is finite-extent, we consider it as a moving frame and fix $x(n)$. In this case, we use $y(n) = \sum_{k=-\infty}^{\infty} x(k)h(n-k)$. There are three regions to consider.

Region 1: $n < n_0$: There is no overlap in this region, thus $y(n) = 0$.

Region 2: $n_0 \le n \le n_0 + N - 1$: In this region, overlap exists between n_0 and n (see Figure 2.2 [top]), hence:

$$y(n) = \sum_{k=n_0}^{n} \alpha^{n-k}\beta^{k-n_0} = \alpha^n\beta^{-n_0} \sum_{k=n_0}^{n} (\beta\alpha^{-1})^k$$
$$= \alpha^n\beta^{-n_0}\left[\frac{(\beta\alpha^{-1})^{n_0} - (\beta\alpha^{-1})^{n+1}}{1-\beta\alpha^{-1}}\right] = \frac{\alpha^{n-n_0+1} - \beta^{n-n_0+1}}{\alpha-\beta}, \text{ for } \alpha \ne \beta$$
$$\text{If } \alpha = \beta \Rightarrow y(n) = \alpha^{n-n_0} \sum_{k=n_0}^{n} 1 = (n - n_0 + 1)\alpha^{n-n_0}$$

Note: we used the finite power series formula $\sum_{k=n}^{m} a^k = \frac{a^n - a^{m+1}}{1-a}$ to get the above result.
Region 3: $n \ge n_0 + N$: In this region, overlap exists between $n - N$ and n (see Figure 2.2 [bottom]), hence:

$$y(n) = \sum_{k=n-N}^{n} \alpha^{n-k}\beta^{k-n_0} = \alpha^n\beta^{-n_0} \sum_{k=n-N}^{n} (\beta\alpha^{-1})^k$$

or

$$y(n) = \alpha^n\beta^{-n_0}\left[\frac{(\beta\alpha^{-1})^{n-N} - (\beta\alpha^{-1})^{n+1}}{1-\beta\alpha^{-1}}\right]$$
$$= \beta^{n-n_0-N}\left(\frac{\alpha^{N+1} - \beta^{N+1}}{\alpha-\beta}\right), \ \alpha \ne \beta$$
$$\text{If } \alpha = \beta \Rightarrow y(n) = \sum_{k=n-N}^{n} \alpha^{n-n_0} = (N+1)\alpha^{n-n_0}$$

Combining these results gives the response $y(n)$ of the system.

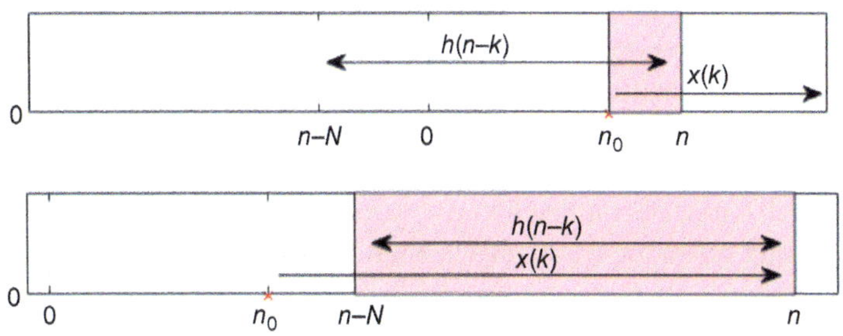

Figure 2.2 Overlap regions 2 and 3 for Example 2.3.

B. Stability of 1-D LSI Systems

Definition 2.2 A 1-D system is said to be bounded-input bounded-output (BIBO) stable [2, 3] *iff* a bounded input results in a bounded output, i.e. $|x(n)| < M < \infty \overset{\text{BIBO}}{\Longrightarrow} |y(n)| < N < \infty, \forall n$.

Using the convolution summation in (2.4), we can arrive at the condition on the impulse response $h(n)$ for BIBO stability of LSI systems. This result is given in the following theorem.

Theorem 2.1 *The necessary and sufficient condition for BIBO stability of a 1-D LSI system is that the impulse response is absolutely summable, i.e.*

$$S = \sum_{k=-\infty}^{\infty} |h(k)| < \infty. \tag{2.5}$$

Proof: The proof can be found in [2, 3].

Example 2.4 *Determine BIBO stability of the system is given by* $y(n) = \sum_{k=-\infty}^{n} x(k)$.

Solution: It is easy to verify that this system is LSI. Thus, we apply Theorem 2.1:

$$h(n) = \sum_{k=-\infty}^{n} \delta(k) = \begin{cases} 1, & \forall n \geq 0 \\ 0, & \text{otherwise} \end{cases}$$

Thus, $\sum_{n=0}^{\infty} |h(n)| = \infty$, i.e. unstable.

2.1.4 Representation of LSI Systems-Difference Equation

A discrete-time LSI system can be represented by the following N^{th} order constant coefficient difference equation:

$$\sum_{k=0}^{N} a_k y(n-k) = \sum_{l=0}^{M} b_\ell x(n-l), \quad a_0 = 1 \tag{2.6}$$

where $y(n)$ is the output, $x(n)$ is the input, and a_ks and b_ls are the coefficients of difference equation. Alternatively, by keeping the current sample of the output $y(n)$ on the left-side and moving the other terms to the right-side, we obtain a more usable form of this difference equation:

$$\underbrace{y(n)}_{\text{present output}} = \underbrace{\sum_{l=0}^{M} b_\ell x(n-l)}_{\text{past \& present inputs}} - \underbrace{\sum_{k=1}^{N} a_k y(n-k)}_{\text{past outputs}}. \tag{2.7}$$

This recursive equation can be used to compute the response of an LSI system to an arbitrary input signal $x(n)$ given the initial conditions. As an example, consider the first-order difference equation $y(n) = ay(n-1) + x(n)$ and let the input be $x(n) = \delta(n)$. Assuming causality $h(n) = 0, \forall n < 0$, using a recursive application of this difference equation, we get $h(0) = ah(-1) + \delta(0) = 1$, $h(1) = a$, $h(2) = a^2, \cdots h(n) = ah(n-1) = a^n$. Thus, the impulse response is $h(n) = a^n u(n)$, which is a stable solution for $|a| < 1$. Although this recursive method for finding the solution of difference equations is simple and suited for actual implementation on software or hardware, it does not provide a closed-form solution for general cases.

It is important to note that the convolution sum, only computes the forced response or particular solution of an LSI system. In contrast, a difference equation representation can be used to find the complete response, which includes both the forced response and the zero-input (or homogeneous) solution resulting from initial conditions. Although time-domain methods [2, 3] exist for solving difference equations, the z-transform method detailed in Appendix A provides a more efficient and systematic approach.

Special Cases:

Case 1 When $N = 0$ and $M \neq 0$ in (2.7), we get:

$$y(n) = \sum_{l=0}^{M} b_l x(n-l),$$

i.e. no recursive terms and hence the system is called *non-recursive*. Comparing to convolution summation:

$$y(n) = \sum_{l=0}^{M} h(l)x(n-l),$$

it is obvious that $h(l) = b_l,\ l \in [0, M]$ i.e. the impulse response is finite duration ($h(l) = 0$ for $n > M$ and $n < 0$) and hence a non-recursive system is also *Finite Impulse Response (FIR)*. Moreover, FIR systems are naturally all-zero systems as their transfer functions contain zeros only, i.e. $H(z) = \mathcal{Z}\{h(n)\} = \sum_{l=0}^{M} b_l z^{-l}$ where $\mathcal{Z}\{.\}$ is the z-transform operator (see Appendix A).

Case 2 When $N \neq 0$ and $M = 0$, in (2.7), we have:

$$y(n) = b_0 x(n) - \sum_{k=1}^{N} a_k y(n-k),$$

where the recursive terms are present but the output $y(n)$ depends only on the present input $x(n)$ and not on the past. In this case, the system is recursive, *infinite impulse response (IIR)*, and all-pole because the transfer function has only poles, i.e. $H(z) = \frac{b_0}{1+\sum_{k=1}^{N} a_k z^{-k}}$.

When both $N \neq 0$ and $M \neq 0$, we have a general case of a recursive or IIR system for which the transfer function contains both poles and zeros.

Note that FIR or non-recursive systems are inherently BIBO stable since condition in Theorem 2.1 is always satisfied. However, the stability of IIR systems must be carefully considered when designing these types of systems.

2.1.5 Frequency Domain Analysis – Discrete-Time Fourier Transform

Discrete-time Fourier transform (DTFT) is used for frequency domain analysis of nonperiodic discrete-time signals both of finite and infinite duration. DTFT is obtained by taking the FT of a sampled continuous-time signal $x(t)$. That is, for the sampling period is T, the sampled signal is:

$$x_s(t) = x(t)\delta_T(t) = \sum_{n=-\infty}^{\infty} x(nT)\delta(t-nT)$$

where $\delta_T(t) = \sum_{n=0}^{\infty} \delta(t-nT)$ is the sampling function. Now, taking the FT of $x_s(t)$ gives:

$$X_s(\omega) = \sum_{n=-\infty}^{\infty} x(nT)\mathcal{F}\{\delta(t-nT)\} = \sum_{n=-\infty}^{\infty} x(nT)e^{-j\omega nT},$$

where $\mathcal{F}\{.\}$ is the FT operator. If we let $x(n) = x(nT)$ and define $\Omega \triangleq \omega T$ to represent the digital frequency, we then obtain:

$$X(e^{j\Omega}) = \sum_{n=-\infty}^{\infty} x(n)e^{-j\Omega n} = \text{DTFT}\{x(n)\}, \tag{2.8}$$

where $e^{-j\Omega n}, \forall n \in \mathbb{Z}$ are the *basis functions* of DTFT that trace the unit circle when $0 \leq \Omega \leq 2\pi$.

2.1.5.1 Properties of DTFT

1. Recall that the complex exponential $e^{-j\Omega n}$ is periodic in Ω with period 2π (see Section 2.1.1). Therefore, DTFT will also be periodic with period 2π. This is the consequence of sampling in the time domain.
2. The basis functions $e^{-j\Omega n}$ are DTFT$\{\delta(k-n)\} = e^{-j\Omega n}$. These two signals are orthonormal both in time and frequency domains, i.e. let $d_n(k) = \delta(k-n)$ and $D_n(e^{j\Omega}) = e^{-j\Omega n}$ then:

$$< d_m(k), d_n(k) > = \sum_{k=-\infty}^{\infty} d_m(k) d_n(k) = \delta(m-n) \quad \text{and}$$

$$< D_m(e^{j\Omega}), D_n(e^{j\Omega}) > = \frac{1}{2\pi}\int_{-\pi}^{\pi} D_m(e^{j\Omega}) D_n^*(e^{j\Omega}) d\Omega = \delta(m-n),$$

where $< .,. >$ stands for the inner product operation and for the discrete case $< a, b >= \sum_m a_m b_m^*$ while for the continuous case $< a, b >= \int a(e^{j\Omega}) b^*(e^{j\Omega}) d\Omega$.

3. Applying the orthonormality property to (2.8), we obtain the inverse DTFT (or IDTFT) as:

$$\frac{1}{2\pi}\int_{-\pi}^{\pi} X(e^{j\Omega}) D_n^*(e^{j\Omega}) d\Omega = \sum_m x(m) \frac{1}{2\pi}\int_{-\pi}^{\pi} D_m(e^{j\Omega}) D_n^*(e^{j\Omega}) d\Omega$$
$$= \sum_m x(m)\delta(m-n) = x(n).$$

That is, signal $x(n)$ can be recovered from $X(e^{j\Omega})$ using IDTFT:

$$x(n) = \frac{1}{2\pi}\int_{-\pi}^{\pi} X(e^{j\Omega}) e^{j\Omega n} d\Omega \tag{2.9}$$

4. DTFT can also be obtained by restricting the z-plane to the unit circle, i.e.

$$X(z)\Big|_{z=e^{j\Omega}} = \sum_{n=-\infty}^{\infty} x(n) e^{-j\Omega n}$$

Example 2.5 *Consider a rectangular pulse signal* $x(n) = \begin{cases} 1, & |n| \le N \\ 0, & \text{elsewhere} \end{cases}$

Find $X(e^{j\Omega})$ *and discuss its properties.*

Solution: Using the DTFT expression, we can write:

$$X(e^{j\Omega}) = \sum_{n=-\infty}^{\infty} x(n) e^{-j\Omega n} = \sum_{n=-N}^{N} e^{-j\Omega n} = \sum_{n=0}^{N} e^{-j\Omega n} + \sum_{n=0}^{N} e^{j\Omega n} - 1$$
$$= \frac{1 - e^{-j\Omega(N+1)}}{1 - e^{-j\Omega}} + \frac{1 - e^{j\Omega(N+1)}}{1 - e^{j\Omega}} - 1 = \frac{e^{-j\Omega N} + e^{j\Omega N} - e^{-j\Omega(N+1)} - e^{j\Omega(N+1)}}{2 - e^{-j\Omega} - e^{j\Omega}}$$
$$= \frac{2\cos(\Omega N) - 2\cos(\Omega(N+1))}{2 - 2\cos(\Omega)} = \frac{\sin(\Omega(N+1/2))\sin(\Omega/2)}{\sin^2(\Omega/2)} = \frac{\sin(\Omega(N+1/2))}{\sin(\Omega/2)}$$

This is the *aliased* version of the continuous-time sinc function. It is aliased because the continuous-time sinc function is not band-limited and hence sampling in the time domain causes aliasing. This aliased function is periodic with period 2π and at $\Omega = 0$ we have $X(e^{j\Omega})|_{\Omega=0} = 2N + 1$. Figure 2.3 shows the plot of this function for $-2\pi \le \Omega \le 2\pi$.

Example 2.6 *Find the response of an FIR system* $y(n) = x(n) + x(n-4)$ *to input signal* $x(n) = \cos\left(\frac{\pi}{2}n\right) + \cos\left(\frac{\pi}{4}n\right)$.

Solution: Taking the DTFT of both sides gives, $Y(e^{j\Omega}) = (1 + e^{-j4\Omega})X(e^{j\Omega})$. Thus, the transfer function is:

$$H(e^{j\Omega}) = \frac{Y(e^{j\Omega})}{X(e^{j\Omega})} = 2e^{-j2\Omega}\cos 2\Omega$$

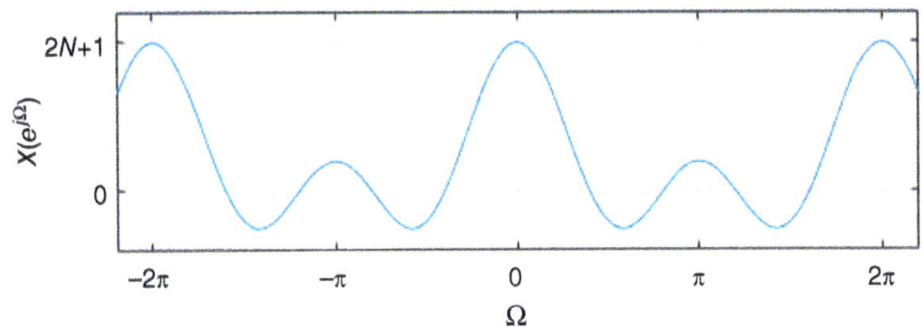

Figure 2.3 Aliased sinc function of Example 2.5.

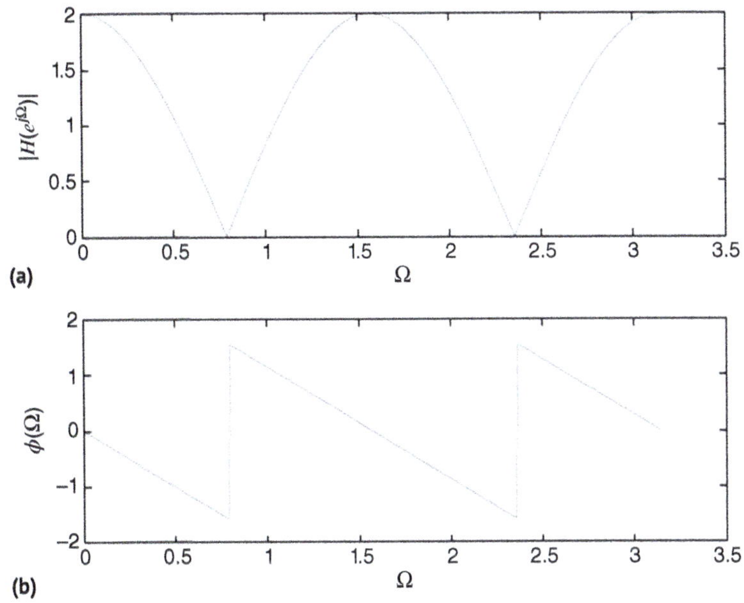

Figure 2.4 (a) Magnitude and (b) phase responses of $H(e^{j\Omega})$ in Example 2.6.

where the magnitude response is $|H(e^{j\Omega})| = 2|\cos 2\Omega|$ and the phase response is $\phi(\Omega) = -2\Omega$ for $\cos 2\Omega > 0$ or $\phi(\Omega) = -2\Omega + \pi$ for $\cos 2\Omega < 0$. That is, the phase is a linear function of Ω. Note that $|H(e^{j\frac{\pi}{2}})| = 2$ and $|H(e^{j\frac{\pi}{4}})| = 0$, therefore, the filter does not pass the signal component $\cos\left(\frac{\pi}{4}n\right)$ of $x(n)$ while passing the component $\cos\left(\frac{\pi}{2}n\right)$ without any attenuation. The plots of the magnitude and phase responses as a function of Ω are shown in Figure 2.4. Alternatively, we can apply this input to the system in the time domain and use the trig-identities to simplify the result:

$$
\begin{aligned}
y(n) &= x(n) + x(n-4) \\
&= \cos\left(\frac{\pi}{2}n\right) + \cos\left(\frac{\pi}{4}n\right) + \cos\left(\frac{\pi}{2}(n-4)\right) + \cos\left(\frac{\pi}{4}(n-4)\right) = 2\cos\left(\frac{\pi}{2}n\right)
\end{aligned}
$$

2.1.6 Frequency Domain Analysis – Discrete Fourier Transform

The DTFT described in the previous section is primarily an analytical tool and is unsuitable for practical DSP. Its frequency variable, Ω, is a continuous variable, making it impractical to implement on digital computers. In contrast, the Discrete Fourier Transform (DFT) [2, 3] enables Fourier representation with discrete frequency variables, which means sampling is performed in both the time and frequency domains.

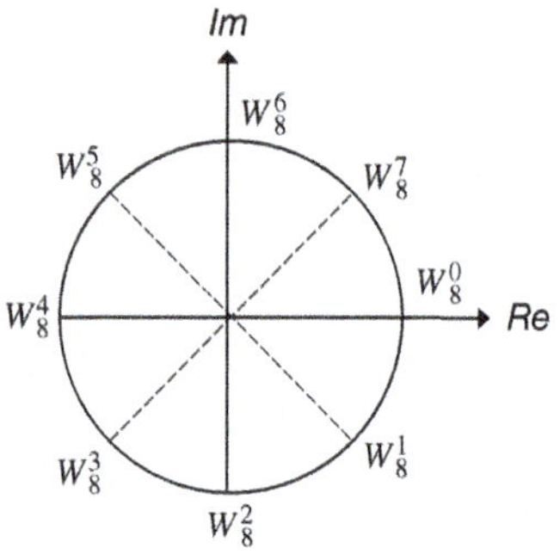

Figure 2.5 Basis functions of DFT are equally spaced samples around the unit circle ($N = 8$).

Consider a finite-duration sequence $x(n)$ of length N, i.e. $x(n) = 0 \;\; n < 0, \;\; n > N-1$. The DTFT of this sequence is:

$$X(e^{j\Omega}) = \sum_{n=0}^{N-1} x(n)e^{-j\Omega n}$$

where $X(e^{j\Omega})$ is periodic with period 2π in Ω. Now, if sampling is applied in the frequency domain such that $\Omega = \frac{2\pi k}{N}$, $k \in [0, N-1]$ i.e. fit exactly N samples in one period we obtain:

$$X(k) = \sum_{n=0}^{N-1} x(n)e^{\frac{-2\pi jkn}{N}} = \sum_{n=0}^{N-1} x(n)W_N^{kn} = \text{DFT}\{x(n)\}, \quad k \in [0, N-1] \tag{2.10}$$

where $W_N := e^{-j\frac{2\pi}{N}}$ and $X(k)$ is the 1-D DFT of $x(n)$. Several important remarks are in order. First, the basis functions of the 1-D DFT, W_N^{kn}, are complex-valued roots of unity of order N, i.e. $W_N^N = 1$ and since they are periodic in both k and n with period N, there are only N distinct roots of unity. Second, as can be observed from (2.10), $X(k)$ is also a finite-duration sequence of size N corresponding to one period of $X(k)$. Third, the basis functions $e^{-j\frac{2\pi kn}{N}}$ represent equally spaced samples around the unit circle as shown in Figure 2.5.

Finally, the normalized basis functions of DFT $W_N^{kn}/\sqrt{N}$ form orthonormal basis, i.e. $< W_N^{kn}, W_N^{km} >= N\delta(m-n)$ since:

$$\begin{aligned} < W_N^{kn}, W_N^{km} > &= \sum_{k=0}^{N-1} W_N^{(m-n)k} \\ &= \sum_{k=0}^{N-1} e^{2\pi j(m-n)k/N} = \begin{cases} N, & m = n \\ \dfrac{1 - e^{2\pi j(m-n)}}{1 - e^{2\pi j(m-n)/N}} = 0, & m \neq n \end{cases} \\ &= N\delta(m-n). \end{aligned} \tag{2.11}$$

Using the orthonormality property, it is easy to show that $x(n)$ can be recovered from its DFT $X(k)$ via the inverse DFT given by:

$$x(n) = \frac{1}{N}\sum_{k=0}^{N-1} X(k)W_N^{-kn} = \text{IDFT}\{X(k)\}, \quad n \in [0, N-1] \tag{2.12}$$

There exist fast algorithms, known as FFT [2, 3], for computing of DFT and IDFT of finite-duration sequences.

Example 2.7 *Find DFT of the signals : (a)* $x(n) = \delta(n - n_0)$, *(b)* $x(n) = a^n, n \in [0, N-1]$, *(c)* $x(n) = e^{j\frac{2\pi}{N}k_0 n}$, $n \in [0, N-1]$, *(d)* $x(n) = \cos\left(\frac{2\pi}{N}k_0 n\right), n \in [0, N-1]$, *and (e)* $x(n) = \begin{cases} 1, & n \text{ even} \\ 0, & n \text{ odd} \end{cases} \quad n \in [0, N-1]$.

Solution:

a. Using property of the Kronecker delta function:

$$X(k) = \sum_{n=0}^{N-1} \delta(n - n_0)e^{-j\frac{2\pi}{N}kn} = e^{-j\frac{2\pi}{N}kn_0} \text{ for } k \in [0, N-1]$$

b. Using series formula $\sum_{n=0}^{N-1} \alpha^n = \frac{1-\alpha^N}{1-\alpha}$, we have:

$$X(k) = \sum_{n=0}^{N-1} a^n e^{-j\frac{2\pi}{N}kn} = \sum_{n=0}^{N-1} \left(ae^{-j\frac{2\pi}{N}k}\right)^n = \frac{1 - a^N}{1 - ae^{-j\frac{2\pi}{N}k}}$$

c. Dual property of (a) gives:

$$X(k) = \sum_{n=0}^{N-1} e^{j\frac{2\pi}{N}k_0 n} e^{-j\frac{2\pi}{N}kn} = N\delta(k - k_0)$$

d. Using Euler formula and result of Part (c) we can write:

$$X(k) = \sum_{n=0}^{N-1} \frac{1}{2}\left[e^{j\frac{2\pi}{N}k_0 n} + e^{-j\frac{2\pi}{N}k_0 n}\right]e^{-j\frac{2\pi}{N}kn} = \frac{N}{2}[\delta(k - k_0) + \delta(k - N + k_0)]$$

e. Assuming N is odd and expanding the DFT, we get:

$$X(k) = \sum_{n=0}^{N-1} x(n)e^{-j\frac{2\pi}{N}kn} = 1 + e^{-j\frac{2\pi}{N}2k} + e^{-j\frac{2\pi}{N}4k} + \ldots + e^{-j\frac{2\pi}{N}(N-1)k}$$

$$= \frac{1 - e^{-j\frac{2\pi}{N}2k\frac{N+1}{2}}}{1 - e^{-j\frac{2\pi}{N}2k}} = \frac{1 - e^{-j\frac{2\pi}{N}k}}{1 - e^{-j4\pi\frac{k}{N}}} = \frac{1}{1 + e^{-j\frac{2\pi}{N}k}}$$

2.1.6.1 Properties of DFT

Many of the DFT properties are similar to those of FT. However, a few properties are different, mainly because of the finite duration and periodicity in both time and frequency domains. These are discussed in this section.

1. **Linearity:** Let $x_1(n)$ and $x_2(n)$ be two finite-duration sequences of sizes N_1 and N_2, respectively, then their linear combination $x(n) = ax_1(n) + bx_2(n)$ is of duration $N = \max[N_1, N_2]$ and hence the DFTs must be computed of size N to generate $X(k) = aX_1(k) + bX_2(k)$. For example, if $N_1 < N_2$ then $x_1(n)$ is augmented by $N_2 - N_1$ zeros, and we compute $X_1(k) = \text{DFT}_{N_2}\{x_1(n)\}$ and $X_2(k) = \text{DFT}_{N_2}\{x_2(n)\}$ and combine them to get $X(k)$.
2. **Circular Shift:** Suppose that $\tilde{x}(n)$ is the periodic extension of the finite-duration signal $x(n)$ and $\tilde{x}_1(n) = \tilde{x}(n + m)$ is the linearly shifted version of $\tilde{x}(n)$ by m samples, then due to the shifting property we have $\tilde{X}_1(k) = W_N^{-km}\tilde{X}(k)$. Now, the shifted version of $x(n)$ is obtained by extracting a period of $\tilde{x}(n + m)$ in the range $0 \leq n \leq N - 1$, i.e. $x_1(n) = \tilde{x}(n + m)R_N(n)$

   $$\text{where } R_N(n) = \begin{cases} 1, & 0 \leq n \leq N-1 \\ 0, & \text{elsewhere.} \end{cases}$$

   Note that while $\tilde{x}_1(n)$ is linearly shifted version of the periodic sequence $\tilde{x}(n)$, $x_1(n)$ is obtained by a rotation of $x(n)$ signal samples around a cylinder or a *circular shift* [2, 3]. For this circularly shifted signal we have $X_1(k) = W_N^{-km}X(k)$ with $X(k) = \tilde{X}(k)R_N(k)$. That is, the shifting property of DFT applies to the circularly shifted version of the finite-duration signal.

   Due to the duality between the time and frequency domains, a similar result holds when a circular shift is applied to the DFT coefficients.
3. **Circular Convolution:** Consider two finite-duration sequences $x(n)$ and $h(n)$ with DFTs that are $X(k)$ and $H(k)$. We wish to determine $\bar{y}(n) = \text{IDFT}\{X(k)H(k)\}$. Specifically, $\bar{y}(n)$ corresponds to one period of $\tilde{y}(n)$ which is the result of linear convolution of the periodic extensions of $x(n)$ and $h(n)$ i.e.

$$\bar{y}(n) = \left[\sum_{m=0}^{N-1} \tilde{x}(m)\tilde{h}(n-m)\right]R_N(n) = x(n) \circledast h(n)$$

where $\circledast$ represents circular convolution operation. Thus, multiplying the DFTs of two finite-duration sequences and then taking the IDFT of the result is equivalent to *circularly convolving* the sequences or linear convolution of their periodic extensions and then extracting the region between $0 \leq n \leq N-1$ by multiplying by $R_N(n)$. The result of this circular convolution, $\bar{y}(n)$, substantially differs from the desired linear convolution result, $y(n)$. The next example illustrates this discrepancy.

Example 2.8 *Consider finite-extent signals* $x(n) = \{1, 2\}$ *and* $h(n) = \{-1, 1\}$ *with their first element corresponding to* $n = 0$. *Both signals are zero outside their respective intervals. Determine:*

a. *linear convolution* $y(n) = x(n) * h(n)$
b. *circular convolution* $\bar{y}(n) = \text{IDFT}\{X(k)H(k)\} = x(n) \circledast h(n)$.

Solution:

a. It is easy to verify that the linear convolution result is $y(n) = \{-1, -1, 2\}$ where the length of $y(n)$ is $M + N - 1 = 3$ with $M = N = 2$ being the sizes of the input and impulse response sequences.
b. Taking the DFTs of $x(n)$

$$X(k) = \sum_{n=0}^{1} x(n)e^{-j\pi kn} = x(0) + x(1)e^{-j\pi k},$$

which yields $X(k) = \{3, -1\}$. Similarly, the DFT of $h(n)$ gives $H(k) = \{0, -2\}$. Thus, we get $\bar{Y}(k) = H(k)X(k) = \{0, 2\}$. Taking the IDFT of $\bar{Y}(k)$ yields $\bar{y}(n) = \frac{1}{2}\sum_{k=0}^{1} \bar{Y}(k)e^{j\pi kn} = \frac{1}{2}(\bar{Y}(0) + \bar{Y}(1)e^{j\pi n})$ or $\bar{y}(n) = \{1, -1\}$.

The discrepancies may be easily seen between these two results $y(n)$ and $\bar{y}(n)$. It is interesting to note that wrapping the sequence $y(n)$ around itself and adding the overlapping first and last terms, we obtain $\bar{y}(n)$ i.e. $\bar{y}(n) = \{-1 + 2, -1\} = \{1, -1\}$. This implies that circular convolution produces the aliased version of the linear convolution result and hence contains wraparound errors.

We will now answer the question of how the DFT can be utilized to produce the desired linear convolution result.

2.1.6.2 Computing Linear Convolution Using DFT

To perform linear convolution using the DFTs of a finite-duration input signal $x(n)$ of length M and an FIR filter with impulse response $h(n)$ of length N (usually $N \ll M$), the following steps are used:

1. Extend both the input signal $x(n)$ and the impulse response $h(n)$ by appending zeros to their tails. The length of the zero-padded sequences must be $L \geq M + N - 1$ where L is typically chosen to be a power of two for more efficient FFT implementation. The zero-padding ensures that the wraparound errors inherent in circular convolution using DFT are eliminated, thus producing the linear convolution result.
2. Compute the L-point DFTs of the zero-padded input signal and impulse response to obtain $X(k)$ and $H(k)$, respectively. Then, multiply these frequency-domain representations point-by-point to get $\bar{Y}(k) = X(k)H(k)$ for $k \in [0, L-1]$.
3. Compute the L-point IDFT of $\bar{Y}(k)$ to obtain $\bar{y}(n) = \text{IDFT}_L\bar{Y}(k)$. The first $M + N - 1$ samples of $\bar{y}(n)$ constitute the desired linear convolution result $y(n)$.

2.2 Discrete-Time Signals and Systems-Random Case

In many DSP applications, we often encounter problems that involve random signals, e.g. channel noise in communication systems or quantization noise in digital systems. We know that r.v.s are obtained by assigning numerical values to the outcome of random events. Random processes, or random signals, on the other hand, are generated by assigning time functions to the outcome of random events. Assigning a discrete-time function, $x(n, s)$, to each outcome s of an experiment results in a discrete random process $X(n, s)$. Here, $x(n, s)$ is called a sample function of the ensemble set, $X(n, s) = \{x(n, s_1), x(n, s_2), \dots , x(n, s_N)\}$. In other words, a random process, $X(n, s)$, can take on one of many possible time functions, $x(n, s_i)$, where $i \in [1, N]$. For simplicity, we can drop the index 's' and simply use $x(n)$ as a sample function of the random process $X(n)$. An example of this is repeating an experiment of collecting speech signals N times, where the presence of noise causes a new signal, $x_i(n)$, to be recorded each time. The set $X(n) = \{x_1(n), x_2(n), \dots , x_N(n)\}$ is a random process (r.p.) representing the ensemble of these speech signals.

2.2.1 Statistical Measures on Random Processes

For a fixed $n = n_0$, $X(n_0)$ can be viewed as a r.v. implying that a r.p. is an *indexed set* of r.v.s where each r.v. can be characterized by its statistical averages. Such averages are called *Ensemble Averages*.

1. **Ensemble Mean:** Ensemble mean of a discrete r.p. $X(n)$ is:

$$\mu_X(n) = E[X(n)] = \sum_{i=1}^{N} x_i(n)P[X(n) = x_i(n)]$$

   where $x_i(n),\ i \in [1, N]$ is the ith sample function, $E[.]$ represents the expectation operation or the ensemble averaging and $P[X(n) = x_i(n)]$ is the *Probability Mass Function* (PMF) of discrete r.v. $X(n)$ at particular time instance n. In general, we have:

$$\mu_{g(X)}(n) = E[g(X(n))] = \sum_{i=1}^{N} g(x_i(n))P[X(n) = x_i(n)]$$

   where g(.) is any scalar function.
2. **Ensemble Auto-Correlation Function (ACF):** Ensemble ACF of r.p. $X(n)$ is a measure of how much $X(n)$ varies in time on the average. Let n_1 and n_2 be two instances in time, then the ACF of $X(n)$ is:

$$\begin{aligned} R_{XX}(n_1, n_2) &= E[X(n_1)X(n_2)] \\ &= \sum_{i=1}^{N}\sum_{j=1}^{M} x_i(n_1)x_j(n_2)P[X(n_1) = x_i(n_1), X(n_2) = x_j(n_2)] \end{aligned}$$

   where $P[X(n_1) = x_i(n_1), X(n_2) = x_j(n_2)]$ is the joint PMF of discrete r.v.s $X(n_1)$ and $X(n_2)$ at particular times instances n_1 and n_2.
3. **Ensemble Auto-Covariance:** Ensemble auto-covariance is a measure of spread from the mean. Let n_1 and n_2 be two instances in time, then the auto-covariance function of $X(n)$ is:

$$C_{XX}(n_1, n_2) = E[(X(n_1) - \mu_X(n_1))(X(n_2) - \mu_X(n_2))] = E[X(n_1)X(n_2)] - \mu_X(n_1)\mu_X(n_2).$$

   Thus, we have the following relation between $C_{XX}(n_1, n_2)$ and $R_{XX}(n_1, n_2)$

$$C_{XX}(n_1, n_2) = R_{XX}(n_1, n_2) - \mu_X(n_1)\mu_X(n_2).$$

4. **Cross-Correlation:** The *cross-correlation* between two r.p.s , $X(n)$ and $Y(n)$, gives a measure of their dependency and is defined as:

$$\begin{aligned} R_{XY}(n_1, n_2) &= E[X(n_1)Y(n_2)] \\ &= \sum_{i=1}^{N}\sum_{j=1}^{M} x_i(n_1)y_j(n_2)P[X(n_1) = x_i(n_1), Y(n_2) = y_j(n_2)] \end{aligned}$$

   where $P[X(n_1) = x_i(n_1), Y(n_2) = y_j(n_2)]$ is the joint PMF of r.v.s $X(n_1)$ and $Y(n_2)$ at particular times instances n_1 and n_2.

5. **Cross-Covariance:** The cross-covariance function of two r.p.s $X(n)$ and $Y(n)$ is defined as:

$$C_{XY}(n_1, n_2) = E[(X(n_1) - \mu_X(n_1))(Y(n_2) - \mu_Y(n_2))]$$
$$= R_{XY}(n_1, n_2) - \mu_X(n_1)\mu_Y(n_2).$$

Remarks:

1. Cross-correlation is not, in general, an interchangeable operation, i.e. $R_{YX}(n_1, n_2) \neq R_{XY}(n_1, n_2)$.
2. If $X(n_1)$ and $Y(n_2)$ are *statistically independent* r.v.s for $\forall n_1, n_2$ then $R_{XY}(n_1, n_2) = E[X(n_1)]E[Y(n_2)] = \mu_X(n_1)\mu_Y(n_2)$, while $C_{XY}(n_1, n_2) = 0$.
3. If $X(n_1)$ and $Y(n_2)$ are *orthogonal* for $\forall n_1, n_2$ then $R_{XY}(n_1, n_2) = 0$.

Example 2.9 *Find the mean and ACF of the pseudo r.p. $X(n) = a\cos(\Omega n) + b\sin(\Omega n)$, where a and b are two independent zero-mean normal (Gaussian) continuous r.v.s with $E[a^2] = E[b^2] = \sigma^2$, and Ω is constant.*

Solution: Clearly r.v.s $X(n_i)$ at different time instances n_i are linear combinations of two normal r.v.s a and b, thus they are jointly normal (why?) and to determine their statistics, it suffices to find the mean and ACF of $X(n)$:

$$E[X(n)] = E[a]\cos(\Omega n) + E[b]\sin(\Omega n) = 0,$$

and

$$R_{XX}(n_1, n_2) = E[X(n_1)X(n_2)]$$
$$= E[\{a\cos(\Omega n_1) + b\sin(\Omega n_1)\}\{a\cos(\Omega n_2) + b\sin(\Omega n_2)\}]$$
$$= E[a^2]\cos(\Omega n_1)\cos(\Omega n_2) + E[b^2]\sin(\Omega n_1)\sin(\Omega n_2)$$
$$= \sigma^2\cos(\Omega n_1)\cos(\Omega n_2) + \sigma^2\sin(\Omega n_1)\sin(\Omega n_2)$$
$$= \sigma^2\cos(\Omega(n_1 - n_2)),$$

where the cross product terms are zero because $E[ab] = 0$. This is due to statistical independence of r.v.s a and b and their zero-mean values.

2.2.2 Wide Sense Stationary Random Processes

The ensemble statistics defined earlier require calculating all first- and second-order statistics at every point in time. Clearly, this is too computationally intensive and impractical for defining a r.p. This requirement can be relaxed by assuming wide-sense stationarity (WSS) defined next.

Definition 2.3 A r.p. $X(n)$ is WSS if

$$E[X(n)] = \mu_X = \text{constant}$$
$$E[X(n_1)X(n_2)] = R_{XX}(n_2 - n_1),$$

or $E[X(n)X(n+m)] = R_{XX}(m)$ i.e. $R_{XX}(m)$ dependent only on lag m and not on time n. That is, for a WSS r.p., the 1st and 2nd order moments are independent of time. In this case, the auto-covariance function becomes:

$$C_{XX}(m) = E[(X(n) - \mu_X)(X(n+m) - \mu_X)] = R_{XX}(m) - \mu_X^2$$

A WSS r.p. $X(n)$ is said to be *white* if $C_{XX}(m) = \sigma_X^2\delta(m)$. This is in analogy to *white light*, in which all frequencies are present (see Section 2.2.4).

2.2.2.1 Time-Domain Properties of WSS Processes

For WSS r.p.s we have the following time-domain properties:

1. Recall that $R_{XX}(m) = E[X(n)X(n+m)]$. Thus, we have:

$$R_{XX}(0) = E[X^2(n)] \text{ mean squared (MS) value}$$

$$C_{XX}(0) = E[(X(n) - \mu_X)^2] = \sigma_X^2 \text{ variance}$$

$$= R_{XX}(0) - \mu_X^2$$

2. Even symmetry, $R_{XX}(m) = R_{XX}(-m)$, since $E[X(n)X(n+m)] = E[X(n+m)X(n)]$
3. The ACF is a decreasing function of m and bounded by $R_{XX}(0)$, that is $|R_{XX}(m)| \leq R_{XX}(0)$
4. If $X(n)$ and $Y(n)$ are WSS and *jointly* WSS r.p.s, then:

$$R_{XY}(n_1, n_2) = R_{XY}(n_2 - n_1)$$

$$C_{XY}(n_1, n_2) = C_{XY}(n_2 - n_1)$$

   For statistically independent WSS r.p.s, $R_{XY}(m) = \mu_X \mu_Y$ and $C_{XY}(m) = 0$.
5. For WSS r.p.s, $X(n)$ and $Y(n)$, we also have $R_{XY}(m) = R_{YX}(-m)$ and $|R_{XY}(m)| \leq [R_{XX}(0)R_{YY}(0)]^{\frac{1}{2}}$ and $|C_{XY}(m)| \leq [C_{XX}(0)C_{YY}(0)]^{\frac{1}{2}}$
6. Many r.p.s become less correlated as they are more separated in time, that is:

$$\lim_{m\to\infty} R_{XX}(m) = \mu_X^2 \quad \text{and} \quad \lim_{m\to\infty} C_{XX}(m) = 0$$

$$\lim_{m\to\infty} R_{XY}(m) = \mu_X \mu_Y \quad \text{and} \quad \lim_{m\to\infty} C_{XY}(m) = 0$$

Example 2.10 *Given statistically independent zero-mean r.p.s $X(n)$ and $Y(n)$ with ACF $R_{XX}(m) = e^{-|m|}$ and $R_{YY}(m) = \cos(2\pi m)$, find: (a) $R_{W_1W_1}(m)$ of $W_1(n) = X(n) + Y(n)$, (b) $R_{W_2W_2}(m)$ of $W_2(n) = X(n) - Y(n)$, and (c) $R_{W_1W_2}(m)$.*

Solution: For Part (a), since $X(n)$ and $Y(n)$ are zero-mean statistically independent r.p.s, we have:

$$R_{W_1W_1}(m) = E[W_1(n)W_1(n+m)] = E[(X(n) + Y(n))(X(n+m) + Y(n+m))]$$

$$= E[X(n)X(n+m)] + E[X(n)Y(n+m)]$$

$$= R_{XX}(m) + R_{YY}(m) = e^{-|m|} + \cos(2\pi m)$$

where $E[X(n)Y(n+m)] = E[Y(n)X(n+m)] = 0$ due to independence.

Using a similar approach for Parts (b) and (c) we get:

$$R_{W_2W_2}(m) = E[W_2(n)W_2(n+m)] = E[(X(n) - Y(n))(X(n+m) - Y(n+m))]$$

$$= R_{XX}(m) + R_{YY}(m) = e^{-|m|} + \cos(2\pi m),$$

and,

$$R_{W_1W_2}(m) = E[W_1(n)W_2(n+m)] = E[(X(n) + Y(n))(X(n+m) - Y(n+m))]$$

$$= E[X(n)X(n+m)] - E[Y(n)Y(n+m)]$$

$$= R_{XX}(m) - R_{YY}(m) = e^{-|m|} - \cos(2\pi m)$$

2.2.3 Time Averages and Ergodicity

The WSS assumption significantly simplifies the statistical representation of r.p.s. However, it still requires many sample functions in the ensemble set to average over. In practice, we would like to deal with only a single signal rather than an ensemble set of signals. That is, we may wish to infer certain averages on the r.p. from the measurements on a single member of the ensemble set. These are known as *time averages*.

Given one signal $x(n)$, its time average mean and ACF are given by:

$$\bar{x} = \langle X(n) \rangle = \lim_{N \to \infty} \frac{1}{2N+1} \sum_{n=-N}^{N} x(n)$$

$$r_{XX}(m) = \langle X(n)X(n+m) \rangle = \lim_{N \to \infty} \frac{1}{2N+1} \sum_{n=-N}^{N} x(n)x(n+m),$$

respectively. If these time averages are computed for each sample function within the ensemble set, these values $\bar{x}$, $r_{XX}(m)$ form r.v.s themselves. Now, it is obvious (due to WSS) that taking $E[\cdot]$ or the ensemble average of these r.v.s yields:

$$E[\bar{x}] = \mu_X \quad \text{and} \quad E[r_{XX}(m)] = R_{XX}(m),$$

i.e. the ensemble averages. Now, if we assume that the deviations of these averages from one sample to another are zero or negligible, then $\bar{x} = \mu_X$ and $r_{XX}(m) = R_{XX}(m)$, that is, the time averages obtained from one signal can be used as surrogates for the ensemble averages.

Definition 2.4 A r.p. is said to be *ergodic* if the time averages computed for any sample function can be used as an approximation to the corresponding ensemble averages. This is a very restrictive form of WSS assumption.

This assumption, although not very practical, greatly simplifies the inference of the statistics, as it allows all the statistics to be measured from only one sample function.

2.2.4 Power Spectrum

Let $X(n)$ be a WSS r.p. with ACF $R_{XX}(m)$, then its Power Spectrum or Spectral Density Function (SDF) is given by:

$$S_{XX}(e^{j\Omega}) = \text{DTFT}\{R_{XX}(m)\} = \sum_{m=-\infty}^{\infty} R_{XX}(m)e^{-j\Omega m}, \tag{2.13}$$

and further using IDTFT we have:

$$R_{XX}(m) = \frac{1}{2\pi} \int_{-\pi}^{\pi} S_{XX}(e^{j\Omega})e^{j\Omega m} d\Omega. \tag{2.14}$$

Thus, there is a one-to-one correspondence between ACF and SDF, which is known as Wiener–Khintchine relation [4].

Example 2.11 *A WSS r.p. $X(n)$ has the ACF given by $R_{XX}(m) = \sigma_X^2 a^{|m|}$ where $a < 1$. Find its power spectrum.*

Solution: Taking DTFT of $R_{XX}(m)$ yields:

$$\begin{aligned}
S_{XX}(e^{j\Omega}) &= \sum_{m=-\infty}^{\infty} R_{XX}(m)e^{-j\Omega m} = \sigma_X^2 \sum_{m=-\infty}^{\infty} a^{|m|}e^{-j\Omega m} \\
&= \sigma_X^2 \sum_{m=-\infty}^{0} a^{-m}e^{-j\Omega m} + \sigma_X^2 \sum_{m=0}^{\infty} a^{m}e^{-j\Omega m} - \sigma_X^2 \\
&= \sigma_X^2 \Big[\sum_{m=0}^{\infty} a^m e^{j\Omega m} + \sum_{m=0}^{\infty} a^m e^{-j\Omega m} - 1 \Big] \\
&= \sigma_X^2 \Big[\frac{1}{1-ae^{j\Omega}} + \frac{1}{1-ae^{-j\Omega}} - 1 \Big] = \sigma_X^2 \Big[\frac{1-a^2}{1+a^2-2a\cos\Omega} \Big]
\end{aligned}$$

2.2.5 Cross Power Spectrum

Consider two r.p.s $X(n)$ and $Y(n)$ with a cross-correlation function $R_{XY}(m)$. We assume that $X(n)$ and $Y(n)$ are individually and jointly WSS. Then, the *cross power spectrum* is defined by:

$$S_{XY}(e^{j\Omega}) = \sum_{m=-\infty}^{\infty} R_{XY}(m)e^{-j\Omega m} \tag{2.15}$$

Also, using IDTFT we have:

$$R_{XY}(m) = \frac{1}{2\pi}\int_{-\pi}^{\pi} S_{XY}(e^{j\Omega})e^{j\Omega m}d\Omega. \tag{2.16}$$

Frequency-Domain Properties of WSS Processes:
For WSS r.p.s we have the following frequency-domain properties:

1. $S_{XX}(e^{j\Omega})$ is real and $S_{XX}(0) \geq 0$ since $R_{XX}(m)$ is symmetric.
2. Since $R_{XX}(m)$ is real, $S_{XX}(e^{-j\Omega}) = S_{XX}(e^{j\Omega})$, i.e. symmetrical.
3. $S_{XX}(e^{j0}) = \sum_{m=-\infty}^{\infty} R_{XX}(m)$.
4. $R_{XX}(0) = \frac{1}{2\pi}\int_{-\pi}^{\pi} S_{XX}(e^{j\Omega})d\Omega$.
5. Recall that for a white r.p., $R_{XX}(m) = \sigma_X^2\delta(m)$. Thus, its SDF is $S_{XX}(e^{j\Omega}) = \sum_{m=-\infty}^{\infty} \sigma_X^2\delta(m)e^{j\Omega m} = \sigma_X^2$, i.e. constant for all frequencies or all frequencies are present with equal contributions in analogy to white light.
6. $S_{XY}(e^{j\Omega}) = S_{YX}(e^{-j\Omega}) = S_{YX}^*(e^{j\Omega})$.
7. The real part of $S_{XY}(e^{j\Omega})$ or $S_{YX}(e^{j\Omega})$ has even symmetry while the imaginary part has odd symmetry

$$\begin{aligned}
\mathrm{Re}\{S_{XY}(e^{j\Omega})\} = \mathrm{Re}\{S_{XY}(e^{-j\Omega})\} &= \mathrm{Re}\{S_{YX}(e^{j\Omega})\} = \mathrm{Re}\{S_{YX}(e^{-j\Omega})\} \\
\mathrm{Im}\{S_{XY}(e^{j\Omega})\} = -\mathrm{Im}\{S_{XY}(e^{-j\Omega})\} &= \mathrm{Im}\{S_{YX}(e^{-j\Omega})\} = -\mathrm{Im}\{S_{YX}(e^{j\Omega})\}
\end{aligned}$$

8. If $X(n)$ and $Y(n)$ are orthogonal, then $S_{XY}(e^{-j\Omega}) = S_{YX}(e^{-j\Omega}) = 0$.
9. $R_{XY}(0) = \frac{1}{2\pi}\int_{-\pi}^{\pi} S_{XY}(e^{j\Omega})d\Omega$.

2.2.6 Response of LTI Systems to Random Inputs

Consider a causal stable LSI system with impulse response $h(n)$ and frequency response $H(e^{j\Omega})$ (or transfer function $H(z)$). The input signal is a WSS r.p. $X(n)$ with mean μ_X and ACF $R_{XX}(m)$. Clearly, the output of this system, $Y(n)$, is also a r.p. and we would like to determine its statistical properties both in time- and frequency-domains. More specifically, given the mean, μ_X, and ACF, $R_{XX}(m)$, of $X(n)$, we want to determine the statistics of $Y(n)$.

To find the mean of $Y(n)$, we take expectation of both sides of the convolution summation:

$$Y(n) = \sum_{k=0}^{\infty} X(k)h(n-k) = \sum_{k=0}^{\infty} h(k)X(n-k),$$

which yields:

$$\mu_Y = E[Y(k)] = \sum_{k=0}^{\infty} h(k)E[X(n-k)] = \mu_X\sum_{k=0}^{\infty} h(k) = \mu_X H(e^{j0}) \tag{2.17}$$

i.e. mean of $Y(n)$ is constant.

Next, we find the cross-correlation and cross-power spectrum between $X(n)$ and $Y(n)$ that will be needed to find the ACF and power spectrum of $Y(n)$. Using convolution sum, cross-correlation function $R_{XY}(m)$ is given by:

$$\begin{aligned} R_{XY}(m) &= E[X(n)Y(n+m)] \\ &= E[X(n)\sum_{k=0}^{\infty} h(k)X(n+m-k)] \\ &= \sum_{k=0}^{\infty} h(k)E[X(n)X(n+m-k)] \\ &= \sum_{k=0}^{\infty} h(k)R_{XX}(m-k) = h(m) * R_{XX}(m). \end{aligned} \tag{2.18}$$

This result implies that the cross-correlation $R_{XY}(m)$ can be obtained by linear convolution of the impulse response with the ACF of $X(n)$. Taking the DTFT of (2.18) gives the cross power spectrum as:

$$S_{XY}(e^{j\Omega}) = H(e^{j\Omega})S_{XX}(e^{j\Omega}) \tag{2.19}$$

or in z-domain $S_{XY}(z) = H(z)S_{XX}(z)$ where $H(z)$ is the transfer function of the system. Now, using the above result, we can easily find the expressions for the ACF and power spectrum of $Y(n)$ using:

$$\begin{aligned} R_{YY}(m) &= E[Y(n)Y(n+m)] \\ &= E[Y(n)\sum_{k=0}^{\infty} h(k)X(n+m-k)] \\ &= \sum_{k=0}^{\infty} h(k)E[Y(n)X(n+m-k)] \\ &= \sum_{k=0}^{\infty} h(k)R_{XY}(-m+k) = h(m) * R_{XY}(-m) \end{aligned} \tag{2.20}$$

or

$$R_{YY}(m) = h(m) * h(-m) * R_{XX}(m) \tag{2.21}$$

since $R_{XX}(-m) = R_{XX}(m)$ due to WSS symmetry property. Finally, taking DTFT of both sides of (2.21), gives the expression for the power spectrum of $Y(n)$:

$$S_{YY}(e^{j\Omega}) = H(e^{j\Omega})H^*(e^{j\Omega})S_{XX}(e^{j\Omega}) = |H(e^{j\Omega})|^2 S_{XX}(e^{j\Omega}) \tag{2.22}$$

or in the z-domain $S_{YY}(z) = H(z)H(z^{-1})S_{XX}(z)$.

The analysis here reveals several key observations. First, since the first- and second-order statistics, $R_{YY}(m)$ and $R_{XY}(m)$, depend only on the time lag m, the output $Y(n)$ is individually and jointly WSS with the input $X(n)$. If the input $X(n)$ is a white process with ACF $R_{XX}(m) = \sigma_X^2\delta(m)$ and a flat power spectrum $S_{XX}(e^{j\Omega}) = \sigma_X^2$, using (2.22) the output power spectrum will be $S_{YY}(e^{j\Omega}) = \sigma_X^2|H(e^{j\Omega})|^2$. Now, because the frequency response, $H(e^{j\Omega})$, is frequency-dependent, the output power spectrum, $S_{YY}(e^{j\Omega})$, is no longer constant. Therefore, the output signal is a *colored* process.

Example 2.12 *A white WSS random process $X(n)$ with uniform PDF (in amplitude),*

$$p_X(x) = \begin{cases} 1, & -\frac{1}{2} \le x \le \frac{1}{2} \\ 0, & \text{elsewhere} \end{cases}$$

is applied to a digital filter with structure shown in Figure 2.6. Find the power spectrum of $Y(n)$.

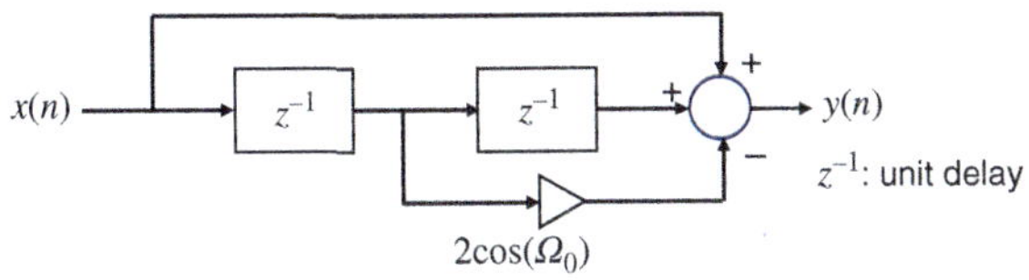

Figure 2.6 Digital filter realization for Example 2.12.

Solution: Since the input is a white process, we have $S_{YY}(e^{j\Omega}) = \sigma_X^2|H(e^{j\Omega})|^2$. Therefore, we need to find σ_X^2 and $H(e^{j\Omega})$. To find σ_X^2, we use the PDF $p_X(x)$:

$$\mu_X = E[X(n)] = \int_{-\frac{1}{2}}^{\frac{1}{2}} x p_X(x)dx = 0 \quad \text{and} \quad \sigma_X^2 = E[X^2(n)] = \int_{-\frac{1}{2}}^{\frac{1}{2}} x^2 p_X(x)dx = \frac{1}{12}$$

From digital filter structure in Figure 2.6, the input–output relation for this FIR filter is:

$$y(n) = x(n-2) - 2\cos(\Omega_0)x(n-1) + x(n)$$

Taking DTFT of this equation yields:

$$H(e^{j\Omega}) = \frac{Y(e^{j\Omega})}{X(e^{j\Omega})} = e^{-j2\Omega} - 2e^{-j\Omega}\cos(\Omega_0) + 1$$

Using this result in the expression for power spectrum of $Y(n)$, we get:

$$\begin{aligned} S_{YY}(e^{j\Omega}) &= \sigma_X^2 H(e^{j\Omega})H(e^{-j\Omega}) \\ &= \frac{1}{12}\left(e^{j2\Omega} + e^{-j2\Omega} + 4\cos^2(\Omega_0) - 4e^{j\Omega}\cos(\Omega_0) - 4e^{-j\Omega}\cos(\Omega_0) + 2\right) \\ &= \frac{1}{12}\left(4\cos^2(\Omega) + 4\cos^2(\Omega_0) - 8\cos(\Omega)\cos(\Omega_0)\right) = \frac{1}{3}(\cos(\Omega) - \cos(\Omega_0))^2 \end{aligned}$$

2.2.7 Spectral Factorization

Recall that the relationship between the output power spectral density, $S_{YY}(z)$, and the input power spectral density, $S_{XX}(z)$, of a LSI system with transfer function $H(z)$ is given in the z-domain as:

$$S_{YY}(z) = H(z)H(z^{-1})S_{XX}(z). \tag{2.23}$$

Spectral factorization is the process of decomposing $S_{YY}(z)$, to determine the system's transfer function, $H(z)$, and the input power spectral density, $S_{XX}(z)$, that produced the output $Y(n)$. Due to the term $H(z)H(z^{-1})$ in (2.23), poles and zeros of $S_{YY}(z)$ occur in reciprocal pairs, that is, if z_i is a pole or zero, then $1/z_i$ will also be a pole or zero of $S_{YY}(z)$. This reciprocal pairing means that a given $S_{YY}(z)$ could correspond to multiple different transfer functions. The idea of spectral factorization is to group all the poles and zeros of $S_{YY}(z)$ in such a way that $H(z)$ captures all those poles and zeros that are within the unit circle, hence $H(z^{-1})$ will contain all the poles and zeros that are outside the unit circle. This process leads to a *minimum-phase* [2, 3] transfer function $H(z)$ that is proper (causal) and stable with poles and zeros that lie inside the unit circle in the z-plane.

Example 2.13 *Consider the power spectrum of a r.p. given by* $S_{YY}(z) = \frac{1}{7/4 + 1/2(z+z^{-1})}$. *Find: (a) the transfer function and difference equation of an LSI system that produced* $Y(n)$ *as its output, and (b) the statistics of the input signal. Is the input a white r.p.?*

Solution: This SDF can be decomposed as:

$$S_{YY}(z) = \frac{1}{(1/2z^{-1} + 1)(1/2z + 1)}$$

where the poles are $z = -1/2$ and $z = -2$. Applying the spectral factorization in (2.23) to $S_{YY}(z)$, gives the system transfer function $H(z) = \frac{1}{(1/2z^{-1}+1)}$ and the input power spectrum $S_{XX}(z) = 1$, which implies that the input is a white process with unity variance. To find the difference equation, we use:

$$H(z) = \frac{Y(z)}{X(z)} = \frac{1}{(1/2z^{-1} + 1)}, \quad \text{or} \quad (1/2z^{-1} + 1)Y(z) = X(z).$$

Taking the inverse z-transform of this equation yields the difference equation:

$$y(n) = -0.5y(n-1) + x(n),$$

which represents a first-order IIR system.

2.3 Problems

Problem 2.1: For each of the following systems find the impulse response and determine whether or not the system is linear, shift invariant, FIR, or IIR. Determine the BIBO stability of these systems.

a. $y(n) = a\,x(n) + b$,
b. $y(n) = n^2 x(n)$,
c. $y(n) = x(n^2)$,
d. $y(n) = \sum_{k=-\infty}^{n} x(k)$,
e. $y(n) = \sum_{k=n}^{\infty} x(k)$,
f. $y(n) = x(n) \sum_{k=-\infty}^{\infty} \delta(n-k)$,
g. $y(n) = \exp\{-|x(n)|\}$.

Problem 2.2: In Section 2.1.2, we used the unit pulse property in (2.1) to arrive at the linear convolution summation.

a. Using a similar approach represents an arbitrary signal $x(n)$ in terms of a linear combination of delayed unit steps.
b. Using the result of part (a) and assuming that the output of the LSI system to a unit step is denoted by $s(n)$, derive the input/output relationship in terms of $s(n)$.
c. What is your conclusion about the result in part (b)? Is this relationship interchangeable with respect to x and s similar to linear convolution sum?

Problem 2.3: Given a discrete-time signal $x(n)$, its even and odd parts can be found using $x_e(n) = \frac{x(n)+x(-n)}{2}$ and $x_o(n) = \frac{x(n)-x(-n)}{2}$, respectively, where we have $x_e(-n) = x_e(n)$ and $x_o(-n) = -x_o(n)$ and further $x(n) = x_e(n) + x_o(n)$.

a. Show that their inner product is zero, i.e. $< x_e(n), x_o(n) >= \sum_{k=-\infty}^{\infty} x_e(k)x_o(k) = 0$.
b. Using the result of part (a), prove $E_x = E_{x_e} + E_{x_o}$ where $E_x = \sum_{n=-\infty}^{\infty} x^2(n)$ is the total energy of the signal $x(n)$ and E_{x_e} and E_{x_o} are the total energies in the even and odd components, respectively.

Problem 2.4: The impulse response of a causal LSI system is $h(n) = a^n u(n)$, where $u(n)$ represents the unit step function.

a. Determine the type of system, i.e. FIR or IIR.
b. Find the unit step response of the system.
c. Using the result of part (b), find the system response to $x(n) = u(n) - u(n-1)$.

Problem 2.5: The impulse response of a causal LSI system is $h(n) = \frac{u(n)-u(n-N)}{N}$ where $u(n)$ represents the unit step function.

a. Determine the type of system, i.e. FIR or IIR.
b. Find the unit step response of the system.

Problem 2.6: Determine the result of linear convolution between sequences $x(n) = a^n u(n)$ where $u(n)$ represents the unit step function and the sequence:

$$h(n) = \sum_{r=-\infty}^{\infty} \delta(n - rN).$$

Problem 2.7: Let $X(e^{j\Omega})$ be the DTFT of infinite-extent signal $x(n)$. Determine the DTFT of: (a) $x(2n + 1)$, (b) $e^{j\pi n/2}x(n + 2)$, (c) $x(n)\cos(0.3\pi n)$, and (d) $x(n) * x(n - 1)$ as a function of $X(e^{j\Omega})$.

Problem 2.8: The transfer function of a first-order IIR filter is given by $H(z) = \frac{1+0.5z^{-1}}{3(1-0.5z^{-1})}$. Determine:

a. Difference equation and impulse response of the system, and
b. magnitude and phase responses.

Problem 2.9: An impulse response of an LSI system is given by $h(n) = \delta(n) + 2\delta(n - 2) + \delta(n - 4)$.

a. Determine the type of system, i.e. FIR or IIR, and the order.
b. Find the frequency response $H(e^{j\Omega})$ and plot its magnitude and phase.
c. Comment on the frequency selectivity (i.e. lowpass, highpass, or bandpass) of this system.

Problem 2.10: The difference equation of an LSI system is given by $y(n) = 0.9y(n - 1) + x(n)$.

a. Determine the type of system, i.e. FIR or IIR.
b. Find the frequency response $H(e^{j\Omega})$ and plot its magnitude and phase.
c. Comment on the frequency selectivity (i.e. lowpass, highpass, or bandpass) of this system.

Problem 2.11: Find the DFT of $x(n) = \sum_{p=0}^{N-1} y(p),\ n \in [0, N - 1]$. Assume $Y(k)$ is known.

Problem 2.12: Given an ergodic r.p. with ACF as shown in Figure 2.7, find (a) $E[X(n)]$, (b) $E[X^2(n)]$, and (c) σ_X^2.

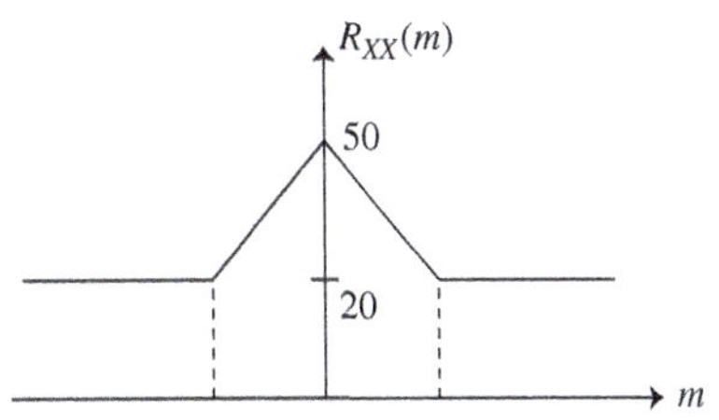

Figure 2.7 Plot of ACF for Problem 2.12.

Problem 2.13: A WSS r.p. is given by $Y(n) = a + bX(n)$ where a, b are constants. Find

a. $R_{YY}(m)$ and $R_{XY}(m)$ in terms of $R_{XX}(m)$.
b. $S_{YY}(e^{j\Omega})$ and $S_{XY}(e^{j\Omega})$ in term of $S_{XX}(e^{j\Omega})$.

Problem 2.14: Consider a r.p. given by $Y(n) = \frac{1}{2N} \sum_{k=n-N}^{n+N} X(k)$ where $X(k)$ is a WSS process with power spectrum $S_{XX}(e^{j\Omega})$.

a. Find $h(n)$ and $H(e^{j\Omega})$.
b. Find the expression for $S_{YY}(e^{j\Omega})$.

Problem 2.15: A zero-mean white r.p. $X(n)$ with variance σ_X^2 is applied to the system in Problem 2.9. Find

a. the ACF of the output $Y(n)$,
b. cross-correlation between $X(n)$ and $Y(n)$, and
c. power spectrum of $Y(n)$.

Problem 2.16: A zero-mean white r.p. $X(n)$ with variance σ_X^2 is applied to the system in Problem 2.10. Find

a. the ACF of the output $Y(n)$,
b. cross-correlation between $X(n)$ and $Y(n)$, and
c. power spectrum of $Y(n)$.

Problem 2.17: A r.p. $Z(n) = X(n) + \Gamma(n)$ consists of signal $X(n)$ and zero-mean white noise $\Gamma(n)$ with variance σ_γ^2. If the ACF of $X(n)$ is $R_{XX}(m) = \sigma_X^2 \rho^{|m|}$, find

a. ACF of $Z(n)$ assuming that signal and noise are uncorrelated,
b. power spectrum of $Z(n)$.

Problem 2.18: Repeat Example 2.12 assuming that the digital filter has a structure shown in Figure 2.8.

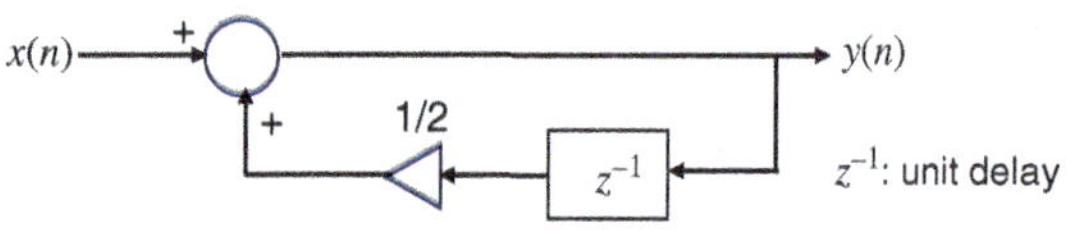

Figure 2.8 Digital filter realization for Problem 2.18.

Problem 2.19: A WSS r.p. with power spectrum $S_{YY}(e^{j\Omega}) = \frac{1}{2+2\cos\Omega}$ is applied to a *whitening* system the output of which is a white process $Z(n)$. Find

a. the frequency response $G(e^{j\Omega})$ of this *inverse* system. Is this system stable?
b. the power spectrum of the whitened output $Z(n)$.

Problem 2.20: Repeat Example 2.13 when the power spectrum of $Y(n)$ is $S_{YY}(z) = \frac{1}{1+\alpha^2+\alpha(z+z^{-1})}$ where $\alpha < 1$.

References

1 T. Matsuura, "*Digital Quantum Information Processing with Continuous Variable Systems*," Springer, 2023.
2 J. Proakis and D. Manolakis, "*Digital Signal Processing: Principles, Algorithms and Applications*," Pearson, 4th Edition, 2006.
3 A. V. Oppenheim and R. W. Schafer, "*Discrete-time Signal Processing*," Pearson, 3rd Edition, 2009.
4 A. Lean-Garcia, "*Probability, Statistics, and Random Processes for Electrical Engineers*," Prentice-Hall, 3rd Edition, 2008.

3

Theoretical Background

This chapter provides theoretical foundations for 2-D signal processing and their use in the Digital Image Processing (DIP) area. We show the extension of the 1-D signal processing tools covered in the previous chapter to the 2-D case. Although in most cases these extensions are straightforward, there are a few instances where they will not be directly applicable. Among such cases are the extension of causality and stability to 2-D systems, which differ from their 1-D counterparts in many ways [1, 2]. This chapter intends to shed light on all such straightforward and more complicated extensions and provide an explanation and understanding of how these concepts can be applied to the DIP area. These extensions also apply to both deterministic and random images, as in the previous chapter.

3.1 Discrete-Space 2-D Signals and Systems-Deterministic Case

3.1.1 Discrete-Space 2-D Signals

Images acquired by sensing systems are typically in a continuous form. A continuous image, $x(u, v)$, is represented by a function of two spatial variables (u, v), $u, v \in \mathbb{R}$, where x is the image intensity at the spatial coordinates (u, v). When the image is sampled in the spatial coordinates, it is represented by $x(m, n)$, where $m, n \in \mathbb{Z}$ represent the sampled spatial variables. For a specific location (m, n), $x(m, n)$ represents the intensity of the pixel at that position. If the intensity of the pixels is also quantized, then each pixel is assigned a value represented by B bits (typically $B = 8$ bits/pixel), resulting in a digital image. Thus, the digitization process involves two steps: sampling of spatial variables and quantization of intensity values. Figure 3.1 illustrates a digital image $x(m, n)$ of size $M \times N$ where the indices $m \in [0, M-1]$ and $n \in [0, N-1]$ correspond to rows and columns, respectively. In this representation, it is assumed that the image is scanned from left to right and top to bottom from the first pixel $x(0, 0)$ to the last pixel $x(M-1, N-1)$. Although other scanning or coordinate assignment schemes are also possible, this is the commonly used one.

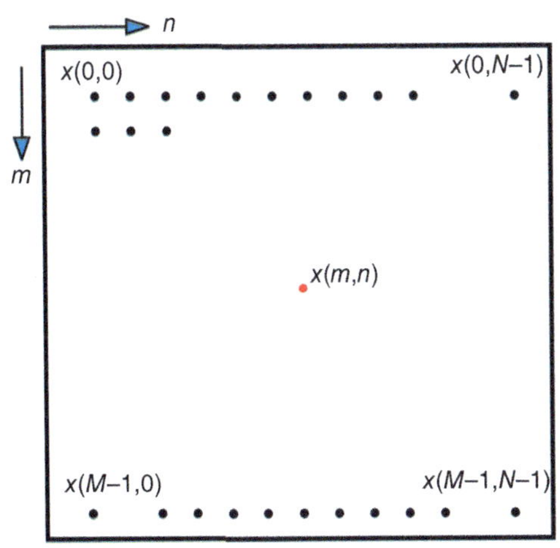

Figure 3.1 A digital image $x(m, n)$ and row and column indexing.

A. Some Important 2-D Signals

The following 2-D signals play an important role in DIP.

1. *2-D Dirac Delta:* A 2-D Dirac delta function is defined as:

$$\delta(u, v) = \begin{cases} \infty, & (u, v) = (0, 0) \\ 0, & \text{Otherwise} \end{cases} \tag{3.1}$$

This function is *separable*, i.e. $\delta(u, v) = \delta(u)\ \delta(v)$ where $\delta(u)$ and $\delta(v)$ represent lines on the v- and u-axis, respectively. Additionally, we have the following properties:

$$\begin{gathered} \iint_{-\epsilon}^{\epsilon} \delta(u, v) du\, dv = 1 \qquad \forall\epsilon > 0 \\ \iint_{-\infty}^{\infty} x(u', v')\delta(u - u', v - v')du'dv' = x(u, v) \end{gathered} \tag{3.2}$$

These properties are directly extended from their 1-D counterparts.

2. *2-D Kronecker Delta:* A 2-D Kronecker delta function is defined as:

$$\delta(m, n) = \begin{cases} 1, & (m, n) = (0, 0) \\ 0, & \text{Otherwise} \end{cases} \tag{3.3}$$

As with the Dirac delta, this function is separable, i.e. $\delta(m, n) = \delta(m)\ \delta(n)$ with $\delta(m)$ and $\delta(n)$ being infinite-extent sequences of impulses along the n- and m-axes, respectively. Similarly, we have the following properties for 2-D Kronecker delta function, i.e.

$$\begin{aligned} \sum\sum_{m,n=-\infty}^{\infty} \delta(m, n) &= 1 \\ \sum_{m'=-\infty}^{\infty}\sum_{n'=-\infty}^{\infty} x(m', n')\delta(m - m', n - n') &= x(m, n) \end{aligned} \tag{3.4}$$

The latter property implies that a 2-D sequence (or array) $x(m, n)$ can be viewed as a collection of points $x(m', n')$ located at $m' = m$ and $n' = n$. This property will be used later to develop the 2-D convolution summation.

B. 2-D Signal Types and Properties

In practice, we are primarily concerned with finite-extent images that are represented by a matrix of size $M \times N$. However, theoretically, we can have the following cases of periodic 2-D signals.

1. *Doubly Periodic 2-D Signals:* Consider 2-D sequence $x(m, n)$ which satisfies:

$$\begin{aligned} x(m, n + N) &= x(m, n) \\ x(m + M, n) &= x(m, n) \end{aligned} \tag{3.5}$$

This sequence is said to be *doubly periodic* [2] with horizontal and vertical periods M and N, respectively. Only MN samples that are within the fundamental period are distinct. Figure 3.2a shows an example [2] of a doubly periodic 2-D sequence with horizontal and vertical periods $M = 3$ and $N = 3$.

2. *General Periodic 2-D Signals:* In general, however, we can have the following definition for periodicity of 2-D sequences:

$$\begin{aligned} x(m + M_1, n + N_1) &= x(m, n) \\ x(m + M_2, n + N_2) &= x(m, n) \end{aligned} \tag{3.6}$$

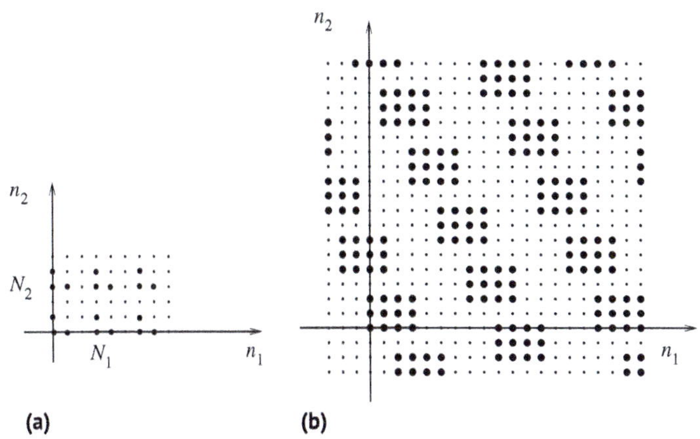

Figure 3.2 (a) A doubly periodic 2-D sequence and (b) a general periodic 2-D sequence [2].

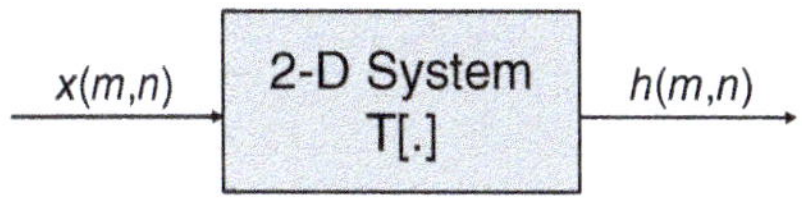

Figure 3.3 A 2-D system with input–output mapping $y(m, n) = T[x(m, n)]$.

In this case $\underline{\mu}_1 = [M_1\ N_1]^T$ and $\underline{\mu}_2 = [M_2\ N_2]^T$ represent the vector periods and $\boldsymbol{D} = \det[\underline{\mu}_1\ \underline{\mu}_2] = M_1N_2 - M_2N_1 \neq 0$ is number of distinct samples in one period where $det[\mathbf{A}]$ stands for the determinant of matrix **A**. Figure 3.2b illustrates an example [2] of a general periodic 2-D sequence with period vectors $\underline{\mu}_1 = [7, 2]^T$ and $\underline{\mu}_2 = [-2, 4]^T$.

3.1.2 Discrete-Space 2-D Systems and Properties

1. **Linearity:** The system in Figure 3.3 defined by the input–output (I/O) mapping $y(m, n) = T[x(m, n)]$, is said to be *linear* iff the operator $T[.]$ satisfies $T[ax_1(m, n) + bx_2(m, n)] = ay_1(m, n) + by_2(m, n)$ where $y_1(m, n) = T[x_1(m, n)]$ and $y_2(m, n) = T[x_2(m, n)]$ and a and b are any arbitrary constants. That is, any linear combination of two or more inputs gives the same combination of their corresponding outputs.
2. **Space Invariance:** The system defined by $y(m, n) = T[x(m, n)]$ is said to be *space invariant* iff for $y(m, n) = T[x(m, n)]$ the operator $T[.]$ satisfies $T[x(m - k, n - l)] = y(m - k, n - l), \forall k, l$.

3.1.3 Representation of 2-D LSI Systems-Convolution Summation

The response of a 2-D system to a 2-D Kronecker delta input is the 2-D impulse response, i.e. $h(m, n; k, l) := T[\delta(m - k, n - l)]$ For an imaging system, it represents the image of an ideal point source. Thus, it is also called the *point spread function (PSF)*. The PSF is real and nonnegative since the input and output of an imaging system represent positive quantities, e.g. the intensity of light. The PSFs of some imaging systems are often separable and circularly symmetric.

For a linear 2-D system, the output can be obtained from its PSF by applying the second equation in (3.4) and using the linearity property:

$$\begin{aligned} y(m, n) = T[x(m, n)] &= T[\sum_k \sum_l x(k, l)\delta(m - k, n - l)] \\ &= \sum_k \sum_l x(k, l)T[\delta(m - k, n - l)] \\ &= \sum_k \sum_l x(k, l)h(m, n; k, l). \end{aligned} \tag{3.7}$$

If the system is also space invariant then:

$$\begin{aligned} h(m, n; k, l) &= T[\delta(m - k, n - l)] \\ &= h(m - k, n - l; 0, 0) = h(m - k, n - l). \end{aligned} \tag{3.8}$$

Thus, for 2-D linear space-invariant (LSI) systems, the input–output relation in (3.7) becomes the convolution sum:

$$y(m,n) = \sum_k \sum_l x(k,l)h(m-k,n-l)$$
$$= x(m,n) ** h(m,n), \tag{3.9}$$

where ** stands for 2-D linear convolution operation.

A. Properties of Linear Convolution

1. As in the 1-D case, 2-D convolution operation has the following important properties [2] that are utilized in 2-D system theory and analysis.
   Commutativity: The order in which we convolve two 2-D signals does not change the result. That is, we have:

$$x(m,n) ** h(m,n) = \sum_k \sum_l x(k,l)h(m-k,n-l)$$
$$= \sum_k \sum_l x(m-k,n-l)h(k,l)$$
$$= h(m,n) ** x(m,n) \tag{3.10}$$

   Distributivity: 2-D linear convolution operation is distributive with respect to addition, i.e.

$$x(m,n) ** [h_1(m,n) + h_2(m,n)] = x(m,n) ** h_1(m,n) + x(m,n) ** h_2(m,n)$$

   Associativity: Grouping of the 2-D signals in the convolution operation doesn't change their linear convolution results, i.e.

$$[x(m,n) ** h(m,n)] ** g(m,n) = x(m,n) ** [h(m,n) ** g(m,n)]$$

2. The result of the linear convolution of two 2-D finite-extent arrays of sizes $(M_1 \times N_1)$ and $(M_2 \times N_2)$ is another 2-D finite-extent array of size $(M_1 + M_2 - 1) \times (N_1 + N_2 - 1)$.
3. The result of 2-D linear convolution can be verified by checking the *area conservation* property $\sum_m \sum_n y(m,n) = (\sum_m \sum_n h(m,n))(\sum_m \sum_n x(m,n))$.

The graphical representation of the 2-D convolution is depicted in Figure 3.4. Using the first 2-D convolution in (3.10), PSF is first rotated about the origin by 180° to produce $h(-k,-l)$ which is then shifted by m, n and overlayed on image $x(k,l)$. The sum of the products of the pixels of $\{x(k,l)\}$ and $\{h(m-k,n-l)\}$ which occupy a common overlap region gives the output pixel at (m,n).

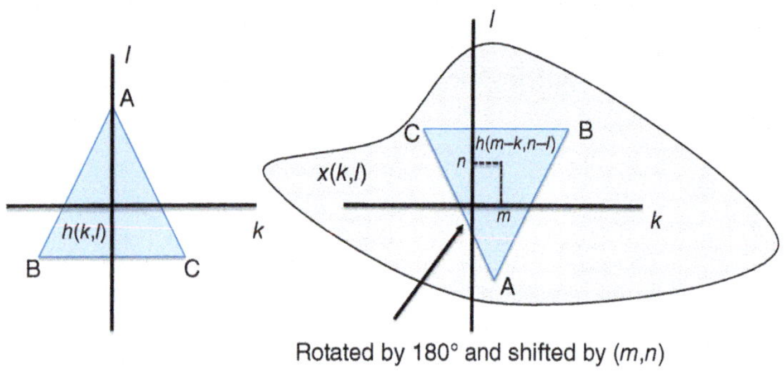

Figure 3.4 Graphical representation of 2-D convolution of $x(m,n)$ and PSF $h(m,n)$.

Example 3.1 *Using the graphical method, find the 2-D linear convolution result of the arrays $h(m, n)$ and $x(m, n)$ (shown in Figure 3.5) of sizes 2×2 and 3×2, respectively. Note that each square box in these arrays represents a pixel.*

Solution: Using the graphical representation, we rotate the PSF $h(k, l)$ by 180° and then shift it by (m, n) one pixel at a time. We then find the sum of the products of the pixels in the overlapped region for each shifted position of the PSF. Figure 3.5 shows the overlap regions for two pixels of the output at $m = 1, n = 1$ and $m = 2, n = 2$ that are indicated by red shaded pixels. The final result, which is of size 3×3 is also shown. Note that using area conservation Property 3, we can verify the final result, i.e. $\sum_m \sum_n h(m, n) = 2$, $\sum_m \sum_n x(m, n) = 16$ and $\sum_m \sum_n y(m, n) = 32$ which verifies the validity of the final result.

B. Stability of 2-D LSI Systems

Definition 3.1 [2] A 2-D system is said to be bounded-input bounded-output (BIBO) stable *iff* a bounded input results in a bounded output, i.e.

$$|x(m, n)| < M < \infty \overset{\text{BIBO}}{\Longrightarrow} |y(m, n)| < N < \infty, \qquad \forall m, n.$$

Theorem 3.1 [2] *The necessary and sufficient condition for BIBO stability of a 2-D LSI system is that the PSF must be absolutely summable, i.e.*

$$\sum_{m=-\infty}^{\infty} \sum_{n=-\infty}^{\infty} |h(m, n)| = P < \infty$$

Proof: The proof can be found in [2].

Note that unlike 1-D signals that could have no limits in time (streaming data), images are often spatially limited. Thus, from a practical viewpoint, it is possible to assess the stability of a system by evaluating the sum over the bounds of the output image and requiring this to be finite for BIBO stability.

C. Causality of 2-D LSI Systems

In processing 2-D signals or images since the independent variables are no longer *temporal* and are rather *spatial*, the causality restriction does not hold and the output can have left and right dependencies on the input image pixels. Nevertheless, when processing images in real time (e.g. raster scan or video), causality still comes to play.

We can evaluate the causality of 2-D systems using the following definitions [1, 2].

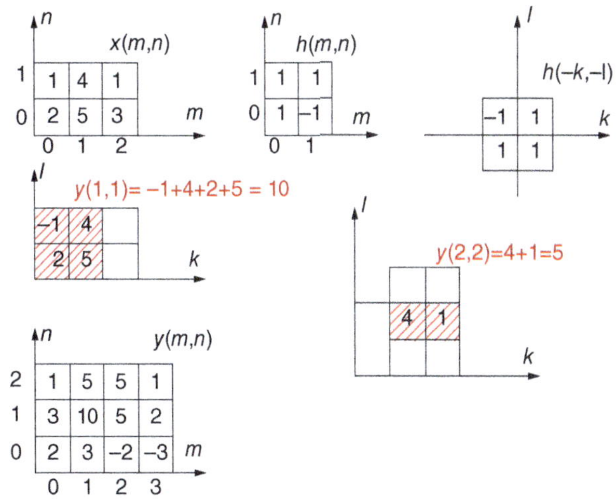

Figure 3.5 2-D linear convolution of $x(m, n)$ and PSF $h(m, n)$ arrays using the graphical method.

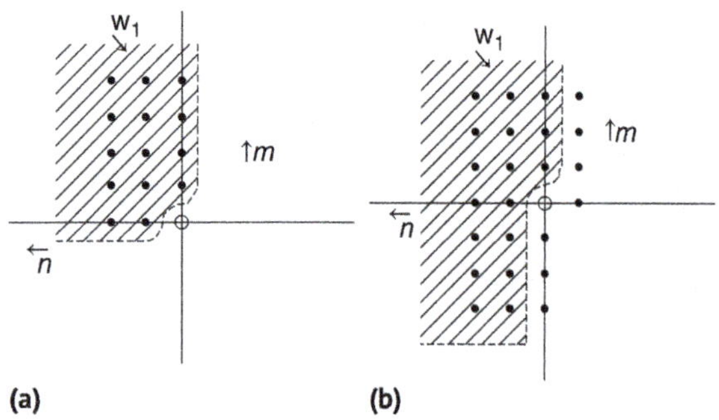

Figure 3.6 (a) Quarter plane ROS, (b) NSHP ROS.

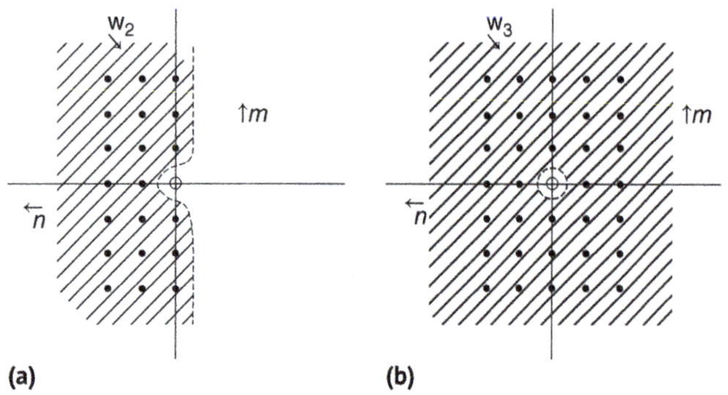

Figure 3.7 (a) Semi-causal ROS, (b) noncausal ROS.

Definition 3.2 (Strongly Causal) A 2-D PSF sequence $\{h(m, n)\}$ is said to be *strongly causal* if $h(m, n) = 0 \; \{m < 0, \forall n\} \cup \{n < 0, \forall m\}$. In this case, the region of support (ROS) of $h(m, n)$ is confined within the *quarter plane (QP)* (see Figure 3.6a).

Definition 3.3 (Weakly Causal) A 2-D sequence $\{h(m, n)\}$ is said to be *weakly causal* if $h(m, n) = 0 \; \{n < 0, \forall m\} \cup \{n = 0, m < 0\}$, i.e. *Non-symmetric Half Plane* (NSHP) ROS (see Figure 3.6b).

Definition 3.4 (Semi-Causal) A 2-D sequence $\{h(m, n)\}$ is said to be *semi-causal* if it is causal in one coordinate and noncausal in the other, i.e. $h(m, n) = 0 \; n < 0, \forall m$, i.e. *Symmetric Half Plane* (SHP) ROS (see Figure 3.7a).

Definition 3.5 (Noncausal) A 2-D sequence $\{h(m, n)\}$ is said to be *noncausal* when the ROS spans the *full plane*, i.e. exists in all four quadrants (see Figure 3.7b).

3.1.4 Representation of 2-D LSI Systems-Difference Equation

1. **Infinite Impulse Response (IIR) or Recursive 2-D Systems**

   A general discrete-space 2-D LSI system can be represented by the following 2-D constant coefficient difference equation:

$$\sum \sum_{k,l \in W_i} a_{k,l} y(m-k, n-l) = \sum \sum_{p,q \in W_j} b_{p,q} x(m-p, n-q) \tag{3.11}$$

   where $a_{k,l}$s and $b_{p,q}$s are the coefficients of the difference equation. Assuming $a_{0,0} = 1$ (normalized) rearranging (3.11) gives:

$$y(m, n) = \sum \sum_{p,q \in W_j} b_{p,q} x(m-p, n-q) - \sum \sum_{k,l \in \hat{W}_i} a_{k,l} y(m-k, n-l) \tag{3.12}$$

   where $W_i = \hat{W}_i \cup (0, 0)$. Due to the recursive term on the right side, this system is called *recursive*. Moreover, it can easily be shown that [2] the PSF of a recursive system is of infinite-extent, hence the system is also referred to as IIR. For recursive systems, $\hat{W}_i$ in (3.12) can only take causal ROS, i.e. QP or NSHP as shown in Figure 3.6, while W_j could take any geometry when postprocessing images. However, when processing is performed while acquiring image data sequentially, both ROSs $\hat{W}_i$ and W_j should be causal (i.e. QP or NSHP).

A special case of recursive systems is when $W_j = (0,0)$ in the 2-D difference equation (3.12), which gives:

$$y(m,n) = b_{0,0}x(m,n) - \sum\sum_{k,l\in\widehat{W}_i} a_{k,l}y(m-k,n-l). \tag{3.13}$$

This case corresponds to 2-D recursive or IIR system of *all-pole* type as in the 1-D case of Section 2.1.4.

2. **Finite Impulse Response (FIR) or Non-Recursive 2-D Systems**
If the ROS $\hat{W}_i = \varnothing$ in the 2-D difference equation (3.12), then we get:

$$y(m,n) = \sum\sum_{k,l\in W_j} b_{k,l}x(m-k,n-l), \tag{3.14}$$

That is, no recursive term in the difference equation. Thus, these systems are called *non-recursive*. Additionally, comparing (3.14) with the 2-D convolution summation:

$$y(m,n) = \sum\sum_{k,l\in W_i} h(k,l)x(m-k,n-l), \tag{3.15}$$

It is clear that $h(k,l) = b_{k,l}$ and $h(k,l) = 0 \ \ (k,l) \notin W_i$. That is, the PSF of a non-recursive system is of finite-extent. Thus, the system is also called 2-D FIR system. For FIR systems, the ROS, W_i, of the PSF can have different geometry, such as the QP, NSHP, SHP, or full plane (FP) shown in Figures 3.6 and 3.7. Note that when processing image data that arrive sequentially (e.g. row-wise or column-wise), only QP or NSHP can be used.

In analogy to the 1-D case (see Section 2.1.4), this system is also *all-zero* system since the transfer function has no denominator polynomial.

Example 3.2 *Consider a 2-D system given by input–output relationship* $y(m,n) = a^{x(m,n)},\ a < 1$ and $\forall m,n \geq 0$. *Determine whether or not the system is linear, space-invariant, and comment on its stability.*

Solution: The system is not linear since:

$$z(m,n) = a^{c_1x_1(m,n)+c_2x_2(m,n)} \neq c_1a^{x_1(m,n)} + c_2a^{x_2(m,n)} = c_1y_1(m,n) + c_2y_2(m,n).$$

The system is space-invariant due to $z(m,n) = a^{x(m-k,n-l)} = y(m-k,n-l)$. The PSF of this system is:

$$h(m,n) = a^{\delta(m,n)} = \begin{cases} a, & (m,n) = (0,0) \\ 1, & \forall m,n \geq 0,\ (m,n) \neq (0,0). \end{cases}$$

Finally, the system is BIBO stable since for any $|x(m,n)| < M \rightarrow y(m,n) = a^{x(m,n)} < N$.

3.1.5 Frequency Domain Analysis – 2-D Continuous Fourier Transform

To extend the frequency domain analysis of the previous chapter to the 2-D signals and systems, we begin with the 2-D continuous Fourier transform (FT) and its properties. For a continuous and absolutely integrable 2-D function $x(u,v)$, the 2-D FT pair [1, 2] is given by:

$$X(\omega_1,\omega_2) = \iint_{-\infty}^{\infty} x(u,v)\exp(-j\omega_1u - j\omega_2v)du dv \tag{3.16}$$

$$x(u,v) = \frac{1}{4\pi^2}\iint_{-\infty}^{\infty} X(\omega_1,\omega_2)\exp(j\omega_1u + j\omega_2v)d\omega_1 d\omega_2. \tag{3.17}$$

where ω_1, ω_2 are related to *spatial frequencies*. The basis functions in the 2-D FT are $e^{-j\omega_1 u}$ and $e^{-j\omega_2 v}$ while those of the inverse 2-D FT are their complex-conjugates. The following properties are essential to the application of 2-D FT in image processing.

3.1.5.1 Remarks and Properties

1. The FT is a complex-valued function with magnitude $|X(\omega_1, \omega_2)|$ and phase $\phi(\omega_1, \omega_2)$, i.e. $X(\omega_1, \omega_2) = |X(\omega_1, \omega_2)|e^{j\phi(\omega_1,\omega_2)}$.

   Since both photographic film and human eye (or any other detector) respond to the magnitude of an image field, it is difficult to directly observe the phase spectrum. Holographic techniques are, however, based upon recording both the magnitude and phase spectra. The phase spectrum contains most of the information about the positions of edges in an image [1].
2. *Spatial Frequencies:* If $x(u, v)$ represents the luminance at spatial coordinates u and v then $\xi_1 = \frac{\omega_1}{2\pi}$ and $\xi_2 = \frac{\omega_2}{2\pi}$ are the spatial frequencies that represent luminance change with respect to spatial distances. The units of ξ_1 and ξ_2 are reciprocal of u and v. If the coordinates u, v are normalized by the viewing distance of the image $x(u, v)$, then the units ξ_1 and ξ_2 are cycles per degree of the viewing angle.
3. *Uniqueness:* The function $x(u, v)$ and its FT $X(\omega_1, \omega_2)$ are unique with respect to one another, i.e. instead of an image its FT can be stored without any loss of information.
4. *Separability:* 2-D FT can be written as a separable transformation in u and v coordinates

$$X(\omega_1, \omega_2) = \int_{-\infty}^{\infty} \left(\int_{-\infty}^{\infty} x(u, v)e^{-j\omega_1 u} du \right) e^{-j\omega_2 v} dv \tag{3.18}$$

   Thus, 2-D FT can be implemented by successive 1-D FTs along rows and columns [1, 2]. This is shown in Figure 3.8 where $\mathfrak{F}\{x(u, ;)\}$ represents 1-D FT of $x(u, v)$ along u coordinate which generates $X(\omega_1, v)$. Then, 1-D FT along v coordinate is applied to $X(\omega_1; , v)$ to produce $X(\omega_1, \omega_2)$, i.e. $\mathfrak{F}\{X(\omega_1; , v)\} = X(\omega_1, \omega_2)$.
5. *2-D Convolution and Correlation:* Any 2-D LSI image formation system can be represented by 2-D convolution integral [1, 2]:

$$y(u, v) = \iint_{-\infty}^{\infty} h(u', v')x(u - u', v - v')du'dv' = h(u, v) ** x(u, v) \tag{3.19}$$

   where $x(u, v)$ and $y(u, v)$ are the input and output images, respectively, $h(u, v)$ is the PSF of the imaging system, and $**$ stands for 2-D convolution operation. Alternatively, in the frequency domain, we have:

$$Y(\omega_1, \omega_2) = H(\omega_1, \omega_2) \cdot X(\omega_1, \omega_2) \tag{3.20}$$

   where $X(\omega_1, \omega_2)$ and $Y(\omega_1, \omega_2)$ are the 2-D FT of the input and output images, respectively and $H(\omega_1, \omega_2)$ is the 2-D frequency response of the system. Thus, the FT of the convolution of two 2-D functions is the product of their FTs.

   The dual property applies to the product of two 2-D functions in the spatial domain, i.e. $y(u, v) = x(u, v) \cdot h(u, v)$, in which case in the frequency domain, we convolve the FTs using:

$$Y(\omega_1, \omega_2) = \frac{1}{4\pi^2}[X(\omega_1, \omega_2) ** H(\omega_1, \omega_2)]. \tag{3.21}$$

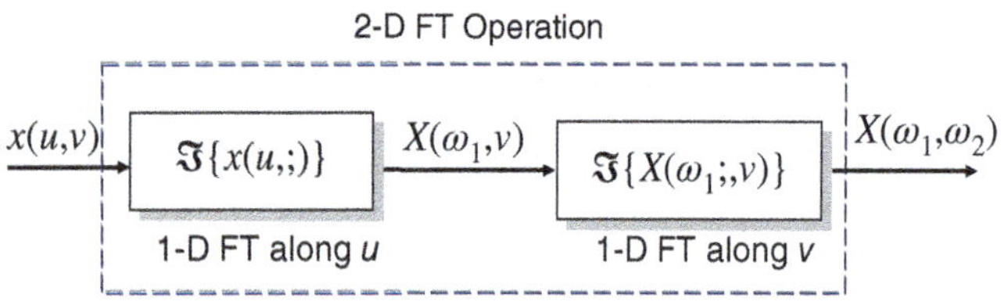

Figure 3.8 2-D FT can be performed using two 1-D FT along rows and columns.

The convolution operation can be used to perform the spatial cross-correlation between two real functions $x(u, v)$ and $t(u, v)$, that is:

$$c(u, v) = \iint_{-\infty}^{\infty} t(u', v')x(u + u', v + v')du'dv' = t(u, v) ** x(-u, -v) \tag{3.22}$$

Alternatively, with change of variables, we get:

$$c(u, v) = \iint_{-\infty}^{\infty} t(u' - u, v' - v)x(u', v')du'dv' = t(-u, -v) ** x(u, v).$$

In the frequency-domain cross-correlation can be carried out using:

$$C(\omega_1, \omega_2) = T(-\omega_1, -\omega_2)X(\omega_1, \omega_2) = T(\omega_1, \omega_2)X(-\omega_1, -\omega_2).$$

Correlation-based methods are often used in image processing for *template matching* in order to find the closest match between a given image $x(u, v)$ and a set of known templates $t(u, v)$. The closest match occurs when a selected template yields the largest cross-correlation value at a particular location within the search image, which corresponds to the location of the detected target.

If $t(u, v) = x(u, v)$ this process becomes *spatial auto-correlation* of a 2-D function, i.e. $a(u, v) = x(u, v) ** x(-u, -v)$ with FT, which is:

$$A(\omega_1, \omega_2) = X(-\omega_1, -\omega_2)X(\omega_1, \omega_2) = |X(\omega_1, \omega_2)|^2.$$

6. *Parseval's Theorem and Inner Product Preservation:* Another important property of FT is that the inner product of two functions in the spatial domain is equal to the inner product of their FTs in the frequency domain, i.e.

$$\iint_{-\infty}^{\infty} x(v, u)y^*(u, v)dudv = \frac{1}{4\pi^2} \iint_{-\infty}^{\infty} X(\omega_1, \omega_2)Y^*(\omega_1, \omega_2)d\omega_1 d\omega_2$$

where $*$ stands for complex conjugate operation. When $x = y$, we obtain the well-known *Parseval's energy conservation* formula, i.e.

$$\iint_{-\infty}^{\infty} |x(u, v)|^2 dudv = \frac{1}{4\pi^2} \iint_{-\infty}^{\infty} |X(\omega_1, \omega_2)|^2 d\omega_1 d\omega_2, \tag{3.23}$$

which implies that the total energy in the spatial domain is the same as that in the frequency domain.

7. *Frequency Response and Eigenfunctions of 2-D LSI Systems:* An eigen-function of a system is defined as an input function that is reproduced at the output with a possible change in amplitude and phase. For a 2-D LSI system eigen-functions are given by $f(u, v) = \exp(j\omega_1 u + j\omega_2 v)$. Using the 2-D convolution integral, the response of a 2-D system with PSF $h(u, v)$ to this input is:

$$g(u, v) = \iint_{-\infty}^{\infty} h(u - u', v - v') \exp[j\omega_1 u' + j\omega_2 v']du'dv'.$$

Now, changing variables $\tilde{u} = u - u'$ and $\tilde{v} = v - v'$ we can easily get:

$$g(u, v) = H(\omega_1, \omega_2) \exp[j\omega_1 u + j\omega_2 v].$$

That is, the output is a complex exponential function with the same frequency as the input signal but its amplitude and phase are changed by the complex frequency response $H(\omega_1, \omega_2)$ of the 2-D system.

Example 3.3 *Determine frequency response, $H(\omega_1, \omega_2)$, of a 2-D system whose PSF is:*

$$h(u, v) = \begin{cases} 1, & |u| \le X, |v| \le Y \\ 0, & \text{elsewhere} \end{cases}$$

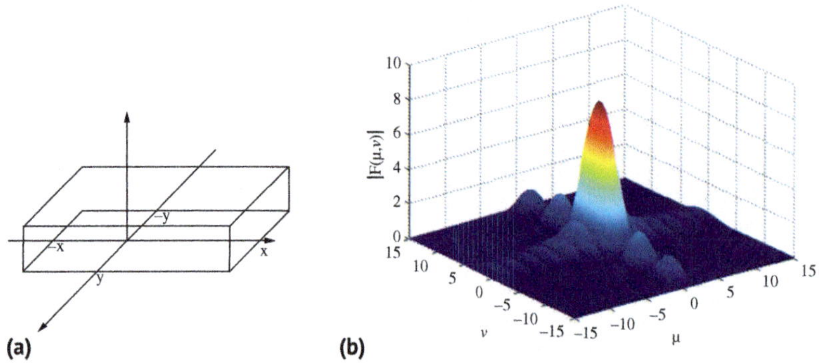

Figure 3.9 (a) 2-D continuous PSF, (b) 2-D sinc frequency response.

Solution: Using the 2-D FT equation (3.16) and its separability property we get:

$$
\begin{aligned}
H(\omega_1, \omega_2) &= \iint_{-\infty}^{\infty} h(u', v') e^{-j(\omega_1 u' + \omega_2 v')} du' dv' \\
&= \int_{-X}^{X} \int_{-Y}^{Y} e^{-j(\omega_1 u' + \omega_2 v')} du' dv' \\
&= \left(\int_{-X}^{X} e^{-j\omega_1 u'} du' \right) \left(\int_{-Y}^{Y} e^{-j\omega_2 v'} dv' \right) \\
&= \left(\frac{e^{j\omega_1 X} - e^{-j\omega_1 X}}{j\omega_1} \right) \left(\frac{e^{j\omega_2 Y} - e^{-j\omega_2 Y}}{j\omega_2} \right) \\
&= 2X \left(\frac{\sin \omega_1 X}{\omega_1 X} \right) 2Y \left(\frac{\sin \omega_2 Y}{\omega_2 Y} \right) \\
&= 4XY \mathrm{sinc}(\omega_1 X) \mathrm{sinc}(\omega_2 Y),
\end{aligned}
$$

which is a separable 2-D sinc function. Figure 3.9a,b shows the rectangular PSF $h(u, v)$ (left) and its 2-D frequency response $H(\omega_1, \omega_2)$ (right), which extends to infinity along both frequency variables, i.e. not band-limited.

3.1.6 Frequency Domain Analysis – 2-D Discrete Space Fourier Transform

Consider the 2-D FT of a continuous image $x(u, v)$ in (3.16). Suppose that the image is sampled uniformly at sampling intervals Δu and Δv (see Chapter 5 for sampling theorem), i.e. $u = m\Delta u$ and $v = n\Delta v$. Then the FT of the sampled image $x(m\Delta u, n\Delta v)$ or simply $x(m, n)$ becomes:

$$X(\omega_1 \Delta u, \omega_2 \Delta v) = \sum_{m=-\infty}^{\infty} \sum_{n=-\infty}^{\infty} x(m, n) e^{-j\omega_1 m\Delta u} e^{-j\omega_2 n\Delta v} \tag{3.24}$$

Alternatively, if we define discrete frequency variables $\Omega_1 = \omega_1 \Delta u$ and $\Omega_2 = \omega_2 \Delta v$, we obtain the 2-D Discrete Space Fourier Transform (DSFT) of sampled image $x(m, n)$ as:

$$X(e^{j\Omega_1}, e^{j\Omega_2}) = \sum_{m=-\infty}^{\infty} \sum_{n=-\infty}^{\infty} x(m, n) e^{-j(\Omega_1 m + \Omega_2 n)} = \text{2-D DSFT}\{x(m, n)\}. \tag{3.25}$$

Note that since the basis functions of 2-D DSFT, $e^{-j\Omega_1 m}$ and $e^{-j\Omega_2 n}$, are periodic with period 2π since $e^{-j\Omega_1 m} = e^{-j(\Omega_1+2k\pi)m}$, the spectrum $X(e^{j\Omega_1}, e^{j\Omega_2})$ will also be doubly periodic with periods 2π, i.e. $X(e^{j\Omega_1}, e^{j\Omega_2}) = X(e^{j(\Omega_1+2\pi k)}, e^{j(\Omega_2+2\pi l)})$. Considering this periodicity property, the inverse DSFT becomes:

$$x(m,n) = \frac{1}{4\pi^2} \iint_{-\pi}^{\pi} X(e^{j\Omega_1}, e^{j\Omega_2}) e^{j(\Omega_1 m+\Omega_2 n)} d\Omega_1 d\Omega_2 \tag{3.26}$$

Equations (3.25) and (3.26) form the 2-D DSFT pair for the sampled image $x(m, n)$. All the properties of 1-D DTFT covered in Section 2.1.5 are also applicable to the 2-D DSFT.

Example 3.4 *Determine frequency response of a discrete-space 2-D system with PSF*

$$h(m,n) = \begin{cases} 1, & |m| < M, |n| < N \\ 0, & \text{otherwise} \end{cases}$$

Solution: Using (3.25) and the definition of the PSF function, we have:

$$\begin{aligned} H(e^{j\Omega_1}, e^{j\Omega_2}) &= \sum_{m=-\infty}^{\infty} \sum_{n=-\infty}^{\infty} h(m,n) e^{-j(\Omega_1 m+\Omega_2 n)} \\ &= \sum_{m=-M+1}^{M-1} \sum_{n=-N+1}^{N-1} e^{-j\Omega_1 m} e^{-j\Omega_2 n} \end{aligned}$$

Now, we use the separability property of the DSFT and the finite power series formula $\sum_{m=-M_1}^{M_2} \alpha^m = \frac{\alpha^{-M_1} - \alpha^{M_2+1}}{1-\alpha}$, to get:

$$\begin{aligned} H(e^{j\Omega_1}, e^{j\Omega_2}) &= \left(\frac{e^{-j\Omega_1(-M+1)} - e^{-j\Omega_1 M}}{1 - e^{-j\Omega_1}} \right) \left(\frac{e^{-j\Omega_2(-N+1)} - e^{-j\Omega_2 N}}{1 - e^{-j\Omega_2}} \right) \\ &= \left(\frac{e^{j\Omega_1\left(M-\frac{1}{2}\right)} - e^{-j\Omega_1\left(M-\frac{1}{2}\right)}}{e^{j\Omega_1/2} - e^{-j\Omega_1/2}} \right) \left(\frac{e^{j\Omega_2\left(N-\frac{1}{2}\right)} - e^{-j\Omega_2\left(N-\frac{1}{2}\right)}}{e^{j\Omega_2/2} - e^{-j\Omega_2/2}} \right) \\ &= \frac{\sin(\Omega_1(M-1/2))}{\sin(\Omega_1/2)} \frac{\sin(\Omega_2(N-1/2))}{\sin(\Omega_2/2)} \end{aligned}$$

Note that the PSF $h(m, n)$ in this example is the sampled version of that in Example 3.2 (Figure 3.9a). Since the 2-D spectrum of the continuous PSF is not band-limited, sampling in the spatial domain leads to aliasing in the frequency domain (see Chapter 5 for sampling theorem). That is, $H(e^{j\Omega_1}, e^{j\Omega_2})$ in this example is the aliased version of the 2-D sinc function $H(\omega_1, \omega_2)$ in Example 3.2 (Figure 3.9b). Figure 3.10 shows this aliased 2-D sinc function for $M = 4$ and $N = 8$. Note that this function is also separable.

Figure 3.10 An aliased 2-D sinc function ($M = 4$ and $N = 8$).

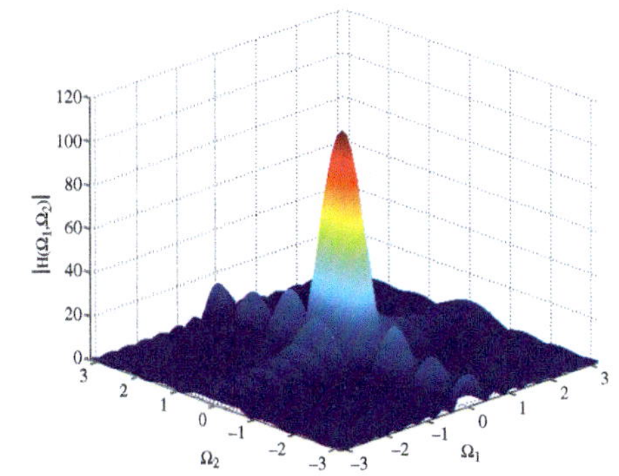

Example 3.5 *Determine the 2-D DSFT of 2-D separate signal* $x(m,n) = a^m u(m)\upsilon(n)$, $|a| < 1$ *where* $u(m)$ *is 1-D unit step function and*

$$\upsilon(n) = \begin{cases} 1, & 0 \le n \le N-1 \\ 0, & \text{elsewhere} \end{cases}$$

Solution: Using the expression for 2-D DSFT in (3.25), we get:

$$\begin{aligned} X(e^{j\Omega_1}, e^{j\Omega_2}) &= \sum_{m,n=-\infty}^{\infty} x(m,n)e^{-j(\Omega_1 m + \Omega_2 n)} \\ &= \sum_{m=-\infty}^{\infty} a^m u(m) e^{-j\Omega_1 m} \sum_{n=-\infty}^{\infty} e^{-j\Omega_2 n}\upsilon(n) \\ &= \sum_{m=0}^{\infty} a^m e^{-j\Omega_1 m} \sum_{n=0}^{N-1} e^{-j\Omega_2 n} \\ &= \left(\frac{1}{1 - ae^{-j\Omega_1}}\right) \times e^{-j\Omega_2\left(\frac{N-1}{2}\right)} \frac{\sin(N\Omega_2/2)}{\sin(\Omega_2/2)} \end{aligned}$$

3.1.7 Frequency Domain Analysis - 2-D DFT

Recall that in Section 2.1.6, the 1-D DFT of a finite-duration sequence $x(n)$, $n \in [0, N-1]$ was obtained by sampling the DTFT $X(e^{j\Omega})$ in the frequency domain. Similarly, the 2-D DFT of a finite-extent digital image $x(m,n)$ of size $M \times N$ is obtained by sampling the doubly periodic 2-D spectrum $X(e^{j\Omega_1}, e^{j\Omega_2})$ in the frequency domain using sampling intervals $\Delta\Omega_1 = \frac{2\pi}{M}$ and $\Delta\Omega_2 = \frac{2\pi}{N}$. That is:

$$\begin{aligned} \tilde{X}(k,l) = X(e^{j\Omega_1}, e^{j\Omega_2})\Big|_{\substack{\Omega_1 = k\Delta\Omega_1 \\ \Omega_2 = l\Delta\Omega_2}} &= \sum_{m=0}^{M-1}\sum_{n=0}^{N-1} x(m,n) W_M^{mk} W_N^{nl}, \quad k \in [0, M-1], l \in [0, N-1] \\ &= \text{2-D DFT}\{x(m,n)\} \end{aligned} \tag{3.27}$$

where $W_M = e^{-\frac{j2\pi}{M}}$ and $W_N = e^{-\frac{j2\pi}{N}}$. Also, as in the 2-D FT and DSFT cases, the basis functions of 2-D DFT are also separable. This allows the operation to be performed using consecutive 1-D DFTs along rows and columns of the digital image. Note that since $X(e^{j\Omega_1}, e^{j\Omega_2})$ is doubly periodic with period 2π, the sampled spectrum $\tilde{X}(k,l)$ is also doubly periodic with periods M and N.

Using the orthogonality of the basis functions of the 2-D DFT, we can arrive at the 2-D IDFT to recover the image from its 2-D DFT:

$$\begin{aligned} \tilde{x}(p,q) &= \frac{1}{MN}\sum_{k=0}^{M-1}\sum_{l=0}^{N-1} \tilde{X}(k,l) W_M^{-kp} W_N^{-lq} \quad p \in [0, M-1], q \in [0, N-1] \\ &= \text{2-D IDFT}\{\tilde{X}(k,l)\} \end{aligned} \tag{3.28}$$

where the basis functions of the 2-D IDFT are the complex conjugate of those of the 2-D DFT. To see how $\tilde{x}(p,q)$ is related to the original finite-extent image $x(m,n)$, we replace $\tilde{X}(k,l)$ from (3.27) into (3.28) and use the definitions of W_M and W_N which after rearranging gives:

$$\tilde{x}(p,q) = \sum_{m,n}\sum \left[\frac{1}{MN}\sum_{k=0}^{M-1}\sum_{l=0}^{N-1} e^{j\frac{2\pi(p-m)k}{M}} e^{j\frac{2\pi(q-n)l}{N}}\right] x(m,n) \tag{3.29}$$

Using the powers series formula $\sum_{k=0}^{M-1} \alpha^k = \frac{1-\alpha^M}{1-\alpha}$, it is easy to show that the term in the square bracket in (3.29) can be simplified to:

$$\frac{1}{MN}\sum_{k=0}^{M-1}\sum_{l=0}^{N-1} e^{j\frac{2\pi(p-m)k}{M}} e^{j\frac{2\pi(q-n)l}{N}} = \begin{cases} 1, & \text{for } m = p + rM,\ n = q + sN \\ 0, & \text{otherwise} \end{cases}$$
$$= \delta(p + rM - m, q + sN - n)$$

Now, using this result in (3.29) yields:

$$\tilde{x}(p,q) = \sum_{r=-\infty}^{\infty}\sum_{s=-\infty}^{\infty} x(p + rM, q + sN), \tag{3.30}$$

which implies that sampling in the frequency domain results in periodic extension of the finite-extent image in the spatial domain where $\tilde{x}(p, q)$ is also doubly periodic with periods M and N. This is the dual of the spatial domain sampling theorem, which is covered in Chapter 5.

The important observation that can be drawn from the above results is that we can view the DFT as two things: (i) Fourier representation for finite-extent signals/images, in which case we remove the $\tilde{x}$ notation; or (ii) Discrete Fourier Series (DFS) expansion for periodic sequences [2]. However, in the 2-D case we rarely deal with periodic images.

Remarks:

1. Since $X(k, l)$ is complex-valued, it may lead to the belief that 2-D DFT increases the dimensionality of the representation. However, this is not true due to conjugate symmetry property, i.e. $X(M - k, N - l) = X^*(k, l)$.
2. For displaying purposes, the quadrant of 2-D DFT $X(k, l)$ can be reordered so that its origin is in the center. This can be done by multiplying the image $x(m, n)$ by the factor $(-1)^{m+n}$ before taking 2-D DFT since:

$$X(k + pM, l + qN) = \sum_{m=0}^{M-1}\sum_{n=0}^{N-1} x(m,n)\exp\left(-2\pi j\left(\frac{mk}{M} + \frac{nl}{N}\right)\right)\exp(-2\pi j(pm + qn))$$

Now, for $p = q = 1/2$, we get:

$$X\left(k + \frac{M}{2}, l + \frac{N}{2}\right) = \sum_{m=0}^{M-1}\sum_{n=0}^{N-1} x(m,n)(-1)^{m+n}\exp\left(-2\pi j\left(\frac{mk}{M} + \frac{nl}{N}\right)\right),$$

which is centered at $\frac{M}{2}, \frac{N}{2}$.

3. Since the dynamic range of the transform coefficients $|X(k, l)|$ is much larger than the exposure range of display systems, it is necessary to compress the coefficients before displaying them using, for example, log operation, $D(k, l) = a\log(1 + b|X(k, l)|)$ where a and b are appropriate constants. If $|X(k, l)| = 0$, then $D(k, l) = 0$, i.e. zero-preserving mapping.

   Figure 3.11 shows the *Peppers* image and its magnitude of 2-D DFT after using *log* and centering operations. As can be seen, a large number of components are insignificant.

Example 3.6 *Find the 2-D DFT of* $x(m, n) = a^{m-n},\ m, n \in [0, N - 1]$.

Solution: Using 2-D DFT equation, we have:

$$X(k,l) = \sum_{m=0}^{N-1}\sum_{n=0}^{N-1} x(m,n) W_N^{mk} W_N^{nl},\ k, l \in [0, N-1]$$
$$= \left\{\sum_{m=0}^{N-1} a^m W_N^{mk}\right\}\left\{\sum_{n=0}^{N-1} a^{-n} W_N^{nl}\right\},\ k, l \in [0, N-1]$$
$$= \frac{1 - a^N}{1 - aW_N^k} \times \frac{1 - a^{-N}}{1 - a^{-1}W_N^l},\ k, l \in [0, N-1]$$

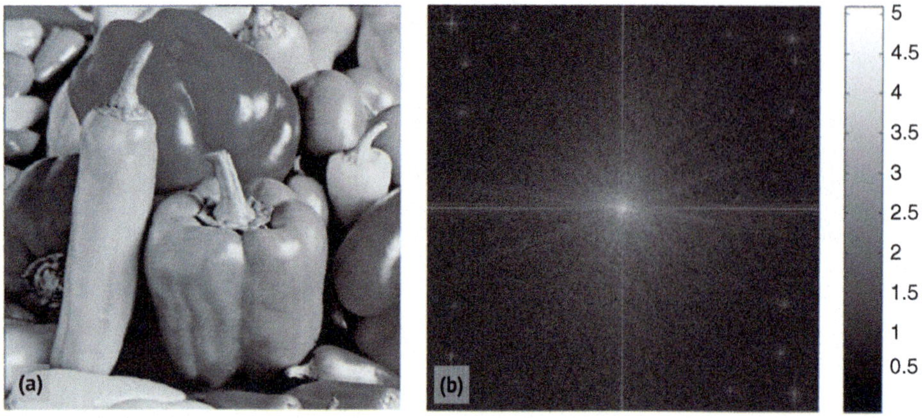

Figure 3.11 (a) Original Peppers image and (b) its *log*-compressed and centered magnitude spectrum using 2-D DFT. *Source:* USC-SIPI Image Database

3.1.7.1 Properties of 2-D DFT

The following properties are the 2-D extensions of those in Section 2.1.6 for 1-D DFT:

1. **Circular Shift:**
Let $y(m, n)$ be the circularly shifted version of $x(m, n)$ by m_0 and n_0, i.e.

$$y(m, n) = x(((m - m_0))_M, ((n - n_0))_N)$$

where $((n))_N$= mod N, i.e. if $p = ((n))_N$, then $0 \leq p \leq N - 1$ and there exists I such that $n = p + IN$. Then the 2-D DFT of $y(m, n)$ is:

$$Y(k, l) = W_M^{m_0 k} W_N^{n_0 l} X(k, l) \quad k, l \in R_{MN}$$

where R_{MN} is a finite-extent region $m \in [0, M - 1],\ n \in [0, N - 1]$. This property implies that the spatial domain shifting property of a finite-extent image applies to its circularly shifted version, i.e.

$$x(((m - m_0))_M, ((n - n_0))_N) \overset{\text{DFT}}{\longleftrightarrow} W_M^{m_0 k} W_N^{n_0 l} X(k, l).$$

Also, due to the duality principle, we have the following circular shift property in the frequency domain:

$$W_M^{-mk_0} W_N^{-nl_0} x(m, n) \overset{\text{DFT}}{\longleftrightarrow} X(((k - k_0))_M, ((l - l_0))_N),$$

That is, multiplication of image $x(m, n)$ by complex exponential functions $W_M^{-mk_0}$ and $W_N^{-nl_0}$ results in a circular shift of its 2-D DFT $X(k, l)$ by k_0 and l_0.

2. **Circular Convolution:**
Consider two images $x(m, n)$ and $h(m, n)$ with ROS in R_{MN} and let $X(k, l)$ and $H(k, l)$ be their corresponding 2-D DFTs. The 2-D IDFT of their product corresponds to the 2-D *circular convolution* of the two images in the spatial domain. That is, we have:

$$\begin{aligned}\hat{y}(m, n) &= \text{2-D IDFT}\{H(k, l)X(k, l)\} \\ &= \sum_{p=0}^{M-1} \sum_{q=0}^{N-1} x(p, q)h(((m - p))_M, ((n - q))_N) \\ &= \sum_{p=0}^{M-1} \sum_{q=0}^{N-1} h(p, q)x(((m - p))_M, ((n - q))_N),\end{aligned}$$

or simply

$$\hat{y}(m, n) = h(m, n) \circledast \circledast x(m, n) = x(m, n) \circledast \circledast h(m, n)$$

where ⊛⊛ symbol stands for 2-D circular convolution operation. As we saw in Section 2.1.6, circular convolution result is totally different from the desired linear convolution result of $x(m, n)$ and $h(m, n)$. This is due to the fact that in circular convolution one image, e.g. $h(m, n)$, is shifted circularly and then convoluted with the other image, i.e. $x(m, n)$. The *wraparound error* caused by the circular shift impacts almost all pixels of the resultant image. However, circular convolution also possesses properties such as commutative, associative, and distributive over addition similar to its linear counterpart.

In many linear DIP tasks, e.g. filtering operation, we are primarily interested in computing the linear convolution results using fast transforms such as 2-D FFT [2]. The following procedure can be used to perform linear convolution using 2-D DFT without aliasing error and with much better efficiency.

3.1.8 Computing 2-D Linear Convolution Using 2-D DFT

Recall that the linear convolution of two finite-extent sequences $x(m, n)$ and $h(m, n)$ with ROS of $R_{M_1N_1}$ and $R_{M_2N_2}$, respectively, produces an image of size $(M_1 + M_2 - 1) \times (N_1 + N_2 - 1)$. Now, to perform linear convolution using DFT we follow the following steps:

- Choose P and Q such that $P \geq M_1 + M_2 - 1, Q \geq N_1 + N_2 - 1$. Note that we often choose $P = 2^p$ and $Q = 2^q$, i.e. powers of 2 for more efficient radix-2 or 4 FFT implementation [2].
- Zero-pad both $h(m, n)$ and $x(m, n)$ to have their ROS extended to R_{PQ}.
- Compute 2-D DFT of size $(P \times Q)$ of $x(m, n)$ and $h(m, n)$ to generate $X(k, l)$ and $H(k, l)$.
- Compute $(P \times Q)$-point 2-D IDFT of the product $Y(k, l) = X(k, l)H(k, l)$. The first $(M_1 + M_2 - 1) \times (N_1 + N_2 - 1)$ portion of the result contains the desired linear convolution result, $y(m, n)$.

3.2 Discrete-space 2-D Signals and Systems-Random Case

This section presents 2-D extension of the 1-D random signal processing in Section 2.2. To this end, we assume that several images $x(m, n)$ from the same scene have been collected under different noise/interference conditions, e.g. radar images of the same scene under varying ground conditions and at different times. The collection of the recorded images forms an *ensemble* set and since noise and interference effects are random, we represent this ensemble set of images by the discrete 2-D random field $X(m, n)$. For a particular (m, n), $X(m, n)$ forms a discrete random variable (r.v.) with possible values that are the pixel intensities $x_i(m, n)$, $i \in [1, N]$ of all images in the ensemble set at the same location. Thus, one way to represent this 2-D random process (r.p.) is to define probability distributions for a huge number of r.v. $\forall(m, n)$ and their joint behavior, which will obviously be impractical.

3.2.1 Statistical Measures on 2-D Random Processes

To simplify statistical representation on 2-D random images, we assume that the random field is *wide-sense stationary* (WSS), which requires that the first- and second-order statistics, namely the ensemble mean and *autocorrelation* function (ACF) be space invariant, i.e.

$$\begin{aligned} \mu_X &= E[X(m, n)] \\ R_{XX}(k, l) &= E[X(m, n)X(m - k, n - l)], \end{aligned} \tag{3.31}$$

where μ_X is the mean, $R_{XX}(k, l)$ is the ACF at lags k, l and $E[.]$ represents the ensemble averaging or expectation operation. As can be seen, these statistics are independent of spatial variables (m, n). The *auto-covariance* function $C_{XX}(k, l)$ of random image $X(m, n)$ is given by:

$$\begin{aligned} C_{XX}(k, l) &= E[(X(m, n) - \mu_X)(X(m - k, n - l) - \mu_X)] \\ &= R_{XX}(k, l) - \mu_X^2. \end{aligned} \tag{3.32}$$

For $(k, l) = (0, 0)$, we have $C_{XX}(0, 0) = \sigma_X^2$, i.e. the variance of the WSS 2-D r.p. X.

The statistical relationship of two 2-D WSS r.p.s X and Y that are also jointly WSS can be represented by their *cross-correlation* or *cross-covariance* functions that are:

$$R_{XY}(k,l) = E[X(m,n)Y(m-k,n-l)] \tag{3.33}$$

$$\begin{aligned} C_{XY}(k,l) &= E[(X(m,n)-\mu_X)(Y(m-k,n-l)-\mu_Y)] \\ &= R_{XY}(k,l) - \mu_X\mu_Y, \end{aligned} \tag{3.34}$$

respectively. Table 3.1 gives the important spatial-domain properties of these statistics for one and two WSS random images. These properties are the 2-D extensions of those in Section 2.2.2. The last two properties in this table imply that the second-order statistics decay as the spatial separation between points of the same r.p. and two different r.p.s increases.

3.2.2 Power Spectra of 2-D Random Processes

The power spectrum or spectral density function (SDF) of the WSS 2-D r.p. $X(m,n)$ is the 2-D DSFT of its auto-correlation function (ACF). That is:

$$S_{XX}(e^{j\Omega_1}, e^{j\Omega_2}) = \sum_{k=-\infty}^{\infty}\sum_{l=-\infty}^{\infty} R_{XX}(k,l)\, e^{-j(\Omega_1 k+\Omega_2 l)} = \text{DSFT}\{R_{XX}(k,l)\}. \tag{3.35}$$

As shown before, this power spectrum is doubly periodic with period 2π. The power spectrum is real $S_{XX}(e^{j\Omega_1}, e^{j\Omega_2}) = S_{XX}^*(e^{j\Omega_1}, e^{j\Omega_2})$ and nonnegative function $S_{XX}(e^{j\Omega_1}, e^{j\Omega_2}) \geq 0, \forall\, \Omega_1, \Omega_2$ due to the fact that the ACF of a WSS r.p. is an even function. Also, $S_{XX}(e^{j\Omega_1}, e^{j\Omega_2}) = S_{XX}(e^{-j\Omega_1}, e^{-j\Omega_2})$, i.e. an even function since the ACF is a real function. The other power spectrum properties in Section 2.2.4 for 1-D signals can also be extended to the 2-D case.

Using the inverse DSFT, we can recover $R_{XX}(k,l)$ from its 2-D power spectrum using:

$$R_{XX}(k,l) = \frac{1}{4\pi^2}\int_{-\pi}^{\pi}\int_{-\pi}^{\pi} S_{XX}(e^{j\Omega_1}, e^{j\Omega_2})\, e^{j(\Omega_1 k+\Omega_2 l)}. \tag{3.36}$$

Table 3.1 Properties of 2-D WSS random processes.

Property	Formulation
Mean-squared	$R_{XX}(0,0) = E[X^2(m,n)]$
Variance	$R_{XX}(0,0) = \sigma_X^2 = E[X^2(m,n)] - \mu_X^2$
Symmetry	Auto-correlation $R_{XX}(-k,-l) = R_{XX}(k,l)$
	Cross-correlation $R_{XY}(-k,-l) = R_{XY}(k,l)$
Bounded 1	Auto-correlation $\|R_{XX}(k,l)\| \leq R_{XX}(0,0)$
	Auto-covariance $\|C_{XX}(k,l)\| \leq \sigma_X^2$
Uncorrelated	$R_{XY}(k,l) = \mu_X\, \mu_Y$ or $C_{XY}(k,l) = 0$
Bounded 2	Cross-correlation $\|R_{XY}(k,l)\| \leq \{R_{XX}(0,0)\, R_{YY}(0,0)\}^{1/2}$
	Cross-covariance $\|C_{XY}(k,l)\| \leq \sigma_X\, \sigma_Y$
Limits 1	$\lim_{k,l\to\infty} R_{XX}(k,l) = \mu_X^2$
	$\lim_{k,l\to\infty} C_{XX}(k,l) = 0$
Limits 2	$\lim_{k,l\to\infty} R_{XY}(k,l) = \mu_X\, \mu_Y$
	$\lim_{k,l\to\infty} C_{XY}(k,l) = 0$

A zero-mean 2-D r.p. is called a *white* process (noise) if its ACF is $R_{XX}(k,l) = \sigma_X^2 \delta(k,l)$ where $\delta(k,l)$ is the Kronecker delta function. Therefore, its 2-D power spectrum will be:

$$S_{XX}(e^{j\Omega_1}, e^{j\Omega_2}) = \sum_{k=-\infty}^{\infty} \sum_{l=-\infty}^{\infty} \sigma_X^2 \delta(k,l)\, e^{-j(\Omega_1 k + \Omega_2 l)} = \sigma_X^2, \tag{3.37}$$

That is, constant over all frequencies. This is in analogy to white light, where all frequencies or color spectral components are present.

For two individually and jointly WSS 2-D r.p.s $X(m,n)$ and $Y(m,n)$ with cross-correlation function $R_{XY}(k,l)$, the cross-power spectrum is defined by:

$$S_{XY}(e^{j\Omega_1}, e^{j\Omega_2}) = \sum_{k=-\infty}^{\infty} \sum_{l=-\infty}^{\infty} R_{XY}(k,l)\, e^{-j(\Omega_1 k + \Omega_2 l)} = \mathrm{DSFT}\{R_{XY}(k,l)\}. \tag{3.38}$$

This 2-D cross-power spectrum is also doubly periodic with period 2π. Moreover, the cross-power spectrum properties in Section 2.2.4 can also be extended to the 2-D case.

The cross-correlation can be recovered from the 2-D cross-power spectrum using IDSFT:

$$R_{XY}(k,l) = \frac{1}{4\pi^2} \int_{-\pi}^{\pi} \int_{-\pi}^{\pi} S_{XY}(e^{j\Omega_1}, e^{j\Omega_2})\, e^{j(\Omega_1 k + \Omega_2 l)} d\Omega_1 d\Omega_2 \tag{3.39}$$

The 2-D power spectrum and cross-power spectrum can alternatively be defined in (z_1, z_2)-transform domain (see Appendix B). This can be done by changing $e^{j\Omega_1} = z_1$ and $e^{j\Omega_2} = z_2$ which yields:

$$S_{XX}(z_1, z_2) = \sum_{k=-\infty}^{\infty} \sum_{l=-\infty}^{\infty} R_{XX}(k,l)\, z_1^{-k} z_2^{-l}, \tag{3.40}$$

and

$$S_{XY}(z_1, z_2) = \sum_{k=-\infty}^{\infty} \sum_{l=-\infty}^{\infty} R_{XY}(k,l)\, z_1^{-k} z_2^{-l}. \tag{3.41}$$

These (z_1, z_2) domain representations are more useful for applications such as 2-D spectral factorization and image modeling (see Chapter 9). In practice, however, the 2-D power spectrum and cross-power spectrum operations are performed using 2-D DFT of size $M \times N$, i.e.

$$S_{XX}(p,q) = \sum_{k=0}^{M-1} \sum_{l=0}^{N-1} R_{XX}(k,l)\, W_M^{kp}\, W_N^{lq}, \quad p \in [0, M-1],\ q \in [0, N-1] \tag{3.42}$$

and

$$S_{XY}(p,q) = \sum_{k=0}^{M-1} \sum_{l=0}^{N-1} R_{XY}(k,l)\, W_M^{kp}\, W_N^{lq}, \quad p \in [0, M-1],\ q \in [0, N-1] \tag{3.43}$$

where M and N represent the extent of the auto-correlation or cross-correlation functions and are usually taken to be the dimension of the original images $X(m,n)$ and $Y(m,n)$. Figure 3.12a,b shows the 2-D ACF and power spectrum of the Peppers image in Figure 3.11, respectively. The power spectrum was computed using 2-D DFT of the ACF.

The 2-D IDFT can be used to recover the auto-correlation and cross-correlation functions from the power spectrum and cross-power spectrum:

$$R_{XX}(k,l) = \frac{1}{MN} \sum_{k=0}^{M-1} \sum_{l=0}^{N-1} S_{XX}(p,q)\, W_M^{-kp}\, W_N^{-lq}, \quad k \in [0, M-1],\ l \in [0, N-1] \tag{3.44}$$

and

$$R_{XY}(k,l) = \frac{1}{MN} \sum_{k=0}^{M-1} \sum_{l=0}^{N-1} S_{XY}(p,q)\, W_M^{-kp}\, W_N^{-lq}, \quad k \in [0, M-1],\ l \in [0, N-1], \tag{3.45}$$

respectively.

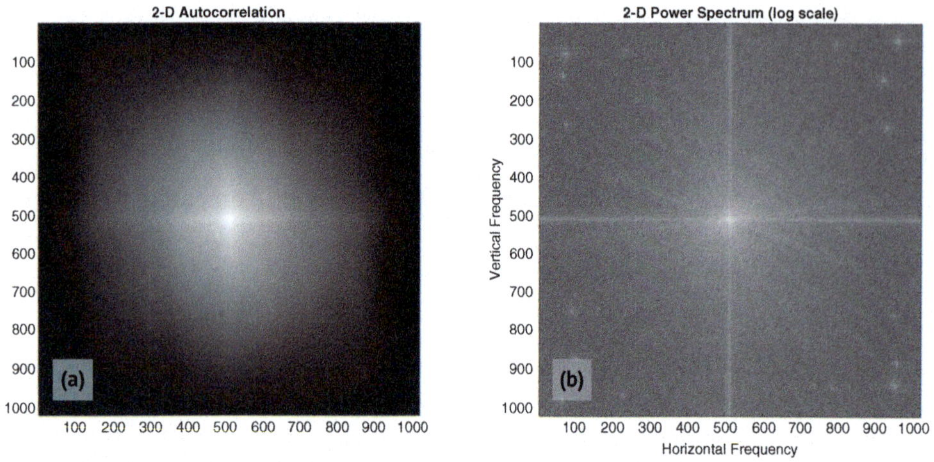

Figure 3.12 (a) 2-D ACF of Peppers image, (b) power spectrum of Peppers image.

3.2.3 Response of 2-D LSI Systems to Random Images

Consider a stable 2-D LSI system with PSF $h(m, n)$ and frequency response $H(e^{j\Omega_1}, e^{j\Omega_2})$. A WSS random image $X(m, n)$ with zero mean and ACF $R_{XX}(k, l)$ is applied to this 2-D system. We would like to determine expressions for the ACF and the power spectrum of its output $Y(m, n)$. Note that since the input is WSS, it is guaranteed that the output r.p. $Y(m, n)$ will also be WSS since the system is LSI.

Assuming causality and using the 2-D convolution summation, $Y(m, n) = \sum_{i=0}^{\infty} \sum_{j=0}^{\infty} h(i, j)X(m - i, n - j)$, we can obtain the expression for the ACF of $Y(m, n)$. However, we first need to derive the expressions for the cross-correlation and cross-power spectrum between $X(m, n)$ and $Y(m, n)$, i.e.

$$\begin{aligned} R_{XY}(k, l) &= E[X(m, n)Y(m - k, n - l)] \\ &= E[X(m, n)\sum_{i=0}^{\infty}\sum_{j=0}^{\infty} h(i, j)X(m - k - i, n - l - j)] \\ &= \sum_{i=0}^{\infty}\sum_{j=0}^{\infty} h(i, j)E[X(m, n)X(m - k - i, n - l - j)] \\ &= \sum_{i=0}^{\infty}\sum_{j=0}^{\infty} h(i, j)R_{XX}(k - i, l - j) = h(k, l) ** R_{XX}(k, l). \end{aligned} \tag{3.46}$$

The corresponding 2-D cross-power spectrum is then obtained by taking 2-D DSFT of (3.46), which gives:

$$S_{XY}(e^{j\Omega_1}, e^{j\Omega_2}) = H(e^{j\Omega_1}, e^{j\Omega_2})\, S_{XX}(e^{j\Omega_1}, e^{j\Omega_2}) \tag{3.47}$$

or in the (z_1, z_2)-domain:

$$S_{XY}(z_1, z_2) = H(z_1, z_2)\, S_{XX}(z_1, z_2) \tag{3.48}$$

where $H(e^{j\Omega_1}, e^{j\Omega_2})$ and $H(z_1, z_2)$ are the 2-D frequency response and transfer function of the 2-D LSI system, respectively. Using a similar approach, we can obtain the expression for the ACF of the output r.p. as:

$$\begin{aligned} R_{YY}(k, l) &= E[Y(m, n)Y(m - k, n - l)] \\ &= h(k, l) ** R_{XY}(-k, -l) \\ &= h(k, l) ** h(-k, -l) ** R_{XX}(k, l). \end{aligned} \tag{3.49}$$

Note that $R_{XX}(k, l) = R_{XX}(-k, -l)$ for WSS r.p.s. Now, taking the 2-D DSFT or (z_1, z_2) transform, we obtain the expressions for the 2-D power spectrum of the output:

$$S_{YY}(e^{j\Omega_1}, e^{j\Omega_2}) = |H(e^{j\Omega_1}, e^{j\Omega_2})|^2\, S_{XX}(e^{j\Omega_1}, e^{j\Omega_2}) \tag{3.50}$$

and

$$S_{YY}(z_1, z_2) = H(z_1, z_2)\, H(z_1^{-1}, z_2^{-1})\, S_{XX}(z_1, z_2), \tag{3.51}$$

respectively. Equation (3.51) is the 2-D version of the spectral factorization method covered in Section 2.2.7 and is widely used for the design of stable 2-D digital systems and also for image modeling applications [3, 4], which will be covered in Chapter 9.

As a special case, let us assume that the input to this 2-D LSI system is a white noise process with ACF $R_{XX}(k, l) = \sigma_X^2\, \delta(k, l)$ or equivalently the 2-D power spectrum $S_{XX}(e^{j\Omega_1}, e^{j\Omega_2}) = \sigma_X^2$. The question is whether or not the output is also white? The answer is no. The reason being, using (3.50), we have:

$$S_{YY}(e^{j\Omega_1}, e^{j\Omega_2}) = \sigma_X^2 |H(e^{j\Omega_1}, e^{j\Omega_2})|^2, \tag{3.52}$$

which implies that the 2-D power spectrum of the output is frequency-dependent and hence it is *colored*.

3.3 Computer Projects

For the following computer projects, use the Python modules *numpy, scipy.signal and matplotlib.pyplot.*

Computer Project 3.1: The goal of this computer project is to study how 2-D DFT can be used to analyze 2-D signals and systems in the frequency domain.

1. Generate the cubic 2-D PSF given by:

$$h(m, n) = \begin{cases} 1, & m, n \in [0, L-1] \\ 0, & m, n \in [L, N-1] \end{cases}$$

   Plot the magnitude response of $|H(k, l)|$, for various pairs of pairs L and N, e.g. $L = 5, 7, 11$ and $N = 32, 64, 128$, and comment on the results. Show how a high density magnitude response can closely represent the 2-D DSFT of this PSF. This can be done by zero-padding the PSF to a larger size, e.g. $N = 512$, before taking its 2-D DFT.
2. Generate the 2-D signal $x(m, n) = \cos(0.48\pi(m + n)) + \cos(0.52\pi(m + n))$. Use the first 10 points of the signal and zero-pad the rest to size 256. Plot the magnitude of the 2-D DFT of the zero-padded signal. Then, generate all the 256 samples and repeat (no zero-padding). Compare your results and describe your observations. Gradually decrease or increase the separation between the frequencies and plot the magnitude of the 2-D DFT. Explain your observations.

Computer Project 3.2: The purpose of this computer assignment is to show how the materials covered in Section 3.2 can be applied to processing random images. For this computer project, we use the Boat image "Boat.mat."

1. Assume that this image is transmitted multiple times through a noisy channel that adds white additive noise to the image every time the same image is transmitted. That is, different realizations of the received images are $x_i(m, n) = x(m, n) + \eta_i(m, n)$, where $\eta_i(m, n),\ i \in [1, P]$ is a zero-mean random noise that is different for every i, and is statistically independent from the original noise-free image $x(m, n)$. If we have P received images containing different independent identically distributed (iid) random noise $\eta_i(m, n)$ with identical variance, we can estimate the noise-free image based on the ensemble averaging over all P measurements. The estimate approaches the true image as the number of measurements increases, i.e.

$$\hat{x}(m, n) = \lim_{P \to \infty} \frac{1}{P} \sum_{i=1}^{P} x_i(m, n)$$

   Compute an estimate of the actual noise-free image based on these received noisy images for $P = 10$ and $P = 25$. Compare the visual quality of your estimates.
2. Assume that the estimate represents your actual noise-free image, then compute the statistics (pdf, mean, and variance) of the noise for both values of P. Is this noise distributed in a known manner? What kind and why?
3. One measure of the quality of a processed image is its signal-to-noise ratio (SNR), which can be computed for statistically independent signal and noise using $\mathrm{SNR}_{dB} = 10 \log_{10} \frac{\sigma_X^2}{\sigma_\eta^2}$ where σ_X^2 and σ_η^2 are the variance of the image and noise, respectively. Select a single sample function (or image) of the ensemble set of received images and compute its SNR for both values of P. What is your conclusion? Use the same sample function for the next part.

4. In this section, we decide to do image estimation by spatial-domain filtering based on only one received image. One possible spatial domain filtering operation that we will discuss in Chapter 8 is the spatial averaging digital filter of order M, N,

$$y(m,n) = \sum_{k=0}^{M-1} \sum_{l=0}^{N-1} b_{k,l}\, x(m-k, n-l)$$

where $b_{k,l}$ are the filter coefficients that are $b_{k,l} = 1/MN,\ k \in [0, M-1], l \in [1, N-1]$. Note that this is a 2-D FIR filter discussed in Section 3.1.4. Plot the frequency response of this filter and comment on its frequency selectivity. Pick the same sample image from the previous part and filter it with this spatial averaging filter of order $M = N = 3$. Compute the SNR of the filtered image using the estimate of the noise-free image for $P = 25$. Comment on your results and its visual quality.
5. Generate the ACF of (a) estimate of the actual image for $P = 25$, (b) the sample image in the previous part, (c) noise in that sample image, and (d) the filtered image $y(m, n)$. Additionally, generate the cross-correlation of the chosen sample image with its corresponding added noise. Provide the plots of these correlation functions. Compare these plots and try to explain your results in the context of the operations that you performed in the previous parts.
6. Using 2-D FFT, generate the plots of 2-D power spectrum and cross-power spectrum for the results generated in the previous part. Provide your detailed analysis on these results.

3.4 Problems

Problem 3.1: In each of the following systems find the PSF and determine whether or not the system is linear, space-invariant, FIR, or IIR. Comment on the BIBO stability of these systems.

a. $y(m,n) = 3\,x(m,n) + 9$
b. $y(m,n) = m^2 n^2 x(m,n)$
c. $y(m,n) = x(m^2, n^2)$
d. $y(m,n) = \sum_{k=-\infty}^{m} \sum_{l=-\infty}^{n} x(k,l)$
e. $y(m,n) = \sum_{k=m}^{\infty} \sum_{l=n}^{\infty} x(k,l)$
f. $y(m,n) = \exp\{-|x(m,n)|\}$

Problem 3.2:

a. All that is known about the 2-D sequences $x(m, n)$ and $h(m, n)$ is the fact that each has a region of support (ROS) that is restricted to the first quadrant, i.e. $x(m, n) = 0$ if $m < 0$ or $n < 0$ and $h(m, n) = 0$ if $m < 0$ or $n < 0$. Show that the ROS of their convolution is also confined to that quadrant.
b. Repeat Part (a) assuming that this time $x(m, n)$ and $h(m, n)$ are each confined to the third quadrant.
c. Suppose that $x(m, n)$ and $h(m, n)$ are each one-quadrant sequence but each has ROS on a different quadrant. What statement, if any, can be made about the ROS of their convolution?

Problem 3.3: Consider two 2-D sequences that are separable, i.e. $x(m, n) = a(m)b(n)$ and $h(m, n) = c(m)d(n)$.

a. Show that their convolution is also a separable 2-D sequence.
b. Express that convolution in terms of a, b, c, and d.

Problem 3.4: In Section 3.1.2, we used the 2-D Kronecker delta property in (3.4) to arrive at the 2-D linear convolution summation. Using a similar approach:

a. Represent an arbitrary 2-D signal $x(m, n)$ in terms of a linear combination of delayed 2-D unit steps.
b. Using the expression obtained in Part (a) and assuming that the output of the LSI system to a 2-D unit step is denoted by $s(m, n)$, derive the input–output relationship of an LSI 2-D system in terms of $s(m, n)$.
c. What is your conclusion about the result in Part (b)? Is this relationship interchangeable with respect to x and s?

Problem 3.5: Determine the result of 2-D linear convolution between 2-D sequence $x(m,n) = a^m b^n u(m,n)$, where $u(m,n)$ represents 2-D unit step function, with the 2-D sequence:

$$h(m,n) = \sum_{r=-\infty}^{\infty} \delta(m - rN, n - rN).$$

Problem 3.6:

a. Determine the result of the 2-D linear convolution of $x(m,n)$ in Example 3.1 with each of the following 2-D arrays. Note that the bold number denotes the location of (0, 0) pixel.

1	−2	3
2	**0**	−5
3	−2	1

,

−1	**4**	1

, and

−1
4
−1

b. Verify your answers in each case via the area conservation property.

Problem 3.7: For the optical imaging system shown in Figure 3.13, determine the output, $y(u,v)$, in terms of the input image $x(u,v)$.

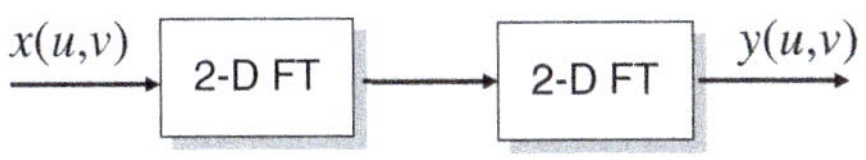

Figure 3.13 Optical imaging system in Problem 3.7.

Problem 3.8: Show that the Fourier transform of a 2-D Gaussian function $x(u,v) = e^{-\pi(u^2+v^2)}$ is another 2-D Gaussian in the frequency domain with variance $\sigma^2 = 1/2\pi$.

Problem 3.9: If the 2-D FT of $x(u,v)$ is given by $X(\omega_1, \omega_2)$ find:

a. FT of $x(au, bv)$,
b. FT of $x(au + bv, cu + dv)$ with $ad - bc = 1$.

Problem 3.10: A bilinear system represents a generalization of the class of linear systems. Although they operate with 1-D inputs and outputs, they are closely related to 2-D systems. The output of a bilinear system can be expressed as:

$$y(n,n) := y(n) = \sum_{m_1} \sum_{m_2} x(m_1)x(m_2)h(n - m_1, n - m_2)$$

If a bilinear system is characterized by the 2-D DSFT of $h(n_1, n_2)$, denoted by $H(e^{j\Omega_1}, e^{j\Omega_2})$, find the response of the system to the input $x(n) = ae^{j\Omega_1 n} + be^{j\Omega_2 n}$.

Problem 3.11: Find the 2-D DSFT and DFT of the following 2-D functions:

a. $x(m,n) = a^m b^{-n} \delta(m+n) u(m)$, $|a|, |b| < 1$, where $u(.)$ represents the 1-D unit step function.
b. $x(m,n) = \sum_{p=0}^{N-1} \delta(m - p, n - p)$,
c. $x(m,n) = \sum_{p=0}^{N-1} \delta(m - pK, n - pL)$, where K and L are fixed integers.

Problem 3.12: Consider a 2-D function of the form $x(m,n) = \sum_{k=0}^{N-1} \delta(m - kn_1, n - kn_2)$.

a. Sketch this 2-D sequence.
b. Determine and sketch the 2-D DSFT of this sequence.

Problem 3.13: The DSFT of the signal $x(m, n) = \alpha^m \beta^n u(m, n)$, where $u(m, n)$ is 2-D unit step function, is sampled at $\Omega_1 = \frac{\pi}{2}k,\ k \in [0, 3]$ and $\Omega_2 = \frac{\pi}{4}l,\ l \in [0, 7]$. The inverse DFT of the sampled spectrum is then taken. Determine the resultant spatial domain image and its relation to the original image $x(m, n)$, if any. Explain the results.

Problem 3.14: Consider the image $x(m, n) = \delta(m) + \delta(n)$.

a. Sketch this function on the (m, n)-plane.
b. Find the 2-D DSFT of this image.
c. Find the 2-D DFT of this image.

Problem 3.15: Consider the image $x(m, n) = \delta(m, n) + \delta(m - 1, n - 1)$.

a. Sketch this function on the (m, n)-plane.
b. Find the 2-D DSFT of this image and comment on its magnitude spectrum.
c. Find the 2-D DFT of this image.

Problem 3.16: Find the frequency response (magnitude and phase) of the following 2-D FIR system and comment on its frequency domain behavior, i.e. the type of filtering it performs.

$$y(m, n) = 1/8[x(m - 1, n) + x(m, n - 1) + 4x(m, n) + x(m, n + 1) + x(m + 1, n)].$$

Problem 3.17: Find the 2-D DFT of $x(m, n) = \sum_{p=0}^{N-1} y(m, p),\quad m, n \in [0, N - 1]$. Assume $Y(k, l)$ is known.

Problem 3.18: Consider a 2-D FIR filter with impulse response $h(m, n) = \rho^m u(n)$.

a. Determine $H(e^{-j\Omega_1}, e^{-j\Omega_2})$ and the filter type.
b. Determine the response of this filter to the following sequences using the frequency domain method.

(i) $\delta(m, n)$, (ii) $\delta(m, n) - \delta(m - 1, n - 1)$, and (iii) $u(m, n)$.

Problem 3.19: Prove the following properties in Table 3.1.

a. Symmetry and
b. Bounded 1 and 2.

Problem 3.20: Consider a 2-D r.p. with ACF $R_{XX}(k, l) = \sigma_X^2 \rho_1^{|k|} \rho_2^{|l|}$ where $\rho_1, \rho_2 < 1$. Find the expression for the corresponding 2-D power spectrum $S_{XX}(e^{j\Omega_1}, e^{j\Omega_2})$.

References

1 A. K. Jain, "*Fundamentals of Digital Image Processing*," Prentice-Hall, 1989.
2 D. E. Dudgeon and R. M. Mersereau, "*Multidimensional Digital Signal Processing*," Prentice Hall, 1984.
3 A. K. Jain, "Advances in mathematical models for image processing," Proceedings of the IEEE, vol. 69, pp. 502–528, 1981.
4 R. King, M. Ahmadi, R. Gorgui-Naguib, A. Kwabwe, and M. R. Azimi-Sadjadi, "*Digital Filtering in One and Two Dimensions*," Plenum Press, 1989.

4

Image Formation and Perception

In this chapter, we discuss some basic principles of image formation and the perception of the human visual system. These principles are central not only to the development and evaluation of various image processing operations that will be covered in subsequent chapters but also to gaining a better understanding of the structure and functional aspects of visual information processing, as well as their limitations. More specifically, visual information processing related to luminance and lighting, spatial vision, and color perception will be briefly discussed. Psychophysical findings [1] that describe some of the limitations of human visual information processing, particularly in detection and discrimination, will also be reviewed. This is especially important in light of the fact that in many such difficult tasks, AI-based solutions have been shown [2] to outperform human visual information processing, while in many other simpler tasks, they do not perform as well.

Since visual information processing and perception begins when light enters the eye, we also begin our coverage with a brief review of the optical and neuronal components of the human eye and their functionalities.

4.1 Human Eye

The human eye is a remarkable biological system responsible for visual formation and perception in the visual cortex. The process begins when light stimuli generated from the outside world enter the eye through the cornea, pupil, and lens, which focus the light onto the retina. The cornea, pupil, lens, and humor collectively form the optical system of the eye, whereas the retina forms the neuronal information processing system [1, 3]. Figure 4.1 illustrates a simple anatomy of the eye and these components. The quality of the image formed and perceived by the retina is directly dependent on the proper function of the optical components. In this section, we briefly describe these optical and neuronal components and their functions for visual information processing.

The *visible light* is referred to the range of electromagnetic radiation spectrum that a human visual system can perceive. This spectral band corresponds to frequencies in the range of $f = 400 - 750$ THz or wavelengths of $\lambda = \frac{c}{f} = 400$ nm (violet) to 750 nm (red), where $c \approx 299,800$ km/s is the speed of light. The first surface that light encounters when entering the eye is the cornea, which is a refractive layer with a nonuniform thickness in the center and along the sides. The cornea is nearly transparent to visible wavelengths. However, for wavelengths below 300 nm, the corneal absorption increases to block exposure, and hence protect the other components.

The space between the corneal surface and the iris and between the lens and the retina are filled with different optically clear media called aqueous and vitreous humors [1, 3], respectively, which form integral parts of the optics of the eye. The lens is a transparent flexible elliptic structure in the sense that its curvature changes to focus objects at different distances. This process, called *accommodation* [3], ensures that image formation on the retina can be accommodated by adjusting the focal length without changing the lens-to-retina distance unlike typical imaging sensors. If the focusing is imperfect due to lens *aberrations* caused by an imperfection of the eye's optical system, a blurred image will form in the retina. The amount of light that can enter the eye is controlled by the iris. The iris muscles change the size of the pupil in response to feedback signals received from the visual cortex to shrink the pupil size when the light intensity is high or to dilate when the intensity is low.

As pointed out in Chapter 3, the point spread function (PSF) of an imaging system completely characterizes its physical properties. Similarly, the PSF of the eye, which is the image formed on the retina in response to a point light source, gives a complete description of the optical system of the eye. However, this PSF is varying depending on the wavelength of

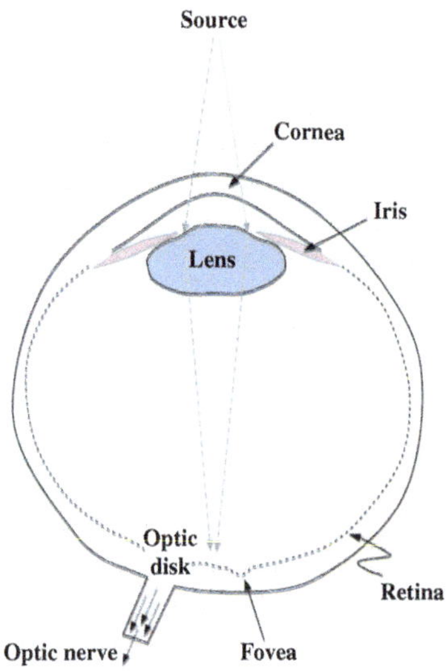

Figure 4.1 Optical and neuronal structures of human eye.

the light i.e. shorter wavelengths lead to narrower PSF while longer wavelengths widen the support of PSF. Dilation and constriction of the pupil also change the PSF. That is, constriction reduces aberration at the cost of increasing distortion due to diffraction.

Once the light reaches the retina, the second visual subsystem, i.e. the neuronal part, takes over the encoding of visual information for image perception. The retina is made up of multiple layers that process incoming light and transmit information to two types of *photoreceptors* namely *cones* and *rods*. Cones are light-sensitive cells responsible for daytime vision called *photoptic*, while rods are those responsible for night vision or *scotopic* vision. The region in between is called *mesoptic* where both the cones and the rods are active. For daytime vision, there are three types of cone with various short, medium, and long wavelength absorption characteristics. The spatial distributions of rods and cones are nonuniform, though they are radially symmetric. That is, the spatial distribution of the cones peaks sharply in the center of the fovea. In contrast, rods are present at high density in most of the retina, with a sharp decline in the fovea [3].

The visual activity detected by the photoreceptors is then received, via the bipolar cells, in the form of electrical signals by the retinal ganglion neurons, which in turn transmit their responses in form of a series of discrete impulses or spikes to the brain via the optic nerve. The experiments carried out by Hubel and Weisel [4] indicated that the response of these cells to a light stimulus decreases as a function of the lateral distance from the center with the maximum response. Figure 4.2 shows the plot of this activity region known as the *receptive field* of a neuron. This region of activity with a small lateral distance tends to be *excitatory* thus promoting neurons firing, while regions outside the receptive field correspond to the *inhibitory* regions that hinder firing activity. These excitatory and inhibitory areas are shown with positive (+) and negative (−) in the activity bubble below the plot in Figure 4.2. The negative region is often referred to as lateral inhibition, which plays an important role in image perception [3, 5]. The responses of ganglion cells in the form of electrical impulses are then transmitted to the brain through the optic nerve, where visual perception of the scene will occur.

The human eye is capable of responding to an enormous range of light intensity, exceeding 10 units on a logarithmic scale (i.e. a minimum-to-maximum intensity variation of over 10-billion times). Inevitably, the eye response to the signal intensity, which determines its apparent intensity, or brightness, is not linear. That is, it is not determined by the nominal change in physical stimulus (light energy), but rather by its change relative to its initial level.

4.2 Elements of Visual Perception

4.2.1 Brightness Sensitivity

The human visual system perceives information about properties of objects within its visual field via the spectral reflectance characteristics that are produced under certain incident lighting conditions. This is done at the expense of losing

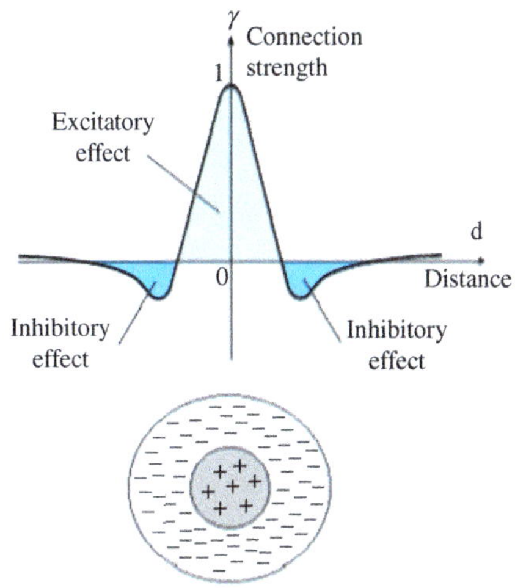

Figure 4.2 Receptive field of retinal cells forming a *Mexican hat* plot and its activity bubble.

information about the quality and quantity of light entering the eye. The light received by the visual system when we observe an illuminated object can be expressed as:

$$\ell(\lambda, u, v) = i(\lambda, u, v).r(\lambda, u, v) \tag{4.1}$$

where u and v are spatial coordinates (locations on the object surface), $0 < i(\lambda, u, v) < \infty$ is the incident or illumination distribution component and $0 \leq r(\lambda, u, v) \leq 1$ is the reflectivity component, which is related to the properties of the object surface.

What are the quantifiable attributes of light that play an important role in human perception? There are many quantifiable attributes of light, including but not limited to luminance, brightness, contrast, and color. Luminance is a measurable physical quantity (not a perceptual quantity) that corresponds to light produced by an illuminating source from some region of space. Its unit is candela per square meter cd/m^2 [6]. Brightness, on the other hand, refers to the perceived amount of light (or luminance) coming from a source or object, e.g. a bright display or TV screen. It is not a physically measurable quantity as it is related to many factors, including psychophysical aspects and the subjectivity of human perception [5, 6]. Consequently, there is no one-to-one correspondence between luminance and perceived brightness. For example, two objects on different backgrounds may appear to have different brightness even though their luminance is identical. In contrast, for two different luminance, it is possible that objects appear to have the same brightness [7]. Brightness is particularly important in the design of critical displays where ambient light is highly variable [8].

The human visual system is more sensitive to the difference in luminance of an object with respect to that of the background than to the absolute luminance. Weber's law [3, 6] gives a threshold for this sensitivity, called just noticeable difference (JND), which is the smallest difference that human eye can perceive. According to Weber's law, this difference ΔI is proportional to the background luminance I of the surrounding. Mathematically, this is given by:

$$\frac{\Delta I}{I} = \kappa \tag{4.2}$$

where $\kappa = 0.01 - 0.02$ is called Weber constant (or ratio), which depends on many factors including the retinal location, stimulus size, duration, wavelength, etc. Thus, Weber's law gives a general rule of thumb for the theoretical limit of the discrimination threshold ΔI as a function of background luminance.

Figure 4.3 shows threshold–intensity (TVI) curve, i.e. a log–log plot of ΔI versus the background luminance I. This plot exhibits four distinct regions. In the first region (very low luminance), the curve is approximately flat, indicating that ΔI is nearly constant. In this dark-light region, visual performance is limited by the neural noise of the retina. As the luminance increases, the second region follows the De Vries–Rose law [3], where the detection threshold varies as $\Delta I \propto \sqrt{I}$. For higher luminance levels, the system enters the Weber region, where Weber's law in (4.2) holds.

At very high luminance levels, ΔI increases faster than I, indicating saturation of the visual system and reduced sensitivity. This behavior led Fechner (Weber's student) to propose a relationship between physical luminance and perceived

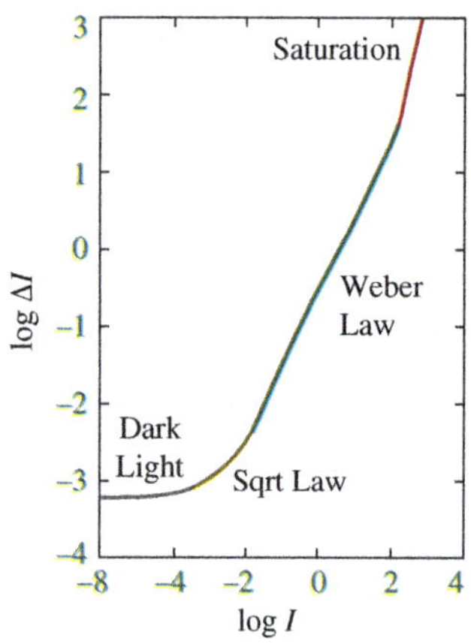

Figure 4.3 Threshold-intensity plot – Four regions corresponding to dark light noise, square root (sqrt) law, Weber's law, and saturation. *Source:* [3] / with permission of Springer Nature.

brightness. Assuming that a just noticeable change in sensation ds is proportional to the relative change in stimulus, we have $ds = k\frac{dI}{I}$. Now, integrating both sides yields

$$s = k \log I + a, \tag{4.3}$$

which is known as the Weber–Fechner law. This relation implies that perceived brightness is a logarithmic function of the luminance I, where k determines the overall sensitivity.

To account for neural noise at low luminance levels, the model can be modified as

$$ds = k\frac{dI}{I + I_0}$$

where I_0 represents the effective noise level of the visual system.

Although Fechner's law holds for a wide range of luminance conditions, it breaks down in some regions. Stevens' law [3] was proposed to provide a more accurate model for these regions. Stevens' law follows $s = \alpha I^{\beta}$ where α is a constant amplitude and $\beta = 1/3$ is the typical choice. It must be pointed out that the human visual system is directionally sensitive, i.e. more sensitive to the light coming in through the center of the eye pupil than to that entering from the periphery of the pupil.

4.2.2 Contrast Sensitivity and Spatial-Temporal Vision

In the previous section, we discussed visual sensitivity for different illumination regions. This section discusses important aspects of spatial and temporal vision [3]. The spatial vision is responsible for still pattern recognition and interpretation by forming spatial configuration of objects, while the temporal vision is responsible for motion detection and perception.

One of the most important aspects of spatial vision is the edge extraction ability of the retinal system that detects abrupt transitions in luminance via a spatial filtering operation. This spatial filtering operation is the result of the lateral inhibition process among neighboring neurons explained in Section 4.1. The main byproduct of lateral inhibition is the contrast sensitivity, which allows us to distinguish edges and fine details of objects within our visual field. Figure 4.4 shows a series of gray scale transition bands from dark (left) to lighter (right) with uniform luminance within each band. However, the perceived brightness response is nonuniform across each band and more pronounced at the transition points from one level to another. More specifically, the transition appears to be brighter on the right-side of a band and darker on the left-side, hence resulting in overshoot and undershoot behavior at these transition points. This optical illusion phenomenon known as Mach bands occurs when transitions from darker to lighter shades are encountered, in which case the perceived brightness appears to be exaggerated at the transition points of luminance change. The impulse response of the Mach band system can be modeled using the Mexican hat plot in Figure 4.2. It should be noted that lateral inhibition can be viewed as the eye's edge detection process that extracts the positions and contrasts of edges in various situations. The edge detector is a high-pass filter that extracts or accentuates high-frequency information in the spatial domain.

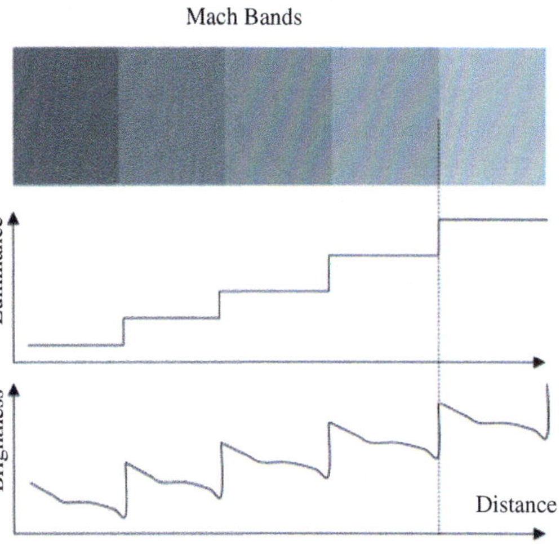

Figure 4.4 Mach bands and luminance versus perceived brightness. *Source*: [9] / with permission of Springer Nature / https://link.springer.com/rwe/10.1007/978-3-642-27851-8_348-1#rightslink/last accessed on Jan 30, 2026.

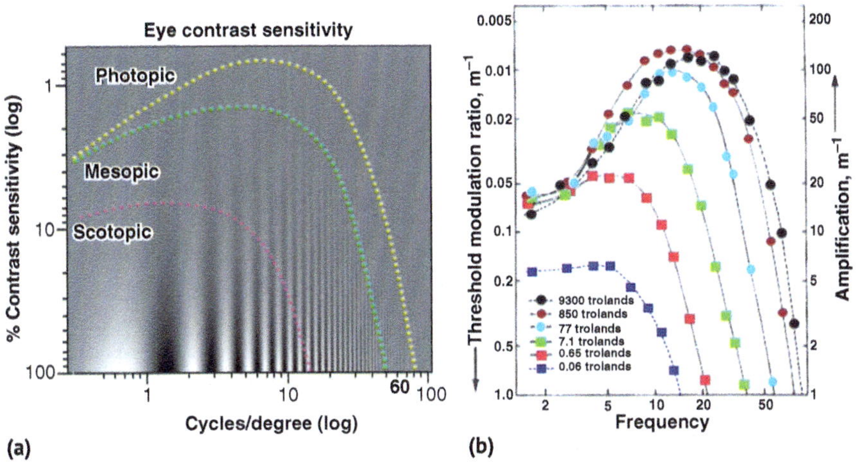

Figure 4.5 (a) Sinusoidal grating stimuli and CSF plots of photopic (yellow), mesoptic (green), and scotopic (red) vision (*Source*: Vladimir Sacek / https://www.telescope-optics.net/aberrations_extended.htm/last accessed on Jan 30, 2026) and (b) contrast sensitivity for various luminance levels as a function of spatial frequency. *Source*: [3] / with permission of Springer Nature.

The ability to perceive luminance variation across space is characterized by the contrast sensitivity function (CSF) [3], which is a measure of the ability of our spatial vision to detect spatial patterns. From a system theory perspective, the CSF represents the spatial frequency response of the eye and can be used as a broad descriptor of visual performance and its links to psychophysical mechanisms [3, 5]. To determine CSF, sinusoidal gratings of varying contrast and spatial frequency are applied as the stimuli, and the threshold (inverse of sensitivity) at which a particular spatial frequency (cycles/degree) is detected is then recorded for various spatial frequencies. Figure 4.5a shows the sinusoidal gratings with different spatial frequencies and the CSF plots for photopic (yellow), mesoptic (green), and scotopic (red) vision. As can be seen, the CSF decreases both at low and high spatial frequencies, hence acting like a band-pass filter. The high frequency roll-off is attributed to the resolution limitation of the visual system or *grating acuity*. The lower roll-off corresponds to stimuli that cover the receptive field uniformly, hence causing lateral inhibition. More recent studies [10] indicated that the CSF is actually a superposition of multiple channels, each tuned to a certain spatial frequency. It should be emphasized that the performance metric of the optical visual system alone is known as the modulation transfer function (MTF) [3, 7], which determines the quality of the image formed on the retina. The MTF is basically the optical transfer function of the eye that precedes the retina. However, unlike the CSF, the MTF has high-pass characteristics. It is important to note that the measured eye CSF is in fact a combined (cascaded) effect of the responses of both the optical and the neuronal systems for spatial vision. In [7], a monochromatic model for the human visual system was presented which consists of the models for the optical system of the eye, its spectral response, and those of the neuronal parts, including the response of the cone receptors and the effects of the Mach band of the cells.

The ability of the eye to form continuous sequences of images even in the presence of a discontinuity of luminance in time is attributed to temporal vision. Perceptual continuity is a function of the temporal frequency (Hz) of moving stimuli or the critical flicker frequency (CFF) [3]. The CFF measures the ability of the brain to resolve temporal visual information to form a flicker-free perception of the motion, e.g. continuous images on display systems. The minimum temporal resolution, or *temporal acuity* is analogous to the grating acuity in spatial vision. The perception of flicker depends on many factors including stimulus size, luminance, and location of the retina. From a physiological point of view, CFF is related to modulation of the firing rates of retinal neurons. As with the CSF for spatial vision, the temporal contrast sensitivity function (TCSF) gives a complete description of temporal vision [3]. The TCSF is generated using a similar approach by presenting a flickering pattern at varying frequency to the subjects and recording the minimum contrast threshold needed to perceive the flicker at each frequency. Figure 4.5b shows the TCSF plots for different luminance levels as a function of frequency. As can be noted, this frequency response is also a band-pass filter with a high temporal frequency cutoff and a low temporal frequency roll-off. The peak temporal contrast sensitivity frequency shifts from approximately 20 Hz to 5 Hz as the luminance decreases, while the cutoff at high temporal frequency goes from around 50 Hz to approximately 10 Hz.

4.2.3 Color Discrimination

The ability of the human eye to distinguish different color spectra is much superior to our ability to discriminate gray levels. To be more specific, the human eye can discriminate a few hundred distinct gray levels, while at the same time we can differentiate over several hundreds of thousand different colors. The neuronal system of the eye and all its photoreceptors (namely cones) as well as many layers of neurons and visual cortex are behind performing this complex task. As a consequence of color discrimination, we can distinguish multiple objects in a scene by performing color-based segmentation. The perception of color in the retinal neural system is attributed to combination of signals from three types of cone sensitive to the red, green, and blue ranges of the spectrum, i.e. long, medium, and short wavelengths, respectively. This implies that the human eye is trichromatic in that we can perfectly match any color by mixing only three wavelengths each with adjustable intensity. The trichromatic system is used to match the color of any single wavelength using mixtures of three primary colors [1, 3].

The word color refers to the effects of light stimuli measured by three attributes that are *Hue*, *Saturation*, and *Intensity* or brightness [3, 8]. The hue depends on the wavelength of the color, e.g. redness, greenishness, etc. while intensity and saturation are representative of brightness and purity of a hue. Purity or saturation refers to how a particular wavelength is diluted by *white* light. High saturation corresponds to colors that are highly pure, i.e. less affected by white light and vice versa. Hue and saturation together provide the chromaticity, or color content of a pixel, while intensity, which is independent of hue and saturation, gives the brightness of the light. These three attributes form a 3-D conical color space referred to as HSI space where the vertical axis represents intensity (I), the horizontal axis represents saturation (S) and the hue (H) corresponds to the azimuth angle usually measured anticlockwise, as shown in Figure 4.6a. Any point in this space represents a specific combination of these three basic components. As can be noted, for the same hue the perceived color can be different due to the other two components. For a fixed intensity, the red, green, and blue locations are indicated in Figure 4.6b.

Colorimetry or color-based science is not only essential for understanding human visual color perception and the related physiological and medical research but also of great importance to many different imaging and printing industries to develop and evaluate better quality systems. The Commission Internationale de l'éclairage (CIE) has been responsible for developing universal standards in this area since the 1930s. The three primary wavelength regions within $\lambda \in [380\text{ nm}, 780\text{ nm}]$ and the associated *spectral sensitivity functions* $s_1(\lambda)$, $s_2(\lambda)$ and $s_3(\lambda)$ defined by the CIE standard for

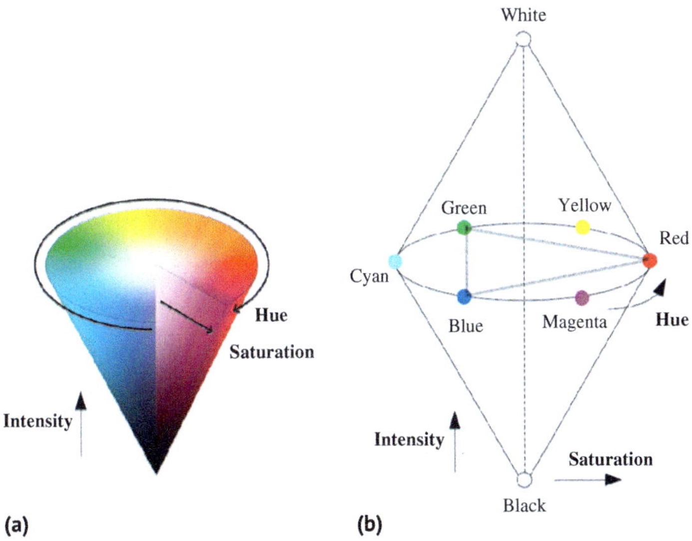

Figure 4.6 (a) HSI conical color space and (b) geometry of the HSI space showing locations of the primary and secondary colors and color triangle. *Source*: Adapted from [11] 2015 / https://www.researchgate.net/publication/282544600_Quantitative_Assessment_of_Flame_Stability_Through_Image_Processing_and_Spectral_Analysis/last accessed on Jan 30, 2026.

Figure 4.7 CIE color matching functions.

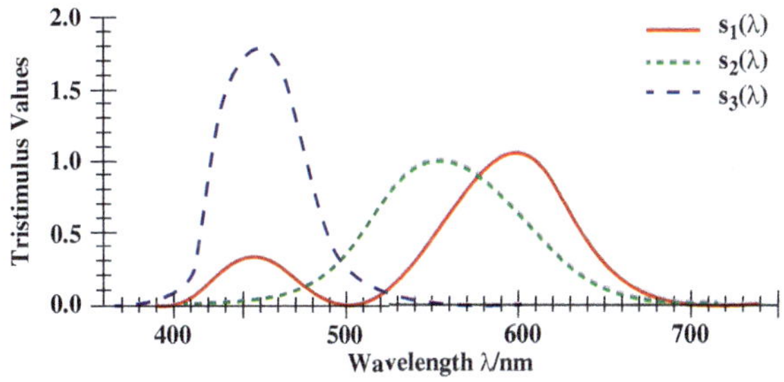

color matching and reproduction [12] are depicted in Figure 4.7. These functions represent the observer's spectral sensitivity functions (or color matching functions) for three cone receptors. The peaks of these functions occur at approximately 610 nm, 546 nm, and 437 nm corresponding to red, green, and blue primary colors. In trichromatic space, these three primary wavelength regions are used to derive three color attributes known as *tristimulus values* $X-Y-Z$, the additive mixture of which can produce any particular color. These tristimulus values for the *reflective* case [3] are given by:

$$
\begin{aligned}
X &= \frac{1}{\rho}\int_{\lambda} r(\lambda)i(\lambda)s_1(\lambda)d\lambda \\
Y &= \frac{1}{\rho}\int_{\lambda} r(\lambda)i(\lambda)s_2(\lambda)d\lambda \\
Z &= \frac{1}{\rho}\int_{\lambda} r(\lambda)i(\lambda)s_3(\lambda)d\lambda
\end{aligned}
\tag{4.4}
$$

where $r(\lambda)$ is the spectral reflectance of the surface under illumination with spectral power distribution $i(\lambda)$, and $\rho = \int_\lambda s_2(\lambda)i(\lambda)d\lambda$ is a normalizing factor. For uniform illumination (CIE-E), we simply use $i(\lambda) = 1$ in (4.4). The normalization factor ρ is used to match a perfectly white surface regardless of illumination (the tristimulus values of the reference white are unity). The trichromatic coefficients used to represent a color are then:

$$\begin{aligned} x &= \frac{X}{X+Y+Z} \\ y &= \frac{Y}{X+Y+Z} \\ z &= \frac{Z}{X+Y+Z} = 1 - x - y \end{aligned} \tag{4.5}$$

where z is dependent on the other two trichromatic coefficients. However, in the CIE system $X - Y - Z$, the tristimulus component Y is a measure of luminance, while X and Z collectively represent *chromaticity* of a color (i.e. hue and saturation). The points in the 2-D CIE chromaticity diagram [8, 12] of Figure 4.8a can be used to determine the wavelength composition of a particular color based on its $x - y$ trichromatic coefficients. The trichromatic coefficient z can subsequently be determined using $z = 1 - x - y$. More specifically, the CIE chromaticity diagram illustrates how a particular color in the diagram can be produced using an additive mixture of two corresponding colors with x and y coordinates. The upper curved boundary of the diagram gives the spectral loci of monochromatic light (fully saturated-pure hue) with their corresponding single wavelengths. These wavelengths are illustrated around the boundary. The lower line boundary that connects red and violet points represents colors that can be generated using additive mixtures of the two monochromatic endpoints. The points on the black curve in the chromaticity diagram of Figure 4.8a correspond to the wavelengths that are less saturated, i.e. highly contaminated with white light. The point of equal energy (point E) on this curve at which $x = y = 0.3333$ represents the *pure white light* standard with equal contributions from all three primary colors (red, green, and blue) and hence zero saturation. The other points on this curve correspond to commonly used reference white, e.g. warm white, cool white, and daylight, used in the lighting industry. To determine dominant wavelength of a particular color using the CIE chromaticity diagram, one can draw a line from the point of equal energy (e.g. point E) to the point corresponding to the color, and extend it until it intersects the spectral boundary. Once this line is drawn, the *purity* (or saturation level) of the color is determined by the ratio of the distance of the specific color from the reference white point to the distance of the intersection point on the spectral boundary from the white point. If the purity index is close to one,

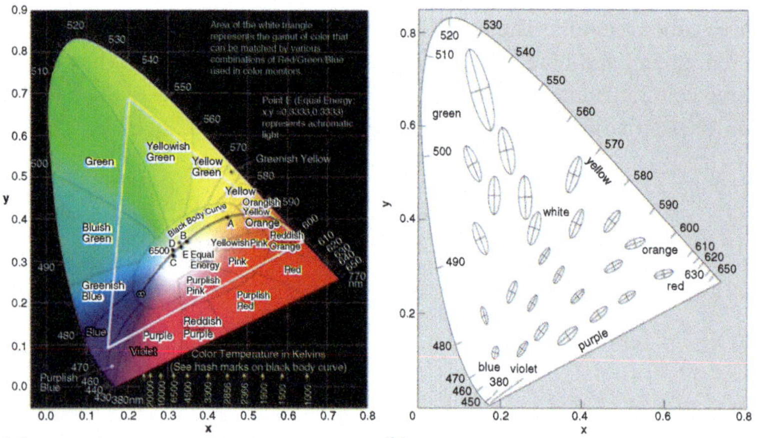

(a) (b)

Figure 4.8 (a) CIE chromaticity diagram. *Source*: LEDtronics, Inc. / https://web.ledtronics.com/last accessed on Jan 30, 2026. (b) Locations of MacAdam ellipses.

the color appears more similar to that associated with the dominant wavelength and hence more saturated. Otherwise, the color appears more white, e.g. pastel colors. The area within the white triangle in the CIE chromaticity diagram of Figure 4.8a corresponds to the color *gamut* that can be reproduced using the primary colors of red, green, and blue on the RGB monitors. Figure 4.8b shows the locations of the so-called MacAdam ellipses [7] in the CIE chromaticity diagram. These ellipses have different sizes and orientations depending on their locations in the diagram. The importance of these ellipses is that all colors within one ellipse are indistinguishable to the human eye.

Example 4.1 *Show that two different monochromatic sources cannot be perceived to have the same color.*

Solution: Let us denote these two different colors by $c_1(\lambda)$ and $c_2(\lambda)$. Now, since they are monochromatic, we have $c_1(\lambda) = P_1\delta(\lambda - \lambda_1)$ and $c_2(\lambda) = P_2\delta(\lambda - \lambda_2)$ where λ_1 and λ_2 are the associated wavelengths and P_1 and P_2 represent their amplitudes. Then, the perceived spectral responses to these colors for cone type i are:

$$g_i(c_1) = \int_\lambda s_i(\lambda)c_1(\lambda)d\lambda = P_1 s_i(\lambda_1) \quad \text{and} \quad g_i(c_2) = \int_\lambda s_i(\lambda)c_2(\lambda)d\lambda = P_2 s_i(\lambda_1)$$

For the two colors to be indistinguishable by the human eye, $g_i(c_1) = g_i(c_2)$, $i = 1, 2, 3$ that is, $P_1 s_i(\lambda_1) = P_2 s_i(\lambda_2)$ or $s_i(\lambda_1) = P_2/P_1 s_i(\lambda_2)$, which can never be satisfied unless both sources are black or white.

4.3 Color Systems

We saw that using an additive mixture of three primary colors, a wide variety of colors can be reproduced for many different devices, including LED televisions and display systems, digital photography, cinematography, printers, etc. [8]. The suitable color gamut for each device varies significantly depending on the specific requirements. Although the RGB color gamut in Figure 4.8a constitutes a large portion of the CIE chromaticity diagram of possible color choices, it is not a universal system. The idea of color systems or spaces is to define different coordinate systems that can be used to represent or map colors for specific applications. These color systems are related to each other via transformations. In this section, we review a few color system standards that are adopted in many different color imaging systems [8].

4.3.1 The RGB System

The RGB color system/space was developed on the basis of the trichromatic nature of the human visual system where there are three types of cones with maximal sensitivity to three different ranges of wavelengths. Moreover, many different colors can be reproduced by using mixtures of three primary colors, and the gamut of such colors is bigger for the RGB system than those of many other systems. The latter characteristic makes the RGB system more popular even though not all possible perceivable colors can be reproduced by the RGB system. A convenient way to represent the RGB system is using a Cartesian coordinate system within a unit cube as shown in Figure 4.9a where Black is at the origin and Red, Green and Blue correspond to the three vertices of the unit cube in R-G-B coordinates (primary colors), and other vertices represent Magenta, Cyan, Yellow, and White (secondary colors). The first three secondary colors lie in the 2-D planes of R-B (Magenta), G-B (Cyan), and R-G (Yellow), while the White color requires all three colors with the same contribution.

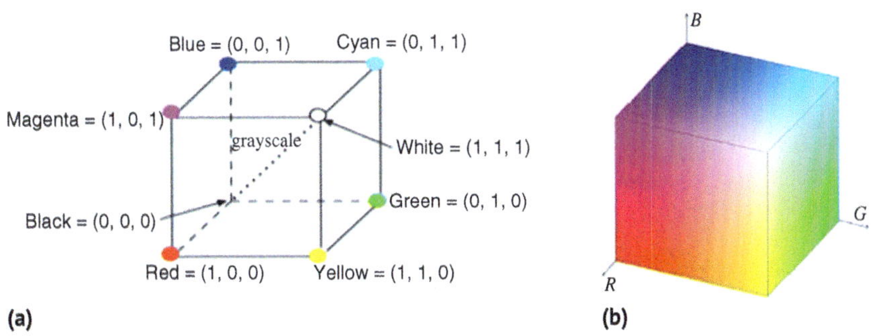

Figure 4.9 (a) Geometry of RGB color space and locations of primary and secondary colors and grayscale line and (b) RGB cubical color space.

The points on the dotted diagonal line connecting Black to White vertices cover different gray levels. The other points inside or on the cube span all possible RGB gamuts as shown in Figure 4.9b. Thus, a pixel at spatial location (m, n) in a full-color image can be represented by a vector of three color components, $\underline{c}(m,n) = [c_r(m,n)\ c_g(m,n)\ c_b(m,n)]^T$, where $c_r(m,n)$, $c_g(m,n)$, and $c_b(m,n)$ are the corresponding coordinates of the color within the RGB unit cube of Figure 4.9b. Clearly, a similar vector representation holds for other color systems that will be studied in the following sections. This implies that a color image must be treated as a *multivariate* spatial process and not as three separate or unrelated univariate spatial processes for each color component/channel. Thus, in several subsequent chapters we will show how different image processing operations can be carried out taking into account this true multivariate nature of the color image data.

4.3.2 The CMYK System

As shown in Figure 4.9a, the three secondary colors at the vertices of the RGB unit cube are Cyan, Magenta, and Yellow, which can be mixed to produce the CMY color gamut. The CMY color gamut is smaller than that of the RGB since unlike the RGB system, which is additive, the CMY mixing is subtractive. Subtractive mixing absorbs radiant power rather than emitting or reflecting the light as in the additive case. Consequently, the CMY system is widely adopted in color printers that use absorbent inks and pigments [8]. The primary absorption spectral regions of Cyan, Magenta, and Yellow correspond to those of Red, Green, and Blue regions, respectively. The RGB system can be mapped to the CMY system (see Figure 4.9a) using:

$$\begin{bmatrix} \mathrm{C} \\ \mathrm{M} \\ \mathrm{Y} \end{bmatrix} = \begin{bmatrix} 1 \\ 1 \\ 1 \end{bmatrix} - \begin{bmatrix} \mathrm{R} \\ \mathrm{G} \\ \mathrm{B} \end{bmatrix}.$$

Figure 4.10 shows the basic idea of subtractive mixing to produce RGB colors and black using the CMY system. However, since generating black color using these three colorants is inefficient and unnecessary, a fourth colorant (black ink) called *key* is added to all color printers. This carbon black key is denoted by the letter K, thus leading to the CMYK system. There are more recent variants of the CMYK system used in higher-quality printers that have a larger color gamut [8].

4.3.3 The HSI System

As stated earlier, the HSI color space is closely related to human perception of color, since we see colors by their hue, saturation, and intensity rather than red, green, and blue. Moreover, for some color image processing operations the HSI system is the preferred choice because unlike the RGB system, which requires multivariate processing of color images, each color component in the HSI system can be processed independently. Figure 4.6 illustrates the geometry of the HSI color

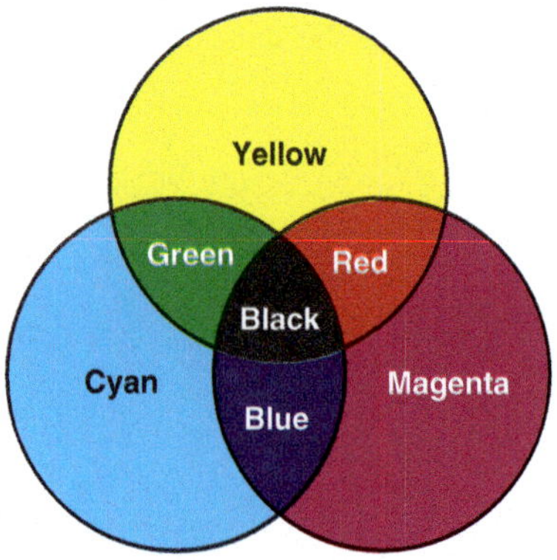

Figure 4.10 Subtractive color mixing used in printing industry to generate various colors.

Table 4.1 Color space conversion formulae.

RGB-to-HSI	HSI-to-RGB
$\phi = \cos^{-1}\left\{\frac{0.5(2R-G-B)}{\sqrt{[(R-G)^2+(R-B)(R-G)]}}\right\}$	For $0^\circ \le H < 120^\circ$,
$H = \begin{cases} \phi, & \text{if } B \le G \\ 360^\circ - \phi, & \text{if } B > G \end{cases}$	Then $R = I\left[1 + \frac{S\cos H}{\cos(60^\circ - H)}\right]$
	$B = I(1-S)$, and $G = 3I - (R+B)$
	For $120^\circ \le H < 240^\circ$, set $H = H - 120^\circ$,
$S = 1 - \frac{3}{R+G+B}[\min(R, G, B)]$	Then $R = I(1-S)$ $G = I\left[1 + \frac{S\cos H}{\cos(60^\circ - H)}\right]$
	and $B = 3I - (R+G)$
	For $240^\circ \le H \le 360^\circ$, set $H = H - 240^\circ$,
$I = \frac{1}{3}(R+G+B)$	Then $G = I(1-S)$, $B = I\left[1 + \frac{S\cos H}{\cos(60^\circ - H)}\right]$
	and $R = 3I - (G+B)$

system which, unlike the unit cube for the RGB system, is conical in shape. In this conical geometry, the central vertical axis connecting black to white vertices corresponds to the grayscale intensity axis where different gray levels lie along this line. The intensity increases when traversing from black to white along this axis. On the other hand, saturation or purity of a color increases as we get closer to the outer surface of the cone. The hue increases anticlockwise as we increase the angle from the red point. The primary RGB colors are 120 degrees apart from each other on the main circumference of the cone in Figure 4.6 where red is at angle $\theta = 0$. The secondary colors, i.e. cyan, magenta, and yellow, are also 120 degrees apart from each other, while 60 degrees apart from the primary ones. The equilateral triangle with vertices at the points R, G, and B is known as the *color triangle*. The colors corresponding to the sides of this triangle are pure colors with no dilution by white light, whereas getting closer to the center point increases the amount of white-light pollution (gray). In RGB space, the plane of the color triangle is perpendicular to the gray line that connects the white to black vertices [12].

Because in the HSI system, intensity is largely decoupled from hue and saturation, one can perform most image processing tasks on the intensity image without significantly affecting hue or saturation, hence preserving the original color presentation. The result can then be converted to the RGB system for display purposes. This is in contrast to processing RGB colors directly, which usually results in adversely affecting the color quality and visually unpleasing effects. Consequently, it will be desirable to develop mapping from the RGB system to the HSI system and reverse. One can then map an RGB image to the HSI space, perform the desired image processing operations on the intensity component, and then apply an inverse mapping to convert the image back to the RGB space. The conversion equations are given in Table 4.1 without derivation. The reader can refer to [12, 13] for details of the derivations. Clearly, the downside is the increased computational cost due to this conversion between RGB and HSI, which involves trigonometric operations as given in Table 4.1.

4.3.4 The YC_bC_r System

This color space and its variants are popular for digital video and DVD/Blu-ray disk players. In the YC_bC_r color system, the colors are mixtures of the luminance component Y (luma) and C_b and C_r, which are chroma components. The luma component is generated using a weighted combination of the R-G-B channels, in which the weights depend on the type of system using this color system. The chroma components represent the difference between blue and luma $C_b = B - Y$ and the difference between red and luma $C_r = R - Y$.

We conclude by pointing out that there are many other useful color spaces [14] that have not been covered in this chapter. These color spaces have been designed specifically for certain types of imaging systems, including digital photography, cinematography, and video.

4.4 Problems

Problem 4.1: Show how Weber–Fechner equation was derived from Weber's law and explain the reasons behind it.

Problem 4.2: Can Stevens' law be derived from a version of Weber–Fechner's law? Which segment of the threshold–intensity plot in Figure 4.3 does this law represent?

Problem 4.3: In [7], a model for the radial MTF of the human visual system was given as:

$$H(\xi_1, \xi_2) = H(\eta) = A\left[\alpha + \left(\frac{\eta}{\eta_0}\right)\right] \exp\left[-\left(\frac{\eta}{\eta_0}\right)^{\beta}\right]$$

where A, α, β, and η_o are constants and $\eta = \sqrt{\xi_1^2 + \xi_2^2}$ represents radial spatial frequency (cycles/deg).

a. Determine the frequency selectivity of this filter in terms of η.
b. Find the expressions for the peak and cut-off frequencies.
c. For $A = 2.6$, $\alpha = 0.0192$, $\eta_0 = 8.772$ and $\beta = 1$, plot this function and determine the values of the frequencies in Part (b).

Problem 4.4: In [15], an analytic formula for the mean radial optical MTF of the human eye was proposed. This simple one-parameter radial MTF is given by:

$$H(\xi_1, \xi_2) = H(\eta) = \left[1 + \left(\frac{\eta}{\eta_1}\right)^2\right]^{-0.62} \sqrt{D(\eta, d, \lambda)}$$

where

$$D(\eta, d, \lambda) = \begin{cases} \frac{2}{\pi}(\cos^{-1}\hat{\eta} - \hat{\eta}\sqrt{1 - \hat{\eta}^2}), & \hat{\eta} < 1 \\ 0, & \hat{\eta} \geq 1 \end{cases},$$

$\hat{\eta} = \frac{\eta}{\eta_0(d,\lambda)}$ with $\eta_0(d, \lambda) = \frac{\pi d 10^6}{\lambda 180}$ and $\eta_1(d) = 21.95 - 5.512d + .3922d^2$. The only parameter is the diameter of the pupil, which can vary $2 \leq d \leq 6$ and $\lambda = 555$.

a. Investigate the effects of each term in $H(\eta)$ by graphing them as functions of the radial spatial frequency $\eta = \sqrt{\xi_1^2 + \xi_2^2}$ for a few choices of d. Comment on the result.
b. Graph $H(\eta)$ and determine its frequency selectivity, i.e. filter type.
c. Using the results in Part (b), find the peak and cutoff frequencies of this filter for various choices of d.

Problem 4.5: The experimental findings have shown simple cells in the retina have a receptive field called Mexican hat, which is often modeled by weights that are Laplacian of Gaussian (LOG) of the form:

$$w(\underline{x}) = \nabla^2\left\{\exp\left\{\frac{-||\underline{x}||^2}{2\sigma^2}\right\}\right\}$$

where $\nabla^2 = \left\{\frac{\partial^2}{\partial u^2} + \frac{\partial^2}{\partial v^2}\right\}$ is the Laplacian operator, $\underline{x} = [u\ v]^T$ and u and v are spatial coordinates.

a. Derive the explicit expression for $w(\underline{x})$ and show that it is radially symmetric, i.e. constant when radial distance $r = \sqrt{u^2 + v^2} =$ constant.
b. Is the area under LOG zero? Why?

Problem 4.6: Cells in the visual pathway have been found to be orientation-selective and therefore more accurately represented by a 2-D Gabor filter:

$$g(\underline{x}) = \exp\left\{-\left(\frac{x^2}{2\sigma_1^2} + \frac{y^2}{2\sigma_2^2}\right)\right\}\cos(\omega_1 u + \omega_2 \upsilon + \phi) \quad \text{even Gabor} \quad \text{or,}$$

$$g(\underline{x}) = \exp\left\{-\left(\frac{x^2}{2\sigma_1^2} + \frac{y^2}{2\sigma_2^2}\right)\right\}\sin(\omega_1 u + \omega_2 \upsilon + \phi) \quad \text{odd Gabor}$$

where $x = u\cos\theta + \upsilon\sin\theta$ and $y = u\sin\theta - \upsilon\cos\theta$ and θ is the orientation of the Gaussian envelope, usually chosen as $\theta = \tan^{-1}\frac{\omega_2}{\omega_1}$.

a. Determine the role of each term in the even or odd versions of the 2-D Gabor filter. Does the Gabor filter have a Mexican hat shape?
b. Assuming $\sigma_1 = \sigma_2$, is this response radially symmetric for the radial distance $r = \sqrt{u^2 + \upsilon^2}$? What about radial spatial frequency $\eta = \sqrt{\xi_1^2 + \xi_2^2}$ where $\omega_1 = 2\pi\xi_1$ and $\omega_2 = 2\pi\xi_2$.

Problem 4.7: Using the CIE standard for the three primary sources (red, green, blue) the spectral energy distributions are $p_i(\lambda) = \delta(\lambda - \lambda_i)$, $i = 1, 2, 3$ where λ_i are the associated wavelengths. Defining the kth cone response to primary source i as $a_{i,k}$, $i, k \in [1, 3]$:

a. Determine $a_{i,k}$ in terms of the spectral sensitivity function of each cone.
b. How does $a_{i,k}$ expression change for a monochromatic color?

Problem 4.8: For Problem 4.7:

a. Find the expressions for the spectral responses of the secondary colors that are obtained by mixing two primary colors (red, green, blue).
b. In the CIE tristimulus $X - Y - Z$ color space, the white point is at $X = Y = Z = 1$. Determine conditions on the spectral sensitivity functions of the cones $s_i(\lambda)$ that must be satisfied at this point.

Problem 4.9: Consider matching a color $c(\lambda) = \delta(\lambda - \lambda_0)$ with wavelength λ_0 using a three-color mixture principle using $c(\lambda) = \sum_{i=1}^{3} \beta_i p_i(\lambda)$ where β_i's are the unknown mixing coefficient and $p_i(\lambda)$, $i = 1, 2, 3$ is the spectral power distribution of the primary color i. Given the spectral sensitivity functions of the three types of cones $s_i(\lambda)$ and the spectral power distribution $p_i(\lambda)$:

a. Develop a method to solve the mixing coefficients.
b. Are the mixing coefficients linear combinations of the spectral sensitivity functions at specific wavelengths?

Problem 4.10: In Problem 4.9, if the same color is to be matched using a different set of three primary colors with spectral power distributions $q_i(\lambda)$, $i = 1, 2, 3$, show that the new mixing coefficients β_i' are linearly related to the previous β_i.

References

1 T. Cornsweet, "*Visual Perception*," New York, Academic Press, 1970.
2 J. E. Korteling, G. C. Van de Boer-Visschedijk, R. A. M. Blankendaal, R. C. Boonekamp, and A. R. Eikelboom, "Human-versus artificial intelligence," *Front Artificial Intelligence*, vol. 4, p. 622364, 2021. https://doi.org/10.3389/frai.2021.622364.
3 M. Ramamurthy and V. Lakshminarayananb, "Human vision and perception," In: R. Karlicek, G. Zissis, and R. Ma, (Editors) *Handbook of Advanced Lighting Technology*, Springer International Pub, 2015.
4 D. H. Hubel and T. N. Wiesel, "Receptive fields, binocular interaction and functional architecture in the cat's visual cortex," *The Journal of Physiology*, vol. 160, pp. 106–154, 1962.

5 T. Norton, D. Corliss, and J. E. Bailey, "*Psychophysical Foundations of Visual Perception*," Butterworth-Heinemann, 2002.

6 J. S. Werner and L. M. Chalupa, "*The New Visual Neurosciences*," Cambridge, MIT Press, 2013.

7 A. K. Jain, "*Fundamentals of Digital Image Processing*," Prentice-Hall, 1989.

8 J. Chen, W. Cranton, and M. Fihn, "*Handbook of Visual Display Technology*," Springer, 2012.

9 S. Zdravković, "Mach bands," In: R. Luo, (editor) *Encyclopedia of Color Scienceand Technology*, Springer, pp. 1–5, 2015.

10 P. J. Kohler and A. D. Clarke, "The human visual system preserves the hierarchy of two-dimensional pattern regularity," Proceedings of the Royal Society B: Biological Sciences, vol. 288, pp. 20211142, 2021.

11 D. Sun, G. Lu, H. Zhou, Y. Yan, and S. Liu, "Quantitative assessment of flamestability through image processing and spectral analysis", *IEEE Transactions on Instrumentation and Measurement*, vol. 64, pp. 3323–3333, 2015.

12 R. C. Gonzalez and R. E. Woods, "*Digital Image Processing*," Pearson, 2018.

13 Unit 8. "*Processing Color Images*," University of Nevada, 2018.

14 M. D. Fairchild, "*Color Appearance Models*," Wiley, 3rd Edition, 2013.

15 A. B. Watson, "A formula for the mean human optical modulation transfer function as a function of pupil size," *Journal of Vision*, vol. 13, pp. 1–11, 2013.

5

Image Sampling and Quantization

The most basic requirement for processing images is that they must be available in digital forms, i.e. 2-D arrays of integer numbers. For digitization, the image is first *sampled* on a discrete grid, and then each pixel is *quantized* to an integer value representing a particular gray level or color component. The entire process is depicted in Figure 5.1 where the acquired continuous image $x(u, v)$ is first sampled to produce the image $x_s(m, n)$ and then quantized to give the digital image $x(m, n)$. These processes are covered in this chapter for both grayscale and color images.

5.1 Image Sampling

5.1.1 2-D Sampling Theorem

Definition 5.1 A continuous image $x(u, v)$ is called *band-limited* if its 2-D frequency spectrum $X(\omega_1, \omega_2)$ is zero outside a bounded region in the frequency domain, i.e. $X(\omega_1, \omega_2) = 0 \quad |\omega_1| > \omega_{10}, |\omega_2| > \omega_{20}$ where ω_{10}, ω_{20} are the band limits of the image [1, 2]. Note that if the spectrum is circularly symmetric, then there is a single band limit $\omega_0 := \omega_{10} = \omega_{20}$.

We sample $x(u, v)$ using an ideal 2-D sampling function, $s_a(u, v)$, defined as:

$$s_a(u, v; \Delta u, \Delta v) := \sum_{m=-\infty}^{\infty} \sum_{n=-\infty}^{\infty} \delta(u - m\Delta u, v - n\Delta v) \tag{5.1}$$

where Δu and Δv are the sampling spacings in the spatial domain. Then, the sampled image is:

$$\begin{aligned} x_s(u, v) &= x(u, v) s_a(u, v; \Delta u, \Delta v) \\ &= \sum_{m=-\infty}^{\infty} \sum_{n=-\infty}^{\infty} x(m\Delta u, n\Delta v)\delta(u - m\Delta u, v - n\Delta v). \end{aligned} \tag{5.2}$$

The choice of the sampling spacings Δu and Δv determine the *spatial resolution* of the sampled image. That is, the smaller the sampling spacings, the higher the spatial resolution. However, the lower bound on the sampling spacings is given by the following result.

Equation (5.2) gives the relationship between the original image $x(u, v)$ and the sampled version $x_s(u, v)$ (or $x_s(m, n)$) in the spatial domain, stating that the sampled image is a collection of pixels from the original image $x(m\Delta u, n\Delta v)$ equally spaced in grids of size $\Delta u \times \Delta v$. To find the corresponding relationship in the frequency domain, we use the result

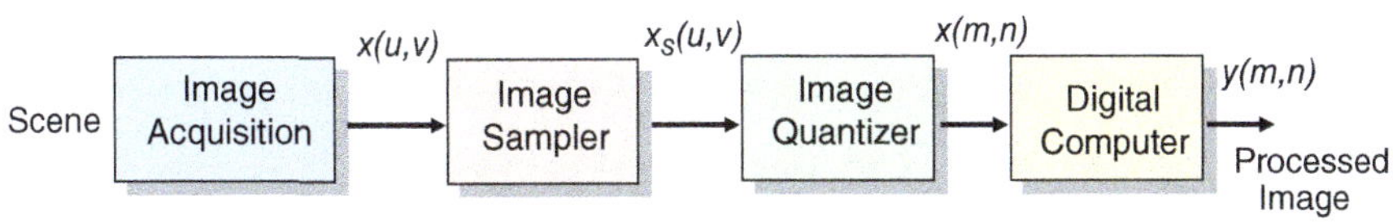

Figure 5.1 Image digitization involves sampling and quantization.

(see Problem 5.3) that the Fourier transform (FT) of the sampling function with spacing Δu and Δv is another sampling function with spacings $\frac{2\pi}{\Delta u}$ and $\frac{2\pi}{\Delta v}$. That is:

$$\mathcal{F}\{s_a(u, v; \Delta u, \Delta v)\} = \frac{4\pi^2}{\Delta u \Delta v} \sum \sum_{k,l=-\infty}^{\infty} \delta\left(\omega_1 - \frac{2\pi k}{\Delta u}, \omega_2 - \frac{2\pi l}{\Delta v}\right)$$
$$= \frac{4\pi^2}{\Delta u \Delta v} s_a\left(\omega_1, \omega_2; \frac{2\pi}{\Delta u}, \frac{2\pi}{\Delta v}\right)$$

where $\mathcal{F}\{.\}$ stands for FT operation. Using the convolution property in the frequency domain (see Property 5 of FT in Section 3.1.5) the 2-D FT of the sampled image is:

$$X_s(\omega_1, \omega_2) = \frac{1}{4\pi^2}\left[X(\omega_1, \omega_2) ** \frac{4\pi^2}{\Delta u \Delta v} s_a\left(\omega_1, \omega_2; \frac{2\pi}{\Delta u}, \frac{2\pi}{\Delta v}\right)\right]$$
$$= \frac{1}{\Delta u \Delta v} \sum_{k=-\infty}^{\infty} \sum_{l=-\infty}^{\infty} X\left(\omega_1 - \frac{2\pi k}{\Delta u}, \omega_2 - \frac{2\pi l}{\Delta v}\right). \tag{5.3}$$

Alternatively, defining the *sampling frequencies* $\omega_{us} = \frac{2\pi}{\Delta u}$ and $\omega_{vs} = \frac{2\pi}{\Delta v}$, we get:

$$X_s(\omega_1, \omega_2) = \frac{1}{4\pi^2} \omega_{us} \omega_{vs} \sum_{k=-\infty}^{\infty} \sum_{l=-\infty}^{\infty} X(\omega_1 - k\omega_{us}, \omega_2 - l\omega_{vs}). \tag{5.4}$$

Now, if we define *discrete* frequency variables $\Omega_1 = \omega_1 \Delta u$ and $\Omega_2 = \omega_2 \Delta v$, then (5.3) can be rewritten as:

$$X_s(\Omega_1, \Omega_2) = \frac{1}{\Delta u \Delta v} \sum_{k=-\infty}^{\infty} \sum_{l=-\infty}^{\infty} X\left(\frac{\Omega_1 - 2\pi k}{\Delta u}, \frac{\Omega_2 - 2\pi l}{\Delta v}\right) \tag{5.5}$$

This result states that the FT of a sampled image is the scaled (by $\frac{1}{\Delta u \Delta v}$ or $\frac{1}{4\pi^2}\omega_{us}\omega_{vs}$), periodic replication of the FT of the original image. That is, $X_s(\omega_1, \omega_2)$ is doubly periodic with periods ω_{us} and ω_{vs}. Thus, *sampling in the spatial domain causes periodicity in the frequency domain*.

Figure 5.2 shows the contour plot of a band-limited 2-D frequency spectrum of an image with band limits ω_{10} and ω_{20} sampled at sampling frequencies ω_{us} and ω_{vs}. As can be seen, to avoid overlap between the periodic replica, the lower bounds on the sampling frequencies should be $2\omega_{10}$ and $2\omega_{20}$, i.e. $\omega_{us} \geq 2\omega_{10}$ and $\omega_{vs} \geq 2\omega_{20}$. These lower bound frequencies are referred to as *Nyquist frequencies* and their reciprocals are called *Nyquist intervals*. From a practical point of view, for images, the sampling frequency must be equal to or greater than twice the highest frequency associated with the finest details in the image, i.e. edges and/or textures.

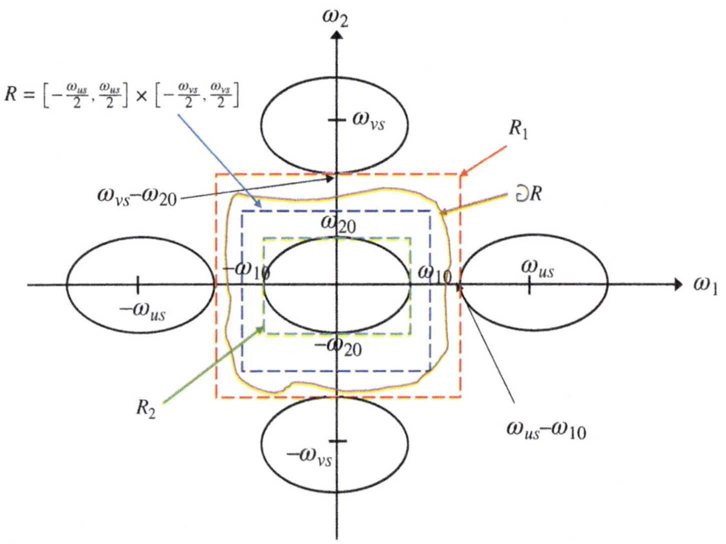

Figure 5.2 2-D spectrum, $X_s(\omega_1, \omega_2)$, of sampled image is periodic replication of $X(\omega_1, \omega_2)$.

5.1.2 Reconstruction

If we sample above the Nyquist frequencies, i.e. $\omega_{us} \geq 2\omega_{10}$, $\omega_{vs} \geq 2\omega_{20}$, then the spectrum $X(\omega_1, \omega_2)$ can be recovered from $X_s(\omega_1, \omega_2)$ by a 2-D low-pass filter (interpolation filter) with frequency response (ideal filter):

$$H(\omega_1, \omega_2) = \begin{cases} \Delta u \Delta v, & \omega_1, \omega_2 \in R \\ 0, & \text{Otherwise} \end{cases} \tag{5.6}$$

where R is any region whose boundary ∂R is contained within the region between R_1 (red) and R_2 (green) rectangles as shown in Figure 5.2. Then, the spectrum of the reconstructed image is:

$$\tilde{X}(\omega_1, \omega_2) = H(\omega_1, \omega_2)X_s(\omega_1, \omega_2) = X(\omega_1, \omega_2) \tag{5.7}$$

That is, exact reconstruction. A typical choice for R is a rectangular passband region (blue) with $R = \left[-\frac{\omega_{us}}{2}, \frac{\omega_{us}}{2}\right] \times \left[-\frac{\omega_{vs}}{2}, \frac{\omega_{vs}}{2}\right]$ for which the PSF is the ideal 2-D interpolation function, i.e. $h(u, v) = \text{sinc}(u\omega_{us})\text{sinc}(v\omega_{vs})$. In this case, the reconstructed image is obtained by convolving the sampled image with this PSF:

$$\begin{aligned} \tilde{x}(u, v) &= h(u, v) ** x_s(u, v) \\ &= \sum_{m=-\infty}^{\infty} \sum_{n=-\infty}^{\infty} x(m\Delta u, n\Delta v)\text{sinc}(\omega_{us}(u - m\Delta u))\text{sinc}(\omega_{vs}(v - n\Delta v)) \end{aligned} \tag{5.8}$$

Note that (5.8) represents an infinite-order interpolation that reconstructs the continuous function $x(u, v)$ from its samples $x(m\Delta u, n\Delta v)$. The basis functions for this representation are $\text{sinc}(\omega_{us}(u - m\Delta u))$ and $\text{sinc}(\omega_{vs}(v - n\Delta v))$ which extend to infinity. Thus, in practice, it is not possible to achieve perfect interpolation and reconstruction. A more practical solution is to use a zero-order or first-order hold device [2], which is obtained by using the Taylor series expansion of the continuous function $x(u, v)$ around a particular point. To see this, let us extend $x(u, v)$ around point (u_0, v_0) using Taylor series expansion:

$$x(u, v) = x(u_0, v_0) + \frac{\partial x(u, v)}{\partial u}\Big|_{u_0,v_0}(u - u_0) + \frac{\partial x(u, v)}{\partial v}\Big|_{u_0,v_0}(v - v_0) + \ldots.$$

For zero-order hold (ZOH), the approximation only includes the first term. Now, let $u_0 = m\Delta u$ and $v_0 = n\Delta v$. Then, ZOH performs the following interpolation:

$$\tilde{x}(u, v) = x(m\Delta u, n\Delta v) \quad \text{for } m\Delta u \leq u < (m + 1)\Delta u, \;\; n\Delta v \leq v < (n + 1)\Delta v$$

That is, ZOH holds each sampled value constant over a rectangular region (pixel cell). In practice, this corresponds to repeating each pixel value along a scan line and then repeating the scan lines. The point spread function (PSF) of the ZOH is a rectangular function of the form $h(u, v) = \text{rect}\left(\frac{u}{\Delta u}, \frac{v}{\Delta v}\right)$, where

$$\text{rect}(u, v) = \begin{cases} 1, & |u| \leq \frac{1}{2}, \quad |v| \leq \frac{1}{2}, \\ 0, & \text{otherwise.} \end{cases}$$

5.1.3 Undersampling and Aliasing Effects

If the sampling frequencies are below the Nyquist frequencies, i.e. $\omega_{us} < 2\omega_{10}$ and/or $\omega_{vs} < 2\omega_{20}$, then the periodic replications of $X(\omega_1, \omega_2)$ will overlap, resulting in a distorted spectrum $X_s(\omega_1, \omega_2)$ from which $X(\omega_1, \omega_2)$ cannot be recovered. The frequencies above half the sampling frequencies, $\frac{\omega_{us}}{2}$ and $\frac{\omega_{vs}}{2}$, are called the *fold-over frequencies*. The overlapping causes these high frequency components in the original image to fold-over and appear as low frequency components (below $\frac{\omega_{us}}{2}$ and $\frac{\omega_{vs}}{2}$) in the sampled image. This phenomenon is called *Aliasing*, which cannot be removed by post-filtering, but can be avoided by prefiltering the image with a low-pass filter prior to sampling so that its band limits are less than one-half of the sampling frequencies. Aliasing in images causes smearing of edges and textures and loss of detail.

To quantify the aliasing effect after the reconstruction, let us rewrite (5.4) as:

$$X_s(\omega_1, \omega_2) = \frac{1}{\Delta u \Delta v}[X(\omega_1, \omega_2) + E(\omega_1, \omega_2)],$$

where $E(\omega_1, \omega_2)$ represents the collective effect of the periodic replica, that is:

$$E(\omega_1, \omega_2) = \sum_{\substack{k,l=-\infty \\ (k,l)\neq(0,0)}}^{\infty} \sum X(\omega_1 - k\omega_{us}, \omega_2 - l\omega_{vs}).$$

After filtering with a 2-D ideal low pass filter

$$H(\omega_1, \omega_2) = \begin{cases} \Delta u \Delta v, & |\omega_1| \leq \frac{\omega_{us}}{2}, |\omega_2| \leq \frac{\omega_{vs}}{2} \\ 0, & \text{otherwise} \end{cases} \tag{5.9}$$

the spectrum of the reconstructed undersampled image is:

$$\tilde{X}(\omega_1, \omega_2) = H(\omega_1, \omega_2)X_s(\omega_1, \omega_2) = X(\omega_1, \omega_2) + \tilde{E}(\omega_1, \omega_2) \tag{5.10}$$

where $\tilde{E}(\omega_1, \omega_2) = H(\omega_1, \omega_2)E(\omega_1, \omega_2)$. Thus, in the spatial domain, we have $\tilde{x}(u, v) = x(u, v) + \tilde{e}(u, v)$ where $\tilde{e}(u, v) = \frac{1}{\Delta u \Delta v}\mathcal{F}^{-1}\{H(\omega_1, \omega_2)E(\omega_1, \omega_2)\}$ represents the aliasing artifacts in the reconstructed image after the low pass filtering (i.e. wrap-around error in passband of the filter):

$$\tilde{e}(u, v) = \frac{1}{4\pi^2} \int_{-\frac{\omega_{us}}{2}}^{\frac{\omega_{us}}{2}} \int_{-\frac{\omega_{vs}}{2}}^{\frac{\omega_{vs}}{2}} E(\omega_1, \omega_2) \exp[j(\omega_1 u + \omega_2 v)] d\omega_1 d\omega_2.$$

Note that here we assume that the reconstruction filter does not impact the image itself.

Example 5.1 *A continuous image $x(u, v) = 2\cos(3u + 4v)$ is sampled at sampling grids $\Delta u = \Delta v = 0.4\pi$. Find the reconstructed image $\tilde{x}(u, v)$ after reconstruction with the ideal low pass filter in (5.9).*

Solution: Using Euler's formula, we expand $\cos(3u + 4v) = (e^{j(3u+4v)} + e^{-j(3u+4v)})/2$, and then use the property $\mathcal{F}\{e^{j\omega_0 t}\} = 2\pi\delta(\omega - \omega_0)$, to find the 2-D spectrum of $x(u, v)$ as:

$$X(\omega_1, \omega_2) = (2\pi)^2[\delta(\omega_1 - 3, \omega_2 - 4) + \delta(\omega_1 + 3, \omega_2 + 4)].$$

As can be seen, this spectrum is band-limited since $X(\omega_1, \omega_2) = 0$ for $|\omega_1| > 3, |\omega_2| > 4$ with band limits $\omega_{10} = 3$ and $\omega_{20} = 4$. The sampling frequencies are $\omega_{us} = \frac{2\pi}{\Delta u} = 5$ and $\omega_{vs} = \frac{2\pi}{\Delta v} = 5$ which are less than the Nyquist frequencies $2\omega_{10} = 6$ and $2\omega_{20} = 8$ hence aliasing is inevitable. Using (5.4), the spectrum of the sampled image is:

$$X_s(\omega_1, \omega_2) = \frac{1}{4\pi^2}\omega_{us}\omega_{vs} \sum_{k=-\infty}^{\infty} \sum_{l=-\infty}^{\infty} X(\omega_1 - k\omega_{us}, \omega_2 - l\omega_{vs})$$

$$= 25 \sum_{k,l=-\infty}^{\infty} \sum [\delta(\omega_1 - 3 - 5k, \omega_2 - 4 - 5l) + \delta(\omega_1 + 3 - 5k, \omega_2 + 4 - 5l)].$$

To reconstruct, we apply the ideal low-pass filter with frequency response:

$$H(\omega_1, \omega_2) = \begin{cases} (0 \cdot 4\pi)^2, & |\omega_1| \leq 2.5, |\omega_2| \leq 2.5 \\ 0, & \text{otherwise} \end{cases}$$

which removes all frequency components in $X_s(\omega_1, \omega_2)$ except the ones inside the passband of the filter, resulting in the spectrum of the reconstructed image:

$$\tilde{X}(\omega_1, \omega_2) = (2\pi)^2[\delta(\omega_1 - 2, \omega_2 - 1) + \delta(\omega_1 + 2, \omega_2 + 1)]$$

Thus, the aliased reconstructed image is $\tilde{x}(u, v) = 2\cos(2u + v)$.

As can be observed from this example, as a consequence of aliasing any frequency component in the original image that is above $\left[\frac{\omega_{us}}{2}, \frac{\omega_{vs}}{2}\right]$ by $(\Delta\omega_u, \Delta\omega_v)$ is reproduced as a lower frequency component at $\left[\frac{\omega_{us}}{2} - \Delta\omega_u, \frac{\omega_{vs}}{2} - \Delta\omega_v\right]$. In this example, the frequency components 3 and 4 are above $\frac{\omega_{us}}{2} = 2.5$ and $\frac{\omega_{vs}}{2} = 2.5$ by $\Delta\omega_u = 0.5$ and $\Delta\omega_v = 1.5$, respectively. Thus, these frequency components are reproduced as $2.5 - 0.5$ and $2.5 - 1.5$, giving $\tilde{x}(u, v) = 2\cos(2u + v)$.

Remark 5.1: Images corrupted by additive wideband noise have frequency spectra with long tails. Thus, sampling based on the band limits of the original image will result in aliasing effects as the tails of the spectrum will fold-over into the band limits of the image. This will result in additional noise in the reconstructed image. To prevent this problem, the image must be prefiltered prior to sampling.

To show the effects of aliasing in images and the importance of using an antialiasing filter prior to sampling, consider the Barbara image in Figure 5.3a. This image, which is of size 512 × 512, is first downsampled by a factor of 2 without any antialiasing filter. Figure 5.3b shows this downsampled image, where the effects of aliasing are clearly noticeable in the textured regions, namely scarf, pants, and tablecloth. Figure 5.3c, on the other hand, illustrates the result of applying antialiasing filtering prior to downsampling, where the aliasing effects are avoided. Finally, Figure 5.3d shows the reconstructed image, obtained by applying ZOH interpolation to the image in Figure 5.3c.

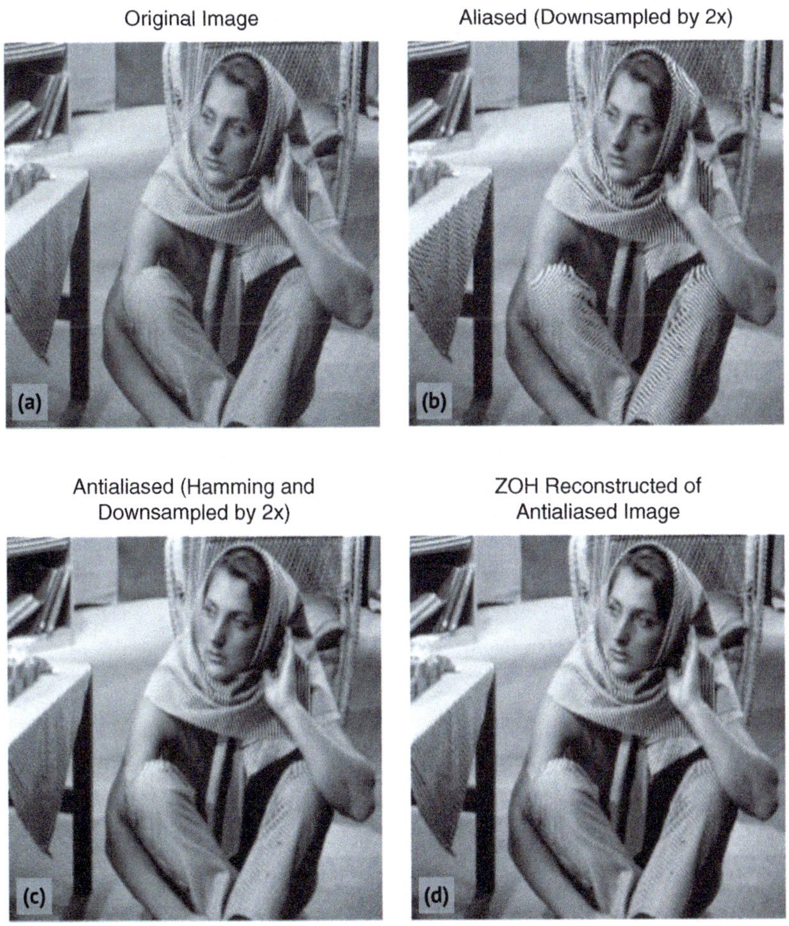

Figure 5.3 (a) Original Barbara image. *Source:* Juang et al. (2012)/John Wiley & Sons/CC BY 3.0. (b) Down-sampled Barbara by 2×, (c) filtered and down-sampled Barbara by 2×, and (d) reconstructed Barbara using ZOH.

5.1.4 Color Image Sampling

Sampling of color images uses essentially similar principles to those covered in the previous sections. Most color imaging sensors use charge-coupled device (CCD) or complementary MOS (CMOS) or N-type MOS (NMOS) to capture images in digital forms. CMOS technologies are used primarily in digital cameras and smartphones because of their lower cost and power consumption. These devices are equipped with millions of *photosites* each containing a photodiode that records a luminance value corresponding to the incident light. A color filter allows only one color, i.e. red (R), green (G), and blue (B), to pass while the others are stopped. This process results in an array of values R, G, and B from which a color image is generated. The most commonly used array is due to Bayer and is therefore known as a Bayer color filter array [3] as shown in Figure 5.4. As can be seen, the most dominant color (by twofold) in this checker-board color pattern array is Green mainly because of the higher sensitivity of the human eye to green versus red and blue. In the Bayer array, each 2 × 2 subarray or *mosaic* of $\begin{matrix} G & R \\ B & G \end{matrix}$ corresponds to a half-resolution pixel. A full-color (visible spectrum) full-resolution image can be generated using the Bayer interpolation or *demosaicing* process or its variants, e.g. the Quad Bayer developed by Sony [4].

As discussed in Section 4.3.3, Hue-Saturation-Intensity (HSI) color space provides a biologically more meaningful full-color representation where the hue component controls the color spectrum (red, green, blue, yellow, etc.), the saturation component regulates color purity, and the intensity component adjusts the brightness. However, as we saw, the HSI model uses an entirely different method to produce a full-color image.

5.2 Image Quantization

A quantizer maps the continuous variable x into a discrete quantity x_q which takes values from a finite set of possible values, denoted as $\{r_1, \dots, r_L\}$ where $r_k \in \mathbb{Z}$. This mapping is generally a staircase function and hence many-to-one, as shown in Figure 5.5.

Quantization Rule:

Define $\{t_k, k = 1, \dots, L+1\}$ as a set of *transition* or *decision* levels with t_1 and t_{L+1} as the minimum and maximum values of x, respectively. Now, if x lies in the interval $[t_k, t_{k+1})$, then $x_q = r_k$, where r_k is called the kth *reconstruction* level.

The precision with which the quantized values x_q capture the information of the original value x is directly proportional to the number of quantization levels L. That is, the higher the value of L, the better the *gray-level or color resolution*. The original and quantized intensities are usually viewed as realizations of continuous and discrete random variables (r.v.s)

Figure 5.4 Bayer color filter array sensor.

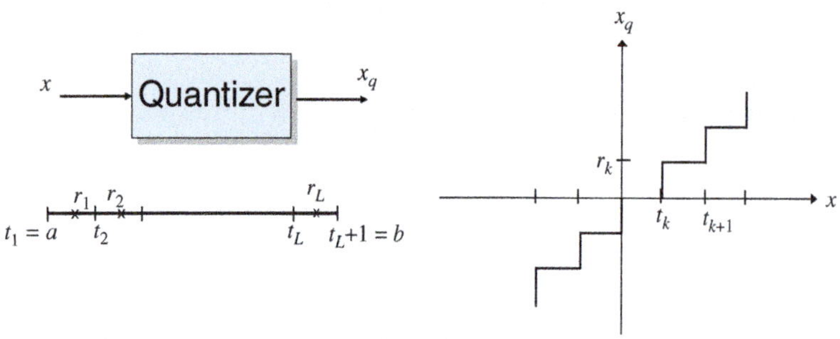

Figure 5.5 A quantizer mapping and staircase function.

X and X_q, respectively where r.v. X takes on values in region $\{X : t_1 \leq x \leq t_{L+1}\}$ and r.v. X_q takes on values in finite set $\{X_q : x_q \in [r_1, r_L]\}$. Thus, quantization is a mapping of a continuous r.v. X to a discrete r.v. X_q.

Properties of Quantization:

1. Quantization mapping is irreversible. That is, for a given quantizer output, the input value cannot be determined or recovered uniquely.
2. Quantization induces distortion $e = x - x_q$. The design of a quantization process is, therefore, a trade-off between simplicity and performance (minimum distortion). For a given L, the quantization error depends on the choices of t_k and r_k.
3. Quantization is a "zero-memory" operation, i.e. output depends on only one input.

5.2.1 Uniform Quantizer

Uniform quantization is the simplest and most commonly used quantizer. Assume that the output of an imaging sensor takes values between 0 and A. If samples are quantized uniformly to L (e.g. $L = 256$) levels, then transition and reconstruction levels for a uniform quantizer are:

$$t_k = \frac{A(k-1)}{L} \quad k = 1, \ldots, L+1$$

$$r_k = t_k + \frac{A/2}{L} \quad k = 1, \ldots, L$$

The interval $q := t_k - t_{k-1} = r_k - r_{k-1}$ is constant for different values of k and is called *quantization interval*.

Example 5.2 *Consider a sampled image whose intensity is represented by a continuous r.v., X, with probability density function (PDF):*

$$p_X(x) = \begin{cases} 1 - |x|, & |x| \leq 1 \\ 0, & \text{elsewhere} \end{cases}$$

Design a four-level ($L = 4$) uniform quantizer and find probability mass function (PMF) of the quantized image X_q.

Solution: For our uniform quantizer, we have $r_k = t_k + A/2L,\ \ k \in [1, L]$ where $A = 2$ and $L = 4$. Thus, transition and reconstruction levels are $t_k = \{-1, -1/2, 0, 1/2, 1\}$ and $r_k = \{-3/4, -1/4, 1/4, 3/4\}$, respectively. Figure 5.6a shows the PDF of this sampled image. All the values of the original image X between $-1 \leq t < -\frac{1}{2}$ are mapped to one $X_q = r_1 = -3/4$ with the associated probability, which is:

$$P(X_q = r_1) = \int_{-1}^{-1/2} (1 + x)dx = 1/8.$$

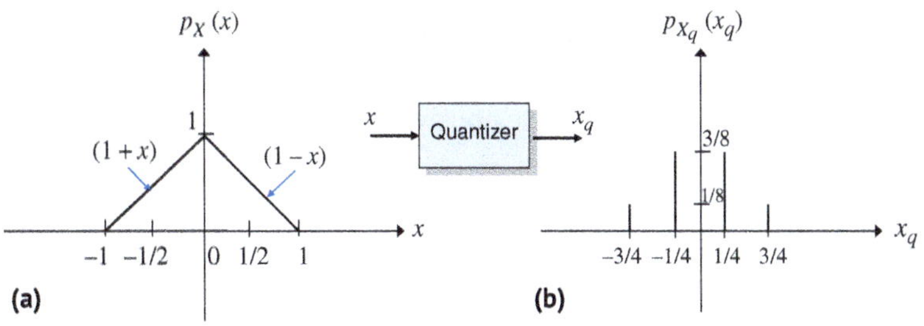

Figure 5.6 (a) PDF of sampled image in Example 5.2; (b) the PMF of the quantized image.

Similarly, for other regions, we have:

$$P(X_q = r_2) = \int_{-1/2}^{0} (1 + x)dx = 3/8$$

$$P(X_q = r_3) = \int_{0}^{1/2} (1 - x)dx = 3/8$$

$$P(X_q = r_4) = \int_{1/2}^{1} (1 - x)dx = 1/8.$$

Figure 5.6b shows the PMF of this sampled and quantized digital image.

5.2.2 Minimum Mean Squared or Lloyd-Max Quantizer

The question here is how should we choose the optimum transition levels, t_k, and reconstruction levels, r_k, to minimize the quantizer's distortion? This quantizer, referred to as Lloyd-Max quantizer [5, 6] minimizes the quantization distortion or the mean squared error (MSE) for a given quantization level L. Let the intensity of pixels in the sampled image, $x(m, n)$ be represented by a real, scalar, and continuous r.v. X with PDF $p_X(x)$, $t_1 \leq x \leq t_{L+1}$.

If the quantization error is $e = x - x_q$, where x_q are realizations of discrete r.v. X_q representing the quantized image, it is desired to find optimum decision levels, t_k, and reconstruction levels, r_k, for an L-level quantizer such that the MSE:

$$\epsilon = E[(X - X_q)^2] = \int_{t_1}^{t_{L+1}} (x - x_q)^2 p_X(x)dx \tag{5.11}$$

is minimized. The MSE in (5.11) can be rewritten as:

$$\epsilon = \sum_{i=1}^{L} \int_{t_i}^{t_{i+1}} (x - r_i)^2 p_X(x)dx, \tag{5.12}$$

since $x_q = r_i$ when $t_i \leq x \leq t_{i+1}$. Minimizing ϵ with respect to unknowns t_k and r_k yields (necessary conditions):

$$\frac{\partial \epsilon}{\partial t_k} = 0 \Rightarrow (t_k - r_{k-1})^2 p_X(t_k) - (t_k - r_k)^2 p_X(t_k) = 0$$

$$\frac{\partial \epsilon}{\partial r_k} = 0 \Rightarrow 2\int_{t_k}^{t_{k+1}} (x - r_k)p_X(x)dx = 0, \quad k \in [1, L],$$

where the first derivative is obtained using the fact that in (5.12), there are two regions that depend on t_k. These are $t_{k-1} \leq x < t_k$ for which $x_q = r_{k-1}$ and $t_k \leq x < t_{k+1}$ for which $x_q = r_k$. Thus, the above equations become:

$$t_k = \frac{r_k + r_{k-1}}{2} \tag{5.13}$$

$$r_k = \frac{\int_{t_k}^{t_{k+1}} x p_X(x)dx}{\int_{t_k}^{t_{k+1}} p_X(x)dx} \tag{5.14}$$

That is, the optimum transition levels lie halfway between the optimum reconstruction levels, which in turn lie at the centroid (center of mass) of the PDF in between the optimum transition levels. Obviously, these two equations are coupled and hence must be solved iteratively given the boundary values t_1 and t_{L+1}.

5.2.2.1 High-Rate Quantization

When the number of quantization levels is large, an approximate solution can be obtained by modeling the smooth PDF $p_X(x)$ as a piecewise constant function [2, 7], i.e.

$$p_X(x) \simeq p_X(\hat{t}_j), \quad t_j \leq x \leq t_{j+1}$$

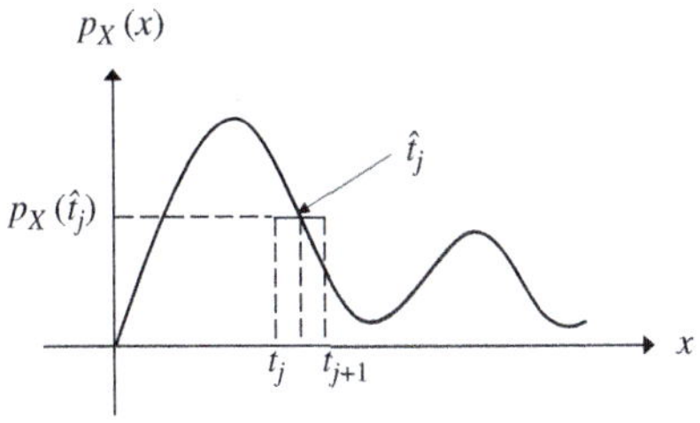

Figure 5.7 Piecewise linear approximation of the image pdf.

where $\hat{t}_j = \frac{t_j+t_{j+1}}{2}$, i.e. the midpoint between t_j and t_{j+1}. Figure 5.7 illustrates this piecewise linear approximation of an arbitrary PDF. Using this approximation and performing the required minimization, the equation for the optimum transition levels becomes:

$$t_{k+1} \approx \frac{(t_{L+1} - t_1)\int_{t_1}^{a_k} [p_X(x)]^{-1/3} dx}{\int_{t_1}^{t_{L+1}} [p_X(x)]^{-1/3} dx} + t_1$$

with $a_k = \frac{k(t_{L+1}-t_1)}{L} + t_1$, $k \in [0, L]$. Now, this together with the optimum reconstruction level $r_k = \frac{t_k+t_{k+1}}{2}$ are two decoupled equations that can be used to find the optimum values of these unknowns. It can be shown [2, 7] that in this case the MSE distortion of the quantizer is given by:

$$\epsilon \approx \frac{1}{12L^2}\left\{\int_{t_1}^{t_{L+1}} [p_X(x)]^{\frac{1}{3}} dx\right\}^3 \tag{5.15}$$

To compare the performance of the optimal Lloyd-Max quantizer with that of the uniform quantizer, we use the original Baboon image in Figure 5.8a, which is quantized to 8 bits/pixel. Figure 5.8b,c shows the results of requantizing this image into 6 bits/pixel using the iterative version of Lloyd-Max quantizer and the uniform quantizer, respectively. The iterative Lloyd-Max implementation starts with initial guesses for the reconstruction levels, then calculates the corresponding transition levels using (5.13), and updates the reconstruction levels using (5.14). These steps are repeated until no significant reduction in distortion is observed. Using the error (difference) images, the MSE values are also computed for each case and presented in the captions of these figures. As can be seen from these figures, the requantized image using the Lloyd-Max quantization exhibits very little smearing compared to the original image in Figure 5.8a. In contrast, the requantized image using the uniform quantization shows very noticeable smearing and patchiness due to loss of gray level resolution. The computed MSEs and SNRs shown in the figure captions also attest to the superiority of the Lloyd-Max quantizer.

Properties of Lloyd-Max Quantizers:

The Lloyd-Max quantizer has several important properties [2] that are discussed next.

1. Unbiased in the mean, i.e. $E[X] = E[X_q]$.

   Proof: Since X_q is a discrete r.v. with values r_k, $k \in [1, L]$, we have:

   $$E[X_q] = \sum_{k=1}^{L} r_k \, p_{X_q}(X_q = r_k)$$

   But using the staircase mapping of the quantizer, the PMF at r_k is $p_{X_q}(X_q = r_k) = \int_{t_k}^{t_{k+1}} p_X(x)dx$. Thus, using (5.14) of the Llyod-Max quantizer we get:

   $$E[X_q] = \sum_{k=1}^{L} r_k \int_{t_k}^{t_{k+1}} p_X(x)dx = \sum_{k=1}^{L} \int_{t_k}^{t_{k+1}} x \, p_X(x)dx,$$
   $$= \int_{t_1}^{t_{L+1}} x \, p_X(x)dx = E[X].$$

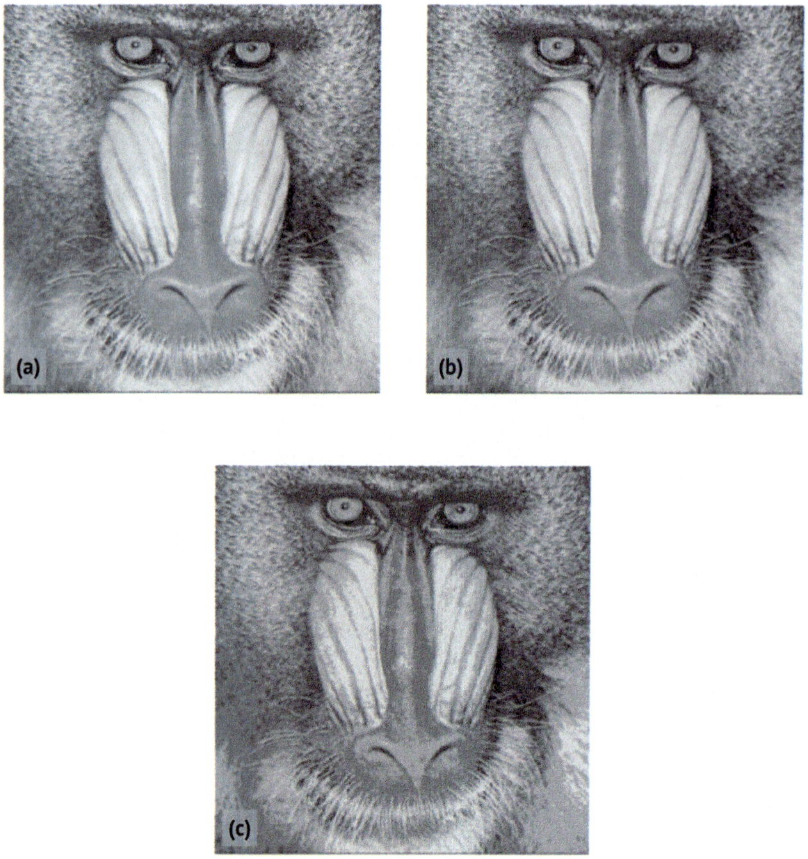

Figure 5.8 Comparison of the Lloyd-Max and uniform quantizers on the Baboon image: (a) original Baboon image, (b) Lloyd-Max quantized-6 bits, MSE = 26 (SNR = 19.9 dB), and (c) uniform quantizer-6 bits, MSE = 238 (SNR=10.7 dB). *Source:* Juang et al. (2012)/ John Wiley & Sons/CC BY 3.0.

2. Orthogonality of the quantization error and quantizer's output, i.e. $E[(X - X_q)X_q] = 0$.

   Proof: It suffices to show $E[XX_q] = E[X_q^2]$. The correlation of r.v.s. X and X_q is:

$$E[XX_q] = \sum_{k=1}^{L} \int_{t_k}^{t_{k+1}} xx_q\, p_{X,X_q}(x, X_q = r_k)dx$$

   When $x \in \Omega_k$ in segment k, i.e. $t_k \leq x < t_{k+1}$, the joint PDF $p_{X,X_q}(x, X_q = r_k)$ can be expressed as $p_{X_q}(X_q = r_k|x \in \Omega_k)\, p_X(x \in \Omega_k) = p_X(x \in \Omega_k)$ since $p_{X_q}(X_q = r_k|x \in \Omega_k) = 1$. Thus:

$$E[XX_q] = \sum_{k=1}^{L} r_k \int_{t_k}^{t_{k+1}} x\, p_X(x)dx$$

   Now, using (5.14) of Llyod-Max quantizer and result in Property 1:

$$E[XX_q] = \sum_{k=1}^{L} r_k \int_{t_k}^{t_{k+1}} x\, p_X(x)dx = \sum_{k=1}^{L} r_k^2\, p_{X_q}(X_q = r_k) = E[X_q^2]$$

3. The result in Property 2 implies that the minimum MSE estimator (MMSE) of r_k is the *conditional mean estimator* (Problem 5.9), i.e.

$$r_k = \int_{t_k}^{t_{k+1}} x\, p_{X_q}(x_q|x \in \Omega_k)dx = E[X_q|X \in \Omega_k].$$

4. If $f(B)$ is the distortion function of the B-bit quantizer for the unity variance input, then the variance of the quantizer's output is $\sigma^2_{X_q} = (1 - f(B))\sigma^2_X$ (Problem 5.10). The distortion function, $f(B)$ is a monotonically decreasing convex function with $f(0) = 1$ and $f(\infty) = 0$.
5. For the special case of a high rate quantizer, each interval makes an approximately equal contribution to the overall MSE, i.e.

$$Pr[t_i \leq X < t_{i+1}]E[(X - X_q)^2 | t_i \leq X < t_{i+1}] \approx Pr[t_j \leq X < t_{j+1}]E[(X - X_q)^2 | t_j \leq X < t_{j+1}], \quad i \neq j.$$

This implies that for the highly probable part of the distribution, conditional MSE is lower and vice versa so that the product remains approximately the same (Problem 5.11).

5.2.2.2 Special Case

It is interesting to note that if a uniformly distributed image is assumed, the Lloyd-Max quantizer reduces to a uniform quantizer. That is, for a uniformly distributed image, the uniform quantizer is indeed the optimal choice [2]. To see this, consider r.v. X with uniform PDF given by:

$$p_X(x) = \begin{cases} \dfrac{1}{t_{L+1} - t_1}, & t_1 \leq x \leq t_{L+1} \\ 0, & \text{otherwise} \end{cases}$$

Then, from (5.14) we get $r_k = \frac{(t^2_{k+1} - t^2_k)}{2(t_{k+1} - t_k)} = \frac{t_{k+1} + t_k}{2}$. Now combining this result with (5.13) yields $t_k = \frac{t_{k+1} + t_{k-1}}{2}$. Thus, $t_k - t_{k-1} = t_{k+1} - t_k = q = \text{constant}$. Also:

$$q = \frac{t_{L+1} - t_1}{L}, \quad t_k = t_{k-1} + q$$

and further $r_k = t_k + q/2$, hence $r_k - r_{k-1} = t_k - t_{k-1} = q$, which implies that all transition and reconstruction levels are equally spaced, i.e. uniform quantizer.

The quantization error $e = x - x_q$ is uniformly distributed over the interval $(-q/2, q/2)$ (Problem 5.12) with the variance (MSE)

$$\epsilon = \frac{1}{q} \int_{-q/2}^{q/2} \alpha^2 d\alpha = \frac{q^2}{12}.$$

If we assume that $t_1 = 0$ and $t_{L+1} = A$, it is simple to show that the variance σ^2_x of image X is $\sigma^2_x = A^2/12$. Now, if the number of quantization level of the B-bit uniform quantizer is $L = 2^B$, the quantization interval will be $q = A/2^B$. This gives the signal-to-quantization error (noise) ratio of:

$$\frac{\sigma^2_x}{\epsilon} = 2^{2B} \quad \text{or} \quad \text{SNR}_{\text{dB}} = 10 \log_{10} 2^{2B} = 6B \text{ dB}.$$

That is, in this special case, every additional bit in the quantization improves SNR by 6 dB.

Remark 5.2: Using a uniform quantizer, smaller intensity values suffer a higher percentage distortion. To see this, assume a sampled image $x(m, n)$ with intensity values in the range of $[0, 255]$ uniformly quantized into $L = 16$ levels, i.e. $t_k = \{0, 15, 31, 47, 63, 79, 95, 111, 127, \dots, 255\}$ and hence $r_k = \{7, 23, 39, 55, 71, 87, 103, 119, 135, \dots, 247\}$. Now compare the following relative changes.

For $x = 12 \in [t_1 = 0, t_2 = 15) \Rightarrow r_1 = 7$ and quantization error $e = x - r_1 = 5$, i.e. 5/12=41.7% change, while for $x = 100 \in [t_7 = 95, t_8 = 111) \Rightarrow r_7 = 103$ and quantization error $e = x - r_7 = -3$, i.e. only 3/100=3% change. A nonlinear dynamic range compressor can *equalize* for these effects that are more visible for low intensities.

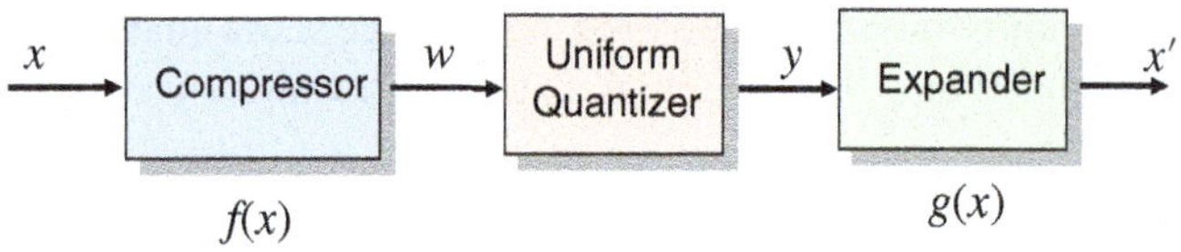

Figure 5.9 A *compandor* consisting of a compressor, a uniform quantizer, and an expander acts as a nonuniform quantizer.

5.2.3 Compandor Design

In the previous section, we showed that for a uniformly distributed image, the Lloyd-Max quantizer degenerates to a uniform quantizer. In other words, in this case, the uniform quantizer is optimum in the MSE sense. Thus, the idea here is to devise a nonlinear mapping that not only compresses the dynamic range (compressor) to overcome the problem stated in Remark 5.1 above but also maps the image to a uniformly distributed one so that a simple uniform quantizer can perform optimally in the mapped domain. The dynamic range of the resultant image is then expanded using another nonlinear transformation (expander). This combination of compressor-expander is called *compandor* [2] and the overall system as shown in Figure 5.9 mimics a nonuniform quantizer. Note that a key benefit of the compandor design over the Lloyd-Max quantizer is that it does not require solving a pair of coupled equations.

The decision and reconstruction levels of the compandor for an even PDF, i.e. $p_X(x) = p_X(-x)$, are:

$$t_k = f^{-1}(kq), \quad t_{-k} = -t_k, \quad k = 0, \dots, L/2$$

$$r_k = g((k - \frac{1}{2})q), \quad r_{-k} = -r_k, \quad k = 1, \dots, L/2$$

The compressor function $f(.)$ and the expander function $g(.)$ can be found [2] to approximate the Lloyd-Max quantizer without any iterative process for finding t_k and r_k. These are:

$$f(x) = 2a\frac{\int_{t_1}^{x} [p_X(u)]^{1/3} du}{\int_{t_1}^{t_{L+1}} [p_X(u)]^{1/3} du} - a$$

$$g(x) = f^{-1}(x)$$

where $[-a, a]$ is the range of $w = f(x)$ over which the uniform quantizer operates. If $p_X(x) = p_X(-x)$, i.e. is even function, then:

$$\begin{aligned} f(x) &= a\frac{\int_{0}^{x} [p_X(u)]^{1/3} du}{\int_{0}^{t_{L+1}} [p_X(u)]^{1/3} du}, \quad x \geq 0 \\ &= -f(-x), \quad x < 0. \end{aligned}$$

For large L, the MSE distortion of the compandor can be approximated as:

$$\varepsilon \approx \frac{1}{12L^2}\left\{\int_{t_1}^{t_{L+1}} [p_X(x)]^{\frac{1}{3}} dx\right\}^3,$$

which is the same as that of the Lloyd-Max quantizer.

Note that the compandor design generally does not make the compressor output w to be uniformly distributed unless the compressor mapping is specifically designed to accomplish this goal. This is shown in the next example.

Example 5.3 *A compandor transformation is:*

$$w = \begin{cases} f(x) = \int_{0}^{x} p_x(\alpha)d\alpha, & x > 0 \\ -f(-x), & x < 0 \end{cases}$$

with the expander mapping $g(.) = f^{-1}(.)$*. Show that this compressor makes* w *(input to the quantizer) uniformly distributed, making the uniform quantizer the optimal choice. If* $p_X(x) = \begin{cases} 1 - |x|, & |x| \le 1 \\ 0, & \text{elsewhere} \end{cases}$

Find t_k *and* r_k *for the overall system when* $L = 4$ *and find the MSE.*

Solution: We realize that the compressor transformation is basically the cumulative distribution function (CDF) of r.v. X, as:

$$f(x) = P_X(x) = \int_0^x p_x(\alpha)d\alpha, \quad \forall x > 0$$

Using this transformation (monotonic), the PDF of compressor's output W becomes:

$$p_W(w) = \left.\frac{p_X(x)}{|f'(x)|}\right|_{x=f^{-1}(w)} = 1$$

That is, uniform. This process is called *histogram equalization* and is also used for contrast enhancement (see Section 8.1.2). Using the above results $f(x) = P_X(x) = \int_0^x (1-\alpha)d\alpha = x - \frac{x^2}{2}$, $x > 0$. Thus:

$$w = \begin{cases} f(x) = x - \dfrac{x^2}{2}, & 0 \le x \le 1 \\ -f(-x) = x + \dfrac{x^2}{2}, & -1 \le x \le 0 \end{cases}$$

Additionally:

$$x = f^{-1}(w) = \begin{cases} 1 - \sqrt{1 - 2w}, & 0 \le w \le 1/2 \\ -1 + \sqrt{1 + 2w}, & -1/2 \le w \le 0 \end{cases}$$

Thus, $w \sim \text{unif}[-1/2, 1/2]$. Now, using a uniform quantizer with $L = 4$ and $q = 1/4$, we get quantizer's transition and reconstruction levels:

$$t_k^w = \{-1/2, -1/4, 0, 1/4, 1/2\} \text{ and } r_k^y = \{-3/8, -1/8, 1/8, 3/8\}$$

The output of the uniform quantizer y is mapped using the expander transformation:

$$x' = g(y) = f^{-1}(y) = \begin{cases} 1 - \sqrt{1 - 2y}, & 0 \le y \le 1/2 \\ -1 + \sqrt{1 + 2y}, & -1/2 \le y \le 0 \end{cases}$$

Now, for the overall system, we need to find transition levels at the input of the compressor x and reconstruction levels at the output of expander x'. Using the transition levels t_k^w and reconstruction levels r_k^y and the inverse functions, we get $t_k^x = \{-1, -0.293, 0, 0.293, 1\}$ and $r_k^{x'} = \{-1/2, -0.134, 0.134, 1/2\}$. We can then use these values to calculate the MSE of this compandor:

$$\epsilon = E[(X - X')^2] = \sum_{i=1}^{4} \int_{t_i^x}^{t_{i+1}^x} (x - r_i^{x'})^2 p_X(x)dx$$

$$\epsilon = \int_{-1}^{-0.293} (x + 0.5)^2(1 + x)dx + \int_{-0.293}^{0} (x + 0.134)^2(1 + x)dx + \int_{0}^{0.293} (x - 0.134)^2(1 - x)dx + \int_{0.293}^{1} (x - 0.5)^2(1 - x)dx = 0.0178$$

It will be interesting to compare this MSE with that of the uniform quantizer in Example 5.2 (Problem 5.5).

5.2.4 Color Image Quantization

In general, all operations used for grayscale images can also be applied to each color channel separately and then combined. Each color component may be quantized at a resolution equivalent to that utilized in a grayscale image, e.g. 8-bits/pixel. The results can then produce a combined 24-bits/pixel true color image. The Lloyd-Max and the uniform quantizers (Figure 5.10) covered earlier can also be applied to the three-channel color images using the same procedures.

For example, we applied these quantizers to the original Finn color image in Figure 5.10a, which was initially quantized to 8 bits/pixel for each color channel. Figure 5.10b shows the results of requantizing the image to 6 bits/pixel per channel using the iterative Lloyd-Max and uniform quantizers, respectively. The corresponding MSE and SNR values are provided for each case. Visual evaluation of the results, along with the computed MSE and SNR values, clearly demonstrates the superiority of the Lloyd-Max quantizer.

Alternatively, one can approach this problem using vector quantization (VQ), a widely used method for quantizing multidimensional data [8–12]. When applied to color image data, the VQ process involves two main stages: (i) *palette* design or *codebook* of representative colors. These colors serve as cluster centroids in the 3-D color space; (ii) mapping that assigns each original pixel's color vector to the nearest color in the designed palette. The core objective of this approach is to significantly reduce the number of unique colors in an image while minimizing visual distortion. Typically, this technique reduces the original 24-bit color depth to an 8-bit representation or even fewer bits/pixel, leading to substantial data compression and efficient storage.

There are generally two classes of color quantization methods, namely, image-independent and image-dependent methods. The latter determines a specific palette depending on the color distribution of the image while the former uses a fixed palette without considering image properties. Various clustering algorithms used for image-dependent color quantization

Figure 5.10 Comparison of the Lloyd-Max and uniform quantizers on the Finn color image: (a) Original Finn, (b) Lloyd-Max quantized-6 bits, MSE = 27 (SNR = 18.4 dB), and (c) uniformly quantized-6 bits, MSE = 225 (SNR = 8.7 dB).

include k-means [8], minmax [9], competitive learning [10], fuzzy c-means [11], and self-organizing maps [12]. The reader is referred to the relevant sources in these areas. In Chapter 17, we present unsupervised learning methods for image compression using VQ.

5.3 Computer Projects

The computer projects in this chapter use Python libraries OpenCV, NumPy, Matplotlib, SciPy, scikit-image.

Computer Project 5.1: A zone plate image of size 512 × 512 is shown in Figure 5.11 which consists of a set of concentric rings, known as Fresnel zones. The rings alternate between being opaque and transparent with a spatial frequency, which varies as a function of the radial distance from the center. The pattern can be considered a type of chirp signal due to the varying frequency/spacing. That is, the rings closer to the center are wider (lower spatial frequency), and the ring width decreases toward the periphery, corresponding to a higher spatial frequency. The equation for this quadratic chirp that is used to generate the zone plate pattern is:

$$z(m, n) = \sin\left(2\pi\left(f_0 R_{m,n} + \frac{k}{2} R_{m,n}^2\right)\right)$$

where the radial distance $R_{m,n}$ is defined as $R_{m,n} = \sqrt{\left(\frac{m}{N} - 0.5\right)^2 + \left(\frac{n}{N} - 0.5\right)^2}$, N is the dimension of the image, m, n are the indices of the point from the origin, and f_0 and k are initial frequency and the sweep rate of the zone plate.

The primary objective of this project is to conduct a comprehensive investigation into the effects of undersampling and aliasing artifacts on a digitally generated zone plate image. The project will evaluate various low-pass antialiasing filters and a ZOH reconstruction filter to quantify their performance in mitigating these effects and restoring image quality. First, generate a high-resolution zone plate image using the above formulations. Downsample the generated image by integer factors of 2 and 4. Analyze and explain the visual effects of aliasing, specifically observing how low- and high-spatial frequency regions of the zone plate are impacted in each case. Document observations regarding the nature and severity of the artifacts introduced (e.g. moiré patterns, stair-stepping, loss of detail). Then, implement and apply at least three different types of low-pass antialiasing filters (e.g. spatial averaging, Gaussian, Hamming/Hanning windowed filters) prior to downsampling at the same factors (2 and 4). Evaluate the effectiveness of each filter in reducing aliasing artifacts. Analyze the results to determine if aliasing effects are totally eliminated, explaining the theoretical reasons (e.g. the Nyquist sampling theorem and filter characteristics) for the observed outcomes. Apply a 2-D ZOH reconstruction filter to all downsampled images without and with prefiltering. Compare the reconstructed images with the original high-resolution image using both visual inspection and quantitative metrics, e.g. Peak Signal-to-Noise Ratio (PSNR) or MSE to objectively assess the quality of reconstruction.

Computer Project 5.2: The goal of this computer project is to study the effects of undersampling and aliasing for a three-channel RGB color image of a dog, "Finn," and his toy in Figure 5.10a. The original high-resolution image is of size 4032 × 3024 × 3 pixels. Apply the same procedures used in Computer Project 5.1 to downsample the image by factors of 4 and 8. The downsampling should be performed both without and with pre-lowpass filtering (antialiasing). Determine

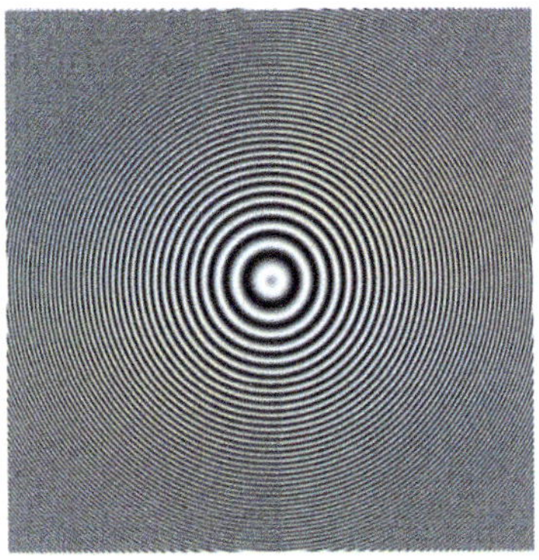

Figure 5.11 A zone plate image.

which spatial filter provides the best overall results in mitigating the aliasing effects and explain the reasons. Reconstruct all downsampled images using a 2-D ZOH interpolation filter. Compare the reconstructed images with the original high-resolution image using visual inspection and other measures (see Computer Project 5.1). To focus the evaluation, concentrate your visual inspection on a small region of interest (ROI) within the image, for example, the toy. What are your observations?

Computer Project 5.3: The purpose of this computer project is to study image quantization using uniform and the optimal Lloyd-Max quantizers and evaluate their performance on the Barbara image (Figure 5.3a). This 512×512 pixel image is originally quantized at 8 bits/pixel. The objective is to requantize this image at B bits/pixel, for $B \in \{4, 5, 6, 7\}$, using both a uniform quantizer and an iterative Lloyd-Max quantizer. The iterative Lloyd-Max quantizer implementation starts with initial guesses for the reconstruction levels, then computes the corresponding transition levels, and repeats this process until no significant reduction in distortion is observed, e.g. using a predefined convergence criterion. Generate the difference images between the original and the requantized images for each case and then compute the MSE and SNR using difference (error) image. Plot these measures versus B and comment on the results. Provide a detailed analysis of these quantizers and state their pros and cons.

Computer Project 5.4: In this computer project, we would like to design a compander (compressor/expander) for the Barbara image in Computer Project 5.3 to mimic the optimal Lloyd-Max quantizer. The core principle is that a uniform quantizer is optimal for a uniformly distributed image. Thus, by using a compressor function that makes the PDF of the image approximately uniform, a subsequent uniform quantizer becomes more effective. As shown in Example 5.3, the compressor function, $f(x)$, that makes the PDF of the output of the compressor approximately uniform is the CDF of image x. However, in this project we replace the CDF by a sigmoidal function of the form $w = f(x) = \frac{1}{1+e^{-x}}$ in order to obtain a simple compressor whose inverse, $f^{-1}(.) = g(.)$ is available in closed form. First, normalize the histogram of the Barbara image to obtain the PDF. Perform the compressor mapping using the sigmoidal function. Apply an $L = 2^B$-level uniform quantizer for $B \in \{4, 5, 6, 7\}$ bits/pixel to the output of the compressor $w = f(x)$. This step is followed by applying the expander function $g(.)$ to the uniformly requantize image y, after the uniform quantization, using the inverse mapping $x' = g(y) = -\ln \frac{1-y}{y}$ where x' represents the requantized image using this nonuniform quantization system. Generate the difference images between the original image, x, and the requantized (output of the expander) image, x', for each value of B and then compute the MSE and SNR using the difference (error) images. To see the benefits of nonuniform quantization in image processing, apply a uniform quantizer directly to the Barbara image without using a compander for the same values of B. Then, using the difference images generate the same plots of the MSE and SNR versus B for this quantizer. Provide a comprehensive evaluation of the performance of both quantization systems. Finally, present a detailed discussion of the pros and cons of each quantization approach, highlighting the benefits of nonuniform quantization in image processing.

5.4 Problems

Problem 5.1: The image $x(u, v) = 2 \cos \alpha u \cos \beta v$ is sampled using $\Delta u = \Delta v = 0.5\pi$. The reconstruction filter is an ideal low-pass filter whose passband lies within $R = \left[-\frac{\omega_{us}}{2}, \frac{\omega_{us}}{2}\right] \times \left[-\frac{\omega_{vs}}{2}, \frac{\omega_{vs}}{2}\right]$.

(a) Find the limits on α and β for perfect reconstruction.
(b) If $\alpha = 3$ and $\beta = 4$, find the reconstructed image.

Problem 5.2: A band-limited image, $x(u, v)$, with band limits $\omega_{10} = \omega_{20} = 1$ is corrupted by additive noise with band limits that extend to 3. Thus, the spectrum of this noisy image is:

$$X(\omega_1, \omega_2) = \begin{cases} 1 + \varepsilon, & |\omega_1, \omega_2| \leq 1 \\ \varepsilon, & 1 < |\omega_1, \omega_2| \leq 3 \\ 0, & \text{elsewhere} \end{cases}$$

where ε represents the effect of noise. This image is sampled at frequencies $\omega_{us} = \omega_{vs} = 2$ and then reconstructed using an ideal 2-D lowpass filter with passband within the frequency region $R = \left[-\frac{\omega_{us}}{2}, \frac{\omega_{us}}{2}\right] \times \left[-\frac{\omega_{vs}}{2}, \frac{\omega_{vs}}{2}\right]$.

(a) Find the spectrum of the reconstructed image.
(b) If the peak signal-to-noise ratio (PSNR) in the noisy image is measured by $1/\varepsilon$ ratio, what is the PSNR in the reconstructed image? Explain your result.

Problem 5.3: Show that the 2-D Fourier transform of sampling function:

$$s_a(u, v; \Delta u, \Delta v) = \sum_{m=-\infty}^{\infty} \sum_{n=-\infty}^{\infty} \delta(u - m\Delta u, v - n\Delta v)$$

is another sampling function:

$$\mathcal{F}\{s_a(u, v; \Delta u, \Delta v)\} = \frac{4\pi^2}{\Delta u \Delta v} s_a\left(\omega_1, \omega_2; \frac{2\pi}{\Delta u}, \frac{2\pi}{\Delta v}\right)$$

This result was used in Section 5.1.

Problem 5.4: Drive the expression for the frequency response of the 2-D ZOH for image reconstruction and comment on its frequency domain behavior.

Problem 5.5: For the uniform quantizer in Example 5.2, compute the MSE and compare your result with that in Example 5.3.

Problem 5.6: Consider a sampled image with PDF:

$$p_X(x) = \begin{cases} \frac{1}{2}x, & 0 \leq x \leq 2 \\ 0, & \text{elsewhere} \end{cases}$$

Design a four-level ($L = 4$) uniform quantizer for this image. Find the PDF of the quantized image and the associated mean squared error (MSE).

Problem 5.7: Repeat Example 5.3 for the PDF in Problem 5.6.

Problem 5.8: Consider a sampled image with a truncated Laplacian PDF:

$$p_X(x) = \begin{cases} c\, e^{|x|}, & -1 \leq x \leq 1 \\ 0, & \text{elsewhere} \end{cases}$$

First, find c for a valid PDF. Then, design a four-level ($L = 4$) uniform quantizer for this image and find the histogram of the quantized image. Determine the quantizer's MSE distortion.

Problem 5.9: Prove Property 3 of the Lloyd-Max quantizer.

Problem 5.10: Prove Property 4 of the Lloyd-Max quantizer.

Problem 5.11: Prove Property 5 of the Lloyd-Max quantizer.

Problem 5.12: Show that for the special case of uniform distribution, the quantization error of the Lloyd-Max quantizer $e = x - x_q$ is also uniformly distributed over the interval $(-q/2, q/2)$. Then, show that the MSE is $\epsilon = \frac{q^2}{12}$.

Problem 5.13: Consider a uniformly distributed image with PDF:

$$p_X(x) = \begin{cases} \frac{1}{\alpha}, & \text{for } x \in [-\alpha/2, \alpha/2] \\ 0, & \text{otherwise} \end{cases}.$$

Using (5.15) find:

(a) The MMSE of the Lloyd-Max quantizer for this image.
(b) The SNR when the noise is due to the quantization distortion.

Problem 5.14: For a zero-mean Gaussian r.v. X with variance σ^2 find the pair of compandor transformations $f(x)$ and $g(y)$.

Problem 5.15: Consider a truncated Laplacian r.v. X with pdf $p_X(x) = ce^{-\alpha|x|}$ where $-A \leq x \leq A$. Find c for a valid pdf and then find the compandor transformation pair $f(x)$ and $g(y)$.

References

1 D. E. Dudgeon and R. M. Mersereau, "*Multidimensional Digital Signal Processing*," Prentice Hall, 1984.
2 A. K. Jain, "*Fundamentals of Digital Image Processing*," Prentice Hall, 1989.
3 B. E. Bayer, "Color Imaging Array," US Patent 3971065A, 1976.
4 SONY, "Sony Releases Stacked CMOS Image Sensor for Smartphones with Industry's Highest 48 Effective Megapixels." Sony Global - Sony Global Headquarters, 2019.
5 J. Max, "Quantizing for minimum distortion," *IEEE Transactions on Information Theory*, vol. 6, no. 1, pp. 7–12, 1960.
6 S. P. Lloyd, "Least squares quantization in PCM," *IEEE Transactions Information Theory*, vol. 28, no. 2, pp. 129–137, 1982.
7 W. K. Pratt, "*Digital Image Processing*," Wiley, 4th Edition, 2007.
8 Y- C. Hu and M-G. Lee, "K-means based color palette design scheme with the use of stable flags," *Journal of Electronic Imaging*, vol. 16, no. 3, pp. 033003, 2007.
9 Z. Xiang, "Color image quantization by minimizing the maximum inter-cluster distance," *ACM Transactions on Graphics*, vol. 16, no. 3, pp. 260–276, 1997.
10 P. Scheunders, "Comparison of clustering algorithms applied to color image quantization," *Pattern Recognition Letters*, vol. 18, no. 11–13, pp. 1379–1384, 1997.
11 D. Ozdemir and L. Akarun, "Fuzzy algorithm for color quantization of images," *Pattern Recognition*, vol. 35, no. 8, pp. 1785–1791, 2002.
12 C.-H. Chang, P. Xu, R. Xiao, and T. Srikanthan, "New adaptive color quantization method based on self-organizing maps," *IEEE Transactions on Neural Networks*, vol. 16, no. 1, pp. 237–249, 2005.
13 Y.-S., Juang, L.-T., Ko, J.-E., Chen, Y.-S., Shieh, T.-Y., & Sung, H.-C. Hsin, "Histogram modification and wavelet transform for high performance watermarking," *Mathematical Problems in Engineering*, Article 164869, 2012, https://doi.org/10.1155/2012/164869

6

Image Transforms

2-D unitary transforms have found numerous applications in digital image processing. Important application areas [1, 2] include: (i) Image data compression using transforms such as 2-D discrete cosine transform (DCT) or 2-D discrete wavelet transform (DWT) to reduce the volume of data by discarding low-magnitude transform coefficients; (ii) Feature extraction to extract salient and representative features from an image for subsequent detection and classification tasks; (iii) Dimensionality reduction to reduce the dimension of the image data by discarding the less informative components; and (iv) Transform-domain filtering for image enhancement and restoration.

In this chapter, we will introduce several important and widely used unitary transforms, including discrete Fourier transform (DFT), DCT, Walsh-Hadamard transform (WHT), and Karhunen–Loeve transform (KLT), and their extensions to the 2-D case for image processing. Important properties of these transforms are also discussed, together with a unified matrix representation which provides better insights into their geometric properties. A specific application of the 2-D KLT for facial pattern representation and classification, along with a step-by-step algorithm, is also presented.

6.1 2-D Discrete Fourier Transform

In Section 3.1.7, we introduced 2-D DFT and many of its important properties, including the separability property, which allows for its efficient implementation using the 1-D DFT. However, the approach presented in this section is based on the matrix representation of the 1-D and 2-D DFT [1]. The main motivations behind the matrix representation framework are to: (i) use a unified and more compact representation that can be applied to all unitary transforms, and (ii) enable much deeper insights into the geometrical properties of these transforms. To this end, we begin by considering the matrix representation of the 1-D DFT and IDFT.

6.1.1 Matrix Representation of 1-D DFT

Consider a finite-extent 1-D sequence $x(n)$, $n \in [0, N-1]$ with 1-D DFT:

$$X(k) = \frac{1}{\sqrt{N}} \sum_{n=0}^{N-1} x(n) W_N^{nk} \quad k \in [0, N-1]$$

where $W_N := e^{-2\pi j/N}$. The IDFT is:

$$x(n) = \frac{1}{\sqrt{N}} \sum_{k=0}^{N-1} X(k) W_N^{-nk} \quad n \in [0, N-1].$$

Note that for the sake of uniformity, we have split the scaling term $\frac{1}{N}$ in the 1-D IDFT between the forward and inverse transforms. Now, if we arrange components of $X(k)$ and $x(n)$ in vector forms, we get the following matrix representation of DFT:

$$\begin{bmatrix} X(0) \\ \vdots \\ X(N-1) \end{bmatrix}_{N\times 1} = \frac{1}{\sqrt{N}} \begin{bmatrix} 1 & 1 & \cdots & 1 \\ 1 & W_N & \cdots & W_N^{N-1} \\ \vdots & \vdots & & \vdots \\ 1 & W_N^{N-1} & \cdots & W_N \end{bmatrix}_{N\times N} \begin{bmatrix} x(0) \\ \vdots \\ x(N-1) \end{bmatrix}_{N\times 1}$$

or in compact form:

$$\underline{X} = \mathbf{W}_N \underline{x} \tag{6.1}$$

where $\mathbf{W}_N$ is the DFT matrix with elements $w_{kn} = \frac{1}{\sqrt{N}} W_N^{kn \bmod N}$, $\underline{x} = [x(0)\, x(1) \cdots x(N-1)]^T$, and $\underline{X} = [X(0)\, X(1) \cdots X(N-1)]^T$ are the signal and DFT coefficient vectors, respectively. Note that since the basis functions of the IDFT are complex conjugate of those of the DFT, the IDFT equation in matrix form becomes:

$$\underline{x} = \mathbf{W}_N^* \underline{X} \tag{6.2}$$

where $\mathbf{W}_N^*$ is the same as $\mathbf{W}_N$ where each element is complex-conjugated. Thus, we have $\mathbf{W}_N^* \mathbf{W}_N = \boldsymbol{I}$ or $\mathbf{W}_N^* = \mathbf{W}_N^{-1}$.

6.1.1.2 Properties of DFT Matrix

a. *Symmetry and Unitary:* The DFT matrix is symmetrical, i.e. $\mathbf{W}_N^T = \mathbf{W}_N$ and unitary since $\mathbf{W}_N^{-1} = \mathbf{W}_N^*$. Combining these two properties yields $\mathbf{W}_N^{-1} = \mathbf{W}_N^H$ where the superscript H denotes the Hermitian or conjugate transpose operation.
b. *Energy Conservation:* The energy is preserved in both domains, i.e. $\|\underline{X}\|^2 = \|\underline{x}\|^2$. Using Property 1 and (6.1) we can write:

$$\|\underline{X}\|^2 \triangleq \sum_{k=0}^{N-1} |X(k)|^2 = \underline{X}^H \underline{X} = \underline{x}^H \mathbf{W}_N^H \mathbf{W}_N \underline{x} = \underline{x}^H \underline{x} = \|\underline{x}\|^2$$

c. Let us rewrite (6.2) as:

$$\underline{x} = \mathbf{W}_N^* \underline{X} = \sum_{k=1}^{N} X(k) \underline{w}_k^* \tag{6.3}$$

where $\underline{w}_k^*$ is the kth column of the IDFT matrix $\mathbf{W}_N^*$ or the kth *basis vector* in this representation. Consequently, (6.3) shows how signal vector $\underline{x}$ can be represented using a linear combination of these basis vectors weighted by their corresponding DFT coefficients $X(k)$. Furthermore, these basis vectors form an orthonormal basis, i.e. $\langle \underline{w}_k^*, \underline{w}_l^* \rangle = \underline{w}_k^H \underline{w}_l^* = \delta(k-l)$. This orthonormality property allows one to uniquely compute the DFT coefficients:

$$X(l) = \underline{w}_l^H \underline{x} = \langle \underline{w}_l, \underline{x} \rangle, \quad \forall l \in [1, N]$$

d. Due to the unitary property, the inner products are preserved in both domains. That is:

$$\langle \underline{x}, \underline{y} \rangle = \langle \mathbf{W}_N^* \underline{X}, \mathbf{W}_N^* \underline{Y} \rangle = \langle \underline{X}, \underline{Y} \rangle.$$

This is known as Parseval's identity.
e. Owing to the unitary property, all the eigenvalues of the DFT matrix $\mathbf{W}_N$ have unity magnitudes, i.e. they lie on the unit circle (Problem 6.5).

6.1.2 Matrix Representation of 2-D DFT

To extend the above matrix representation to the 2-D DFT of a finite-extent image $x(m, n)$ with region of support (ROS) in R_{MN}, one option is to treat the image as a vector (row or column arrangement) of size $MN \times 1$ and then apply the same results as in (6.1) and (6.2). However, this leads to $MN \times MN$ DFT and IDFT matrices, which become computationally prohibitive. Alternatively, we can take advantage of the separability property of the 2-D DFT and apply the 1-D DFT to rows and columns consecutively [1]. That is, if $\mathbf{x}$ and $\mathbf{X}$ are matrices of size $M \times N$ representing the original image and its 2-D DFT image, respectively, the separable 2-D DFT and IDFT pair can be written as:

$$\mathbf{X} = \mathbf{W}_M \mathbf{x} \mathbf{W}_N \tag{6.4}$$

and

$$\mathbf{x} = \mathbf{W}_M^* \mathbf{X} \mathbf{W}_N^* \tag{6.5}$$

respectively. In (6.4), pre-multiplying image matrix $\mathbf{x}$ by the matrix $\mathbf{W}_M$ performs column-wise 1-D DFT operations of length M. The resultant matrix is then post-multiplied by the DFT matrix $\mathbf{W}_N$, which performs row-wise 1-D DFT operations of length N. The same operations are performed on the transformed image $\mathbf{X}$ in (6.5) using the inverse DFT matrices $\mathbf{W}_M^*$ and $\mathbf{W}_N^*$. In the subsequent sections, we will show that other image transforms such as 2-D DCT, WHT, and KLT also admit similar matrix representations.

Let $\underline{w}_k^*$ be the kth column of $\mathbf{W}_M^*$ and $\underline{w}_l^H$ be the lth row of $\mathbf{W}_N^*$, then the $(k, l)^{\text{th}}$ *basis image*, which is also a matrix of size $M \times N$, is:

$$\mathcal{W}^*(k, l) = \underline{w}_k^* \underline{w}_l^H.$$

Now, the 2-D IDFT in (6.5) can be expressed in terms of these basis images using:

$$\mathbf{x} = \sum_{k=0}^{M-1} \sum_{l=0}^{N-1} X(k, l) \mathcal{W}^*(k, l) \tag{6.6}$$

that is, the image is decomposed onto a space spanned by the MN *basis images*. The coefficients (or projection of the image onto the basis images) in this representation are given by:

$$X(k, l) = \langle \mathbf{x}, \mathcal{W}^*(k, l) \rangle \tag{6.7}$$

where the inner product of two matrices $\mathbf{A}$ and $\mathbf{B}$ is defined as $\langle \mathbf{A}, \mathbf{B} \rangle = \sum_{m=0}^{M-1} \sum_{n=0}^{N-1} a(m, n) b^*(m, n)$, where $a(m, n)$ and $b(m, n)$ are the $(m, n)^{\text{th}}$ elements of matrices $\mathbf{A}$ and $\mathbf{B}$, respectively.

6.2 2-D Discrete Cosine Transform

Recall that the DFS of any real, even-symmetric signal contains only real coefficients corresponding to the *cosine* terms. This can be extended to the DFT of a symmetrically extended signal or image (Problem 6.9). There are many ways to symmetrically extend a signal or an image, leading to a variety of DCT types. In this section, we present DCT-II, which is the most widely used version [1–3]. DCT is used in a variety of applications, including image data compression (e.g. in JPEG [4]), video compression (e.g. in MPEG [5]), and feature extraction [6, 7].

6.2.1 Matrix Representation of 1-D DCT

We begin with the scalar form of 1-D DCT-II for a finite-extent signal $x(n)$, $n \in [0, N-1]$:

$$X(k) = \begin{cases} \dfrac{1}{\sqrt{N}} \displaystyle\sum_{n=0}^{N-1} x(n), & k = 0 \\ \sqrt{\dfrac{2}{N}} \displaystyle\sum_{n=0}^{N-1} x(n) \cos\left[\dfrac{(2n+1)k\pi}{2N}\right], & k \in [1, N-1] \end{cases} \tag{6.8}$$

where, unlike the basis functions of DFT, which are complex-valued, the real-valued cosine functions, $\cos\left[\frac{(2n+1)k\pi}{2N}\right]$, form the basis functions of 1-D DCT. This is one of the main reasons for its popularity in signal and image compression. The inverse 1-D DCT or IDCT:

$$x(n) = \frac{1}{\sqrt{N}}X(0) + \sqrt{\frac{2}{N}}\sum_{k=1}^{N-1} X(k)\cos\left[\frac{(2n+1)k\pi}{2N}\right], \quad n \in [0, N-1] \tag{6.9}$$

can be used to recover the signal $x(n)$ from its DCT coefficients $X(k)$. It can be observed from (6.9) that the 1-D IDCT and DCT share identical basis functions. This shared characteristic offers an advantage over transforms such as the DFT, where the basis functions for the forward and inverse transforms are different.

Now, arranging the signal $x(n),\ n \in [0, N-1]$ and DCT coefficients $X(k),\ k \in [0, N-1]$ in vector forms, we get the following matrix representation of (6.8):

$$\underline{X} = \mathbf{C}_N \underline{x} \tag{6.10}$$

where $\underline{x}$ and $\underline{X}$ are the signal and the DCT vectors defined similar to those in the DFT case, and $\mathbf{C}_N$ is the 1-D DCT matrix with elements $c(k, n)$ given by:

$$c(k,n) = \begin{cases} 1/\sqrt{N}, & k = 0, n \in [0, N-1] \\ \sqrt{\frac{2}{N}}\cos\left[\frac{(2n+1)k\pi}{2N}\right], & k \in [1, N-1], n \in [0, N-1] \end{cases}$$

Repeating this process for the inverse DCT in (6.9), we get 1-D IDCT in matrix form as:

$$\underline{x} = \mathbf{C}_N^T \underline{X} \tag{6.11}$$

The properties of the DCT matrix $\mathbf{C}_N$ are discussed next.

6.2.1.1 Properties of DCT Matrix

1. The DCT matrix is real, i.e. $\mathbf{C}_N = \mathbf{C}_N^*$, and *orthogonal* [8] since $\mathbf{C}_N^{-1} = \mathbf{C}_N^T$ or equivalently, $\mathbf{C}_N^T\mathbf{C}_N = \mathbf{C}_N\mathbf{C}_N^T = \mathbf{I}_N$, where $\mathbf{I}_N$ is an identity matrix of dimension N.
2. The DCT matrix also preserves the inner product property due to orthogonality of matrix $\mathbf{C}_N$.
3. Since $\mathbf{C}_N$ is an orthogonal matrix, the basis vectors of the 1-D DCT (i.e. the columns of $\mathbf{C}_N$), are mutually orthonormal. That is, we have $\underline{c}_k^T\underline{c}_l = \delta(k-l)$. This property allows us to compute the DCT coefficients using $X(l) = \langle \underline{c}_l, \underline{x} \rangle = \underline{c}_l^T\underline{x}$.
4. The energy of the signal is highly concentrated in a small number of the DCT coefficients $X(k)$. This *energy compaction* property [1, 2] is a primary reason why the DCT is so effective for data compression applications. The property will be illustrated in an example involving image data in subsequent sections.
5. Note that it is possible to perform circular convolution between two finite-extent signals using the relationship between the DCT of the original signal and the DFT of its symmetrically extended version [9, 10].

Example 6.1 *Show that the $N \times N$ DCT matrix $\mathbf{C}_N$ is an orthogonal matrix, i.e. prove $\mathbf{C}_N\mathbf{C}_N^T = \mathbf{I}_N$.*

Solution: Consider the $(k, l)^{\text{th}}$ (for $k, l \neq 0$) element of $\mathbf{C}_N\mathbf{C}_N^T$ and use the trig identities and Euler's formula to get:

$$[\mathbf{C}_N\mathbf{C}_N^T]_{k,l} = \frac{2}{N}\sum_{n=0}^{N-1}\cos\left[\frac{(2n+1)k\pi}{2N}\right]\cos\left[\frac{(2n+1)l\pi}{2N}\right]$$

$$= \frac{1}{2N}\left[e^{\frac{-j(k+l)\pi}{2N}}\sum_{n=0}^{N-1}e^{\frac{-jn(k+l)\pi}{N}} + e^{\frac{j(k+l)\pi}{2N}}\sum_{n=0}^{N-1}e^{\frac{jn(k+l)\pi}{N}} + e^{\frac{-j(k-l)\pi}{2N}}\sum_{n=0}^{N-1}e^{\frac{-jn(k-l)\pi}{N}} + e^{\frac{j(k-l)\pi}{2N}}\sum_{n=0}^{N-1}e^{\frac{jn(k-l)\pi}{N}}\right]$$

where $\sum_{n=0}^{N-1} e^{\frac{\pm jn(k+l)\pi}{N}} = \delta(k+l-N)$ and $\sum_{n=0}^{N-1} e^{\frac{\pm jn(k-l)\pi}{N}} = \delta(k-l)$. Why? Thus, the above equation simplifies to:

$$[\mathbf{C}_N\mathbf{C}_N^T]_{k,l} = \cos\left[\frac{(k+l)\pi}{2N}\right]\delta(k+l-N) + \cos\left[\frac{(k-l)\pi}{2N}\right]\delta(k-l)$$

where the first term is always zero (since for $k + l - N \neq 0 \Rightarrow \delta(k + l - N) = 0$ and for $k + l - N = 0 \Rightarrow \cos\left[\frac{(k+l)\pi}{2N}\right] = \cos\left[\frac{\pi}{2}\right] = 0$) and the second term is also zero except when $k = l$. Thus:

$$[\mathbf{C}_N\mathbf{C}_N^T]_{k,l} = \begin{cases} 1, & \text{for } k = l \\ 0, & \text{otherwise} \end{cases}$$

For $k, l = 0$, $[\mathbf{C}_N\mathbf{C}_N^T]_{0,0} = \sum_{n=0}^{N-1} \frac{1}{N} = 1.$

Hence, $\mathbf{C}_N\mathbf{C}_N^T = \mathbf{I}_N$, i.e. orthogonal transformation matrix.

6.2.2 Matrix Representation of 2-D DCT

Let us first extend 1-D DCT and IDCT in (6.8) and (6.9) to the 2-D case. For an image $x(m, n)$ of size $N \times N$ the 2-D DCT is:

$$X(k,l) = \begin{cases} \frac{1}{N}\sum_{m=0}^{N-1}\sum_{n=0}^{N-1} x(m,n), & k,l = 0,0 \quad m,n \in [0, N-1] \\ \frac{2}{N}\sum_{m=0}^{N-1}\sum_{n=0}^{N-1} x(m,n)\cos\left[\frac{(2m+1)k\pi}{2N}\right]\cos\left[\frac{(2n+1)l\pi}{2N}\right], & k,l \neq 0,0 \begin{cases} k,l \in [0, N-1], \\ m,n \in [0, N-1] \end{cases} \end{cases} \tag{6.12}$$

which suggests that the 2-D DCT is also a separable transform and can be performed using consecutive 1-D DCTs applied to columns and rows of the image. The 2-D IDCT is:

$$x(m,n) = \frac{1}{N}X(0,0) + \frac{2}{N}\sum_{\substack{k=0,\,l=0 \\ k,l \neq (0,0)}}^{N-1}\sum^{N-1} X(k,l)\cos\left[\frac{(2m+1)k\pi}{2N}\right]\cos\left[\frac{(2n+1)l\pi}{2N}\right], \tag{6.13}$$

which allows recovering $x(m, n)$ from its 2-D DCT coefficients $X(k, l)$.

As in Section 6.1.2, to arrive at the matrix representation for 2-D DCT, we use the separability property of the basis functions. Thus, for an image matrix $\mathbf{x}$ of size $N \times N$, the 2-D DCT in matrix form becomes:

$$\mathbf{X} = \mathbf{C}_N\mathbf{x}\mathbf{C}_N^T \tag{6.14}$$

where pre-multiplying the image matrix $\mathbf{x}$ by the DCT matrix $\mathbf{C}_N$ performs column-wise 1-D DCT operations, while post-multiplying the results by $\mathbf{C}_N^T$ performs row-wise 1-D DCT transformation. Using the orthogonality of the DCT matrix $\mathbf{C}_N$, the inverse 2-D DCT in matrix form can be derived as:

$$\mathbf{x} = \mathbf{C}_N^T\mathbf{X}\mathbf{C}_N \tag{6.15}$$

In (6.15), let us denote the kth column of $\mathbf{C}_N^T$ by $\underline{c}_k$ and the lth row of $\mathbf{C}_N$ by $\underline{c}_l^T$. Then, the $(k, l)^{\text{th}}$ *basis image* of 2-D DCT is $\mathsf{C}(k, l) = \underline{c}_k\underline{c}_l^T$, $k, l \in [0, N - 1]$ and hence (6.15) can be rewritten as:

$$\mathbf{x} = \sum_{k=0}^{N-1}\sum_{l=0}^{N-1} X(k,l)\mathsf{C}(k,l). \tag{6.16}$$

That is, the image $\mathbf{x}$ is decomposed onto a space spanned by these N^2 *basis images* with coefficients that are $X(k, l) = \langle \mathbf{x}, \mathsf{C}(k, l)\rangle$.

Figure 6.1 displays all 64 basis images of the 2-D DCT for $N = 8$. The images are arranged from the lowest spatial frequencies in the upper left corner $(k, l) = (0, 0)$ to the highest frequencies in the lower right corner $(k, l) = (N - 1, N - 1)$. Each basis image effectively represents different possible spatial patterns that can occur within an 8×8 image block, including a plain background (upper left), various edge structures, and different possible coarse and fine texture details (lower right). A key property of the DCT is its ability to compact most of the image energy into a few low-frequency

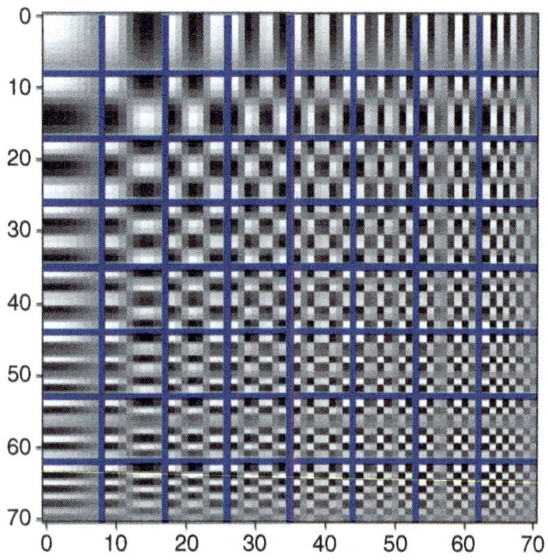

Figure 6.1 Basis images of 2-D DCT for $N = 8$.

coefficients. Thus, any arbitrary 8×8 image block can be efficiently approximated by a linear combination of only a small number (e.g. $P \times Q$) of these basis images, weighted by their corresponding 2-D DCT coefficients $X(k, l)$, that is, $\hat{\mathbf{x}} = \sum_{k=0}^{P-1} \sum_{l=0}^{Q-1} X(k, l)\mathrm{C}(k, l)$ where $P, Q < N$. To illustrate this property, consider the original Peppers image shown in Figure 6.2a. We apply a 2-D DCT to the entire image and then perform a truncation by retaining only the top 64×64 low-frequency DCT coefficients and setting all others to zero (i.e. an ideal low-pass filtering operation in the frequency domain). The image is then reconstructed using the inverse 2-D DCT from these truncated coefficients. Figure 6.2b,c displays the resulting DCT coefficients and the reconstructed image, respectively. The reconstructed image exhibits noticeable smearing and ringing artifacts. The ringing artifacts (Gibbs phenomena) occur because of the sharp cutoff frequency of the ideal filter.

In practice, however, the original image is typically partitioned into blocks of size $L \times L$, and 2-D DCT coefficients are computed for each block. Figure 6.3a displays the top 2×2 DCT coefficients that were retained for each 16×16 block of the original image shown in Figure 6.2a. The remaining DCT coefficients in each block were set to zero. This blocking procedure significantly improves the compression rate because only a small fraction (<1.6%) of the DCT coefficients is kept for each image block. The reconstructed image is obtained by applying the 2-D IDCT to the truncated coefficients of each block, followed by a block-by-block reconstruction. Figure 6.3b illustrates the reconstructed image for this case, demonstrating significantly better reconstruction quality, while simultaneously achieving a superior compression rate compared to the previous (unblocked) case. The SNR for the reconstructed image was found to be 23 dB. As can be observed, smearing and ringing artifacts are significantly reduced in this reconstructed image.

6.3 2-D Walsh-Hadamard Transform

6.3.1 Walsh Transform

We first briefly introduce the Walsh transform for 1-D signals and then discuss how this transform is related to the "Walsh-Hadamard" transform (WHT) [1, 2]. The Walsh transform gives a complete orthonormal signal space to represent any continuous-time signal $x(t)$ defined over interval $0 \leq t \leq 1$. The basis functions of this transform, $\mathrm{wal}(k, t)$, are generated recursively over $0 \leq t \leq 1$ using:

$$\mathrm{wal}(k, t) = \begin{cases} \mathrm{wal}\left(\left\lfloor \frac{k}{2} \right\rfloor, 2t\right), & 0 < t < \frac{1}{2} \\ (-1)^k \mathrm{wal}\left(\left\lfloor \frac{k}{2} \right\rfloor, 2t - 1\right), & \frac{1}{2} \leq t \leq 1 \end{cases} \tag{6.17}$$

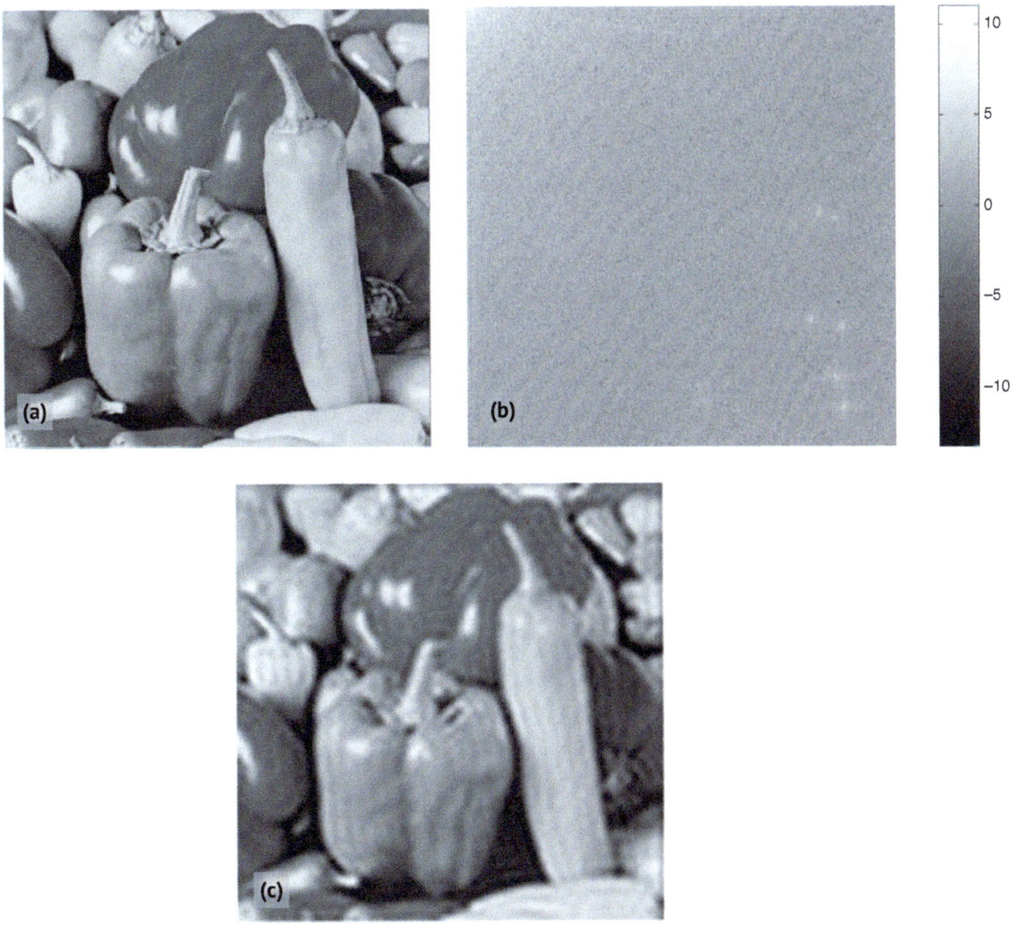

Figure 6.2 (a) Original Peppers image, (b) 2-D DCT coefficients of the entire Peppers image, and (c) reconstructed Peppers image from truncated 2-D DCT coefficients.

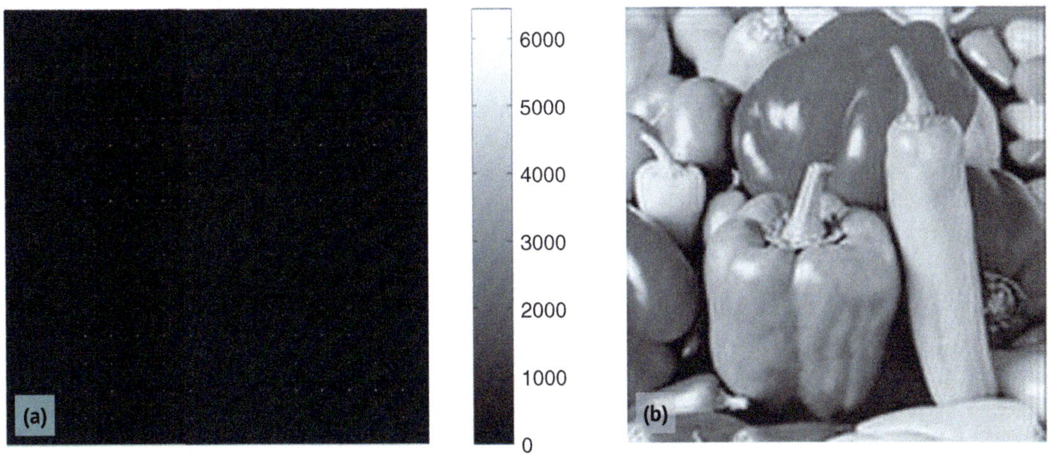

Figure 6.3 (a) Image formed of the retained 2-D DCT coefficients in each block and (b) reconstructed Peppers image.

where $\left\lfloor \frac{k}{2} \right\rfloor$ is the integer part of $\frac{k}{2}$ and by definition $\text{wal}(0, t) = 1$ and $\text{wal}(1, t) = \begin{cases} 1, & 0 < t < \frac{1}{2} \\ -1, & \frac{1}{2} < t < 1 \end{cases}$.

Figure 6.4 shows the plot of the first eight Walsh basis functions $\text{wal}(n,t)$, $n \in [0,7]$ as a function of time t. Walsh basis functions form a complete orthonormal basis over $[0,1]$:

$$\langle \text{wal}(m,t), \text{wal}(n,t)\rangle = \int_0^1 \text{wal}(m,t)\text{wal}(n,t)dt = \delta(m-n) \tag{6.18}$$

Thus, a continuous-time signal $x(t)$ can be represented as a linear combination of these basis functions:

$$x(t) = \sum_{n=0}^{\infty} \phi_n \text{wal}(n,t) \tag{6.19}$$

where due to the orthonormality, we have $\phi_n = \int_0^1 x(t)\text{wal}(n,t)dt$.

6.3.1.1 Properties of Walsh Transform

1. The product of two Wash functions is another Walsh function:

$$\text{wal}(k,t)\text{wal}(l,t) = \text{wal}(m,t) \tag{6.20}$$

with $m_i = k_i \oplus l_i$ where k_i, l_i, and m_i are the ith bit in binary representation of k, l, and m integers and $\oplus$ stands for *Exclusive OR* operation. For example, $\text{wal}(2,t)\text{wal}(3,t) = \text{wal}(1,t)$ since $k_0 = 0, k_1 = 1, l_0 = 1, l_1 = 1$ and hence $m_0 = 0 \oplus 1 = 1$ and $m_1 = 1 \oplus 1 = 0$.
2. The number of zero crossings (or sign change) of a Walsh function is referred to as its *sequency,* in a manner similar to frequency of a sinusoidal function. The recursive order of generation of Walsh functions ensures a sequency i for $\text{wal}(i,t)$.
3. The Walsh function, $\text{wal}(i,t)$, can be synthesized from the binary representation of integer i. That is, let $i < 2^n$ with binary representation $i = \sum_{k=0}^{n-1} b_k 2^k$, $b_k = 0, 1$. If $b_k = 1$, then $\text{wal}(i,t)$ has transitions at dyadic points $r_k = \left\{t_j = \frac{2j+1}{2^{k+1}},\ 1 - t_j,\ j = [0, 2^k - 1]\right\}$. Otherwise, for $b_k = 0$ there is no transition. For example, for $\text{wal}(5,t)$, we have $b_0 = 1, b_1 = 0$ and $b_2 = 1$. Thus, this function has transitions at $t_j = 1/2, 1/8, 3/8, 5/8$, and $7/8$, i.e. the transitions occur only at dyadically rational points in interval $[0,1]$ (see Figure 6.4).
4. For $i \in [0, 2^n - 1]$, the Walsh functions $\text{wal}(i,t)$ sampled at $t = \frac{j}{2^n}$, $j \in [0, 2^n - 1]$ (where $\text{wal}(i,t) = \text{wal}(i,t^+)$ at transitions), yields a complete orthonormal discrete basis, leading to the WHT for discrete time signal of size $N = 2^n$.

6.3.2 Matrix Representation of 1-D WHT

For a finite-extent discrete-time signal $x(n)$ of length $N = 2^n$, WHT pair in scalar form is given by:

$$X(k) = \frac{1}{\sqrt{N}} \sum_{m=0}^{N-1} x(m)(-1)^{b(k,m)}, \quad k \in [0, N-1] \tag{6.21}$$

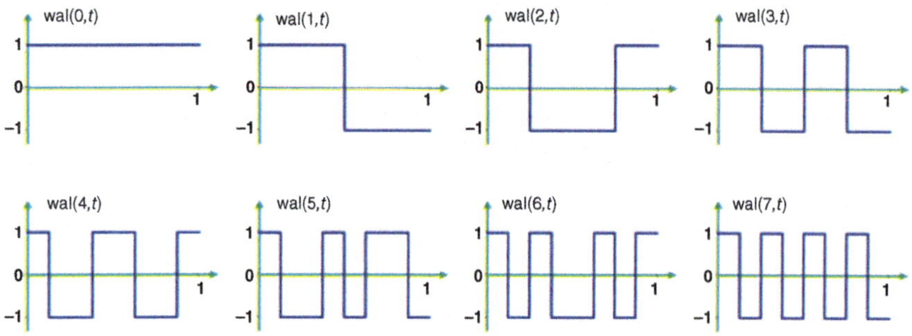

Figure 6.4 The Walsh basis functions $\text{wal}(k,t)$ for $k \in [0,7]$.

and

$$x(m) = \frac{1}{\sqrt{N}} \sum_{k=0}^{N-1} X(k)(-1)^{b(k,m)}, \quad m \in [0, N-1] \tag{6.22}$$

where $b(k, m) = \sum_{i=0}^{n-1} k_i m_i$ with $n = \log_2 N$, and $k_i, m_i = 0, 1$ being the ith bits in binary representation of k and m, respectively. From (6.21) and (6.22), it is evident that the basis functions for both forward and inverse WHT are $(-1)^{b(k,m)}$, $k, m \in [0, N-1]$.

As in previous transforms, arranging the signal $x(n)$ and WHT coefficients $X(k)$ in vector forms leads to the matrix representations:

$$\underline{X} = \mathbf{H}_n \underline{x} \tag{6.23}$$

and

$$\underline{x} = \mathbf{H}_n \underline{X}, \tag{6.24}$$

where $\mathbf{H}_n$ is the Walsh-Hadamard matrix of size $N \times N$ used in both forward and inverse WHTs. For $N = 2^n$, the WHT matrices can be generated [1, 2] recursively using:

$$\mathbf{H}_n = \mathbf{H}_{n-1} \otimes \mathbf{H}_1 = \frac{1}{\sqrt{2}} \begin{bmatrix} \mathbf{H}_{n-1} & \mathbf{H}_{n-1} \\ \mathbf{H}_{n-1} & -\mathbf{H}_{n-1} \end{bmatrix}, \tag{6.25}$$

where $\otimes$ represents Kronecker product of two matrices [8] and $\mathbf{H}_1$ is an *elementary* matrix given by, $\mathbf{H}_1 = \frac{1}{\sqrt{2}} \begin{bmatrix} 1 & 1 \\ 1 & -1 \end{bmatrix}$. As an example, for $n = 2$ using (6.25) we get:

$$\mathbf{H}_2 = \frac{1}{2} \begin{bmatrix} 1 & 1 & 1 & 1 \\ 1 & -1 & 1 & -1 \\ 1 & 1 & -1 & -1 \\ 1 & -1 & -1 & 1 \end{bmatrix}.$$

Note that the number of zero crossings, known as the WHT *sequency,* for rows 1–4 in $\mathbf{H}_2$ are 0,3,1, and 2, respectively, which correspond to the sampled wal$(0, t)$, wal$(3, t)$, wal$(1, t)$, and wal$(2, t)$. Similarly, for $\mathbf{H}_3$, the sequency of rows 1–8 is 0,7,3,4,1,6,2 and 5, respectively. Note that the basis functions in the WHT equations in (6.21) and (6.22) correspond to the elements of the kth row of $\mathbf{H}_n$ for $n = \log_2 N$.

An alternative scalar representation exists for the WHT in which the transform coefficients are ordered by their sequency. This ordering arranges the rows of the WHT matrix such that the sequency increases as a function of the increasing row index, k. In this case, we have $b(k, m) = \sum_{i=0}^{n-1} g_i(k) m_i$ where $g_i(k)$ is generated using $g_i(k) = k_{n-i} \oplus k_{n-i-1}$ with $g_0(k) = k_{n-1}$, and $m_i, k_i = 0, 1$ were defined as before. This ordering arranges the rows of the WHT matrix such that the sequency increases with the row index k, analogous to the frequency ordering in DFT and DCT matrices.

6.3.2.1 Properties of WHT Matrix

1. The WHT matrix is real, symmetric, and orthogonal, i.e. $\mathbf{H}_n = \mathbf{H}_n^* = \mathbf{H}_n^{-1} = \mathbf{H}_n^T$.
2. Note that $\mathbf{H}_1 = \mathbf{W}_2$, where $\mathbf{W}_2$ is the 2×2 DFT matrix highlighting a connection between the two transforms for $N = 2$. As a consequence, one can perform a $N = 2$-point circular convolution with WHT.

3. The 1-D WHT can be implemented efficiently using $N \log_2 N = Nn$ additions or subtractions. This follows from the factored structure of $\mathbf{H}_n$, that is, $\mathbf{H}_n = \widehat{\mathbf{H}}^n$ where $\widehat{\mathbf{H}}$ is:

$$\widehat{\mathbf{H}} = \frac{1}{\sqrt{2}} \begin{bmatrix} 1 & 1 & 0 & 0 & \cdots & 0 \\ 0 & 0 & 1 & 1 & 0 & \cdots \\ \cdot & & & & & \\ \cdot & & & & & \\ 0 & 0 & \cdots & 0 & 1 & 1 \\ - & - & - & - & \cdots & - \\ 1 & -1 & 0 & 0 & \cdots & 0 \\ 0 & 0 & 1 & -1 & 0 & \cdots \\ \cdot & & & & & \\ \cdot & & & & & \\ 0 & 0 & \cdots & 0 & 1 & -1 \end{bmatrix}. \tag{6.26}$$

This matrix contains only two nonzero elements along each row. Thus, multiplying $\widehat{\mathbf{H}}$ by an $N \times 1$ vector requires only N additions or subtractions. Repeating this over n factors yields a total of $Nn = N \log_2 N$ additions or subtractions.
4. There is no need to store the entire Hadamard matrix $\mathbf{H}_n$. Instead, one can generate the $(k, m)^{\text{th}}$ element using $\mathbf{H}_n(k, m) = (-1)^{b(k,m)}$ where $b(k, m)$ was defined before and similarly for the ordered Hadamard matrix.
5. The WHT has a relatively good energy compaction ability, particularly for correlated data. However, unlike the DCT, the WHT basis functions are piecewise constant, which lead to poorer energy compaction for smoothly varying signals.
6. The WHT is used extensively in quantum computing in Hadamard gates to rotate the qubit states [11].

6.3.3 Matrix Representation of 2-D WHT

For a finite-extent image $x(p, q)$ of size $N \times N$, $N = 2^n$, the 2-D WHT and inverse WHT pair are:

$$X(k, l) = \frac{1}{N} \sum_{p=0}^{N-1} \sum_{q=0}^{N-1} x(p, q)(-1)^{b(k,l;p,q)}, \quad k, l \in [0, N-1] \tag{6.27}$$

and

$$x(p, q) = \frac{1}{N} \sum_{k=0}^{N-1} \sum_{l=0}^{N-1} X(k, l)(-1)^{b(k,l;p,q)}, \quad p, q \in [0, N-1] \tag{6.28}$$

where $b(k, l; p, q) = \sum_{i=0}^{n-1} (k_i p_i + l_i q_i)$, and k_i, l_i p_i, and q_i represent the ith bits in binary representations of k, l, p, and q, respectively, and + represents binary (or modulo 2) addition. It can be shown (Problem 6.15) that $b(k, l; p, q) = b(k, p) + b(l, q)$ where $b(k, p)$ was defined as before. This implies that the 2-D WHT basis functions are separable. Using this separability property, the 2-D WHT pair in (6.27) and (6.28) in matrix forms are:

$$\mathbf{X} = \mathbf{H}_n \mathbf{x} \mathbf{H}_n \tag{6.29}$$

and

$$\mathbf{x} = \mathbf{H}_n \mathbf{X} \mathbf{H}_n \tag{6.30}$$

The basis images of the 2-D inverse WHT can be generated in a manner similar to those of the 2-D DFT and DCT. That is, the $(k, l)^{\text{th}}$ basis image of the 2-D inverse WHT in (6.30) is $\mathsf{H}(k, l) = \underline{h}_k \underline{h}_l^T$, $k, l \in [0, N-1]$ where $\underline{h}_k$ is the kth column of $\mathbf{H}_n$ and $\underline{h}_l^T$ is the lth row of $\mathbf{H}_n$. Then, we can alternatively express the image matrix $\mathbf{x}$ in (6.30) as:

$$\mathbf{x} = \sum_{k=0}^{N-1} \sum_{l=0}^{N-1} X(k, l) \mathsf{H}(k, l) \tag{6.31}$$

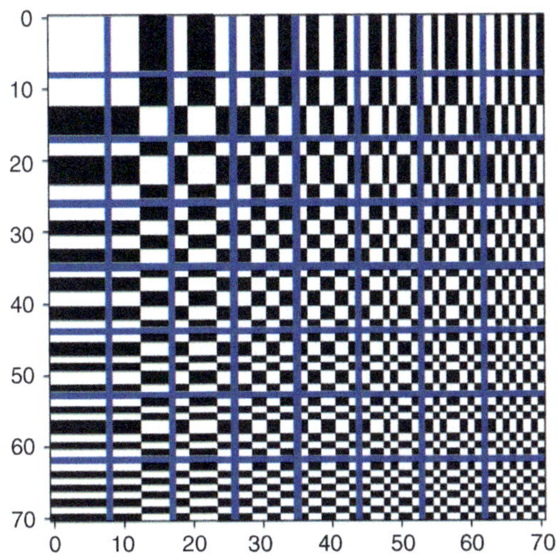

Figure 6.5 Basis images of 2-D ordered WHT for $N = 8$.

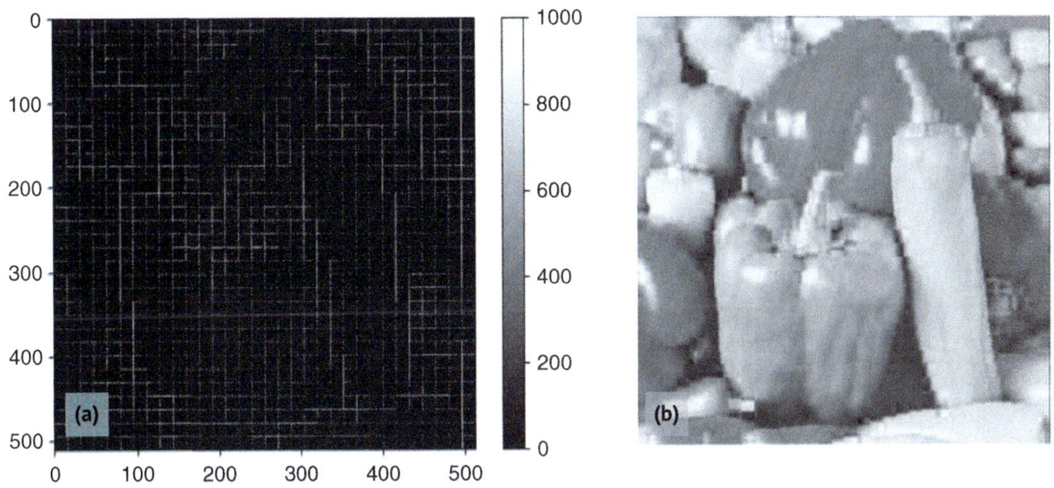

Figure 6.6 (a) Image formed of the retained 2-D WHT coefficients in each block and (b) reconstructed image.

where the coefficients are obtained using $X(k, l) =< \mathbf{x}, \mathrm{H}(k, l) >$ as in the 2-D DCT case.

Figure 6.5 displays all 64 basis images of 2-D WHT for $N = 8$. Note that these basis images are sequency-ordered to facilitate comparison with those of the 2-D DCT in Figure 6.1. As can be viewed, these basis images can be regarded as binary versions of those of the 2-D DCT. Also, they have similar structural properties that can be used to capture various combinations of edges and textures.

To examine the usefulness of 2-D WHT for image data reduction applications, the same block-based procedure adopted in Section 6.2.2 for 2-D DCT was applied here. The original Peppers image in Figure 6.2a was used in this experiment for the purpose of comparing the result with that of the 2-D DCT. After applying 2-D WHT to each 16×16 nonoverlapping block of this image, the top 2×2 WHT coefficients are retained in each block, while the remaining coefficients are set to zero. The image formed using these retained coefficients is shown in Figure 6.6a. The 2-D inverse WHT is then applied to each filtered image block, and the reconstructed image is generated block-by-block. This image is shown in Figure 6.6b, which exhibits considerable blocking artifacts. The SNR of the reconstructed image is approximately 11.7 dB. This result clearly demonstrates that, compared to 2-D DCT, 2-D WHT performs poorly for image data reduction applications. This is mainly due to the better energy compaction property of the DCT and its continuous and smooth basis functions.

6.4 2-D Karhunen–Loeve Transform

Unlike the other transforms covered so far, this transform applies to random signals and images and has wide applications in data reduction, visualization, and decorrelation applications. For discrete data, the KLT is also known as principal component analysis (PCA) [1].

6.4.1 1-D KLT of Random Signals

Let $X(n),\ n \in [0, N-1]$ be a finite-extent 1-D wide-sense stationary (WSS) random process (r.p.) with zero mean and autocovariance function $R_{XX}(m) = E[X(n)X(n-m)]$ where $E[\cdot]$ represents the expectation operator (ensemble average). If $\underline{X}$ is the vector arrangement of $\{X(n)\},\ n \in [0, N-1]$, i.e. $\underline{X} = [X(0) \dots X(N-1)]^T$, then the covariance matrix of this random vector $\underline{X}$ is given by:

$$\mathbf{R}_X = E[\underline{X}\underline{X}^T] = \begin{bmatrix} R_{xx}(0) & R_{xx}(-1) & \cdots & R_{xx}(-N+1) \\ R_{XX}(1) & R_{XX}(0) & \cdots & R_{XX}(-N+2) \\ \vdots & & & \vdots \\ R_{XX}(N-1) & R_{XX}(N-2) & \cdots & R_{XX}(0) \end{bmatrix},$$

which is a Toeplitz matrix [8]. To find the KLT of $\underline{X}$, we diagonalize matrix $\mathbf{R}_X$ using:

$$\mathbf{\Psi}^H \mathbf{R}_X \mathbf{\Psi} = \mathbf{\Gamma} = \text{Diag}[\gamma_0, \cdots, \gamma_{N-1}] \tag{6.32}$$

where matrix $\mathbf{\Psi} = [\underline{\xi}_0, \cdots, \underline{\xi}_{N-1}]$ consists of eigenvectors $\underline{\xi}_k$ of matrix $\mathbf{R}_X$ as its columns, and $\mathbf{\Gamma}$ is a diagonal matrix consisting of the corresponding eigenvalues $\gamma_0 > \gamma_1 > \cdots > \gamma_{N-1}$ along its diagonal. Using (6.32), for each pair of eigenvalue γ_k and eigenvector $\underline{\xi}_k$ we have:

$$\mathbf{R}_X \underline{\xi}_k = \gamma_k \underline{\xi}_k, \qquad k \in [0, N-1]$$

where these eigenvectors form an orthonormal basis for the KLT, satisfying the condition $\underline{\xi}_l^H \underline{\xi}_k = \delta(k-l)$. As a result, transformation matrix $\mathbf{\Psi}$ is an orthonormal (or unitary) matrix, which means it satisfies the condition $\mathbf{\Psi}\mathbf{\Psi}^H = \mathbf{I}$.

Now, the 1-D KLT of $\underline{X}$ can be generated using:

$$\underline{\mathcal{X}} = \mathbf{\Psi}^H \underline{X}. \tag{6.33}$$

Using the orthonormal property of matrix $\mathbf{\Psi}$, the inverse KLT becomes:

$$\underline{X} = \mathbf{\Psi}\underline{\mathcal{X}}. \tag{6.34}$$

As can be seen, the KLT matrix $\mathbf{\Psi}^H$ depends on the second-order statistics of the data.

6.4.1.1 Properties of KLT Matrix

The KLT has many desirable properties [1] that make it optimal for many signal and image processing applications. Some of these important properties are listed below.

1. *Decorrelation:* The KLT coefficients $\{\mathcal{X}(k), k \in [0, N-1]\}$ are uncorrelated, i.e.

   $$E[\mathcal{X}(k)\mathcal{X}^*(l)] = \gamma_k \delta(k-l).$$

   It suffices to show that the correlation matrix of $\underline{\mathcal{X}}$, i.e. $E[\underline{\mathcal{X}}\underline{\mathcal{X}}^H]$, is diagonal, i.e.

   $$E[\underline{\mathcal{X}}\underline{\mathcal{X}}^H] = \mathbf{\Psi}^H E[\underline{X}\underline{X}^T]\mathbf{\Psi} = \mathbf{\Psi}^H \mathbf{R}_X \mathbf{\Psi} = \mathbf{\Gamma} = \text{Diag}[\gamma_0, \cdots, \gamma_{N-1}].$$

   This implies that each eigenvalue γ_i is the variance (or energy) of the corresponding KLT coefficient $\mathcal{X}(i)$ along the eigenvector $\underline{\xi}_i$. Thus, the KLT components represent the contributions of the data along the relevant coordinates. In other words, KLT rotates the coordinate system so that the new orthogonal coordinates, $\underline{\xi}_i$, maximize the energy or variance, γ_i, of the data. The projection of each data point onto these coordinates gives the KLT coefficients or principal components (PCs) of the data sample.

   Note that the matrix $\mathbf{\Psi}$ is not a unique matrix with respect to the decorrelation property, that is, there could be many matrices (unitary or non-unitary) that would decorrelate the data in the transformed domain.

2. *Optimality in Data Reduction:* Consider the scenario depicted in Figure 6.7 where the vector $\underline{X}$ is mapped to vector $\underline{\mathcal{X}}$ using a unitary matrix $\mathbf{A}$ of dimension $N \times N$. In the transformed domain, the first m elements of $\underline{\mathcal{X}}$ are retained, while the rest are set to zero. This reduced-dimensional vector is denoted by $\underline{Z}$, i.e.

$$Z(k) = \begin{cases} \mathcal{X}(k), & k \in [0, m-1] \\ 0, & k \geq m \end{cases}$$

Thus, $\underline{Z}$ lies in an $(m \leq N)$-D dimensional subspace, while both $\underline{X}$ and $\underline{\mathcal{X}}$ are in N-D space. The vector $\underline{Z}$ is then used to reconstruct $\hat{\underline{X}}$ using inverse unitary matrix $\mathbf{B}$ of dimension $N \times N$. The average MSE between the original signal $X(n)$ and the reconstructed signal $\hat{X}(n)$ is:

$$J_m = \frac{1}{N} E\left[\sum_{n=0}^{N-1} |X(n) - \hat{X}(n)|^2\right] \tag{6.35}$$

where the expectation $E[.]$ is over an ensemble set of signal.

Now, the question is what are the optimal choices of the matrices $\mathbf{A}$ and $\mathbf{B}$ are such that J_m is minimized for each and every choice of $m \in [1, N]$? It can be shown [1] that the MSE J_m is minimized for every choice of m when $\mathbf{A} = \mathbf{\Psi}^H$, $\mathbf{B} = \mathbf{\Psi}$, and $\mathbf{AB} = \mathbf{I}$, i.e. the KLT and inverse KLT matrices, respectively.

Using the energy preservation property of unitary transforms, (6.35) can also be rewritten as:

$$J_m = \frac{1}{N} E\left[\sum_{k=0}^{N-1} |\mathcal{X}(k) - Z(k)|^2\right] = \frac{1}{N} E\left[\sum_{k=m}^{N-1} |\mathcal{X}(k)|^2\right] = \frac{1}{N} \sum_{k=m}^{N-1} \gamma_k. \tag{6.36}$$

That is, the MSE J_m is equal to the total energy of the discarded components in the transformed domain or the sum of their corresponding eigenvalues. This result leads to the following procedure for applying KLT for data reduction.

6.4.1.2 KLT-Based Data Reduction Algorithm

a. Form covariance matrix $\mathbf{R}_X$ using many independent zero-mean data samples $\underline{x}$ of multivariate r.p. $\underline{X}$. Diagonalize matrix $\mathbf{R}_X$ to find its eigenvectors $\underline{\xi}_i$, $i \in [0, N-1]$ and corresponding eigenvalues γ_i, $i \in [0, N-1]$ where $\gamma_0 > \gamma_1 > \cdots > \gamma_{N-1}$.
b. Find m such that the first m eigenvalues capture a desired percentage of the total energy, e.g. $\eta = \frac{\sum_{i=0}^{m-1} \gamma_i}{\sum_{i=0}^{N-1} \gamma_i} \geq 97\%$.
c. Form the reduced $N \times m$ transformation matrix:

$$\mathbf{\Psi}_{\text{red}} = [\underline{\xi}_0, \underline{\xi}_1, \cdots, \underline{\xi}_{m-1}]$$

and transform each data vector $\underline{x}$ (a realization of $\underline{X}$) into its corresponding $m \times 1$ KLT vector:

$$\underline{\mathrm{x}}_{\text{red}} = \mathbf{\Psi}_{\text{red}}^H \underline{x}.$$

d. To reconstruct, use the reduced-dimensional KLT matrix $\mathbf{\Psi}_{\text{red}}$, i.e.

$$\underline{\hat{x}}_{\text{rec}} = \mathbf{\Psi}_{\text{red}} \underline{\mathrm{x}}_{\text{red}}.$$

Clearly, if $m = N$, *perfect reconstruction* is achieved, i.e. $\underline{\hat{x}}_{\text{rec}} = \underline{x}$.

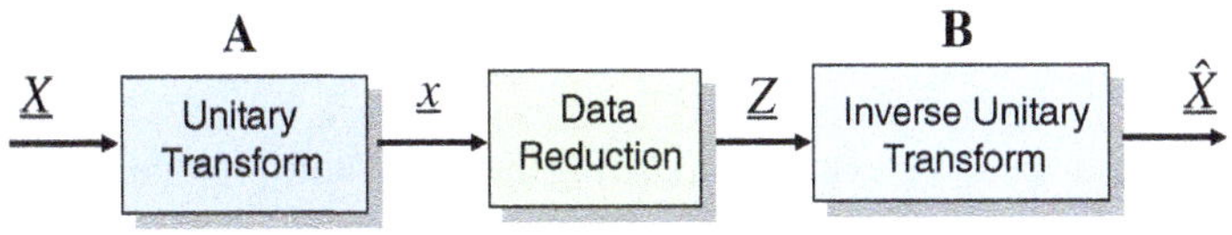

Figure 6.7 Optimal data reduction in transform domain.

3. *Distribution of Variance:*
   Among all unitary transforms, the KLT packs the maximum average energy into the first $m \leq N$ elements of $\underline{\mathcal{X}}$. That is, if $\mathbf{A}$ is any other unitary transform and $\mathbf{\Psi}^H$ is the KLT matrix, then for any $m \in [1, N]$:

   $$S_m(\mathbf{\Psi}^H) \geq S_m(\mathbf{A})$$

   where $S_m(\mathbf{A})$ is the energy captured by the first m transformed coefficients, defined as $S_m(\mathbf{A}) = \sum_{k=0}^{m-1} \sigma_k^2$ and $\sigma_k^2 = E[|\mathcal{X}(k)|^2]$, with $\sigma_0^2 \geq \sigma_1^2 \cdots \geq \sigma_{N-1}^2$. The reader is referred to [1] for the proof.

6.4.2 2-D KLT of Random Images

Consider a finite-extent image $x(m, n),\ m, n \in [0, N-1]$, which is a sample of a zero-mean 2-D r.p. $X(m, n)$. To extend the KLT to the 2-D case, we have two possible approaches. In the first approach, we reshape each sample image into a vector of dimension $N^2 \times 1$ and then treat the problem as the 1-D by finding the covariance matrix of size $N^2 \times N^2$ and performing the diagonalization procedure. Clearly, this approach, in its original form, is not only computationally prohibitive but also may become ill-conditioned when the number of samples is limited. Alternatively, we can perform the 2-D KLT using separable 1-D transforms:

$$\mathbf{X} = \mathbf{\Psi}_1^H \mathbf{x} \mathbf{\Psi}_2^* \tag{6.37}$$

where as before $\mathbf{x}$ and $\mathbf{X}$ are matrices of image $x(m, n)$ and its 2-D KLT, respectively, and $\mathbf{\Psi}_1^H$ and $\mathbf{\Psi}_2^*$ are 1-D KLT matrices applied to the columns and rows of the image matrix $\mathbf{x}$. The inverse 2-D KLT in matrix form is:

$$\mathbf{x} = \mathbf{\Psi}_1 \mathbf{X} \mathbf{\Psi}_2^T. \tag{6.38}$$

The basis images (or eigen-images) of the 2-D KLT can similarly be defined as $\mathsf{K}(k, l) = \underline{\xi}_{1k} \underline{\xi}_{2l}^T,\ k, l \in [0, N-1]$ where $\underline{\xi}_{1k}$ is the kth column of $\mathbf{\Psi}_1$ and $\underline{\xi}_{2l}^T$ is l^{th} row of $\mathbf{\Psi}_2^T$. Then, in (6.38) the image matrix $\mathbf{x}$ can be expressed as a linear combination of these eigen-images with weights that are the 2-D KLT coefficients, $X(k, l)$, i.e.

$$\mathbf{x} = \sum_{k=0}^{N-1} \sum_{l=0}^{N-1} X(k, l) \mathsf{K}(k, l).$$

Although this matrix representation approach was found to be advantageous for other transforms, here it requires finding two different 1-D KLT mapping matrices, $\mathbf{\Psi}_1^H$ and $\mathbf{\Psi}_2^*$. These matrices must be found by solving two separate eigenvalue problems with two different data covariance matrices, which are formed based on column-wise and row-wise arrangements of the image data. An alternative and more efficient method was proposed in [12] for computing 2-D KLT and eigen-images with applications to facial pattern representation and recognition. A slightly modified version of this algorithm is presented next.

6.4.3 An Efficient Algorithm for Eigen-Image and PCA of Images

Let $\{\mathbf{x}_i\}_{i=1}^M$ be a set of $N \times N$ training images. Each image matrix $\mathbf{x}_i$ is first reshaped into a vector, $\underline{x}_i \in \mathbb{R}^{N^2}$. Then, the following steps are performed.

1. Find the sample mean vector of the training dataset using $\underline{\mu} = \frac{1}{M} \sum_{i=1}^M \underline{x}_i$ and subtract it from each training sample $\underline{\bar{x}}_i = \underline{x}_i - \underline{\mu}$. Then, form data matrix $\mathbf{\Upsilon} = \frac{1}{\sqrt{M}} [\underline{\bar{x}}_1 \cdots \underline{\bar{x}}_M] \in \mathbb{R}^{N^2 \times M}$.
2. Find the sample covariance matrix of the training dataset $\bar{\mathbf{R}}_X = \frac{1}{M} \sum_{i=1}^M \underline{\bar{x}}_i \underline{\bar{x}}_i^t$ or $\bar{\mathbf{R}}_X = \mathbf{\Upsilon}\mathbf{\Upsilon}^T \in \mathbb{R}^{N^2 \times N^2}$.
3. Assuming that we can find the principal eigenvectors of $\bar{\mathbf{R}}_X$ by solving the eigenvalue problem:

   $$\bar{\mathbf{R}}_X \underline{\xi}_k = \gamma_k \underline{\xi}_k \tag{6.39}$$

   the eigen-images can then be found by reshaping these eigenvectors $\underline{\xi}_k$. However, solving the eigenvalue problem (6.39) of N^2 dimension is not only computationally impractical, but also unnecessary since matrix $\bar{\mathbf{R}}_X$ is rank-deficient with rank at most $M-1$. Instead, we can form matrix $\mathbf{\Upsilon}^T\mathbf{\Upsilon}$ of size $M \times M$ and solve the following eigenvalue problem with the same eigenvalues:

$$\mathbf{\Upsilon}^T\mathbf{\Upsilon}\underline{\zeta}_k = \gamma_k \underline{\zeta}_k \tag{6.40}$$

where $\underline{\zeta}_k$ is the kth eigenvector of this problem associated with eigenvalue γ_k. To see the relation between these two eigenvalue problems, pre-multiply (6.40) by $\mathbf{\Upsilon}$:

$$(\mathbf{\Upsilon}\mathbf{\Upsilon}^T)\mathbf{\Upsilon}\underline{\zeta}_k = \gamma_k \mathbf{\Upsilon}\underline{\zeta}_k \tag{6.41}$$

or equivalently

$$\bar{\mathbf{R}}_X\mathbf{\Upsilon}\underline{\zeta}_k = \gamma_k \mathbf{\Upsilon}\underline{\zeta}_k. \tag{6.42}$$

This result implies that while the two eigenvalue problems in (6.39) and (6.41) have identical eigenvalues, the eigen-images of the former can be obtained from those of the latter using:

$$\underline{\xi}_k = \mathbf{\Upsilon}\underline{\zeta}_k = \frac{1}{\sqrt{M}}\sum_{l=1}^{M} \zeta_{l,k}\underline{\bar{x}}_l \tag{6.43}$$

where $\zeta_{l,k}$ is the lth element of vector $\underline{\zeta}_k$. We keep $P \leq M$ eigen-images associated with the largest eigenvalues. For visualization purposes, each eigenvector $\underline{\xi}_k$ can be reshaped into an $N \times N$ image.

4. To reconstruct the ith training image using the top P PCs and the corresponding eigen-vectors we apply:

$$\underline{\hat{x}}_i = \sum_{k=1}^{P} X_i(k)\underline{\xi}_k + \underline{\mu} \tag{6.44}$$

where $X_i(k) = \underline{\xi}_k^H \underline{\bar{x}}_i/\gamma_k$ is its kth PC. Note that normalization by γ_k is needed since even though $\underline{\zeta}_k$ are orthonormal, $\underline{\xi}_k$ are orthogonal but not normal. This is due to $\underline{\xi}_l^H\underline{\xi}_k = \underline{\zeta}_l^H\mathbf{\Upsilon}^T\mathbf{\Upsilon}\underline{\zeta}_k = \underline{\zeta}_l^H\gamma_k\underline{\zeta}_k = \gamma_k\delta(k-l)$.

Any new image $\underline{\bar{y}}$ (mean subtracted by $\underline{\mu}$) with similar second-order statistics can also be represented by the same eigen-images fairly accurately using:

$$\underline{y} = \sum_{k=1}^{P} Y(k)\underline{\xi}_k + \underline{\mu} \tag{6.45}$$

where $Y(k) = \underline{\xi}_k^H\underline{\bar{y}}/\gamma_k$ is the kth PC of image $\underline{y}$.

5. The extracted top P PCs for any image $\underline{y}$ can be used to form a reduced-dimensional feature vector $\underline{Y} = [Y(0), \cdots, Y(P-1)]^t$ that represents this image. This feature vector can then be applied to a classifier to classify the particular image. A (*minimum distance classifier*) is a simple linear classifier (see Chapter 16) that can be implemented to perform image discrimination using these PC feature vectors. Using this classifier, making a decision about the class membership of an unknown sample $\underline{Y}$ can be done using:

$$j = \underset{k}{\text{argmin}}\ ||\underline{Y} - \underline{X}_k||^2 \qquad \Longrightarrow \qquad \underline{y} \in C_j \quad \text{i.e. class j} \tag{6.46}$$

where $\underline{X}_k$ is the PC vector of the kth training sample and j is the index of the class of the training sample that has the shortest distance or closest match to the unknown pattern $\underline{Y}$. Clearly, there are more sophisticated classifiers that can be trained using the same feature sets to perform more accurate discrimination. These will be covered in Chapters 16 and 18.

6.4.4 Face Representation Using Eigen-Faces

Turk & Pentland showed [12] that with only a few eigen-faces (*standardized face ingredients*) extracted from a set of training images, any *other face* can be represented with good accurately. This method has been used not only for face recognition but also for applications such as handwriting analysis, lip reading, voice recognition, sign language and hand gesture interpretation, and medical imaging.

Figure 6.8 shows a very small ($M = 15$) subset of the Yalefaces image database [13] used as the training set to find the eigen-faces. There are 14 male and 1 female subjects in this training set. Several male subjects have facial hair, and some wear transparent glasses. The original Yalefaces database consists of six facial expressions (normal, sad, happy, sleepy, surprised, and wink) and three lighting conditions (center, left, and right). All faces have been centered as much as possible. In this experiment, only images with neural expressions and centered lighting are used for training Figure 6.9 shows the first $P = 10$ eigen-faces (out of 14) obtained using the procedure outlined in Section 6.4.3. Figure 6.10 shows fairly accurate reconstruction of all the training images using these 10 eigen-faces and the associated PCs.

To assess the generalization capability of the eigen-faces to represent facial images with differing expressions and lighting conditions, six test images were randomly chosen. Figures 6.11 and 6.12 show the original test images and their reconstructed versions using all 14 eigen-faces, respectively. The results indicate poor reconstruction performance, even when all eigen-faces are used. This failure is primarily attributed to the limited size and diversity of the training set, which does not capture the variations in expression and illumination present in the test images. Computer Project 6.3 is designed to address these issues by repeating the experiment using a larger and more diverse training dataset, thereby improving the robustness and generalization capability of the eigen-face representation.

6.4.5 PCA and Singular Value Decomposition (SVD)

The data matrix $\mathbf{\Upsilon}$ is typically a very tall and skinny matrix, i.e. $N^2 \gg M$. If the rank of this matrix is $r \leq M$, the *thin* SVD of this matrix [8, 14] is given by:

$$\mathbf{\Upsilon} = \mathbf{U}\mathbf{\Sigma}\mathbf{V}^H \tag{6.47}$$

where $\mathbf{U} \in \mathbb{R}^{N^2 \times r}$ and $\mathbf{V} \in \mathbb{R}^{r \times r}$ contain the left and right singular vectors of $\mathbf{\Upsilon}$ as their columns, respectively, and $\mathbf{\Sigma} \in \mathbb{R}^{r \times r}$ is a diagonal matrix with diagonal elements $\sigma_1, \sigma_2, \cdots, \sigma_r$ that are singular values of $\mathbf{\Upsilon}$. The matrices $\mathbf{U}$ and $\mathbf{V}$ are orthogonal, that is, $\mathbf{U}^H\mathbf{U} = \mathbf{I}_r$ and $\mathbf{V}^H\mathbf{V} = \mathbf{I}_r$. Now, using this thin SVD of $\mathbf{\Upsilon}$, the sample covariance matrix of the data can be expressed as:

$$\bar{\mathbf{R}} = \mathbf{\Upsilon}\mathbf{\Upsilon}^H = \mathbf{U}\mathbf{\Sigma}^2\mathbf{U}^H \quad \text{or} \quad \bar{\mathbf{R}}\mathbf{U} = \mathbf{U}\mathbf{\Sigma}^2. \tag{6.48}$$

Figure 6.8 Training facial images $M = 15$. *Source*: A. S. Georghiades et al. (2001)/Yale Center for Computational Vision and Control.

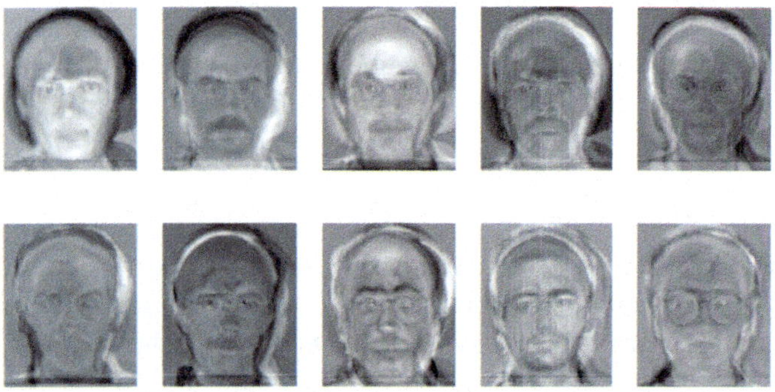

Figure 6.9 First 10 eigen-faces obtained using the training dataset.

Figure 6.10 Reconstructed training images using $P = 10$ eigen-faces.

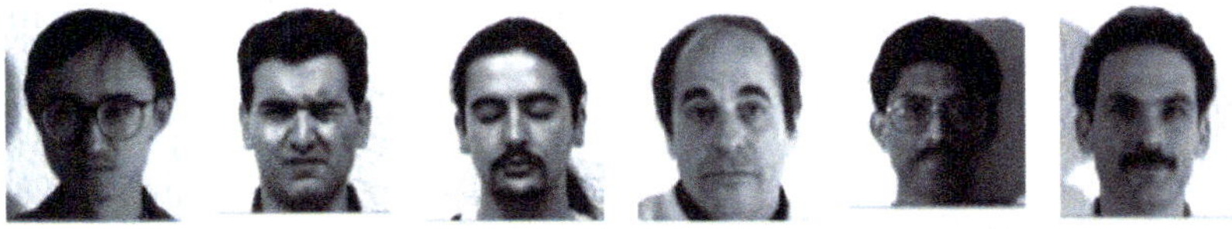

Figure 6.11 Original test images with different pose, gesture, and lighting conditions. *Source*: A. S. Georghiades et al. (2001)/Yale Center for Computational Vision and Control.

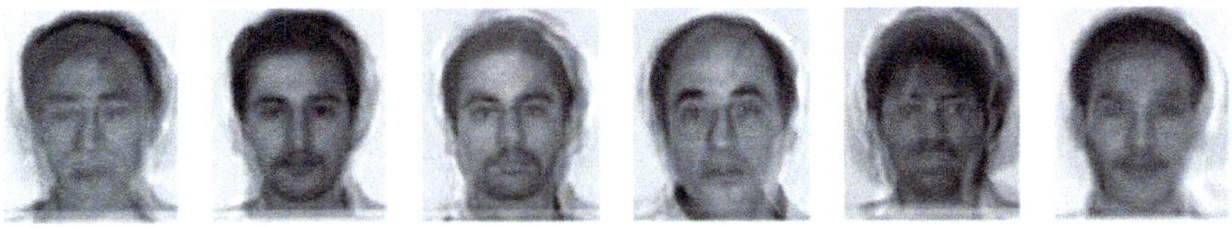

Figure 6.12 Reconstructed test images using all 14 eigen-faces.

Comparing this result to (6.32), it is clear that $\mathbf{\Gamma} = \mathbf{\Sigma}^2$, i.e. $\gamma_k = \sigma_k^2,\ k \in [1, r]$, and moreover $\underline{\xi}_k = \underline{u}_k\ k \in [1, r]$ where $\underline{u}_k$ is the kth column of $\mathbf{U}$. On the other hand, matrix $\mathbf{\Upsilon}^H\mathbf{\Upsilon}$ can also be expressed using the thin SVD of $\mathbf{\Upsilon}$ as:

$$\mathbf{\Upsilon}^H\mathbf{\Upsilon} = \mathbf{V}\mathbf{\Sigma}^2\mathbf{V}^H \quad \text{or} \quad \mathbf{\Upsilon}^H\mathbf{\Upsilon}\mathbf{V} = \mathbf{V}\mathbf{\Sigma}^2. \tag{6.49}$$

Comparing this equation with the eigenvalue problem in (6.40), we deduce that $\underline{\zeta}_l = \underline{v}_l,\ l \in [1, r]$. Thus, we can first find matrices $\mathbf{V}$ and $\mathbf{\Sigma}$ using the SVD in (6.49) and then use them to find $\mathbf{U} = \mathbf{\Upsilon}\mathbf{V}\mathbf{\Sigma}^{-1}$ using (6.47). This algorithm is basically the same as that in Section 6.4.3 although using the thin SVD framework.

6.5 Computer Projects

The computer projects in this chapter use Python libraries OpenCV, NumPy, matplotlib, SciPy, scikit-image, SymPy.

Computer Project 6.1: Write a Python code to generate all the basis images of 2-D DCT and 2-D WHT for $N = 8$. Note that to compare each individual basis image of 2-D WHT with that of the 2-D DCT one needs to order the sequency of the basis vectors for the 2-D WHT.

Computer Project 6.2: This computer project explores the application of the 2-D DCT to image compression using the Baboon image. The primary objective is to investigate the trade-offs between compression ratio and image quality for different block sizes and data reduction strategies. Partition the Baboon image into nonoverlapping blocks and apply 2-D DCT to each block. Use two different blocks sizes, for example, 8×8 and 16×16. Devise and implement a scheme to reduce data in the DCT domain by selectively setting a percentage of coefficients to zero. This zeroing operation should be applied to coefficients that contribute least to the overall image energy (e.g. those with the smallest magnitudes, often corresponding to high-frequency components), thereby achieving compression. Among the meaningful criteria that can be used for zeroing process are magnitude or energy-based and frequency-based, which works by keeping low frequency components (filtering in transform domain). For each block size and data reduction criterion, reconstruct the image block-by-block from the retrained coefficients and display the results. Compare the results of your reconstructions using performance measures such as signal-to-noise ratio (SNR) for different percent reduction. The SNR is defined by $\mathrm{SNR}_{dB} = 10 \log_{10} \frac{\sigma_s^2}{\sigma_n^2}$ where σ_s^2 and σ_n^2 are the variances of the signal and noise, respectively. Note that in this case, noise is the difference image between the original image and the reconstructed version. Generate and display plots of SNR versus percentage of zeroed coefficients (or compression ratio) and determine the optimal operating point corresponding to a point at which significant data reduction is achieved with acceptable image quality.

Computer Project 6.3: Repeat the steps in Computer Project 6.2 for the RGB color dog (Finn) image.

Computer Project 6.4: The goal of this computer project is to study PCA for the purpose of image data reduction and facial pattern representation. The same Yalefaces image database used in Section 6.4.4 is used in this project. This dataset [13] contains 165 images of 15 male and female subjects taken under 3 different lighting conditions (center, left, and right); and 8 different facial expressions including: normal, happy, sad, sleepy, surprised, wink, with glasses, and without glasses. Repeat the procedure in Section 6.4.4 using an extended training dataset that constitutes 2/3 of the images randomly selected from the Yalefaces database. Ensure an equal number of cases are chosen for each facial expression and lighting condition to avoid possible bias. Compute the eigen-faces using this training dataset and display the dominant ones, selected based on their corresponding eigenvalues. Randomly select 15 images from the training dataset and reconstruct the faces using the dominant eigen-faces and the corresponding PCs of the images. Comment on the reconstruction ability both visually and using SNR measure. Test the reconstruction ability of your trained system on at least 11 randomly picked test cases that were not part of the training data and comment on the reconstruction quality and generalization performance. You should also try to test the capabilities of the trained system on a new face (e.g. your own) taken under similar conditions. Finally, design and train a minimum distance classifier to classify facial patterns using their PC features. You may choose the particular classes of interest, such as female versus male, happy versus sad, or any other meaningful categories, and report the classification accuracy.

6.6 Problems

Problem 6.1: Consider the following coordinate mapping:

$$\begin{bmatrix} x \\ y \end{bmatrix} = \begin{bmatrix} \cos\theta & \sin\theta \\ -\sin\theta & \cos\theta \end{bmatrix} \begin{bmatrix} u \\ v \end{bmatrix}.$$

a. Comment on the properties of this mapping matrix.
b. Find the inverse mapping matrix and comment on its properties.
c. Find the basis vectors of this representation.

Problem 6.2: In Problem 6.1, if x and y represent realizations of r.v.s X and Y find θ so that:

a. Mean squared value of u is maximized.
b. u and v become uncorrelated.

Problem 6.3: For an image $\mathbf{x} = \begin{bmatrix} a & b \\ c & d \end{bmatrix}$ and orthogonal transformation matrix $\mathbf{A} = \frac{1}{\sqrt{2}} \begin{bmatrix} 1 & 1 \\ 1 & -1 \end{bmatrix}$.

a. Find the basis images for the transform.
b. Find the transformed image.
c. Show the inverse mapping yields the original image.

Problem 6.4: If $\underline{w}_k$ represents the kth column of the DFT matrix $\mathbf{W}_N$, show that these vectors form an orthonormal basis, i.e. $\langle \underline{w}_k, \underline{w}_l \rangle = \underline{w}_k^H \underline{w}_l = \delta(k-l)$.

Problem 6.5: Show that a unitary matrix $\mathbf{U}$ has eigenvalues that all have unity magnitudes.

Problem 6.6: Suppose $\underline{X}$ is a zero-mean random vector with covariance matrix $\mathbf{C}_X = \sigma^2\mathbf{I}$. Show that the DFT of $\underline{X}$ has the same covariance matrix, that is, remains invariant under unitary transformation. Can this result be extended to a random vector with a general covariance matrix?

Problem 6.7: Consider a digital image $x(m, n)$ with ROS in $m, n \in [-N, N]$. Show that:

a. $X(-k, -l) = X^*(k, l)$ where $X(k, l)$ is the 2-D DFT of $x(m, n)$,
b. if the image is symmetrical around the origin, then $Im\{X(-k, -l)\} = 0$,
c. for $Re\{X(-k, -l)\} = 0$ the image must be antisymmetric around the origin.

Problem 6.8: Consider a digital image $x(m, n)$ with ROS in $m, n \in [0, N-1]$. Given that the image is defined as:

$$x(m,n) = \begin{cases} c, & m = m_1, m = m_2, \quad n \in [0, N-1] \\ 0, & \text{elsewhere} \end{cases}.$$

where intensity $0 \leq c \leq 255$ and $m_1, m_2 \in [0, N-1]$, and $m_1 \neq m_2$.

a. Plot the image and find its 2-D DFT. Compare the original image and its transform.
b. Repeat part (a) for the image

$$x(m,n) = \begin{cases} c, & m = n, \quad m, n \in [0, N-1] \\ 0, & \text{otherwise} \end{cases}.$$

Problem 6.9: Consider the symmetrically extended version of finite-extent signal $x_N(n), n \in [0, N-1]$ defined as:

$$\hat{x}_{2N}(n) = x_{2N}(n) + x_{2N}(-n-1)$$

where $x_{2N}(n) = \begin{cases} x_N(n), & n \in [0, N-1] \\ 0, & n \in [N, 2N-1] \end{cases}.$

Show that the $2N$-point DFT of $\hat{x}_{2N}(n)$ is related to the N-point DCT-II of $x_N(n)$.

Problem 6.10: Let $x(m, n)$ be a finite-extent image with ROS, R_{NN}. The symmetrically extended version of this image is given by:

$$\hat{x}_{2N}(m,n) = \begin{cases} x_N(m,n) + x_N(2N-1-m,n) + x_N(m,2N-1-n) + \\ x_N(2N-1-m, 2N-1-n) \quad \text{for} \quad m, n \in [0, N-1] \\ 0, \quad \text{Otherwise,} \end{cases}.$$

Show the relationship between the 2-D DFT of this symmetrically image and the 2-D DCT of the original image.

Problem 6.11: The 1-D discrete sine transform (DST) of a finite-extent signal $x(n), n \in [0, N-1]$ is given by:

$$X(k) = \sqrt{\frac{2}{N+1}} \sum_{n=0}^{N-1} x(n) \sin\left[\frac{(n+1)(k+1)\pi}{N+1}\right], \quad k \in [1, N-1].$$

Find the DST transformation matrix $\mathbf{S}$ and its properties.

Problem 6.12: In Problem 6.11, show that the DST matrix $\mathbf{S}$ is an orthogonal matrix.

Problem 6.13: Prove Property 1 in Section 6.3.2 for the WHT matrix.

Problem 6.14: Prove Property 2 in Section 6.3.2 for the WHT matrix.

Problem 6.15: Show that the basis functions of 2-D WHT are separable, i.e. $b(k, l; p, q) = b(k, p) + b(l, q)$ leading to a separable transform.

Problem 6.16: Consider a flat 4×4 image:

$$x(m, n) = a, \qquad m, n \in [0, 3]$$

Find the 2-D WHT of this image and explain your results.

Problem 6.17: Consider two 2×2 images $\mathbf{x} = \begin{bmatrix} 1 & 0 \\ 2 & 1 \end{bmatrix}$ and $\mathbf{h} = \begin{bmatrix} 1 & -1 \\ 0 & 1 \end{bmatrix}$.

a. Find their 2-D WHT images $\mathbf{X}$ and $\mathbf{H}$ and the element-wise product $\hat{\mathbf{Y}} = [\hat{Y}(k, l)]$ where $\hat{Y}(k, l) = X(k, l)H(k, l)$.
b. Find the inverse 2-D WHT of image $\hat{\mathbf{Y}}$ and show that the result is related to the circular convolution of images $x(m, n)$ and $h(m, n)$.
c. Explain your results. Can we extend this observation to larger size images?

Problem 6.18: Let $\underline{X}$ be a zero-mean random vector with covariance matrix $\mathbf{R}_X = E[\underline{X}\underline{X}^T]$ where $E[\cdot]$ represents the expectation operator. Determine a unity norm vector $\underline{\xi}$ such that the projection of $\underline{X}$ onto the subspace spanned by $\underline{\xi}$ captures maximum energy of $\underline{X}$. What is this subspace?

Problem 6.19: Let $X(n)$ be a zero-mean wide sense stationary (WSS) r.p. with auto-covariance function $R_X(m) = E[X(n)X(n-m)] = \sigma^2\rho^{|m|}$, and let $\underline{X} = [X(n)\, X(n-1)]^T$ be a 2×1 random vector of this r.p. Determine:

a. Covariance matrix of $\underline{X}$ and its eigenvalues and eigenvectors.
b. PCs of this vector process and comment on the results.

Problem 6.20: In Problem 6.19, assume that the signal is corrupted by noise, i.e. $Y(n) = X(n) + \eta(n)$, where η is a zero-mean white Gaussian noise with unit variance. Find:

a. Covariance matrix of $\underline{X}$ and its eigenvalues and eigenvectors.
b. PCs of this vector process and comment on the results.

References

1 A. K. Jain, "*Fundamentals of Digital Image Processing*," Prentice Hall, 1989.
2 R. C. Gonzalez and R. E. Woods, "*Digital Image Processing*," Pearson, 4th Edition, 2018.
3 N. Ahmed, T. Natarajan, and K. R. Rao, "Discrete cosine transform," *IEEE Transactions on Computers*, vol. 23, no. 1, pp. 90–93, 1974.
4 G. K. Wallace, "The JPEG still picture compression standard," *Communications of the ACM*, vol. 34, no. 4, pp. 30–44, 1991.
5 D. J. Le Gall, "The MPEG video compression algorithm," *Signal Processing: Image Communication*, vol. 4, no. 2, pp. 129–140, 1992.
6 C. Podilchuk and X. Zhang, Face recognition using DCT based feature vectors, *Proceedings of IEEE International Conference on Acoustics, Speech, and Signal Processing Conference*, 1996.
7 X. Y. Jing and D. Zhang, "A face and palmprint recognition approach based on discriminant DCT feature extraction," *IEEE Transactions on Systems, Man and Cybernetics*, vol. 34, pp. 2405–2415, 2004.
8 G. H. Golub and C. F. Van Loan, "*Matrix Computations*," Johns Hopkins Studies in Mathematical Sciences, 4th Edition, 2013.
9 V. G. Reju, S. N. Koh, and I. Y. Soon, "Convolution using discrete sine and cosine transforms," *IEEE Signal Processing Letters*, vol. 14, pp. 445–448, 2007.

10 I. Ito, “Convolution using discrete cosine transforms for improving performance of convolutional neural networks,” *Proceedings of 2022 30th European Signal Processing Conference (EUSIPCO)*, 2022.

11 A. B. Suksmono and Y. Minato, “Quantum computing formulation of some classical Hadamard matrix searching methods and its implementation on a quantum computer,” Scientific Reports, vol. 12, p. 197, 2022.

12 M. Turk and A. Pentland, “Eigenfaces for recognition,” *International Journal of Cognitive Neuroscience*, vol. 3, no. 1, pp. 71–86, 1991.

13 Yale University Facial Database (Yalefaces) http://cvc.cs.yale.edu/cvc/projects/yalefaces/yalefaces.html.

14 J. J. Gerbrands, “On the relationships between SVD, KLT and PCA,” Pattern Recognition, vol. 14, pp. 375–381, 1981.

7

Wavelet Transform

The wavelet theory is old and yet relatively new. The first work in this area appeared in a paper by A. Haar in 1910 [1]. However, the construction of more general wavelets was investigated in 1980s along with efficient algorithms to compute the expansion. Pioneering work in this area was developed independently by many individuals in different disciplines, including but not limited to Calderon's work in harmonic analysis [2], coherent states in quantum mechanics in 1968 by Aslaksen and Klauder [3], and seismic data analysis by Morlet in 1983 [4, 5]. The connection between wavelets and filter banks and applications in signal and image processing is due to many individuals including Daubechies [6], Vetterli [7, 8], Meyer [9], Mallat [10, 11], Donoho [12], Coifman [13] and many others.

The simplicity and elegance of wavelet theory led to extraordinary growth in application areas over the past twenty years. Among such applications are image compression and subband coding standards [14], feature extraction for pattern classification, discrete multi-tone modulation transceivers, and others.

In this chapter, we intend to demonstrate the applications of this powerful tool in signal and image processing. To this end, our discussion begins by outlining the limitations of traditional Fourier analysis for many signal and image applications. We then review an intermediate solution to some of these issues using the Short-Time Fourier Transform (STFT) [15, 16]. The lack of multiresolution ability of the STFT leading to the continuous wavelet transform (CWT) will also be discussed [8]. The implementation of wavelet transform for discrete-time signals using subband decomposition with filter banks [8] yields the 1-D discrete wavelet transform (DWT). We will then exploit the separability property to extend the 1-D DWT to the 2-D case for various image processing applications.

7.1 Short-Time Fourier Transform

Although the Fourier bases have many useful properties, e.g. eigenfunction and convolution for linear time-invariant systems, they also have several shortcomings including:

1. Inability to perform time- and frequency-domain analysis, e.g. it cannot localize a particular note in a given piece of music in the frequency domain. The lack of time-frequency localization is an important drawback in detecting and isolating events in the time or frequency domains.
2. Inefficient for representing discontinuities or sharp changes, that is, requires a large number of Fourier components to represent discontinuities.
3. Does not provide a multiresolution look for signal analysis. That is, it doesn't allow simultaneous ability to examine a signal with fine time resolution for high-frequency analysis and fine frequency resolution for low-frequency analysis.

These deficiencies were first identified by Gabor in 1946 [16] who introduced the *time-localization* principle for extracting local spectral information from a signal. Time localization is achieved by windowing the continuous-time signal $x(t)$ prior to taking the FT. This leads to the STFT as:

$$X_{\text{STFT}}(\tau, \omega) = \int_{-\infty}^{\infty} x(t)g(t-\tau)e^{-j\omega t}dt, \tag{7.1}$$

where $g(t)$ is the window function and τ is the center of the window or the shift variable. This windowing in the STFT introduces a time dependency in the spectral representation, resulting in a 2-D function $X_{\text{STFT}}(\tau, \omega)$ of time (or shift) τ and frequency ω.

Closer observation of the expression in (7.1) reveals two different interpretations of the STFT. First, we can view STFT as the FT (over all frequencies) of the windowed signal around every possible shift τ. Alternatively, comparing this equation with the convolution integral $y(t) = \int_{-\infty}^{\infty} x(\tau)h(t-\tau)d\tau$ where $h(t)$ is the impulse response of a linear time-invariant (LTI) system, one can deduce that $h(t-\tau) = g(t-\tau)e^{-j\omega t}$. For a given frequency ω_0, the effect of multiplying by $e^{-j\omega_0 t}$ is equivalent to modulating $g(t)$ or shifting its FT $G(\omega)$ to the center frequency ω_0. Thus, another way to interpret STFT is the output of a bandpass filter whose frequency response is centered at ω_0. But since ω can vary over the entire frequency range, the STFT involves filtering the signal $x(t)$ with a bank of bandpass filters, each has the same frequency response centered at a different frequency. Each filter has an impulse response that is the window function modulated to that frequency. Thus, the STFT can alternatively be viewed as a modulated filter bank [8, 15].

A typical choice for the window function $g(t)$ is to use a Gaussian window $g_\alpha(t) = \frac{1}{2\sqrt{\pi\alpha}}e^{\frac{-t^2}{4\alpha}}$ with variance $\sigma^2 = 2\alpha$ in which case the STFT becomes the Gabor transform:

$$X_{GT}(\tau, \omega) = \int_{-\infty}^{\infty} x(t)g_\alpha(t-\tau)e^{-j\omega t}dt. \tag{7.2}$$

For this choice of the window function, the FT $G(\omega)$ will also be Gaussian, with variance inversely proportional to that of $g_\alpha(t)$. Additionally, because for a Gaussian PDF we have $\int_{-\infty}^{\infty} g_\alpha(t-\tau)d\tau = \int_{-\infty}^{\infty} g_\alpha(\tau)d\tau = 1$, we can easily arrive at:

$$\int_{-\infty}^{\infty} X_{GT}(\tau, \omega)d\tau = X(\omega) \tag{7.3}$$

that is, collection of localized Gabor transform recovers the global FT of the signal. This result also implies that the STFT is invertible since from (7.3) and using inverse FT we get:

$$x(t) = \frac{1}{2\pi}\int_{-\infty}^{\infty}\int_{-\infty}^{\infty} X_{GT}(\tau, \omega)e^{j\omega t}d\tau d\omega \tag{7.4}$$

7.1.1 Time-Frequency Resolutions

The resolution in the frequency domain can be defined by the bandwidth of the bandpass filter $G(\omega)$ measured by:

$$\Delta\omega = \frac{\left(\int_{-\infty}^{\infty} \omega^2|G(\omega)|^2 d\omega\right)^{\frac{1}{2}}}{\left(\int_{-\infty}^{\infty} |G(\omega)|^2 d\omega\right)^{\frac{1}{2}}}.$$

Similarly, the time resolution is given by the spread of the window $g(t)$ in the time domain, that is:

$$\Delta t = \frac{\left(\int_{-\infty}^{\infty} t^2|g(t)|^2 dt\right)^{\frac{1}{2}}}{\left(\int_{-\infty}^{\infty} |g(t)|^2 dt\right)^{\frac{1}{2}}}$$

Due to the uncertainty principle or *Heisenberg inequality* [6, 15] we have:

$$\text{Time} - \text{Bandwidth Product} = \Delta t \Delta\omega \geq 1/2 \quad \text{Lower bound}$$

that is, resolution in time and frequency cannot be made arbitrarily small. More specifically, to resolve two sinusoids with close frequencies in the frequency domain, $\Delta\omega$ should be arbitrarily small; in which case, we lose the ability to resolve two closely spaced events in the time domain that are less than Δt apart and vice versa. Note that the lower bound (equality) is achieved when a Gabor window is used. Another important observation is that once Δt is chosen, $\Delta\omega$ is also decided, i.e. both time and frequency resolutions are fixed. Figure 7.1a shows the frequency subband decomposition for the STFT where the filter bank consists of modulated filters with constant bandwidth $\Delta\omega$ (Figure 7.1).

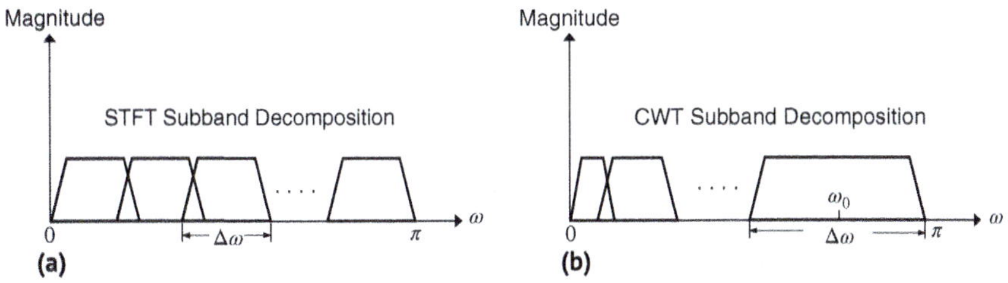

Figure 7.1 Frequency subband decomposition of (a) the STFT and (b) the CWT.

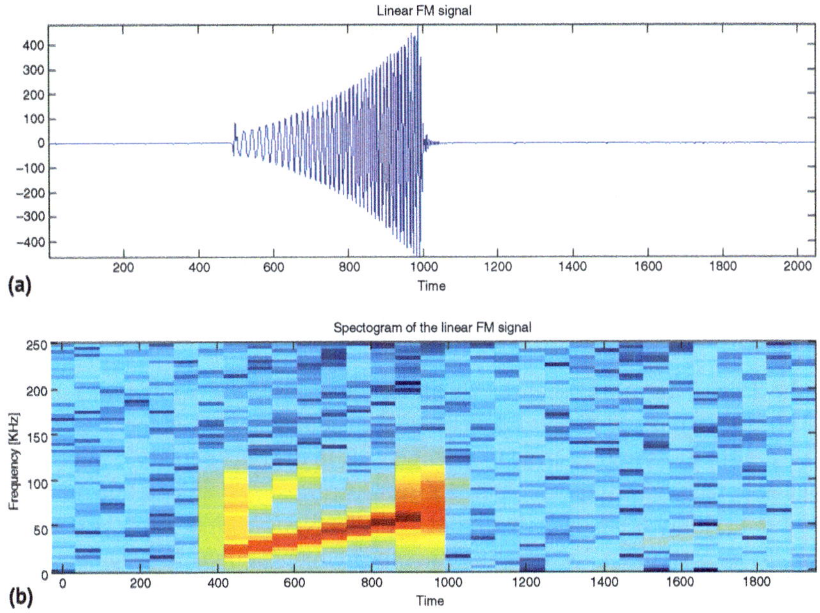

Figure 7.2 (a) A LFM upsweep signal and (b) its spectrogram.

Remark 7.1:

1. The plot of $|X_{\text{STFT}}(\tau, \omega)|^2$ in the (τ, ω)-plane is referred to as *Spectrogram* [15], which is a useful tool in signal analysis, as it provides a distribution of the signal in time-frequency plane. Examples of spectrograms for two different signals, namely a linear frequency modulated (LFM) and one with transient events, are shown in Figures 7.2 and 7.3, respectively. As can be seen in Figure 7.2, the LFM upsweep is clearly observable together with the location of high-frequency transitions at the beginning and end of the signal. The spectrogram in Figure 7.3 also clearly exhibits the locations of the impulsive events, as well as the low frequency content of the signal.
2. For a discrete-time signal, $x(n)$, the STFT is expressed in terms of the DTFT of the windowed signal:

$$X_{\text{STFT}}(m, \Omega) = \sum_{n=-\infty}^{\infty} x(n)g(n-m)e^{-j\Omega n} \tag{7.5}$$

where $g(n-m)$ is the window function centered at m and Ω is the digital frequency.

7.2 Continuous Wavelet Transform

Although the STFT remedies the lack of time-frequency localization of the FT, the resolutions in time and frequency are fixed. That is, good resolution in time and frequency cannot be achieved simultaneously.

The wavelet transform offers multiresolution analysis by allowing variation of time and frequency resolutions. This multiresolution representation is performed using an *analysis* filter bank consisting of bandpass filters with *constant Q* [15]

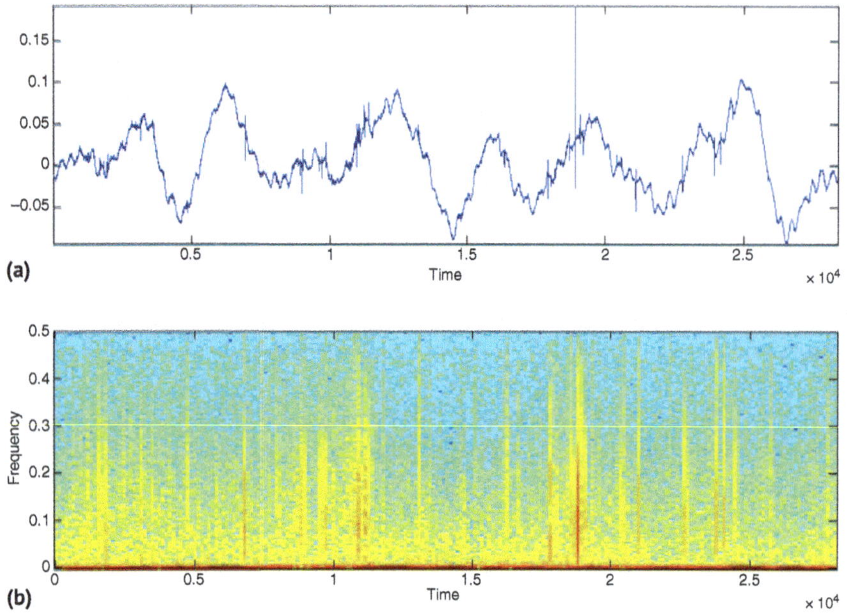

Figure 7.3 (a) A low frequency signal with multiple superimposed impulsive events and (b) its spectrogram.

instead of the constant bandwidth filters in the STFT. This means that the center frequency over the bandwidth (quality factor) ratio of the filter is fixed, i.e. $Q = \frac{\omega_0}{\Delta\omega}$. Thus, as we move to higher frequencies, the bandwidth $\Delta\omega$ can increase, offering a smaller time resolution Δt. This provides the ability to resolve close transient events in the time domain. Conversely, separating sinusoidal signals with close frequencies requires arbitrarily small $\Delta\omega$ in the frequency domain, while the time resolution Δt can be large. This nice *multiresolution* ability is biologically similar to the frequency response of the cochlea in the human inner ear, which yields a great auditory perception. Figure 7.1b shows the frequency subband decomposition for the CWT, where the filter bank consists of constant $-Q$ filters.

The CWT achieves this by using basis functions that are scaled (i.e. stretched or compressed) and shifted versions of a prototype function, $\psi(t)$, referred to as the *mother wavelet*:

$$\psi_{a,\tau}(t) = \frac{1}{\sqrt{a}}\psi\left(\frac{t-\tau}{a}\right),$$

where $\psi_{a,\tau}(t)$ is called the *daughter* wavelet, $a > 0$ is the scaling or *dilatation* parameter, τ is the shift parameter, and the factor $\frac{1}{\sqrt{a}}$ is used for energy normalization purposes. Small values of $a < 1$ correspond to compressed wavelets, providing fine time resolution for detecting high-frequency activities (i.e. *zoom-in*) while large values of $a > 1$ widens the wavelets, providing coarse time resolution but finer frequency resolution for low-frequency analysis (i.e. *zoom-out*). The CWT is then defined as:

$$X_{\mathrm{CWT}}(a,\tau) = \frac{1}{\sqrt{a}}\int_{-\infty}^{\infty} x(t)\psi^*\left(\frac{t-\tau}{a}\right)dt = \int_{-\infty}^{\infty} x(t)\psi_{a,\tau}^*(t)dt. \tag{7.6}$$

Alternatively, using a change of variable $t \to at$ in (7.6) we get:

$$X_{\mathrm{CWT}}(a,\tau) = \sqrt{a}\int_{-\infty}^{\infty} x(at)\psi^*\left(t-\frac{\tau}{a}\right)dt \tag{7.7}$$

in which the signal is scaled while the window width is fixed. Thus, a large contraction (for large a) of the signal, $x(at)$, is seen through a fixed width window (global view) whereas a large expansion of the signal (for very small a) gives the detailed view.

7.2.1 Properties of CWT

1. Scale change of continuous-time signals (e.g. $x(at)$) does not change resolution since it can be reversed, while in discrete-time signals increasing scale involves downsampling (e.g. $x(an)$ for $a > 1$) and reduction of resolution, which is irreversible. Moreover, downsampling discrete-time signals is not a time-invariant operation either. In contrast, decreasing scale (upsampling) is reversible, and hence it does not change the resolution.
2. The function $\psi(t)$ must satisfy the following conditions to qualify as a mother wavelet.
   a. $\psi(t)$ should be absolutely integrable and square integrable (finite energy).

   $$\int_{-\infty}^{\infty} |\psi(t)|dt < \infty$$

   and

   $$\int_{-\infty}^{\infty} |\psi(t)|^2 dt < \infty$$

   b. If $\Psi(\omega)$ is FT of $\psi(t)$ the *admissibility* condition, required for inverse operation, should also be satisfied, i.e.

   $$C_\psi = \int_{-\infty}^{\infty} \frac{|\Psi(\omega)|^2 d\omega}{|\omega|} < \infty.$$

   Since $\Psi(\omega)$ is a continuous function, the finiteness of this function implies that $\Psi(0) = \int_{-\infty}^{\infty} \psi(t)dt = 0$ (the reason $\psi(t)$ is called a *wavelet*). This implies that the wavelet must have a bandpass-like spectrum. Figure 7.4 shows four mother wavelet functions for Haar, Meyer, Morlet, and db4 wavelets [6, 8], respectively.
3. If the above conditions are satisfied, the inverse CWT becomes:

$$x(t) = \frac{1}{C_\psi} \int_0^{\infty} \int_{-\infty}^{\infty} X_{\text{CWT}}(a, \tau)\psi_{a,\tau}^*(t) \frac{da d\tau}{a^2} \tag{7.8}$$

where C_ψ is a constant that depends on choice of mother wavelet function $\psi(t)$. This *synthesis* equation shows that any signal $x(t)$ can be reconstructed as a superposition of the shifted and scaled wavelets.

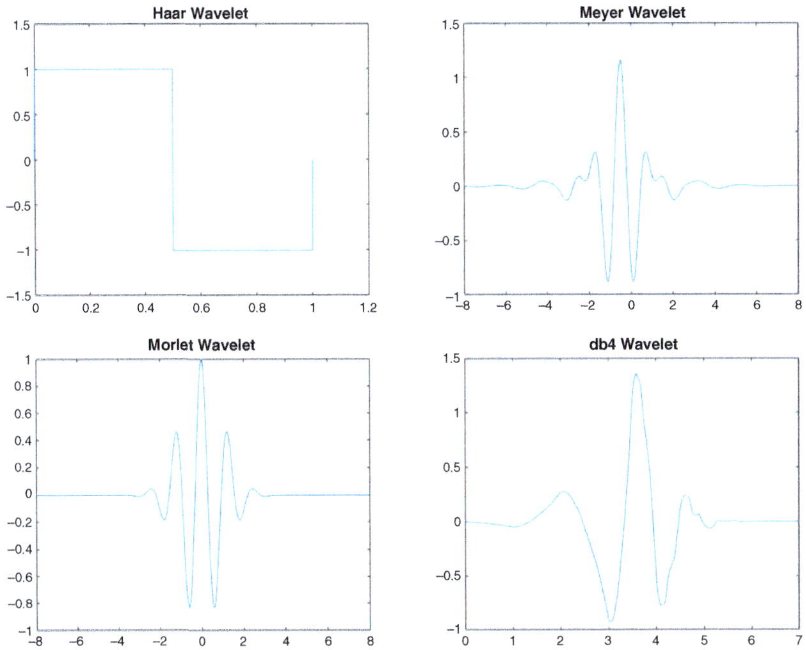

Figure 7.4 Several different mother wavelet functions.

4. *Energy Preservation:* CWT preserves energy in both domains, i.e.

$$\frac{1}{C_\psi}\int_0^\infty\int_{-\infty}^\infty |X_{\mathrm{CWT}}(a,\tau)|^2\frac{da\,d\tau}{a^2} = \int_{-\infty}^\infty |x(t)|^2dt, \tag{7.9}$$

where $|X_{\mathrm{CWT}}(a,\tau)|^2$ is called the *Scalogram,* representing the energy distribution of the signal in time-scale plane. In contrast to the spectrogram, the energy of the signal is distributed at multiple resolutions.

5. *Shifting:* If $X_{\mathrm{CWT}}(a,\tau)$ is the CWT of signal $x(t)$, then the CWT of the shifted signal $y(t) = x(t-\tau')$ is:

$$Y_{\mathrm{CWT}}(a,\tau) = X_{\mathrm{CWT}}(a,\tau-\tau'),$$

that is, invariance to shift.

6. *Scaling:* If $X_{\mathrm{CWT}}(a,\tau)$ is the CWT of signal $x(t)$, then the CWT of the scaled signal $y(t) = (1/\sqrt{b})x(t/b)$ is:

$$Y_{\mathrm{CWT}}(a,\tau) = X_{\mathrm{CWT}}\left(\frac{a}{b},\frac{\tau}{b}\right),$$

that is, scaling the signal results in scaling the scale and shift parameters.

7. *Time Localization:* Let $x(t) = \delta(t-t_0)$ then CWT is:

$$X_{\mathrm{CWT}}(a,\tau) = \frac{1}{\sqrt{a}}\int \psi\left(\frac{t-\tau}{a}\right)\delta(t-t_0)dt = \frac{1}{\sqrt{a}}\psi\left(\frac{t_0-\tau}{a}\right),$$

which is the scaled mother wavelet reversed in time and located at t_0. This shows perfect localization ability of the CWT.

Example 7.1

a. *Represent the CWT of signal $x(t)$ using the FT of the mother wavelet function and $X(\omega)$.*
b. *Using the result in Part (a) prove the energy preservation property stated above.*

Solution:

a. From (7.6) and using Parseval's theorem we can write $X_{\mathrm{CWT}}(a,\tau)$ as:

$$X_{\mathrm{CWT}}(a,\tau) = \int_{-\infty}^\infty x(t)\psi^*_{a,\tau}(t)dt = \frac{1}{2\pi}\int_{-\infty}^\infty X(\omega)\Psi^*_{a,\tau}(\omega)d\omega,$$

where $\Psi_{a,\tau}(\omega)$ is the Fourier transform of the daughter wavelet $\psi_{a,\tau}(t)$. Using the shifting and scaling properties, this FT can be related to that of the mother wavelet, i.e. $\Psi_{a,\tau}(\omega) = \sqrt{a}e^{-j\tau\omega}\Psi(a\omega)$. This gives:

$$X_{\mathrm{CWT}}(a,\tau) = \frac{\sqrt{a}}{2\pi}\int_{-\infty}^\infty X(\omega)\Psi^*(a\omega)e^{j\tau\omega}d\omega.$$

b. To prove energy preservation property, we use the result in Part (a) in the left side of (7.9), i.e.

$$\frac{1}{C_\psi}\int_0^\infty\int_{-\infty}^\infty |X_{\mathrm{CWT}}(a,\tau)|^2\frac{da\,d\tau}{a^2} = \frac{1}{C_\psi}\int_0^\infty\left[\int_{-\infty}^\infty\left|\frac{\sqrt{a}}{2\pi}\int_{-\infty}^\infty X(\omega)\Psi^*(a\omega)e^{j\tau\omega}d\omega\right|^2 d\tau\right]\frac{da}{a^2}.$$

Now, let us denote $P(\omega) = X(\omega)\Psi^*(a\omega)$. Using inverse FT and invoking Parseval's theorem, the right side of the above equation becomes:

$$\begin{aligned}\frac{1}{C_\psi}\int_0^\infty\left[\int_{-\infty}^\infty\left|\frac{1}{2\pi}\int_{-\infty}^\infty P(\omega)e^{j\tau\omega}d\omega\right|^2 d\tau\right]\frac{da}{|a|} &= \frac{1}{C_\psi}\int_0^\infty\left[\int_{-\infty}^\infty |p(\tau)|^2d\tau\right]\frac{da}{a}\\ &= \frac{1}{C_\psi}\int_0^\infty\left[\frac{1}{2\pi}\int_{-\infty}^\infty |P(\omega)|^2d\omega\right]\frac{da}{a}.\end{aligned}$$

Thus, we can write:

$$\frac{1}{C_\psi}\int_{-\infty}^{\infty}\int_{-\infty}^{\infty}|X_{\mathrm{CWT}}(a,\tau)|^2\frac{da d\tau}{a^2}=\frac{1}{2\pi C_\psi}\int_{-\infty}^{\infty}|X(\omega)|^2\left[\int_0^{\infty}\frac{|\Psi(a\omega)|^2}{a}da\right]d\omega.$$

Applying the admissibility condition to the inner integral and using Parseval's theorem again yields the final result

$$\frac{1}{C_\psi}\int_0^{\infty}\int_{-\infty}^{\infty}|X_{\mathrm{CWT}}(a,\tau)|^2\frac{da d\tau}{a^2}=\frac{1}{2\pi}\int_{-\infty}^{\infty}|X(\omega)|^2 d\omega=\int_{-\infty}^{\infty}|x(t)|^2 dt.$$

7.2.1.1 Wigner–Ville Distribution

An alternative to the spectrogram for nonstationary signal analysis is the Wigner–Ville distribution (WVD) [15, 17, 18]. It provides a quadratic time-frequency representation with many desirable properties. For a continuous time signal $x(t)$ the WVD is given by:

$$\mathrm{WV}_x(\tau,\omega)\equiv\int_{-\infty}^{\infty}x\left(\tau+\frac{t}{2}\right)x^*\left(\tau-\frac{t}{2}\right)e^{-j\omega t}dt. \tag{7.10}$$

For nonstationary signals, WVD shows how the signal energy is distributed in the time-frequency plane.

7.2.1.1.1 Properties of WVD

The following are some of the important properties of the WVD.

1. WVD is real-valued, i.e. $\mathrm{WV}_x^*(\tau,\omega)=\mathrm{WV}_x(\tau,\omega)$.
2. It preserves time and frequency shifts.
3. The frequency and time marginals of WVD correspond to the signals instantaneous power and its spectral energy density, respectively, i.e.

$$\frac{1}{2\pi}\int_{-\infty}^{\infty}\mathrm{WV}_x(\tau,\omega)d\omega=|x(\tau)|^2$$

$$\int_{-\infty}^{\infty}\mathrm{WV}_x(\tau,\omega)d\tau=|X(\omega)|^2$$

4. The WVD has time-frequency shift property, i.e. if $y(t)=x(t-\tau_0)e^{j\omega_0 t}$, then $\mathrm{WV}_y(\tau,\omega)=\mathrm{WV}_x(\tau-\tau_0,\omega-\omega_0)$.
5. The WVD and CWT are related by:

$$|X_{\mathrm{CWT}}(\tau,a)|^2=\iint\mathrm{WV}_x(t,\xi)\mathrm{WV}_\Psi\left(\frac{t-\tau}{a},a\xi\right)dtd\xi$$

that is, 2-D correlation between WVDs of the signal and the wavelet.

6. Moyal Formula: This property is central to the design of time-frequency processing algorithms and is given by:

$$\frac{1}{2\pi}\int_{-\infty}^{\infty}\int_{-\infty}^{\infty}\mathrm{WV}_x(\tau,\omega)\mathrm{WV}_y(\tau,\omega)d\tau d\omega=\left|\int_{-\infty}^{\infty}x(t)y^*(t)dt\right|^2.$$

For other useful properties of WVD the reader is referred to [19].

Example 7.2 *If continuous-time signals $x(t)$ and $y(t)$ have WVDs $WV_x(\tau,\omega)$ and $WV_y(\tau,\omega)$, respectively, prove that the WVD of their product $z(t)=x(t)y(t)$ is:*

$$\mathrm{WV}_z(\tau,\omega)=\frac{1}{2\pi}\int_{-\infty}^{\infty}\mathrm{WV}_x(\tau,\xi)\mathrm{WV}_y(\tau,\omega-\xi)d\xi$$

that is, convolution of their WVDs.

Solution: Start with WVD of $z(t)$

$$\begin{aligned}
\mathrm{WV}_z(\tau,\omega) &= \int_{-\infty}^{\infty} z\left(\tau+\frac{t}{2}\right)z^*\left(\tau-\frac{t}{2}\right)e^{-j\omega t}dt \\
&= \int_{-\infty}^{\infty} x\left(\tau+\frac{t}{2}\right)x^*\left(\tau-\frac{t}{2}\right)y\left(\tau+\frac{t}{2}\right)y^*\left(\tau-\frac{t}{2}\right)e^{-j\omega t}dt \\
&= \int_{-\infty}^{\infty} x\left(\tau+\frac{t}{2}\right)x^*\left(\tau-\frac{t}{2}\right)dt \int_{-\infty}^{\infty} y\left(\tau+\frac{t_1}{2}\right)y^*\left(\tau-\frac{t_1}{2}\right)\delta(t-t_1)e^{-j\omega t_1}dt_1.
\end{aligned}$$

Using FT properties we know that $\delta(t-t_1)=\frac{1}{2\pi}\int_{-\infty}^{\infty} e^{-j\xi(t-t_1)}d\xi$. Now, substituting this result gives:

$$\begin{aligned}
WV_z(\tau,\omega) &= \frac{1}{2\pi}\int_{-\infty}^{\infty} x\left(\tau+\frac{t}{2}\right)x^*\left(\tau-\frac{t}{2}\right)dt \int_{-\infty}^{\infty} y\left(\tau+\frac{t_1}{2}\right)y^*\left(\tau-\frac{t_1}{2}\right)e^{-j\omega t_1}dt_1 \int_{-\infty}^{\infty} e^{-j\xi(t-t_1)}d\xi \\
&= \frac{1}{2\pi}\int_{-\infty}^{\infty}\left[\int_{-\infty}^{\infty} x\left(\tau+\frac{t}{2}\right)x^*\left(\tau-\frac{t}{2}\right)e^{-j\xi t}dt \int_{-\infty}^{\infty} y\left(\tau+\frac{t_1}{2}\right)y^*\left(\tau-\frac{t_1}{2}\right)e^{-j(\omega-\xi)t_1}dt_1\right]d\xi \\
&= \frac{1}{2\pi}\int_{-\infty}^{\infty} WV_x(\tau,\xi)WV_y(\tau,\omega-\xi)d\xi.
\end{aligned}$$

7.3 Digitization of Shift-Scale Parameters and Wavelet Series

In practice, the shift-scale parameters τ and a cannot take continuous values and must be discretized for computational purposes even though the signal is continuous-time. This is particularly the case for the inverse formula, which involves double integrals with respect to these parameters. A typical way to discretize these parameters is to use $a=a_o^j$ and $\tau = k\,a_o^jT$ with $j,k\in\mathbb{Z}$, $a_0>1$. The idea behind this sampling is that narrow windows are translated by small steps in order to catch high frequency details, whereas wider (low frequency) wavelets are translated by larger steps. The sampled daughter wavelets become:

$$\psi_{j,k}(t)=a_o^{\frac{-j}{2}}\,\psi(a_o^{-j}t-kT),$$

resulting in wavelet series coefficients [8, 15],

$$X_{j,k}=\int_{-\infty}^{\infty} x(t)\psi_{j,k}^*(t)dt, \tag{7.11}$$

which is known as the *Analysis* equation. To observe small details (i.e. magnification), a_o^{-j} must be large, which requires large negative j. On the other hand, global view requires small positive j. A typical choice of digitization process is known as the *dyadic sampling* [8] for which $a_o=2$ and $T=1$. In this case, there exist certain choices of $\psi(t)$ such that $\{\psi_{j,k}\}$ forms an orthonormal basis, hence facilitating the reconstruction from the transform coefficients. The reconstruction or the *synthesis* equation in this case becomes:

$$x(t)=\sum_j\sum_k X_{j,k}\psi_{j,k}(t). \tag{7.12}$$

7.3.1 Wavelet Frames

The theory of wavelet frames provides a balance between redundancy (or sampling density) and restrictions on $\psi(t)$ for stable reconstruction. That is, large redundancy (i.e. over-sampling) leads to mild conditions on the basis functions while small redundancy (critical sampling) causes basis functions to be very constrained.

The family of wavelet functions is called *frame* if the energy of the wavelet series coefficients lies between two positive *frames bounds* A and B, i.e.

$$A||x||^2\le\sum_j\sum_k|X_{j,k}|^2\le B||x||^2$$

where $||x||^2 = \int |x(t)|^2 dt$. This is indeed the condition for having a numerically stable reconstruction of the signal from its wavelet series coefficients. The frame bounds $0 < A \leq B < \infty$ can be calculated from a_0, T and $\psi(t)$ [6, 8]. These limits determine the accuracy of the signal reconstruction, that is:

$$x(t) \approx \frac{2}{A+B} \sum_j \sum_k X_{j,k} \psi_{j,k}(t).$$

The closer A and B are, the more accurate the reconstruction. If $A = B$ and $|\psi_{j,k}| = 1, \forall j, k$ we have the so-called *tight frames*. If we have tight frame, the case $A = B = 1$ (critical sampling) constitutes an orthonormal basis, i.e.

$$\int \psi_{j,k} \psi^*_{j',k'}(t)dt = \delta(j - j')\delta(k - k')$$

Thus, in this case, an arbitrary signal can be represented exactly as a weighted sum of the basis functions using the following synthesis equation:

$$x(t) = \sum_j \sum_k X_{j,k} \psi_{j,k}(t) \tag{7.13}$$

7.4 Filter Banks and Discrete Wavelet Series

Here, our goal is to achieve frequency subband decomposition, similar to that in Figure 7.1b, using filter bank implementation for discrete-time signals. The analysis presented in this section is solely based on the time domain, as opposed to the z-transform and polyphase approaches [8, 11, 20]. We will then extend this analysis to the spatial domain for images.

Let us consider a discrete-time signal $x(n)$ that is filtered using a half-band low-pass filter (LPF) and a half-band high-pass filter (HPF) with impulse responses $g(n)$ and $h(n)$, respectively. Using these half-band filters, the frequencies above $\frac{\pi}{2}$ for the low-pass filtered signal and those below $\frac{\pi}{2}$ for the high-pass filtered signal are suppressed. Thus, the signal spectrum is decomposed into two subbands as shown in the left side of Figure 7.5. For each subband signal, the bandwidth is reduced by a factor of two due to filtering, although the two signals together contain all the information of the original signal. According to the Nyquist criterion, each subband signal can now be downsampled or decimated (drop every odd indexed sample) by a factor of two without losing any information. This downsampling, denoted by $\downarrow 2$, doubles the scale while the resolution remains unchanged. The downsampling operation serves another purpose in preventing an increase in the dimensionality of the data representation because the total length of the two filtered signals is more than twice the size of the input. This is due to the fact that the linear convolution of a finite-extent signal of size L with each half-band filter impulse response of order N produces an output of length $L + N - 1$ samples. Finally, downsampling by $\downarrow 2$ stretches the spectrum by a factor of two, which typically causes aliasing effects in absence of the half-band filters.

The low-pass and high-pass filtered and downsampled outputs, $X_0(k)$ and $X_1(k)$, in Figure 7.5, are called the *low-pass approximation* and *added detail*, respectively and can be expressed as:

$$\begin{aligned} X_0(k) &= \sum_n x(n)g(2k - n) = \langle x(n), g(2k - n) \rangle \\ X_1(k) &= \sum_n x(n)h(2k - n) = \langle x(n), h(2k - n) \rangle \end{aligned} \tag{7.14}$$

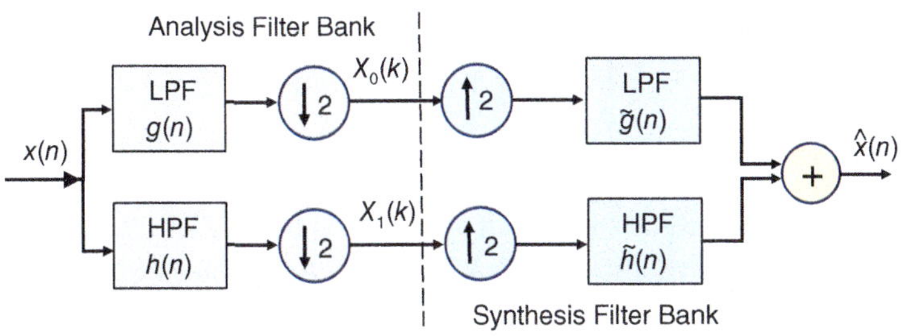

Figure 7.5 Analysis and synthesis filter banks of one-level DWS and subband decomposition.

where $\langle a(n), b(n)\rangle$ is the inner product operation. The analysis filter bank (left-hand side of Figure 7.5) performs one-level discrete wavelet series (DWS) for implementing the multi-resolution subband decomposition of Figure 7.1b. The half-band HPF, $h(n)$, can be created from the half-band LPF, $g(n)$ using the *alternating flip* operation, i.e. $h(n) = (-1)^n g(N-1-n)$. This is shown in the following example.

Example 7.3 *Show that alternating flip* $h(n) = (-1)^n g(N-1-n)$ *maps a half-band LPF* $g(n)$ *to a half-band HPF* $h(n)$.

Solution: Since $G(z) = \sum_{n=0}^{N-1} g(n)z^{-n}$ is the LPF transfer function, it satisfies $G(z)|_{z=-1} = 0$ and $G(z)|_{z=1} = 1$. The transfer function of the corresponding HPF is:

$$H(z) = \sum_{n=0}^{N-1} (-1)^n g(N-1-n)z^{-n} = \sum_{n=0}^{N-1} (-1)^{N-1-n} g(n)z^{-(N-1-n)}$$
$$= (-z)^{-(N-1)} \sum_{n=0}^{N-1} g(n)(-z^{-1})^{-n} = (-z)^{-(N-1)} G(-z^{-1})$$

Using this relationship, it is clear that $H(z)|_{z=-1} = 1$ and $H(z)|_{z=1} = 0$, i.e., $H(z)$ is a transfer function of a HPF.

The synthesis filter bank illustrated in the right-hand side of Figure 7.5 involves upsampling by a factor of two before using half-band reconstruction filters. These upsamplers are illustrated by symbol $\uparrow 2$. The upsampling process involves inserting a 0 sample between every pair of samples to bring the signals back to their original sizes. Thus, in contrast to downsampling, upsampling operation is reversible. Moreover, upsampling compresses the spectrum so that the synthesis half-band filters can reconstruct the corresponding low-pass and high-pass information, the combination of which gives the reconstructed signal $\hat{x}(n)$. Let us denote the synthesis half-band LPF and HPF impulse responses by $\tilde{g}(n)$ and $\tilde{h}(n)$, respectively. Then, the reconstructed version of $x(n)$ is:

$$\hat{x}(n) = \sum_k [X_0(k)\tilde{g}(n-2k) + X_1(k)\tilde{h}(n-2k)] \tag{7.15}$$

that is, weighted sum of the impulse responses where weights are the inner products of the signal with impulse responses as shown in (7.14). We will show later in this chapter that when the impulse responses of the synthesis filters are the *time-revered* versions of the corresponding analysis filters, that is, $\tilde{g}(n) = g(N-1-n)$ and $\tilde{h}(n) = h(N-1-n)$, an orthogonal filter bank is constructed which leads to perfect reconstruction (PR), i.e. $\hat{x}(n) = x(n)$. Thus, the entire filter bank can be designed from a single prototype filter $g(n)$.

7.4.1 Matrix Representation of DWS

Assuming that the impulse response $g(n)$ of a low-pass FIR filter is of length N, the linear convolution in (7.14) can be represented in matrix form as:

$$\underline{X}_0 = \mathbf{G}\underline{x} \tag{7.16}$$

where $\mathbf{G}$ is an infinite matrix (streaming data) defined as:

$$\mathbf{G} = \begin{bmatrix} \vdots & \vdots & \vdots & \vdots & \vdots & \vdots & \vdots & \vdots & \vdots \\ \cdots & g(N-1) & g(N-2) & g(N-3) & \cdots & g(0) & 0 & 0 & \cdots \\ \cdots & 0 & 0 & g(N-1) & \cdots & g(2) & g(1) & g(0) & \cdots \\ \vdots & \vdots & \vdots & \vdots & \vdots & \vdots & \vdots & \vdots & \vdots \end{bmatrix},$$

$\underline{x} = [\cdots\ x(-1)\ x(0)\ x(1)\ \cdots]^T$ is the infinite-dimensional input vector and the output vector $\underline{X}_0$ is defined similarly. Note that downsampling results in eliminating every other rows of the convolution matrix which leads to the *double-shift* to the right from one row to the next row as evident in matrix $\mathbf{G}$.

The same procedure can be repeated for the high-pass filtered and downsampled signal:

$$\underline{X}_1 = \mathbf{H}\underline{x} \tag{7.17}$$

where matrix **H** is defined in a similar manner to **G** but in terms of the elements of the impulse response, $h(n)$, of the same order N. These two matrices together form the *Analysis Filter Bank* matrix **W**, i.e.

$$\begin{bmatrix} \underline{X}_0 \\ \underline{X}_1 \end{bmatrix} = \begin{bmatrix} \mathbf{G} \\ \mathbf{H} \end{bmatrix} \underline{x} = \mathbf{W}\underline{x}.$$

Assuming that the impulse response $h(n)$ is the alternating flipped version of $g(n)$, we get $\mathbf{GH}^T = \mathbf{0}$ (for N even). That is, another useful byproduct of the alternating flip operation is that it makes the rows of **G** orthogonal to those of **H**. The proof follows easily by observing the structures of these matrices (Problem 7.11).

Now, the question is, what are the basis functions of this signal representation? To answer this question, we first interleave $\underline{X}_0$ and $\underline{X}_1$ to get $\underline{X}$ defined as:

$$\underline{X} = [\dots X_0(0) X_1(0) X_0(1) X_1(1) \dots]^T.$$

Note that interleaving $\underline{X}_0$ and $\underline{X}_1$ is equivalent to upsampling and adding them together, that is, $\underline{X} = \underline{X}_0 \uparrow 2 + \underline{X}_1 \uparrow 2$. Then, the analysis filter bank equations in (7.14) can be rewritten as:

$$\begin{aligned} X(2k) = X_0(k) &= \sum_n x(n)\phi_{2k}(n) \\ X(2k+1) = X_1(k) &= \sum_n x(n)\phi_{2k+1}(n) \end{aligned} \tag{7.18}$$

where $\phi_{2k}(n) = g(2k-n)$ and $\phi_{2k+1}(n) = h(2k-n)$ are the dual basis functions for the analysis filter bank. From (7.15), it is clear that the dual basis for the synthesis (or reconstruction) filter bank are $\tilde{\phi}_{2k}(n) = \tilde{g}(n-2k)$ and $\tilde{\phi}_{2k+1}(n) = \tilde{h}(n-2k)$.

Similar to the analysis filter bank, the equations for the synthesis filter bank can also be written in vector-matrix form using:

$$\begin{aligned} \underline{u}_0 &= \tilde{\mathbf{G}}\underline{X}_0 \\ \underline{u}_1 &= \tilde{\mathbf{H}}\underline{X}_1 \end{aligned} \tag{7.19}$$

where $\tilde{\mathbf{G}}$ is an infinite matrix defined in terms of $\tilde{g}(n)$ as:

$$\tilde{\mathbf{G}} = \begin{bmatrix} \vdots & \vdots & \vdots & \vdots & \vdots & \vdots \\ \cdots & \tilde{g}(0) & 0 & 0 & \cdots & 0 \\ \cdots & \tilde{g}(1) & 0 & \cdots & \cdots & 0 \\ \cdots & \tilde{g}(2) & \tilde{g}(0) & 0 & \cdots & 0 \\ \cdots & \vdots & \vdots & \vdots & \vdots & \vdots \\ \vdots & \tilde{g}(N-1) & \vdots & \vdots & \vdots & \vdots \\ \cdots & 0 & \tilde{g}(N-2) & \vdots & \vdots & \vdots \\ \cdots & 0 & \tilde{g}(N-1) & \vdots & \vdots & \vdots \\ \vdots & \vdots & \vdots & \vdots & \vdots & \vdots \end{bmatrix}$$

and $\underline{u}_0 = [\cdots u_0(-1)\, u_0(0)\, u_0(1) \cdots]^T$ is an infinite dimensional vector representing the reconstructed version of the low-pass approximation. Matrix $\tilde{\mathbf{H}}$ and vector $\underline{u}_1$ are also similarly defined in terms of $\tilde{h}(n)$ and $u_1(n)$, respectively. Then, the reconstruction equation in (7.19) in matrix-vector form becomes:

$$\underline{\hat{x}} = \underline{u}_0 + \underline{u}_1 = \tilde{\mathbf{G}}\underline{X}_0 + \tilde{\mathbf{H}}\underline{X}_1 \tag{7.20}$$

Substituting (7.16) and (7.17) for $\underline{X}_0$ and $\underline{X}_1$ into (7.20) yields:

$$\underline{\hat{x}} = (\tilde{\mathbf{G}}\mathbf{G} + \tilde{\mathbf{H}}\mathbf{H})\underline{x}. \tag{7.21}$$

Hence, the condition for PR [8, 20], i.e. $\underline{\hat{x}} = \underline{x}$ is:

$$\tilde{\mathbf{G}}\mathbf{G} + \tilde{\mathbf{H}}\mathbf{H} = \tilde{\mathbf{W}}^T\mathbf{W} = \mathbf{I} \tag{7.22}$$

where $\tilde{\mathbf{W}}^T = [\tilde{\mathbf{G}}\ \tilde{\mathbf{H}}]$ is the synthesis filter bank matrix. This condition requires that the dual basis functions of the analysis and synthesis filter banks satisfy the so-called *biorthogonality* condition, i.e. $\langle \phi_k(n), \tilde{\phi}_l(n) \rangle = \delta(k-l)$. In other words, we have the following properties between the dual analysis and synthesis filters:

$$\begin{aligned} \langle g(2k-n), \tilde{g}(n-2l) \rangle &= \delta(k-l) \\ \langle h(2k-n), \tilde{h}(n-2l) \rangle &= \delta(k-l) \end{aligned} \tag{7.23}$$

Note that for this biorthogonal filter bank there are two prototype filters, i.e. $g(n)$ and $\tilde{g}(n)$ since $\tilde{h}(n)$ can be obtained from $\tilde{g}(n)$ using the alternating flip operation.

7.4.2 Orthonormal Filter Banks and DWT

Now, if the filter $g(n)$ is designed so that the combined analysis filter bank matrix $\mathbf{W}$ is an orthogonal matrix [20, 21], that is, $\mathbf{W}\mathbf{W}^T = \mathbf{I}_1$ and $\mathbf{W}^T\mathbf{W} = \mathbf{I}_2$, we will get an orthogonal filter bank with many useful properties. Note that since the infinite matrix $\mathbf{W}$ is not square (due to downsampling), $\mathbf{I}_1$ and $\mathbf{I}_2$ are two identity matrices of different dimensions. The wavelet representation in this case is referred to as DWT [8].

As we saw in Chapter 6, the orthogonality (or unitary) property of the transformation matrices allows us to perfectly recover the original signal from its transformed version using an inverse (or synthesis) mapping matrix $\tilde{\mathbf{W}}^T = \mathbf{W}^T$ or $\tilde{\mathbf{G}} = \mathbf{G}^T$ and $\tilde{\mathbf{H}} = \mathbf{H}^T$. Considering the structures of $\mathbf{G}$ and $\mathbf{G}^T$ one can easily deduce the fact that the condition $\tilde{\mathbf{G}} = \mathbf{G}^T$ will be met when the impulse response of synthesis filter $\tilde{g}(n)$ is the *time-revered* version of the corresponding analysis filter, i.e. $\tilde{g}(n) = g(N-1-n)$ and similarly for $\tilde{h}(n) = h(N-1-n)$. This was stated without proof in the last paragraph of the previous section. Therefore, for an orthonormal filter bank, only the LPF in the analysis filter bank needs to be designed since all other filters are defined in terms of $g(n)$.

The PR condition for the orthonormal case becomes:

$$\mathbf{W}^T\mathbf{W} = \mathbf{G}^T\mathbf{G} + \mathbf{H}^T\mathbf{H} = \mathbf{I}_2 \tag{7.24}$$

Note that the complementary condition $\mathbf{W}\mathbf{W}^T = \mathbf{I}_1$ yields:

$$\mathbf{W}\mathbf{W}^T = \begin{bmatrix} \mathbf{G}\mathbf{G}^T & \mathbf{G}\mathbf{H}^T \\ \mathbf{H}\mathbf{G}^T & \mathbf{H}\mathbf{H}^T \end{bmatrix} = \begin{bmatrix} \mathbf{I} & \mathbf{0} \\ \mathbf{0} & \mathbf{I} \end{bmatrix} \tag{7.25}$$

which implies that $\mathbf{G}$ and $\mathbf{H}$ are semi-orthogonal matrices [21] with $\mathbf{G}\mathbf{G}^T = \mathbf{I}$ and $\mathbf{H}\mathbf{H}^T = \mathbf{I}$ but $\mathbf{G}^T\mathbf{G} \neq \mathbf{I}$ and $\mathbf{H}^T\mathbf{H} \neq \mathbf{I}$. In other words, the rows of $\mathbf{G}$ and $\mathbf{H}$ form orthonormal sets. Thus, we have:

$$\begin{aligned} \langle g(n-2k), g(n-2l) \rangle &= \delta(k-l) \\ \langle h(n-2k), h(n-2l) \rangle &= \delta(k-l). \end{aligned} \tag{7.26}$$

1. **Haar Filter Bank**
   Haar filter bank is an example of an extreme and yet simple filter bank with many interesting and useful properties. The FIR filters in the Haar analysis filter bank have impulse responses $g(n) = \{g(0), g(1)\} = \left\{\frac{1}{\sqrt{2}}, \frac{1}{\sqrt{2}}\right\}$ and $h(n) = \{h(0), h(1)\} = \left\{\frac{1}{\sqrt{2}}, -\frac{1}{\sqrt{2}}\right\}$ which perform moving average and difference operations, respectively. It can easily be shown that the frequency responses of these filters are:

$$G(e^{j\Omega}) = \sum_n g(n)e^{-j\Omega n} = \sqrt{2}\cos\frac{\Omega}{2}e^{-j\frac{\Omega}{2}}$$
$$H(e^{j\Omega}) = \sum_n h(n)e^{-j\Omega n} = \sqrt{2}j\sin\frac{\Omega}{2}e^{-j\frac{\Omega}{2}}. \tag{7.27}$$

Thus, Haar filters have linear phase characteristics with $\theta_G(\Omega) = -\frac{\Omega}{2} + k\pi,\ k = 0, 1$ and $\theta_H(\Omega) = -\frac{\Omega}{2} + \pi/2$. Furthermore, it is simple to show that the dual basis $\phi_{2k}(n) = g(2k - n)$ and $\phi_{2k+1}(n) = h(2k - n)$ are orthonormal. The output of the analysis filter bank are:

$$X(2k) = X_0(k) = \frac{1}{\sqrt{2}}(x(2k) + x(2k-1))$$
$$X(2k+1) = X_1(k) = \frac{1}{\sqrt{2}}(x(2k) - x(2k-1)) \tag{7.28}$$

Comparing these equations with a two-point DFT of $\underline{x}$ we can see that Haar expansion is indeed related to two-point DFT of $\underline{x}$. Adding and subtracting the above equations yield the reconstructed versions of the low-pass and high-pass signals:

$$x(2k) = \frac{1}{\sqrt{2}}(X(2k) + X(2k+1))$$
$$x(2k-1) = \frac{1}{\sqrt{2}}(X(2k) - X(2k+1)), \tag{7.29}$$

respectively. Comparing these synthesis equations in (7.29) with those of the analysis in (7.28) reveals the fact that the dual bases for the synthesis filter bank are $\tilde{g}(n) = \left\{\frac{1}{\sqrt{2}}, \frac{1}{\sqrt{2}}\right\}$ and $\tilde{h}(n) = \left\{-\frac{1}{\sqrt{2}}, \frac{1}{\sqrt{2}}\right\}$. That is, they are the *time-reversed* versions of those of the analysis filter bank, as expected. It is important to note that although the Haar filter bank is simple, it is the only one that offers both linear phase and orthonormality. This is not shared by the other DWT orthonormal filter banks.

2. **Daubechies' Family of Orthonormal Filters**
Daubechies [6, 8, 11, 20] designed a family of FIR half-band filters with maximally flat frequency response that provide orthonormal basis functions. The main purpose was to ensure that they lead to a good approximation to continuous-time wavelet and scaling functions when the filter bank is iterated several times. To satisfy these requirements, the filter transfer function should contain many zeros at $z = -1$ (or $\Omega = \pi$). An example of this family of half-band filters is Daubechies' db4, which is a short length 4 (third order) FIR filter bank with LPF coefficients:

$$g(n) = \frac{1}{4\sqrt{2}}\{(1+\sqrt{3}), (3+\sqrt{3}), (3-\sqrt{3}), (1-\sqrt{3})\}.$$

The normalization factor $\frac{1}{4\sqrt{2}}$ is added to ensure $\sum_n g(n)^2 = 1$. The transfer function of this LPF is:

$$G(z) = \frac{1}{4\sqrt{2}}(1+z^{-1})^2((1+\sqrt{3}) + (1-\sqrt{3})z^{-1}),$$

which has two zeros at $z = -1$ as desired. The corresponding half-band HPF obtained by using the alternating flip operation is:

$$h(n) = \frac{1}{4\sqrt{2}}\{(1-\sqrt{3}), (-3+\sqrt{3}), (3+\sqrt{3}), (-1-\sqrt{3})\}.$$

Synthesis filters have impulse responses that are the time-reversed version of those of the corresponding analysis filter bank, as shown before. Although this filter bank offers orthonormality, which is an important property for signal and image processing applications, the filters do not have a linear phase characteristic. This is evident from the fact that the impulse responses of the filters do not possess any symmetry (or anti-symmetry) property, which is required to have FIR filters with linear phase characteristic.

3. **Biorthogonal Filter Bank**
Unlike the orthogonal filter bank family, these filters provide linear phase characteristics while the synthesis filters are not related to those of the analysis filter bank hence requiring two pairs of filters. Nevertheless, as we saw before, they still satisfy the PR condition. They are good choices for many signal processing applications (e.g. audio or acoustics) and data compression. An example of biorthogonal case is the 5/3 filter bank with the following analysis and synthesis FIR filters that have short lengths of 5 and 3. The impulse responses and transfer functions of the analysis and synthesis LPF are:

$$g(n) = \frac{1}{8}\{-1, 2, 6, 2, -1\}, \quad G(z) = \frac{1}{8}(1 + z^{-1})^2(-1 + 4z^{-1} - z^{-2})$$

and

$$\tilde{g}(n) = \frac{1}{4}\{1, -2, 1\}, \quad \tilde{G}(z) = \frac{1}{4}(1 + z^{-1})^2,$$

respectively, both of which have two zeros at $z = -1$ as desired. The other filters are obtained by the following mappings that map low-pass to high-pass:

$$H(z) = \tilde{G}(-z) = \frac{1}{4}(1 - z^{-1})^2$$

and

$$\tilde{H}(z) = -G(-z) = \frac{1}{8}(1 - z^{-1})^2(1 + 4z^{-1} + z^{-2}).$$

As can be seen, the alternating flip is no longer satisfied here, so the analysis filters are not orthogonal to each other. However, the biorthogonality conditions in (7.23) are satisfied (Problem 7.16) to provide a PR filter bank.

4. **Quadrature Mirror Filter (QMF) Bank**
The quadrature mirror filter (QMF) bank, introduced by Croisier, Esteban, and Galand [22], employs a specific structural symmetry (mirror) between its low-pass and high-pass filters. In this configuration, the HPF in the analysis bank is derived from the LPF by applying an alternating sign, rather than an alternating flip operation. This relationship is expressed in the time domain as $h(n) = (-1)^n g(n)$, or in the z-transform domain $H(z) = G(-z)$. Furthermore, the synthesis filters are directly related to their analysis counterparts through the following relations:

$$\tilde{G}(z) = G(z), \text{ and } \tilde{H}(z) = -H(z) = -G(-z).$$

Despite these constraints, QMF filter banks can be designed to satisfy the PR condition [8, 22].

7.4.3 Iterated Filter Banks

The subband decomposition can be iterated on the *low-pass approximations* to produce a multi-level filter bank. This is often called *octave-band* filter bank since each successive high-pass filtered output contains an octave of the input bandwidth. A three-level filter bank for subband decomposition is shown in Figure 7.6, which results in one *lowest order approximation*, $X_0^{(3)}(m)$ and three *added details* $X_1^{(3)}(m)$, $X_1^{(2)}(l)$, and $X_1^{(1)}(k)$. The superscript denotes the level in the subband decomposition. The added detail, $X_1^{(1)}(k)$, mainly contains high-frequency information of the original signal since it is the result of high-pass filtering followed by downsampling. The second added detail, $X_1^{(2)}(l)$, contains the mid-frequency information extracted from the second-level filter bank. This added detail is obtained by filtering and downsampling the signal using the first-level LPF followed by a second-level HPF. Similarly, the third-level added detail, $X_1^{(3)}(m)$, was generated by filtering and downsampling the signal using two LPF in the first- and second-levels and one HPF in the third-level of decomposition. Finally, the lowest-order approximation, $X_0^{(3)}(m)$, corresponds to the narrowest-bandwidth subband in the low-frequency region of the spectrum. This subband decomposition implements the desired subband decomposition in Figure 7.1b.

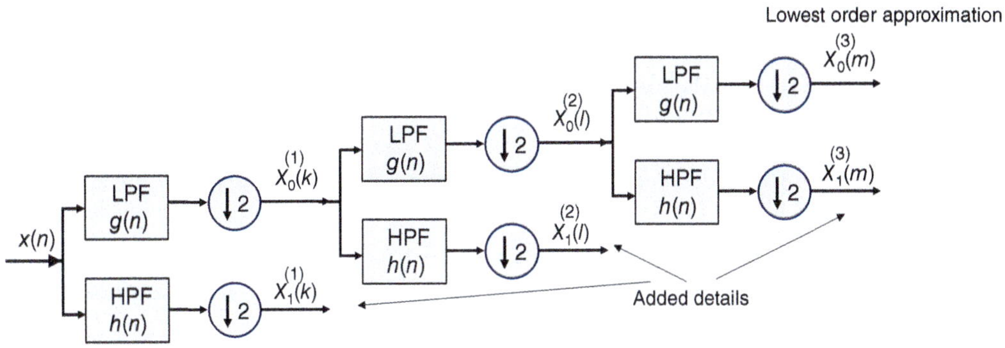

Figure 7.6 Three-level 1-D DWT analysis filter bank.

In the general case for L-level filter bank, we have one lowest order approximation and L added details given by:

$$X_0^{(L)}(p) = \langle g_L(2^L p - n), x(n)\rangle = \sum_n x(n) g_L(2^L p - n)$$
$$X_1^{(j)}(m) = \langle h_j(2^j m - n), x(n)\rangle = \sum_n x(n) h_j(2^j m - n), \quad j \in [1, L] \tag{7.30}$$

where $g_L(n) = \sum_i g_{L-1}(i) g(n-2i)$ is the impulse response of the LPF associated with the lowest order approximation, while $h_j(n) = \sum_i g_{j-1}(i) h(n-2i)$ is the jth band-pass filter (BPF) for $j \in [2, L]$ or the HPF for $j = 1$, corresponding to the added details.

Assuming an orthonormal filter bank, the reconstruction equation for this multi-level DWT becomes:

$$x(n) = \sum_{j=1}^{L} \sum_i X_1^{(j)}(i) h_j(n - 2^j i) + \sum_p X_0^{(L)}(p) g_L(n - 2^L p). \tag{7.31}$$

Now, let us associate continuous-time functions $\varphi^{(L)}(t)$ and $\psi^{(j)}(t)$ to the impulse responses, $g_L(n)$ and $h_j(n)$, respectively. That is

$$\varphi^{(L)}(t) = 2^{\frac{L}{2}} g_L(n), \quad \text{for} \quad \frac{n}{2^L} \le t < \frac{n+1}{2^L},$$
$$\psi^{(j)}(t) = 2^{\frac{j}{2}} h_j(n), \quad \text{for} \quad \frac{n}{2^j} \le t < \frac{n+1}{2^j}. \tag{7.32}$$

As can be observed from their definitions, these functions are piecewise constant over each interval. The interval decreases as the lengths of these filters increase by increasing the number of level (or iterations) L in the filter bank. This is due to the fact that the length of $g_L(n)$ is $N^{(L)} = (2^L - 1)(N-1) + 1$ (L consecutive linear convolutions). Hence, when $L \to \infty$ the function $\varphi^{(L)}(t)$ becomes compactly supported over the interval $[0, N-1]$. The factor $2^{\frac{L}{2}}$ ensures that the functions have unit norms.

Now, the key questions are: (a) Do these piecewise constant functions converge? (b) And if they do, to which functions do they converge when $L \to \infty$? The answer to the first question is the issue of *regularity*, which is not addressed here. The reader is referred to [6, 8] for further reading of the regularity condition. Assuming convergence, the above functions converge to continuous-time *scaling* (or father wavelet) and *wavelet* (or mother wavelet) functions, i.e.

$$\lim_{L\to\infty} \varphi^{(L)}(t) = \varphi(t),$$
$$\lim_{L\to\infty} \psi^{(L)}(t) = \psi(t). \tag{7.33}$$

The wavelet function $\psi^{(J)}(t)$ is related to the scaling function. To show this, let us use the expressions $g_k(n) = \sum_i g_{k-1}(i)g(n-2i)$ and $h_j(n) = \sum_i g_{j-1}(i)h(n-2i)$ together with those in (7.32), which yield the following two-scale recursive equations for generating $\varphi^{(L)}(t)$ and $\psi^{(J)}(t)$ from the scaling functions at the previous iteration (or levels), i.e.

$$\begin{aligned} \varphi^{(L)}(t) &= \sqrt{2}\sum_i g(i)\varphi^{(L-1)}(2t-i) \\ \psi^{(J)}(t) &= \sqrt{2}\sum_i h(i)\varphi^{(J-1)}(2t-i). \end{aligned} \tag{7.34}$$

Assuming regularity condition (relates to the number of zeros of the filters at $z = -1$) is met [6, 8], at convergence we get:

$$\begin{aligned} \varphi(t) &= \sqrt{2}\sum_i g(i)\varphi(2t-i) \\ \psi(t) &= \sqrt{2}\sum_i h(i)\varphi(2t-i). \end{aligned} \tag{7.35}$$

These equations, which are also called *dilation* and *wavelet* equations, respectively, provide a mechanism for building continuous-time wavelet bases from the discrete iterated filters.

Figure 7.7 shows the scaling and wavelet functions of db4 filter bank for $L = 1, 2, 4,$ and 8 levels or iterations, respectively. As can be seen, by increasing L the wavelet function becomes closer to the corresponding continuous-time wavelet function in Figure 7.4.

The orthonormal subband decomposition described in this chapter can be iterated on both the low-pass approximation and the added details to form any octave-band tree structure with different interesting properties [11, 13]. Clearly, the full-tree structure yields a subband decomposition similar to that of the STFT. These tree structures form a family of orthonormal bases known as *wavelet packets* (WP) from which the best one can be chosen according to some criterion, e.g. the rate distortion function for image data compression [23].

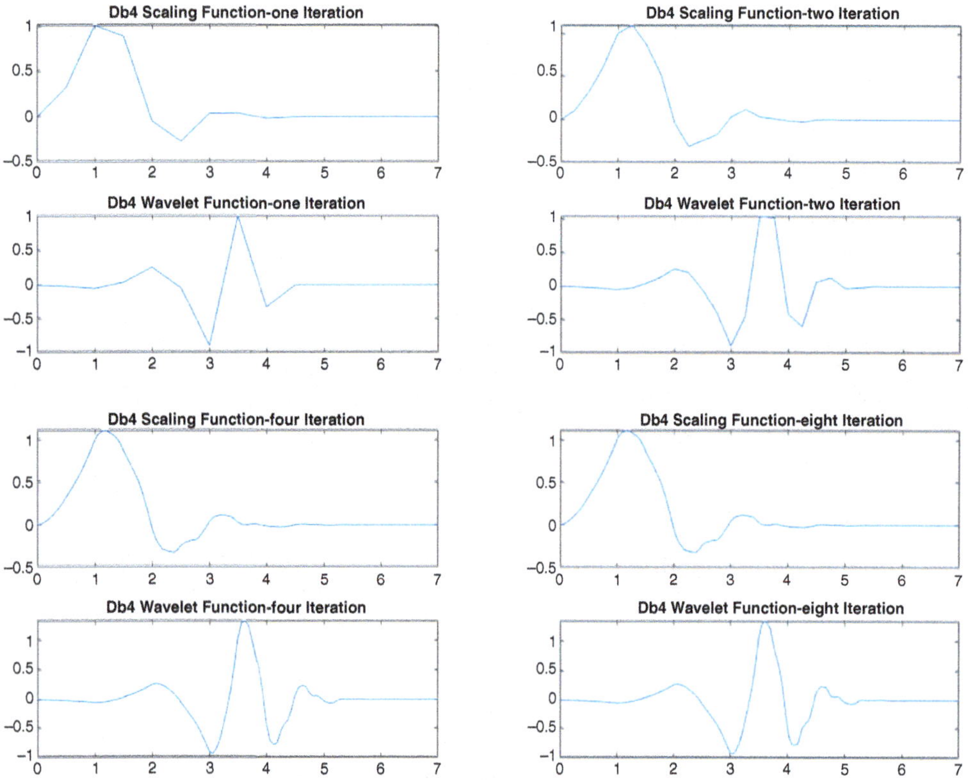

Figure 7.7 Scaling and wavelet functions of db4 at various filter bank iterations.

7.5 2-D DWT and Filter Banks

The wavelet representation can be extended to the 2-D for many image processing applications [23, 24]. As in the 1-D case, there exists a unique scaling function $\varphi(u, v)$ whose dilation and translation:

$$\varphi_{i;k,l}(u, v) = 2^{-i}\varphi(2^{-i}u - k, 2^{-i}v - l)$$

form an orthonormal basis for the lowest order approximation. Note that, for reasons that will be evident shortly, we have a slight change of notation in that the first subscript i represents the number of iterations (or levels) in the 2-D case. Additionally, without loss of generality, we consider the separable multi-resolution case [10, 14] for which $\varphi(u, v) = \varphi(u)\varphi(v)$. Thus, the orthonormal basis also are separable, i.e.

$$\varphi_{i;k,l}(u, v) = 2^{-\frac{i}{2}}\varphi(2^{-i}u - k)2^{-\frac{i}{2}}\varphi(2^{-i}v - l) = \varphi_k^i(u)\varphi_l^i(v).$$

The wavelet functions in this case consist of three parts with orthonormal basis associated with the horizontal $\psi^h_{i;k,l}(u, v) = \varphi_k^i(u)\psi_l^i(v)$, vertical $\psi^v_{i;k,l}(u, v) = \psi_k^i(u)\varphi_l^i(v)$, and diagonal $\psi^d_{i;k,l}(u, v) = \psi_k^i(u)\psi_l^i(v)$. Clearly, these lead to one lowest order approximation and three horizontal, vertical, and diagonal added details for an image $x(m, n)$ for just one-level 2-D wavelet decomposition. These subimages are given by:

$$\begin{aligned}
X_0(k, l) &= \langle \varphi_{i;k,l}(u, v), x(u, v)\rangle, \\
X_1^{\mathbf{h}}(k, l) &= \langle \psi^h_{i;k,l}(u, v), x(u, v)\rangle, \\
X_1^{\mathbf{v}}(k, l) &= \langle \psi^v_{i;k,l}(u, v), x(u, v)\rangle, \\
X_1^{\mathbf{d}}(k, l) &= \langle \psi^d_{i;k,l}(u, v), x(u, v)\rangle.
\end{aligned} \tag{7.36}$$

For digital images, the separability property allows efficient implementation of the 2-D DWT by successive application of the 1-D filters to the rows and columns of an image. For an image of size $M \times M$, applying a 1-D DWT along the rows leads to two subimages each of size $M \times \frac{M}{2}$. This is followed by another 1-D DWT along the columns of these two subimages, resulting in four subimages each of size $\frac{M}{2} \times \frac{M}{2}$. Figure 7.8 shows the associated analysis filter bank for one-level 2-D DWT.

The first subimage that is obtained by low-pass filtering and downsampling (by 2) along rows and columns gives the low-order approximation, the second one is obtained by low-pass filtering and downsampling along rows and high-pass filtering (edge extraction) and downsampling along columns giving the first added detail corresponding to the vertical edge features, the third and fourth added details similarly give the horizontal and diagonal edge features. Reconstruction from

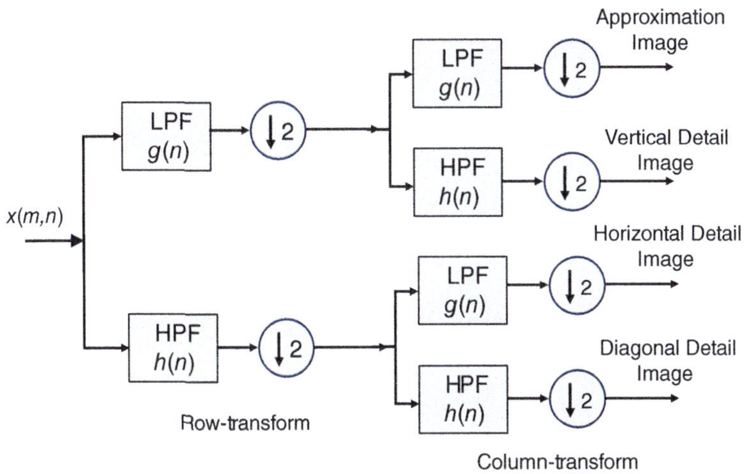

Figure 7.8 One-level 2-D DWT analysis filter bank and four subimages corresponding to low order approximation and vertical, horizontal and diagonal added details.

these subimages or their manipulated versions, e.g. histogram thresholded for denoising or compression, can be done using the corresponding synthesis filter bank configured in a manner similar to the 1-D case.

This process can be iterated on the low-pass filtered and downsampled subimage several times as in the 1-D case to perform multi-level 2-D DWT. Subsequent iterations will, however, provide subimages that contain finer details such as textural and edge-based features.

Figure 7.9a and b shows 2-D DWT for one- and two-level decomposition of the Peppers image using the orthonormal db4 wavelet, respectively. As can be seen, the detail subimages mostly contain very low intensities pixels and those of edge/texture details. To compress the image data, the histograms of the added detail subimages can be thresholded to extract only the useful edge and texture coefficients and set those with intensities smaller than the chosen thresholds to zero. The threshold for each added detail subimage is determined based its own histogram. While each pixel in the lowest-order approximation subimage can be encoded using, e.g. 8 bits/pixel, only those remaining coefficients (after the thresholding) need to be encoded in the added detail subimages. This yields substantial reduction in the total number of bits required to represent an image. This reduction in the bit rate is even more substantial for the 2-D DWT compared to the 2-D DCT compression rate, while providing much better quality reconstruction of the compressed images. For these reasons, 2-D DWT is adopted in the new JPEG and MPEG standards.

To compare the image data compression performance of these two methods, the full-color Finn image in Figure 5.10a is used for this experiment. The 2-D DCT is applied to each nonoverlapping image blocks of size 16×16 and low-pass filtering is performed in the transformed domain to discard approximately 95% of the low-magnitude DCT coefficients in each block. The image is then reconstructed block by block using the inverse 2-D DCT. The reconstructed Finn image using this method is shown in Figure 7.10a. The corresponding error image is also shown in Figure 7.10b. The SNR obtained using the error image is found to be approximately 24.6 dB. Next, we applied a two-level 2-D DWT using db4 filters to the original Finn image. The histograms of the added detail subimages (six) are thresholded using their own chosen thresholds to provide approximately the same percentage reduction of the DWT coefficients. The corresponding synthesis filter bank is then applied to reconstruct the image using the lowest-order approximation subimage together with the thresholded added details. This reconstructed image is shown in Figure 7.10c. Visual evaluations of this reconstructed image and its error image in Figure 7.10d indicate much better quality reconstruction for the 2-D DWT in comparison with 2-D DCT using the same percent reduction. The SNR for this image is 27.4 dB, which also attests to this better performance. Computer Project 7.2 further investigates the applications of 2-D DCT and DWT methods in image data compression and provides a more comprehensive performance evaluation of these methods in terms of compression rate versus SNR.

Another application domain for DWT is denoising for both signals and images [8, 12]. The idea is that noise manifests itself as fine-grained structure in an image, and the DWT provides a scale-based decomposition. Consequently, noise tends to be mostly captured by the subbands associated with the finer scales (or added details). Proper thresholding [12] these subimages would naturally result in filtering out the noise. However, one must be careful not to distort edge-related features

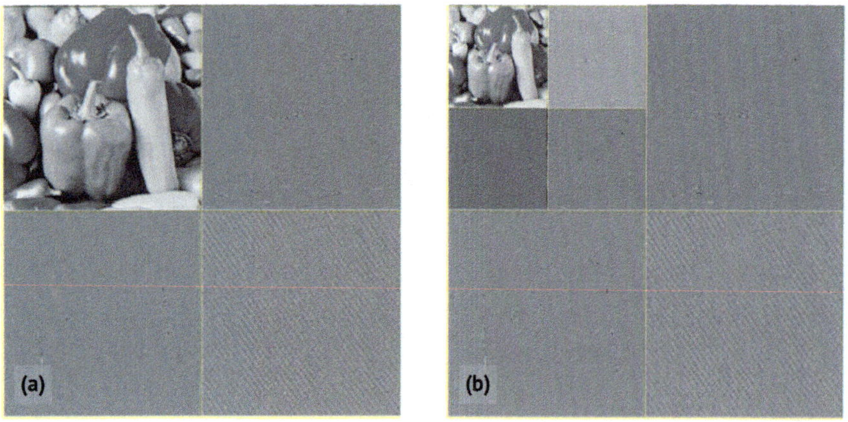

Figure 7.9 (a) One- and (b) two-level 2-D DWT filter bank decompositions of the Peppers image.

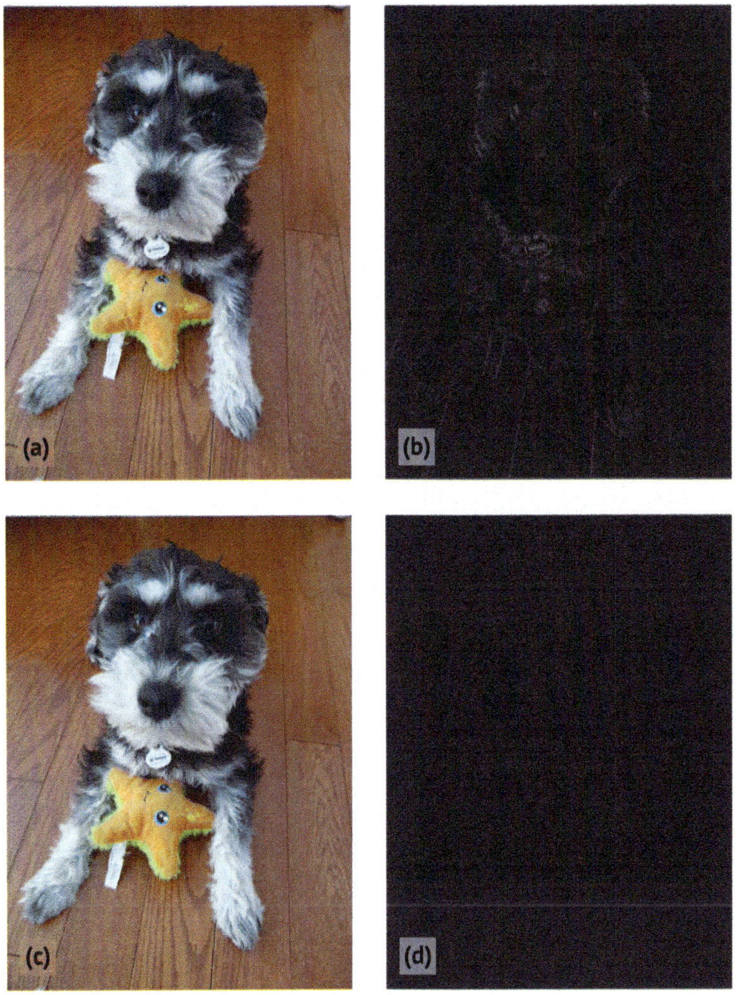

Figure 7.10 Reconstructed Finn images from the compressed versions using (a) 2-D DCT and (c) 2-D DWT. Error images of (b) 2-D DCT and (d) 2-D DWT.

when choosing the thresholds. There are several approaches for choosing proper threshold for each subband. A common approach is to decide the threshold based on the histogram of each subimage as stated before and then use a *soft thresholding*, which leads to a less severe distortion of the features of interest. A possible soft thresholding scheme is given as:

$$x_{\text{th}} = \begin{cases} \text{sign}(x)(|x| - Tr), & \text{for} \quad |x| > Tr \\ 0, & \text{for} \quad |x| \leq Tr \end{cases},$$

where Tr is the chosen threshold. In the next chapter, we provide an example of image denoising using 2-D DWT.

Two-dimensional wavelet transform can also be used for a wide variety of other image processing applications, including textural feature extraction [25], image restoration [24], image classification [26], image fusion, etc. Figure 7.11a and b shows a pair of Geostationary Satellite (GOES) images corresponding to the visible and one of the infrared (IR) channels of GOES-8, respectively. The visible channel provides useful textural and other features complementing those of the IR features which depend on ground and cloud top temperature profiles. Note that the IR image in Figure 7.11b is, in fact, inverted in intensities for better comparison with the visible-channel image. The textural features of the visible channel can help in the classification of different types of clouds when used together with those of the IR channels [26]. Image fusion in this case involves capturing discriminatory information that would enable cloud typing from just one image that contains the essential features of both channels. To this effect, we can first decompose both visible and IR images using two-level

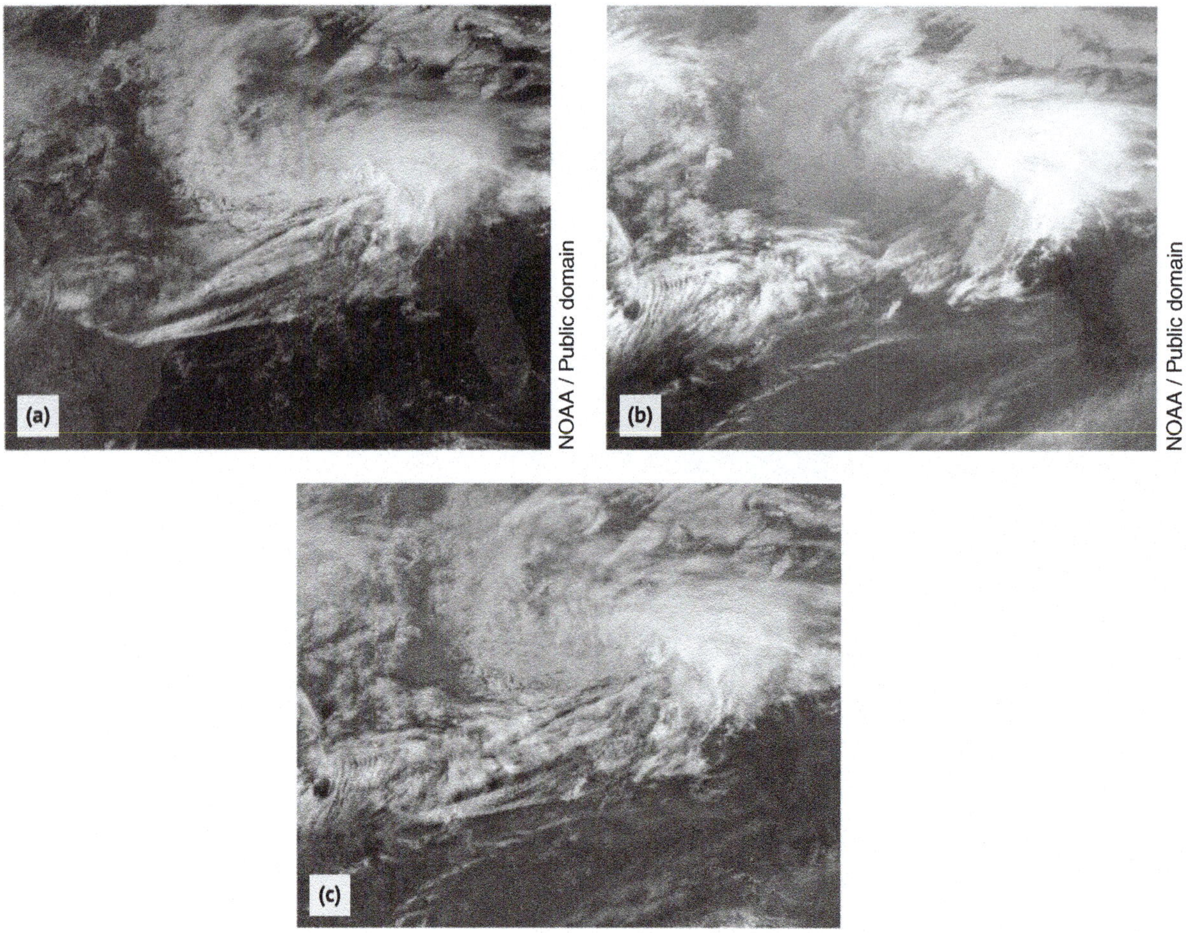

Figure 7.11 A pair of (a) visible, (b) IR GOES-8 images, and (c) fused image using 2-D DWT.

2-D DWT with db4 filters. The fused image in Figure 7.11c is then generated by keeping the lowest-order approximation from the IR images and added details that contain textural information from the visible image. As can be seen, the fused image retains almost all essential information of both channels and hence can be used for subsequent cloud type classification.

7.6 Computer Projects

The computer projects in this chapter use Python libraries OpenCV, NumPy, matplotlib, SciPy, pywt, os. and math.

Computer Project 7.1: Generate the impulse responses of the orthogonal db4, sym4, and biorthogonal 5/3 using the Python wavelet library and validate all the important properties of the analysis and synthesis filters for each case. Then, investigate the frequency response of each filter and elaborate on its magnitude and phase properties. Iterate the corresponding filter bank for each filter type for $L = 1, 2, 4, 8, 16$ iterations. Experimentally determine the convergence to the continuous-time scaling and wavelet functions. Describe your results and observations.

Computer Project 7.2: The goal of this computer project is to study and compare the data compression capabilities of two different filter banks for the 2-D DWT namely the db4 Daubechies filter and the biorthogonal 5/3 filter. The project uses the original Baboon image. The compression methodology involves applying two- and three-level decomposition using the two filter banks. Compression is achieved by thresholding the DWT coefficients in the added detail subimages, specifically by discarding low-magnitude coefficients to introduce a desired percentage of zero coefficients within each subimage. Perform the 2-D DWT on the Baboon image using both the db4 and biorthogonal 5/3 filters. Select thresholds for each added detail subimage based on its histogram to achieve specific levels of coefficient sparsity (a desired percentage of zero coefficients). For each nonzero added detail coefficient, assume an 8-bit codeword. The pixels in the lowest-order approximation subimage are also assigned 8 bits/pixel. Use the synthesis filter bank to reconstruct the images from the retained coefficients and the lowest order approximation subimage. Generate the error image (difference between original and reconstructed images) and produce plots of the SNR versus bits/pixel (BPP) for each case and percent coefficient reduction by varying the thresholds. Then, determine the best filter bank based on the results of the first part. Refer to the results in Figure 7.10 and the description of the corresponding experiment. Provide a detailed analysis of your results, and state the pros and cons of this data compression method.

Computer Project 7.3: This project evaluates the image denoising capabilities of the 2-D DWT using both the db4 and biorthogonal 5/3 wavelets, for the two- and three-level filter banks. In this project, we use the original Barbara image. First, introduce additive white noise to the original image to make the SNR of the noisy image approximately 5 dB. Apply wavelet-based denoising filter to this noisy image by proper thresholding the added detail subimages. The objective is to remove the effects of noise as much as possible while preserving essential edge and fine image details, hence balancing effective noise removal with high-frequency information retention. Reconstruct the filtered images for each case (db4 and biorthogonal 5/3, with two- and three-level filter banks) using the lowest-order approximation and the thresholded added detail subimages. Compare the noise removal efficacy across all implemented methods. The comparison should include both a qualitative visual assessment and a quantitative SNR performance analysis of the final filtered images.

Computer Project 7.4: This computer project aims to develop and evaluate a robust image fusion methodology for multispectral satellite data from the GOES-R system. The core objective is to generate a single, highly informative fused image from a visible and multiple infrared (IR) channels, with enhanced capability for ground and cloud-type discrimination. The project investigates the performance of the 2-D DWT, using both db4 and biorthogonal 5/3 filter banks, for two- and three-level decompositions. The effectiveness of different channel combinations and various added detail subimage selection rules will be comprehensively evaluated. The analysis should focus on the ability of the fused images to preserve and enhance relevant features for improved visual interpretation and discrimination of ground and cloud structures.

7.7 Problems

Problem 7.1: Given a continuous-time signal $x(t)$, express its STFT, $X_{\text{STFT}}(\tau, \omega)$, for two extreme choices of the window functions namely $g(t) = \delta(t)$ and $g(t) = u(t)$. Provide explanations of the result in each case.

Problem 7.2: For a continuous-time signal $x(t)$ with CWT $X_{\text{CWT}}(a, \tau)$ prove (a) shifting and (b) scaling properties in Section 7.2.

Problem 7.3: In template matching, we often search for components in a signal (or objects in an image) which are shifted versions of a template or prototype function. If these are viewed as bases functions, in essence we would like to know if a particular signal or image can be represented using linear combinations of these bases. That will tell us if we have a potential match or not. Now, consider a finite-extent signal $x(n)$ of length N and a template $t(n)$ of the same length. Assume that we choose our basis functions according to $\phi_k(n) = t(((n-k))_N)$, i.e. circularly shifted versions of the template (finite-extent case). Show:

a. How $x(n)$ or its vector version $\underline{x}$ can be expressed in terms of these bases functions,
b. Based on the results in part (a) devise a simple hypothesis testing scheme to decide if a match is found in the signal or not.

Problem 7.4: The Haar wavelet function is given by $\psi(t) = \begin{cases} 1, & -\frac{1}{2} \le t \le 0 \\ -1, & 0 \le t \le \frac{1}{2} \\ 0, & \text{else} \end{cases}$. Using this wavelet, find the CWT $X_{\text{CWT}}(a, \tau)$ for the following signals:

a. Unit step $u(t) = \begin{cases} 1, & t \geq 0 \\ 0, & t < 0 \end{cases}$.

b. Shifted unit step $x(t) = u(t - t_0)$.

c. The Haar scaling signal $x(t) = \varphi(t) = \begin{cases} 1, & 0 \leq t \leq 1 \\ 0, & \text{else} \end{cases}$.

Problem 7.5: Repeat Problem 7.4 for the following signals (assume $\tau = 0$).

a. Unit ramp $r(t) = \begin{cases} t, & t \geq 0 \\ 0, & t < 0 \end{cases}$.

b. Signal $x(t) = \begin{cases} t^n, & t \geq 0 \;\; n = 0, 1, 2, \ldots \\ 0, & t < 0 \end{cases}$.

Problem 7.6: Find the Wigner–Ville distribution (WVD) of the following typical signals and explain your results.

a. Impulse $\delta(t)$.
b. Complex exponential $x(t) = e^{j\omega_0 t}$.
c. Linear chirp $x(t) = e^{j\omega_0 t^2}$.
d. Rectangular (gate) $x(t) = \begin{cases} 1, & \text{for } t \leq \left|\frac{1}{2}\right| \\ 0, & \text{else} \end{cases}$.

Problem 7.7: If $\mathrm{WV}_x(\tau, \omega)$ is the WVD of signal $x(t)$, prove the following properties.

a. Time-shifting, i.e. if $y(t) = x(t - \tau_0)$ then $\mathrm{WV}_y(\tau - \tau_0, \omega)$.
b. Modulation (frequency shift), i.e. if $y(t) = e^{j\omega_0 t}x(t)$ then $\mathrm{WV}_y(\tau, \omega - \omega_0)$.
c. Scaling, i.e. if $y(t) = \sqrt{a}\, x(at)$ $a > 0$ then $\mathrm{WV}_y\left(a\tau, \frac{\omega}{a}\right)$.

Problem 7.8: If $\mathrm{WV}_x(\tau, \omega)$ is the WVD of signal $x(t)$, prove the following *recovery* properties.

$$\frac{1}{2\pi}\int_{-\infty}^{\infty} WV_x\left(\frac{t}{2}, \omega\right)e^{j\omega t}\, d\omega = x(t)x^*(0)$$

$$\int_{-\infty}^{\infty} WV_x\left(t, \frac{\omega}{2}\right)e^{j\omega t}\, dt = X(\omega)X^*(0)$$

Problem 7.9: If $\mathrm{WV}_x(\tau, \omega)$ is the WVD of signal $x(t)$, prove the following properties.

a. Convolution Property:

$$\text{If } \; y(t) = \int_{-\infty}^{\infty} x(t-\tau)h(\tau)\, d\tau \; \text{ then } \; WV_y(t, \omega) = \int_{-\infty}^{\infty} WV_x(\rho, \omega)WV_h(t - \rho, \omega)\, d\rho.$$

b. Cross-correlation Property:

$$\text{If } \; y(t) = \int_{-\infty}^{\infty} x(t+\tau)h^*(\tau)\, d\tau \; \text{ then } \; WV_y(t, \omega) = \int_{-\infty}^{\infty} WV_x(\rho, \omega)WV_h(-t + \rho, \omega)\, d\rho.$$

Problem 7.10: Find the wavelet series coefficients for the following signals using the dyadically sampled Haar wavelet function.

a. Unit step $u(t)$.
b. Shifted unit step $u(t - t_0)$
c. The Haar continuous scaling function.

Problem 7.11: Show that the alternating flip results in an analysis filter bank with orthogonal properties, i.e. $\mathbf{GH}^T = \mathbf{0}$ (for N even) or alternatively $\langle g(n)h(n)\rangle = 0, \ \forall k, l$ where $\langle a(n), b(n)\rangle = \sum_n a(n)b^*(n)$, i.e. the inner product operation.

Problem 7.12: Prove that the condition to have linear phase FIR filter of order N is that the impulse response possesses symmetry or anti-symmetry property. Prove this property for both N odd and N even cases.

Problem 7.13: Show that the dual basis of the Haar filter bank, i.e. $\phi_{2k}(n) = g(2k - n)$ and $\phi_{2k+1}(n) = h(2k - n)$ are orthonormal.

Problem 7.14: Show that for the 5/3 biorthogonal filter bank the following biorthogonality properties hold:

$$\langle g(2k - n), \tilde{g}(n - 2l)\rangle = \delta(k - l)$$

$$\left\langle h(2k - n), \tilde{h}(n - 2l)\right\rangle = \delta(k - l)$$

Problem 7.15: Derive:

a. Time domain input–output relationships for a downsampler $\downarrow 2$ and an upsampler $\uparrow 2$.
b. Using the results in part (a) develop the same relationships in the z-domain.
c. Using results in part (b) analyze their frequency domain effects and then explain the roles of the analysis and synthesis filters in a one-level filter bank.

Problem 7.16: Extend the result in Problem 7.15 for a two-level filter bank. What conclusions can be drawn from these results?

Problem 7.17: Consider a one-level filter bank such as the one in Figure 7.5. Develop the PR condition in the z-transform domain using the results in part (b) of Problem 7.15. Note that a delay between the input signal and the reconstructed signal is still acceptable for the PR condition. Did you arrive at any other condition in this development? Can you explain the importance of meeting this condition?

Problem 7.18: Derive the expression for the FT, $\Phi(\omega)$, of the scaling function, $\varphi(t)$, in terms of the frequency response of the LPF $g(n)$ using the two-scale equation (7.35). Explain the importance of the result.

Problem 7.19: Repeat Problem 7.18 for the FT, $\Psi(\omega)$, of the wavelet function, $\psi(t)$, in terms of the frequency response of the HPF $h(n)$ using the two-scale equation (7.34). Explain the importance of the result.

Problem 7.20: Using the same approach presented in Section 7.4.2 for orthogonal filter banks, develop the PR condition for one-level 2-D orthogonal DWT filter bank shown in Figure 7.8.

References

1 A. Haar, "Zur Theorie der orthogonalen Funktionen systeme," *(Erste Mitteilung), Mathematische Annalen*, vol. 69, pp. 331–371, 1910.
2 A. Calderon, "Intermediate spaces and interpolation, the complex method," *Studia Mathematica*, vol. 24, pp. 113–190, 1964.

3 E. W. Aslaksen and J. R. Klauder, "Unitary representations of the affine group," *Journal of Mathematical Physics*, vol. 9, pp. 206–211, 1968.

4 J. Morlet, G. Arens, E. Fourgeau, and D. Giard, "Wave propagation and sampling theory- Part I: Complex signal and scattering in multilayered media," *Geophysics*, vol. 47, pp. 203–221, 1982.

5 J. Morlet, G. Arens, E. Fourgeau, and D. Giard, "Wave propagation and sampling theory-Part II: Sampling theory and complex waves," *Geophysics*, vol. 47, pp. 222–236, 1982.

6 I. Daubechies, "Ten lectures on wavelets," CBMS-NSF Regional Conference Series in Applied Mathematics, Series Number 61, 1992.

7 M. Vetterli and C. Herley, "Wavelets and filter banks: Theory and design," *IEEE Transactions on Signal Processing*, vol. 40, pp. 2207–2232, 1992.

8 M. Vetterli and J. Kovačević, "*Wavelets and Subband Coding*," Prentice Hall, 1995.

9 Y. Meyer, "*Wavelets, Algorithms and Applications*," Philadelphia, SIAM, 1993.

10 S. G. Mallat, "A theory for multiresolution signal decomposition: The wavelet representation," *IEEE Transactions on Pattern Analysis and Machine Intelligence*, vol. 11, pp. 674–693, 1989.

11 S. G. Mallat, "*A Wavelet Tour of Signal Processing: The Sparse Way*," Academic Press, 2009.

12 D. Donoho, "Unconditional bases are optimal bases for data compression and statistical estimation," *Applied Computational Harmonic Analysis*, vol. 1, pp. 100–115, 1993.

13 R. R. Coifman and M. V. Wickerhauser, "Entropy-based algorithms for best basis selection," *IEEE Transactions on Information Theory, Special Issue on Wavelet Transforms and Multiresolution Signal Analysis*, vol. 38, pp. 713–718, 1992.

14 M. Antonini, M. Barlaud, P. Mathieu, and I. Daubechies, "Image coding using wavelet transform," *IEEE Transactions on Image Processing*, vol. 1, pp. 205–220, 1992.

15 O. Rioul and M. Vetterli, "Wavelets and signal processing," *IEEE Signal Processing Magazine*, 1991.

16 D. Gabor, "Theory of communication," *Journal of the Institution of Electrical Engineers*, vol. 93, pp. 429-457, 1946.

17 L. Cohen, "Time-frequency distributions: A review," *Proceedings of IEEE*, vol. 77, pp. 941–981, 1989.

18 A. Flandrin, "*Time-frequency/Time-scale Analysis*," Academic press, 1999.

19 B. V. Rao, "Two-dimensional Wigner-Ville transforms and their properties," PhD Dissertation, University of Texas-Pan American, 2009.

20 G. Strang and T. Nguyen, "*Wavelets and Filter Banks*," Wellesley-Cambridge Press, 1996.

21 G. H. Golub and C. F. Van Loan, "*Matrix Computations*," Johns Hopkins Studies in Mathematical Sciences, 4th Edition, 2013.

22 A. Croisier, D. Esteban, and C. Galand, "Perfect channel splitting by use of interpolation/decimation tree decomposition techniques," First International Conference on Sciences and Systems, Patras, pp. 443–446, 1976.

23 K. Ramchandran and M. Vetterli, "Best wavelet packet bases in a rate-distortion sense," *IEEE Transactions on Image Processing*, vol. 2, pp. 160–175, 1993.

24 M. A. T. Figueiredo and R. D. Nowak, "An EM algorithm for wavelet-based image restoration," *IEEE Transactions on Image Processing*, vol. 12. pp. 906–916, 2003.

25 S. Arivazhagan and L. Ganesan, "Texture classification using wavelet transform," *Pattern Recognition Letters*, vol. 24, pp. 1513–1521, 2003.

26 B. Tian, M. R. Azimi-Sadjadi, M. Shaike, T. Vonder-Haar, and D. Reinke, "A study of cloud classification with neural networks using textural and spectral features," *IEEE Transactions on Neural Networks*, vol. 10, pp. 138–151, 1999.

8

Image Enhancement

Images acquired by various imaging sensors often require processing before they are suitable for further evaluation by humans or machines. Numerous factors can degrade the quality of captured images, including operating and environmental conditions during the acquisition process, the physical characteristics of the objects being imaged, and the inherent properties and imperfections of the imaging sensors themselves. For example, poor lighting conditions during photography frequently lead to low-contrast images.

The goal of image enhancement is to design a set of operations that manipulate acquired images to improve their visual appearance or quality, thereby preparing them for high-level processing and analysis by an automated system or a human expert. Improving "quality" can imply various things, such as: removing the effects of undesirable noise and interference; enhancing certain features or facilitating their extraction; improving contrast; and assigning color codes to different grayscale areas to create a pseudo-color image. Preprocessing using image enhancement is an important step in any image processing system, since the success of subsequent middle- and high-level processing steps heavily relies on it. The methods in this area are often ad hoc or heuristic, in the sense that they do not typically consider any prior knowledge about the imaging systems or the environment. An acceptable result is usually obtained after several trials that may involve fine-tuning the parameters of a particular algorithm or even trying different algorithms. The criterion for converging to an acceptable result is often based upon the subjective assessment of a human observer, which can obviously vary depending on an individual's preferences and design requirements.

The methods used in this area can be categorized into three general categories [1] depending on the type of operations performed on the images. These include: point operations, neighborhood or spatial operations, and transform-based operations. The following sections describe different commonly used algorithms in each category.

8.1 Point Operations

These memoryless operations are primarily used to enhance image contrast by mapping the pixel intensities regardless of their spatial locations. There are two general types of point operations, namely grayscale mapping and histogram modeling, which are covered in this section together with their main differences in the way they operate on an image.

8.1.1 Grayscale Mapping

The grayscale mapping transforms the original image pixel intensity $x \in \{x_k, k \in [0, K-1]\}$, which is a finite set of discrete intensity values, x_k, into the mapped image pixel intensity $y \in \{y_l, l \in [0, L-1]\}$ using the transformation $y = f(x)$ where $f(\cdot)$ is the mapping function. This function can be linear or nonlinear and is typically monotonic, depending on the type of operation. Using this transformation, all pixels with intensity x are mapped to pixels with intensity y. Since digital images have a finite number of gray levels (e.g. $K = 256$), these mapping operations can easily be performed using a lookup table where x is the input or index to the lookup table and y is the resultant mapped version. The same idea can be applied to color images where a separate lookup table is used for each RGB color channel.

Given an image, the main goal of grayscale mapping is to devise a specific transformation $f(\cdot)$ that improves contrast in certain intensity regions. This choice depends on the properties of the image and, in particular, on its histogram, which represents the frequency of occurrences of each gray level in the image. Let us illustrate this by considering a simple example. Figure 8.1 shows a low contrast image and its corresponding histogram. The highest intensity in this image is $x_K = 51$, which explains why the image is predominantly dark. To improve its contrast, we can apply histogram stretching using a simple linear mapping $y = ax$ with $a = 5$, which is chosen to make the mapping one-to-one and avoid any intensity saturation. The effect of this simple mapping on the image and its histogram is shown in Figure 8.2. As can be seen, the mapped image clearly has a better visual appearance and its histogram is stretched into a wider range of gray levels, although there are a large number of histogram bins that have no values.

A piecewise linear mapping is often used for many contrast mapping applications. A three-segment linear mapping with slopes a, b and c can be expressed as:

$$y = \begin{cases} ax, & 0 \leq x < x_{k_1} \\ b(x - x_{k_1}) + l_1, & x_{k_1} \leq x < x_{k_2} \\ c(x - x_{k_2}) + l_2, & x_{k_2} \leq x \leq x_{K-1}, \end{cases}$$

where x_{K-1} is the maximum intensity in the original image. The number of line segments is decided on the basis of the shape and the number of *modes* in the histogram where each mode corresponds to an intensity region in the image. For example, the image in Figure 8.1 has a unimodal histogram, while the geostationary satellite (GOES) image in Figure 8.3 has three modes in its histogram, as can be seen by the three distinguishable intensity regions, although the number of pixels (area under each mode) in the third mode is small. The transition points x_{k_1} and x_{k_2} are typically placed at the valleys of the histogram or the boundaries between the modes. Line segments with slopes greater than one perform contrast stretching in the corresponding region, while those with slopes less than one lead to compression of the intensity region.

Suppose that we would like to enhance the first mode corresponding to the lower intensity region in the GOES image of Figure 8.3. The GOES series of satellites [2] provides multiple spectral bands that capture the amount of reflected or emitted energy at different wavelengths to obtain information about land, oceans, and clouds. In the visible channel, darker intensity regions typically correspond to land and ocean areas, while brighter values correspond to clouds. This is illustrated by the first two dominant modes in the histogram of Figure 8.3b. To selectively apply contrast enhancement to the first mode, we devise a two-segment piecewise linear mapping where the first segment has a slope > 1 and the second segment has a slope < 1. Figure 8.4a shows this mapping and its effect on the histogram of the GOES image in Figure 8.3b. The

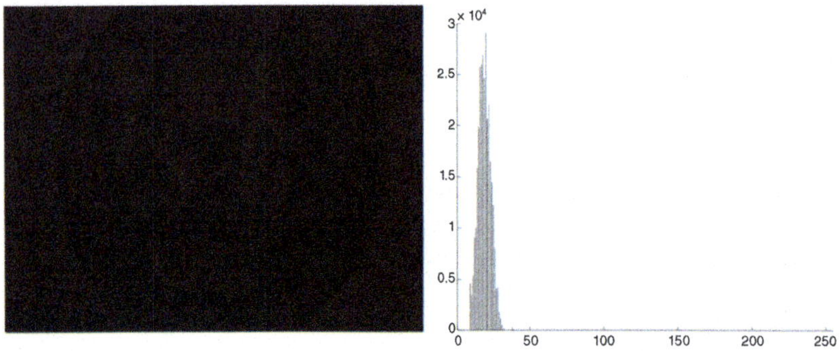

Figure 8.1 A low contrast image (left) and its histogram (right).

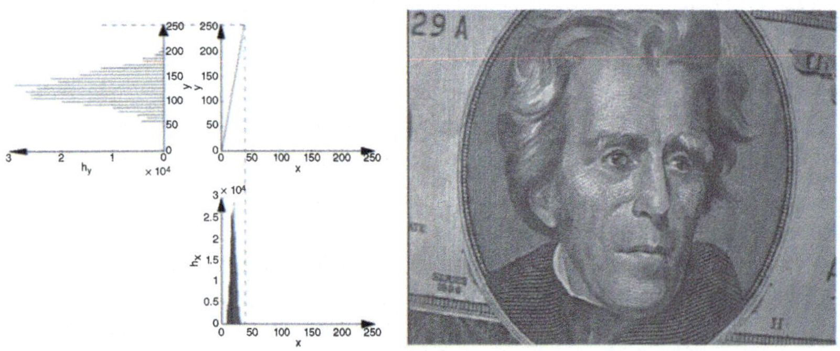

(a) (b)

Figure 8.2 (a) Linear grayscale mapping and its effects on histogram of Figure 8.1; (b) resultant image after grayscale mapping.

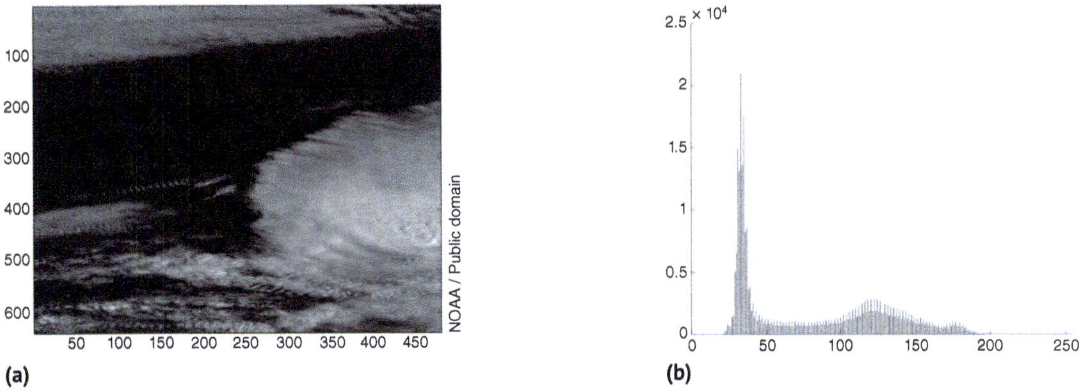

Figure 8.3 (a) GOES 8 satellite image (visible channel) and (b) its bimodal histogram.

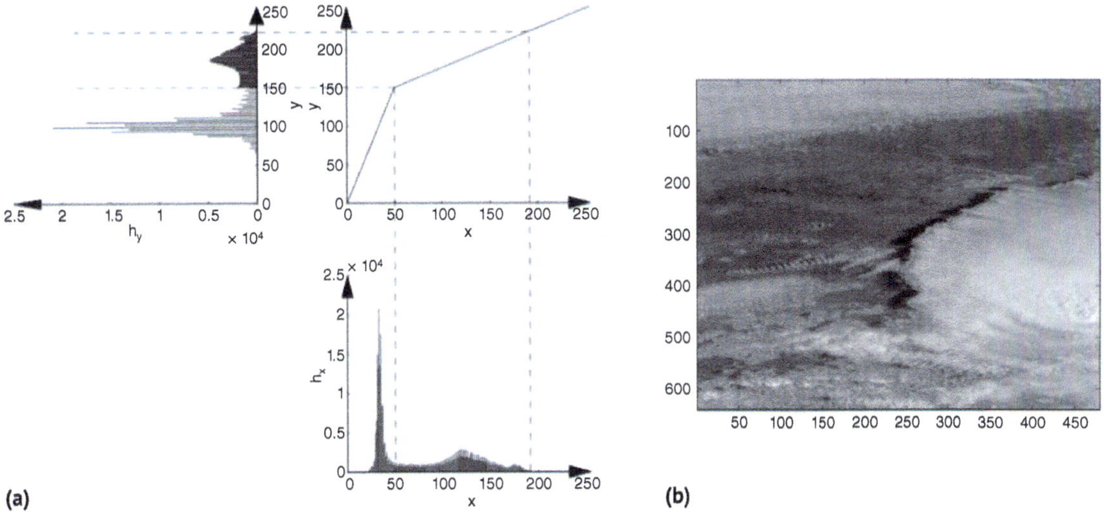

Figure 8.4 (a) Piecewise linear grayscale mapping and (b) contrast-enhanced image.

contrast-enhanced image is shown in Figure 8.4b, which exhibits a much better definition of cloud shadow areas and low-intensity ground/ocean features. Note that this improvement is achieved at the price of compressing the second histogram mode, which results in poorer contrast in the cloud regions.

The relationship between the histograms of the original image, $h_X(x)$, and the mapped image, $h_Y(y)$, can be determined considering a monotonic one-to-one mapping function $y = f(x)$. In this case, if the original image, X, has histogram:

$$h_X(x) = \sum_{k=0}^{K-1} N_k \delta(x - x_k) \tag{8.1}$$

where N_k is the number of pixels with intensity x_k and $\sum_{k=0}^{K-1} N_k = N$ is the total number of pixels in the image, then the histogram of the mapped image will be:

$$h_Y(y) = \sum_{l=0}^{L-1} M_l \delta(y - y_l) \tag{8.2}$$

where $y_l = f(x_k)$ is the mapped version of x_k with exactly the same number (for one-to-one mapping) of occurrences, i.e. $M_l = N_k$.

The pixel intensities, x_k and y_l, in the original and mapped images can be viewed as possible values of two discrete random variables (r.v.s) X and Y, with probability mass functions (PMFs) that are $P(X = x_k) = \frac{N_k}{N}$ and $P(Y = y_l) = \frac{M_l}{N}$ [3]. That is, normalizing the frequency of gray levels occurrences in the histogram by the total number of pixels N gives the PMF of the r.v. If this mapping is one-to-one, the probabilities are equal since $M_l = N_k$, although their locations change from x_k to y_l. Thus, for monotonic one-to-one mapping functions, the shape of the histograms and number of bins remain unchanged while the spacing between the histogram bins will change. However, if the mapping becomes many-to-one, then contrast compression or clipping occurs in the affected regions, as the number of histogram bins reduces in the compressed regions.

8.1.1.1 Special Cases

1. **Thresholding:** Figure 8.5a shows the mapping for the thresholding operation where the threshold value is x_{k_t}. Pixel intensities below x_{k_t} are assigned a value of 0 (black), while those above this threshold are assigned a value of 255 (white) or any other desired intensity value. The choice of x_{k_t} is made based on examining the histogram of the original image. Thresholding operation is used in many applications, e.g. handwriting character recognition, signature, and fingerprint identification. Figure 8.5b,c shows a fingerprint image and its binary version obtained by thresholding this image at $x_{k_t} = 80$, respectively.
2. **Segmentation:** Figure 8.6a shows one possible mapping for segmenting a region of pixel intensities between x_{k_1} and x_{k_2} from the rest of the image. This mapping is useful when the desired features are known to lie within this intensity range. Examples of its applications include background removal and object detection and extraction. Although the histogram-based segmentation methods are simple, they suffer from many shortcomings. In Chapter 12, we present more reliable segmentation methods for practical scenarios.
3. **Digital Negative:** Figure 8.6b shows a grayscale mapping used for digital negative operation. The result of this mapping on the Lena image of Figure 8.7a is illustrated in Figure 8.7b. This type of mapping is used in printing industry.
4. **Nonlinear Mapping:** Nonlinear grayscale transformations may also be employed for specific contrast enhancement applications. Figure 8.8a shows the effect of a nonlinear sigmoidal mapping on the histogram of the GOES image in Figure 8.3a. In this case, the mapping resulted in larger separation of the modes or intensity regions, though within each intensity region the contrast is reduced. The result of this nonlinear mapping is shown in the image of Figure 8.8b. Another useful example of nonlinear transformation is the logarithmic operation:

$$y = a \log_{10}(b + |x|)$$

where a and b are constants. This mapping compresses the dynamic range of the image by reducing high-intensity values, while boosting lower-intensity values. A typical choice is $b = 1$ to ensure zero-preserving mapping.

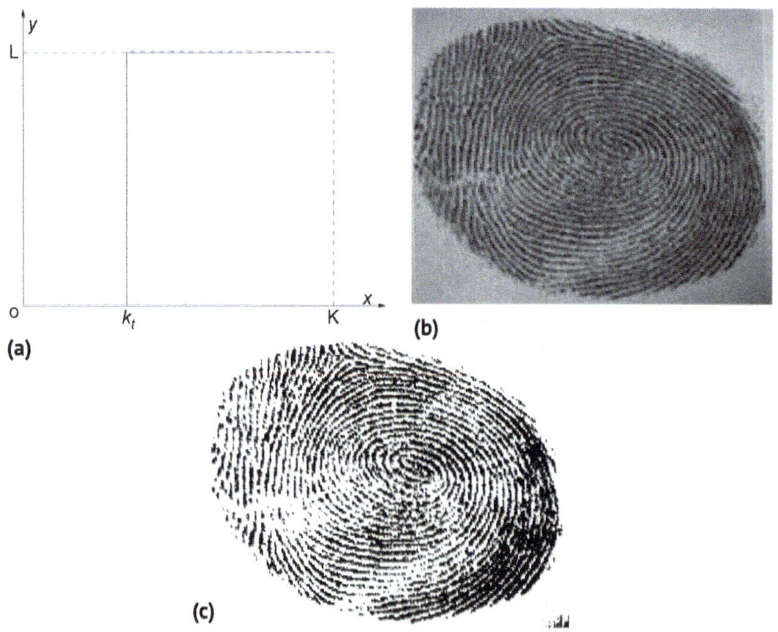

Figure 8.5 (a) Binary thresholding, (b) fingerprint image, and (c) thresholded fingerprint image.

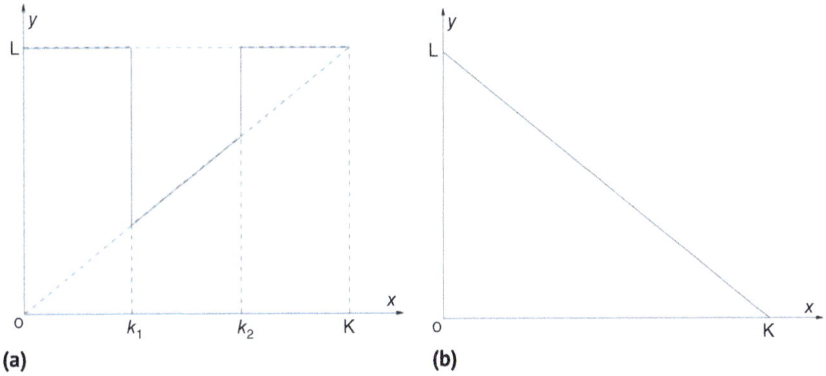

Figure 8.6 (a) Grayscale mapping for segmentation and (b) grayscale mapping for digital negative.

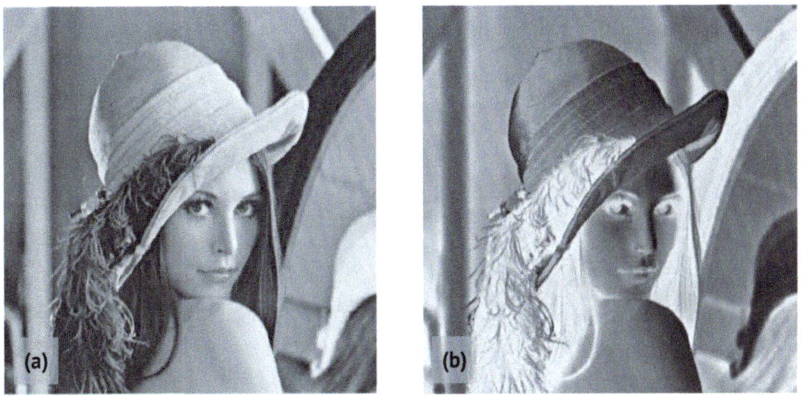

Figure 8.7 (a) Original Lena image. *Source*: [4] / John Wiley & Sons / CC BY 3.0. (b) its digital negative. *Source*: [5] / MDPI / CC BY 4.0.

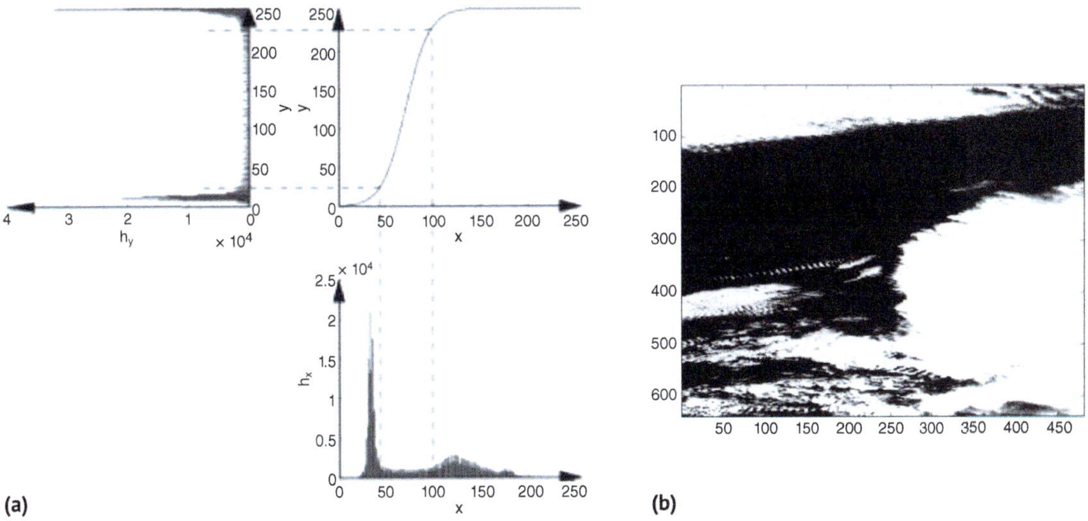

Figure 8.8 (a) Nonlinear sigmoidal mapping and (b) grayscale mapped GOES image.

8.1.2 Histogram Modeling

Histogram modeling methods differ from grayscale mapping methods in many ways. Unlike grayscale mapping, where the user explicitly devises a mapping function, histogram modeling aims to reshape the histogram to a predetermined *desired* histogram. These operations generally use nonlinear transformations that may not be explicitly visible to the user. However, the user sees the outcome based on the desired specified histogram. There are two types of histogram modeling methods: histogram equalization and histogram specification; the latter is a more general case of the former.

8.1.2.1 Histogram Equalization

The goal of histogram equalization is to apply a mapping that yields an image with a uniformly distributed histogram. For an image with continuous-valued pixel intensity (i.e. sampled but not quantized) the explicit solution to this problem exists. That is, if we denote the pixel intensity by a continuous r.v. X with PDF (or the normalized histogram) $p_X(x)$, the cumulative distribution function (CDF) of the image given by:

$$y = f(x) = P_X(x) = \int_0^x p_X(\alpha)d\alpha \tag{8.3}$$

serves as the mapping function that produces a uniformly distributed mapped image. To show this, consider the relationship between the PDF of a r.v., X, and the PDF of its mapped version, $Y = f(X)$, where $f(\cdot)$ is monotonic [3]:

$$p_Y(y) = \left.\frac{p_X(x)}{|f'(x)|}\right|_{x=f^{-1}(y)} = 1, \quad y \in [0, 1] \tag{8.4}$$

where $f'(x) = \frac{df(x)}{dx}$. Thus, in this case, since CDF is used as a mapping function $f(x) = P_X(x)$ and $f'(x) = p_X(x)$, we achieve the desired uniformly distributed image $Y \sim unif\ [0, 1]$. Note that CDF is single-valued and monotonically increasing function in the range $0 \le P_X(x) \le 1$, hence satisfying all the necessary requirements for a mapping function.

In practice, digital images are quantized, and pixel intensities are represented by a discrete r.v., X with values x_k and a CDF of stair-step form. That is, if the PMF of r.v. X is $P(X = x_k)$, then the CDF is given by:

$$P_X(x) = \sum_{k=0}^{K-1} P(X = x_k)u(x - x_k) = \sum_{k=0}^{K-1} \frac{N_k}{N}u(x - x_k) \tag{8.5}$$

where $u(x)$ represents the unit step function and N_k and N were defined before. This CDF starts from 0 and gets to its final value of 1, since $\sum_{k=0}^{K-1} \frac{N_k}{N} = 1$. The amplitude of the step at $x = x_i$ is $P(X = x_i) = \frac{N_i}{N}$. At $x = x_i$, the mapped intensity is the sum of all step heights up to the ith step, i.e.

$$y_i' = f(x_i) = P_X(x_i) = \sum_{k=0}^{i} \frac{N_k}{N}. \tag{8.6}$$

However, since $P(X = x_i) = P(Y' = y_i')$, the histogram (or PDF) of the mapped image will be the scaled version of that of the original image with the same number of bins and frequency of occurrences, and only the spacing between the bins changes. Consequently, this mapping in its original form fails to equalize a digital image. Additionally, since for CDF we have $0 \le P_X(x) \le 1$, the mapped values need to be converted to actual gray-level values. These problems can be overcome by performing a requantization after the equalization process. A typical choice for this requantization process [1] is of the form:

$$y_l = \text{Int}\left[\frac{(y_l' - y'_{\min})}{1 - y'_{\min}}(y_L - y_{L_0}) + y_{L_0} + 0.5\right] \tag{8.7}$$

where $y'_{\min}$ is the smallest value of y_l', y_{L_0}, and y_L are the smallest and largest intensity levels in the mapped image and Int$[x]$ represents integer part of quantity x. The number of bins in the histogram of the image Y is determined by $(y_L - y_{L_0})$, which is the only parameter that can be varied in this method. By decreasing the number of bins, the histogram of the mapped image becomes closer to a uniform one. However, this is achieved at the cost of reduced gray-level resolution, which cannot be afforded in images with great details. The appropriate number of bins is typically determined experimentally by observing the visual appearance of the equalized images.

Figure 8.9a–d shows equalized versions of the "Dollar" image in Figure 8.1a along with their histograms for 60 and 8 bins, respectively. The improvement in visual appearance of these images over the original one is clearly noticeable. The histogram of the equalized image in Figure 8.9d obtained for 8 bins looks closer to the uniformly distributed histogram compared to that in Figure 8.9b using 60 bins. Nevertheless, the loss of gray-level resolution resulted in some patchiness or clipping in the mapped image. Note that unlike grayscale mapping, histogram equalization does not require explicit user involvement in designing the mapping function or fine-tuning its parameters. While this non-interactive nature simplifies implementation, it also limits flexibility, as the user cannot experiment with different mappings to achieve the most visually pleasing result.

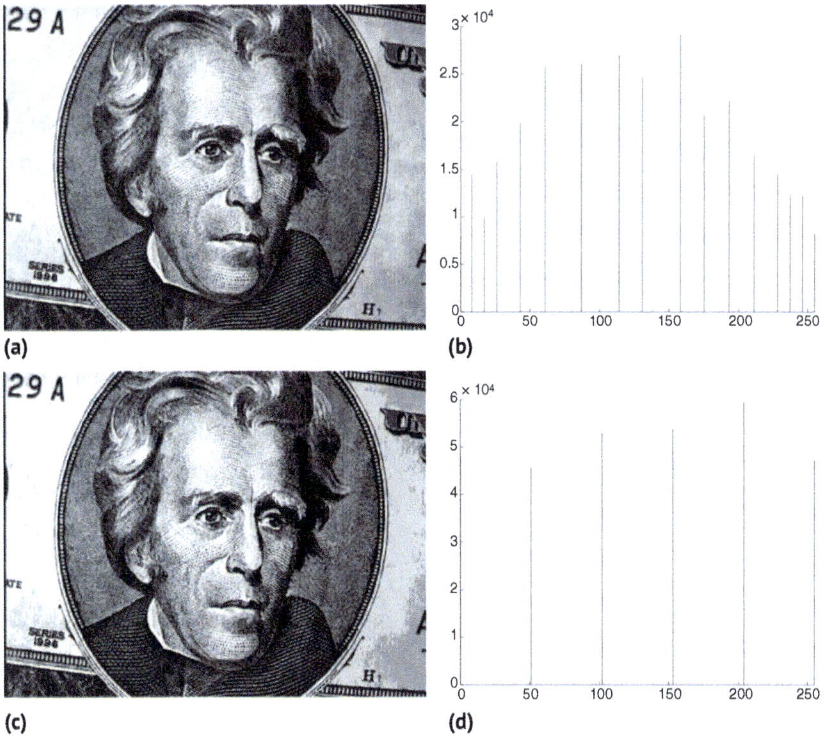

Figure 8.9 (a) and (c) Histogram equalized images and (b) and (d) their histograms for 60 and 8 number of bins.

8.1.2.2 Histogram Specification

In this method, the user can specify a desired histogram and interactively examines its effects on the image. The choice of desired histogram is dependent on the characteristics of the histogram of the original image and the specific goals of the enhancement task.

Again, let us begin with the continuous case where X and Y are two continuous r.v.s representing the pixel intensities of the original and histogram-specified images, respectively. Now, given the original image, X, and its PDF, $p_X(x)$, and the desired PDF, $p_Y(y)$, our goal is to map the original image into a new image, Y, with the specified PDF $p_Y(y)$.

If we hypothetically assume that the desired mapped image, Y, is available, then both the original and desired image can be histogram-equalized using their corresponding CDFs, i.e.

$$w = f(x) = P_X(x) = \int_0^x p_X(\alpha)d\alpha,$$

and

$$z = g(y) = P_Y(y) = \int_0^y p_Y(\beta)d\beta.$$

Now, since w and z are realizations of the equalized r.v.s W and Z with uniformly distributed PDFs over $[0, 1]$, and further they are related via an unknown transformation, we can assume that $z = w$ or $g(y) = f(x)$. As a result, the desired image y can be obtained using the inverse transformation $y = g^{-1}(f(x))$ or $y = P_Y^{-1}(P_X(x))$. Since $g(\cdot)$ is single-valued and monotonic, the above transformation gives the unique mapped values. Of course, this solution relies on the existence of an analytical inverse function $g^{-1}(\cdot)$, which may not be available for certain functions. However, this is not an issue for the discrete case.

For the discrete case, let us represent the pixel intensity of the original and mapped digital images by discrete r.v.s X and Y, respectively. Given their histograms or PMFs, as in the continuous case, we first find their respective CDFs:

$$w = P_X(x) = \sum_{k=0}^{K-1} \frac{N_k}{N} u(x - x_k)$$

and

$$z = P_Y(y) = \sum_{l=0}^{L-1} \frac{M_l}{N} u(y - y_l)$$

where N_k and M_l are the number of pixels with intensities x_k and y_l in images X and Y, respectively, and w and z are the possible values of their equalized versions W and Z. These CDFs are also stair-step functions with jumps that are located at $x = x_i$ and $y = y_i$. The heights of the steps at the transition points x_i and y_i are $w_i = \sum_{k=0}^{i} \frac{N_k}{N}$ and $z_i = \sum_{l=0}^{i} \frac{M_l}{N}$, respectively. Now, the equivalent of the inverse mapping in the continuous case can be performed using a simple procedure [1]. For each equalized gray level w_i, let j be the smallest possible index for which $z_j \geq w_i$, then map all N_i pixels with gray level x_i to the intensity level y_j in the mapped image. It is important to note that this mapping is typically many to one, implying that several intensity levels in the original image may be mapped to one level in the histogram-specified image.

Figure 8.10a shows the histogram-specified version of the "Dollar" image in Figure 8.1a, where the desired histogram is a 1/2 cycle sine wave shown in Figure 8.10b. The histogram of the resultant image in Figure 8.10a is given in Figure 8.10c, which closely resembles the shape of the desired histogram. Visual comparison of the histogram-specified image in Figure 8.10a with those of the histogram-equalized ones in Figure 8.9 shows much better contrast enhancement performance without introducing any artifacts.

Example 8.1 *Histogram of a continuous image, X, is given by* $h_X(x) = Ae^{-x}$ *where* $x \in [0, b]$ *and* A *is a normalization factor. We would like to map this image to a new image, Y, with the desired histogram* $h_Y(y) = Bye^{-y^2}$ *with range of values* $y \in [0, b]$ *and B is the normalization factor. Find the transformation function* $y = T(x)$ *that achieves this task [1].*

Solution: First, normalize the histograms by the areas under the histograms to generate PDFs. Then, find the corresponding CDFs by integrating the PDFs:

$$w = f(x) = P_X(x) = \frac{1}{AR_1} \int_0^x Ae^{-\alpha} d\alpha = \frac{A}{AR_1}(1 - e^{-x})$$

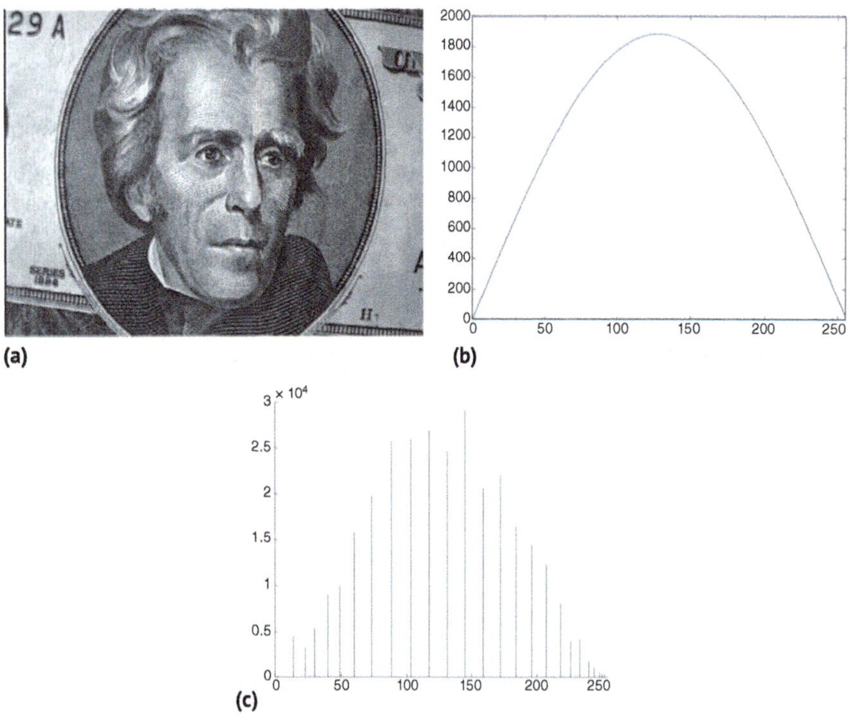

Figure 8.10 (a) Histogram-specified dollar image, (b) desired histogram, and (c) reshaped histogram of the histogram-specified image.

where AR_1 is the area under $h_X(x)$, i.e. $AR_1 = A\int_0^b e^{-\alpha} d\alpha = A(1 - e^{-b})$. Thus, we get:

$$w = f(x) = \frac{(1 - e^{-x})}{(1 - e^{-b})}.$$

Also, for the other image we have:

$$z = g(y) = P_Y(y) = \frac{1}{AR_2}\int_0^y B\alpha e^{-\alpha^2} d\alpha$$

Let us define $\alpha^2 = \beta$, then we get:

$$z = g(y) = \frac{1}{2AR_2}\int_0^{y^2} Be^{-\beta} d\beta = \frac{B}{2AR_2}(1 - e^{-y^2}),$$

where $AR_2 = \frac{B}{2}\int_0^{b^2} e^{-\beta} d\beta = \frac{B}{2}(1 - e^{-b^2})$. Thus, we obtain:

$$z = g(y) = \frac{(1 - e^{-y^2})}{(1 - e^{-b^2})}$$

Now, using $w = z$, we can find $y = T(x) = g^{-1}(f(x))$. In this case, $g(\cdot)$ has an inverse function, which is:

$$y = g^{-1}(z) = \sqrt{ln\frac{1}{1 - z(1 - e^{-b^2})}}$$

Now, substituting $z = w = \frac{(1-e^{-x})}{(1-e^{-b})}$ in the above equation gives

$$y = T(x) = \sqrt{ln\frac{1}{1 - \frac{(1-e^{-x})(1-e^{-b^2})}{(1-e^{-b})}}}.$$

8.2 Spatial Operations

Spatial operations involve defining a local neighborhood, referred to as a mask (or window), surrounding a given pixel. The intensity of the center pixel is then replaced by the result of performing some type of linear or nonlinear spatial domain operation on all the pixels within the mask. The mask is swept across the image, one pixel at a time, in a regular raster scan fashion until all the pixels have been processed. The choice of the geometry of the mask is somewhat arbitrary, though square windows with odd dimensions (e.g. 3×3 or 5×5) are commonly used where the spatial operations are performed for the center pixel.

The spatial operations introduced in this section are generally *ad hoc* in nature, since they do not rely on prior knowledge of the properties of the image or the imaging system. As a result, they can be applied to many different types of images irrespective of their properties or the type of imaging sensors that captured them.

In this section, we study different spatial operators of both linear and nonlinear types and examine their effectiveness in different applications. As far as linear spatial operations are concerned, the process can be represented as the linear convolution of the image with the PSF of a 2-D finite impulse response (FIR) filter, whose coefficients correspond to the mask values. These coefficients can be chosen to perform a variety of image processing tasks, including noise reduction or smoothing, edge detection and sharpening, template matching, and target detection.

8.2.1 Noise Reduction and 2-D Low-Pass Filtering

Any image is subject to noise and interference from various sources, such as sensor noise, film grain noise [1], channel noise, and speckle noise in synthetic aperture radar/sonar (SAR/SAS) [6, 7]. Noise can corrupt an image additively or multiplicatively [8], depending on the source.

Spatial low-pass filtering is a simple and effective method for reducing additive white Gaussian noise in images. Using a 2-D linear FIR filter, each pixel intensity in an image is replaced by the weighted sum of those within its local neighborhood

1/9	1/9	1/9
1/9	1/9	1/9
1/9	1/9	1/9

(a)

0	1/8	0
1/8	1/2	1/8
0	1/8	0

(b)

0	1/5	0
1/5	1/5	1/5
0	1/5	0

(c)

Figure 8.11 Three typical spatial low-pass filtering masks of size 3 × 3. (a) Square spatial averaging mask, (b) cross-shaped weighted averaging mask, and (c) cross-shaped spatial averaging mask.

in a mask. The weights correspond to the filter coefficients used for the spatial operation. This 2-D linear filtering can be represented by a 2-D linear convolution sum:

$$y(m,n) = \sum\sum_{k,l \in W} h(k,l)x(m-k,n-l) \tag{8.8}$$

where $x(m,n)$ is the input image, $y(m,n)$ is the output image, $h(k,l)$ represents the filter coefficients or the weights, and W is the support of the mask. To avoid changing the intensity of the pixels in uniform background areas, the filter coefficients should add up to a unity value, i.e. $\sum\sum_{k,l \in W} h(k,l) = 1$. A common choice of the coefficients is $h(k,l) = \frac{1}{N_W}, \forall (k,l) \in W$, where N_W represents the number of pixels in W. In this case, the filter performs spatial averaging of the pixel values within the mask, W.

Figure 8.11a–c shows three 3 × 3 low-pass filter masks with different choices of weights. Mask 2 places more emphasis on the center pixel while the other two are spatial averaging masks. The choices of the filter coefficients and mask size present a trade-off between noise reduction ability of the filter and the edge smearing artifacts caused due to loss of high-frequency components.

To illustrate the effects of changing filter coefficients on the smoothed image, consider the noise-free "Peppers" image in Figure 8.12a and its histogram in Figure 8.12b. This image is then corrupted by adding white noise to produce the signal-to-noise ratio of SNR ≈ 6.6 dB. The noisy image and its histogram are shown in Figure 8.12c,d, respectively. As can be seen from this image and its histogram, the additive noise severely degrades the image and the gray-level distribution.

The three low-pass filtering masks in Figure 8.11a–c are then applied to the noisy image in Figure 8.12c. The filtered images are shown in Figure 8.13a–c for Masks 1–3, respectively. Clearly, in all the cases, this simple spatial low-pass filtering substantially reduced the amount of noise. The corresponding SNR values for the filtered images are found to be $SNR_1 =$ 16 dB, $SNR_2 = 11.5$ dB and $SNR_3 = 13.5$ dB, respectively. Although the noise reduction ability of Mask 1 appears to be better than those of the other two masks, as evident from the filtered images and the SNR values, some smearing artifacts are also more noticeable in the filtered image of this mask. The smearing artifacts are caused by excessive attenuation of high-frequency components corresponding to sharp edges and textures in the image.

This can be explained by comparing the magnitude responses of these filters in Figure 8.14a–c, which indicates that Mask 1 has a smaller bandwidth and therefore better noise suppression, at the price of increased attenuation of high-frequency components, and hence more smearing artifacts. In contrast, the bandwidth of the Mask 2 is the largest among all the three filters, thereby preserving more high-frequency content. As a result, it produces less noise reduction but better retention of edges and fine structures. The histogram of the filtered image obtained using Mask 2 in Figure 8.13d shows the fact that this filter actually recovered most of the characteristics of the histogram of the original image in Figure 8.12b from that of the noisy image in Figure 8.12d.

In summary, when using spatial low-pass filtering, the filter coefficients must be properly selected to maintain a balance between noise suppression and retention of essential high-frequency information corresponding to edges and fine details of the image.

Example 8.2 *Consider the two 5-point masks in Figure 8.11b,c with input–output equations:*

$$y_1(m,n) = \frac{1}{5}[x(m,n) + x(m-1,n) + x(m+1,n) + x(m,n-1) + x(m,n+1)]$$

$$y_2(m,n) = \frac{1}{8}[4x(m,n) + x(m-1,n) + x(m+1,n) + x(m,n-1) + x(m,n+1)]$$

Compare their frequency responses in terms of their frequency selectivity and noise reduction ability.

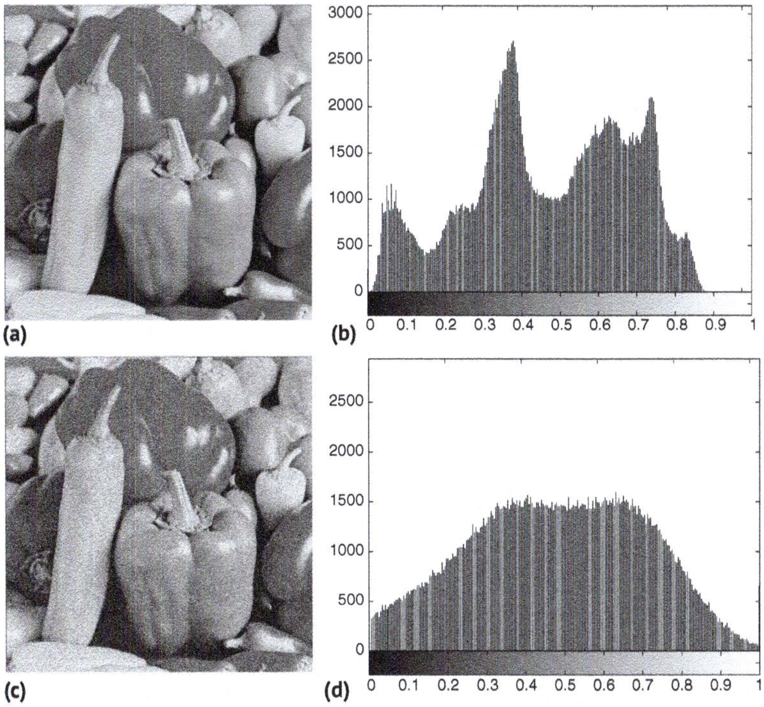

Figure 8.12 (a) Original Peppers image, (b) histogram of original image, (c) noisy Peppers image, and (d) histogram of noisy image. *Source*: [5] / MDPI / CC BY 4.0.

Solution: Using the 2-D DSFT and shifting property, the frequency response of the first mask is:

$$H_1(e^{j\Omega_1}, e^{j\Omega_2}) = \frac{Y_1(e^{j\Omega_1}, e^{j\Omega_2})}{X(e^{j\Omega_1}, e^{j\Omega_2})} = \frac{1}{5}[1 + e^{-j\Omega_1} + e^{j\Omega_1} + e^{-j\Omega_2} + e^{j\Omega_2}]$$

$$= \frac{1}{5}[1 + 2\cos\Omega_1 + 2\cos\Omega_2]$$

Similarly, for Mask 2 we get $H_2(e^{j\Omega_1}, e^{j\Omega_2}) = \frac{1}{8}[4 + 2\cos\Omega_1 + 2\cos\Omega_2]$. To compare their frequency selectivity, consider the diagonal cross-sections obtained by setting $\Omega_1 = \Omega_2 = \Omega$. This yields:

$$H_1(\Omega) = \frac{1}{5}[1 + 4\cos\Omega], \quad H_2(\Omega) = \frac{1}{2}[1 + \cos\Omega]$$

Figure 8.15 shows the plots of these magnitude responses as a function of Ω. Comparison of these plots clearly explains why the first filter (Mask 3) gave a higher SNR = 13.5 dB compared to the second one (11.5 dB), at the cost of increased smearing.

8.2.1.1 Effects of Mask Size on Performance

To determine the effects of mask size on the noise removal performance of spatial filters, consider a noisy image $z(m, n)$ given by:

$$z(m, n) = x(m, n) + \eta(m, n),$$

where $x(m, n)$ is the noise-free image and $\eta(m, n)$ is the additive white noise with zero mean ($\mu_\eta = 0$) and variance σ_η^2. The filtered image after spatial averaging is given by:

$$y(m, n) = \frac{1}{N_W} \sum\sum_{k,l \in W} z(m - k, n - l) = \overline{x}(m, n) + \overline{\eta}(m, n) \tag{8.9}$$

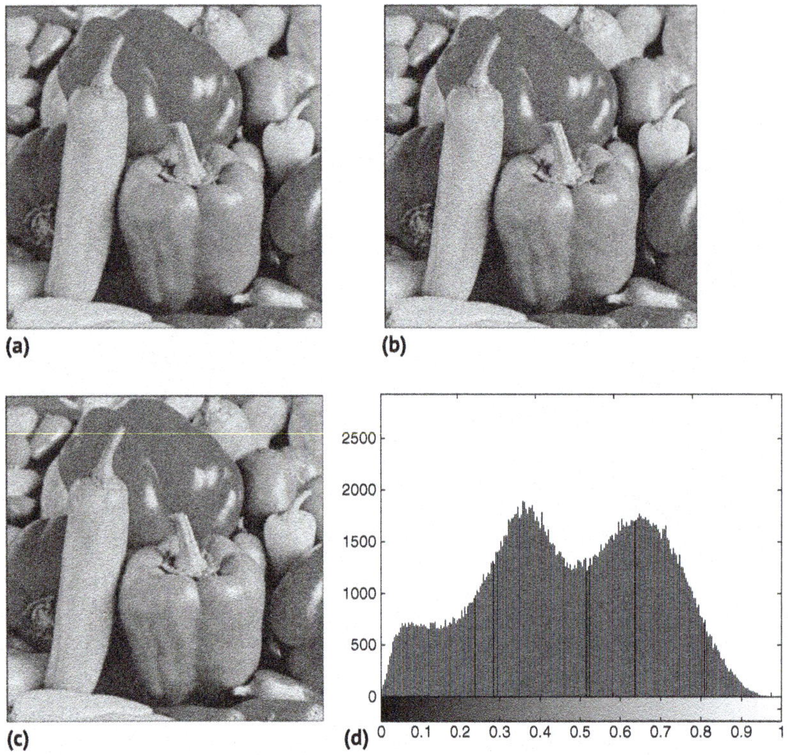

Figure 8.13 (a–c) Filtered images using Masks 1–3; (d) histogram of filtered image using Mask 2.

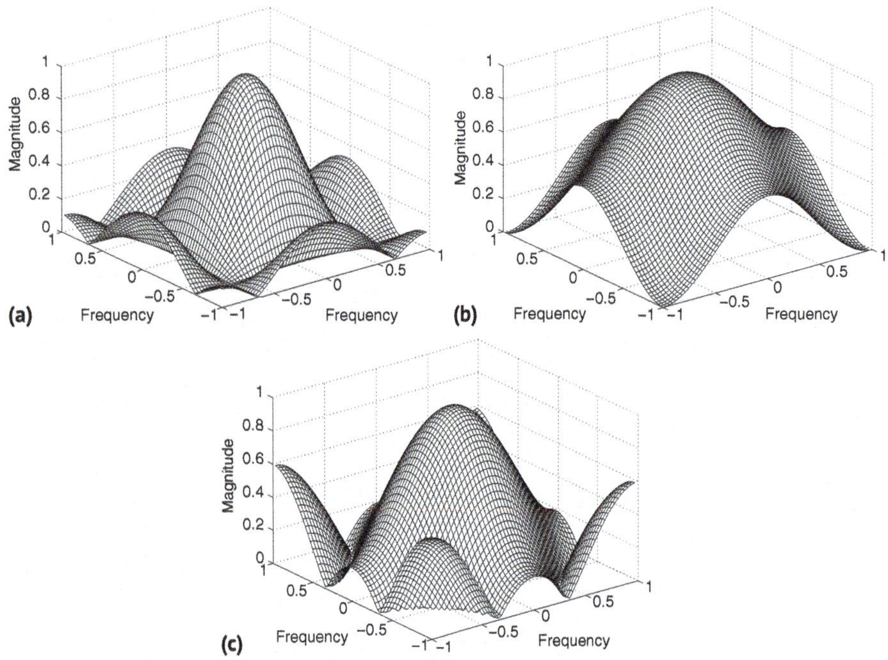

Figure 8.14 (a)–(c) Magnitude responses of spatial filters for Masks 1–3, respectively.

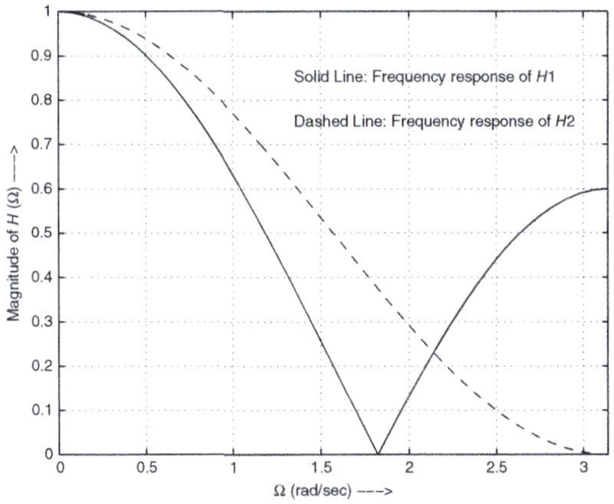

Figure 8.15 Diagonal cross-section of magnitude responses of spatial filters in Example 8.2.

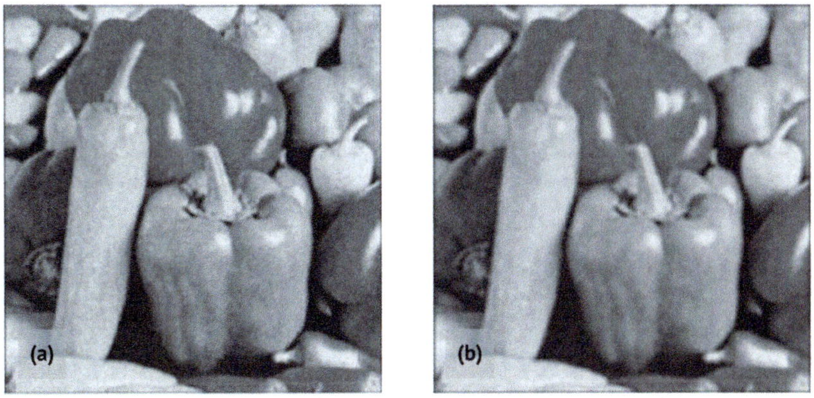

Figure 8.16 Spatial averaging results using (a) 5 × 5 and (b) 7 × 7 mask sizes.

where $\overline{\eta}(m,n) = \frac{1}{N_W} \sum\sum_{k,l \in W} \eta(m-k, n-l)$ is the filtered version of $\eta(m,n)$ (or residual noise) in the processed image and $\overline{x}(m,n) = \frac{1}{N_W} \sum\sum_{k,l \in W} x(m-k, n-l)$ is the filtered version of the original image $x(m,n)$.

It can be shown (Problem 8.12) that $\overline{\eta}(m,n)$ has zero mean and variance $\sigma_{\overline{\eta}}^2 = \frac{\sigma_\eta^2}{N_W}$, that is, the variance of noise is reduced by a factor of N_W. This suggests that increasing the mask size improves the SNR performance of the filter. This conclusion is based on the assumption that spatial averaging does not impact the noiseless image $x(m,n)$, and hence its variance $\sigma_{\overline{x}}^2$ remains unchanged. However, this assumption is not valid as filtering does indeed impact the image itself as can be observed from (8.9) since increasing the mask size also increases the edge smearing artifacts as a result of reducing the bandwidth of the spatial filter. Nevertheless, this assumption is valid for areas where the noiseless image $x(m,n)$ has a uniform intensity within the mask. Figure 8.16a,b shows the results of applying spatial averaging filters with mask sizes of 5 × 5 and 7 × 7, respectively. The increased level of blurring is clearly noticeable in these results, especially for the 7 × 7 mask.

8.2.1.2 Edge-Preserving Spatial Filters

These types of filters preserve the sharpness of the edges while suppressing the effects of noise in the non-edge areas. This is achieved by detecting the presence of edges within a local neighborhood (mask) and adjusting the filter coefficients accordingly. That is, in regions where edges are detected, smoothing is either reduced or all together avoided, whereas in regions with uniform intensity more smoothing is applied.

There are several recent edge-preserving filters [9, 10] that can be used to protect the edges from undesirable smearing artifacts. The simplest approach is to apply a gradient-based scheme (see Section 8.2.3) to detect the presence of edges. A threshold is then specified, and only those masks for which the gradient values are smaller than the predefined threshold, i.e. no edge, will undergo the spatial averaging. The challenge, of course, is the appropriate choice of the threshold for a given image. Alternatively, a directional averaging filter can be devised where the spatial averages, $\bar{y}_{i,j}(m,n)$ are calculated in several directions using:

$$\bar{y}_{i,j}(m,n) = \frac{1}{N_{S_{i,j}}} \sum\sum_{i,j \in S_{i,j}} x(m-i, n-j) \tag{8.10}$$

where $S_{i,j}$ represents a set of pixel indices within the mask along a specific direction, and $N_{S_{i,j}}$ is the number of pixels in that set. For example, for the 3×3 mask, there are four different sets of triple pixels in four directions. That is, $S_{i,j} = \{(1,0),(0,0),(-1,0)\}, \{(0,1),(0,0),(0,-1)\}, \{(-1,1),(0,0),(1,-1)\}$, and $\{(-1,-1),(0,0),(1,1)\}$. These correspond to four lines centered at $x(m,n)$ along the horizontal, 45° orientation, vertical, and 135° orientation, respectively. For each direction, the difference between the center pixel and the directional average is computed. The output pixel value is then chosen as:

$$y(m,n) = \bar{y}_{i^*,j^*}(m,n), \quad \text{where} \quad (i^*, j^*) = \arg\min_{(i,j)} \left| x(m,n) - \bar{y}_{i,j}(m,n) \right|.$$

This selection criterion favors the direction with the least intensity variation, thereby preserving edge structures while still providing noise reduction. Computer Assignment 8.6 explores the application of this method for removing additive noise while minimizing edge smearing.

8.2.2 Median Filtering

Median filtering is a nonlinear filtering operation mainly used to remove impulsive or "salt & pepper" type noise. As in spatial filtering, the median filtering involves sliding a mask encompassing an odd number of pixels. The center pixel in the window is then replaced by the median of the pixel values within the mask, i.e.

$$y(m,n) = \text{median}\{x(m-k, n-l), k, l \in W\}. \tag{8.11}$$

Finding the median value requires arranging pixel intensities within the mask in increasing or decreasing order and picking the middle value. Generally, the mask size, N_W, is chosen to be an odd number to ensure a unique median. Some of the key properties of the median filter are listed below.

8.2.2.1 Properties of Median Filter

1. Median filter is a nonlinear filter, i.e.

   $$\text{median}\{x_1(m) + x_2(m)\} \neq \text{median}\{x_1(m)\} + \text{median}\{x_2(m)\}.$$

2. It does not perform well on additive Gaussian noise.
3. It performs poorly when the number of noise pixels within a mask $\geq \frac{N_W}{2}$, i.e. 50% or more contaminated pixels.
4. Unlike spatial filtering, median filtering generally does not cause edge smearing artifacts. Nevertheless, as the mask size increases, the median gets closer to the local mean and hence more smearing and edge distortion will be noticeable.
5. When N_W is odd, search for the median requires $(N_W - 1) + (N_W - 2) + \cdots + \frac{N_W - 1}{2} = \frac{3(N_W^2 - 1)}{8}$ comparisons. For example, for 3×3 and 5×5 mask sizes, the number of comparisons are 30 and 224, respectively. This number can be drastically reduced using more recent algorithms, including those in [11, 12].
6. To protect sharp corners and thin lines in the images, other shaped masks (e.g. cross-shaped) are preferred.

To show the effectiveness of the median filtering, the original "Baboon" image in Figure 8.17 was corrupted by adding salt & pepper noise with a density varying from 10 to 100%. Different mask sizes were tested to determine the noise removal ability as well as the smearing artifacts in each case. Figure 8.18a shows the corrupted Baboon image with the noise density of 40%. Figure 8.18b–d shows the median filtered images for the mask sizes of 3×3, 5×5, and 7×7, respectively. As can

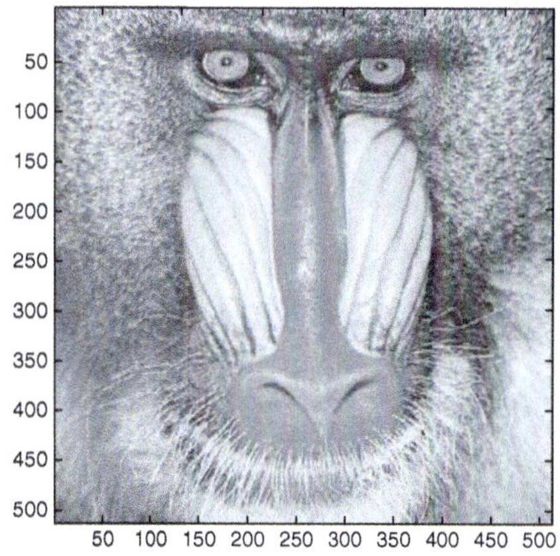

Figure 8.17 Original Baboon image. *Source:* [4] / John Wiley & Sons / CC BY 3.0.

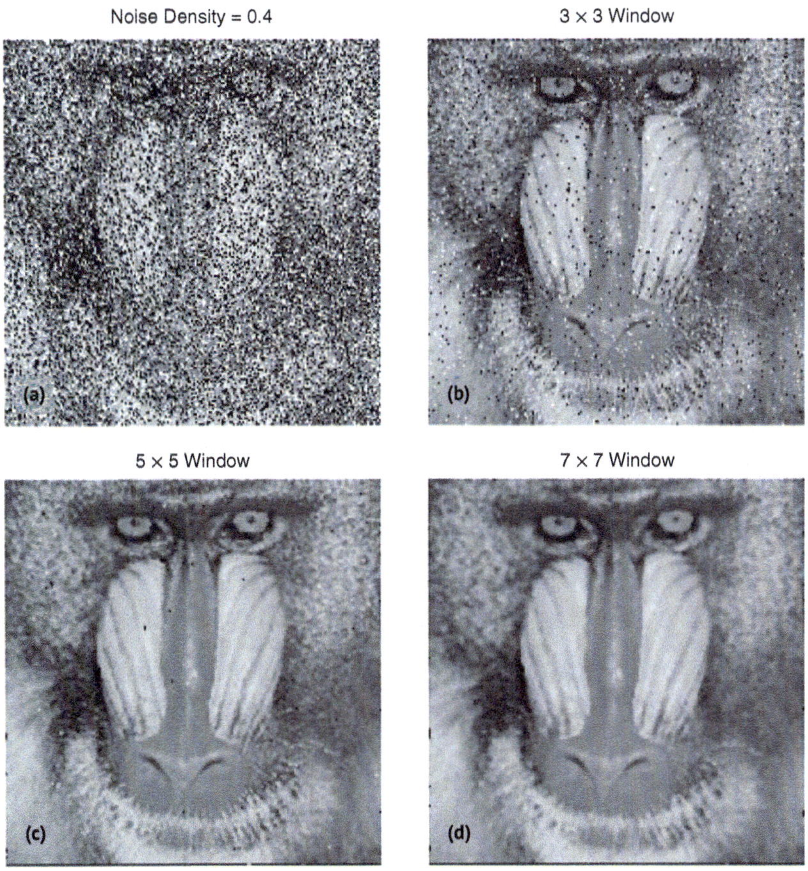

Figure 8.18 (a) Noise-corrupted Baboon with 40% density of salt & pepper noise and (b–d) median filtered images for 3×3, 5×5, and 7×7 mask sizes, respectively.

be observed from these results, even for this high density of salt & pepper noise, the median filter performs very well. Although the 3×3 median filter leaves some residual noise pixels, due to relatively high density of the noise, it causes the least smearing artifacts comparing to the other mask sizes. The smearing is more noticeable in Baboon's hair and whiskers, particularly for the larger mask sizes. The performance plot, which is the plot of the variance of the residual noise in the processed image versus the noise density is shown in Figure 8.19 for different mask sizes. As expected, when the noise density increases, the filter's ability to remove the noise deteriorates. However, this deterioration is more pronounced for smaller masks.

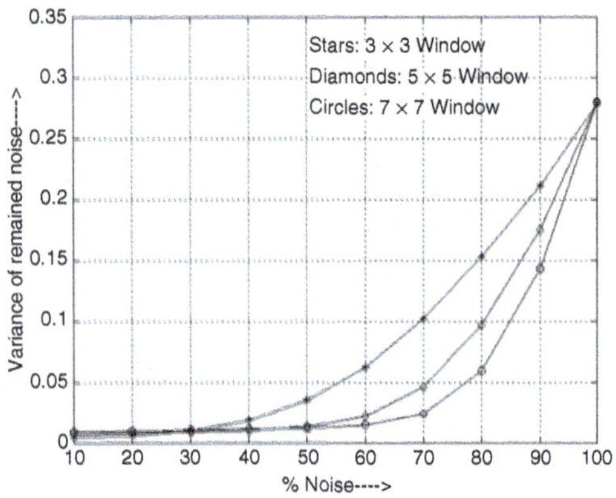

Figure 8.19 Performance plots for median filter for different mask sizes.

Example 8.3 *Consider a discrete-time signal* $x(m)$, $m \in [1, M]$. *Show that the objective function* $\zeta(\hat{x})$ *given below is minimized when* $\hat{x} = \text{median}\{x(m),\ m \in [1, M]\}$.

$$\zeta(\hat{x}) = \sum_{m=1}^{M} |x(m) - \hat{x}| = \sum_{m=1}^{M} |e(m)| = ||\underline{e}||_1$$

where $e(m) = x(m) - \hat{x}$ *and* $||\underline{e}||_1$ *denotes* ℓ_1 *norm of vector* $\underline{e} = [e(1)e(2)...e(M)]^T$.

Solution: Let S_1 and S_2 be two subsets of indices m with $S_1 \bigcup S_2 = \{1, ..., M\}$ and cardinality M_1 and M_2 respectively, such that:

$$\zeta(\hat{x}) = \sum_{m \in S_1} (x(m) - \hat{x}) + \sum_{m \in S_2} (-x(m) + \hat{x})$$

Now, taking partial derivative with respect to $\hat{x}$ and setting the result to zero gives $-M_1 + M_2 = 0$, i.e. $M_1 = M_2$. That is, the minimum of $\zeta(\hat{x})$ is achieved when $\hat{x}$ is in the midpoint or median of these data points. In other words, the median filter is the minimizer of the ℓ_1-norm of the error vector $\underline{e}$.

8.2.3 Edge Extraction and 2-D High-Pass Filtering (HPF)

Edge extraction and sharpening are often used to improve the visual appearance of images with less prominent edges. This is because an image with accentuated edges is subjectively more pleasing to human eyes than an exact reproduction. This procedure, which is also referred to as *unsharp masking*, is widely used in the printing industry to crisp the edges of a picture. Edge extraction is also an important part of typical image analysis systems that rely on contours or other edge-based features to discriminate different objects.

The basic idea is to detect edges of an image using a high-pass filter or a 2-D gradient operator, since the gradient takes relatively small values in the regions that are fairly smooth and larger values where edges are located. To sharpen or emphasize edges, a fraction of this gradient or high-pass filtered image is then added to the original image. The gradient or high-pass image can also be generated by subtracting an unsharp or low-pass filtered image from the original image. This is due to the fact that a 2-D high-pass filter can be produced from the corresponding 2-D low-pass filter using:

$$H_{\text{HP}}(e^{j\Omega_1}, e^{j\Omega_2}) = 1 - H_{\text{LP}}(e^{j\Omega_1}, e^{j\Omega_2})$$

or in the spatial domain $h_{\text{HP}}(m, n) = \delta(m, n) - h_{\text{LP}}(m, n)$.

In general, edge sharpening operation can be represented as:

$$y(m, n) = x(m, n) + \lambda g(m, n) \tag{8.12}$$

where $g(m, n)$ represents a suitable gradient computed for the mask centered at $x(m, n)$ and $\lambda > 0$ is a proportionality constant. The presence of an edge is declared if $g(m, n)$ exceeds a prespecified threshold. This threshold can typically be decided by examining the histogram of the image $g(m, n)$.

The commonly used edge detection operators can be categorized into two general categories. Those that are based on computing the first-order derivatives include: Roberts, Prewitt, and Sobel operators and those that use the second-order derivatives (zero-crossing detector) including 2-D discrete Laplacian and its variants. These operators are briefly described next.

1. Roberts Operator

This simple edge operator uses two 2 × 2 diagonal gradients $g_1(m,n) = x(m,n) - x(m+1,n+1)$ and $g_2(m,n) = x(m+1,n) - x(m,n+1)$ and combining the results using either the absolute value:

$$g(m,n) \equiv |g_1(m,n)| + |g_2(m,n)| \tag{8.13}$$

or the square root operation:

$$g(m,n) = \{g_1^2(m,n) + g_2^2(m,n)\}^{1/2}. \tag{8.14}$$

The corresponding masks are shown in Figure 8.20. The edge orientation with respect to the horizontal axis is determined by using:

$$\theta(m,n) = \tan^{-1}\frac{g_1(m,n)}{g_2(m,n)}. \tag{8.15}$$

The square root combination was shown [13] to provide slightly better results. Figure 8.21b shows the result of applying the Roberts operator to the original "Airplane" image in Figure 8.21a. The major problem with this operator is its susceptibility

0	0	0
0	1	0
0	0	−1

0	0	0
0	0	1
0	−1	0

Figure 8.20 Spatial masks for Roberts gradient operator.

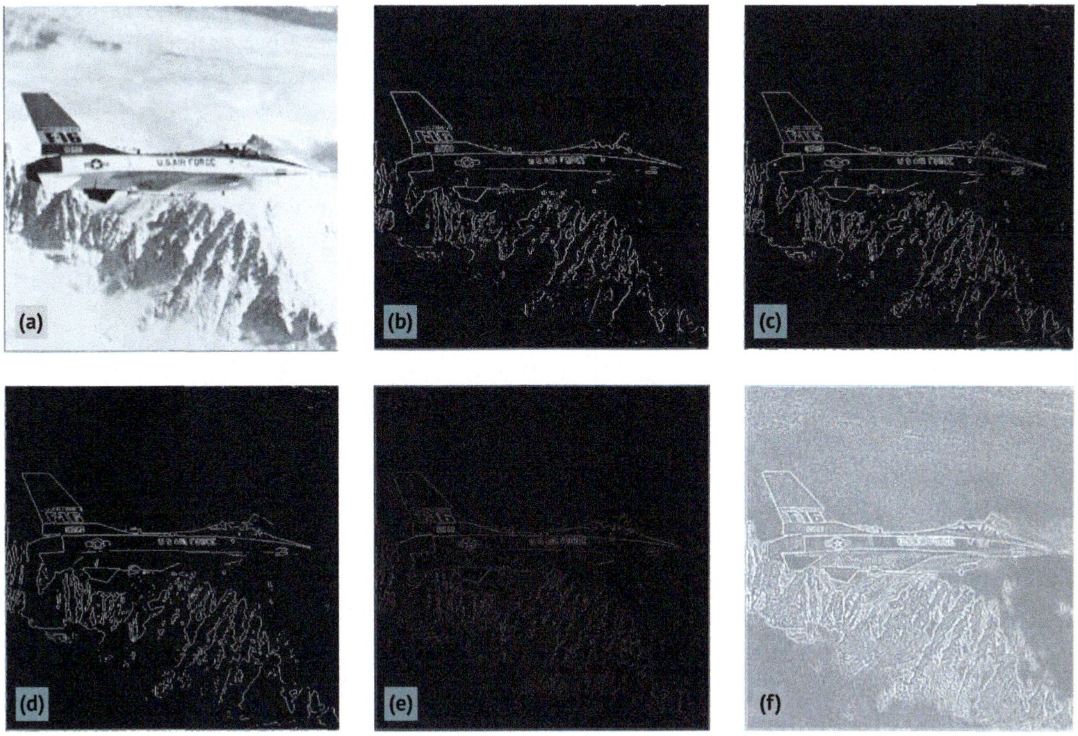

Figure 8.21 (a) Original Airplane image and edge detected image using (b) Roberts, (c) Sobel, (d) Prewitt, (e) Laplacian, and (f) Laplacian operator with histogram equalization. *Source*: U.S. Air Force. F-16 Fighting Falcon (public-domain photograph).

1	0	−1
1	0	−1
1	0	−1

(a)

−1	−1	−1
0	0	0
1	1	1

(b)

1	0	−1
2	0	−2
1	0	−1

(c)

−1	−2	−1
0	0	0
1	2	1

(d)

0	−1	0
−1	4	−1
0	−1	0

(e)

Figure 8.22 Spatial masks for Prewitt (a, b), Sobel (c, d), and Laplacian (e) operators.

to noise and small fluctuations in image intensity. Additionally, it may fail to detect certain edges, particularly those that are weak or not aligned with the diagonal directions. These limitations restrict its effectiveness in practical applications compared to more robust operators.

2. Prewitt and Sobel Operators

The Prewitt and Sobel operators are similar to the Roberts operators in the sense that they are based on approximating the first-order spatial derivatives to detect edges with different orientations. The results are typically combined using the square root operation in (8.14). The 3×3 horizontal and vertical masks associated with these two operators are shown in Figure 8.22. As can be seen, the Sobel masks in Figure 8.22c,d differ from those of the Prewitt in Figure 8.22a,b only by the way the weights of the north, south, west, and east pixels are chosen. Figure 8.21c,d shows the edge-detected Airplane images using the Sobel and Prewitt operators, respectively, which are visually better than that of Roberts in Figure 8.21b. The reason being the size of the masks for these operators is slightly larger resulting in less sensitivity to noise and small perturbations. As can be seen, the Sobel operator performs slightly better than the Prewitt owing to the differences in the weights that resulted in some noise suppression within the mask.

It should be noted that for all high-pass filters used for edge detection, the coefficients, or weights of the masks should add up to 0, i.e. $\sum\sum_{k,l\in W} h(k,l) = 0$. This ensures that the high-pass filter masks produce nearly zero or very small values when placed in flat regions of an image while responding to edge presence within the masks.

3. Laplacian Operator

The Laplacian operator is based on a 2-D discrete approximation of the second-order derivatives of the image function. The idea is that although the first-order derivative peaks where the transition or edges are located, the second-order derivative will have zero-crossings at these locations. The 3×3 mask in Figure 8.22e can be used to perform the 2-D Laplacian operation. The input–output relationship for this edge operator is given by:

$$g(m,n) \equiv 4x(m,n) - [x(m-1,n) + x(m+1,n) + x(m,n-1) + x(m,n+1)] \tag{8.16}$$

Figure 8.21e shows the result of applying this operator to the original Airplane image in Figure 8.21a. As can be observed, the 2-D Laplacian operator is highly sensitive to noise and small spatial variations even more than the other operators. To see this noise sensitivity, Figure 8.21f shows the histogram-equalized version of the 2-D Laplacian results of Figure 8.21e. The effects of noise and other perturbations are clearly noticeable in this result, although it did a nice job on the edges as well. Moreover, because of the second-order derivative, this operator tends to have a double-response to the edges.

Example 8.4 *Derive the 3×3 Laplacian operator from the Laplacian of a continuous 2-D function $x(u, v)$.*

Solution: The 2-D Laplacian operator of a continuous image $x(u, v)$ is given by:

$$\nabla^2(x(u,v)) = \frac{\partial^2 x(u,v)}{\partial u^2} + \frac{\partial^2 x(u,v)}{\partial v^2}$$

where ∇^2 represents the 2-D Laplacian operator. Using the forward difference method for derivatives, discretizing $u = m\triangle u$ and $v = n\triangle v$, and choosing $\triangle u = \triangle v = 1$, we get:

$$\frac{\partial^2 x(u,v)}{\partial u^2} \approx x(m+1,n) - 2x(m,n) + x(m-1,n)$$

and

$$\frac{\partial^2 x(u,v)}{\partial v^2} \approx x(m,n+1) - 2x(m,n) + x(m,n-1)$$

Now, combining these two expressions yields the discrete version of the 2-D Laplacian operation:

$$\nabla^2(x(u,v)) \approx x(m+1,n) + x(m,n+1) - 4x(m,n) + x(m-1,n) + x(m,n-1).$$

4. LoG Operator

To reduce the sensitivity of the 2-D Laplacian operator to noise and small perturbations, and to improve the detection of zero-crossing points, Marr-Hildreth [14] proposed a procedure that involves smoothing the image prior to edge detection using the Laplacian operator. They used a Gaussian filter due to its smoothness and localization ability in both the spatial and frequency domains (see Section 7.1). This is because the Fourier transform of a 2-D Gaussian is another Gaussian.

To derive the LoG edge detection operator, consider a continuous image $x(u,v)$ where u and v are the spatial coordinates. Let the PSF of the 2-D Gaussian smoothing filter be $h(u,v) = \frac{1}{2\pi\sigma^2} e^{-\frac{u^2+v^2}{2\sigma^2}}$ where σ^2 is the smoothing parameter or the variance. This filter is first applied to the original image and then the Laplacian of the resultant smoothed image $y(u,v)$ is taken, i.e.

$$\nabla^2 y(u,v) = \nabla^2[x(u,v) ** h(u,v)]. \tag{8.17}$$

Now, using the interchangeability property of the convolution integral, we can write:

$$\nabla^2 y(u,v) = x(u,v) ** \nabla^2 h(u,v) = x(u,v) ** \left[\frac{\partial^2 h(u,v)}{\partial u^2} + \frac{\partial^2 h(u,v)}{\partial v^2}\right]$$

$$= x(u,v) ** g(u,v) \tag{8.18}$$

where $g(u,v)$ is the LoG edge detection operator given by:

$$g(u,v) = \frac{1}{\pi\sigma^4}\left(\frac{u^2+v^2}{2\sigma^2} - 1\right)e^{-\frac{u^2+v^2}{2\sigma^2}}. \tag{8.19}$$

This result implies that we can perform LoG operation in one-step by convolving the image $x(u,v)$ with the LoG PSF $g(u,v)$ using (8.18) rather than first applying a Gaussian smoother $h(u,v)$ followed by the Laplacian operator as in (8.17). The plot of the function $g(u,v)$ in Figure 8.23 indicates that this function represents an inverted 2-D Mexican hat with a

Figure 8.23 Plot of PSF $g(u,v)$ of LoG operator for $\sigma = 1$.

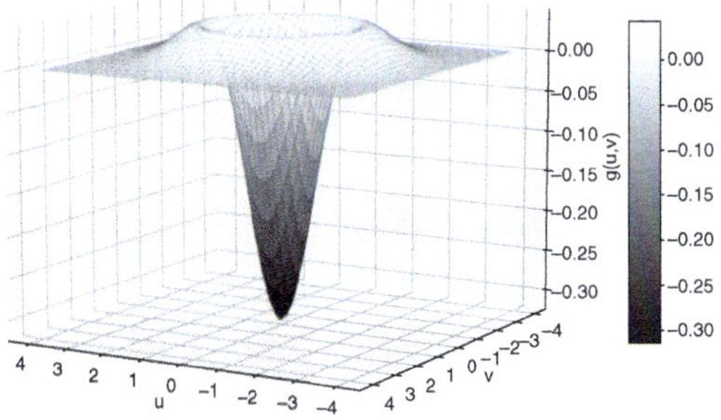

zero-crossing point that satisfies $u^2 + v^2 = 2\sigma^2$ in the spatial domain. It can be shown that the Fourier transform of the LoG function is also another LoG function in the frequency domain, i.e.

$$G(\omega_1, \omega_2) = -(\omega_1^2 + \omega_2^2)e^{-\frac{\sigma^2(\omega_1^2+\omega_2^2)}{2}}, \tag{8.20}$$

with the peak at $\omega_1^2 + \omega_2^2 = \frac{2}{\sigma^2}$. Thus, the LoG is isotropic in both spatial and frequency domains and exhibits band-pass filter characteristics, emphasizing mid-frequency components.

For processing digital images, this continuous operator must be truncated and approximated by a discrete mask of an appropriate size [15]. Figure 8.24 shows the 5×5 mask for LoG operator [16]. The mask coefficients are obtained by sampling the continuous LoG operator on a discrete grid and ensuring that they add up to 0 to avoid any bias. Gunn [15] proposed a method for determining the appropriate mask size in order to reduce the effects of the zero-crossing detection error. Additionally, a discrete zero-crossing detector was developed that guarantees closure of the extracted contours and hence provides a robust implementation of the LoG operator. A commonly used heuristic for determining the mask size is three times the distance between zero-crossings in the spatial domain.

Figure 8.25a shows the result of applying this LoG edge detection operator to the original Airplane image in Figure 8.21a. Compared to the results of the Laplacian operator in Figure 8.21e, it can be seen that the edges are detected reliably and with much better definition. The resolution in the edge detection ability can be determined by adjusting the σ parameter of the Gaussian. For example, a large σ yields an image with global edge maps. This is due to the over-smoothing as a result of using a Gaussian filter with large variance (width) in the spatial domain. The location of the zero-crossing is not affected by this choice. On the other hand, small values of σ produce images with finer edge details.

After extracting the edges, if edge sharpening is required, a fraction of the edge extracted image can be added to the original image using (8.12). Figure 8.25b shows the edge-enhanced Airplane image using the LoG edge operator with $\lambda = 0.05$.

5. Canny Operator

Unlike the other commonly used edge extraction algorithms, the Canny edge detection operator [16, 17] is specifically designed to meet three design criteria to provide an optimal or suboptimal approach with respect to these criteria. Canny's three design criteria are:

1. **Detection:** Low probability of error, that is, both low false negatives to detect real edges and low false positives to falsely mark non-edge points. Clearly, the presence of noise in the vicinity of edges impacts these error probabilities.
2. **Localization:** The points marked as edge points by the detector should be as close to the center of the true edge as possible.
3. **Single Response:** Although implicitly speaking this criterion is directly related to the false positive rate of the detector, the mathematical form of the first criterion does not prevent the occurrence of the multiple edge responses.

Canny [17] showed that the filter operator that meets these criteria can be approximated fairly closely using an operator corresponding to the first directional derivative of a Gaussian. The reason is that there are efficient ways to perform filtering operation in the spatial domain if it can be represented as some derivative of a Gaussian, as we showed for the LoG operator.

Let us see how this works for the continuous case where $h(u, v) = e^{-\frac{u^2+v^2}{2\sigma^2}}$ is a 2-D smoothing Gaussian function and $x(u, v)$ is the original image. The first directional derivative of this Gaussian along the unit vector $\underline{n}$ is:

$$h_n(u, v) = \frac{\partial h}{\partial \underline{n}} = \underline{n}.\vec{\nabla} h(u, v)$$

0	0	−1	0	0
0	−1	−2	−1	0
−1	−2	16	−2	−1
0	−1	−2	−1	0
0	0	−1	0	0

Figure 8.24 Spatial mask for 5 × 5 LoG operator.

Figure 8.25 (a) Edge detected image using LoG operator, (b) edge-enhanced Airplane using LoG, and (c) Edge detected image using Canny operator.

where $\vec{\nabla}h(u,v) = \left[\frac{\partial h}{\partial u} \; \frac{\partial h}{\partial v}\right]^T$. This operator is then applied to image $x(u,v)$ using 2-D convolution $h_n(u,v) ** x(u,v)$. Alternatively, the operation can be carried out more efficiently using the associative property of convolution, i.e.

$$\frac{\partial h}{\partial \underline{n}} ** x(u,v) = \frac{\partial}{\partial \underline{n}} y(u,v) = \vec{\nabla} y(u,v) \tag{8.21}$$

where $y(u,v) = h(u,v) ** x(u,v)$ is the smoothed image. This equation implies that we first convolve the image $x(u,v)$ with a 2-D smoothing Gaussian $h(u,v)$ and then compute the directional gradient of $y(u,v)$ along $\underline{n}$. Note that we do not need to know the direction $\underline{n}$ before convolution since ideally at solution, $\underline{n}$ should be normal to *edge direction* (direction of the tangent to the edge contour). The solution is the local maximum of (8.21) at which point we have $\frac{\partial^2}{\partial \underline{n}^2} y(u,v) = 0$, i.e. the presence of an edge point is declared. At an edge point, $\underline{n}$ satisfies:

$$\underline{n} = \frac{\vec{\nabla} y(u,v)}{|\vec{\nabla} y(u,v)|} = 1 \tag{8.22}$$

or $\vec{\nabla} y(u,v) = |\vec{\nabla} y(u,v)|$ which gives an indication of edge strength.

The method developed for continuous images can easily be extended to the digital case by first convolving the image $x(m,n)$ with a mask of an appropriate size associated with the discretized 2-D symmetric Gaussian. Any of the first-order derivative masks, e.g. Sobel, can then be used to implement (8.21) and compute magnitude and orientation using similar equations to those in (8.14) and (8.15), respectively. Figure 8.25c shows the results of the Canny operator on the Airplane image of Figure 8.21a, which provided an excellent edge detection result with better edge definition and reduced false detections.

There are other steps in Canny algorithms that include a non-maximum suppression scheme, thresholding, and edge linking [16]. The function of non-maximum suppression is to guarantee the third criterion stated above and thin out the thick edges after the gradient operation. There are many algorithms that can be utilized to accomplish this task. A typical non-maximum suppression works by partitioning every 3×3 mask centered at pixel (m,n) into several (e.g. eight) segments of possible edge directions and finding the closest direction (or segment) to the edge orientation estimated using (8.15). Then, the gradient magnitude of the center pixel is compared against the two neighboring pixels in the positive and negative directions of the estimated edge orientation and if its magnitude is the largest, the center pixel is kept. Otherwise, it is set to zero, i.e. effectively thinning the edge to a single-pixel line.

The output of the non-maximum suppression step may still have some spurious edge points and noise. Double thresholding is usually adopted in Canny's algorithm using high τ_H and low τ_L thresholds to remove or reduce these false edge points. Pixels with values larger than τ_H are considered strong edge points, while those smaller than τ_L are set to zero. Pixels whose values are between the two thresholds are considered uncertain cases. To determine whether an uncertain case is an actual edge point or not, we use an edge-linking algorithm. This algorithm determines the connectivity of an uncertain point to strong neighboring edge points. If a certain number of connectivity is confirmed, then the uncertain point is declared as an actual edge. Otherwise, it will be set to zero, i.e. a non-edge point. This procedure is also known as thresholding for hysteresis [16].

8.2.4 Correspondence and 2-D Matched Filtering

Two-dimensional template matching plays an important role in a number of applications such as motion and stereo correspondence and object detection. In such applications, it is often required to search for known objects or structures in the observed images. This process involves defining templates that closely represent the objects to be detected and localized. The original image is then spatially cross-correlated with the template. A match is found at the locations where the cross-correlation peaks and the amplitudes of the peaks exceed a chosen threshold. Since the spatial cross-correlation operation is related to the convolution, the entire operation can be carried out in the spatial domain using spatial filtering with the template as a mask, hence the term *matched filter*.

Let us assume that a target template is denoted by $t(m, n)$, $m, n \in W$ with W being the region of support (ROS) of the template. Then, this template is spatially shifted by (k, l) and cross-correlated with the original image to yield:

$$c_{x,t}(k, l) = \sum_m \sum_n x(m, n) t(m - k, n - l) \tag{8.23}$$

$$= x(k, l) ** t(-k, -l) \tag{8.24}$$

where $c_{x,t}(k, l)$ represents the cross-correlation at the spatial location k, l, and $**$ denotes 2-D convolution operation. At locations where this function exceeds a predetermined threshold, potential matches are declared. Clearly, the highest cross-correlation is achieved when the unknown object in the image is a translated version of the template, that is, $\alpha\, t(m - k, n - l)$ where α is an intensity-dependent factor. However, in practice, the objects may be translated, scaled, rotated, and distorted versions of the templates. In such cases, the standard template matching presented here becomes less effective, since the cross-correlation results may very well be less than the specified threshold. A better approach is to employ feature-based matching methods where the extracted features are invariant to the translation, scaling, and rotation of the templates. Such feature extraction methods will be discussed in Chapter 14.

To demonstrate the application of matched filtering for object/target detection and localization, consider the red blood cell image in Figure 8.26a. Our goal is to detect and locate every red blood cell from the background using the template matching. The result can subsequently be used to count the number of red blood cells in a sample image, a crucial task for many medical diagnostics. To detect those partially obscured cells near the image boundaries, the image is first symmetrically padded around the boundaries of the image as shown in Figure 8.26b. If the template size is $d_1 \times d_2$, the top and bottom padding each has a height of $d_1 - 1$, while the left and right padding each has a width of $d_2 - 1$. Template matching is then applied to the padded image and the result is extracted within the support of the original image to eliminate spurious or repeated detections caused by the symmetric extension around the boundaries. Figure 8.26c shows the result of this template matching process where the center of each red box corresponds to the spatial location of the detected peak in the 2-D cross-correlation map. The green box outlines the chosen template. As can be observed, this simple method effectively

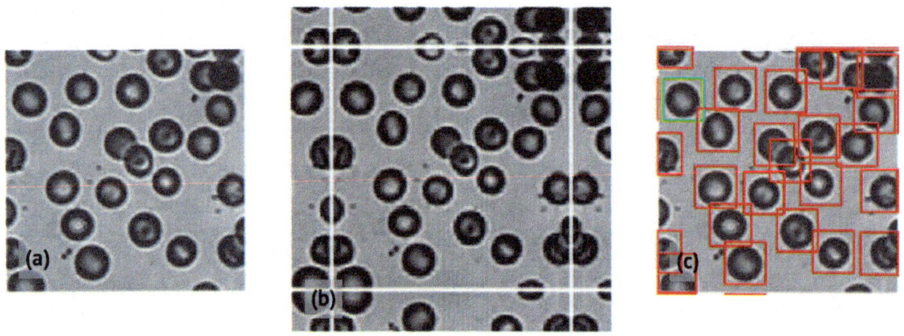

Figure 8.26 (a) Original red blood cell image *Source*: The Northwestern University / http://www.ece.northwestern.edu/local-apps/matlabhelp/toolbox/images/linfilt6.html / last accessed Aug. 13, 2026 Linear Filtering and Filter Design (Image Processing Toolbox). (b) symmetrically padded version, and (c) matched filtered result.

detected and localized most of the cells, with the exception of those that were heavily obscured. The selection of an appropriate threshold presents a trade-off between maximizing the detection rate and minimizing false alarms. It is worth noting that this template matching approach gave favorable results because the image mostly contains cells of uniform size and minimal distortion. As discussed previously, this method typically fails when encountering significant variations in object size, orientation, or shape within the image.

8.2.5 Hybrid Operations

Certain image processing tasks require the combination of multiple types of operations. In particular, the combination of point and spatial operations can be useful in numerous image processing applications. In this section, we provide an example of such hybrid operation for multiplicative noise removal.

In several image generation systems, the source of noise that corrupts the image is multiplicative in nature [1, 18, 19]. This is in contrast to typical additive noise sources that we considered in the previous sections. In such cases, to reduce the effects of multiplicative noise, one can use spatial low-pass filters discussed in Section 8.2.1, after converting the multiplicative noise to additive noise by applying a log to the noisy image. Once spatial filtering is completed, an exponential mapping is applied to convert the filtered image back to the image domain. This procedure constitutes a hybrid operation, combining point operations (logarithm and exponential) with spatial low-pass filtering.

To illustrate the performance of this hybrid operation on images with multiplicative noise, the original "Boat" image in Figure 8.27a is corrupted by a white Gaussian multiplicative noise with mean 1 and variance 0.1 to produce the corrupted image in Figure 8.27b. Then, this hybrid filtering is implemented using the spatial averaging Mask 1 in the log domain followed by the inverse exponential mapping. The result of this hybrid operation is shown in Figure 8.27c.

To evaluate performance of this multiplicative noise filtering operation, a reasonable measure that can be used is the ratio of the local mean to the local standard deviation:

$$p(m,n) = \frac{\mu(m,n)}{\sigma(m,n)} \tag{8.25}$$

Figure 8.27 (a) Original Boat image. *Source*: [20] / John Wiley & Sons / CC BY 3.0. (b) Boat with multiplicative noise, and (c) filtered image using Mask 1 in log-domain.

where $\mu(m,n)$ and $\sigma(m,n)$ are typically computed over a window, W, using:

$$\mu(m,n) = \frac{1}{N_W} \sum\sum_{k,l \in W} x(m-k, n-l)$$

and

$$\sigma(m,n) = \left[\frac{1}{N_W} \sum\sum_{k,l \in W} [x(m-k, n-l) - \mu(m,n)]^2 \right]^{\frac{1}{2}}. \quad (8.26)$$

This measure is closely related to the local peak signal-to-noise ratio (PSNR), especially in flat areas of the image with minimal spatial variations. For quantitative evaluation, windows of size 50 × 100 were selected in certain areas with relatively uniform intensity and the PSNR measures were calculated and averaged. For the noisy image, this measure was found to be $p = \frac{\mu}{\sigma} = 4.3$. After this hybrid filtering operation using the spatial low-pass filter Masks 1–3, the PSNR values increased to $p = \frac{\mu}{\sigma} = 6.3, 4.9$, and 5.3, respectively. These results indicate that Mask 1 provides the most effective noise reduction, albeit at the expense of increased smearing artifacts.

8.3 Transform Domain Operations

This section introduces another category of image enhancement methods that use transform domain operations. Figure 8.28 shows the block diagram of a typical transform domain system with its components. The central idea is to map the image, using a unitary transformation, to the transform domain where the image features or attributes are better represented and localized. The filtering operation is carried out in the transform domain using simple algebraic or point operations instead of the spatial operations discussed in the previous section. A "zonal filter" is typically used that extracts or enhances the desired image attributes while suppressing or removing the others. The inverse unitary mapping is then applied to convert the image from the transform domain back to the spatial domain. There are several key advantages of transform domain operations. First, they provide deeper insight about the features, spectral or otherwise, that need to be removed or enhanced in the image. In addition, they tend to compact the image attributes into localized regions in the transform domain for more effective processing. In other words, unlike the other category of image enhancement methods covered so far, transform domain operations are more *objective* in the sense that the characteristics of the image in the transformed domain are taken into account when designing the zonal filters.

This section presents several 2-D transform domain operations for additive noise removal, matched filtering, cepstral domain filtering, and homomorphic filtering. Several examples are also provided to illustrate the applications of these methods.

8.3.1 2-D Filtering in Transform Domain

As illustrated in the block diagram of Figure 8.28, a unitary transform (e.g. the 2-D DFT or DCT) represented by the operator $\mathcal{A}\{.\}$, is first applied to decompose the image into several components that tend to isolate certain features of the image. For example, high-frequency components in the 2-D DFT or DCT of an image are representative of edges or textural structures. The filtering is then performed in the transform domain using algebraic or point operations:

$$Y(k,l) = H(k,l)X(k,l) \quad (8.27)$$

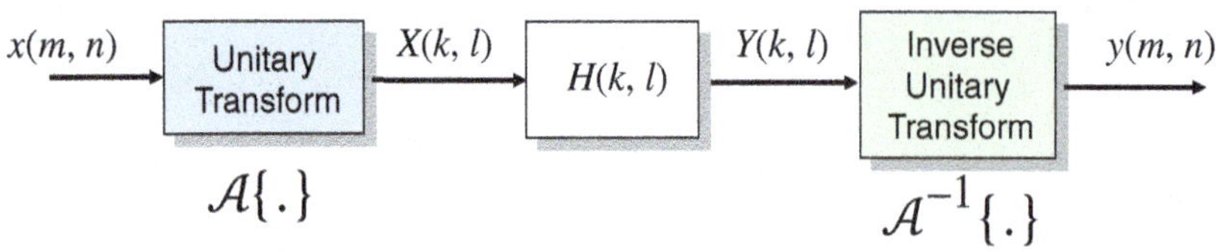

Figure 8.28 Block diagram of typical transform domain operation.

where $X(k,l) = \mathcal{A}\{x(m,n)\}$ and $Y(k,l) = \mathcal{A}\{y(m,n)\}$ are the 2-D unitary transformed versions of the original image, $x(m,n)$, and the filtered image, $y(m,n)$, respectively, and $H(k,l) = \mathcal{A}\{h(m,n)\}$ represents the transform domain response or the unitary transform of the PSF, $h(m,n)$, of the 2-D filter. This is then followed by the inverse unitary transform:

$$y(m,n) = \mathcal{A}^{-1}\{Y(k,l)\} \tag{8.28}$$

to bring back the filtered image to the spatial domain. Note that to avoid wrap-around error due to the circular convolution property of the DFT, the zero-padding procedure in Section 3.1.7 must be performed prior to taking DFT and transform domain filtering.

Different types of zonal filters can be designed for a particular image processing task that vary in terms of their simplicity, effectiveness, and artifacts they produce. The simplest filter is an ideal 2-D filter with a sharp frequency response. For example, for an ideal 2-D low-pass filter, the frequency response is given by:

$$H_I(e^{j\Omega_1}, e^{j\Omega_2}) = \begin{cases} 1, & D(\Omega_1, \Omega_2) \le D_0 \\ 0, & \text{otherwise} \end{cases}$$

where D_0 represents the cutoff frequency of the filter and $D(\Omega_1, \Omega_2)$ describes the passband shape. For example, for *circularly symmetric* ideal filters, $D(\Omega_1, \Omega_2) = (\Omega_1^2 + \Omega_2^2)^{\frac{1}{2}}$, which is the radial distance of the point (Ω_1, Ω_2) from the origin in the frequency domain. The choice of cut-off frequency D_0 presents a trade-off between the noise removal ability of the filter and the smearing artifacts caused by the filter.

Figure 8.29a shows the sampled frequency response of an ideal circularly symmetric 2-D LPF:

$$H_I(k,l) = H_I(e^{j\Omega_1}, e^{j\Omega_2})\Big|_{\Omega_1 = \frac{2\pi k}{N_1},\, \Omega_2 = \frac{2\pi l}{N_2}} \tag{8.29}$$

where N_1 and N_2 are the number of samples in one period of $H_I(e^{j\Omega_1}, e^{j\Omega_2})$. This frequency response is doubly periodic with period 2π in both variables. Now, taking the 2-D IDFT of $H_I(k,l)$ of size $N_1 \times N_2$ yields the impulse response or the filter coefficients $h_I(m,n) = 2 - D\ IDFT\{H_I(k,l)\}$, for the corresponding 2-D FIR filter of order $N_1 \times N_2$. It should be noted that this approach for designing 2-D FIR filters is known as the *frequency sampling* method [21–23]. Thus, the frequency sampling method involves: (i) specifying a desired frequency response for the filter over the range of $[0, 2\pi]$ in both Ω_1 and Ω_2, (ii) sampling Ω_1 and Ω_2 in the frequency domain by fitting N_1 and N_2 samples in one period, and (iii) taking the 2-D IDFT of the sampled frequency response to produce the 2-D FIR filter coefficients. Figure 8.29b shows the contour plot of the PSF of the designed filter using the frequency sampling method. Note that since sampling is carried out in the frequency domain, this PSF is the aliased version of the original infinite-extent 2-D sinc PSF.

The bandwidth D_0 can be chosen either empirically or by specifying the percentage of the energy, β, retained in the filtered image relative to the total energy of the image. This is given by:

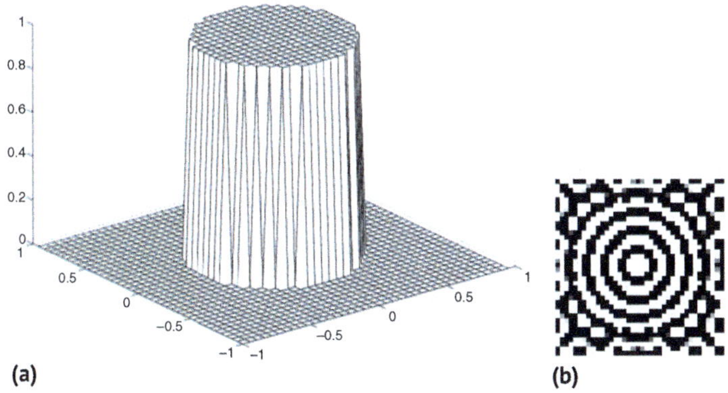

Figure 8.29 (a) Sampled frequency response and (b) PSF contour of a circularly symmetric ideal filter.

$$\beta = \left[\left(\sum\sum_{k,l \in S_0} E(k,l)\right) / \left(\sum_{k=0}^{N-1}\sum_{k=0}^{N-1} E(k,l)\right)\right] \times 100\% \qquad (8.30)$$

where $E(k, l) = |X(k, l)|^2$ and S_0 is the set of (k, l) values within the passband region of the filter. Note that it is assumed that $N_1 = N_2 = N$.

To show the effects of changing β on the filtered images, the original noise-free Peppers image in Figure 3.11a is used here. The centered 2-D magnitude spectrum of this image is shown in Figure 3.11b. This noise-free image is then filtered using a 2-D ideal low-pass filter with increasing D_0 to determine the effects of increasing bandwidth of the filters on the quality of the filtered image. The results in Figure 8.30a–f show the filtered images and the corresponding 2-D magnitude spectra for $D_0 = 20$, 50, and 70% of the maximum radius $D_0 = \frac{\sqrt{2}N}{2}$. The corresponding values of β computed using (8.30) are also

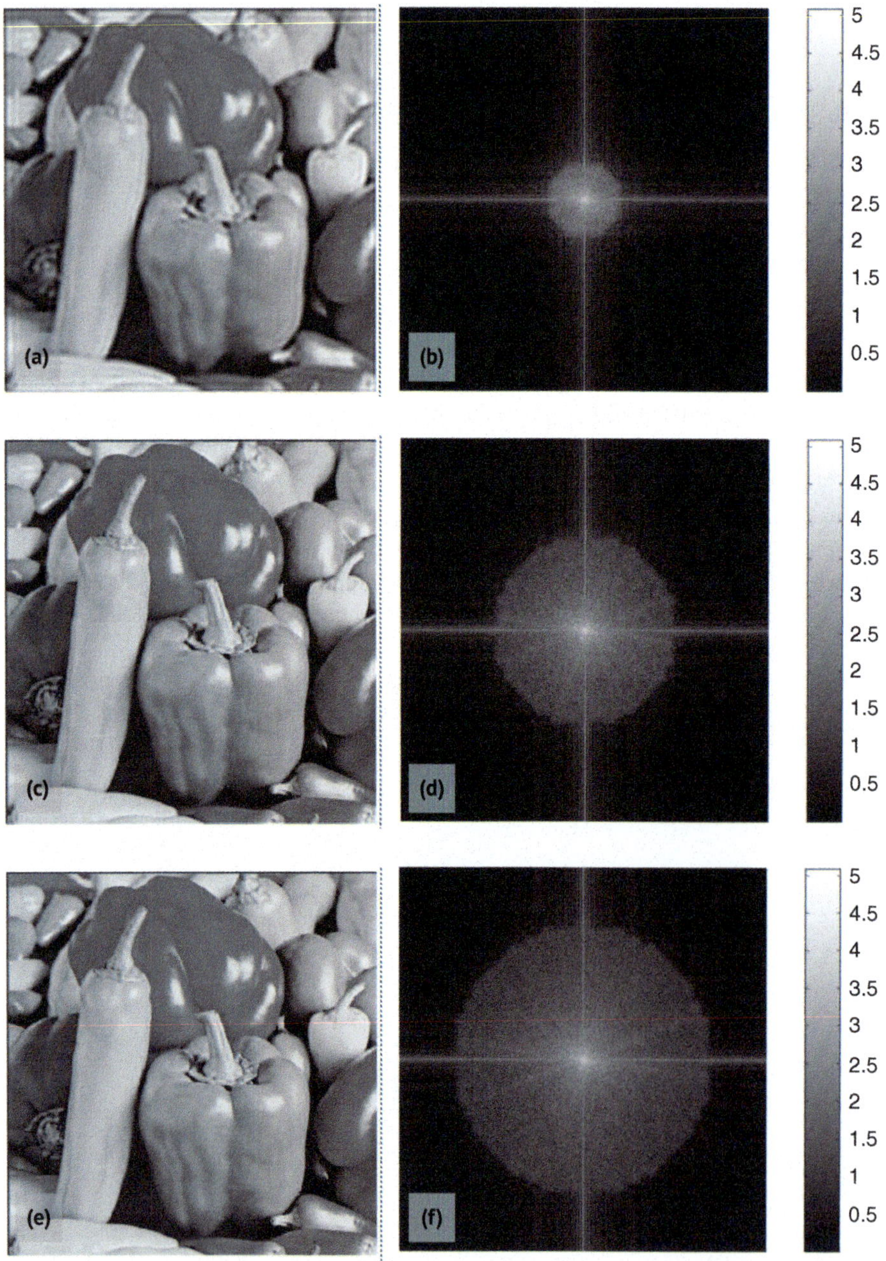

Figure 8.30 Filtered images (a, c, e) and corresponding 2-D magnitude spectrum (b, d, f) for ($D_0 = 20\%$, $\beta = 99.12\%$), ($D_0 = 50\%$, $\beta = 99.73\%$) and ($D_0 = 70\%$, $\beta = 99.84\%$), respectively.

given. Closer evaluation of these images reveals several interesting observations. Although for $D_0 = 20\%$ more than $\beta = 99\%$ of the total energy is retained, the filtered image in Figure 8.30a exhibits noticeable smearing artifacts as a consequence of the loss of high-frequency components. As the bandwidth increases, the amount of smearing is reduced as well. In addition to the smearing effects, the 2-D ideal filter introduces "ringing" artifacts due to the Gibbs phenomenon arising from the sharp cutoff of the ideal filter. As D_0 increases, the number of rings in a given region increases, thus producing more finely spaced rings that are harder to visualize in Figure 8.30c,e even though they are still prevalent.

Ringing artifacts can be overcome by designing filters that exhibit smooth transition regions between the passband and stopband. This can be achieved by using the *windowing method* [21–23], which involves multiplying the infinite-extent PSF of an ideal filter by a 2-D window function with finite support. The windowing is another method used to design FIR filters. Figure 8.31a,b shows the frequency response of a circularly symmetric 2-D non-ideal FIR filter and the contour plot of its PSF. This filter was designed by applying a Hamming window [21] to the PSF of the ideal filter for $D_0 = 20\%$. As can be observed, the filter possesses a smooth transition region. Figure 8.32a,b shows the resulting filtered Peppers image and its magnitude spectrum, respectively. Although the smearing effect is still noticeable, due to small bandwidth, no ringing is evident in the processed image compared to the result in Figure 8.30a.

Other simple non-ideal filters that can be used to reduce the ringing artifacts include: 2-D Butterworth and trapezoidal filters. The frequency response of 2-D low-pass Butterworth filter is:

$$H(e^{j\Omega_1}, e^{j\Omega_2}) = \frac{1}{1 + (\sqrt{2} - 1)[D(\Omega_1, \Omega_2)/D_0]^{2n}}$$

where n is the order of the filter that determines the sharpness of the frequency response. At $D(\Omega_1, \Omega_2) = D_0$ we have $H(e^{j\Omega_1}, e^{j\Omega_2}) = 1/\sqrt{2}$ corresponding to the cutoff point.

For 2-D trapezoidal low-pass filters, the frequency response is:

$$H(e^{j\Omega_1}, e^{j\Omega_2}) = \begin{cases} 1, & D(\Omega_1, \Omega_2) \le D_0, \\ \dfrac{D_1 - D(\Omega_1, \Omega_2)}{D_1 - D_0}, & D_0 < D(\Omega_1, \Omega_2) \le D_1, \\ 0, & \text{otherwise.} \end{cases}$$

Example 8.5 *Consider the problem of approximating an ideal 2-D LPF with frequency response:*

$$H_1(e^{j\Omega_1}, e^{j\Omega_2}) = \begin{cases} 1, & |\Omega_1| < a < \pi, |\Omega_2| < b < \pi \\ 0, & \text{otherwise.} \end{cases}$$

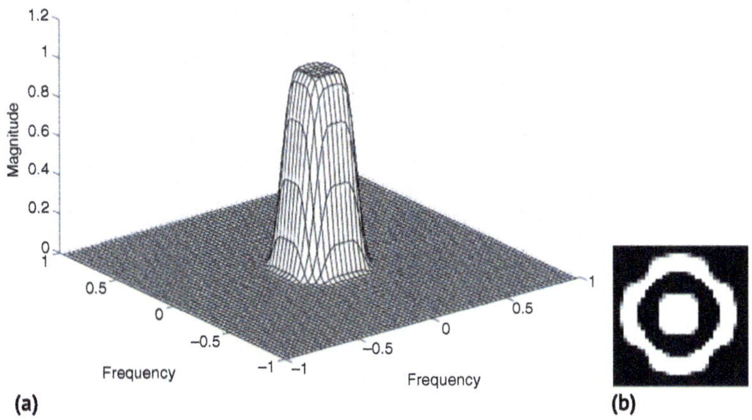

Figure 8.31 (a) Frequency response and (b) PSF contour of a circularly symmetric non-ideal filter designed using windowing method.

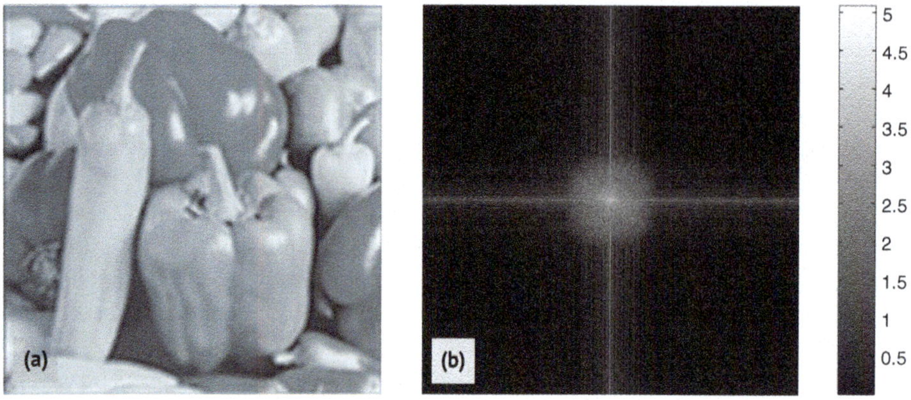

Figure 8.32 (a) Low-pass filtered Peppers image using non-ideal filter and (b) its magnitude spectrum.

by a filter whose PSF is

$$h_D(m,n) = \begin{cases} A, & m = n = 0, \\ B, & m = \pm 1, \quad n = 0, \\ C, & m = 0, \quad n = \pm 1, \\ 0, & \text{otherwise.} \end{cases}$$

Find the values of unknown A, B, and C that minimize the Least Squares (LS) error:

$$\varepsilon = \frac{1}{4\pi^2} \int_{-\pi}^{\pi} \int_{-\pi}^{\pi} |H_I(e^{j\Omega_1}, e^{j\Omega_2}) - H_D(e^{j\Omega_1}, e^{j\Omega_2})|^2 d\Omega_1 d\Omega_2$$

Solution: From the definition of $h_D(m,n)$, we can write:

$$h_D(m,n) = A\delta(m,n) + B[\delta(m-1,n) + \delta(m+1,n)] + C[\delta(m,n-1) + \delta(m,n+1)]$$

Taking the 2-D DSFT gives the frequency response:

$$H_D(e^{j\Omega_1}, e^{j\Omega_2}) = A + 2B\cos\Omega_1 + 2C\cos\Omega_2$$

Now, substituting $H_I(e^{j\Omega_1}, e^{j\Omega_2})$ and $H_D(e^{j\Omega_1}, e^{j\Omega_2})$ in the LS objective function, taking partial derivatives of ε with respect to the unknowns A,B, and C and setting the results to zero, yields:

$$\frac{\partial \varepsilon}{\partial A} = 0 \Rightarrow A = \frac{\int_{-a}^{a}\int_{-b}^{b} d\Omega_1 d\Omega_2}{\int_{-\pi}^{\pi}\int_{-\pi}^{\pi} d\Omega_1 d\Omega_2} = \frac{ab}{\pi^2}$$

$$\frac{\partial \varepsilon}{\partial B} = 0 \Rightarrow B = \frac{\int_{-a}^{a}\int_{-b}^{b} \cos\Omega_1 d\Omega_1 d\Omega_2}{2\int_{-\pi}^{\pi}\int_{-\pi}^{\pi} \cos^2\Omega_1 d\Omega_1 d\Omega_2} = \frac{b\sin a}{\pi^2}$$

$$\frac{\partial \varepsilon}{\partial C} = 0 \Rightarrow C = \frac{\int_{-a}^{a}\int_{-b}^{b} \cos\Omega_2 d\Omega_1 d\Omega_2}{2\int_{-\pi}^{\pi}\int_{-\pi}^{\pi} \cos^2\Omega_2 d\Omega_1 d\Omega_2} = \frac{a\sin b}{\pi^2}.$$

Note that the above results are obtained using the fact that area under cosine is zero in $[-\pi, \pi]$. This example provides yet another method, referred to as the method of *Least Squares*, for optimal FIR filter design [21–23].

8.3.2 2-D Filtering Using Other Unitary Transforms

Other unitary transforms such as 2-D DCT or 2-D DWT may also be used for the mapping operator $\mathcal{A}(\cdot)$ in Figure 8.28. To illustrate this, we first consider the application of the 2-D DCT for filtering the original noise-free Peppers image in Figure 3.11a, and compare the results with those of the 2-D DFT in the previous section. The 2-D DCT of this image is taken and low-pass filtering is performed in the DCT domain using an ideal 2-D filter with rectangular passband of size 120×120. All DCT components outside of this passband region were filtered out (zeroed). The 2-D inverse DCT is then applied to obtain the filtered image in the spatial domain. Figure 8.33 shows the result of this DCT-based low-pass filtering. Again, ringing and smearing artifacts are visible in this filtered image, though the smearing artifacts are less pronounced compared to the 2-D DFT-based result in Figure 8.30. This improvement is primarily due to the superior energy packing property of the DCT. Consequently, the discarded DCT coefficients did not carry that much information pertaining to the edge and other higher-frequency image attributes.

In Chapter 7, we discussed how 2-D DWT can be used to perform low-pass filtering and noise removal. Here, we provide an example to demonstrate the application of the 2-D DWT for image denoising. The Peppers image is corrupted by adding white noise to produce SNR of approximately 8.6 dB. The noisy image is shown in Figure 8.34a. A one-level 2-D DWT using db4 wavelet is then applied to this image, leading to four subimages shown in Figure 8.34b. As can be observed, the noise manifests itself mostly in the added detail subimages. Thus, a simple filtering scheme in the DWT domain entails thresholding the added details based on their histograms and reconstructing the image based upon the lowest order approximation subimage and the thresholded added details. This operation is effectively equivalent to low-pass filtering, since the high- and mid-frequency components in the subbands are attenuated. Figure 8.34c shows the reconstructed or the DWT-filtered Peppers image. The SNR was improved to approximately 13.3 dB with minimal smearing or ringing artifacts compared to the other transform domain low-pass filtering discussed earlier.

8.3.3 2-D Matched Filtering in Transform Domain

In Section 8.2.4, we showed how 2-D correspondence and matched filtering can be performed using 2-D cross-correlation or convolution. This operation can also be performed in the transform domain more efficiently. Using 2-D DFT, the transform domain version of (8.23) becomes:

$$C_{x,t}(k, l) = X(k, l)T^{*}(k, l), \tag{8.31}$$

where $C_{x,t}(k, l)$ is the cross-power spectrum and $T(k, l)$ is the 2-D DFT of the template $t(p, q)$. Using (8.31), the 2-D cross-correlation operation is converted to a simple multiplication in the transform domain. To avoid circular convolution problems, both $c_{x,t}(k, l)$ and $t(p, q)$ must be zero-padded appropriately before taking their 2-D DFTs. The 2-D IDFT of the result will then yield the desired 2-D cross-correlation in the spatial domain.

Figure 8.33 Filtered Pepper image using 2-D DCT ($\beta = 99.48\%$).

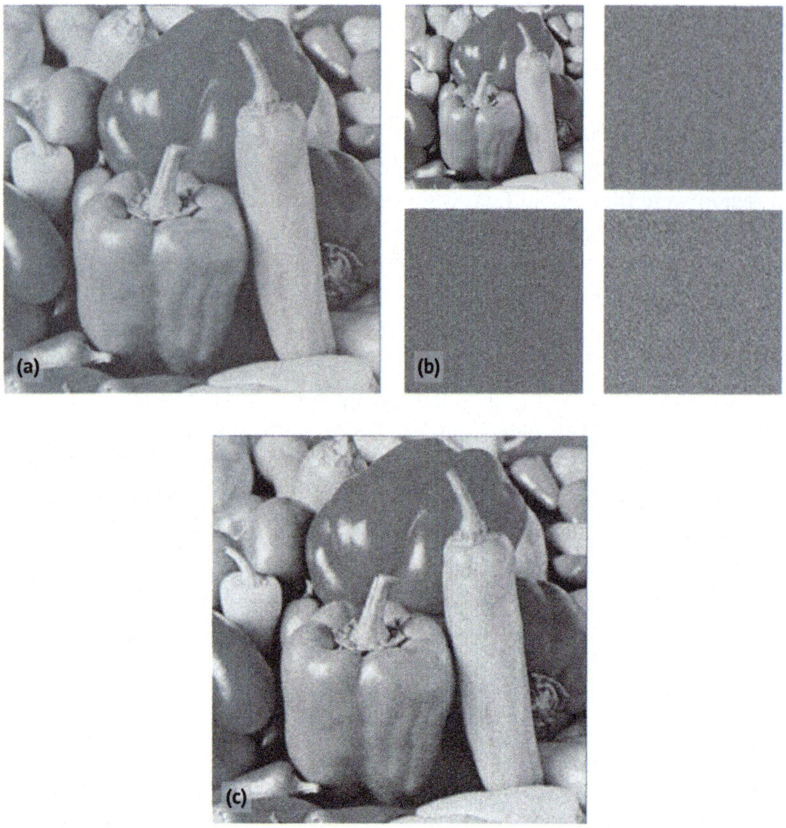

Figure 8.34 (a) Noisy Peppers image, (b) one-level 2-D DWT subimages of noisy Peppers, and (c) DWT-denoised Peppers image (SNR = 13.3 dB).

Figure 8.35 Root filtering operation.

8.3.4 Root Filtering and Cepstral Domain Processing

The idea behind root filtering and Cepstral (juxtaposition of the first four letters of the word spectral) is to apply a nonlinear mapping to the magnitude spectrum of an image in order to redistribute its energy in the transform domain while keeping the phase is unchanged. Several types of nonlinear mapping functions [1] can be employed for different tasks. In this section, we present two specific nonlinear mappings.

1. **Root Filter**
   In root filtering, the magnitude component is raised to a power $\alpha < 1$ while the phase component is retained, i.e.

$$Y(k, l) = |X(k, l)|^{\alpha} \exp(j\theta(k, l)), \qquad 0 \le \alpha \le 1. \tag{8.32}$$

   This operation, shown in Figure 8.35, de-emphasizes large magnitude frequency components while boosting the smaller ones. As a consequence, high-frequency components that correspond to edges and textural features and generally have small magnitudes are accentuated in comparison with those of the lower frequencies. This will yield an image with more pronounced edges. Figure 8.36a–d shows the original Baboon image and the root-filtered versions for α= 0, 0.5 and 0.75, respectively. As evident from these results, the high frequency features in the hair and whiskers are emphasized in the root-filtered images. Note that for $\alpha = 0$, the image in Figure 8.36b is reconstructed using the phase-only information.

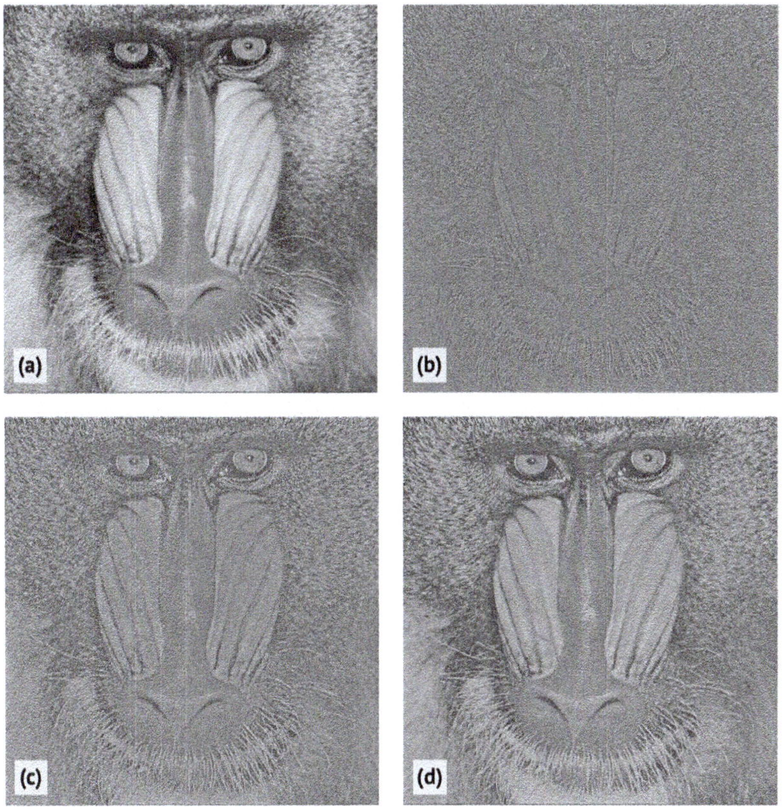

Figure 8.36 (a) Original Baboon image and (b–d) its root-filtered resultant image for $\alpha = 0$, 0.5, and 0.75, respectively.
Source: [4] / John Wiley & Sons / CC BY 3.0.

2. **Generalized Homomorphic Filter and Cepstral Domain Processing**
Another popular choice of nonlinear transformation consists of taking the natural log of the magnitude spectrum while keeping the phase unchanged, i.e.

$$C(k, l) = [\log |X(k, l)|] \exp(j\theta(k, l)). \tag{8.33}$$

However, in practice to avoid numerical issues of log operation, we use:

$$C(k, l) = [\log(a + b|X(k, l)|)] \exp(j\theta(k, l)) \tag{8.34}$$

where a and b control the shape of the log function. Typically $a = 1$ ensures that $X(k, l) = 0$ maps to $C(k, l) = 0$.

Now, the inverse unitary transform of $C(k, l)$, denoted by $c(m, n)$ is called the *generalized cepstrum* or *generalized Homomorphic transform* of the image $x(m, n)$. Figure 8.37a shows the entire operation, indicated by the operator $\mathcal{H}$, to map image $x(m, n)$ to its generalized cepstrum $c(m, n)$. As with the root filter, this operation tends to reduce the dynamic range of the components in the transform domain and increases the dynamic range in the spatial domain upon reconstruction. Consequently, it provides a certain amount of high-frequency enhancement since natural images tend to have small magnitudes at high spatial frequencies. Figure 8.37b shows the inverse generalized cepstrum to reproduce image $x(m, n)$ from its generalized cepstrum image $c(m, n)$. This inverse cepstral mapping is denoted by the operator $\mathcal{H}^{-1}$. Many operations including point or spatial operations can be applied in the cepstral domain for various specific image processing tasks. Figure 8.37c shows the block diagram of a typical cepstral-domain image processing system that consists of generalized cepstral transform, $\mathcal{H}$, inverse generalized cepstral transform, $\mathcal{H}^{-1}$, and a particular point or spatial operation in the cepstral domain.

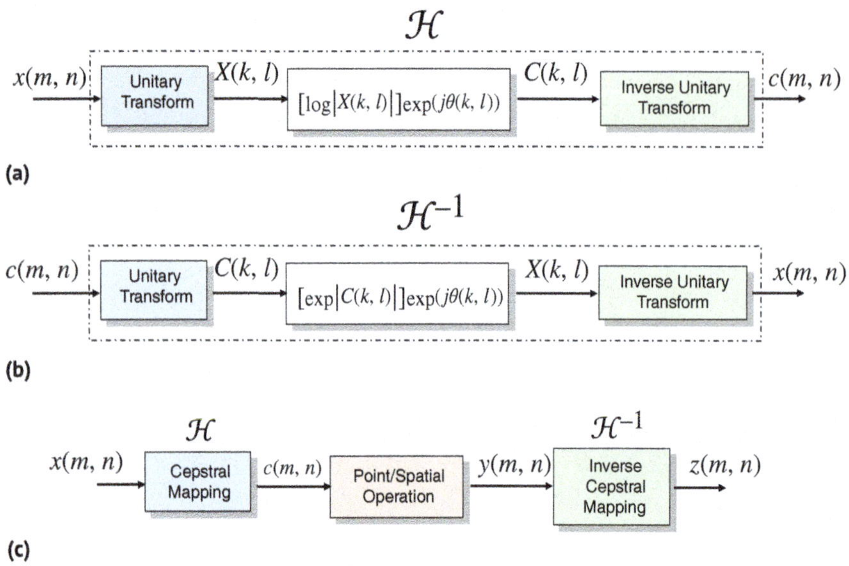

Figure 8.37 (a) Generalized cepstral transform, (b) inverse generalized cepstral transform, and (c) cepstral-domain image processing.

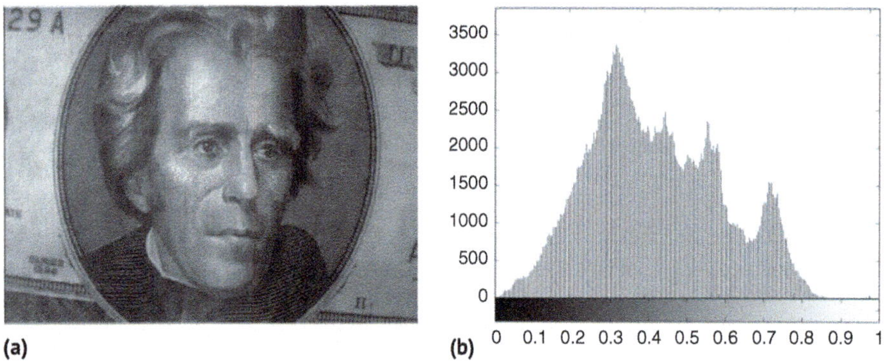

Figure 8.38 (a) Contrast-enhanced Dollar image using generalized homomorphic filtering, and (b) its histogram.

Here, we illustrate one particular application of the generalized cepstrum or Homomorphic operation for contrast enhancement. Unlike histogram manipulation methods, contrast enhancement using this method can be done without loss of gray level resolution. Figure 8.38a shows the result of applying contrast enhancement using linear grayscale mapping in the cepstral domain to the "Dollar" image in Figure 8.1. The histogram of the enhanced image is shown in Figure 8.38b. Comparing this result with those obtained using the grayscale mapping or histogram modeling methods in Section 8.1, clearly shows significantly better quality contrast enhancement.

8.4 Computer Projects

The computer projects in this chapter use Python libraries OpenCV, NumPy, Matplotlib, scikit-image, SciPy, PyWavelets, os. and math.

Computer Project 8.1: Generate the 8×8 image matrix shown in (8.35) that has eight gray levels. It is desirable to improve the relative contrast of the rectangular objects in this image with respect to the background.

$$\begin{bmatrix} 4 & 4 & 4 & 4 & 4 & 4 & 4 & 4 \\ 4 & 5 & 5 & 6 & 6 & 7 & 4 & 4 \\ 4 & 5 & 5 & 6 & 6 & 7 & 4 & 5 \\ 4 & 5 & 4 & 7 & 6 & 6 & 4 & 5 \\ 4 & 5 & 5 & 5 & 5 & 5 & 4 & 5 \\ 4 & 5 & 6 & 6 & 6 & 7 & 4 & 5 \\ 4 & 4 & 7 & 7 & 6 & 6 & 4 & 4 \\ 4 & 4 & 4 & 4 & 4 & 4 & 4 & 4 \end{bmatrix}. \tag{8.35}$$

Apply histogram equalization to this image, display the histograms of the original and equalized images, and comment on the results. Develop a histogram specification scheme that blends the background areas together while enhancing the contrast of the rectangular objects. Then, use an appropriate grayscale mapping to achieve the same goal and compare your results with those of the histogram modeling methods and discuss the advantages and limitations of each contrast enhancement method.

Computer Project 8.2: Scan a scratched or torn photograph, and display the image and its histogram. It is desirable to remove the effects of the scratch or tear as much as possible using point operations. Develop and implement a scheme to accomplish this task. Display the enhanced image and its histogram and comment on the results.

Computer Project 8.3: In this computer project, the objective is to design and implement different contrast enhancement operations in Section 8.1 to improve the visual appearance of the "Pepsi" image. Design a suitable contrast mapping and histogram specification procedure to improve the contrast of the region containing the Pepsi can while ensuring that the other parts of the image are not adversely affected. Evaluate the effectiveness of each approach, compare their performance, and discuss the advantages and disadvantages of each method.

Computer Project 8.4: The nonlinear transformation:

$$b_i = \text{Int}\left[\frac{x}{2^{p-i}}\right] - 2\,\text{Int}\left[\frac{x}{2^{p-i+1}}\right],$$

where Int$[a]$ is the integer part of quantity a and $L = 2^p$, extracts the ith bit in the binary representation of an image x. Using this equation, develop and implement an algorithm to extract all the 8 bit planes in the binary representation of the Baboon image. Determine which bit-planes are visually the more important ones.

Computer Project 8.5: The goal of this computer project is to study the effects of various spatial low-pass filtering operations to remove the effects of additive white noise from the "Peppers" image. Add white noise with mean zero and an appropriate variance to the original Peppers image to produce SNR of approximately 5 dB, where $\text{SNR}_{dB} = 10\log_{10}\frac{\sigma_x^2}{\sigma_n^2}$ and σ_x^2 and σ_n^2 are the variances of the original noise-free image and noise, respectively. Use this noisy Peppers image as the input to different 2-D low-pass filters with various mask sizes and filter coefficients to reduce the effects of additive noise. Compare the results of these filters in terms of SNR improvement and comment on the frequency characteristics of each filter.

Computer Project 8.6: Investigate the effectiveness of the median filter for removing salt & pepper noise from the corrupted version of the Peppers image. Add salt & pepper noise with varying density of 10%–70% in increments of 10%. Then, apply 2-D median filter to each corrupted image using 3×3, 5×5, and 7×7 mask sizes. Plot the variance of the residual noise in the processed images versus noise density (see Figure 8.19) for each mask size. Experimentally determine the maximum noise level that can be tolerated without producing significant artifacts.

Computer Project 8.7: Generate arbitrary 8×8 images in order to experimentally validate the following properties of the 2-D median filter.

1. Effects on thin lines and sharp corners.
2. Separability.
3. Nonlinearity.
4. Repeated application of the median filter leads to the root-filter result that is unchanged by further filtering. Explain the reasons.

Computer Project 8.8: Apply different edge extraction masks covered in Section 8.2.3 to the 8×8 images generated in Computer Project 8.7 that contain different edge structures. Evaluate and compare the performance of different high-pass filters in terms of their edge detection accuracy and noise sensitivity. You may add a small amount of noise to the images to investigate the behavior of each filter in presence of noise, particularly nearby the edges. Repeat your experiment using another image e.g. "Barbara." Compare the results of different edge extraction operators and comment on their performance.

Computer Project 8.9: Generate an 8×8 image that contains an arbitrary-shaped object. Use this object as a fixed target template and develop a 2-D matched filter to perform the following tasks.

1. Scale (in size) the object for different scaling factors and insert the scaled object in an image with uniform background. Study the effect of the scaling on the performance of the matched filter.
2. Repeat Part 1 by rotating the object in the image for different rotation angles, and then study the effect of rotation on the performance matched filter.
3. Provide a comprehensive evaluation of the spatial-domain matched filtering for object detection and localization, including its advantages and disadvantages.

Computer Project 8.10: In this computer project, we investigate the performance of homomorphic filtering for removing multiplicative speckle noise from the synthetic aperture radar (SAR) Farm image. This image was formed by mapping the complex-valued reflection coefficients to an intensity image. Due to the long tail of the distribution, a slight compression had to be made to fit the intensity levels to 256 gray levels. First, use spatial domain low-pass filters, e.g. spatial averaging filters, in the log domain and evaluate the results both visually and also quantitatively in terms of the peak SNR (PSNR) improvements (see Section 8.2.5). Then, repeat by applying homomorphic filtering using 2-D DFT. Compare the results and comment on the advantages and limitations of each method.

Computer Project 8.11: This computer project investigates the effectiveness of 2-D KL transform for noise removal. Partition the noisy Peppers image in Computer Project 8.5 into 8×8 nonoverlapping blocks and arrange each block into a 64-dimensional vector. Using the procedure in Section 6.4.1, compute the covariance matrix of the data and its eigenvalues and eigenvectors. Depending on the eigenvalues, construct a reduced KL transformation matrix $\boldsymbol{\Psi}_{\text{red}}^{H}$ and use it to map each data vector to its reduced dimensional feature space. Then, reconstruct the image block-by-block from the reduced dimensional features and using the 2-D inverse KL transform matrix $\boldsymbol{\Psi}_{\text{red}}$. Compute the SNR of the filtered image and comment on the noise removal ability of this transform domain approach. What are the advantages and disadvantages of the KL transform-based noise filtering comparing to the other unitary transforms?

Computer Project 8.12: Using the generalized cepstral-domain operation presented in Section 8.3.4, improve the contrast of the Pepsi image in Computer Project 8.3. Follow the procedure used in the experiment of Figure 8.38. Comment on the results and state your findings.

8.5 Problems

Problem 8.1: Explain why the contrast mapping function $f(\cdot)$ has to be monotonic. Can a non-monotonic function be used for contrast mapping?

Problem 8.2: A continuous image, X, has a Gaussian histogram with mean μ_X and variance σ_X^2. This image is mapped using a linear grayscale mapping $y = ax$.

a. Show that the histogram of the transformed image is also Gaussian and find the mean and variance of this histogram as a function of those of image X and parameter a.
b. Explain the effects of the parameter a on the mapped histogram.

Problem 8.3: A continuous image with bimodal histogram can be modeled by a mixture of two Gaussian functions with mean values μ_1 and μ_2 and variances σ_1^2 and σ_2^2, respectively. It is desirable to stretch the first mode of the histogram while compressing the second one. Devise a piecewise linear transformation function for this operation and determine the mean and variance associated with the two modes after the mapping.

Problem 8.4: A continuous image has a uniform histogram in the grayscale range of 0 to A. Determine a grayscale transformation function that maps this image to a new image with a uniform histogram in the range 0 to B.

Problem 8.5: The histogram of a continuous image, X, is given by $h_X(x) = Ae^{-x}$ where $x \in [0, b]$ and A is a normalization factor. We would like to map this image to a new image, Y, with the desired histogram $h_Y(y) = B \cos y$ where $y \in [0, b]$ and B is the normalization factor. Find the transformation function $y = f(x)$ that achieves this task.

Problem 8.6: It is known that the response of the human eye can be modeled as a logarithmic function of form $z = \alpha ln(x + \beta)$. If we want the eye to perform histogram equalization on an image, X, with histogram $h_X(x)$, determine the mapping function that should be applied to the image prior to viewing it.

Problem 8.7: A digital image with gray levels in the range $x_k \in [0, K - 1]$ has a uniform histogram. Determine a grayscale transformation function that maps this image to a new image with a uniform histogram in the range $x_l \in [0, L - 1]$. Consider both cases where $K \gtrless L$. What are the main differences between this case and the continuous case in Problem 8.4?

Problem 8.8: A digital image with K (even) levels of gray, $x_k \in [0, K - 1]$, is re-quantized to an image with $K/2$ levels of gray using the mapping $\{x_{2k}, x_{2k+1}\} \rightarrow x_{2k}$.

a. Determine the type of mapping function and histogram of the mapped image.
b. Find the expressions for the mean and variance of the mapped image.

Problem 8.9: Explain, both using theory and intuition, why the histogram equalization for digital images does not lead to a perfect uniform histogram.

Problem 8.10: Explain why the algorithm for the histogram specification yields the desired inverse mapping operation.

Problem 8.11: Suppose that image X is uniformly quantized into p bits. Show that the nonlinear point operation:

$$b_i = \text{Int}\left[\frac{x}{2^{p-i}}\right] - 2\,\text{Int}\left[\frac{x}{2^{p-i+1}}\right],$$

where Int$[a]$ represents integer part of quantity a and $L = 2^p$, extracts the ith bit in the binary representation of this image.

Problem 8.12: Show that the filtered residual white noise, $\overline{\eta}(m, n)$, in (8.9) has zero mean and variance $\sigma_{\overline{\eta}}^2 = \frac{\sigma_\eta^2}{N_W}$ where σ_η^2 is the variance of the noise $\eta(m, n)$ and N_W is the number of pixels in mask W.

Problem 8.13: The 5-point spatial averaging filter using Mask 3 in Figure 8.11 (see Example 8.2) is applied to an image $x(m, n)$ twice.

a. Find the coefficients of the equivalent filter that yields the same result in one pass. Is this a symmetric mask? What can be said about its noise removal ability?
b. Determine the magnitude response of this filter. How does the bandwidth compare to that of the original filter? Note: you can perform this comparison for the cross-section of the frequency response with plane $\Omega_1 = \Omega_2 = \Omega$ as in Example 8.2.

Problem 8.14: Repeat Problem 8.13 for 5-point Mask 2 in Figure 8.11.

Problem 8.15: Assume that the 2-D FIR filter in (8.8) has a symmetrical mask, i.e. $h(k, l) = h(-k, -l)$. Devise a scheme to perform the filtering operations more efficiently with almost half of the number of operations.

Problem 8.16: For each of the following 2-D sequences, determine the results of a 3×3 spatial averaging and 2-D median filters.

a. $\delta(m, n)$,
b. $u(m, n)$,
c. $\delta(m, n) - \delta(m - 1, n - 1)$.

Problem 8.17: Does the nonlinear property of 1-D median filter hold for two independent uniformly distributed signals $x_1(m)$ and $x_2(m)$ in the range $x_1 \in [0, A_1]$ and $x_2 \in [0, A_2]$, respectively? Hint: use the property that the PDF of the sum of two random variables is the convolution of their individual PDFs. Can this result be extended to the general case of arbitrary PDFs?

Problem 8.18: Determine if the 2-D median filter is a separable filter. That is, can it be implemented using two consecutive 1-D median filters?

Problem 8.19: Prove the validity of the number of comparisons in the median filter i.e. $\frac{3(N_W^2-1)}{8}$ for the mask containing N_W pixels (odd mask size).

Problem 8.20: Consider a 1-D signal $x(m)$, $m \in [1, M]$. Find the estimate $\hat{x}$ that minimizes the sum squared error or ℓ_2 norm of error vector $\underline{e}$ given by:

$$\varepsilon(\hat{x}) = \sum_{m=1}^{M} (x(m) - \hat{x})^2 = \sum_{m=1}^{M} e^2(m) = \underline{e}^T\underline{e} = ||\underline{e}||_2,$$

where $e(m) = x(m) - \hat{x}$ and $\underline{e} = [e(1)e(2)...e(M)]^T$ is the error vector. What is your conclusion about the result of this minimization and how do you compare that with the result in Example 8.3?

Problem 8.21: The standard median filter is a non-recursive operation since the output of the filter (median) is determined based on only the input data points. For a 1-D signal, this non-recursive median filter is given by:

$$y(m) = \text{median}\{x(m - N), ...x(m), ...x(m + M)\}$$

A recursive median filter can be defined as:

$$y(m) = \text{median}\{y(m - L), ...y(m - 1), x(m - N), ...x(m), ...x(m + M)\}$$

where $y(m - i)$ are the previously computed outputs. Determine the performance of the recursive median filter for impulsive noise removal in comparison with the non-recursive one.

Problem 8.22: Using approximation of the first-order derivatives, derive the expression for the 2-D discrete gradient operator from the 2-D gradient function:

$$|\nabla x(u, v)| = \sqrt{\left(\frac{\partial x(u, v)}{\partial u}\right)^2 + \left(\frac{\partial x(u, v)}{\partial v}\right)^2}.$$

Problem 8.23: Using the backward difference approximation, derive the expression for the 2-D discrete Laplacian from the 2-D Laplacian function. How does that differ from the results in Example 8.4?

$$\nabla^2 x(u, v) = \frac{\partial^2 x(u, v)}{\partial u^2} + \frac{\partial^2 x(u, v)}{\partial v^2}$$

Problem 8.24: Consider the following two 3×3 masks. Find the frequency spectra of these masks, plot the diagonal

−1	0	1
0	0	0
1	0	−1

0	1	0
−1	0	−1
0	1	0

cross-sections of their magnitude responses as functions of $\Omega = \Omega_1 = \Omega_2$ and comment on their frequency domain characteristics.

Problem 8.25:

a. Find the frequency spectra of the Robert's operators and plot the diagonal cross-sections of their magnitude response as a function of $\Omega = \Omega_1 = \Omega_2$. Explain the spatial and spectral characteristics of these operators.
b. Repeat Part (a) for Prewitt operators.
c. Repeat Part (a) for Sobel operators.

Problem 8.26: Consider a continuous image $x(u, v)$ and assume that this image is rotated by angle θ using the transformation:

$$\begin{bmatrix} u \\ v \end{bmatrix} = \begin{bmatrix} \cos\theta & -\sin\theta \\ \sin\theta & \cos\theta \end{bmatrix} \begin{bmatrix} u' \\ v' \end{bmatrix}$$

where u, v are the original coordinates and u', v' are the new mapped coordinates after the rotation. Show that the following operators are rotation invariant.

a. The gradient operator in $|\nabla x(u, v)|$ in Problem 8.22.
b. The Laplacian $\nabla^2 x(u, v)$ in Example 8.4.
c. Explain the importance of the rotation invariance property.

Problem 8.27: Derive the 2-D Fourier transform of the LoG operator $g(u, v)$ in Eq. (8.20).

Problem 8.28: Consider a signal $x(m)$ that contains a rectangular component, $t(m) = u(m) - u(m - M)$, of width M in additive white noise $\eta(m)$,

$$x(m) = t(m - m_0) + \eta(m).$$

The noise has a zero mean and the variance σ_η^2. Matched filtering with template $t(m)$ is then applied to detect and localize the known signal.

a. Determine the cross-correlation function $c_{x,t}(k)$.
b. Describe the effects of decreasing M on the result. Can you still detect the location of the signal when $M = 1$, i.e. $t(m) = \delta(m)$? Why?

Problem 8.29: In Problem 8.28, if the signal to be detected is a scaled and shifted version of the template in additive noise, that is, $x(m) = t\left(\frac{m-m_0}{a}\right) + \eta(m)$ where a is the scaling factor, determine the cross-correlation function $c_{x,t}(k)$ for some choices of the parameter a. What conclusion can you draw?

Problem 8.30: Show that the location at which the maximum of the cross-correlation in (8.23) is achieved corresponds to the minimum of the sum squared mismatching error:

$$\epsilon(k, l) = \sum_m \sum_n [x(m, n) - t(m - k, n - l)]^2.$$

Problem 8.31: Assume that an image $x(m, n)$ contains a 3×3 square-shaped object at a spatial location (m_0, n_0) within the image with additive white noise:

$$z(m, n) = t(m - m_0, n - n_0) + \eta(m, n)$$

where $t(m, n)$ represents the square template and $\eta(m, n)$ is a zero mean additive white noise with variance σ_η^2. Matched filtering is then applied to detect and localize the object.

a. Determine the cross-correlation function $c_{x,t}(k, l)$ and plot the diagonal cross-section of this function for $k = l$. Where is the peak of this function located?
b. Find the frequency response of this matched filter and comment on the performance.

Problem 8.32: Consider an ideal 2-D low-pass filter with the following frequency response:

$$H(e^{j\Omega_1}, e^{j\Omega_2}) = \begin{cases} 1, & |\Omega_1| \le \beta\pi, \ |\Omega_2| \le \beta\pi \\ 0, & \text{otherwise} \end{cases}$$

where $\beta < 1$ represents the bandwidth.

a. Determine the expression for the 2-D impulse response of this filter.
b. If before taking 2-D DFT the frequency response is sampled in the region $[-\pi, \pi] \times [-\pi, \pi]$ using:

$$H(k, l) = H(e^{j\Omega_1}, e^{j\Omega_2})|_{\Omega_1 = \frac{2\pi k}{N}, \Omega_2 = \frac{2\pi l}{N}}$$

Determine the impulse response of the frequency-sampled filter. Explain the relation between the results in Parts (a) and (b).
c. A noisy image corrupted with additive white noise is filtered using the filter in part (a). If the SNR of the original noisy image was 0 dB, determine the SNR after the filtering. Note that it is assumed that the filtering does not affect signal.

Problem 8.33: Consider an ideal 2-D low-pass filter with the following frequency response:

$$H(e^{j\Omega_1}, e^{j\Omega_2}) = \begin{cases} 1, & |\Omega_1| \le \frac{\pi}{4}, \ |\Omega_2| \le \frac{\pi}{4} \\ 0, & \frac{\pi}{4} < |\Omega_1| \le \pi, \ \frac{\pi}{4} < |\Omega_2| \le \pi \end{cases}$$

This filter is then used as a prototype to construct new filters by manipulating the PSF $h(m, n)$. Determine the frequency response of the following filters whose PSF is:

a. $h_1(m, n) = h(2m, 2n)$,
b. $h_2(m, n) = \begin{cases} h(\frac{m}{2}, \frac{n}{2}), & \text{for} \quad m = 2p, \ n = 2q \quad p, q \in \mathbb{Z} \\ 0, & \text{otherwise,} \end{cases}$,
c. $h_3(m, n) = (-1)^{m+n} \, h(m, n)$.

Problem 8.34: Find the expression for the frequency response $H_1(e^{j\Omega_1}, e^{j\Omega_2})$ of a 2-D FIR filter with symmetric PSF, i.e. $h_1(m, n) = h(-m, -n)$. How can this property be exploited in the transform domain filtering? If the original filter is low-pass, what type of filter does this mapping generate?

Problem 8.35: Repeat Problem 8.34 when the PSF possesses:

a. Alternate symmetry $h_1(m, n) = (-1)^{m+n} h(-m, -n)$,
b. $h_2(m, n) = (-1)^m h(-m, n)$,
c. $h_3(m, n) = (-1)^n h(m, -n)$.

Problem 8.36: Consider an image $x(m, n)$ corrupted by an array of line structures, i.e.

$$z(m, n) = x(m, n) + \zeta(m, n)$$

where $\zeta(m, n)$ represents an array of lines with slope a and distance d, i.e.

$$\zeta(m, n) = \sum_{p=-\infty}^{\infty} \delta(m - a, n - p\,d\,a)$$

a. Sketch the sequence $\zeta(m, n)$.
b. Determine the DSFT of $\zeta(m, n)$. What kind of conclusion can you draw from this result?
c. Design a 2-D filter that removes the effects of the line structures from the image.

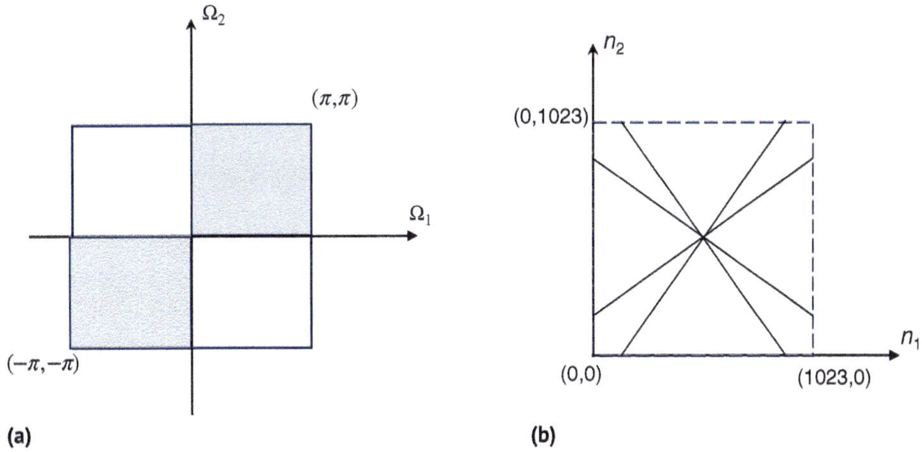

Figure 8.39 (a) Frequency response of the filter and (b) input image in Problems 8.37 and 8.38.

Problem 8.37: Consider a 2-D sequence of the form [21]:

$$x(m,n) = \sum_{p=0}^{N-1} \delta(m - p\,d_1, n - p\,d_2)$$

a. Sketch this sequence.
b. Determine and sketch the DSFT of this sequence.
c. Now consider a 2-D LSI digital filter whose frequency response is shown Figure 8.39a where the shaded region has a value 1 and the blank region has a value 0. The input to this filter is a sampled version of image in Figure 8.39b where white samples are 0, while black samples on the lines are 1. Using the results in Parts (a) and (b), determine and sketch the output of the filter.

Problem 8.38: Repeat Problem 8.37, Part (c), if the positions of the blank and shaded regions in the frequency response of Figure 8.39a are swapped.

Problem 8.39: Repeat Problem 8.28 in the frequency domain by determining $T(k)$, $X(k)$, and the cross-power spectrum using DFT. How do you choose the DFT sizes?

Problem 8.40: Repeat Problem 8.31 in the frequency domain by determining $T(k,l)$, $X(k,l)$, and the cross-power spectrum using 2-D DFT. How do you choose the 2-D DFT sizes?

References

1 A. K. Jain, "*Fundamentals of Digital Image Processing*," Prentice Hall, 1989.
2 GOES Environmental Satellites, EW-2015-04-049-JPL.
3 S. Miller and D. Childers, "*Probability and Random Processes: With Applications to Signal Processing and Communications*," Academic Press, 2nd Edition, 2012.
4 Y.-S. Juang, L.-T. Ko, J.-E. Chen, Y.-S. Shieh, T.-Y. Sung, and H.-C. Hsin, "Histogram modification and wavelet transform for high performance watermarking." *Mathematical Problems in Engineering*, Article 164869, 2012. https://doi.org/10.1155/2012/164869.
5 D. Battikh, S. El Assad, T. M. Hoang, B. Bakhache, O. Deforges, and M. Khalil, "Comparative study of three steganographic methods using a chaotic system and their universal steganalysis based on three feature vectors." *Entropy*, vol. 21, no. 8, p. 748, 2019. https://doi.org/10.3390/e21080748.
6 F. Argenti, A. Lapini, L. Alparone, and T. Bianchi, "A tutorial on speckle reduction in synthetic aperture radar images," *IEEE Geoscience and Remote Sensing Magazine*, pp. 6–35, 2013.

7 A. P. Lyons, D. A. Abraham, and S. F. Johnson, "Modeling the effect of seafloor ripples on synthetic aperture sonar speckle statistics," *IEEE Journal of Oceanic Engineering*, vol. 35, pp. 242–249, 2010.

8 M. Tur, K. C. Chin, and J. W. Goodman, "When is speckle noise multiplicative?" *Applied Optics*, vol. 21, pp. 1157–1159, 1982.

9 P. Jain and V. Tyagi, "A survey of edge-preserving image denoising methods," *Information System Frontier*, vol. 18, pp. 159–170, 2016.

10 F. Zhu, Z. Liang, X. Jia, L. Zhang, and Yizhou Yu, "A benchmark for edge-preserving image smoothing," *IEEE Transactions on Image Processing*, vol. 28, pp. 3556–3570, 2019.

11 S. Perreault and P. Hebert, "Median filtering in constant time," *IEEE Transactions on Image Processing*, vol. 16 pp. 2389–2394, 2007.

12 B. Weiss, "Fast median and bilateral filtering," *ACM Transactions on Graph*, vol. 25, no. 3, pp. 519–526, 2006.

13 W. K. Pratt, "*Digital Image Processing*," Wiley-Interscience, 4th Edition, 2007.

14 D. Marr and E. Hildreth, "Theory of edge detection," *Proceedings of the Royal Society of London (Series B), Biological Sciences*, vol. 207, pp. 187–217, 1980.

15 S. R. Gunn, "On the discrete representation of the Laplacian of Gaussian," *Pattern Recognition*, vol. 32, pp. 1463–1472, 1999.

16 R. C. Gonzalez and R. E. Woods, "*Digital Image Processing*," 4th Edition, Pearson, 2018.

17 J. Canny, "A computational approach to edge detection," *IEEE Transactions on Pattern Recognition and Machine Intelligence (PAMI)*, vol. 8, pp. 679–698, 1986.

18 V. S. Frost, J. A. Stiles, K. S. Shanmugan, and J. C. Holtzman, "A model for radar images and its application to adaptive digital filtering of multiplicative noise," *IEEE Transactions on Pattern Analysis and Machine Intelligence*, vol. 4, pp. 157–166, 1982.

19 D. E. Dudgeon and R. M. Mersereau, "*Multidimensional Digital Signal Processing*," Prentice-Hall, 1984.

20 Y. Hu, C. Zhu, and Z. Wang, "An improved piecewise linear chaotic map based image encryption algorithm." *The Scientific World Journal*, Article 275818, 2014. https://doi.org/10.1155/2014/275818.

21 D. R. Andrews, "*Ultrasonics and Acoustics-multiplicative and Additive Noise*," Third Edition, Encyclopedia of Physical Science and Technology, 2003.

22 R. King, M. Ahmadi, R. Naguib, A. Kwabwe, and M.R. Azimi-Sadjadi, "*Digital Filtering in One and Two Dimensions-Design and Applications*," Plenum Press, 1989.

23 J. S. Lim, "*Two-dimensional Signal and Image Processing*," Prentice-Hall, 1990.

9

Image Modeling

In many image processing applications, it is often of interest to represent an ensemble of images with similar first- and second-order statistics using a linear model of appropriate order. Two-dimensional (2-D) finite-order auto-regressive (AR) or auto-regressive moving average (ARMA) models are typically employed for parametric representations of image fields. These models have found numerous applications in digital image processing, including image restoration [1–4], image encoding [5], and texture analysis [6, 7]. Using these models, the objective is to represent the image field as the output of a 2-D linear system with a white or colored random process as input. The 2-D linear system is described by a linear difference equation (i.e. an ARMA model), whose coefficients are obtained using the image covariance function or, equivalently, the spectral density function (SDF). The procedure in the 2-D case differs from that in the 1-D case in two major ways. First, in the 1-D case, where time is normally the independent variable, causality is a rigid requirement for the physical realizability of the model, while in the 2-D case, where variables are spatial, the region of support (ROS) of the model can be causal, semicausal or noncausal, leading to initial-value, initial-boundary-value, or boundary-value problems [1], respectively. Second, in contrast to the 1-D case where any rational SDF of finite order can be factored into a causal and stable transfer function, in the 2-D case, there is in general no spectral factorization of the SDF that produces a causal and stable transfer function and corresponding 2-D linear difference equation.

The aim of this chapter is to study the problem of image modeling and related issues of causality, stability, and spectral factorization. To this end, we first review 1-D causal and noncausal models and their properties. Then, several 2-D AR and ARMA models with different prediction geometries are introduced. Examples are included to illustrate the applications of these models to image processing. Most of the technical materials of this chapter were extracted from Chapter 10 in [8].

9.1 1-D Linear Models

The parametric approach to spectral estimation involves three steps. In the first step, an appropriate parametric time-series model, such as an auto-regressive (AR), moving average (MA) or combined auto-regressive moving average (ARMA) model, is to be selected to represent the measured data. The choice of one of the three models is dictated by the shape and characteristics of the data frequency spectrum. The AR model is appropriate for spectra with sharp peaks but no deep valleys. The MA model is used for spectra with deep valleys but no sharp peaks. Finally, ARMA models are used to represent spectra with peaks and valleys. Step two involves finding the estimates of the model parameters. In the third step, the estimated parameters are used to match the given SDF (validation step). The computational burden associated with estimating AR model parameters is often significantly lower than that of the MA and ARMA models. Moreover, the principle of parsimony precludes the use of a model that has a large number of parameters. Consequently, low-order AR models are often preferred. In this section, these three models are introduced for modeling 1-D random processes.

9.1.1 1-D Causal Representation

Let $y(k)$ represent a sample function of a discrete-time, zero-mean wide-sense stationary (WSS) process $Y(k)$. Then, a causal (one-sided) representation of the form:

$$y(k) = \sum_{i=1}^{M} a_i y(k-i) + \sum_{i=0}^{N} b_i x(k-i) \tag{9.1}$$

is called an ARMA model with coefficients a_i and b_i. The driving process $X(k)$, with sample function $x(k)$, is a white process with zero mean, i.e. $E[X(k)] = 0$, and autocorrelation $R_{XX}(k-l) = E[X(k)X(l)] = \sigma_X^2 \delta(k-l)$, where σ_X^2 is the variance and $\delta(\cdot)$ represents the Kronecker delta function.

The model parameters a_i define the AR portion of the ARMA model, while the parameters b_i correspond to the moving average portion. The notation ARMA(M, N) is often used to indicate an ARMA model of the form in (9.1).

The system transfer function, which is obtained by taking z-transform (see Appendix A) of (9.1), has a rational form, $H(z) = \frac{B(z)}{A(z)}$, where the polynomials are $A(z) = 1 - \sum_{i=1}^{M} a_i z^{-i}$ and $B(z) = \sum_{i=0}^{N} b_i z^{-i}$ with $b_0 = 1$. The transfer function $H(z) = 1 + \sum_{i=1}^{\infty} h(i) z^{-i}$ is the z-transform of the impulse response sequence $h(i)$. This impulse response is the solution to the equation $A(z)H(z) = B(z)$, or $h(n) = \sum_{i=1}^{M} a_i h(n-i) + \delta(n) + \sum_{i=1}^{M} b_i \delta(n-i)$. Note that for $n > M$, this solution is $h(n) - \sum_{i=1}^{M} a_i h(n-i)$, so the impulse response is AR on its tails. Both $A(z)$ and $B(z)$ are assumed to have all of their roots within the unit circle in the z-plane to ensure that $H(z)$ is a BIBO stable and minimum phase [9]. As shown in Chapter 2, since $Y(k)$ is assumed to be a WSS process, its SDF, $S_{YY}(z) = \sum_{n=-\infty}^{\infty} R_{YY}(n) z^{-m}$ can be expressed as:

$$S_{YY}(z) = S_{XX}(z)H(z)H(z^{-1}) = S_{XX}(z)\frac{B(z)B(z^{-1})}{A(z)A(z^{-1})} \tag{9.2}$$

where $R_{YY}(n) = E[Y(k)Y(k-n)]$ is the autocorrelation function (ACF) of $Y(k)$, and SDF $S_{XX}(z) = \delta_X^2$ since $X(k)$ is a white process.

Let us consider two special cases.

i. **MA Model:** If all AR parameters are zero except $a_0 = 1$, then (9.1) becomes a MA process of order N, or simply MA(N):

$$y(k) = \sum_{i=0}^{N} b_i x(k-i). \tag{9.3}$$

This model is also called all-zero since the transfer function, $H(z) = B(z)$ has no poles. The corresponding SDF in (9.2) becomes:

$$S_{YY}(z) = S_{xx}(z)B(z)B(z^{-1}). \tag{9.4}$$

ii. **AR Model:** If all MA parameters are zero except $b_0 = 1$, then we have:

$$y(k) = \sum_{i=1}^{M} a_i y(k-i) + x(k) \tag{9.5}$$

which is an AR(M) process. This model is also called an all-pole model, since $H(z) = 1/A(z)$ has no zeros. The SDF for this model is:

$$S_{YY}(z) = \frac{S_{XX}(z)}{A(z)A(z^{-1})}. \tag{9.6}$$

We note that in (9.2) and (9.6) the denominator $A(z)A(z^{-1})$ is a polynomial in z of order $2M$. The zeros of this polynomial occur in reciprocal pairs, i.e. if z_k is a root of $A(z)$, then $1/z_k$ is a root of $A(z^{-1})$. Thus, given the SDF of the data, spectral factorization (see Section 2.2.7) aims to decompose the poles and zeros of the SDF into those that are within the unit circle, and those that are outside, leading to minimum-phase and maximum-phase transfer functions, respectively. The minimum phase transfer function, $H(z)$, is chosen to represent the data, yielding a stable, causal 1-D ARMA or AR model. The methods commonly used to accomplish this include the Wiener-Doob method and the linear prediction method [10, 11]. The linear prediction method fits successively higher-order models to the given autocorrelation by solving a finite set of Toeplitz equations. Under some conditions, the SDFs realized by the models can be shown to converge uniformly to the given SDF.

9.1.1.1 Parameter Estimation for Causal Models

Consider the ARMA(M,N) model in (9.1) where M and N are known. Multiplying both sides by $Y(k-n)$ and taking the expectation yields:

$$R_{YY}(n) = \sum_{i=1}^{M} a_i R_{YY}(n-i) + \sum_{i=0}^{M} b_i R_{XY}(n-i) \tag{9.7}$$

where $R_{XY}(n)$ is the cross-correlation function between $X(k)$ and $Y(k)$, and can be expressed in terms of impulse response $h(k)$ using the convolution sum, i.e.

$$\begin{aligned} R_{XY}(n) = E[X(i+n)Y(i)] &= E\left[X(i+n)\left[\sum_{k=0}^{\infty} h(k)X(i-k)\right]\right] \\ &= \sum_{k=0}^{\infty} h(k)R_{XX}(n+k), \qquad h(0)=1. \end{aligned} \tag{9.8}$$

Since $X(k)$ is assumed to be a white process $R_{XX}(n+k) = \sigma_X^2 \delta(n+k)$, then:

$$R_{XY}(n) = \begin{cases} 0, & n > 0 \\ \sigma_X^2, & n = 0 \\ \sigma_X^2 h(-n), & n < 0 \end{cases} \tag{9.9}$$

Substituting into (9.7), the ACF $R_{YY}(n)$ becomes:

$$R_{YY}(n) = \begin{cases} \sum_{i=1}^{M} a_i R_{YY}(n-i), & n > N \\ \sum_{i=1}^{M} a_i R_{YY}(n-i) + \sigma_X^2 \sum_{i=n}^{N} b_i h(i-n), & 0 \le n \le N \\ R_{YY}(-n), & n < 0 \end{cases} \tag{9.10}$$

The AR parameters can be estimated separately from the MA parameters by solving a system of linear equations for $N+1 \le n \le N+M$ known as the *Yule-Walker* equation [10, 11]:

$$\begin{bmatrix} R_{YY}(N) & R_{YY}(N-1)\cdots & R_{YY}(N-M+1) \\ R_{YY}(N+1) & R_{YY}(N)\cdots & R_{YY}(N-M+2) \\ \vdots & & \vdots \\ R_{YY}(N+M-1) & R_{YY}(N+M-2)\cdots & R_{YY}(N) \end{bmatrix} \begin{bmatrix} a_1 \\ a_2 \\ \vdots \\ a_M \end{bmatrix} = \begin{bmatrix} R_{YY}(N+1) \\ R_{YY}(N+2) \\ \vdots \\ R_{YY}(N+M) \end{bmatrix}, \tag{9.11}$$

or simply $\mathbf{R}_Y \underline{a} = \underline{r}_Y$ where $\mathbf{R}_Y$ is the correlation matrix which is positive-definite (PD) and Toeplitz, $\underline{a} = [a_1\ a_2\ ...\ a_M]^T$ is the parameter vector, and $\underline{r}_Y = [R_{YY}(N+1)\ R_{YY}(N+2)\ ...\ R_{YY}(N+M)]^T$ is a vector of auto-correlations. There exist computationally efficient algorithms [12] for solving the Toeplitz-based system of the linear equation in (9.11). In practice, the sample auto-correlation $r_{YY}(n), n \in [N+1, N+M]$ is computed from the data samples:

$$r_{YY}(n) = \frac{1}{P} \sum_{i=1}^{P} y(i) y(i-n)$$

and used in (9.11) instead of the theoretical auto-correlations $R_{YY}(n)$. The number of data samples P is assumed to be large in order to avoid bias in the estimates. Once AR parameters are estimated from the Yule-Walker equation, the unknown variance of the driving process σ_X^2 can be estimated by multiplying the AR model by $Y(n)$, taking the expectation, and using (9.9), i.e. $r_{XY}(0) = \hat{\sigma}_X^2$ which gives:

$$\hat{\sigma}_X^2 = r_{YY}(0) - [r_{YY}(1) \cdots r_{YY}(M)]\ \underline{\hat{a}}. \tag{9.12}$$

The parameters of the MA part can be estimated using an innovation algorithm or, more efficiently, using maximum likelihood (ML) and least square estimations [11, 13]. However, since AR models are used more extensively in signal/image processing applications, we develop our results primarily for these models. For AR models, the ACF in (9.10) simplifies to:

$$R_{YY}(n) = \begin{cases} \sum_{i=1}^{M} a_i R_{YY}(n-i), & n > 0 \\ \sum_{i=1}^{M} a_i R_{YY}(n-i) + \sigma_X^2, & n = 0 \\ R_{YY}(-n), & n < 0, \end{cases} \tag{9.13}$$

which can be arranged in matrix form as:

$$\begin{bmatrix} R_{YY}(0) & R_{YY}(-1) & \cdots & \cdots & R_{YY}(-M) \\ R_{YY}(1) & R_{YY}(0) & \cdots & \cdots & R_{YY}(-M+1) \\ \vdots & & & & \vdots \\ & & \ddots & & \\ \vdots & & & \ddots & \vdots \\ R_{YY}(M) & R_{YY}(M-1) & \cdots & \cdots & R_{YY}(0) \end{bmatrix} \begin{bmatrix} 1 \\ -a_1 \\ \vdots \\ \vdots \\ -a_M \end{bmatrix} = \begin{bmatrix} \sigma_X^2 \\ 0 \\ 0 \\ \vdots \\ 0 \end{bmatrix} \tag{9.14}$$

This matrix equation is known as the AR Yule-Walker equation [11, 13]. Note that although σ_X^2 is also unknown, we intentionally kept the Toeplitz structure of matrix $\mathbf{R}_Y$ intact for efficient computation of $\mathbf{R}_Y^{-1}$. Now, assuming that $\mathbf{R}_Y^{-1}$ is computed, the solutions are given by:

$$\hat{\sigma}_X^2 = \frac{1}{[\mathbf{R}_Y^{-1}]_{1,1}}, \quad \text{and} \quad \hat{a}_i = -\hat{\sigma}_X^2\ [\mathbf{R}_Y^{-1}]_{i+1,1}, \quad i \in [1, M] \tag{9.15}$$

where $[\mathbf{R}_Y^{-1}]_{i,j}$ is the $(i, j)^{\text{th}}$ element of the matrix $\mathbf{R}_Y^{-1}$. Again, it must be mentioned that the sample auto-correlations $r_{YY}(n)$ are used instead of $R_{YY}(n)$ in the Yule-Walter equation to compute estimates of $\hat{\sigma}_X^2$ and AR parameters $\hat{a}_i,\ i \in [1, M]$.

9.1.1.2 Convergence Property

An important consequence of the above result is that any given SDF can be closely approximated by a finite-order AR spectrum [14]. In other words, if the SDF, $S_{YY}(z)$, is *positive analytic*[1] and if $\hat{a}_i(M)$, $i \in [1, M]$, and $\hat{\sigma}_X^2(M)$ represent the solution of (9.14) for an AR model of order M then:

$$\lim_{M\to\infty}\left[\frac{\hat{\sigma}_X^2(M)}{A_M(z)A_M(z^{-1})}\right] = S_{YY}(z) \tag{9.16}$$

where $A_M(z)$ is the same as $A(z)$ defined before but with a_i replaced by $\hat{a}_i(M)$. That is, as $M \to \infty$ the solutions of (9.14) provide exact spectral factorization of $S_{YY}(z)$. In general, (9.14) can be used to find a rational all-pole spectral approximation of an arbitrary positive SDF.

Example 9.1 *Consider a zero-mean WSS process with a two-sided ACF $R_{YY}(n) = E[Y(i+n)Y(i)] = \sigma_Y^2\rho^{|n|}$, $\sigma < \rho < 1$. Find an AR model that represents this process.*

Solution: The SDF of $Y(k)$ is given by:

$$\begin{aligned}S_{YY}(z) &= \sum_{n=-\infty}^{\infty} R_{YY}(n)z^{-n}\\ &= \sigma_Y^2\left(\frac{1}{1-\rho z^{-1}} + \frac{1}{1-\rho z} - 1\right) = \frac{\sigma_y^2(1-\rho^2)}{(1-\rho z^{-1})(1-\rho z)}.\end{aligned}$$

Applying the spectral factorization in (9.6) we obtain the model transfer function $H(z) = \frac{Y(z)}{X(z)} = \frac{1}{(1-\rho z)}$. The stable first-order AR model corresponding to this transfer function is:

$$y(k) = \rho y(k-1) + x(k) \qquad \text{and} \qquad \sigma_X^2 = \sigma_Y^2(1-\rho^2).$$

This first-order model exactly represents the SDF.

9.1.1.3 Causal Minimum Variance Prediction

The conditional mean estimate of $Y(k)$, conditioned on the past $Y(i)$, $i \in [k-M, k-1]$, is

$$\hat{Y}(k) \equiv E\Big[Y(k)|Y(i), i \in [k-M, k-1]\Big] = \sum_{i=1}^{M} a_i Y(k-i). \tag{9.17}$$

This is the best mean-squared predictor [13] of $Y(k)$ based on the past M samples. It corresponds to the minimum variance predictor that minimizes the mean-squared error (MSE):

$$\epsilon = E\left[\left(Y(k) - \hat{Y}(k)\right)^2\right]. \tag{9.18}$$

To minimize (9.18), we take partial derivative with respect to a_l and set the result to zero. This gives $E[(Y(k)-\hat{Y}(k))Y(k-l)] = E[X(k)Y(k-l)] = 0$ which is known as the orthogonality principle. That is, the prediction error $(Y(k)-\hat{Y}(k))$ or $X(k)$ is orthogonal to the past samples $Y(k-i)$. Since $\hat{Y}(k)$ is expressed linearly in terms of $Y(k-i)$ via relation in (9.17), it follows that $E[X(k)\hat{Y}(k)] = 0$, i.e. the prediction error is also orthogonal to the predicted signal $\hat{Y}(k)$. As a consequence of this orthogonality principle, we get the following normal equations:

$$R_{YY}(k) - \sum_{i=1}^{M} a_i R_{YY}(k-i) = \sigma_X^2\delta(k), \qquad 0 \le k \le M, \tag{9.19}$$

1 A positive analytic SDF is a nonnegative function in frequency that has a convergent power series expansion.

which are equivalent to the AR Yule-Walker equations in (9.14). This important result indicates that the Yule-Walker equation provides the minimum variance solution for the AR model parameters, and hence gives the best linear predictor in the MSE sense.

9.1.2 1-D Noncausal Representation

In this section, some results on 1-D noncausal representation [14–16] are presented together with related issues such as stability and correlation matching. The intent is to provide insight into noncausal representation, which commonly arises in image modeling. It must be mentioned that the 1-D noncausal models are important in their own right, as they also have applications in image restoration and data compression.

A noncausal AR(M) representation for a zero-mean discrete-time r.p. $Y(k)$ is:

$$y(k) = \sum_{\substack{i=-M \\ i\neq 0}}^{M} a_i y(k-i) + x(k) \tag{9.20}$$

where the autoregression terms involve all past and future samples for $i \in [-M, M]$, $i \neq 0$ and $x(k)$ is a realization of the prediction error process $X(k)$. The associated SDF of $Y(k)$ can be written as:

$$S_{YY}(z) = \frac{S_{XX}(z)}{A(z)A(z^{-1})} \tag{9.21}$$

where $A(z) = 1 - \sum_{\substack{i=-M \\ i\neq 0}}^{M} a_i z^{-i}$. However, since $A(z^{-1}) = A(z)$ and $S_{XX}(z) = \sigma_X^2 A(z)$ [14, 15], then (9.21) becomes simply $S_{YY}(z) = \frac{\sigma_X^2}{A(z)}$. This result indicates the fact that noncausal models do not require spectral factorization. This will become clear in Example 9.2.

9.1.2.1 Parameter Estimation for Noncausal Models

Using the orthogonality principle $E[Y(k)X(k-n)] = 0$ for $n \neq 0$, we obtain the normal equations:

$$R_{YY}(n) = \sum_{\substack{i=-M \\ i\neq 0}}^{M} a_i R_{YY}(n-i) + \sigma_X^2 \delta(n), \quad |n| \leq M. \tag{9.22}$$

Plugging $n = 0, \cdots 2M$, (9.22) can be arranged in matrix form:

$$\begin{bmatrix} R_y(0) & R_y(-1)\cdots & R_y(-2M) \\ R_{YY}(1) & R_{YY}(0)\cdots & R_{YY}(-2M+1) \\ \vdots & & \vdots \\ R_{YY}(2M) & R_{YY}(2M-1)\cdots & R_{YY}(0) \end{bmatrix} \begin{bmatrix} -a_{-M} \\ -a_{-M+1} \\ \vdots \\ -a_{-1} \\ 1 \\ -a_1 \\ \vdots \\ -a_M \end{bmatrix} = \begin{bmatrix} 0 \\ 0 \\ \vdots \\ 0 \\ \sigma_X^2 \\ 0 \\ \vdots \\ 0 \end{bmatrix} \tag{9.23}$$

or in compact form $\mathbf{R}_Y \underline{a} = \sigma_X^2 \underline{1}_{M+1}$ where $\underline{1}_{M+1} = [0 \ \cdots \ 0\,1\,0 \ \cdots \ 0]^T$ is a unit vector with a 1 at the $(M+1)^{\text{th}}$ element, $\underline{a} = [-a_{-M} \ \cdots \ -a_{-1}\ 1\ -a_1 \ \cdots \ -a_M]^T$ and $\mathbf{R}_Y$ is the correlation matrix of the data. If $\mathbf{R}_Y$ is positive definite, there is a unique solution. Note that here $a_i = a_{-i}$, $i \in [1, M]$. This is due to the fact that $\mathbf{R}_Y^{-1}$ is persymmetric (symmetric about the cross diagonal) and $\underline{a}$ is proportional to the middle column of $\mathbf{R}_Y^{-1}$. As in the causal case, assuming that $\mathbf{R}_Y^{-1}$ is computed, we have:

$$\hat{\sigma}_X^2 = \frac{1}{[\mathbf{R}_Y^{-1}]_{M+1,M+1}}, \quad \text{and} \quad \hat{a}_i = -\hat{\sigma}_X^2\,[\mathbf{R}_Y^{-1}]_{i+M+1,M+1}, \quad \hat{a}_{-i} = \hat{a}_i \tag{9.24}$$

where $[\mathbf{R}_Y^{-1}]_{i,j}$ is the $(i, j)^{\text{th}}$ element of the matrix $\mathbf{R}_Y^{-1}$.

9.1.2.2 Convergence Property

Let $\hat{a}_i(M)$, $i \in [-M, M]$ and $\hat{\sigma}_X^2(M)$ be the solutions of (9.23), then for the r.p. $Y(k)$ with positive analytic SDF the polynomial $A_M(z) = 1 - \sum_{\substack{i=-M \\ i\neq 0}}^{M} \hat{a}_i(M)z^{-i}$ converges uniformly in some neighborhoods of $|z| = 1$ to an analytic limit $\lim_{M\to\infty} A_M(z) = A(z)$. Moreover, the variance of the error also converges to a positive lower limit σ_X^2, where $lim_{M\to\infty}\hat{\sigma}_X^2(M) = \sigma_X^2 = \left[\frac{1}{2\pi}\int_{-\pi}^{\pi}\frac{1}{S_{YY}(e^{j\Omega})}d\Omega\right]^{-1}$. The derivations of these results are given in [15].

Example 9.2 *Consider the ACF in Example 9.1. We would like to develop a noncausal representation for its SDF.*

Solution: Let us rewrite the SDF in Example 9.1 as:

$$S_{YY}(z) = \frac{\sigma_y^2(1-\rho^2)}{1+\rho^2-\rho(z+z^{-1})} = \frac{\beta^2}{1-\alpha(z+z^{-1})}$$

where $\alpha = \rho/(1+\rho^2)$ and $\beta^2 = \sigma_y^2(1-\rho^2)/(1+\rho^2)$. Now, denoting $A(z) = 1-\alpha(z+z^{-1})$ and multiplying the denominator and numerator by $A(z^{-1})$ gives the following.

$$S_{YY}(z) = \frac{\beta^2 A(z^{-1})}{A(z)A(z^{-1})}$$

Comparing this equation with the general expression in (9.6), we obtain $H(z) = \frac{Y(z)}{X(z)} = \frac{1}{1-\alpha(z+z^{-1})}$ and $S_{XX}(z) = \beta^2 A(z^{-1}) = \beta^2[1-\alpha(z+z^{-1}]$. From the transfer function and using inverse z-transform (see Appendix A) we get the following noncausal realization for the same SDF:

$$y(k) = \alpha\big(y(k+1) + y(k-1)\big) + x(k)$$

In addition, taking the inverse z-transform of $S_{XX}(z)$ yields the ACF of $X(k)$:

$$R_{XX}(n) = \beta^2\delta(n) - \beta^2\alpha\big[\delta(n-1)+\delta(n+1)\big],$$

which implies that the driving process $X(k)$ is no longer white (i.e. colored). This example illustrates two important properties of noncausal models: (i) they do not require spectral factorization; (ii) they are driven by colored noise. Thus, for a general noncausal model, the ACF of the driving process is:

$$R_{XX}(n) = \sigma_X^2\delta(n) - \sigma_X^2\sum_{i\neq 0} a_i\delta(n-i),$$

that is, $X(k)$ is a MA process.

9.1.2.3 Noncausal Minimum Variance Prediction

A r.p. $Y(k)$ is called M^{th}-order Markov [14], if all the information needed for minimum variance noncausal prediction of $Y(k)$ is contained in its M immediate past and future samples, i.e. $E\big[Y(k)|Y(k-i), \forall i\neq 0\big] = E\big[Y(k)|Y(k-i), 1\leq|i|\leq M\big]$. In this case, M^{th}-order noncausal MVR predictor of $y(k)$ is:

$$\hat{y}(k) = \sum_{\substack{i=-\infty \\ i\neq 0}}^{M} \hat{a}_i(M)y(k-i) \tag{9.25}$$

where coefficients $\hat{a}_i(M)$ are obtained by solving the noncausal Yule-Walker equation in (9.23). Then, the corresponding noncausal AR model represents an exact realization of the Mth-order Markov process since the SDF generated by this model, $S_{YY}^{(M)}(z) = \hat{\sigma}_X^2(M)/A_M(z)$ where $A_M(z) = 1 - \sum_{\substack{i=-M \\ i\neq 0}}^{M} \hat{a}_i(M)z^{-i}$ matches exactly the SDF of a finite-order process $S_{YY}(z)$. In practice, however, one needs to approximate an infinite-order process by a finite-order MVR prediction model, in which case by choosing large $M > M_0$, the difference between the actual covariances and those generated by the model can be made arbitrarily small. Clearly, the correlation match is exact, i.e. $S_{YY}^{(M)}(z) = S_{YY}(z)$ either when $M \to \infty$ or for $M = M_0$ if $Y(k)$ is indeed a M_0^{th} order process.

9.2 2-D Linear Image Models

As mentioned earlier, a major difference between 1-D and 2-D models is that, for the latter case, causality is not a requirement. Moreover, it may not be possible to achieve a finite-order causal realization for a given rational SDF due to the lack of general factorization for 2-D polynomials. In general, the prediction or support region of 2-D models can take three types of geometry, leading to causal, semicausal, and noncausal models [14–17].

Most of the prior work on image modeling and related issues has focused on causal models. In recent years, however, semicausal and noncausal models have found applications in a wide variety of image processing problems. Semicausal models are used in image data compression, image filtering, and spectral estimation. Noncausal models have been found useful in developing block processing algorithms with applications in image data compression using block coders. In this section, these models and their properties are studied.

Consider a 2-D WSS random field $Y(i, j)$ with zero mean and ACF $R_{YY}(k, l) = E[Y(i + k, j + l)Y(i, j)]$. As before, we denote a sample image of this 2-D r.p. by $y(i, j)$. A 2-D AR model that represents this process is:

$$y(i, j) = \sum\sum_{m,n\in W_r} a_{m,n} y(i - m, j - n) + x(i, j) \tag{9.26}$$

where $x(i, j)$ is a sample image of the driving r.p. $X(i, j)$, $a_{m,n}$ are the 2-D AR model coefficients, and W_r is the ROS of the model. The geometry of W_r determines the type of MVR predictor, namely causal, semicausal, and noncausal. As discussed in Chapter 3, adopting a scanning scheme which scans the image line-by-line, top-to-bottom, and left-to-right on each line, the three prediction regions or ROS associated with these models are defined as follows:

Causal NSHP Model:

$$W_1 = \left\{(m, n) : \{n \geq 1, \forall m\} \cup \{n = 0, m \geq 1\}\right\} \tag{9.27}$$

Semicausal Model:

$$W_2 = \left\{(m, n) : \{n \geq 1, \forall m\} \cup \{n = 0, \forall m \neq 0\}\right\} \tag{9.28}$$

Noncausal Model:

$$W_3 = \left\{(m, n) : \{\forall (m, n) \neq (0, 0)\}\right\} \tag{9.29}$$

These ROS geometries are shown in Figure 9.1a,c,d, respectively. We note that W_1 corresponds to a general causal nonsymmetric half-plane (NSHP) model with larger support compared to that of the quarter-plane in Figure 9.1b. In practice, since only finite-order models are of interest, a finite number of nearest-neighbor pixels are included in the ROS. We refer to these ROS by $\breve{W}_r \subset W_r$, defined as:

$$\breve{W}_1 = \left\{(m, n) : 1 < n \leq N, -M \leq m \leq M \text{ and } n = 0, 1 \leq m \leq M\right\} \tag{9.30}$$

$$\breve{W}_2 = \left\{(m, n) : 0 \leq n \leq N, -M \leq m \leq M, (m, n) \neq (0, 0)\right\} \tag{9.31}$$

$$\breve{W}_3 = \left\{(m, n) : -N \leq n \leq N, -M \leq m \leq M, (m, n) \neq (0, 0)\right\} \tag{9.32}$$

Figure 9.2a–c illustrates examples of finite-extent ROS for three first-order 2-D models of causal quarter-plane, semicausal, and noncausal types, respectively.

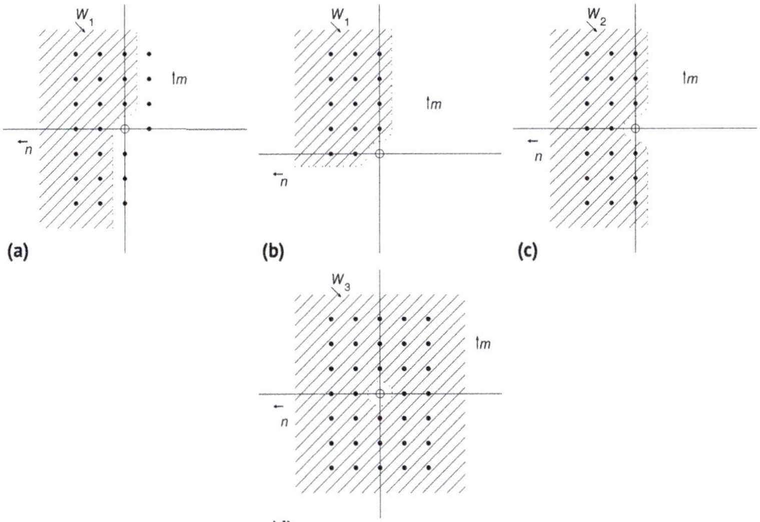

Figure 9.1 ROS for (a) causal nonsymmetric half plane, (b) causal quarter-plane, (c) semicausal, and (d) noncausal finite order 2-D AR models.

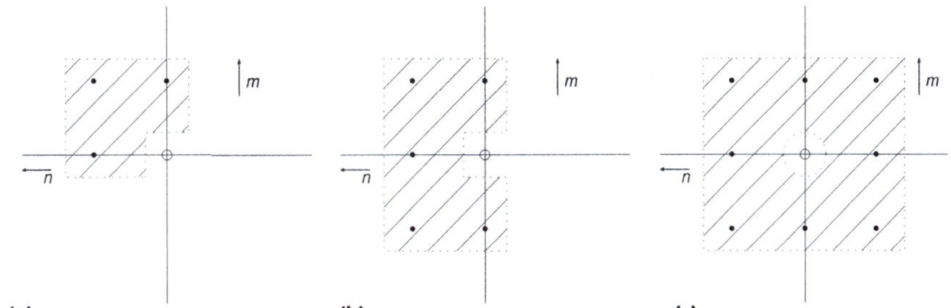

Figure 9.2 Finite-extent ROS of (a) causal quarter-plane, (b) semicausal, and (c) noncausal geometries.

9.2.1 2-D Causal Representation

In general, for a given positive analytic SDF there exists a unique causal realization of infinite order. That is, for a 2-D r.p. $Y(i, j)$, the SDF is:

$$S_{yy}(z_1, z_2) = \frac{S_{XX}(z_1, z_2)}{A(z_1, z_2)A(z_1^{-1}, z_2^{-1})} \tag{9.33}$$

where $S_{XX}(z_1, z_2)$ is the SDF of the driving process (prediction error) $X(i, j)$, and for infinite-order 2-D causal AR models $A(z_1, z_2)$ is:

$$A(z_1, z_2) = 1 - \sum_{m,n \in W_1} \sum a_{m,n} z_1^{-m} z_2^{-n}. \tag{9.34}$$

In general, $X(i, j)$ may be modeled as a MA process, so that (9.33) is a rational function expressed as a ratio of two 2-D polynomials in z_1 and z_2 (see Appendix B). Since 2-D causal models are recursive in both dimensions, their realization from the SDF $S_{YY}(z_1, z_2)$ requires factorization in both z_1 and z_2 variables. Furthermore, it can be shown [14] that causal models are driven by white noise. These observations will also be made through Example 9.3.

It is possible, however, to find a finite-order causal representation that would closely approximate the SDF in (9.33) corresponding to the infinite-order causal model. The corresponding SDF for this finite-order realization is:

$$S_{YY}^{(M,N)}(z_1, z_2) = \frac{\sigma_X^2(M, N)}{A_{M,N}(z_1, z_2)A_{M,N}(z_1^{-1}, z_2^{-1})} \tag{9.35}$$

where $A_{M,N}(z_1, z_2) = 1 - \sum\sum_{m,n \in \breve{W}_1} a_{m,n}(M,N)\, z_1^{-m} z_2^{-n}$. The coefficients $a_{m,n}(M,N)$ and the variance $\sigma_X^2(M,N)$ of the driving process can be computed using a 2-D extension of the Yule-Walker method, which is given next.

9.2.1.1 Parameter Estimation of 2-D Causal AR Model

As in the 1-D case, the MVR predictor of $y(i,j)$ given by $\hat{y}(i,j) = \sum\sum_{m,n \in \breve{W}_r} a_{m,n} y(i-m, j-n)$ leads to the normal equation for an NSHP causal model of order (M,N). Using this normal equation, we obtain:

$$
\begin{aligned}
R_{YY}(k,l) = {} & \sum_{m=1}^{M} a_{m,0}(M,N)\, R_{YY}(k-m,l) + \sum_{m=-M}^{M}\sum_{n=1}^{N} a_{m,n}(M,N)\, R_{YY}(k-m,l-n) \\
& + \sigma_X^2(M,N)\delta(k,l),\ (k,l) \in \breve{W}_1 \cup \{(0,0)\}.
\end{aligned}
\tag{9.36}
$$

Arranging this normal equation in matrix form yields the 2-D extension of the Yule-Walker equation as:

$$
\mathbf{R}_Y\, \underline{a}_{M,N} = \sigma_X^2(M,N)\, \underline{1}_{1,1} \tag{9.37}
$$

where

$$
\mathbf{R}_Y = \begin{bmatrix}
\rho'_0 & \rho'_{-1} & \rho'_{-2} & \cdots & \cdots & \rho'_{-N} \\
\rho'_1 & \rho_0 & \rho_{-1} & & & \rho_{-N+1} \\
\vdots & & \ddots & & & \vdots \\
\vdots & & & \ddots & & \rho_{-1} \\
\rho'_N & \rho_{N-1} & & \cdots & \cdots & \rho_0
\end{bmatrix}. \tag{9.38}
$$

The block matrices ρ_k and ρ'_k are defined, respectively, as

$$
\rho_k = \begin{bmatrix}
R_{YY}(0,k) & R_{YY}(-1,k) & & \cdots & R_{YY}(-2M,k) \\
R_{YY}(1,k) & & & & \vdots \\
\vdots & & \ddots & & \\
\vdots & & & \ddots & \vdots \\
R_{YY}(2M,k) & \cdots & & R_{YY}(1,k) & R_{YY}(0,k)
\end{bmatrix} = \rho_{-k}^T \tag{9.39}
$$

and

$$
\rho'_0 = \begin{bmatrix}
R_{YY}(0,0) & R_{YY}(-1,0) & \cdots & R_{YY}(-M,0) \\
R_{YY}(1,0) & R_{YY}(0,0) & \cdots & R_{YY}(-M+1,0) \\
\vdots & & \ddots & \vdots \\
R_{YY}(M,0) & R_{YY}(M-1,0) & \cdots & R_{YY}(0,0)
\end{bmatrix}
$$

$$
\rho'_{-k} = \begin{bmatrix}
R_{YY}(M,-k) & R_{YY}(M-1,-k) & \cdots & R_{YY}(-M,-k) \\
R_{YY}(M+1,-k) & R_{YY}(M,-k) & \cdots & R_{YY}(-M+1,-k) \\
\vdots & & \ddots & \vdots \\
R_{YY}(2M,-k) & R_{YY}(2M-1,-k) & \cdots & R_{YY}(0,-k)
\end{bmatrix},\ k \in [1,N] \tag{9.40}
$$

$$
\rho'_k = \rho'^T_{-k}.
$$

Moreover, we have:

$$\underline{a}_{M,N}^T = [\underline{a}_0^T \; \underline{a}_1^T \; \cdots \; \underline{a}_N^T]$$
$$\underline{a}_0^T = [1 \; - a_{1,0}(M,N) \; \cdots \; - a_{M,0}(M,N)], \qquad a_{00}(M,N) = 1$$
$$\underline{a}_i^T = [-a_{-M,i}(M,N) \; \cdots \; - a_{0,i}(M,N) \; \cdots \; - a_{M,i}(M,N)], \quad i \in [1,N] \tag{9.41}$$

$$\underline{1}_{1,1}^T = [\underline{1}_1^T \; 0^T \; \cdots \; 0^T], \quad \text{and} \quad \underline{1}_1^T = [1\, 0 \; \cdots \; 0] \tag{9.42}$$

As in the case, assuming that $\mathbf{R}_Y^{-1}$ is available, $\hat{\sigma}_X^2(M,N)$ and $\hat{a}_{m,n}(M,N)$ can be found using:

$$\hat{\sigma}_X^2(M,N) = 1/[\mathbf{R}_Y^{-1}]_{1,1}$$
$$\hat{a}_{m,0}(M,N) = -\hat{\sigma}_X^2(M,N)[\mathbf{R}_Y^{-1}]_{k,1}, \quad k = m+1, \quad m \in [1,M] \tag{9.43}$$
$$\hat{a}_{m,i}(M,N) = -\hat{\sigma}_X^2(M,N)[\mathbf{R}_Y^{-1}]_{k,1}, \quad k = i(2M+1) + m + 1, \quad m \in [-M,M], \; i \in [1,N]$$

where $[\mathbf{R}_Y^{-1}]_{i,j}$ denotes the $(i,j)^{\text{th}}$ element of matrix $\mathbf{R}_Y^{-1}$.

Note that for the special case of 2-D causal model with quarter-plane ROS, the correlation matrix $\mathbf{R}_Y$ becomes doubly Toeplitz, i.e. a block Toeplitz matrix with Toeplitz blocks.

Example 9.3 *Consider a 2-D zero-mean WSS process with a separable ACF:*

$$R_{YY}(k,l) = E\left[Y(i+k, j+l)Y(i,j)\right] = \sigma_Y^2 \rho^{|k|+|l|}, \quad 0 < \rho < 1$$

where σ_Y^2 is the variance of $Y(i,j)$. We seek a causal representation for this 2-D random process.

Solution: For this separable ACF, the SDF is:

$$S_{YY}(z_1,z_2) = \sum_{m=-\infty}^{\infty} \sum_{n=-\infty}^{\infty} R_{YY}(m,n) z_1^{-m} z_2^{-n}$$
$$= \sigma_Y^2 \left(\sum_{m=-\infty}^{\infty} \rho^{|m|} z_1^{-m} \right) \left(\sum_{n=-\infty}^{\infty} \rho^{|n|} z_2^{-n} \right)$$

Using a similar approach as in Example 9.1, we obtain:

$$S_{YY}(z_1,z_2) = \frac{\sigma_y^2 (1-\rho^2)^2}{(1-\rho z_1^{-1})(1-\rho z_2^{-1})(1-\rho z_1)(1-\rho z_2)}$$

For a causal model, spectral factorization must be applied in both variables. Using (9.33), we obtain:

$$A(z_1,z_2) = (1-\rho z_1^{-1})(1-\rho z_2^{-1}) \quad \text{and} \quad S_{XX}(z_1,z_2) = \sigma_Y^2 (1-\rho^2)^2.$$

Thus, the transfer function of the 2-D causal system is $H(z_1,z_2) = \frac{Y(z_1,z_2)}{X(z_1,z_2)} = \frac{1}{A(z_1,z_2)}$. Now, taking the inverse 2-D z-transform gives the 2-D causal model:

$$y(i,j) = \rho y(i-1,j) + \rho y(i,j-1) - \rho^2 y(i-1,j-1) + x(i,j)$$

This 2-D causal model, with quarter-plane ROS shown in Figure 9.2a, is an exact realization of the above SDF. From the SDF, $S_{XX}(z_1,z_2)$, it is evident that the driving process, $X(i,j)$, is white with $\mu_X = 0$ and $R_{XX}(k,l) = \beta_1^2 \delta(k,l)$ where $\beta_1^2 = \sigma_Y^2 (1-\rho^2)^2$.

9.2.1.2 Convergence Property

Given a positive analytic SDF, the polynomials $A_{M,N}(z_1, z_2)$ in (9.35):

$$A_{M,N}(z_1, z_2) = 1 - \sum_{m=1}^{M} a_{m,0}(M, N)z_1^{-m} - \sum_{m=-M}^{M} \sum_{n=1}^{N} a_{m,n}(M, N)z_1^{-m}z_2^{-n}$$

with coefficients generated using the Yule-Walker equation, converge uniformly [16] to $A(z_1, z_2)$ in the region $\Gamma_1 = \{|z_1| = 1, |z_2| \geq 1\} \cup \{|z_1| \geq 1, z_2 = \infty\}$. Moreover, the variance of the driving process converges to

$$\lim_{M,N\to\infty} \hat{\sigma}_X^2(M, N) = \sigma_X^2 = \exp\left[\frac{1}{4\pi^2}\int_{-\pi}^{\pi}\int_{-\pi}^{\pi} \log S_{YY}(e^{j\Omega_1}, e^{j\Omega_2})d\Omega_1 d\Omega_2\right]$$

In summary, the following properties are valid for 2-D causal models [15]:

a. Their realization from $S_{YY}(z_1, z_2)$ requires spectral factorization in both z_1 and z_2 variables;
b. They are driven by white noise, i.e. $R_{XX}(k, l) = \sigma_X^2\delta(k, l)$;
c. They have quarter-plane ROS for separable ACF, although the converse is not true; and
d. They are recursive in both spatial directions.

9.2.2 2-D Semicausal Representation

Semicausal models [14–17] are recursive in one dimension corresponding to the causal direction and non-recursive in the other associated with the noncausal direction. Furthermore, unlike causal models, which require spectral factorization in both variables, semicausal models require factorization of SDF $S_{YY}(z_1, z_2)$ only in one of the variables that corresponds to the causal direction. In addition, semicausal models are not white noise driven.

As in the causal case, it is possible to find finite-order semicausal models that approximate a given positive analytic SDF arbitrarily closely. The corresponding SDF of such a finite-order 2-D semicausal model [15] is:

$$S_{YY}^{(M,N)}(z_1, z_2) = \frac{S_{XX}^{(M,N)}(z_1, z_2)}{A_{M,N}(z_1, z_2)A_{M,N}(z_1^{-1}, z_2^{-1})}, \tag{9.44}$$

where $A_{M,N}(z_1, z_2) = \sum\sum_{m,n\in \check{W}_2 \cup (0,0)} a_{m,n}(M, N)z_1^{-m}z_2^{-n}$ with ROS $\check{W}_2$ defined in (9.30), and $a_{m,n}(M, N)$ being the model coefficients. In addition, the numerator of the SDF in (9.44) is:

$$S_{XX}^{(M,N)}(z_1, z_2) = \sigma_X^2(M, N)A_0(z_1), \tag{9.45}$$

where $A_0(z_1) = 1 + \sum_{m=-M}^{M} a_{m,0}(M, N)z_1^{-m}$, which implies that the driving process $X(i, j)$ is a MA (colored) process along the spatial variable i, and white along the spatial variable j. It can be seen from (9.44) and (9.45) that the finite-order semicausal models realize SDF functions that contain both numerator and denominator polynomials. However, the numerator polynomial is one-dimensional, corresponding to the noncausal direction, i.e. $A_0(z_1) = A(z_1, \infty)$. Additionally, to ensure the symmetry property of the SDF, the coefficients must satisfy $a_{m,0} = a_{-m,0}$.

The corresponding finite-order 2-D semicausal model with ROS $\check{W}_2$ that realizes the SDF in (9.44) is:

$$y(i, j) = \sum\sum_{m,n\in\check{W}_2} a_{m,n}(M, N)\, y(i - m, j - n) + x(i, j), \tag{9.46}$$

where parameters $a_{m,n}(M, N)$ are obtained using the parameter estimation in the next section. Also, taking the inverse 2-D z-transform of (9.45) yields the ACF of the driving process $X(i, j)$:

$$R_{XX}(k, l) = \sigma_X^2(M, N)\left[\delta(k, l) + \sum_{m=-M}^{M} a_{m,0}(M, N)\delta(k - m, l)\right]. \tag{9.47}$$

Thus, the semicausal model in (9.46) is driven by a colored process along the spatial variable k.

9.2.2.1 Parameter Estimation of 2-D Semicausal AR Model

The normal equations for the 2-D semicausal model in (9.46) can be written as:

$$R_{YY}(k,l) = \sum_{\substack{m=-M \\ m\neq 0}}^{M} a_{m,0}(M,N)R_{YY}(k-m,l) + \sum_{m=-M}^{M}\sum_{n=1}^{N} a_{m,n}(M,N)R_{YY}(k-m,l-n) + \sigma_X^2(M,N)\delta(k,l), \quad k,l \in \breve{W}_2 \cup \{(0,0)\} \tag{9.48}$$

Since $a_{0,0}(M,N) = 1$, this normal equation can be arranged [15] in matrix form:

$$\mathbf{R}_Y \underline{a}_{M,N} = \sigma_X^2(M,N)\underline{1}_{M+1,1} \tag{9.49}$$

where in this case matrix $\mathbf{R}_Y$ is doubly Toeplitz given by:

$$\mathbf{R}_Y = \begin{bmatrix} \boldsymbol{\rho}_0 & \boldsymbol{\rho}_{-1} & \boldsymbol{\rho}_{-2} & \cdots & \boldsymbol{\rho}_{-N} \\ \boldsymbol{\rho}_1 & \boldsymbol{\rho}_0 & & & \boldsymbol{\rho}_{-N+1} \\ \vdots & & \ddots & & \vdots \\ \vdots & & & \ddots & \vdots \\ \boldsymbol{\rho}_N & \cdots & & \cdots & \boldsymbol{\rho}_0 \end{bmatrix} \tag{9.50}$$

where the block matrices $\boldsymbol{\rho}_k$ defined in (9.39) are also Toeplitz. Also, we have:

$$\begin{aligned} \underline{a}_{M,N} &= [\underline{a}_0^T \ \underline{a}_1^T \ \cdots \ \underline{a}_N^T]^T, \\ \underline{1}_{M+1,1} &= [\underline{1}_{M+1}^T \ \underline{0}^T \ \cdots \ \underline{0}^T]^T, \end{aligned} \tag{9.51}$$

where

$$\begin{aligned} \underline{a}_k^T &= [-a_{-M,k}(M,N) \ -a_{-M+1,k}(M,N) \ \cdots \ -a_{0,k}(M,N) \ \cdots \ -a_{M,k}(M,N)], \\ \underline{1}_{M+1}^T &= [0 \ \cdots \ 0\,1\,0 \ \cdots \ 0], \end{aligned} \tag{9.52}$$

with 1 appearing in the $(M+1)^{\text{th}}$ position of the vector $\underline{1}_{M+1}$. As in the causal case, estimates of $\sigma_X^2(M,N)$ and $a_{m,n}(M,N)$ can be found using:

$$\hat{\sigma}_X^2(M,N) = 1/[\mathbf{R}_Y^{-1}]_{M+1,M+1} \tag{9.53}$$

and

$$\hat{a}_{m,n}(M,N) = -\hat{\sigma}_X^2(M,N)[\mathbf{R}_Y^{-1}]_{k,M+1}, \quad k = n(2M+1) + m + M + 1 \tag{9.54}$$

Example 9.4 *Consider the problem of finding a 2-D semicausal realization for the separable ACF in Example 9.3.*

Solution: Let us rewrite the SDF in Example 9.3. as:

$$\begin{aligned} S_{YY}(z_1,z_2) &= \frac{\sigma_Y^2(1-\rho^2)^2}{[1+\rho^2-\rho z_1^{-1}-\rho z_1](1-\rho z_2)(1-\rho z_2^{-1})} \\ &= \frac{\sigma_Y^2(1-\rho^2)^2/(1+\rho^2)}{[1-\alpha_1(z_1^{-1}+z_1)](1-\rho z_2)(1-\rho z_2^{-1})} \end{aligned}$$

where $\alpha_1 = \rho/(1+\rho^2)$. Multiplying both the numerator and the denominator by term $[1-\alpha_1(z_1+z_1^{-1})]$ gives the following.

$$S_{YY}(z_1,z_2) = \frac{\sigma_Y^2(1-\rho^2)^2[1-\alpha_1(z_1+z_1^{-1})]/(1+\rho^2)}{[1-\alpha_1(z_1^{-1}+z_1)][1-\alpha_1(z_1+z_1^{-1})](1-\rho z_2)(1-\rho z_2^{-1})}$$

Applying spectral factorization along z_2 gives:

$$A(z_1, z_2) = [1 - \alpha_1(z_1^{-1} + z_1)](1 - \rho z_2^{-1})$$

and

$$S_{XX}(z_1, z_2) = \sigma_Y^2 (1 - \rho^2)^2 [1 - \alpha_1(z_1 + z_1^{-1})]/(1 + \rho^2).$$

Then, the transfer function of the 2-D linear system that realizes this SDF exactly is:

$$H(z_1, z_2) = \frac{Y(z_1, z_2)}{X(z_1, z_2)} = \frac{1}{A(z_1, z_2)} = \frac{1}{(1 - \rho z_2^{-1})(1 - \alpha_1 z_1^{-1} - \alpha_1 z_1)}$$

which after taking inverse 2-D z-transform gives the semicausal model:

$$y(i, j) = \alpha_1 \big(y(i-1, j) + y(i+1, j)\big) - \rho\alpha_1 \big(y(i+1, j-1) + y(i-1, j-1)\big) + \rho y(i, j-1) + x(i, j).$$

This model is recursive along the j direction only. The ACF of the driving process, $X(i, j)$, can be determined from the SDF, $S_{XX}(z_1, z_2)$, as:

$$R_{XX}(k, l) = \beta_2^2 \left[\delta(k, l) - \alpha_1 \delta(k-1, l) - \alpha_1 \delta(k+1, l) \right]$$

where $\beta_2^2 = \sigma_Y^2 (1 - \rho^2)^2 / (1 + \rho^2)$. This ACF indicates the fact that $X(i, j)$ is an MA process along the i variable and white along the j variable.

As can be observed from the poles of the transfer function $H(z_1, z_2)$, the pole at $z_2 = \rho$ is inside the unit circle in the z_2-plane corresponding to a stable causal part along variable j, while one of the poles $z_1 = \frac{1}{2\alpha_1}(1 \pm \sqrt{1 - 4\alpha_1^2})$ is outside the unit circle in the z_1-plane corresponding to the noncausal part along variable i.

The above transfer function can be decomposed and realized by two separable 2-D AR models:

$$y(i, j) = \rho y(i, j-1) + \upsilon(i, j) \quad \text{(causal part)}$$

$$\upsilon(i, j) = \alpha_1(\upsilon(i+1, j) + \upsilon(i-1, j)) + x(i, j) \quad \text{(noncausal part)}$$

These equations form an initial-boundary value problem, with the initial value problem in the j variable and the boundary value problem in the i variable. Thus, the recursive solution of the second equation along the forward direction (i.e. starting from $i = 0$) is unstable. On the other hand, if the boundary conditions $\upsilon(i, 0)$ and $\upsilon(i, N+1)$ are known, the solution for each j is obtained by solving this equation as a boundary value problem by applying a Riccati transformation [18].

9.2.2.2 Convergence Property

Given a positive analytic SDF, the polynomial:

$$A_{M,N}(z_1, z_2) = 1 - \sum\sum_{(m,n)\in \tilde{W}_2} \hat{a}_{m,n}(M, N) z_1^{-m} z_2^{-n}$$

converges uniformly to the analytic semi-minimum phase [16] $A(z_1, z_2)$ in the region $\Gamma_2 = \{|z_1| = 1, |z_2| \geq 1\}$.

The variance of the driving process converges to a positive lower limit σ_X^2:

$$\lim_{M,N\to\infty} \hat{\sigma}_X^2(M, N) = \sigma_X^2 = \left[\frac{1}{2\pi} \int_{-\pi}^{\pi} D^{-1}(e^{j\Omega_1}) d\Omega_1 \right]^{-1}$$

where $D(e^{j\Omega_1}) = \sigma_x^2 A_0^{-1}(e^{j\Omega_1})$ and $A_0(z_1)$ was defined in (9.45). Moreover, by increasing (M, N), the difference between the actual correlations and the correlations generated by the model can be made arbitrarily small.

From the results in this section and those in Example 9.4, the following important observations can be made for 2-D semicausal models:

a. They are driven by colored noise. More specifically, the driving process $X(i, j)$ is white along variable j and correlated with MA along variable i;
b. They are recursive (or causal) in only one variable, namely j. Any attempt to re-index the variable i to obtain a 2-D causal model leads to an unstable model;
c. For a given semicausal model, a causal representation may be found under some mild restrictions [15–17]. However, the causal realization may have a higher (or even infinite) order; and
d. The numerator of the SDF is a 1-D polynomial in z_1 corresponding to the noncausal direction.

9.2.3 2-D Noncausal Representation

Noncausal models [14–17] are nonrecursive in both directions. They do not require spectral factorization in any variable and are driven by colored noise. A positive analytic SDF can be realized arbitrarily closely by a stationary finite-order noncausal model with ROS $\breve{W}_3$ defined in (9.30). Consider the positive analytic SDF:

$$S_{YY}^{(M,N)}(z_1, z_2) = \frac{S_{XX}^{(M,N)}(z_1, z_2)}{A_{M,N}(z_1, z_2)A_{M,N}(z_1^{-1}, z_2^{-1})} \tag{9.55}$$

where $A_{M,N}(z_1, z_2) = \sum\sum_{m,n\in\breve{W}_3\cup\{(0,0)\}} a_{m,n}(M, N)z_1^{-m}z_2^{-n}$, and $a_{m,n}(M, N)$ are the model parameters. The numerator of (9.55) gives the SDF of the driving input $X(i, j)$:

$$S_{XX}^{(M,N)}(z_1, z_2) = \sigma_X^2(M, N)A_{M,N}(z_1, z_2). \tag{9.56}$$

From (9.55) and (9.56), yields:

$$S_{YY}^{(M,N)}(z_1, z_2) = \frac{\sigma_X^2(M, N)}{A_{M,N}(z_1, z_2)}. \tag{9.57}$$

Thus, given a rational SDF with zeroth-order numerator polynomial, there exists a unique finite-order noncausal model to represent it. Furthermore, similar to the 1-D noncausal models, we have symmetry property $a_{-m,-n} = a_{m,n}$.

The corresponding finite-order 2-D noncausal model with ROS $\breve{W}_3$ that realizes the SDF in (9.55) or (9.57) is:

$$y(i, j) = \sum_{m,n\in\breve{W}_3}\sum a_{m,n}(M, N)\, y(i - m, j - n) + x(i, j) \tag{9.58}$$

where the model parameters $a_{m,n}(M, N)$ are obtained using the parameter estimation in the next section.

Using (9.56), it can be easily shown that the ACF of the driving process $X(i, j)$ is:

$$R_{XX}(k, l) = \sigma_X^2(M, N)\left[\delta(k, l) + \sum_{m,n\in\breve{W}_3}\sum a_{m,n}(M, N)\delta(k - m, l - n)\right] \tag{9.59}$$

implying that the noncausal models are driven by colored noise.

9.2.3.1 Parameter Estimation of 2-D Noncausal AR Model

The normal equation for the noncausal model in (9.58) is given by:

$$R_{YY}(k, l) = \sum_{m,n=\in\breve{W}_3}\sum a_{m,n}(M, N)R_{YY}(k - m, l - n) + \sigma_X^2(M, N)\delta(k, l), \quad k, l \in \breve{W}_3 \cup \{(0, 0)\}. \tag{9.60}$$

Arranging this normal equation for various values of (k, l) gives the following Yule-Walker equation:

$$\mathbf{R}_Y \underline{a}_{M,N} = \sigma_X^2(M, N)\underline{1}_{M+1,N+1} \tag{9.61}$$

where $\mathbf{R}_Y$ is a symmetric doubly Toeplitz matrix given by:

$$\mathbf{R}_Y = \begin{bmatrix} \rho_0 & \rho_{-1} & \cdots & \rho_{-2N} \\ \rho_1 & \rho_0 & & \rho_{-2N+1} \\ \vdots & & \ddots & \vdots \\ \rho_{2N} & \rho_{2N-1} & \cdots & \rho_0 \end{bmatrix} \tag{9.62}$$

with ρ_k defined in (9.39) and

$$\begin{aligned} \underline{a}_{M,N} &= [\underline{a}_{-N}^T \ \cdots \ \underline{a}_0^T \ \cdots \ \underline{a}_N^T]^T \\ \underline{1}_{M+1,N+1} &= [\underline{0}^T \cdots \underline{0}^T \ \underline{1}_{M+1}^T \ \underline{0}^T \cdots \underline{0}^T]^T \end{aligned} \tag{9.63}$$

where 1_{M+1}^T appears in the $(N+1)^{\text{th}}$ position of $\underline{1}_{M+1,N+1}$,

$$\underline{a}_k^T = [-a_{-M,k}(M,N) \cdots - a_{0,k}(M,N) \cdots - a_{M,k}(M,N)]$$

with $a_{00}(M,N) = 1$, and further $\underline{1}_{M+1}^T = [0\,0 \ \cdots \ 1\,0 \ \cdots \ 0]$ with 1 at the $(M+1)^{\text{th}}$ element.

Now, if $\mathbf{R}_Y$ is positive definite, there is a unique solution for the model coefficients $a_{m,n}(M,N)$ and the variance of the driving process $\sigma_X^2(M,N)$ which are given by:

$$\hat{\sigma}_X^2(M,N) = 1/[\mathbf{R}_Y^{-1}]_{k,k} \qquad k = N(2M+1) + M + 1 \tag{9.64}$$

and

$$\hat{a}_{m,n}(M,N) = -\hat{\sigma}_X^2(M,N)[\mathbf{R}_Y^{-1}]_{l,k} \qquad l = (n+N)(2M+1) + m + M + 1, \ \ k \in [-N,N] \tag{9.65}$$

Example 9.5 *For the ACF in Example 9.3, develop a 2-D noncausal model.*

Solution: Let us rewrite the SDF in Example 9.3 in the following form:

$$S_y(z_1,z_2) = \frac{\sigma_Y^2(1-\rho^2)^2[1-\alpha_1(z_1^{-1}+z_1)][1-\alpha_1(z_2^{-1}+z_2)]/(1+\rho^2)^2}{[1-\alpha_1(z_1^{-1}+z_1)]^2[1-\alpha_1(z_2^{-1}+z_2)]^2}$$

Comparing this to (9.55)–(9.59), yields:

$$A(z_1,z_2) = [1-\alpha_1(z_1^{-1}+z_1)][1-\alpha_1(z_2^{-1}+z_2)]$$

and

$$S_{XX}(z_1,z_2) = \sigma_Y^2(1-\rho^2)^2[1-\alpha_1(z_1+z_1^{-1})][1-\alpha_1(z_2^{-1}+z_2)]/(1+\rho^2)^2$$

where $\alpha_1 = \rho/(1+\rho^2)$. The transfer function of the 2-D system that realizes this SDF exactly is:

$$\begin{aligned} H(z_1,z_2) = \frac{Y(z_1,z_2)}{X(z_1,z_2)} &= \frac{1}{A(z_1,z_2)} \\ &= \frac{1}{[1-\alpha_1(z_1^{-1}+z_1)][1-\alpha_1(z_2^{-1}+z_2)]} \end{aligned}$$

which after taking inverse 2-D z-transform gives the following 2-D noncausal model:

$$\begin{aligned} y(i,j) = {} & \alpha_1\big(y(i-1,j) + y(i+1,j) + y(i,j-1) + y(i,j+1)\big) \\ & -\alpha_1^2\big(y(i-1,j-1) + y(i+1,j-1) + y(i-1,j+1) + y(i+1,j+1)\big) + x(i,j). \end{aligned}$$

The ACF of the driving process, $X(i, j)$, can be determined from the SDF, $S_{XX}(z_1, z_2)$, as:

$$\begin{aligned} R_{XX}(k, l) = \beta_3^2[&\delta(k, l) - \alpha_1\delta(k-1, l) - \alpha_1\delta(k+1, l) - \alpha_1\delta(k, l-1) \\ &- \alpha_1(k, l+1) + \alpha_1^2\delta(k-1, l-1) + \alpha_1^2\delta(k+1, l-1) \\ &+ \alpha_1^2\delta(k-1, l+1) + \alpha_1^2\delta(k+1, l+1)] \end{aligned}$$

where $\beta_3^2 = \sigma_Y^2(1-\rho^2)^2/(1+\rho^2)$. Clearly, this ACF indicates that $X(i, j)$ is a colored process along both spatial variables.

9.2.3.2 Convergence Property

For a 2-D random field $Y(i, j)$ with a positive analytic SDF [16], the polynomial:

$$A_{M,N}(z_1, z_2) = 1 - \sum_{(m,n)\in \bar{W}_3}\sum \hat{a}_{m,n}(M, N)z_1^{-m}z_1^{-n}$$

converges uniformly on the unit bidisc $|z_1| = |z_2| = 1$ to $A(z_1, z_2)$ as the model orders (M, N) increase, i.e.

$$\lim_{M,N\to\infty} A_{M,N}(z_1, z_2) = A(z_1, z_2), \qquad |z_1| = |z_2| = 1$$

The variance of the driving process converges to a positive lower limit σ_X^2:

$$\lim_{M,N\to\infty} \hat{\sigma}_X^2(M, N) = \sigma_X^2 = \left[\frac{1}{4\pi^2}\int_{-\pi}^{\pi}\int_{-\pi}^{\pi} S_{YY}^{-1}(e^{j\Omega_1}, e^{j\Omega_2})d\Omega_1 d\Omega_2\right]^{-1}$$

The results in this section together with those in Example 9.5 point to the following properties of the noncausal models [15]:

a. They do not require spectral factorization in any direction;
b. They are driven by colored noise;
c. They are not recursive in any direction; and
d. The numerator of their SDF is a 2-D polynomial in z_1 and z_2.

9.2.4 Numerical Example

Here, we use the Lena image in Figure 9.3 to compute the coefficients of 2-D AR models with causal NSHP, semicausal, and noncausal ROS. Both first- and second-order models are considered. First, the sample autocorrelations are computed from the mean subtracted image to form the sample correlation matrix $\mathbf{R}_Y$ for each model and order. Then, the corresponding Yule-Walker parameter estimation methods covered in Sections 9.2.1, 9.2.2, and 9.2.3 are used to estimate the model

Figure 9.3 Lena image. *Source:* Juang et al. 2012 / John Wiley & Sons / CC BY 3.0.

parameters and the variance of the driving process. The estimated model coefficients along with the variance of the driving process for the first- and second-order causal NSHP, semicausal, and noncausal models are given in Figures 9.4a–c and 9.5a–c, respectively. Several important observations can be made from these results. For a model order, the variance of the driving process decreases substantially as the support of the model extends from causal to semicausal and from semicausal to noncausal ROS. Moreover, for a particular choice of an AR model, increasing the model order only minimally decreases the variance of the prediction error. This observation suggests that increasing the AR model order beyond the second order, has diminishing returns in modeling accuracy. Thus, to avoid increasing computational requirements, we will use the first-order 2-D causal NSHP AR model in the subsequent chapters, for example, in image restoration using 2-D Kalman filtering in Section 10.2.5 and also for predictive encoding in Section 11.3.1.

9.3 Multivariate or Vector Linear Models

In the preceding sections, only scalar or univariate linear prediction models have been considered. In these models, each pixel is related to its neighboring pixels using a causal, semicausal, or a noncausal ROS. As a result, if these models are used for prediction or image restoration, at each stage of the processing, only one pixel estimate is produced. With the growing emphasis on parallel processing, it is desirable to develop vector image models that inherently support parallelism. In vector or multivariate models, the image field is represented by a 1-D vector AR process that captures not only intra-dependence among pixels in one vector but also inter-dependence among pixels across neighboring vectors within the AR model support region.

The goal of this section is to introduce vector models and highlight their relations to the scalar models in the previous sections. To this end, only a general vector modeling framework is presented. For more specific work, the reader is referred to [19, 20].

Consider an image of size $N \times N$ that is vector-scanned horizontally in overlapping strips of width W [3, 19], with an overlap typically equal to $W/2$. The direction of scanning is assumed to be from left-to-right to top-to-bottom, where each strip is processed independently. The image is assumed to be a vector Markov process of order M and hence can be modeled

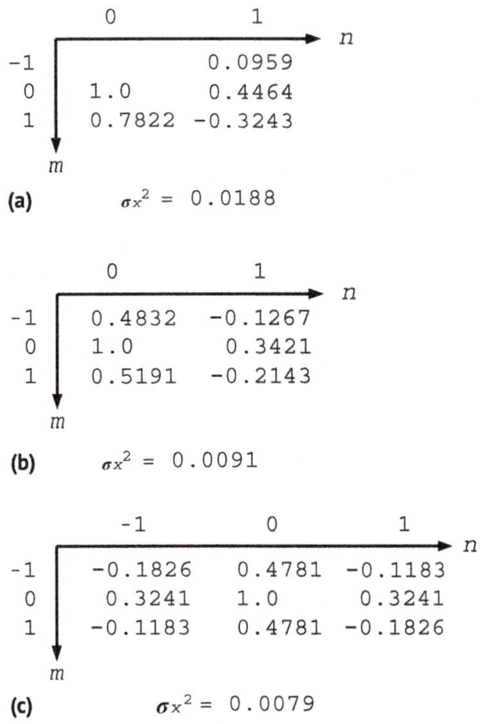

Figure 9.4 First-order 2-D (a) causal NSHP, (b) semicausal, and (c) noncausal AR model parameters and estimated σ_X^2 for Lena image.

(a)

m \ n	0	1	2
-2		0.1475	0.6090
-1		0.1319	-0.1453
0	1.0	-0.4314	0.0124
1	0.8577	-0.1552	0.1206
2	-0.1306	-0.2081	-0.1602

σ_x^2 = 0.0141

(b)

m \ n	0	1	2
-2	-0.1168	-0.0529	0.0577
-1	0.5973	-0.0415	-0.0592
0	1.0	0.2867	-0.0134
1	0.6000	-0.1578	-0.0321
2	-0.1215	0.0202	0.0324

σ_x^2 = 0.0080

(c)

m \ n	-2	-1	0	1	2
-2	0.0329	-0.0014	-0.0779	-0.0654	0.0460
-1	-0.027	-0.1101	0.5113	-0.0212	-0.037
0	-0.0322	0.2832	1.0	0.2832	-0.0322
1	-0.0376	-0.0212	0.5113	-0.110	-0.0275
2	0.0460	-0.0654	-0.0779	-0.0014	0.0329

σ_x^2 = 0.0070

Figure 9.5 Second-order 2-D (a) causal NSHP, (b) semicausal, and (c) noncausal AR model parameters and estimated σ_X^2 for Lena image.

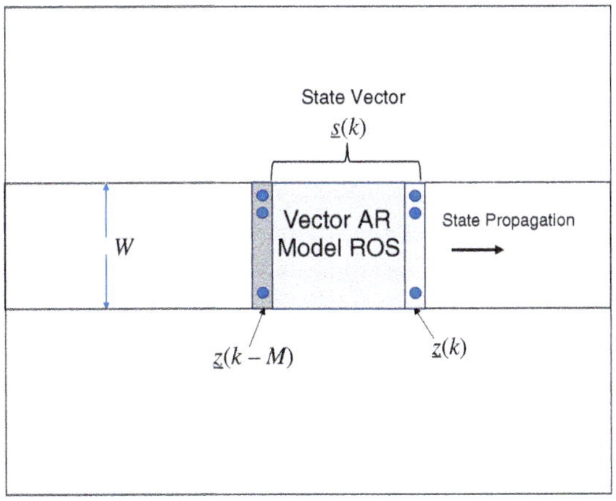

Figure 9.6 ROS of a vector AR model within an image strip.

within each strip by an Mth order vector AR process with causal ROS as shown in Figure 9.6. This 1-D multivariate or vector AR process is given by:

$$\underline{z}(k) = \mathbf{A}_1^T \underline{z}(k-1) + \cdots + \mathbf{A}_M^T \underline{z}(k-M) + \underline{x}(k) \tag{9.66}$$

where $\underline{z}(k)$ is a $W \times 1$ vector of pixels in the ith strip, i.e.

$$\underline{z}(k) = [y((i-1)W, k)\; y((i-1)W+1, k) \cdots y(iW-1, k)]^T,$$

and $\underline{x}(k)$ is a sample vector of the driving random vector process $\underline{X}(k)$. The matrices $\mathbf{A}_i$ are the model coefficient matrices of size $W \times W$. The first- and second-order statistics of the driving process $\underline{X}(k)$ are $E[\underline{X}(k)] = 0$ and $E[\underline{X}(k)\underline{X}^T(l)] = \mathbf{Q}_{\underline{X}}\delta(k-l)$ where $\mathbf{Q}_{\underline{X}}$ is the covariance matrix. Thus, this driving process is white across columns (inter-vector) and colored within each vector (intra-vector), since the elements of vector $\underline{X}(k)$ are mutually correlated. This property can be more evident by expressing (9.66) in scalar form where each pixel in $\underline{z}(k)$ is represented by linear combination of pixels within the ROS in Figure 9.6 and the corresponding scalar driving process. This relationship corresponds to a semicausal representation in which the noncausality occurs along the vertical direction. As we showed earlier, semicausal finite-order models are driven by processes that are white along the causal direction and colored along the noncausal direction. Thus, the elements of $\underline{X}(k)$ are mutually correlated along the noncausal (vertical) direction, whereas the elements of $\underline{X}(k)$ and $\underline{X}(l)$ for all $k \neq l$ within the ROS are uncorrelated, since these are along the causal (horizontal) direction.

It is also important to note that the assumption of WSS within each strip is generally not a valid for vector AR processes. Instead, a more appropriate assumption is column-wise WSS within each strip [19].

9.3.1 Parameter Estimation of Vector AR Models

Consider the MVR vector representation:

$$\underline{z}(k) = \hat{\underline{z}}(k) + \underline{x}(k) \tag{9.67}$$

where the linear prediction vector is $\hat{\underline{z}}(k) = \sum_{i=1}^{M} \mathbf{A}_i^T \underline{z}(k-i)$. As before, we can show that minimizing the variance of the error vector $\underline{e}(k) = \underline{z}(k) - \hat{\underline{z}}(k)$ leads to the orthogonality principle, i.e.

$$E[\underline{Z}(k-l)\underline{X}^T(k)] = \mathbf{Q}_{\underline{X}}\delta(l)$$

and hence $E[\hat{\underline{Z}}(k)\underline{X}^T(k)] = \mathbf{0}$. This orthogonality principle is used to arrive at the vector AR Yule-Walker equation to compute the model parameter matrices.

To estimate the model parameter matrices $\mathbf{A}_i$, $i \in [1, M]$, transpose (9.66), premultiply both sides by $\underline{z}(k-l)$ and then take expectation. This yields:

$$\rho_l = E[\underline{z}(k-l)\underline{z}^T(k)] = \rho_{l-1}\mathbf{A}_1 + \rho_{l-2}\mathbf{A}_2 + \cdots + \rho_{l-M}\mathbf{A}_M + E[\underline{z}(k-l)\underline{x}^T(k)]. \tag{9.68}$$

Plugging $l = 0, 1, \ldots M$ into this normal equation gives the following vector Yule-Walker system of equations, which must be solved for the unknown parameter matrices and also $\mathbf{Q}_{\underline{X}}$, i.e.

$$\begin{bmatrix} \rho_0 & \rho_l^T & \cdots & \cdots & \rho_M^T \\ \rho_l & \rho_0 & \cdots & \cdots & \rho_{M-1}^T \\ \vdots & & & & \vdots \\ & & \ddots & & \\ \vdots & & & \ddots & \vdots \\ \rho_M & \rho_{M-1} & \cdots & \cdots & \rho_0 \end{bmatrix} \begin{bmatrix} \mathbf{I}_W \\ -\mathbf{A}_1 \\ \vdots \\ \vdots \\ -\mathbf{A}_M \end{bmatrix} = \begin{bmatrix} \mathbf{Q}_{\underline{X}} \\ \mathbf{0} \\ \mathbf{0} \\ \vdots \\ \mathbf{0} \end{bmatrix}, \tag{9.69}$$

where $\rho_{-m} = \rho_m^T$. In compact form, we can write (9.69) as:

$$\mathbf{R}_{\underline{Z}} \begin{bmatrix} \mathbf{I}_W \\ -\mathbf{A}_1 \\ \vdots \\ \vdots \\ -\mathbf{A}_M \end{bmatrix} = \begin{bmatrix} \mathbf{Q}_{\underline{X}} \\ \mathbf{0} \\ \mathbf{0} \\ \vdots \\ \mathbf{0} \end{bmatrix}, \tag{9.70}$$

which gives the following solutions for $\mathbf{Q}_{\underline{X}}$ and parameter matrices $\mathbf{A}_i$:

$$\mathbf{Q}_{\underline{X}} = \left[[\mathbf{R}_{\underline{Z}}^{-1}]_{1,1} \right]^{-1} \tag{9.71}$$

and

$$\mathbf{A}_i = -[\mathbf{R}_{\underline{Z}}^{-1}]_{i+1,1}\mathbf{Q}_{\underline{X}}, \quad i \in [1, M] \tag{9.72}$$

where in this case, $[\mathbf{R}_{\underline{Z}}^{-1}]_{i,j}$ is the $(i, j)^{\text{th}}$ block matrix of $\mathbf{R}_{\underline{Z}}^{-1}$ of size $W \times W$. In practice, the sample covariance matrices of the image data within each strip are computed using:

$$\hat{\rho}_l = \frac{1}{N-l} \sum_{k=l}^{N-1} \underline{z}(k)\underline{z}^T(k-l) \tag{9.73}$$

and subsequently used in the vector Yule-Walker equation (9.69) to estimate the model parameters. Reestimating these parameter matrices for every new strip enables model adaptability to the local statistics within each strip.

9.4 Computer Projects

The computer projects in this chapter use Python libraries NumPy, Matplotlib, and SciPy.

Computer Project 9.1: The goal of this computer project is to verify the theoretical results presented in the previous sections for 2-D AR models. In this experiment, a 2-D covariance model is considered that is often used to closely approximate the true covariance function of images. This 2-D covariance model is an exponential isotropic function [15] of the form, $R_{YY}(i, j) = \rho^{\sqrt{i^2+j^2}}$ with $\rho = 0.9$. First, generate and display a 256×256 portion of this covariance function (positive quadrant only). Note that this covariance model is symmetric across all quadrants. Compute and plot the SDF, $S_{YY}(k, l)$, by taking the 2-D DFT of size 256×256 of the covariance array. Comment on the shape of this function. Using the Yule-Walker equations, estimate first-order 2-D AR model parameters with causal NSHP, semicausal, and noncausal ROS. You need to construct the covariance matrix $\mathbf{R}_Y$ for each case depending on the model. Verify your results by comparing them to those reported in [15] for this covariance model. Provide a detailed analysis of your results and comment on the modeling accuracy of each configuration.

Computer Project 9.2: In this computer project, we repeat the procedure in Section 9.2.4 using the "Baboon" image. Use both first- and second- order AR models and compute sample autocovariances from the mean subtracted image to form the covariance matrix $\mathbf{R}_Y$ for each case. Then, use the corresponding Yule-Walker equations for each type of the model to generate the corresponding parameters. Compare these model parameters and the estimate of driving noise variance $\hat{\sigma}_X^2$ with those obtained in Section 9.2.4. Evaluate the accuracy of your results by forming the error image, i.e. the difference between the original and linear MVR predicted images and find the variance of this error image. Provide a comprehensive evaluation of each 2-D model in terms of modeling accuracy, computational complexity, and implementation issues.

Computer Project 9.3: The goal of this computer project is to develop a strip vector AR model for the "Lena" image using the method in Section 9.3. First, divide the image into strips of width $W = 8$ and compute the sample covariance matrices using (9.73) from the mean subtracted (vector-wise) data. Using equations (9.69), (9.71), and (9.72), estimate the model parameter matrices, $\mathbf{A}_i$, and the driving noise error covariance matrix $\mathbf{Q}_{\underline{X}}$ for a second-order $M = 2$ vector model. Perform this estimation for at least three different strips in the image. To compare the accuracy of the modeling in each strip, the Frobenious norm of matrix $\mathbf{Q}_{\underline{X}}$, i.e. $||\mathbf{Q}_{\underline{X}}||_F = \sqrt{tr[\mathbf{Q}_{\underline{X}}\mathbf{Q}_{\underline{X}}^T]}$ may be used. Evaluate the accuracy of your results by forming the error image in each chosen strip and finding its covariance matrix. The error image can be computed by forming the difference between the original image and the linear predicted image within the strip. The linear predicted image is generated using the model with the estimated parameter matrices within each strip. Discuss advantages and disadvantages of the strip vector models in terms of modeling capability, computational cost, and implementation complexity.

9.5 Problems

Problem 9.1 Consider a 1-D MA(1) process $Y(k) = X(k) - b_1X(k-1)$ where $X(k)$ is a zero-mean white process with variance σ_X^2. Determine the SDF, $S_{YY}(z)$, and ACF, $R_{YY}(n)$, of the r.p. $Y(k)$. Is $Y(k)$ a white r.p.?

Problem 9.2 Consider a 1-D first-order AR(1) process $Y(k) = a_1 Y(k-1) + X(k)$ where $X(k)$ is a zero-mean white process with variance σ_X^2.

a. Determine the SDF $S_{YY}(z)$.
b. Time-domain solution of this AR(1).
c. Expression for ACF, $R_{YY}(n)$ of the process $Y(k)$. Is $Y(k)$ a white r.p.?

Problem 9.3 A zero-mean stationary process $y(k)$ has the following frequency-domain SDF:

$$S_{YY}(e^{j\Omega}) = \frac{8.5 - 4\cos\Omega}{5 - 4\cos\Omega}$$

a. Convert this SDF to the z domain and then determine a causal realization for this process using the spectral factorization method.
b. It is desired to whiten this process by passing it through a linear system. Determine the transfer function of this whitening system.

Problem 9.4 Given an ARMA(1,1), $Y(k) - a_1 Y(k-1) = X(k) + b_1 X(k-1)$ where $|a_1| < 1$ and $X(k)$ is a zero-mean white process with variance σ_X^2. Find the expression for SDF, $S_{YY}(z)$, and ACF, $R_{YY}(n)$.

Problem 9.5 Consider the r.p. in Example 9.1. Using the Yule-Walker method, fit a first-order causal AR model to this process and identify the parameters. Verify your results by comparing them with those obtained in Example 9.1.

Problem 9.6 Repeat Problem 9.5 for the first-order noncausal AR model in Example 9.2.

Problem 9.7 A zero-mean WSS image, $Y(i, j)$, is characterized by a separable ACF given by $R_{YY}(i, j) = \sigma_Y^2 \rho_1^{|i|} \rho_2^{|j|}$, $0 < \rho_1, \rho_2 < 1$.

a. Using the spectral factorization method in Section 9.2.1, fit a 2-D causal AR model to this image.
b. Repeat Part (a) for 2-D semicausal AR model (see Section 9.2.2).
c. Repeat Part (a) for 2-D noncausal AR model (see Section 9.2.3). Comment on your results.

Problem 9.8 Consider a first-order causal NSHP model given by:

$$Y(i, j) = a_{1,0} Y(i-1, j) + a_{0,1} Y(i, j-1) + a_{1,1} Y(i-1, j-1) + a_{-1,1} Y(i+1, j-1) + X(i, j)$$

where $X(i, j)$ is a white driving process with variance σ_X^2. The raster scanning is adopted along the direction j. Determine the 2-D SDF of this process. Is it possible to convert this model to a causal quarter-plane model?

Problem 9.9 Consider the following 2-D transfer functions:

$$\text{(i) } H(z_1, z_2) = \frac{1}{1 + 0.5z_1^{-1} - 0.5z_2^{-1} - 0.25z_1^{-1}z_2^{-1}} \tag{9.74}$$

$$\text{(ii) } H(z_1, z_2) = \frac{1}{(1 + z_1^{-1})(1 + z_1)(1 + z_2^{-1})(1 + z_2)}$$

$$\text{(iii) } H(z_1, z_2) = \frac{1}{z_1 + z_1^{-1} + z_1^{-1}z_2^{-1}}$$

Find the corresponding AR model for each and determine if they are causal, semicausal, or noncausal. Is it possible to convert these to initial value and/or boundary value problems? Note that the scanning is vertical and left-to-right.

Problem 9.10 Repeat Problem 9.9 for the following 2-D transfer functions:

$$\text{(i)}\ H(z_1, z_2) = \frac{1}{1 - a_{1,0}z_1^{-1} - a_{0,1}z_2^{-1} - a_{1,1}z_1^{-1}z_2^{-1}} \tag{9.75}$$

$$\text{(ii)}\ H(z_1, z_2) = \frac{1}{1 - a_{1,0}z_1^{-1} - a_{-1,0}z_1 - a_{0,1}z_2^{-1}}$$

$$\text{(iii)}\ H(z_1, z_2) = \frac{1}{1 - a_{1,0}z_1^{-1} - a_{-1,0}z_1 - a_{0,1}z_2^{-1} - a_{0,-1}z_2}$$

Problem 9.11 Repeat Problem 9.9 for the following 2-D transfer functions:

$$\text{(i)}\ H(z_1, z_2) = \frac{z_1 + z_1^{-1}}{1 - a_{1,0}z_1^{-1} - a_{0,1}z_2^{-1} - a_{1,1}z_1^{-1}z_2^{-1}} \tag{9.76}$$

$$\text{(ii)}\ H(z_1, z_2) = \frac{z_1^{-1} - z_2^{-1}}{1 - a_{1,0}z_1^{-1} - a_{-1,0}z_1 - a_{0,1}z_2^{-1}}$$

$$\text{(iii)}\ H(z_1, z_2) = \frac{1 + z_1^{-1} + z_1^{-2}}{1 - a_{1,0}z_1^{-1} - a_{-1,0}z_1 - a_{0,1}z_2^{-1} - a_{0,-1}z_2}$$

Problem 9.12 For the 2-D causal model in Example 9.3, develop a causal vector representation and identify all its parameters.

Problem 9.13 In Problem 9.12, find the covariance matrix $\mathbf{R}_{\underline{Y}}$ of the vector model in terms of the ACF of the scalar 2-D model.

Problem 9.14 Consider the semicausal representation obtained in Example 9.4, i.e.

$$y(i, j) = \alpha_1\big(y(i-1, j) + y(i+1, j)) - \rho\alpha_1(y(i+1, j-1) + y(i-1, j-1)\big)$$
$$+ \rho y(i, j-1) + x(i, j)$$

where $x(i, j)$ is a sample image of a random field $X(i, j)$ that is white along the causal direction j and correlated and MA process along the noncausal direction i. We would like to arrange this model in vector form such that the propagation of the vector is along the causal direction. Develop the corresponding model and identify all its parameters. Is this model stable?

Problem 9.15 Consider the noncausal representation obtained in Example 9.5, i.e.

$$y(i, j) = \alpha_1\big(y(i-1, j) + y(i+1, j) + y(i, j-1) + y(i, j+1)\big)$$
$$- \alpha_1^2\big(y(i-1, j-1) + y(i+1, j-1) + y(i-1, j+1) + y(i+1, j+1)\big) + x(i, j).$$

where $x(i, j)$ is a sample image of a random field $X(i, j)$ that is colored along both directions i and j. Is it possible to arrange this model in an AR or ARMA vector model that is causal? Develop a vector model for this noncausal representation.

References

1. A. K. Jain and J. R. Jain, "Partial difference equations and finite difference methods in image processing-Part 2: image restoration," *IEEE Transactions on Automatic Control*, vol. 23, pp. 817–833, 1978.
2. J. W. Woods and C. H. Radewan, "Kalman filtering in two dimensions," *IEEE Transactions on Information Theory*, vol. 23, pp. 473–482, 1977.
3. B. R. Suresh and B. A. Shenoi, "New results in two-dimensional Kalman filtering with applications to image restoration," *IEEE Transactions on Circuits and Systems*, vol. 28, pp. 307–319, 1981.
4. M. R. Azimi-Sadjadi and P. W. Wong, "Two-dimensional block Kalman filtering for image restoration," *IEEE Transactions on Acoustics, Speech, and Signal Processing*, vol. 35, pp. 1736–1749, 1987.

5 E. J. Delp, R. L. Kashyap, and O. R. Mitchell, "Image data compression using autoregressive time series models," *Pattern Recognition*, vol. 11, pp. 313–323, 1979.

6 R. Chellappa and R. L. Kashyap, "Texture synthesis using 2-D noncausal autoregressive models," *IEEE Transactions on Acoustics, Speech, and Signal Processing*, vol. 33, pp. 194–202, 1985.

7 T. E. Hall and G. B. Giannakis, "Bispectral analysis and model validation of texture images," *IEEE Transactions on Image Processing*, vol. 4, pp. 996–1009, 1995.

8 R. King, M. Ahmadi, R. Naguib, A. Kwabwe, and M. R. Azimi-Sadjadi, "*Digital Filtering in One and Two Dimensions-design and Applications*," Plenum Press, 1989.

9 J. G. Proakis and D. G. Manolakis, "*Digital Signal Processing, Principles, Algorithms, and Applications*," Pearson, 5th Edition, 2021.

10 S. Lawrence Marple, "*Digital Spectral Analysis with Applications*," Prentice-Hall, 1987.

11 P. Stoica, "*Spectral Analysis of Signals*," Prentice-Hall, 2005.

12 R. Kumar, "A fast algorithm for solving a Toeplitz system of equations," *IEEE Transactions on Acoustics, Speech, and Signal Processing*, vol. 33, pp. 254–267, 1985.

13 P. J. Brockwell and R. A. Davis, "*Introduction to Time Series and Forecasting*," Springer-Verlag, 3rd Edition, 2016.

14 A. K. Jain, "Advances in mathematical models for image processing," *Proceedings of the IEEE*, vol. 69, pp. 502–528, 1981.

15 S. Ranganath and A. K. Jain, "Two-dimensional linear prediction models-Part 1: Spectral factorization and realization." *IEEE Transactions on Acoustic, Speech, and Signal Processing*, vol. 33, pp. 280–299, 1985.

16 R. L. Kashyap, "Characterization and estimation of two-dimensional ARMA models," *IEEE Transactions on Information Theory*, vol. 30, pp. 736–745, 1984.

17 A. K. Jain, "A semicausal model for recursive filtering of two-dimensional images," *IEEE Transactions on Computers*, vol. 26, pp. 343–350, 1977.

18 L. Dieci, M. R. Osborne, and R. D. Russell, "A Riccati transformation method for solving linear BVPs. I: Theoretical aspects," *SIAM Journal on Numerical Analysis*, vol. 25, pp. 1055–1073, 1988.

19 M. R. Azimi-Sadjadi, "New results in strip alman filtering," *IEEE Transactions on Circuits and Systems*, vol. 36, pp. 893–897, 1989.

20 E. A. Robinson, "*Multichannel Time Series Analysis With Digital Computer Programs*," Houston, TX: Goose Pond Press, 2nd Edition, 1983.

10

Image Restoration

Any image acquired by an imaging sensor is subject to noise and other degradations caused by both stochastic and deterministic sources. Stochastic sources include additive and multiplicative noise. Examples of additive noise are sensor or detector noise, channel noise, thermal noise, and quantization noise. A common form of multiplicative noise is speckle noise, which arises in coherent imaging systems such as synthetic aperture radar (SAR) [1], synthetic aperture sonar (SAS) [2], ultrasonic imaging [3], and laser imaging [4]. Deterministic degradation is primarily associated with the blurring effects [5], which result from predictable physical phenomena such as motion between the camera and the object, defocusing, atmospheric turbulence (which alters the refractive index of media in satellite imaging), and the finite aperture size in radar or sonar arrays.

The goal of any image restoration system [5] is to recover or estimate an image that has been degraded by these deterministic and stochastic sources. Image restoration can be viewed as an *inverse process*, where the effects of degradation are mitigated, using certain prior knowledge of degradation processes, to recover the original image. In this sense, image restoration methods are more objective compared with those of image enhancement. The idea is depicted in Figure 10.1 where the image formation system produces a degraded continuous image $y(u, v)$ from the original scene image $x(u, v)$. This continuous image is digitized to produce $y(m, n)$, where $m \in [0, M-1]$, $n \in [0, N-1]$. A restoration filter is then applied to obtain an estimate, $\hat{x}(m, n)$, of the original image.

10.1 Image Degradation Models

Random noise is represented by its statistical characteristics, such as first- and second-order moments. Care must be taken in choosing an appropriate distribution function by considering the physical phenomenon [6, 7] that generated the noise. A typical additive noise model used in image restoration is white Gaussian noise (WGN), which is mainly caused by thermal sources in charge-coupled devices (CCDs) widely used in digital cameras. Another source of noise or uncertainty that affects the electron counts in CCD arrays is the *readout noise*, which is independent of exposure time. This noise is modeled using a Poisson distribution [8]. Quantization noise, on the other hand, is represented by a uniform distribution as discussed in Chapter 5. As for the multiplicative sources of noise, speckle [9] is of primary concern in coherent imaging systems such as in SAR [1] employed on many airborne platforms. Speckle appears as granular intensity fluctuations in the image. The source of this noise is due to random interference of waves scattered by microscopic fluctuations on the object's surface within a single resolution cell. These interactions lead to random variations in the phase and amplitude of the received images. Constructive and destructive combinations of these phase components result in high- and low-intensity spots, causing the

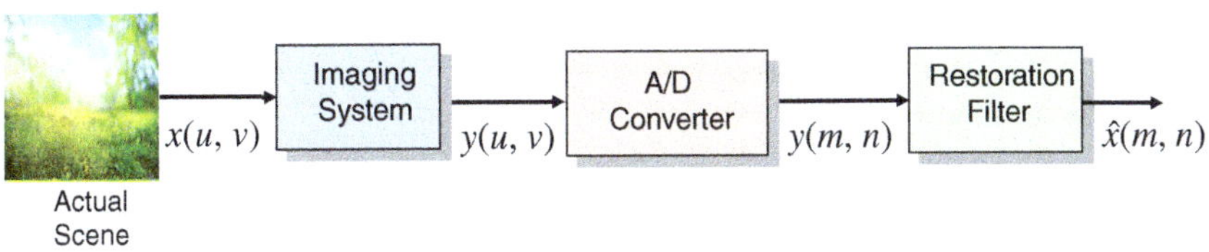

Figure 10.1 Image formation process and restoration filter.

Table 10.1 Different LSI blur, PSF $\boldsymbol{h(u, v)}$, and frequency response, $\boldsymbol{H(\omega_1, \omega_2)}$.

Type	$h(u, v)$	$H(\omega_1, \omega_2)$
Horizontal motion blur	$\frac{1}{\alpha_0}\text{rect}\left[\frac{u}{\alpha_0} - \frac{1}{2}\right]\delta(v)$	$e^{-j\frac{\omega_1\alpha_0}{2}}\text{sinc}\left(\frac{\omega_1\alpha}{2\pi}\right)$
Atmospheric turbulence	$e^{-\pi\alpha^2(u^2+v^2)}$	$\frac{1}{\alpha^2}e^{\left(-\frac{\omega_1^2+\omega_2^2}{4\pi\alpha^2}\right)}$
Rectangular scanning aperture	$\text{rect}\left[\frac{u}{\alpha}, \frac{v}{\beta}\right]$	$\alpha\beta\text{sinc}\left(\frac{\alpha\omega_1}{2\pi}\right)\text{sinc}\left(\frac{\beta\omega_2}{2\pi}\right)$
CCD local interactions	$\sum\sum_{k,l=-1}^{1} \alpha_{k,l}\delta(u - k\Delta, v - l\Delta)$	$\sum\sum_{k,l=-1}^{1} \alpha_{k,l}e^{-j(\omega_1 k+\omega_2 l)\Delta}$
Diffraction-limited (coherent)	$ab\ \text{sinc}(au)\ \text{sinc}(bv)$	$\text{rect}\left[\frac{\omega_1}{2\pi a}, \frac{\omega_2}{2\pi b}\right]$
Diffraction-limited (incoherent)	$\text{sinc}^2(au)\ \text{sinc}^2(bv)$	$\text{tri}\left[\frac{\omega_1}{2\pi a}, \frac{\omega_2}{2\pi b}\right]$

speckle effects. Speckle noise has the characteristics of multiplicative, non-Gaussian, nonstationary, and image-dependent noise, making it particularly difficult to remove effectively [6, 7].

In contrast, blur is a deterministic source of degradation with a known or unknown point spread function (PSF) that convolves with the original image. Before designing an image restoration algorithm, one must identify the specific physical phenomenon that resulted in blurring effects. In some cases, it may be possible to derive an analytic expression for the PSF if sufficient prior information regarding the physical nature of the blurring phenomenon is available. An example of this case [5] is presented in the next example. Table 10.1 provides additional examples of LSI blur and their corresponding PSFs and 2-D frequency responses, derived based on the underlying physical models of the blurring process [5]. However, in many realistic situations, the parameters of the PSF must be estimated directly from the observed image. Among the methods used for these situations include those in [10, 11].

Example 10.1 [5] *An object $x(u, v)$ being imaged moves uniformly along u-direction at velocity v. If the exposure time is T, we would like to find the PSF $h(u, v)$ of the horizontal motion blur.*

Solution: The image intensity at time t is $x_t(u, v) = x(u - vt, v)$. Thus, the observed image over the exposure time T seconds becomes:

$$y(u, v) = \frac{1}{T}\int_0^T x(u - vt, v)dt.$$

If we define $vt = \alpha$ and $vT = \alpha_0$, then we get:

$$y(u, v) = \frac{1}{\alpha_0}\int_0^{\alpha_0} x(u - \alpha, v)d\alpha,$$

which can be expressed as:

$$y(u, v) = \frac{1}{\alpha_0}\int\int_{-\infty}^{\infty} \text{rect}\left[\frac{\alpha}{\alpha_0} - \frac{1}{2}\right]\delta(\beta)x(u - \alpha, v - \beta)d\alpha d\beta$$

Now, we compare this equation with the 2-D convolution integral:

$$y(u, v) = \int\int h(\alpha, \beta)x(u - \alpha, v - \beta)d\alpha d\beta$$

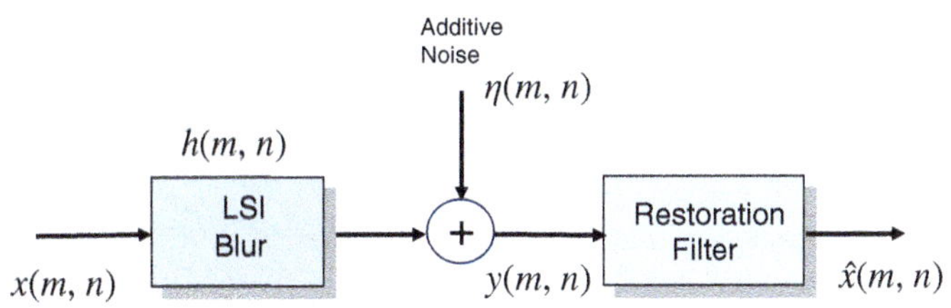

Figure 10.2 A simple degradation model and restoration filter.

which gives the PSF of the horizontal motion blur as:

$$h(u, v) = \frac{1}{\alpha_0}\text{rect}\left(\frac{u}{\alpha_0} - \frac{1}{2}\right)\delta(v).$$

A typical image formation system consisting of additive white noise, $\eta(u, v)$, multiplicative noise, $\gamma(u, v)$, and LSI blur is given by:

$$y(u, v) = \gamma(u, v)[x(u, v) ** h(u, v)] + \eta(u, v) \tag{10.1}$$

where $h(u, v)$ is the PSF of LSI blur, $x(u, v)$ is the original image, and $y(u, v)$ is the observed image. The estimate of the original image, $\hat{x}(u, v)$, must be generated using the restoration filter. Since the sources that produced the additive and multiplicative noise are independent of each other, we can safely assume that they are mutually uncorrelated. Additionally, we can assume that they are uncorrelated with the original image.

In the absence of multiplicative noise, the digitized version of the observation model in (10.1) becomes:

$$y(m, n) = x(m, n) ** h(m, n) + \eta(m, n) \tag{10.2}$$

where $y(m, n)$ is the observed digital image. This simple degradation model is shown in Figure 10.2. Now, given the observed image $y(m, n)$, the goal of image restoration is to design a filter to recover or estimate the original image $\hat{x}(m, n)$ from $y(m, n)$, taking into account the properties of degradation phenomena in terms of the statistics of the noise and the PSF of the blur.

10.2 Image Restoration Filters

There are two general classes of restoration filters that can be employed depending on (i) the degradation model, (ii) the prior knowledge about the degradation sources and their specific characteristics, and (iii) the design requirements and constraints. These are listed below.

1. **Deterministic-based**
   These methods tend to ignore the effects of additive and/or multiplicative noise, as well as the statistical properties of the image. Thus, they are used primarily for *deconvolution* purposes to undo the effects of blur, especially when noise is negligible or when no prior statistical information is available. The most commonly used deterministic filters are: direct inverse and pseudo-inverse filters and Least Squares (LS) filters.
2. **Stochastic-based**
   Unlike their deterministic counterparts, in addition to the blur PSF, these filters incorporate the statistical information about both the noise and original image to design the restoration filters. Therefore, they simultaneously perform deblurring (deconvolution) and noise filtering (image estimation). Among commonly used conventional filters in this category are the 2-D Wiener filter [5] and the 2-D Kalman filter [12–15] that will be covered later in this chapter.

   More recently, neural network-based approaches have been developed that can perform the same tasks with minimal explicit knowledge of the degradation processes. A brief review of such methods is given in [16], and a specific neural network architecture is presented in Chapter 19.

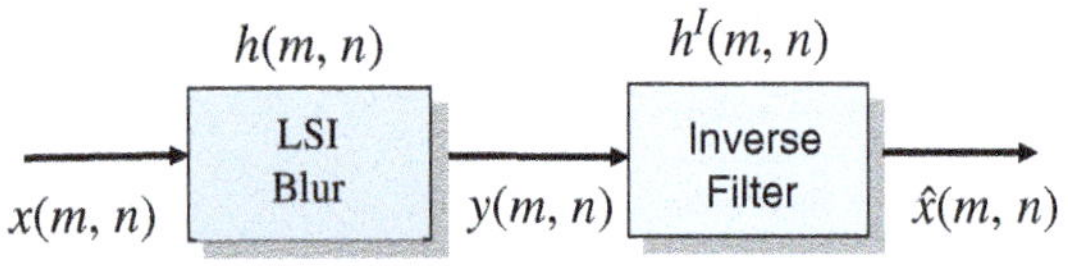

Figure 10.3 Inverse filtering for deconvolution.

10.2.1 Deterministic Filters

10.2.1.1 Direct Inverse Filter

The standard inverse filter attempts to recover the original image from the observed blurred image using an inverse system with PSF, $h^I(m, n)$, as shown in Figure 10.3. Assuming noise-free observation, the observed image is:

$$y(m, n) = h(m, n) ** x(m, n). \tag{10.3}$$

Applying the inverse filter yields:

$$\hat{x}(m, n) = y(m, n) ** h^I(m, n) \tag{10.4}$$

where $\hat{x}(m, n)$ is the restored or deblurred image. Substituting (10.3) into (10.4) gives:

$$\hat{x}(m, n) = x(m, n) ** h(m, n) ** h^I(m, n) \tag{10.5}$$

Perfect recovery requires $\hat{x}(m, n) = x(m, n)$, i.e.

$$h(m, n) ** h^I(m, n) = \delta(m, n) \tag{10.6}$$

which in frequency domain gives

$$H^I(k, l) = \frac{1}{H(k, l)}. \tag{10.7}$$

That is, the inverse filter frequency response, $H^I(k, l)$, is the reciprocal of the blur process frequency response. Then, the recovered image is obtained using:

$$\hat{X}(k, l) = H^I(k, l)Y(k, l) = \frac{Y(k, l)}{H(k, l)} \tag{10.8}$$

where $Y(k, l) = \text{2-D DFT}\{y(m, n)\}$. However, there are several issues that render this operation useless in its original form. Although from (10.8) it appears that the inverse filter yields perfect recovery, in practice, the direct inverse filter suffers from *singularity* and *ill-conditioning* problems at frequencies where $H(k, l) \simeq 0$. In the presence of even small amounts of noise in the observed image, e.g. due to quantization, the inverse filter amplifies its effects since

$$\hat{X}(k, l) = X(k, l) + \frac{N(k, l)}{H(k, l)} \tag{10.9}$$

where the first term on the right-hand side is the 2-D DFT of the original recovered image, while the second term represents the effect of additive noise after the inverse filtering. This leads to a filtered image dominated by noise [5].

10.2.1.2 Regularized Inverse Filter

To address the afore-mentioned problems of the direct inverse filter, we can slightly modify the transfer function of the inverse filter leading to *regularized inverse* filters. The frequency response of a commonly used regularized inverse filter is:

$$H^I(k, l) = \frac{H^*(k, l)}{|H(k, l)|^2 + \varepsilon} \tag{10.10}$$

where ε is a positive quantity, usually chosen as a fraction of $\max_{k,l} |H(k, l)|^2$. In Section 10.3.2, we will show that this regularized inverse filter may actually be regarded as a deterministic version of the Wiener filter.

Alternatively, we can use the following pseudo-inverse filter transfer function:

$$H^{+}(k,l) = \begin{cases} \dfrac{1}{H(k,l)}, & |H(k,l)| \geq \epsilon \\ 0, & |H(k,l)| < \epsilon \end{cases} \tag{10.11}$$

where the higher frequency components associated with noise that have magnitude less than ϵ are zeroed out.

It is important to mention that when implementing these filters using the DFT, proper zero-padding must be employed to avoid circular convolution issues.

10.2.1.3 The Least Squares (LS) Filter

The LS filter estimates the original image by minimizing the sum squared errors between the observed image, $y(m, n)$, and the *estimated observed* image, $\hat{y}(m, n) = h(m, n) ** \hat{x}(m, n)$, hence the name LS. This minimization is with respect to the unknown or estimate of the original image, $\hat{x}(m, n)$. The sum squared error cost function for this minimization is:

$$J(\hat{x}) = \sum_{m}\sum_{n} |y(m,n) - \hat{y}(m,n)|^2 \tag{10.12}$$

where summations are over all the pixels within the compact support of the image. Clearly, (10.12) is the total energy of the error image $e(m, n) = y(m, n) - \hat{y}(m, n)$ in the spatial domain. Using the Parseval's theorem, this can be expressed in the frequency domain as:

$$J(\hat{X}) = \sum_{k}\sum_{l} |Y(k,l) - \hat{Y}(k,l)|^2 \tag{10.13}$$

where $\hat{Y}(k, l) = H(k, l)\hat{X}(k, l)$ is the 2-D DFT of $\hat{y}(m, n)$. Taking the derivative of J with respect to $\hat{X}(k, l)$ and setting the result to zero gives:

$$\frac{\partial J}{\partial \hat{X}(k,l)} = 0 \Rightarrow -H^{*}(k,l)(Y(k,l) - H(k,l)\hat{X}(k,l)) = 0$$

which leads to the same transfer function as the direct inverse filter, i.e.

$$\hat{X}_{\mathrm{LS}}(k,l) = Y(k,l)/H(k,l) \tag{10.14}$$

An interesting question is that, what would you change in the above LS cost function (Problem 10.2) to obtain the regularized inverse filter in (10.5).

Based upon the results in this section, we can conclude that deterministic-based solutions lead to the direct inverse filter or a type of regularized inverse filter.

10.2.2 Stochastic Filters – 1-D Wiener Filter

Wiener filter takes into account the first- and second-order statistics of the noise and image to construct the filter transfer function. It provides the best linear estimate of the original image by minimizing the mean squared error (MSE), and is thus referred to as a Minimum Mean Squared Error (MMSE) estimator. It assumes global wide-sense stationarity (WSS) of the image field, which is not a valid assumption for most images. However, the locally adaptive version of this filter circumvents this shortcoming at the expense of higher computational requirements. In this section, we begin with the development of the 1-D Wiener filter, which can then be extended to the 2-D case in Section 10.2.3.

Consider the scenario depicted in Figure 10.4, where the original signal $x(n)$ is degraded by a known channel with impulse response $h(n)$ and additive zero-mean white noise $\eta(n)$ with variance σ_η^2, resulting in the observed signal $y(n)$. This observed signal is then used as input to the Wiener filter to provide the optimum estimate (in the MMSE sense) $z(n)$ of a desired signal $d(n)$. The error signal $e(n) = z(n) - d(n)$ drives the Wiener filter toward the optimal solution. For our specific application, since we require that the output of the filter provide $z(n) = \hat{x}(n)$, the desired signal will be $d(n) = x(n)$. However, the following development gives a more general Wiener filter transfer function that can be applied to many different estimation problems.

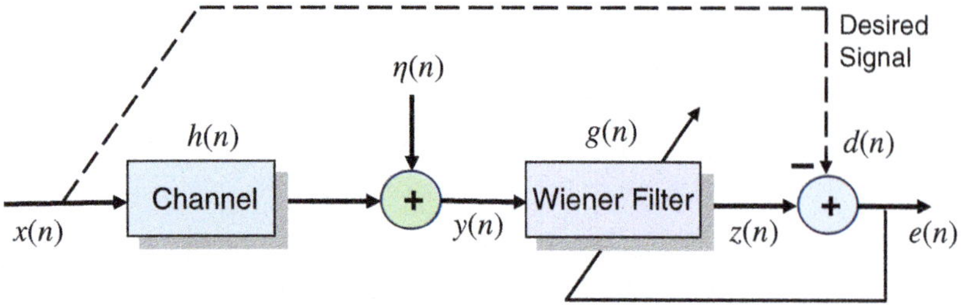

Figure 10.4 1-D Wiener filtering process.

Our goal is to find the finite impulse response (FIR) filter coefficients $g(n)$ so that the output of the filter $z(n)$ is as close as possible to a desired signal $d(n)$ in the MMSE sense. That is, we would like to minimize:

$$J(g(n)) = E[e^2(n)] = E[(z(n) - d(n))^2] \tag{10.15}$$

with respect to $g(j),\ j \in [0, N]$ where $E[.]$ stands for expectation operation. Substituting $z(n) = \sum_{i=0}^{N} g(i)y(n-i) = g(n) * y(n)$ into (10.15), taking partial derivative of $J(g(n))$ with respect to $g(j)$, and setting the result to zero gives:

$$\frac{\partial J(g)}{\partial g(j)} = 0 \Rightarrow 2E[y(n-j)(\sum_{i=0}^{N} g(i)y(n-i) - d(n))] = 0, \quad j \in [0, N] \tag{10.16}$$

Rearranging this equation and interchanging summation and expectation operations produces:

$$\sum_{i=0}^{N} g(i)E[y(n-i)y(n-j)] = E[d(n)y(n-j)].$$

Assuming that random sequences $d(n)$ and $y(n)$ are jointly WSS with cross-correlation function $R_{dy}(j) = E[d(n)y(n-j)]$ and $R_{yy}(j) = E[y(n)y(n-j)]$ is the auto-correlation of $y(n)$, the above equation becomes:

$$\sum_{i=0}^{N} g(i)R_{yy}(j-i) = R_{dy}(j) \quad \text{or} \quad g(j) * R_{yy}(j) = R_{dy}(j). \tag{10.17}$$

Note that in this chapter, to simplify the notation, we use lowercase letters to represent both the r.p.s and their sample functions.

Equation (10.17) is the normal equation to solve for $g(j)$ in the time domain. More specifically, if we plug in $j \in [0, N]$, we obtain the following system of linear equations:

$$\mathbf{R}_y\, \underline{g} = \underline{r}_{dy}, \tag{10.18}$$

where $\underline{g} = [g(0) \cdots g(N)]^T$, $\underline{r}_{dy} = [R_{dy}(0) \cdots R_{dy}(N)]^T$ is the cross-correlation vector, and $\mathbf{R}_y$ is the correlation matrix of data $y(n)$:

$$\mathbf{R}_y = \begin{bmatrix} R_{yy}(0) & R_{yy}(-1) & \cdots & R_{yy}(-N) \\ R_{yy}(1) & R_{yy}(0) & \cdots & R_{yy}(-N+1) \\ \vdots & & \ddots & \vdots \\ R_{yy}(N) & R_{yy}(N-1) & \cdots & R_{yy}(0) \end{bmatrix},$$

which is symmetric and Toeplitz due to WSS, i.e. $\mathbf{R}_y^T = \mathbf{R}_y$. The linear system of equations in (10.18) provides a simple design method in the time domain to find the unknown impulse response coefficients of the Wiener filter. Once the Wiener filter is designed, the implementation phase involves applying the signal $y(n)$ and producing the filtered signal $z(n)$ in the time domain. However, in many situations where the correlation matrix $\mathbf{R}_y$ is of a large dimension, an alternative approach for the design and implementation of the Wiener filter is to perform the operations in the frequency domain. Taking the DFT of both sides of (10.17) yields the general expression for the frequency response of the Wiener filter as:

$$G(k) = \frac{S_{dy}(k)}{S_{yy}(k)} \tag{10.19}$$

where $G(k) = \text{DFT}\{g(n)\}$, $S_{dy}(k) = \text{DFT}\{R_{dy}(j)\}$ is the cross-power spectrum between the desired signal $d(n)$ and the observed signal $y(n)$, and $S_{yy}(k) = \text{DFT}\{R_{yy}(j)\}$ is the power-spectrum of $y(n)$. The transfer function in (10.19) is general in the sense that it can be used to derive the specific transfer function of the Wiener filter for different observation models. We will show how this transfer function can be used for two different observation models: one corresponding to the additive white noise model depicted in Figure 10.4 and another involving both additive and multiplicative noise.

In additive noise case, the observation model is $y(n) = h(n) * x(n) + \eta(n)$ and the goal is to find the optimal (in the MMSE sense) estimate, $\hat{x}(n)$, of the original signal $x(n)$ given the observation signal $y(n)$, the channel impulse response, $h(n)$, and the statistics of the signal $x(n)$ and noise $\eta(n)$. We assume that the signal and noise are uncorrelated. Furthermore, we assume that representative realizations of signal $x(n)$, with the same first- and second-order statistics are available and can be used as the desired signal during the training or design phase, i.e. $d(n) = x(n)$. Thus, the dashed lines in Figure 10.4 are present only during the training phase. Once the Wiener filter is designed, these connections are no longer needed during implementation.

To derive the frequency response of the Wiener filter for this observation model using (10.19), we need to find expressions for the numerator $S_{dy}(k)$ and denominator $S_{yy}(k)$ in terms of the channel frequency response $H(k)$, and the power spectra of the noise and signal. Since $d(n) = x(n)$, we have $R_{dy}(j) = R_{xy}(j)$, which can be expressed as

$$R_{xy}(j) = E[x(n)y(n-j)] = E\left[x(n)\left(\sum_i h(i)x(n-j-i) + \eta(n-j)\right)\right].$$

Interchanging the summation and expectation and using the assumption that $x(n)$ and $\eta(n)$ are uncorrelated yields:

$$\begin{aligned} R_{xy}(j) &= \sum_i h(i)E[x(n)x(n-j-i)] = \sum_i h(i)R_{xx}(i+j) = h(-j) * R_{xx}(-j) \\ &= h(-j) * R_{xx}(j) \end{aligned} \tag{10.20}$$

where $R_{xx}(-j) = R_{xx}(j)$ due to the WSS assumption. In the frequency domain (10.20) becomes:

$$S_{xy}(k) = H^*(k)S_{xx}(k) \tag{10.21}$$

Using a similar approach for the denominator yields:

$$\begin{aligned} R_{yy}(j) &= E[y(n)y(n-j)] = E\left[\left(\sum_i h(i)x(n-i) + \eta(n)\right)y(n-j)\right] \\ &= \sum_i h(i)R_{xy}(j-i) + R_{\eta\eta}(j) \end{aligned} \tag{10.22}$$

Using the result of (10.20) in (10.22) produces the expression for $R_{yy}(j)$ as:

$$R_{yy}(j) = h(j) * h(-j) * R_{xx}(j) + R_{\eta\eta}(j), \tag{10.23}$$

or in the frequency domain:

$$S_{yy}(k) = |H(k)|^2 S_{xx}(k) + S_{\eta\eta}(k). \tag{10.24}$$

Now, substituting the expressions in (10.21) and (10.24) into (10.19) gives the expression for the frequency response of the 1-D Wiener filter for this additive noise observation model:

$$G(k) = \frac{H^*(k)S_{xx}(k)}{|H(k)|^2 S_{xx}(k) + S_{\eta\eta}(k)} \tag{10.25}$$

Once the Wiener filter is designed using (10.25), the filtering operation can simply be implemented in the DFT domain by $\hat{X}(k) = G(k)Y(k)$ and then taking IDFT of the result, i.e. $\hat{x}(n) = \text{IDFT}\{\hat{X}(k)\}$. It must be pointed out that the training phase involves computing the sample autocovariances of noise and signal from the training data with the same statistics and then using them to compute $S_{xx}(k)$ and $S_{\eta\eta}(k)$ in (10.25). However, prior to taking DFT, all the signals must be zero-padded to avoid circular convolution problems.

10.2.2.1 Properties and Remarks

1. For the noise-only case with no channel effects, i.e. $h(n) = \delta(n)$ and $H(k) = 1$, the Wiener filter frequency response becomes:

$$G(k) = \frac{S_{xx}(k)}{S_{xx}(k) + S_{\eta\eta}(k)}$$

If $\eta(n)$ is a white noise process, its power spectrum is $S_{\eta\eta}(k) = \sigma_\eta^2$. In this case, we can easily determine the power spectrum, $S_{xx}(k)$, and the noise variance, σ_η^2 , directly from the observed spectrum $S_{yy}(k) = S_{xx}(k) + \sigma_\eta^2$. Since the signal spectrum $S_{xx}(k)$ is typically band-limited, the observed spectrum becomes approximately flat at high frequencies, where it is dominated by noise, i.e. $S_{yy}(k) \approx \sigma_\eta^2$. Figure 10.5 shows a band-limited power spectrum superimposed on the flat power spectrum of white noise.
Moreover, in this special case, the frequency response of the Wiener filter can alternatively be expressed as:

$$G(k) = \frac{S_{xx}(k)}{S_{xx}(k) + S_{\eta\eta}k)} = \frac{\mathrm{SNR}(k)}{\mathrm{SNR}(k) + 1}$$

where $\mathrm{SNR}(k) = \frac{S_{xx}(k)}{S_{\eta\eta}(k)}$ is the signal-to-noise ratio (SNR) at frequency index k. For frequencies for which $\mathrm{SNR}(k) >> 1$, the filter gain $G(k) \approx 1$, and these frequency components are in the passband, while for frequencies where $\mathrm{SNR}(k) << 1$, we have $G(k) \approx \mathrm{SNR}(k)$, i.e. all such frequencies are attenuated in proportion to their SNR. This shows the *adaptive* nature of the Wiener filter, which adjusts its gain depending on the SNR at each frequency.
2. If the noise is absent, i.e. $S_{\eta\eta}(k) = 0$, the Wiener filter degenerates to the direct inverse filter, i.e. $G(k) = \frac{1}{H(k)}$.
3. Since autocovariances are real and even, the power spectra $S_{xx}(k)$ and $S_{\eta\eta}(k)$ will also be real and nonnegative. Therefore, the phase of $G(k)$ is:

$$\Phi_G(k) = \Phi_{H^*}(k) = -\Phi_H(k),$$

which is the same as the phase of the direct inverse filter.
4. Wiener filter does not, in general, restore the power spectrum of the original signal, i.e. $S_{\hat{x}\hat{x}}(k) \neq S_{xx}(k)$. To see this, let us first find $R_{\hat{x}\hat{x}}(j)$ using $\hat{x}(n) = g(n) * y(n)$ and the results obtained earlier in this section, i.e.

$$R_{\hat{x}\hat{x}}(j) = g(j) * g(-j) * R_{yy}(j) \tag{10.26}$$

Taking the DFT and using $S_{yy}(k) = |H(k)|^2 S_{xx}(k) + S_{\eta\eta}(k)$, the power spectrum of the filter output becomes:

$$\begin{aligned} S_{\hat{x}\hat{x}}(k) &= |G(k)|^2 S_{yy}(k) \\ &= \frac{|H(k)|^2 S_{xx}^2(k)}{|H(k)|^2 S_{xx}(k) + S_{\eta\eta}(k)} \end{aligned} \tag{10.27}$$

That is, only in the absence of noise, i.e. when $S_{\eta\eta}(k) = 0$, we get $S_{\hat{x}\hat{x}}(k) = S_{xx}k)$. Otherwise, $S_{\hat{x}\hat{x}}(k) \neq S_{xx}k)$.
The power spectrum of the estimation error $e(n) = x(n) - \hat{x}(n)$ can also be found as:

$$\begin{aligned} S_{ee}(k) &= |1 - G(k)H(k)|^2 S_{xx}(k) + |G(k)|^2 S_{\eta\eta}(k) \\ &= \frac{S_{xx}(k) S_{\eta\eta}(k)}{|H(k)|^2 S_{xx}(k) + S_{\eta\eta}(k)} \end{aligned} \tag{10.28}$$

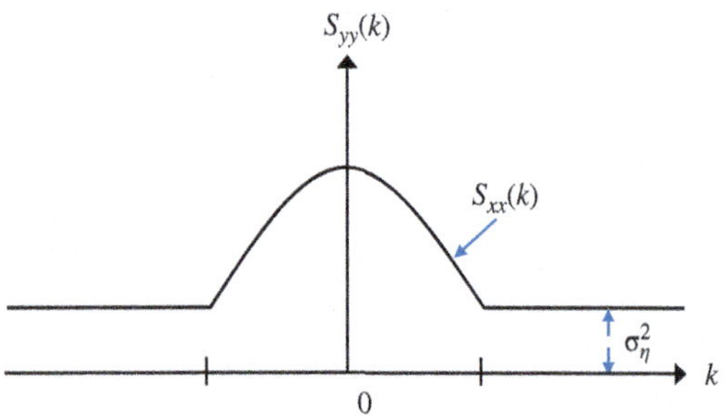

Figure 10.5 Estimating signal power spectrum and noise variance from power spectrum of the observed signal.

Example 10.2 *Geometric mean filter [5] is obtained by geometric mean combination of the frequency responses of the Wiener filter and pseudo-inverse filter in (10.11) as:*

$$G_s(k) = (H^+(k))^s \left[\frac{H^*(k)S_{xx}(k)}{|H(k)|^2 S_{xx}(k) + S_{\eta\eta}(k)} \right]^{1-s}.$$

Show that unlike the Wiener filter, the geometric mean filter restores the power spectrum of the original signal when $s = 1/2$, that is, $S_{\hat{x}\hat{x}} = S_{xx}$.

Solution: Using the geometric mean filter frequency response for $s = 1/2$ in (10.27) gives:

$$S_{\hat{x}\hat{x}}(k) = |G_s(k)|^2 S_{yy}(k) = |H^+(k)\, H(k)| S_{xx}(k)$$

However, since:

$$H^+(k) = \begin{cases} \dfrac{1}{H(k)} & |H(k)| \neq 0 \\ 0 & |H(k)| = 0, \end{cases}$$

we get $S_{\hat{x}\hat{x}}(k) = S_{xx}(k)$ for $|H(k)| \neq 0$.

10.2.3 Stochastic Filters – 2-D Wiener Filter

The extension of the 1-D Wiener filtering for image restoration is straightforward under the assumption of 2-D WSS. For a degradation process consisting of LSI blur with PSF $h(m, n)$ and additive noise $\eta(m, n)$, the 2-D Wiener filter transfer function is:

$$G(k, l) = \frac{H^*(k, l)S_{xx}(k, l)}{|H(k, l)|^2 S_{xx}(k, l) + S_{\eta\eta}(k, l)} \tag{10.29}$$

where $H(k, l) = \text{2-D DFT}\{h(m, n)\}$, and $S_{xx}(k, l)$ and $S_{\eta\eta}(k, l)$ are the power spectra of the original image $x(m, n)$ and additive noise, $\eta(m, n)$, respectively. For practical implementation using the 2-D DFT, it is typically assumed that $h(m, n)$ and $R_{xx}(i, j)$ and $R_{\eta\eta}(i, j)$ are of finite support. All the properties of the 1-D Wiener filter in the previous section extend directly to the 2-D case.

It is instructive to express the transfer function in (10.29) as:

$$G(k, l) = \frac{H^*(k, l)}{|H(k, l)|^2 + S_{\eta\eta}(k, l)/S_{xx}(k, l)}$$

and compare it with the regularized inverse filter in (10.5). The ratio $S_{\eta\eta}(k, l)/S_{xx}(k, l) > 0$ in the above expression can be viewed as a frequency-dependent regularization parameter, analogous to the constant ε used in regularized inverse filter. Thus, the Wiener filter does not have the singularity problems of the direct inverse filter. Moreover, for frequencies where $S_{\eta\eta}(k, l) >> S_{xx}(k, l)$, the filter response $G(k, l)$ becomes small, thereby de-emphasizing the effects of noisy observations at those frequencies. Recall that we had the opposite effects in the inverse filter, i.e. the Wiener filter does not have the ill-conditioning problems either. However, a key limitation of the standard 2-D Wiener filter in image restoration is the WSS assumption, which is not valid for real images. As a result, the restored images can exhibit edge smearing artifacts.

Figure 10.6a,b shows the original "Baboon" image and its blurred version obtained using horizontal motion blur generated by spatial averaging with a 1×5 mask. The blurred image is corrupted by adding white noise to produce a SNR ≈ 7 dB. This blurred and noisy image, shown in Figure 10.6c, is then applied to a 2-D Wiener filter, which produced the restored image illustrated in Figure 10.6d. The effectiveness of the restoration process is clearly evident throughout the filtered image especially in the Baboon's hair and whiskers where subtle details were restored. However, as can be seen, some smearing artifacts are also noticeable mainly because of the global WSS assumption.

To show the effectiveness of the 2-D Wiener filter for color image restoration, consider the original color image "Finn" in Figure 5.10a. A horizontal motion blur image was first applied using a spatial averaging mask of size 1×64. Subsequently, white noise was added to the blurred image to produce an SNR ≈ 4.4 dB. The blurred and noisy version of Finn, shown in Figure 10.7a, was then used as the input to the standard 2-D Wiener filter, the result of which is presented in

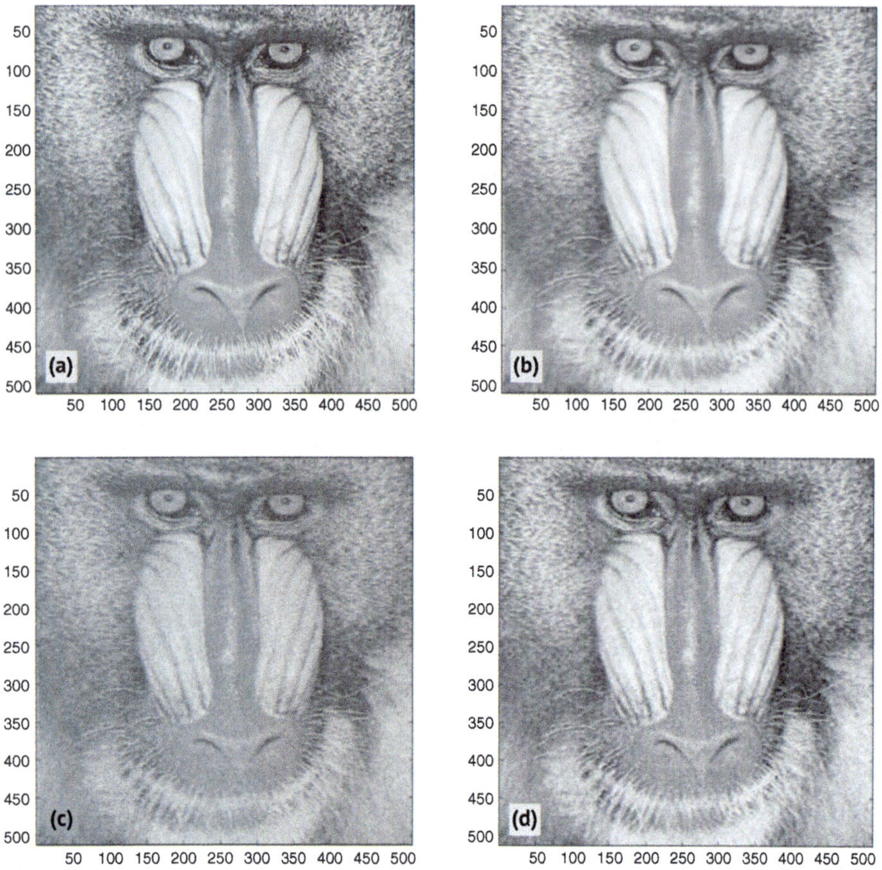

Figure 10.6 (a) Original Baboon image. *Source:* Juang et al. 2012 / John Wiley & Sons / CC BY 3.0. (b) Baboon with motion blur, (c) Baboon with blur and additive noise, and (d) 2-D Wiener filtered result.

Figure 10.7b. The restored image visually reveals the effectiveness of the 2-D Wiener filter for simultaneous noise removal and deblurring. The SNR of the restored image was measured to be approximately 16 dB, which also attests to this significant improvement. The bottom row figures in 10.7 provide a close-up evaluation of the performance in a region of the image around the right eye. Figure 10.7c illustrates the original region, Figure 10.7d shows its blurred and noisy version, whereas Figure 10.7e is the restored result. As can be seen, the Wiener filter successfully restores most of structural details in this region, especially in the eye and hair portions. The smearing artifacts inherent in the standard filtering can be significantly reduced using locally adaptive Wiener filtering [17], which relies on estimating the local sample covariances of noise in relatively flat areas of the image, while the covariances of the image are computed in regions containing edges and textures. These local statistics are then used to build the transfer function of a spatially adaptive Wiener filter. Alternatively, the Kalman filter, in Sections 10.2.4 and 10.2.5, can be used for improved restoration performance.

10.2.3.1 2-D Wiener Filter – Multiplicative and Additive Noise Model

Recall that the general Wiener filter frequency response in (10.19) can be applied to different observation models. Here, we will show how the 2-D version:

$$G(k,l) = \frac{S_{xy}(k,l)}{S_{yy}(k,l)}$$

can be used to derive 2-D Wiener filter for an observation model consisting of LSI blur and both multiplicative and additive noise:

$$y(m,n) = \gamma(m,n)[h(m,n) ** x(m,n)] + \eta(m,n) \tag{10.30}$$

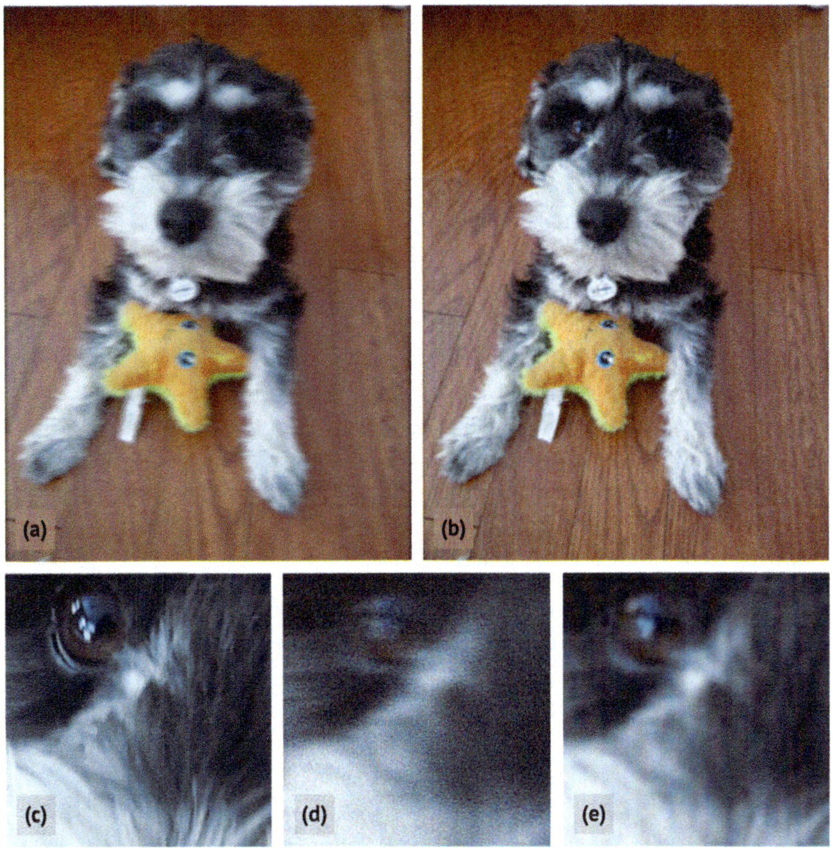

Figure 10.7 (a) Blurred and noisy Finn (SNR 4.38 dB), (b) 2-D Wiener filtered result (SNR 16.09 dB), (c) Original eye region, (d) blurred and noisy eye (SNR 4.38 dB), and (e) restored eye (SNR 16.09 dB).

where $\gamma(m, n)$ and $\eta(m, n)$ are mutually uncorrelated with mean values μ_γ and μ_η, and ACFs $R_{\gamma\gamma}(i, j)$ and $R_{\eta\eta}(i, j)$, respectively. Moreover, they are assumed to be uncorrelated with the original image $x(m, n)$, which has zero-mean and ACF $R_{xx}(i, j)$.

Applying the same approach used for the 1-D additive observation model, the cross-covariance between $x(m, n)$ and $y(m, n)$ is:

$$R_{xy}(i, j) = E\big[x(m + i, n + j)[\gamma(m, n)[h(m, n) ** x(m, n)] + \eta(m, n)]\big]$$
$$= \mu_\gamma R_{xx}(i, j) ** h(-i, -j)$$

where the additive-noise term vanishes because $x(m, n)$ and $\eta(m, n)$ are uncorrelated and $x(m, n)$ has zero mean. Therefore, in the frequency domain,

$$S_{xy}(k, l) = \mu_\gamma S_{xx}(k, l)H^*(k, l) \tag{10.31}$$

For the denominator term, the ACF of the observed image is:

$$R_{yy}(i, j) = R_{\gamma\gamma}(i, j)[h(i, j) ** h(-i, -j) ** R_{xx}(i, j)] + R_{\eta\eta}(i, j)$$

To find the expression for the power spectrum, $S_{yy}(k, l)$, we use the convolution in the frequency domain property, since the above ACF is the product of two terms plus the power spectrum of noise. That is, we have:

$$S_{yy}(k, l) = S_{\gamma\gamma}(k, l) ** [|H(k, l)|^2 S_{xx}(k, l)] + S_{\eta\eta}(k, l), \tag{10.32}$$

where $**$ denotes 2-D convolution in the frequency domain. Thus, the 2-D Wiener filter frequency response for this observation model becomes:

$$G(k,l) = \frac{\mu_y S_{xx}(k,l)H^*(k,l)}{S_{\gamma\gamma}(k,l) ** [|H(k,l)|^2 S_{xx}(k,l)] + S_{\eta\eta}(k,l)}. \tag{10.33}$$

10.2.4 Stochastic Filters – Strip Kalman Filter

The Kalman filter is a *state estimator* in which the state variables are dynamical random variables governed by a differential or difference equation driven by a random process, together with an appropriate set of initial conditions. The concept of *state* is central to the linear system theory and is carried over to the case of stochastic differential and difference equations used in Kalman filtering. In the stochastic case, the state can be defined as the minimum set of information about past and present estimates needed to determine an optimal causal estimate of future states, given future observations.

To apply Kalman filter for image restoration, the state variables correspond to several pixels of the original image within a ROS of an image model. The Kalman filter then produces an estimate of the state vector recursively for every new blurred and noisy observation. Unlike the global Wiener filter, the Kalman filter produces local estimates of the original image within the ROS of the model. As a result, smearing artifacts are reduced compared to the global Wiener filter. To perform image restoration with Kalman filter, the following steps should be implemented:

1. Choose an appropriate 1-D or 2-D AR model that accurately represents the original image random field. The methods in Chapter 9 can be used to find the parameters of 2-D AR models with causal NSHP, semicausal, or noncausal ROS and those of the 1-D vector AR models.
2. Construct the corresponding state and observation equations with state variables being image pixels to be estimated in the ROS of the model, and
3. Implement the recursive Kalman filtering equations to update the state estimates for every new observation.

In this section, we present a 1-D strip Kalman filter [13, 15], which uses a 1-D vector AR model in every image strip to represent the original image. The idea of the strip Kalman filtering is depicted in Figure 10.8, where the image is partitioned into overlapping strips of width W, typically with 50% overlap between adjacent strips. The figure also illustrates the ROS of the vector AR model that consists of M present and past image vectors $\underline{z}(k), \underline{z}(k-1), \ldots \underline{z}(k-M+1)$. Thus, as discussed in Chapter 9, this 1-D vector AR model of order M can be expressed as:

$$\underline{z}(k) = \mathbf{A}_1^T \underline{z}(k-1) + \cdots + \mathbf{A}_M^T \underline{z}(k-M) + \underline{e}(k) \tag{10.34}$$

where $\underline{z}(k) = [x((i-1)W, k)\; x((i-1)W+1, k) \cdots x(iW-1, k)]^T$ is a vector of the original noise-free image pixels, $x((i-1)W+l, k)$, in the ith strip of dimension W, $\underline{e}(k)$ is the driving noise vector, and $A_i \in \mathbb{R}^{W\times W}$ are model parameter matrices. We assume that $E[\underline{e}(k)] = 0$ and $E[\underline{e}(k)\underline{e}^T(l)] = \mathbf{Q}_{\underline{e}}\delta(k-l)$ with $Q_{\underline{e}} \in \mathbb{R}^{W\times W}$ is the covariance matrix of $\underline{e}(k)$. As described in Section 9.3, this driving process is white inter-column-wise while the elements of the vector $\underline{e}(k)$ are mutually correlated within each column (intra-column-wise colored).

In Section 9.3.1, we showed how the Yule-Walker method can be extended to find the parameter matrices, $\mathbf{A}_i$, $i \in [1, M]$, as well as the covariance matrix of the driving process $\mathbf{Q}_{\underline{e}}$. Assuming that these matrices have been computed, the next step is to construct the state equation. We define the *state* vector as:

$$\underline{s}(k) = [\underline{z}^T(k), \underline{z}^T(k-1), \cdots \underline{z}^T(k-M+1)]^T \in \mathbb{R}^{WM}.$$

As shown in Figure 10.8, the state vector, $\underline{s}(k)$, is composed of vector $\underline{z}(k)$ and all $M-1$ previous vectors $\underline{z}(k-l)$, $l \in [1, M-1]$ within the ROS of the model, excluding $\underline{z}(k-M)$. Now, the state equation in the *canonical* form [18] can be derived as:

$$\underline{s}(k) = \mathbf{F}\underline{s}(k-1) + \mathbf{G}\underline{e}(k) \tag{10.35}$$

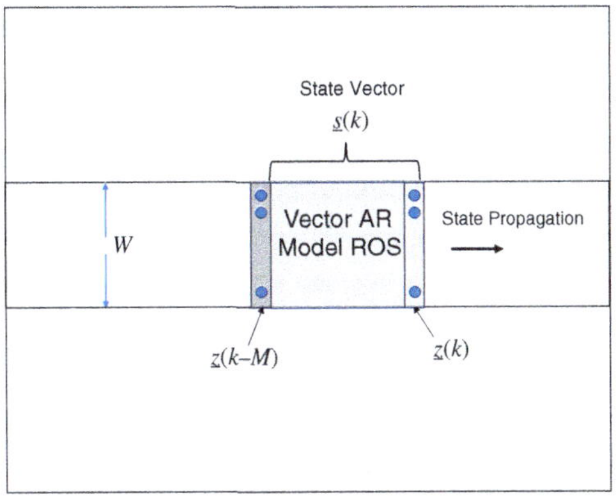

Figure 10.8 Strip vector AR model support region and state vector.

where **F** and **G** matrices are defined as:

$$\mathbf{F} = \begin{bmatrix} \mathbf{A}_1^T & \mathbf{A}_2^T & \cdots & \cdots & \mathbf{A}_M^T \\ \mathbf{I}_W & \mathbf{0} & \cdots & \cdots & \mathbf{0} \\ \vdots & & & & \vdots \\ & & \ddots & & \\ \vdots & & & \ddots & \vdots \\ \mathbf{0} & \mathbf{0} & \cdots & \cdots & \mathbf{I}_W \end{bmatrix}, \quad \mathbf{G} = \begin{bmatrix} \mathbf{I}_W \\ \mathbf{0} \\ \vdots \\ \vdots \\ \mathbf{0} \end{bmatrix}. \tag{10.36}$$

Let the observation vector of the degraded image be defined as $\underline{y}(k) = [y((i-1)W, k),\ y((i-1)W+1, k), \dots, y(iW-1, k)]^T$. Then, to produce estimate of the state vector $\hat{\underline{s}}(k)$ recursively, we need to relate this observation vector to the state vector, PSF of the LSI blur, and the additive measurement noise via an observation equation. Assuming that the LSI blur is modeled by a vector MA process with the same ROS as the AR model, the observation equation becomes:

$$\underline{y}(k) = \mathbf{H}_0\underline{z}(k) + \mathbf{H}_1\underline{z}(k-1) + \cdots + \mathbf{H}_{M-1}\underline{z}(k-M+1) + \underline{n}(k) \tag{10.37}$$

where $\underline{n}(k) \in \mathbb{R}^W$ represents the additive measurement noise vector with $E[\underline{n}(k)] = \underline{0}$ and $E[\underline{n}(k)\underline{n}^T(j)] = \mathbf{R}_{\underline{n}}\delta(k-j)$ with $\mathbf{R}_{\underline{n}}$ being the covariance matrix of $\underline{n}(k)$. The matrices $\mathbf{H}_i \in \mathbb{R}^{W\times W}$ lower triangular Toeplitz matrices defined in terms of the PSF of the 2-D LSI (MA) blur of order M, i.e.

$$\mathbf{H}_i = \begin{bmatrix} h(0,i) & 0 & 0 & 0 & \cdots & 0 \\ h(1,i) & h(0,i) & 0 & 0 & \cdots & 0 \\ \vdots & \vdots & \ddots & \vdots & \vdots & \vdots \\ h(M-1,i) & \cdots & h(1,i) & h(0,i) & 0 & \cdots \\ \ddots & \ddots & \ddots & \ddots & \ddots & 0 \\ 0 & \cdots & h(M-1,i) & \cdots & h(1,i) & h(0,i) \end{bmatrix}, \quad i \in [0, M-1]$$

where it is assumed that $M < W$. The observation equation in (10.37) can be arranged in terms of the state vector $\underline{s}(k)$ as:

$$\underline{y}(k) = \mathbf{H}\underline{s}(k) + \underline{n}(k) \tag{10.38}$$

where $\mathbf{H} = [\mathbf{H}_0 \cdots \mathbf{H}_{M-1}] \in \mathbb{R}^{W\times WM}$.

Given the observation vector $\underline{y}(k)$, the goal of the Kalman filter is to provide recursive estimates of the state vector using all available observations up to time k. Note that here, the index k plays the role of a spatial progression variable (analogous to time), corresponding to the current position of the observation vector within an image strip. We denote this estimate of the state vector by $\hat{\underline{s}}(k|k) = E[\underline{s}(k)|\mathcal{Y}(k)]$, where $\mathcal{Y}(k) = \{\underline{y}(1), \cdots, \underline{y}(k)\}$ is a set of all observations up to the present time k. Then, assuming that the previous state estimate at time $k-1$ is available, the Kalman filter equations are implemented in the following order:

1. *Initial State Estimate (prediction):* $\hat{\underline{s}}(k|k-1) = \mathbf{F}\hat{\underline{s}}(k-1|k-1)$. This step generates an initial (before the updating) estimate of the state vector at time k based on all observations up to $k-1$, i.e. $\hat{\underline{s}}(k|k-1) = E[\underline{s}(k)|\mathcal{Y}(k-1)]$.
2. *A Priori Error Covariance Matrix:* $\mathbf{P}(k|k-1) = \mathbf{F}\mathbf{P}(k-1|k-1)\mathbf{F}^T + \mathbf{G}\mathbf{Q}_{\underline{e}}\mathbf{G}^T$ where $\mathbf{P}(k|k-1) = E[(\underline{s}(k) - \hat{\underline{s}}(k|k-1))(\underline{s}(k) - \hat{\underline{s}}(k|k-1))^T]$. This step computes the error covariance matrix of the a priori (before the updating) state estimate using the error covariance matrix of the updated estimate at iteration $k-1$.
3. *Kalman Gain Matrix:* $\mathbf{K}(k) = \mathbf{P}(k|k-1)\mathbf{H}^T[\mathbf{H}\mathbf{P}(k|k-1)\mathbf{H}^T + \mathbf{R}_{\underline{n}}]^{-1}$. This step computes the Kalman gain matrix needed for the update step.
4. *Update State Estimate:* $\hat{\underline{s}}(k|k) = \hat{\underline{s}}(k|k-1) + \mathbf{K}(k)[\underline{y}(k) - \mathbf{H}\hat{\underline{s}}(k|k-1)]$. This step updates the state estimate using the *innovation* process, $\underline{\xi}(k) = \underline{y}(k) - \mathbf{H}\hat{\underline{s}}(k|k-1)$, and the Kalman gain matrix.
5. *A Posteriori Error Covariance Matrix:* $\mathbf{P}(k|k) = [\mathbf{I} - \mathbf{K}(k)\mathbf{H}]\mathbf{P}(k|k-1)$ where $\mathbf{P}(k|k) := E[(\underline{s}(k) - \hat{\underline{s}}(k|k))(\underline{s}(k) - \hat{\underline{s}}(k|k))^T]$. This step prepares the a posteriori (after the updating) error covariance matrix needed for the next iteration when the state estimation moves to time $k+1$.

10.2.4.1 Important Remarks

- To initialize the Kalman filter, we use $\hat{\underline{s}}(0|0) = \mu_X\underline{1}$ and $\mathbf{P}(0|0) = \sigma_X^2\mathbf{I}$ where μ_X is a user-specified initial guess for the mean value, σ_X^2 is a measure of confidence in μ_X, i.e. larger σ_X^2 for lesser confidence, and $\underline{1} = [1, \cdots 1]^T$, i.e. a vector of all ones.
- The Kalman filter is driven by the *innovation process* $\underline{\xi}(k) = \underline{y}(k) - \hat{\underline{y}}(k|k-1)$ defined in Step 4, where $\hat{\underline{y}}(k|k-1) = \mathbf{H}\hat{\underline{s}}(k|k-1)$ is the estimate of the observation vector based on the initial state estimate.
- Equations in Steps 2, 3, and 5 are not data dependent and therefore can be iterated offline until convergence is reached for $\mathbf{K}(k)$ to a steady-state value $\bar{\mathbf{K}}$. This steady-state gain can then be used online in the state updating process of Step 4. It is also important to note that the convergence of the gain is directly dependent on the modeling accuracy of the image and degradation processes, specifically, the state and observation models. That is, a poor model fit leads to poor convergence and performance.
- The Kalman gain matrix decreases as $\mathbf{Q}_{\underline{e}}$ decreases, which happens when the image model and hence the initial state estimate are accurate in representing the image field. In this case, the Kalman gain reduces to put less emphasize on the observation. The Kalman gain matrix also decreases when $\mathbf{R}_{\underline{n}}$ increases, since in this case the observations are less reliable due to heavy noise, hence more reliance on the model and initial estimate.
- When $\mathbf{Q}_{\underline{e}} \approx \mathbf{0}$, the image model is highly accurate, in which case the Kalman gain approaches zero, causing the filter to rely primarily on predictions as it becomes decoupled from the observations. To avoid data saturation as a result of this decoupling, it is recommended [19] to use $\mathbf{Q}_{\underline{e}} = \varepsilon\mathbf{I}$ where ε is a small positive value.

The original Baboon image in Figure 10.6a was corrupted by adding white noise with a large variance to make SNR of approximately −1.7 dB. This severely degraded image, shown in Figure 10.9a, was then used to test the effectiveness of the strip Kalman filter. A first-order vector AR model was employed, where the parameter matrix $\mathbf{A}_1$ and the covariance matrix of the driving process $\mathbf{Q}_{\underline{e}}$ were obtained using the vector Yule-Walker method in Section 9.3.1 based on a small portion of the noise-free image. The strip width was chosen to be $W = 64$, with no overlap between adjacent strips. Since no blur is present, observation matrix becomes $\mathbf{H} = [\mathbf{I}\ \mathbf{0} \cdots \mathbf{0}]$. The restored image, illustrated in Figure 10.9b, indicates impressive performance of this filtering even in presence of severe noise. The SNR after the filtering was found to be SNR = 2.9 dB which exhibits ≈4.6 dB improvements. The restoration effects are more noticeable in the hair and whiskers areas, which were severely degraded by heavy noise. Some edge effects are also noticeable because no overlap was used between adjacent strips, highlighting the importance of overlapping in such strip-based processing.

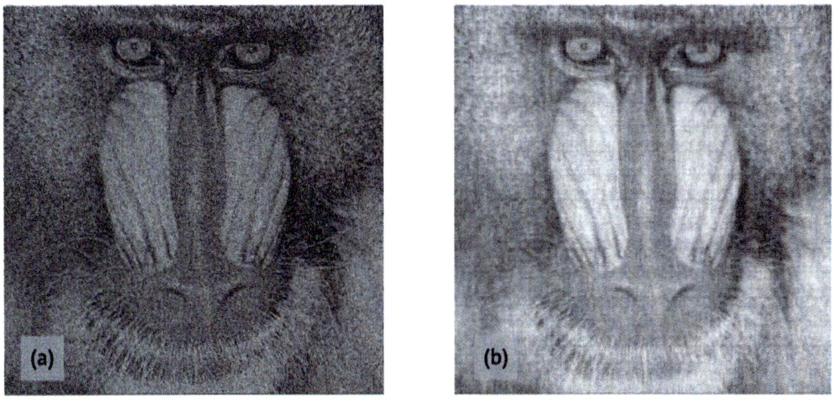

Figure 10.9 (a) Noisy Baboon and (b) restored image using strip Kalman filter with $M = 1$ and $W = 64$.

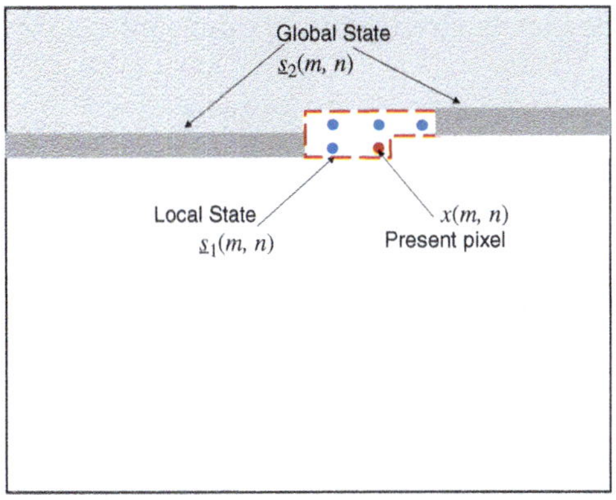

Figure 10.10 Global and local states and NSHP support.

10.2.5 Stochastic Filters – Reduced Update Kalman Filter (RUKF)

There are many 2-D extensions of the 1-D Kalman filter for image restoration, including those in [12, 14, 20]. In this section, we introduce the reduced update Kalman filtering (RUKF) method in [12] that overcomes some of the problems that had plagued the 2-D extension of Kalman processing.

The RUKF framework is based on defining the *local* and *global* states. The local state is defined within a nonsymmetric half-plane (NSHP) ROS, and propagates one pixel at a time (location) in a raster scan fashion. It consists of the *present* pixel being processed and all those pixels within the ROS that lie in the *past*. Figure 10.10 illustrates the local state for a first-order NSHP ROS (red dashed lines) consisting of four past pixels (blue) and the present pixel (red). Pixels not yet scanned are considered to be in the *future*. The global state acts as storage for the previously processed pixels needed for subsequent local state estimation when it moves to a new position along the same row or when it advances to the next row only after completing the entire row of pixels. The global state for this first-order model, shown in dark gray in Figure 10.10, consists of $N - 3$ pixels where N is the dimension of the image.

The original image, $x(m, n)$, is modeled using a causal 2-D AR model of order M with a NSHP ROS:

$$x(m,n) = \sum_{i,j \in \widehat{W}_1} \sum a_{i,j} x(m-i, n-j) + e(m,n) \tag{10.39}$$

where $e(m, n)$ is a zero-mean white driving process with variance σ_e^2 and $\breve{W}_1$ is the ROS of the model, i.e. $\breve{W}_1 = \{(i, j) : i = 0,\ 1 \le j \le M \text{ and } 1 < i \le M, -M \le j \le M\}$. The model parameters $a_{i,j}$ and the driving noise variance σ_e^2 are assumed to be generated using the Yule-Walker method in Section 9.2.1. For an Mth order 2-D NSHP model, the local state vector, $\underline{s}_1(m, n)$, consists of $2M(M + 1) + 1$ pixels in the ROS:

$$\begin{aligned}\underline{s}_1(m,n) = [&x(m,n)\, x(m,n-1) \ldots x(m,n-M)\ \vdots\ x(m-1,n+M) \ldots \\ &x(m-1,n-M)\ \vdots\ \ldots \vdots\ x(m-M,n+M) \ldots x(m-M,n-M)]^T.\end{aligned} \tag{10.40}$$

The global state, which serves as storage, consists of $M(N - 2M - 1)$ pixels:

$$\begin{aligned}\underline{s}_2(m,n) = [&x(m-1,n+M+1) \ldots x(m-1,N-1)\ \vdots \ldots\ \vdots\ x(m-M,n+M+1) \ldots x(m-M,N-1) \\ &\vdots\ x(m,n-M-1) \ldots x(m,0) \ldots \vdots\ x(m-M+1,n-M-1) \ldots x(m-M+1,0)]^T.\end{aligned} \tag{10.41}$$

Augmenting these two state vectors forms the entire state vector:

$$\underline{s}(m,n) = [\underline{s}_1^T(m,n)\, \underline{s}_2^T(m,n)]^T \in \mathbb{R}^{MN+M+1}. \tag{10.42}$$

The state update takes place only within the ROS of the model and hence does not require significant computations since $M \ll N$.

Using the above definitions of the local and global state vectors, the state equation for the RUKF becomes:

$$\underline{s}(m,n) = \mathbf{C}\underline{s}(m,n-1) + \underline{e}(m,n) \tag{10.43}$$

where $\underline{e}(m,n) = [e(m,n)\, 0 \cdots 0]^T$ and the matrix $\mathbf{C}$ is partitioned as:

$$\mathbf{C} = \begin{bmatrix} \mathbf{C}_{11} & \mathbf{C}_{12} \\ \mathbf{C}_{21} & \mathbf{C}_{22} \end{bmatrix}. \tag{10.44}$$

Thus, (10.43) can be rewritten as:

$$\begin{aligned}\underline{s}_1(m,n) &= \mathbf{C}_{11}\underline{s}_1(m,n-1) + \mathbf{C}_{12}\underline{s}_2(m,n-1) + \underline{e}_1(m,n) \\ \underline{s}_2(m,n) &= \mathbf{C}_{21}\underline{s}_1(m,n-1) + \mathbf{C}_{22}\underline{s}_2(m,n-1)\end{aligned} \tag{10.45}$$

where $\underline{e}_1(m,n) = [e(m,n)\, 0 \ldots 0]^T \in \mathbb{R}^{2M(M+1)+1}$ and matrices $\mathbf{C}_{11}$ and $\mathbf{C}_{12}$ are defined as:

$$\mathbf{C}_{11} = \begin{bmatrix}
a_{0,1} & \cdots & a_{0,M} & 0 & a_{1,-M+1} & \cdots a_{1,M} & 0 & \cdots a_{M,-M+1} & \cdots a_{M,M} & 0 \\
1 & 0 & \cdots & 0 & .. & \cdots & 0 & \cdots & \cdots & 0 \\
\vdots & \ddots & \vdots & \vdots & .. & \vdots & \vdots & \vdots & \vdots & \vdots \\
0 & \cdots & 1 & 0 & .. & \cdots & 0 & \cdots & \cdots & 0 \\
0 & \cdots & 0 & 0 & .. & \cdots & 0 & \cdots & \cdots & 0 \\
0 & \cdots & 0 & \cdots & 1 & \cdots & 0 & \cdots & \cdots & 0 \\
\vdots & \vdots & \vdots & \vdots & \vdots & \ddots & \vdots & \vdots & \vdots & \vdots \\
0 & \cdots & 0 & \cdots & 0 & \cdots & 1 & 0 & \cdots & 0 \\
0 & \cdots & 0 & 0 & .. & \cdots & 0 & \cdots & \cdots & 0 \\
0 & \cdots & 0 & \cdots & .. & 0 & 0 & 1 & \cdots & 0 \\
\vdots & \vdots & \vdots & \vdots & \vdots & \vdots & \vdots & \vdots & \ddots & \vdots \\
0 & \cdots & 0 & \cdots & .. & \cdots & 0 & \cdots & 1 & 0
\end{bmatrix}_{(2M(M+1)+1)\times(2M(M+1)+1)},$$

and

$$
\mathbf{C}_{12} = \begin{bmatrix}
a_{1,-M} & 0 & \cdots & a_{2,-M} & 0 & \cdots & .. & a_{M,-M} & 0 & \cdots \\
0 & \cdots & 0 & 0 & .. & \cdots & 0 & \cdots & \cdots & 0 \\
\vdots & \vdots & \vdots & \vdots & .. & \vdots & \vdots & \vdots & \vdots & \vdots \\
0 & \cdots & 0 & 0 & .. & \cdots & 0 & \cdots & \cdots & 0 \\
1 & \cdots & 0 & 0 & .. & \cdots & 0 & \cdots & \cdots & 0 \\
0 & \cdots & 0 & 0 & .. & \cdots & 0 & \cdots & \cdots & 0 \\
\vdots & \vdots & \vdots & \vdots & .. & \vdots & \vdots & \vdots & \vdots & \vdots \\
0 & \cdots & 0 & 0 & .. & \cdots & 0 & \cdots & \cdots & 0 \\
0 & \cdots & 0 & \cdots & .. & 0 & 0 & 1 & \cdots & 0 \\
0 & \cdots & 0 & \cdots & .. & \cdots & 0 & 0 & \cdots & 0 \\
0 & \cdots & 0 & 0 & .. & \cdots & 0 & \cdots & \cdots & 0 \\
\vdots & \vdots & \vdots & \vdots & \vdots & \vdots & \vdots & \vdots & \ddots & \vdots
\end{bmatrix}_{(2M(M+1)+1)\times(M(N-2M-1))}, \tag{10.46}
$$

respectively. The local state vector $\underline{s}_1(m, n)$ contains all pixels from the previous local state $\underline{s}_1(m, n-1)$ except those in its rightmost column, i.e. pixels $x(m-M, n+M)$, $x(m-M+1, n+M) \ldots , x(m-1, n+M)$. These pixels are retrieved from $\underline{s}_2(m, n-1)$ and weighted by the corresponding coefficients of the AR model in the first row of the matrix $\mathbf{C}_{12}$. The remaining unit entries in these matrices represent shift operations. The matrices $\mathbf{C}_{21}$ and $\mathbf{C}_{22}$ mainly implement shift operations for updating the storage in the global state and are given by:

$$
\mathbf{C}_{21} = \left[\begin{array}{cccccc}
0 & 0 & \cdots & \cdots & 0 & 0 \\
\vdots & \vdots & \vdots & \vdots & \vdots & \vdots \\
0 & 0 & \cdots & \cdots & 0 & 0 \\
\hline
0 & 1 & 0 & \cdots & \cdots & 0 \\
0 & 0 & 0 & \cdots & \cdots & 0 \\
\vdots & \vdots & \vdots & \vdots & \vdots & \vdots
\end{array}\right]_{(M(N-2M-1))\times(2M(M+1)+1)} \tag{10.47}
$$

and

$$
\mathbf{C}_{22} = \left[\begin{array}{cccccccc}
0 & 1 & 0 & \cdots & \cdots & \cdots & 0 & 0 \\
0 & 0 & 1 & 0 & 0 & \cdots & \cdots & 0 \\
\vdots & \vdots & \vdots & \ddots & \vdots & \vdots & \vdots & \vdots \\
0 & \cdots & \cdots & 0 & 1 & 0 & \cdots & 0 \\
0 & 0 & 0 & \cdots & 0 & \cdots & 0 & 0 \\
\hline
0 & 0 & \cdots & 0 & 0 & 1 & 0 & \cdots \\
\vdots & \vdots & \vdots & \vdots & \vdots & \ddots & \vdots & \vdots \\
0 & 0 & \cdots & \cdots & 0 & 0 & 0 & 1
\end{array}\right] = \begin{bmatrix} \mathbf{S}_u & \mathbf{O} \\ \mathbf{O} & \mathbf{I} \end{bmatrix}_{(M(N-2M-1))\times(M(N-2M-1))}, \tag{10.48}
$$

where $\mathbf{S}_u$ is a *shifting up* matrix whose dimension depends on the column index n of the current pixel:

$$\mathbf{S}_u = \begin{bmatrix} 0 & 1 & 0 & \cdots & 0 & 0 \\ 0 & 0 & 1 & 0 & \cdots & 0 \\ \vdots & \vdots & \vdots & \ddots & \vdots & \vdots \\ 0 & \cdots & \cdots & 0 & 0 & 1 \\ 0 & \cdots & \cdots & 0 & 0 & 0 \end{bmatrix}_{(M(N-M-n-1))\times(M(N-M-n-1))}.$$

Pre-multiplication by $\mathbf{S}_u$ shifts rows upward by one position and inserts a row of zeros at the bottom row. The identity matrix in (10.48) is also of variable dimension $M(n - M)$.

Assuming that the PSF of the LSI blur is also confined within a causal ROS (quarter plane or NSHP), the observation equation becomes:

$$y(m, n) = \underline{h}^T \underline{s}_1(m, n) + \eta(m, n) \tag{10.49}$$

where $\underline{h}^T$ is a row vector containing the PSF coefficients of the blur and $\eta(m, n)$ is a zero-mean white noise process with variance σ_η^2 representing the additive measurement noise.

Now, using the state and observation equations in (10.45) and (10.49), and given all previous estimates up to $\hat{\underline{s}}(m, n-1)$, the RUKF equations [12] are implemented as follows:

1. ***Initial State Estimate (Prediction):***

   $\hat{\underline{s}}_1(m, n|n-1) = \mathbf{C}_{11}\hat{\underline{s}}_1(m, n-1|n-1) + \mathbf{C}_{12}\hat{\underline{s}}_2(m, n-1|n-1).$

2. ***A Priori Error Covariance Matrix:***

   $\mathbf{P}(m, n|n-1) = \mathbf{C}\mathbf{P}(m, n-1|n-1)\mathbf{C}^T + \sigma_e^2 \underline{g}\underline{g}^T$

   where $\underline{g} = [1\,0 \ldots 0]^T$ and $\mathbf{P}(m, n|n-1) = E[(\underline{s}(m, n) - \hat{\underline{s}}(m, n|n-1))(\underline{s}(m, n) - \hat{\underline{s}}(m, n|n-1))^T]$.

3. ***Kalman Gain Matrix:***

   $\mathbf{K}(m, n) = \mathbf{P}_{11}(m, n|n-1)\underline{h}\left[\underline{h}^T\mathbf{P}_{11}(m, n|n-1)\underline{h} + \sigma_\eta^2\right]^{-1}.$

4. ***Update State Estimate:***

   $\hat{\underline{s}}_1(m, n|n) = \hat{\underline{s}}_1(m, n|n-1) + \mathbf{K}(m, n)[y(m, n) - \underline{h}^T\hat{\underline{s}}_1(m, n|n-1)].$

5. ***A Posteriori Error Covariance Matrix:***

   $\mathbf{P}_{11}(m, n|n) = [\mathbf{I} - \mathbf{K}(m, n)\underline{h}^T]\mathbf{P}_{11}(m, n|n-1)$ where

   $\mathbf{P}_{11}(m, n|n) = E[(\underline{s}_1(m, n) - \hat{\underline{s}}_1(m, n|n))(\underline{s}_1(m, n) - \hat{\underline{s}}_1(m, n|n))^T]$

All the important properties outlined in the previous section for the strip Kalman filter are also applicable for the RUKF. Below are some additional remarks that are specific to this filtering process.

10.2.5.1 RUKF Remarks

1. If there is no blurring process, then $\underline{h}^T = [1\,0 \cdots 0]$ and in Step 3, the term $\underline{h}^T\mathbf{P}_{11}(m, n|n-1)\underline{h} = [\mathbf{P}_{11}(m, n|n-1)]_{1,1}$, i.e. the first element of matrix $\mathbf{P}_{11}(m, n|n-1)$.

2. The error covariance matrix $\mathbf{P}(m, n|n-1)$ in Step 2 is partitioned similar to the matrix $\mathbf{C}$, i.e. $\mathbf{P}(m, n|n-1) = \begin{bmatrix} \mathbf{P}_{11}(m, n|n-1) & \mathbf{P}_{12}(m, n|n-1) \\ \mathbf{P}_{21}(m, n|n-1) & \mathbf{P}_{22}(m, n|n-1) \end{bmatrix}$ where $\mathbf{P}_{21}(m, n|n-1) = \mathbf{P}_{12}^T(m, n|n-1)$. Although only $\mathbf{P}_{11}(m, n|n-1)$ is required for computing the Kalman gain and the a posteriori covariance in Steps 3 and 5, evaluating the full error covariance matrix in Step 2 is computationally expensive. However, due to the sparsity of the block matrices in $\mathbf{C}$, efficient implementations can be developed to avoid computing $\mathbf{P}(m, n|n-1)$ in its full form.
3. The state propagation occurs only along the horizontal direction n. Thus, RUKF is inherently a 1-D Kalman filtering.
4. To avoid artifacts near image boundaries, appropriate boundary conditions were suggested in [12].
5. It was shown [12] that the RUKF is optimal in the sense that it minimizes the MSE over the class of linear, recursive, spatially variable filters of the same order. This implies that the RUKF is better suited for nonstationary image fields when compared with the Wiener filter.

Since the first-order 2-D AR model with NSHP ROS is more commonly used, we illustrate how the local and global state vectors are formed for this case. The local and global state vectors are:

$$\underline{s}_1(m, n) = [x(m, n)\ x(m, n-1)\ x(m-1, n+1)\ x(m-1, n)\ x(m-1, n-1)]^T \tag{10.50}$$

and

$$\underline{s}_2(m, n) = [x(m-1, n+2)\ x(m-1, n+3) \cdots x(m-1, N-1)\ x(m, n-2)\ x(m, n-3) \cdots x(m, 0)]^T. \tag{10.51}$$

Furthermore, the matrices $\mathbf{C}_{11}$, $\mathbf{C}_{12}$, $\mathbf{C}_{21}$, and $\mathbf{C}_{22}$ are defined as:

$$\mathbf{C}_{11} = \begin{bmatrix} a_{0,1} & 0 & a_{1,0} & a_{1,1} & 0 \\ 1 & 0 & 0 & 0 & 0 \\ 0 & 0 & 0 & 0 & 0 \\ 0 & 0 & 1 & 0 & 0 \\ 0 & 0 & 0 & 1 & 0 \end{bmatrix}_{5\times 5}, \tag{10.52}$$

$$\mathbf{C}_{12} = \begin{bmatrix} a_{1,-1} & 0 & \cdots & \cdots & 0 & 0 \\ 0 & 0 & \cdots & \cdots & 0 & 0 \\ 1 & 0 & \cdots & \cdots & 0 & 0 \\ 0 & 0 & \cdots & \cdots & 0 & 0 \\ 0 & 0 & \cdots & \cdots & 0 & 0 \end{bmatrix}_{5\times(N-3)}, \tag{10.53}$$

$$\mathbf{C}_{21} = \begin{bmatrix} 0 & 0 & 0 & 0 & 0 \\ \vdots & \vdots & \vdots & \vdots & \vdots \\ 0 & 0 & 0 & 0 & 0 \\ \hline 0 & 1 & 0 & 0 & 0 \\ 0 & 0 & 0 & 0 & 0 \\ \vdots & \vdots & \vdots & \vdots & \vdots \end{bmatrix}_{(N-3)\times 5},$$

and

$$\mathbf{C}_{22} = \left[\begin{array}{cccccccc} 0 & 1 & 0 & \cdots & \cdots & \cdots & 0 & 0 \\ 0 & 0 & 1 & 0 & 0 & \cdots & \cdots & 0 \\ \vdots & \vdots & \vdots & \ddots & \vdots & \vdots & \vdots & \vdots \\ 0 & \cdots & \cdots & 0 & 1 & 0 & \cdots & 0 \\ 0 & 0 & 0 & \cdots & 0 & \cdots & 0 & 0 \\ \hline 0 & 0 & \cdots & 0 & 0 & 1 & 0 & \cdots \\ \vdots & \vdots & \vdots & \vdots & \vdots & \ddots & \vdots & \vdots \\ 0 & 0 & \cdots & \cdots & 0 & 0 & 0 & 1 \end{array}\right]_{(N-3)\times(N-3)}.$$

To demonstrate the performance of the RUKF for removing additive noise, the original "Lena" image in Figure 10.11a was corrupted by adding white noise to produce an SNR = 5.2 dB. This noisy image is shown in Figure 10.11b. A first-order 2-D AR model with NSHP ROS was used to model the original image. The coefficients of this model, and the variance of the driving noise, were obtained in Section 9.2.4 to be $a_{1,-1} = 0.0959$, $a_{0,1} = 0.4464$, $a_{1,0} = 0.7822$, $a_{1,1} = -0.3234$, and $\sigma_e^2 = 0.0188$. The AR model coefficients are then used to build the matrices $\mathbf{C}_{11}$ and $\mathbf{C}_{12}$, together with the local and global state vectors in (10.50) and (10.51), respectively. The RUKF was then applied recursively to each row of the noisy image. The restored image with an SNR value of approximately SNR = 7.7 dB is shown in Figure 10.11c, providing an improvement of about 2.5 dB. The Wiener filter in (10.29) with $H(k, l) = 1$ was also applied to this noisy image, the result of which is shown in Figure 10.11d. The SNR of this image was found to be SNR = 11.9 dB, i.e. more than 4 dB improvements over that of the RUKF. Visual evaluation of these results reveals that both filters resulted in some smearing artifacts. However, these artifacts are less noticeable for the RUKF, though the Wiener filter provided better noise removal. The blurring artifact is primarily due to the WSS assumption and the global nature of the standard Wiener filter. In contrast, the RUKF is a local filtering method, which is more suited for processing nonstationary data.

10.2.5.2 RUKF for Speckle Noise Removal

As pointed out in Section 10.1, speckle noise commonly occurs in coherent imaging systems and is characterized as multiplicative noise with a negative exponential distribution. For fully developed speckle [9], this exponential distribution has a mean equal to its standard deviation. In general, the higher the average intensity in an area, the noisier it will appear due to the multiplicative nature of speckle. There are certain conditions that must be satisfied to ensure the validity of the exponential distribution and multiplicative characteristics of speckle. These are:

a. the surface is sufficiently rough compared to the wavelength of coherent imaging system;
b. a large number of independent scatterers are present within each resolution cell, so that the central limit theorem applies; and
c. the detector elements are small relative to the speckle size. With square-law detector, the recorded intensity then follows an exponential distribution [9].

In [20], a 2-D block Kalman filter was derived for observation models involving multiplicative speckle noise, LSI blur, and additive white noise. Here, we adopt a similar approach to develop a modified RUKF for such observation models. This approach assumes that the speckle noise $\gamma(m, n)$ is white with nonzero mean μ_γ and variance σ_γ^2 and is uncorrelated with both the additive noise $\eta(m, n)$, and the original image $x(m, n)$. The observation equation in presence of these degradations can be expressed as:

$$y(m, n) = \gamma(m, n)\,\underline{h}^T \underline{s}_1(m, n) + \eta(m, n) \tag{10.54}$$

Under these assumptions, the modified RUKF is implemented as follows:

1. ***Initial Estimate (Prediction):***

$$\hat{\underline{s}}_1(m, n|n-1) = \mathbf{C}_{11}\hat{\underline{s}}_1(m, n-1|n-1) + \mathbf{C}_{12}\hat{\underline{s}}_2(m, n-1|n-1) \tag{10.55}$$

Figure 10.11 (a) Original Lena image. *Source*: Juang et al. 2012 / John Wiley & Sons / CC BY 3.0. (b) Noisy Lena image, (c) RUKF processed image, and (d) 2-D Wiener filtered image.

2. ***A Priori Error Covariance Matrix:***

$$\mathbf{P}(m, n|n-1) = \mathbf{C}\mathbf{P}(m, n-1|n-1)\mathbf{C}^T + \sigma_e^2 \underline{g}\underline{g}^T \tag{10.56}$$

3. ***Kalman Gain Matrix:***

$$\mathbf{K}(m, n) = \mu_\gamma \mathbf{P}_{11}(m, n|n-1)\underline{h}\left[\mu_\gamma^2 \underline{h}^T \mathbf{P}_{11}(m, n|n-1)\underline{h} + \sigma_\gamma^2 \underline{h}^T \mathbf{R}_x(m, n)\underline{h} + \sigma_\eta^2\right]^{-1} \tag{10.57}$$

where $\mathbf{R}_x(m, n) = E[\underline{\mathbf{s}}_1(m, n)\underline{\mathbf{s}}_1^T(m, n)]$, i.e. correlation matrix of the original image.

4. ***Update Estimate:***

$$\underline{\hat{s}}_1(m, n|n) = \underline{\hat{s}}_1(m, n|n-1) + \mathbf{K}(m, n)[y(m, n) - \mu_\gamma \underline{h}^T \underline{\hat{s}}_1(m, n|n-1)] \tag{10.58}$$

5. ***A Posteriori Error Covariance Matrix:***

$$\mathbf{P}_{11}(m, n|n) = [\mathbf{I} - \mu_\gamma \mathbf{K}(m, n)\underline{h}^T]\mathbf{P}_{11}(m, n|n-1) \tag{10.59}$$

$$\mathbf{P}_{12}(m, n|n) = [\mathbf{I} - \mu_\gamma \mathbf{K}(m, n)\underline{h}^T]\mathbf{P}_{12}(m, n|n-1) \tag{10.60}$$

10.2.5.3 Special Case

In the absence of blurring process $\underline{h}^T = [1\ 0 \cdots 0]$ and hence $\underline{h}^T \mathbf{R}_x(m, n)\underline{h} = (\sigma_x^2(m, n) + \mu_x^2(m, n))$ and the Kalman gain equation in Step 3 simplifies to:

$$\mathbf{K}(m, n) = \mu_\gamma \mathbf{P}_{11}(m, n|n-1)\underline{h}\left[\mu_\gamma^2[\mathbf{P}_{11}(m, n|n-1)]_{1,1} + \sigma_\gamma^2[\sigma_x^2(m, n) + \mu_x^2(m, n)] + \sigma_v^2\right]^{-1} \tag{10.61}$$

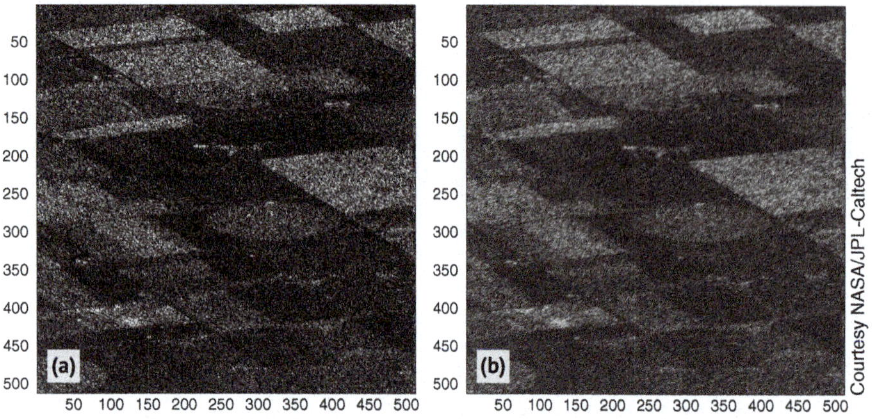

Figure 10.12 (a) Original SAR image of a farm and (b) modified RUKF speckle filtering result.

where $[\mathbf{P}_{11}(m, n|n-1)]_{1,1}$ is the first element of matrix $\mathbf{P}_{11}(m, n|n-1)$ as defined before. In this case, the local mean, $\mu_x(m, n)$, and variance, $\sigma_x^2(m, n)$, of the original image can be computed from those of the observed image $y(m, n)$ within the ROS of the model using:

$$\mu_x(m,n) = \frac{\mu_y(m,n)}{\mu_\gamma}$$
$$\sigma_x^2(m,n) = \frac{\sigma_y^2(m,n) - \sigma_\eta^2 - \mu_x^2(m,n)\sigma_\gamma^2}{\sigma_\gamma^2 + \mu_\gamma^2} \tag{10.62}$$

This multiplicative noise version of the RUKF was applied to a single-look SAR image of a Farm, shown in Figure 10.12a, with resolution 512 × 512 and 256 gray levels. The speckle noise has variance $\sigma_\gamma^2 = 1$ and mean $\mu_\gamma = 1$. The same 2-D AR(1) model parameters obtained in the previous experiment was also used here. A steady-state Kalman gain vector was computed after 10 iterations and used throughout the entire processing. The processed Farm image using this modified RUKF filter is shown in Figure 10.12b. To evaluate the performance of this filter in removing speckle noise, the local *contrast ratio* $\frac{\mu}{\sigma}$ was computed in several windows of size 50×50 pixels. The average value of this contrast ratio improved from 1.27 to 2.04, corresponding to about 4 dB improvement. Visual comparison of the two images also confirms the speckle noise removal ability of this modified RUKF, though some smearing can also be noticed.

10.3 Computer Projects

The computer projects in this chapter use Python libraries OpenCV, NumPy, Matplotlib, scikit-image, and SciPy.

Computer Project 10.1: In this computer project, we study the performance of 2-D inverse filtering methods covered in Section 10.2.1 for de-blurring applications. Use the color image "Finn" for this project. First, convolve the image (each row) with 1-D spatial averaging filters of order 16 and 32 to simulate the effects of the horizontal motion blur. Then, apply the three types of inverse filter, including the direct inverse filter, to the blurred images. Choose an appropriate value of ε for the regularized inverse filter transfer function. Compare the de-blurring performance of each inverse filter for both horizontal motion blur PSFs and provide a detailed discussion on the effectiveness and limitations of each method.

Computer Project 10.2: The goal of this computer project is to investigate the effectiveness of the 2-D Wiener filter covered in Section 10.2.3 for simultaneous de-blurring and noise removal. Use the color image "Finn" for this project. To investigate the use of 2-D Wiener filter for de-blurring purposes, generate a blurred version of the image by introducing horizontal motion blur with a small ROS as described in Computer Project 10.1. Then, apply a 2-D frequency domain Wiener filter to this blurred image and comment on the performance of this filter for removing the effects of horizontal motion blur. Note: choose an arbitrary small positive value for the variance of noise to avoid any singularity problems. Next, generate a noisy version of the Finn image by adding white Gaussian noise to produce an SNR = 5 dB. Use this image as the input to the 2-D Wiener filter to investigate the noise removal ability of this filter. Compare the result, both in terms of SNR and visual appearance, with the results obtained using spatial averaging filters. If the variance of the white additive noise is not known a priori, how would you estimate it from the observed noisy image? The last part of this project involves testing the Wiener filter's capability for simultaneous noise removal and de-blurring. Add white Gaussian noise to the blurred image you generated previously to produce SNR = 5 dB and then apply the 2-D Wiener filter in the frequency domain. Comment on the performance of the Wiener filter in this case. Provide a detailed discussion on the effectiveness and limitations of the 2-D Wiener filter for image restoration.

Computer Project 10.3: This project investigates the applicability of the 2-D Wiener filter to restore images degraded by multiplicative noise. Use the Wiener filter frequency response for the multiplicative noise in (10.33) to remove the effects of speckle noise in the SAR image "Farm." Assume that there is no additive noise or blur. The multiplicative speckle noise is assumed to have a variance $\sigma_\gamma^2 = 1$ and mean $\mu_\gamma = 1$. Apply this modified Wiener filter transfer function to the Farm image and evaluate its performance in several windows of size 50×50 by computing the contrast ratio $\frac{\mu}{\sigma}$ in each window and averaging the results. Perform this evaluation for both the original and Wiener filtered results to determine the improvements resulted from the filtering operation. Compare the performance of the modified Wiener filter with that of the modified RUKF in Section 10.2.5. Provide a comprehensive evaluation of the Wiener filter abilities for multiplicative noise removal. What are the simplifying assumptions made in the derivation of the frequency response of (10.33) that, in your opinion, do not make practical sense?

Computer Project 10.4: The purpose of this computer project is to study and implement a locally adaptive 2-D Wiener filter for noise removal in the absence of blur. Generate a noisy version of the "Lena" image by adding white Gaussian noise, with variance selected to produce an SNR = 5 dB. The locally adaptive Wiener filter is designed to improve the performance of image restoration by reducing the smearing artifacts that result from the WSS assumption inherent in the standard Wiener filter. It achieves this by computing the local statistics of the image and the noise, and subsequently adjusting the filter gain based on the estimated local statistics. Modify the 2-D Wiener filter algorithm by processing the noisy image in sliding windows around each pixel and computing the local mean and variance of the image and the noise in each window. It is assumed that the image and noise are uncorrelated. The transfer function of the local Wiener filter should be adjusted based on the presence or absence of edges and textures. More specifically, in relatively flat areas where the local variance is low, the filter gain should be increased to do more smoothing to effectively remove the noise; while in areas where the local variance is large, indicative of presence of edges or textures, the filter applies less smoothing to preserve structural details and reduce smearing. Evaluate the performance of your locally adaptive Wiener filter on the noisy Lena image and state your observations. More specifically, discuss edge preservation, smearing artifacts, and computational requirements of this filter.

Computer Project 10.5: In this project, we develop a strip Kalman filter and investigate its effectiveness in removing additive noise from the "Lena" image. In Computer Project 9.3, a vector AR model for this image was developed using the method in Section 9.3. Using the same Python code in Colab, estimate the parameter matrices of a second-order vector AR model and the covariance matrix of the driving process $\mathbf{Q}_{\underline{e}}$ in nonoverlapping strips of width $W = 32$. Then, the estimated matrices are averaged over all the strips to arrive at only one set of these matrices for the entire image. Using these estimated parameter matrices, construct the state equation in (10.35) with the driving process $\underline{e}(k)$. Then, generate a noisy version of the image by adding white Gaussian noise and choosing the noise variance to produce SNR = 5 dB. Use (10.38) to build the observation equation for the noisy image (no blur). The observation and state equations are then used to implement the strip Kalman filter in each strip. Consecutive strips should be overlapped by 8 pixels. Additionally, pad the image around the boundaries by an adequate number of rows and columns with pixel values equal to the mean of the image to prevent any boundary effects. The strip Kalman filter should be initialized using the procedure suggested in the first remark of Section 10.2.4. Provide a detailed discussion on the effectiveness and limitations of the strip Kalman filter for image restoration.

10.4 Problems

Problem 10.1: An object $x(u, v)$ being imaged moves uniformly along the $u = v$ line at a speed of ν. If the exposure time is T, find the PSF $h(u, v)$ of the motion blur. Is this motion blur LSI?

Problem 10.2: Consider the LS cost function in (10.12) or (10.13). Add a meaningful constraint term to this cost function so that the LS solution coincides with the regularized inverse filter frequency response in (10.5).

a. Provide justifications for the use of this constraint.
b. Show that minimizing the cost function with your constraint leads to (10.5).

Problem 10.3: The standard image degradation model consisting of LSI blur with PSF $h(m, n)$ and additive noise is:

$$y(m,n) = \sum_{k=0}^{N-1}\sum_{l=0}^{N-1} x(k,l)h(m-k,n-l) + \eta(m,n).$$

a. Show that this model can alternatively be represented as:

$$\underline{y} = \mathbf{H}\underline{x} + \underline{\eta}$$

where $\underline{x} = [\underline{x}_0^T \underline{x}_1^T \cdots \underline{x}_{N-1}^T]^T$ with $\underline{x}_i = [x(i,0)x(i,1)\cdots x(i,N-1)]^T$ and similarly for $\underline{y}$ and $\underline{\eta}$. The matrix $\mathbf{H}$ contains the 2-D PSF coefficients of the LSI blur with finite support ($M \ll N$).

b. Given the observed image $\underline{y}$ and knowledge of the degradation in form of matrix $\mathbf{H}$, obtain an estimate $\hat{\underline{x}}$ of original image $\underline{x}$ such that:

$$\varepsilon = ||\underline{y} - \hat{\underline{y}}||^2$$

where $\hat{\underline{y}} = \mathbf{H}\hat{\underline{x}}$ is minimized.

Problem 10.4: Assuming that the observation model is arranged in vector form as in part (a) of Problem 10.3, minimize the following constrained objective function with respect to $\underline{x}$ and find the solution.

$$\varepsilon(\underline{x}, \gamma) = ||\underline{y} - \mathbf{H}\underline{x}||^2 + \gamma(||\underline{x}||^2 - c)$$

where the second term is the constant norm (energy) $||\underline{x}||^2 = c$ constraint and γ is the Lagrangian multiplier.

Problem 10.5: Repeat Problem 10.4 by changing the constraint term to $||\mathbf{Q}\underline{x}||^2$ where $\mathbf{Q}$ is a positive definite (PD) matrix (i.e. for any $\underline{u}$, $\underline{u}^T\mathbf{Q}\underline{u} > 0$). This makes the constrained objective function of the form:

$$\varepsilon(\underline{x}, \gamma) = ||\underline{y} - \mathbf{H}\underline{x}||^2 + \gamma||\mathbf{Q}\underline{x}||^2$$

where γ is the Lagrangian multiplier.

Problem 10.6: An important constraint that can be used in the objective function of Problem 10.4 is the statistical entropy of the original image, which is $H(\underline{x}) = -\underline{x}^T \ln \underline{x}$ where $\ln \underline{x}$ refers to component-wise logarithm operation. Here we would like to maximize $H(\underline{x})$ (or minimize the $-H(\underline{x})$) with the constraint of $||\underline{y} - \mathbf{H}\underline{x}||^2$. Define the constrained objective function, and find the solution of this constraint LS for $\hat{\underline{x}}$.

Problem 10.7: An image is observed N times via a noisy channel to produce an observed image:

$$y_i(m,n) = x(m,n) + \eta_i(m,n),\ i \in [1, N]$$

at every observation time i where the noise images $\eta_i(m, n)$ are zero-mean mutually uncorrelated white processes with a common variance σ_η^2. The original image $x(m, n)$ is assumed to be deterministic, but unknown.

a. Determine the observation equation in vector form by arranging all N observed images in a vector, i.e. $\underline{y}(m,n) = [y_1(m,n)\cdots y_N(m,n)]^T$ and similarly for $\eta_i(m, n)$.
b. Develop an LS approach to find $\hat{\underline{x}}(m, n)$. Explain why this approach makes practical sense.
c. What is the effect of increasing the number of observations N on the accuracy of the result?

Problem 10.8: In Problem 10.7, let us assume that $\eta_i(m,n)$ are independent identically distributed (iid) Gaussian with zero mean and variance σ_η^2, i.e. $\eta_i(m,n) \sim \mathcal{N}(0,\sigma_\eta^2), \forall i \in [1,N]$. An alternative approach to find the solution of this problem, $\hat{x}(m,n)$, is to use the Maximum Likelihood (ML) method that maximizes the likelihood function (or conditional pdf) $\hat{x}(m,n) = \text{argmax}_{\hat{x}} p\left\{y_1(m,n) \cdots y_N(m,n) | x(m,n)\right\}$ with respect to the unknown image.

a. Find the expression for this likelihood function.
b. Find the ML solution of $\hat{x}(m,n)$. What is your observation?

Problem 10.9: In Problem 10.8, assume that the variance of the noise, σ_η^2, is also unknown. Using the same likelihood function in Problem 10.8, find the ML estimate of the noise variance. Can you justify the practical validity of this result?

Problem 10.10: In Problem 10.8, let us assume that the observation model is:

$$y_i(m,n) = \alpha x(m,n) + \eta_i(m,n), i \in [1,N]$$

where α is an unknown scalar parameter. All other assumptions are still valid here.

a. Find the expression for the likelihood function in this case.
b. Find the ML estimates for both α and $x(m,n)$.

Problem 10.11: Assume a sequence of degraded images $y_i(m,n)$ are observed, i.e.

$$y_i(m,n) = x(m,n)\gamma_i(m,n), \quad i \in [1,N]$$

where $\gamma_i(m,n)$, $\forall\, i$ are iid processes with right-sided negative exponential pdf of the form $p_\Gamma(\gamma) = \alpha e^{-\alpha\gamma}u(\gamma)$ where $u(.)$ is a 1-D unit step function. Assuming that the image $x(m,n)$ is deterministic but unknown:

a. Find the expression for the likelihood function $p\left\{y_1(m,n) \cdots y_N(m,n) | x(m,n)\right\}$ in this case.
b. Find the expression for the ML estimate $\hat{x}(m,n)$.

Problem 10.12: A zero-mean image is degraded by horizontal motion blur (see Example 10.1) and additive zero-mean white noise with variance σ_η^2. If the ACF of the original image can be represented by $R_{xx}(i,j) = \sigma_x^2 e^{-\alpha|i|-\beta|j|}$, find the 2-D Wiener filter frequency response using 2-D DSFT.

Problem 10.13: Consider an image degraded by additive noise:

$$y(m,n) = x(m,n) + \eta(m,n)$$

where $x(m,n)$ is the original zero-mean image and $\eta(m,n)$ is a zero-mean white noise with variance σ_η^2. Assuming that the ACF of the original image can be represented by $R_{xx}(i,j) = \rho^{|i|+|j|}, \quad 0 < \rho < 1$.

a. Derive the Wiener filter's frequency response using DSFT, as functions of ρ and σ_η^2.
b. Develop a method to estimate ρ using the ACF of the observed image $y(m,n)$.

Problem 10.14: In Problem 10.13, assume that the ACF of the original image is represented by $R_{xx}(i,j) = \sigma_x^2 \rho_1^{|i|} \rho_2^{|j|}$, $0 < \rho_1, \rho_2 < 1$

a. Derive the Wiener filter frequency response using DSFT, as functions of ρ_1, ρ_2, σ_x^2, and σ_η^2.
b. Develop a method for estimating ρ_1 and ρ_2, using the ACF of the observed image $y(m,n)$.

Problem 10.15: Show that for the geometric mean filter of Example 10.2, the power spectrum of the estimation error $S_{ee}(k)$ is greater than that of Wiener filter. Explain the importance of this result.

Problem 10.16: A variant of 2-D Wiener filter transfer function for additive noise removal is:

$$G(e^{j\Omega_1}, e^{j\Omega_2}) = \left[\frac{S_{xx}(e^{j\Omega_1}, e^{j\Omega_2})}{S_{xx}(e^{j\Omega_1}, e^{j\Omega_2}) + S_{\eta\eta}(e^{j\Omega_1}, e^{j\Omega_2})}\right]^{\beta}$$

where $S_{xx}(e^{j\Omega_1}, e^{j\Omega_2})$ is the power spectrum of the original image, which decreases as a function of frequencies Ω_1 and Ω_2 and $S_{\eta\eta}(e^{j\Omega_1}, e^{j\Omega_2}) = \sigma_\eta^2$ is the power spectrum of noise. When $\beta = 1$ we have the standard Wiener filter while $\beta = \frac{1}{2}$ leads to the *power spectrum* filter. Determine:

a. which filter has better noise removal ability.
b. which filter causes more blurring artifacts.

Problem 10.17: An image $x(m, n)$ of size $N \times N$ is corrupted by an additive white noise $\eta(m, n)$ with zero mean and variance σ_η^2. If the original image is modeled by a 2-D causal quarter-plane AR model:

$$x(m, n) = x(m-1, n) + x(m, n-1) - x(m-1, n-1) + e(m, n)$$

where $e(m, n)$ is a zero-mean white driving process with variance σ_e^2, find:

a. power spectrum of the original image as a function of Ω_1 and Ω_2,
b. power spectrum of the noisy image $y(m, n)$,
c. transfer function of the Wiener filter.

Problem 10.18: An image $x(m, n)$ is degraded by LSI blur and additive white noise with zero mean and variance σ_η^2. If the ACF of the original image can be represented by $R_{xx}(i, j) = \sigma_x^2 \rho_1^{|i|} \rho_2^{|j|}, \quad 0 < \rho_1, \rho_2 < 1$ and the PSF of the blur is given by $h(m, n) = e^{-\alpha(m+n)} u(m, n), \quad \alpha > 0$ where $u(m, n)$ represents a 2-D unit step function, find:

a. 2-D Wiener filter frequency response using the 2-D DSFT (assume $\mu_x = 0$).
b. Find the power spectrum of the error, i.e. $S_{ee}(e^{j\Omega_1}, e^{j\Omega_2})$ (see Property 4 in Section 10.2.2).

Problem 10.19: Consider the image $x(m, n)$ in Problem 10.17 which is represented by a 2-D causal AR model driven by white noise $e(m, n)$ with zero mean and variance σ_e^2.

a. Represent this 2-D AR model using an appropriate state equation and drive the observation equation for the additive noise degradation (assume no blur).
b. Devise a Kalman filter process for this state estimation problem.

Problem 10.20: A zero-mean random image of size $N \times N$ is represented by a separable ACF $R_{xx}(i, j) = \sigma_x^2 \rho_1^{|i|} \rho_2^{|j|}$, $0 < \rho_1, \rho_2 < 1$. In Chapter 9, we showed that the image with such ACF can be modeled as an output of a 2-D AR process:

$$x(m, n) = \alpha(x(m-1, n) + x(m+1, n)) - \rho_2 \alpha(x(m+1, n-1) + x(m-1, n-1)) + \rho_2 x(m, n-1) + e(m, n)$$

where $\alpha = \frac{\rho_1}{(1+\rho_1^2)}$ and ACF of the driving process $e(m, n)$ which is $R_{ee}(i, j) = \beta^2[\delta(i, j) - \alpha\delta(i-1, j) - \alpha\delta(i+1, j)]$ with $\beta^2 = \frac{\sigma_x^2(1-\rho_1^2)(1-\rho_2^2)}{(1+\rho_1^2)}$.

a. By proper state assignment, arrange this 2-D AR model driven by colored noise into a state equation driven by a white process. The image is assumed to be zero outside the boundaries.
b. If the degradation model consists of additive zero-mean white noise with variance σ_η^2, develop the relevant Kalman filter equations that generate the estimate of the original image.
c. Explain the advantages and disadvantages of this filter.

References

1 F. Argenti, A. Lapini, L. Alparone, and T. Bianchi, "A tutorial on speckle reduction in synthetic aperture radar images," *IEEE Geoscience and Remote Sensing Magazine*, pp. 6–35, 2013.

2 A. P. Lyons, D. A. Abraham, and S. F. Johnson, Modeling the effect of seafloor ripples on synthetic aperture sonar speckle statistics," *IEEE Journal of Oceanic Engineering*, vol. 35, pp. 242–249, 2010.

3 M. Karaman, M.A. Kutay and G. Bozdagi, "An adaptive speckle suppression filter for medical ultrasonic imaging," *IEEE Transactions on Medical Imaging*, vol. 14, pp. 283–292, 1995.

4 P. Zakharov, A. Völker, A. Buck, B. Weber, and F. Scheffold, "Quantitative modeling of laser speckle imaging," *Optics Letters*, vol. 31, pp. 3465–3467, 2006.

5 A. K. Jain, "*Fundamentals of Digital Image Processing*," Prentice-Hall, 1989.

6 C. Boncelet, "Image noise models." In Alan C. Bovik, *Handbook of Image and Video Processing*, Academic Press, 2005.

7 A. K. Boyat and B. K. Joshi, "A review paper: Noise models in digital image processing," *Signal & Image Processing: An International Journal (SIPIJ)*, vol. 6, pp. 63–75, 2015.

8 A. Lean-Garcia, "*Probability, Statistics, and Random Processes for Electrical Engineers*," Prentice-Hall, 3rd Edition, 2008.

9 M. Tur, K. C. Chin, and J. W. Goodman, "When is speckle noise multiplicative?" *Applied Optics*, vol. 21, pp. 1157–1159, 1982.

10 A. Tekalp, H. Kaufman and J. Woods, "Identification of image and blur parameters for the restoration of noncausal blurs," *IEEE Transactions on Acoustics, Speech, and Signal Processing*, vol. 34, pp. 963–972, 1986.

11 Y. Yitzhaky and N. S. Kopeika, "Identification of blur parameters from motion blurred images," *Graphical Models Image Processing*, vol. 59, pp. 310–320, 1997.

12 J. W. Woods and C. H. Radewan, "Kalman filtering in two dimensions," *IEEE Transactions on Information Theory*, vol. 23, pp. 473–482, 1977.

13 B. R. Suresh and B. A. Shenoi, "New results in two-dimensional Kalman filtering with applications to image restoration," *IEEE Transactions on Circuits and Systems*, vol. 28, pp. 307–319, 1981.

14 M. R. Azimi-Sadjadi and P. W. Wong, "Two-dimensional block Kalman filtering for image restoration," *IEEE Transactions on Acoustics, Speech, and Signal Processing*, vol. 35, pp. 1736–1749, 1987.

15 M. R. Azimi-Sadjadi, "New results in strip Kalman filtering," *IEEE Transactions on Circuits and Systems*, vol. 36, pp. 893–897, 1989.

16 J. Su, B. Xu, and H. Yin "A survey of deep learning approaches to image restoration," *Neurocomputing*, vol. 487, pp. 46–65, 2022.

17 J. S. Lim, "*Two-dimensional Signal and Image Processing*," Prentice-Hall, 1990.

18 J. M. Mendel, "*Lessons in Estimation Theory for Signal Processing, Communications, and Control*," Prentice Hall, 1995.

19 A. P. Sage and J. L. Melsa, "*Estimation Theory With Applications to Communications and Control*," McGraw-Hill, 1971.

20 M. R. Azimi-Sadjadi and S. Bannour, "Two-dimensional adaptive block Kalman filtering of SAR imagery," *IEEE Transactions on Geoscience and Remote Sensing*, vol. 29, pp. 742–753, 1991.

11

Image Compression and Encoding

In the modern digital technology era, an enormous volume of data is continuously generated in form of signals, text, documents, and images that need to be transmitted or stored efficiently using existing devices with limited bandwidth and capacity. As an example, Geostationary satellite series (GOES-16) [1] has 16 spectral channels that capture images at a spatial resolution exceeding 5000×5000 pixels at 12 bits/pixel, every 10 minutes. The volume of data that needs to be transmitted and/or stored in a single day exceeds 100 GB. Clearly, this puts significant strain on the resource-constrained systems that transmit, receive, or store this massive data. Thus, image data compression and encoding have become integral components of modern digital imaging systems. Applications of image data compression and encoding in image transmission are found in remote sensing and surveillance, wireless networks, multimedia, meteorological satellite imaging, military applications, radar and sonar imaging, etc. Applications in image storage and retrieval are found in large database managements for high-resolution medical images, such as MRI or CT scans, weather maps, geographical surveys, and office automation, among others.

The main objective of image data compression is to reduce the number of bits required to represent an image without sacrificing its *useful* properties or compromising its visual quality. This is accomplished by exploiting and removing various forms of *redundancies* in images. In general, four types of redundancy can be exploited in image compression [2–5]. First, psycho-visual redundancy of human visual systems can be described as information that can be discarded without significantly impairing the quality of image perception. This redundancy is exploited in the design of quantizers (see Chapter 5), where resolutions beyond approximately 10 bits/pixel will be indistinguishable to human eyes. Second, inter-pixel redundancy exists among neighboring pixels due to strong spatial correlations in small neighborhoods within an image. In most natural images, pixel intensities vary smoothly, especially in background regions, leading to significant statistical dependence among adjacent pixels. This correlation typically decreases as the spatial distance between pixels increases. Third, spatial or temporal inter-frame redundancy occurs in stereo and video images, respectively. Spatial inter-frame redundancy arises in images captured from a scene by two or more disparate cameras, while temporal inter-frame redundancy corresponds to motionless parts in consecutive video frames. Finally, the fourth is coding redundancy, which refers to the redundancy (or inefficiency) in assigning codes to gray level. Coding redundancy is directly related to the assignment of code lengths to symbols and can be minimized through the use of optimal encoding schemes. In this chapter, we will focus on the second and fourth types of redundancy in images.

From an information-theoretic perspective, an image can be represented as a discrete *source* of scalar random variable X that generates a sequence of symbols corresponding to pixel intensity values, $\{x_i\}_{i=1}^{L}$ from a finite *alphabet* of symbols (gray levels) as shown in Figure 11.1. The sequence of symbols is drawn from a probability distribution function (pdf) or the image histogram. In some cases, an image must be considered as a multivariate source of random vector (r.v.) $\underline{X}$, e.g. in RGB color system. A coding process then maps each symbol to a codeword, or string of bits, that uniquely represents that symbol. The goal of an image data compression and encoding system is to devise efficient lossless or lossy methods to reduce the intra- and inter-frame redundancies, and design optimal encoding strategies to minimize coding redundancy.

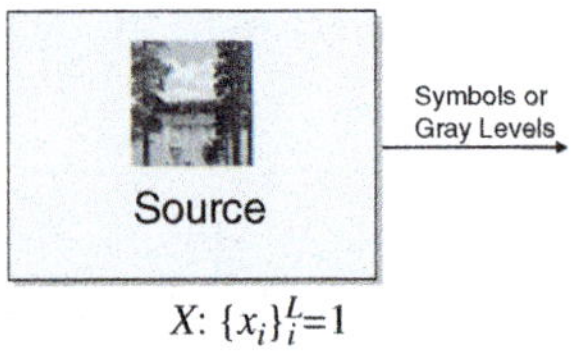

Figure 11.1 An information source (image) produces a sequence of symbols (gray levels) from a finite alphabet of symbols. *Source*: Hu et al., 2014 / John Wiley & Sons / CC BY 3.0.

11.1 A Review of Information Theory

As mentioned earlier, data compression and encoding systems exploit image redundancies to reduce the number of bits required to represent them while preserving image quality and relevant information for human perception or machine analysis. Now, the questions that arise are: how can redundancy be quantified, and how is it related to the number of bits required for representation? These questions are addressed through the concepts of information and entropy defined next.

11.1.1 Measure of Information and Entropy

Consider a discrete *memoryless* source (e.g. an image) $X = \{x_i\}_{i=1}^{L}$ that generates messages comprising of symbols (e.g. gray levels), x_i, with associated probabilities P_k, $k \in [1, L]$, where L is the number of symbols in the alphabet of gray levels. Note that a memoryless source [6] is an information source for which successive symbols are statistically independent, that is, each symbol is selected without any relation to the previous one.

Then, *information* associated with symbol x_k is:

$$I_k = -\log_2 P_k = \log_2 \frac{1}{P_k} \qquad \text{bits.}$$

That is, an unlikely symbol carries more information than a likely symbol. Note that since the base of the log operation is 2, the unit of information is in bits. For example, for a symbol x_k with $P_k = 1$, then $I_k = 0$, i.e. no information is conveyed (or zero bits), while for equally likely symbols with $L = 2^B$, we have $P_k = \frac{1}{L} = \frac{1}{2^B} \rightarrow I_k = B$ bits, i.e. information can be transmitted as a B-bit binary number implying that each symbol can be represented by B bits.

Entropy of the discrete *memoryless* source X is the *average information* associated with all its symbols x_k with probabilities P_k, i.e.

$$H(X) = \sum_{k=1}^{L} P_k I_k = -\sum_{k=1}^{L} P_k \log_2 P_k = -E[\log_2 p_X(X = x)] \qquad \text{Average bits/pixel} \tag{11.1}$$

where $E[.]$ denotes expectation and $p_X(X = x)$ is the probability mass function (pmf) of X. Clearly, we have $\sum_{k=1}^{L} P_k = 1$. The entropy reaches its maximum value when all symbols are equally likely, i.e. $P_k = \frac{1}{L}$, $k \in [1, L]$ with $L = 2^B$, then:

$$H_{\max}(X) = -\sum_{k=1}^{L} \frac{1}{L} \log_2 \frac{1}{L} = \log_2 L = B \qquad \text{bits} \tag{11.2}$$

which corresponds to the least redundant case. In contrast, let us consider the case where $P_i = 1$ and $P_k = 0\ k \neq i\ \ k \in [1, L]$, then $H_{\min}(X) = 0$ which corresponds to the most redundant. Thus, entropy is a measure of redundancy and further $0 \leq H(X) \leq H_{\max}$.

As an example, consider a discrete source of r.v. X with $L = 8$ symbols (or gray levels) and associated probabilities $\{P_k\}_{k=1}^{8} = \left\{\frac{1}{2}, \frac{1}{4}, \frac{1}{8}, \frac{1}{16}, \frac{1}{64}, \frac{1}{64}, \frac{1}{64}, \frac{1}{64}\right\}$. The entropy of this source is:

$$H(X) = -\sum_{k=1}^{8} P_k \log_2 P_k = 2 \qquad \text{bits.}$$

What does this mean? Suppose that we want to transmit the source messages as a sequence of symbols through a communication channel. Since the alphabet has $L = 8 = 2^3$ symbols, we can obviously use 3-bit strings (or codewords) to send each symbol (or pixel). However, since the probabilities are unequal, we can assign shorter bit strings for more probable symbols and longer codewords for less probable ones. One possible set of codewords is: 0, 10, 110, 1110, 111100,

111101, 111110, and 111111, which gives the average code length of 2 bits equal to the entropy. There is no other choice of codewords that can provide a lower than average of 2 bits/symbol for this source [6].

11.1.2 Entropy and Coding

Entropy represents the lower bound on the average number of bits required to encode a source. That is, for a set of coder input symbols $x_i, i \in [1, L]$, with associated probabilities P_k, it is impossible to code them using fewer than $H(X)$ bits/symbol on average. This result is due to Shannon's first theorem of source coding [6], which established the foundation of information theory. We first present this theorem and then show how it can be applied to our specific problem of constructing codewords for the symbols or gray levels.

11.1.2.1 Shannon's Source Coding Theorem

Let X be a memoryless source with entropy $H(X)$ that generates a sequence of L independent and identically distributed (iid) symbols $S = \{X_1, X_2, \cdots, X_L\}$. Let the expected (or average) codeword length per symbol be R_L then:

$$H(X) \leq R_L < H(X) + \frac{1}{L}, \tag{11.3}$$

where $R_L = E[\beta(S)]$ and $\beta(S)$ is the length of codeword assigned to the sequence S. There are many important observations resulting from this theorem. First, as the length of the sequence of symbols (or block) L increases, we can achieve expected code length/symbol arbitrarily close to the entropy (lower bound). Additionally, for finite-length sequences, lossless encoding is achievable if the average codeword length satisfies the right-hand inequality.

What happens if the symbols $X_i, i \in [1, L]$ are not iid? It has been shown [6] that Shannon's first theorem in this case becomes:

$$H(X_1, X_2, \cdots X_L) \leq R_L < H(X_1, X_2, \cdots X_L) + 1 \tag{11.4}$$

where $H(X_1, X_2, \cdots X_L)$ is the joint entropy of the discrete r.v.s $\{X_1, X_2, \cdots X_L\}$ with the joint pdf $p_{X_1,X_2,\cdots X_L}(x_1, x_2, \cdots x_L)$, i.e.

$$\begin{aligned} H(X_1, X_2, \cdots X_L) &= -E[\log_2 p_{X_1,X_2,\cdots X_L}(x_1, x_2, \cdots x_L)] \\ &= -\sum_{x_1 \in \mathcal{X}_1} \cdots \sum_{x_L \in \mathcal{X}_L} p_{X_1,X_2,\cdots X_L}(x_1, x_2, \cdots x_L) \log_2 p_{X_1,X_2,\cdots X_L}(x_1, x_2, \cdots x_L) \end{aligned} \tag{11.5}$$

where $\mathcal{X}_i$ is a set of possible values of symbol X_i. Clearly, for the iid case $p_{X_1,X_2,\cdots X_L}(x_1, x_2, \cdots x_L) = \prod_{i=1}^{L} p_{X_i}(x_i) = [p_X(x)]^L$ and hence:

$$H(X_1, X_2, \cdots X_L) = -E[\log_2 p_X(x)^L] = -LE[\log_2 p_X(x)] = LH(X),$$

which gives the same result as in Shannon's theorem when divided by L.

Now, let us consider our main problem of assigning codewords to encoder inputs $x_i, i \in [1, L]$ as shown in Figure 11.2. If we design a code with codewords c_i, $i \in [1, L]$ of lengths β_i, the average code length required by the coder is $R_L = \sum_{i=1}^{L} \beta_i P_i$. Then, Shannon's source coding theorem gives $H(X) \leq R_L < H(X) + \frac{1}{L}$, i.e. it is possible to encode without loss (or distortion) a source X with entropy $H(X)$ using an average of less than $H(X) + \frac{1}{L}$ bits/symbol. The optimality of the encoder depends on how close R_L is to $H(X)$.

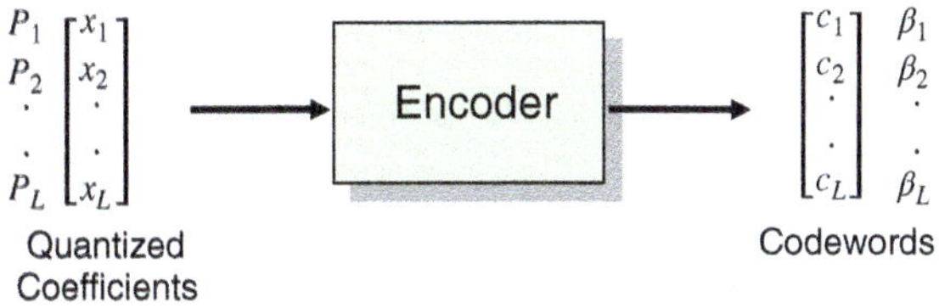

Figure 11.2 Encoder assigning codewords $\{c_i\}$ with lengths $\{\beta_i\}$ to quantized coefficients $\{x_i\}$.

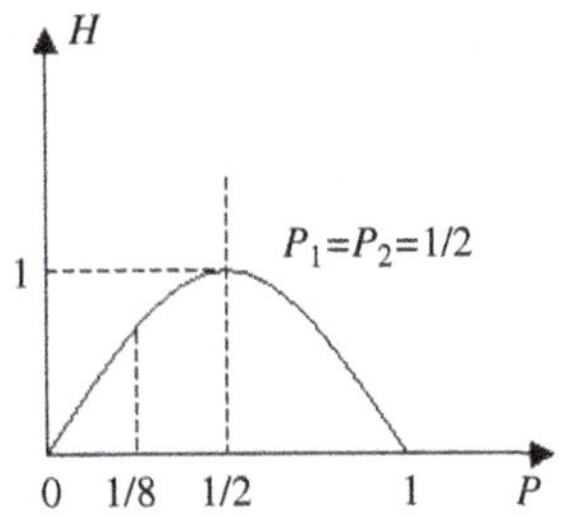

Figure 11.3 Plot of entropy, $H(X)$, in Example 11.1 as a function of probability p.

From the above result, the maximum achievable compression, C, which is the ratio of the average bit rate of the original raw data, B to the average bit rate of the encoded data R_L, satisfies:

$$\frac{B}{H+\epsilon} \leq C \leq \frac{B}{H}, \qquad \epsilon = 1/L. \tag{11.6}$$

Example 11.1 [6] *For a binary source X with L = 2 and probabilities $P_1 = p$ and $P_2 = 1 - p$,* where $0 \leq p \leq 1$, *find the expression for the entropy $H(X)$ and plot it as a function of p.*

Solution: The entropy is $H(X) = -p \log_2 p - (1-p)\log_2(1-p)$. Figure 11.3 shows plot of $H(X)$ as a function of p. Since the source is binary, we can use $B = 1$ bit/pixel corresponding to $H_{\max} = 1$ at $p = 1/2$. For a skewed distribution, e.g. $p = 1/8$, the entropy decreases to $H \approx 0.2$, i.e. more redundancy. Then, it is possible to construct a coding scheme that uses an average of 0.2 bits/symbol.

11.1.3 Rate Distortion Function

In the previous section, we discussed lossless or distortion-free data compression where the achievable rate is lower-bounded by the source entropy. However, since certain distortion may be inevitable in any image compression and encoding, then the question becomes what is the minimum average distortion achievable at a particular rate? Or conversely, what is the minimum rate (number of bits) necessary to encode the image while allowing a certain level of distortion? More specifically, let $D(X, \hat{X})$ be a fixed distortion between the actual r.v., X and its reproduced version, $\hat{X}$ after decompression and decoding. Then, the question is, allowing for a fixed distortion $D(X, \hat{X})$, what is the minimum number of bits required to encode the data? The answer is given in the following rate distortion function theorem [6, 7].

1. Rate Distortion Function Theorem (IID Gaussian Source):

Theorem: *The rate distortion function for an iid Gaussian source $X \sim \mathcal{N}(0, \sigma_X^2)$ with distortion D is:*

$$R(D) = \begin{cases} \frac{1}{2}\log_2 \frac{\sigma_X^2}{D}, & 0 \leq D \leq \sigma_X^2 \\ 0, & D > \sigma_X^2 \end{cases} = \mathrm{Max}\left[0, \frac{1}{2}\log_2 \frac{\sigma_X^2}{D}\right] \tag{11.7}$$

where $D(X, \hat{X}) = \mathrm{E}[(X - \hat{X})^2]$ is the MSE distortion and $\hat{X}$ is a reproduced version of X. The proof of this theorem can be found in [6]. The plot of this rate distortion function $R(D)$ that gives the number of bits required for distortion D is shown in Figure 11.4. As can be seen, for distortion $D \geq \sigma_X^2$, $R(D) = 0$, i.e. the source can be represented with zero bits. For non-iid Gaussian sources, the threshold at which the rate becomes zero will be different.

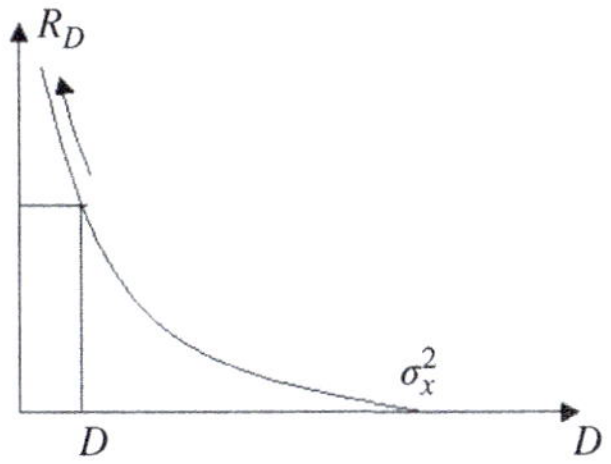

Figure 11.4 Rate distortion function $R(D)$ as a function of distortion D for a Gaussian source.

Remark:

1. Using (11.7) we can express D as a function of R, i.e. $D(R) = \sigma_X^2 2^{-2R}$. Hence, each extra bit in R reduces the distortion by a factor of 4, i.e. if $R \to R+1$ then $D(R+1) = \sigma_X^2 2^{-2(R+1)} = D(R)/4$
2. A more general expression for the rate distortion function $R(D)$ for a source X with pdf $p_X(x)$ and distortion measure $D(X, \hat{X})$ is given [6] by

$$R(D) = \min_{p_{X,\hat{X}}(x,\hat{x})} I(X; \hat{X}), \tag{11.8}$$

where $I(X;\hat{X})$ is the mutual information between X and its reproduced version $\hat{X}$, and the minimization is over all joint distribution $p_{X\hat{X}}(x, \hat{x})$ that satisfy the expected distortion constraint. The mutual information $I(X;\hat{X})$ can be expressed [6] in terms of the source entropy and conditional entropy of the source given the reproduced version:

$$I(X;\hat{X}) = \sum_{x,\hat{x}} p_{X,\hat{X}}(x, \hat{x}) \log \frac{p_{X,\hat{X}}(x, \hat{x})}{p_X(x) p_{\hat{X}}(\hat{x})} = H(X) - H(X|\hat{X}) \tag{11.9}$$

where $H(X|\hat{X})$ is the conditional entropy $H(X|\hat{X}) = -E[\log p_{X|\hat{X}}(x|\hat{x})]$.

2. Rate Distortion Function Theorem (Multivariate Gaussian Source):
Consider a multivariate Gaussian source $\underline{X} = [X_1, X_2, \cdots X_m]^T$, where the elements $X_1, ..., X_m$ are independent but not identically distributed Gaussian r.v.s, i.e. $X_i \sim \mathcal{N}(0, \sigma_{X_i}^2)$. Given a total bit budget R bits, to be assigned to all the elements of the random vector $\underline{X}$, the question is how to optimally allocate bits to each component to minimize the overall squared error distortion function $D(\underline{X}, \underline{\hat{X}}) = \sum_{i=1}^{m} (X_i - \hat{X}_i)^2$? The rate distortion function for this *parallel Gaussian source* [6] case is:

$$R(D) = \sum_{i=1}^{m} \frac{1}{2} \log_2 \frac{\sigma_{X_i}^2}{D_i} \tag{11.10}$$

where

$$D_i = \begin{cases} \lambda, & \text{if } \lambda < \sigma_{X_i}^2, \\ \sigma_{X_i}^2 & \text{otherwise,} \end{cases} \tag{11.11}$$

and λ is chosen so that $\sum_{i=1}^{m} D_i = D$. This result implies that components of $\underline{X}$ for which $\sigma_{X_i}^2 \leq \lambda$ are not allocated any bits. However, for those with $\sigma_{X_i}^2 > \lambda$, bit allocation will be based on the ratio of $\frac{\sigma_{X_i}^2}{\lambda}$, subject to the total bit budget constraint.

11.2 Image Compression and Encoding Methods – Lossless

In general, any image data compression and encoding system consists of two main subsystems, an encoder and a decoder, together referred to as CODEC as shown in Figure 11.5a,b. The encoder subsystem first maps the image to the transform domain, where interpixel or spatial redundancies are reduced by discarding insignificant coefficients as much as possible

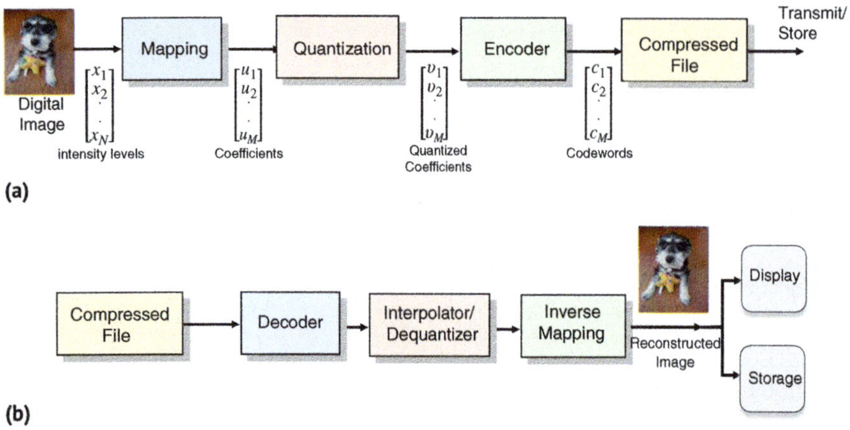

Figure 11.5 A typical image compression with coder and decoder subsystems: (a) compressor and encoder; (b) decoder and reconstruction.

and reducing the dimensionality of the data. The mapping must be invertible to enable data decompression and reconstruction processes at the decoder. The transformed coefficients are then quantized to which codewords are assigned at the encoder. Lloyd-Max quantizer can be employed to reduce the loss or distortion inherent in the quantization process. At the decoder, the codewords are first decoded and then converted to real-valued coefficients by a dequantizer. Using these real-valued coefficients, the image can be reconstructed using the inverse transformation.

There are two general categories of image compression methods, namely lossless and lossy data compression [2–5]. In lossless case, the reconstruction from the compressed and encoded data (decompression and decoding) leads to perfect reproduction of the original image, whereas lossy compression allows some distortion in exchange for higher compression efficiency. The choice of one category over the other is highly dependent on the specific application and the required compression rate. In this section, we investigate various lossless image compression and encoding methods and discuss their optimality as well as advantages and limitations. Lossy compression methods will be discussed in Section 11.3. Lossless image compression and encoding methods rely on encoding pixel values without introducing distortion. These methods typically exploit statistical redundancy through efficient coding design. Among such methods are Pulse-Code Modulation (PCM), Huffman encoding, arithmetic encoding, and run-length encoding [2, 6, 7].

11.2.1 Pulse-Code Modulation

This simple method [7, 8] has many applications in speech, audio, and image encoding in which analog data is sampled and quantized into fixed length codewords using A/D converters. Typically, uniform quantization is employed in this process. The fixed-length codewords simplify both the encoding and the decoding processes. However, despite all the advantages of PCM, it does not produce optimal codes and any attempt to reduce the length of codewords by simply truncating the least-significant bits for each pixel value leads to distortion in the encoding process.

11.2.2 Huffman Encoding

This method, proposed by Huffman in 1952, provides an optimal lossless encoding with minimum redundancy using a simple procedure [6, 7]. Its optimality stems from using variable-length codewords, where more probable symbols are assigned shorter codewords than those with a low probability of occurrence. Huffman coding is a *prefix-free* coding in the sense that the bit string representing a symbol is not a prefix of a different symbol, hence one can uniquely decode a sequence of bits without ambiguity. Without this property, a bit string could be interpreted in multiple ways.

The algorithm for building codewords consists of the following steps [6, 7]:

1. Arrange symbols' probabilities, P_k, in a decreasing order and consider them as *leaf nodes* of a tree.
2. Merge the two nodes with the smallest probabilities to form a new node whose probability is the sum of those of the two merged nodes. Go to Step 1 and repeat until only two *root nodes* are left.
3. Arbitrarily assign 1s and 0s to each pair of branches merging into a node.
4. Read sequentially from the root node to the leaf nodes to form the associated code for each symbol.

Example 11.2 *Consider the following 8×8 image with $L = 8$ gray levels. Construct Huffman codes for the pixel values of this image [7].*

4	4	4	4	4	4	4	0
4	5	5	5	5	5	4	0
4	5	6	6	6	5	4	0
4	5	6	7	6	5	4	0
4	5	6	6	6	5	4	0
4	5	5	5	5	5	4	0
4	4	4	4	4	4	4	0
4	4	4	4	4	4	4	0

Solution: The codewords c_k are obtained by constructing the binary tree shown in Figure 11.6 and assigning 0 and 1 to the upper and lower branches. The codewords together with their lengths β_k are given in Table 11.1. The average bits/pixel and source entropy are:

$$R_L = \sum_{k=1}^{8} \beta_k P_k = 1.923 \text{ bits/pixel}$$

$$H(X) = -\sum_{k=1}^{8} P_k \log_2 P_k = 1.852 \text{ bits/pixel.}$$

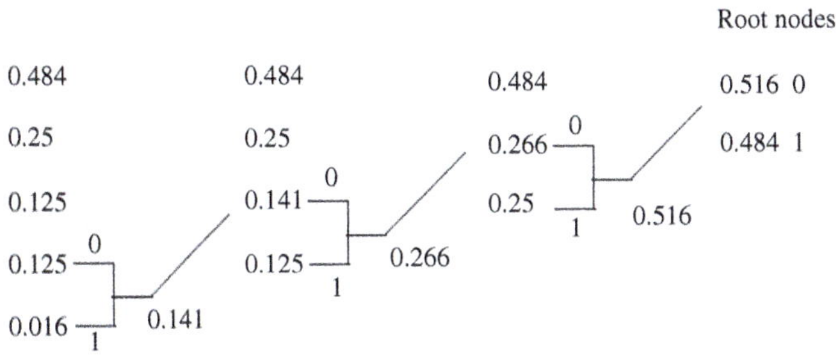

Figure 11.6 Tree structure for Huffman encoding.

Table 11.1 Huffman encoding.

Gray levels	No. of occurrences	P_k	c_k	β_k	$P_k\beta_k$	$-P_k \log_2 P_k$
0	8	0.125	0000	4	0.5	0.375
1	0	0	–	0	–	–
2	0	0	–	0	–	–
3	0	0	–	0	–	–
4	31	0.484	1	1	0.484	0.507
5	16	0.25	01	2	0.5	0.5
6	8	0.125	001	3	0.375	0.375
7	1	0.016	0001	4	0.064	0.095
	64	1			R_L=1.923	$H(X)$=1.852

That is, an average of 1.923 bits/pixel (instead of 3 bits/pixel using PCM) can be used to encode this image. Additionally,

$$H(X) = 1.852 < R_L = 1.923 < H(X) + \frac{1}{L} = 1.977$$

confirming that Huffman code is optimal. It must be pointed out that there are many optimal codes, e.g. complementing all bits of an optimal code will produce another optimal code. Huffman encoding produces one such code. In [6], it has been shown that optimal codes satisfy the following properties:

1. If $P_j > P_k$, then $\beta_j \leq \beta_k$,
2. The two longest codes corresponding to the least likely symbols have the same lengths and they differ only in the last bit.

Huffman codes are uniquely decipherable, and can be efficiently decoded from left to right using a lookup table. For example, in the previous example, the binary string 0000001101 corresponds to four-pixel intensities 0,6,4, and 5. Due to its optimality and simplicity of the encoding and decoding processes, it is a widely adopted lossless compression method in many standards, e.g. JPEG and MPEG [9, 10].

11.2.3 Arithmetic Encoding

Arithmetic encoding [6] also produces variable-length codewords much like Huffman encoding. However, unlike Huffman encoding where each symbol is assigned a specific codeword, in arithmetic encoding, a codeword is assigned to a sequence or block of source symbols from dictionary $\mathcal{X} = \{x_1, x_2, ...x_n\}$. To determine the codeword associated with a specific message that is a combination of the symbols, e.g. "$x_2, x_5, x_2, ... x_j$" in the collection, all the possible symbols $x_1, \ldots, x_n$ are represented in an interval of $[0, 1)$. This interval is then divided into smaller disjoint intervals in proportion to probability P_i, of the symbol x_i. The lower limit of the interval S_i for the x_i symbol is closed at $P(X \leq x_{i-1}) = \sum_{k=1}^{i-1} P_k$, i.e. the cumulative density function (CDF) at step $i-1$ while the upper limit is open at $P(X \leq x_i) = \sum_{k=1}^{i} P_k$, i.e. CDF at step i, hence $S_i \in \left[\sum_{k=1}^{i-1} P_k, \sum_{k=1}^{i} P_k\right)$. The first symbol in a specific message, for example, x_2 in the above combination, selects one of these intervals or segments, i.e. S_2 with probability P_2. This interval is then subdivided, in the same way, in proportion to the symbol probabilities. The next symbol, for example x_5 in the above combination, then chooses the next interval with probability P_2P_5. This procedure is repeated until we reach the last symbol in the message. As the number of symbols in a message increases, the intervals become progressively smaller at each level. However, a symbol with low probability narrows the interval more than a symbol with higher probability. A binary codeword is then assigned to the entire message based on the final interval. This binary codeword corresponds to any number within the final interval, and its length is inversely related to the size of the final interval. That is, smaller intervals correspond to longer codewords.

Arithmetic encoding performance asymptotically approaches the Shannon entropy limit as the number of symbols in a message increases. However, a mechanism is needed to unambiguously determine the end of the message to distinguish one message from the next. The main advantage of arithmetic encoding compared to Huffman encoding is its better compression efficiency, especially for sources with small alphabets or alphabets with highly skewed probabilities. Nevertheless, this benefit comes with many issues. First, arithmetic encoding assigns a codeword to the entire message consisting of several symbols rather than one codeword for each symbol as in Huffman encoding. Consequently, it is more prone to transmission errors, i.e. a single bit error corrupts the entire message. Second, the decoding process is more complex than that of the Huffman encoding. Let us illustrate the coding and decoding processes by giving an example.

Example 11.3 *Reconsider the* 8×8 *image in Example 11.2 with* $L = 8$ *gray levels. Assume that we code every* 2×2 *block of this image arranged in a 1-D sequence of 4 symbols using a row-wise arrangement. Arbitrarily choose two blocks and encode them using both Huffman and arithmetic encoding.*

Solution: We choose the first two 2×2 blocks from the two top rows going from left to right. These two blocks arranged in 1-D sequences are $\{4, 4, 4, 5\}$ and $\{4, 4, 5, 5\}$. The schematic diagram in Figure 11.7 shows the entire process.

Encoding Process: We start with the interval $[0, 1)$ to represent the alphabet $\{0, 4, 5, 6, 7\}$. Subsequently, this interval is partitioned into disjoint segments based on the symbol probabilities in Table 11.1. These segments correspond to $[0, 0.016)$, $[0.016, 0.141)$, $[0.141, 0.266)$, $[0.266, 0.516)$ and $[0.516, 1)$. Now, since the first symbol in the message is 4, the interval $[0.516, 1)$ is selected and in turn partitioned in proportion to the same probabilities. Note that the lower closed limit, i.e. 0.516, corresponds to the CDF value $P_7 + P_6 + P_0 + P_5 = 0.016 + 0.125 + 0.125 + 0.25 = 0.516$. The next symbol is also 4,

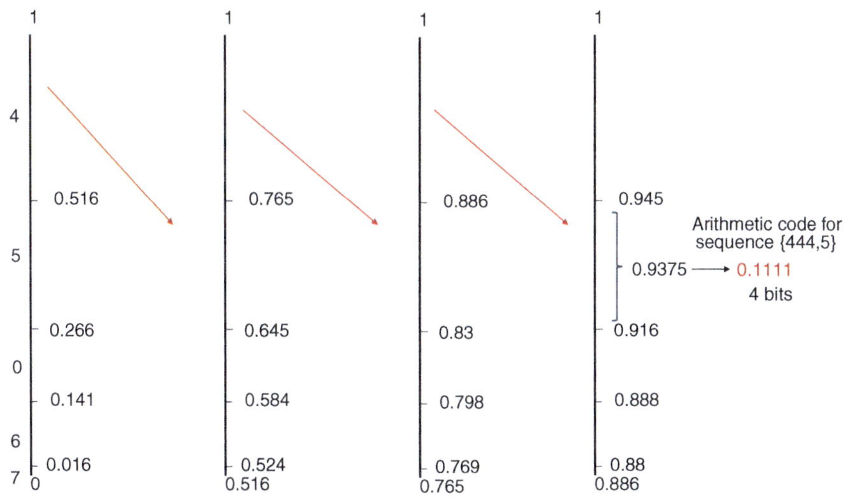

Figure 11.7 Arithmetic encoding for first block of image in Example 11.2.

which selects the interval [0.765, 1) and partitions it into sub-intervals in proportion to the probabilities. This process is repeated until the last symbol in the message is 5, which leads to the interval [0.916, 0.945). Any number within this interval can be selected as the final code to represent the entire sequence and then converted to a binary string. Choosing 0.9375 yields a binary code 0.1111, i.e. 4 bits/block excluding the end of the message bit (usually zero). For this sequence, Huffman code, according to the codeword assignment in Example 11.2, will be 11101, i.e. 5 bits/block. If we repeat the same process for the second block sequence {4, 4, 5, 5}, the final interval will be [0.845, 0.858) and the corresponding binary arithmetic code will be 0.1101101 (or 0.8515625 decimal), i.e. 7 bits/block versus the Huffman code, which is 110101, i.e. 6 bits/block. Thus, as we can see, the efficiency of arithmetic code depends on the length of the message and also the skewness of the distribution of the symbols in the alphabet. For the first message, since the final interval was larger, the arithmetic code gave better bit/message compression.

Decoding Process: The decoder also has the same symbol probabilities and uses the same partitioning process as with the encoder (see Figure 11.7). Once the decoder receives the binary code 0.1111, it associates with the last interval [0.516, 1), since its decimal number 0.9375 lies in that interval, thereby identifying the first symbol 4. Now, the interval for 4 is partitioned according to the probabilities as in Figure 11.7. Again, since the value 0.9375 is in the interval [0.765, 1), the next symbol is also 4. The same partitioning process is repeated for the interval [0.765, 1) according to probabilities as shown in the third level of the diagram in Figure 11.7. The last sub-interval [0.885, 1) associated with symbol 4 will be selected since 0.9375 lies in this segment. This sub-interval is partitioned again in the same way. However, this time 0.9375 lies in the segment [0.916, 0.945), which corresponds to symbol 5. Therefore, the entire decoded message will be {4, 4, 4, 5}.

11.2.4 Run-Length Encoding

Run-length coding [6, 7] is simple and effective lossless method that is particularly suited for binary images, e.g. text, finger prints, facsimile, etc. Its efficiency relies on having large areas within an image with uniform intensity, which is typically true in these applications. As a result, it is not efficient for images with a large number of gray levels or textured images. The process involves scanning the image horizontally or vertically and while scanning assign a group of pixels with the same intensity to a pair (g_i, l_i), where g_i is the gray level and l_i is the length of the *run*. These pairs of numbers will be encoded instead of the actual pixels in each group. Run-length encoding can also be used to detect thin edges and boundaries of regions in an image.

As an example, if we consider the image in Example 11.2 and apply a vertical continuous top-down scanning (lawn mower pattern), we obtain the run length codes shown in Table 11.2. For this scanning mode, there is a total of 20 pairs of (g_i, l_i) where using PCM every intensity, g_i, needs 3 bits while the run length, l_i, require 5 bits to encode, i.e. 2.5 bits/pixel. The horizontal scanning, on the other hand, would have led to 34 pairs (68 numbers), which is more than the total number of pixels (i.e. 64).

Table 11.2 Pair of $\{(g_i, l_i)\}$ for vertical scan.

(4,9)	(5,5)	(4,3)	(5,1)	(6,3)
(5,1)	(4,3)	(5,1)	(6,1)	(7,1)
(6,1)	(5,1)	(4,3)	(5,1)	(6,3)
(5,1)	(4,3)	(5,5)	(4,10)	(0,8)

The number of bits/pixel may further be reduced by preprocessing the original image using reversible mappings to make runs longer. For example, the added details subimages produced using 2-D DWT contain mostly zeros (after thresholding) and hence are suitable for run-length encoding. Variants of run length encoding are used in modern standards such as JPEG XS [11], where modifications are introduced to control run-length variability and reduce encoding complexity.

Example 11.4 *The transition probabilities for a run-length encoding of a binary image (0: black and 1: white) are* $P_0 = P(0|1)$ *and* $P_1 = P(1|0)$ *with* $P(0|0) = 1 - P(1|0)$ *and* $P(1|1) = 1 - P(0|1)$. *Assuming that all runs are independent, find (a) average run lengths, (b) entropy of white and black runs, and (c) compression ratio.*

Solution: A run of length $l \geq 1$ can be modeled by a geometric r.v. X_i with pmf $P(X_i = l) = p_i(1-p_i)^{l-1}, l \geq 1, i = 0, 1$ which corresponds to the probability of the first occurrences of 0 (or 1) after l independent trials of $l - 1$ ones. The average length of run is:

$$\mu_{X_i} = \sum_{l=1}^{\infty} lP(X_i = l) = \sum_{l=1}^{\infty} lp_i(1-p_i)^{l-1}, \quad i = 0, 1$$

Now, using series formula $\sum_{n=1}^{\infty} na^{n-1} = \frac{1}{(1-a)^2}$, we get $\mu_{X_i} = \frac{1}{p_i}, i = 0, 1$, i.e. the average length of the runs is inversely proportional to the transition probability p_i. That is, low transition probabilities imply longer runs. The entropy for r.v. X_i is given by:

$$\begin{aligned} H_{X_i} &= -\sum_{l=1}^{\infty} P(X_i = l) \log_2 P(X_i = l) \\ &= -p_i \sum_{l=1}^{\infty} (1-p_i)^{l-1} [\log_2 p_i + (l-1) \log_2 (1-p_i)], \quad i = 0, 1 \end{aligned}$$

Using the same series formula, we can get:

$$H_{X_i} = -\frac{1}{p_i} [p_i \log_2 p_i + (1-p_i) \log_2 (1-p_i)], \quad i = 0, 1$$

The achievable compression ratio is:

$$C_r = \frac{H_{X_0} + H_{X_1}}{\mu_{X_0} + \mu_{X_1}} = \frac{H_{X_0} P_0}{\mu_{X_0}} + \frac{H_{X_1} P_1}{\mu_{X_1}}$$

where $P_i = \frac{p_i}{p_0 + p_1}$ are the *prior* probabilities of black and white pixels.

For example, if $p_0 = 0.4$ and $p_1 = 0.5$, we get $\mu_{X_0} = 2.5$ and $\mu_{X_1} = 2$. Also, the entropies will be $H_{X_0} = 2.4275$ and $H_{X_1} = 2$, and the achievable compression ratio becomes is $C_r \approx 0.984$.

11.3 Image Compression and Encoding Methods – Lossy

The achievable compression rates of pixel-based encoding methods in the previous section are fundamentally limited by Shannon's entropy encoding theorem. Now, the questions are: (1) Is it possible to achieve better compression beyond this limit? (2) If so, how and at what cost? The answer to the first question is yes, it is possible to achieve higher compression rates using lossy image compression and encoding methods. These methods work by either (a) removing local inter-pixel dependencies among a group of pixels in a small neighborhood within an image prior to quantization and encoding processes, or (b) transforming the image to a domain in which its essential information can be represented using significantly fewer coefficients. The obvious cost of achieving higher compression is the loss or distortion that occurs due to data reduction and quantization. However, owing to its psycho-visual redundancy, the human visual system is often incapable of discerning such distortions in the images. Quantization exploits these perceptual limitations. In what follows, we will discuss several lossy image compression methods, including predictive encoding, transform-based encoding, and vector quantization (VQ). The block diagram in Figure 11.5 shows all the key components of lossy image compression systems of the first two types. Vector quantization, however, follows a different structure, which will be discussed later.

11.3.1 Predictive Encoding

Predictive encoding aims to remove mutual redundancy among neighboring pixels within a region of support (ROS) of a model, and then encodes only the innovation or *new information* [7]. This is achieved by predicting the value of a pixel to be encoded based on the previously processed pixels within its ROS and then encoding the difference (or residual) between the actual and predicted values. This residual or prediction error mainly contains the *new information* of the pixel, while the predicted value captures the *old information* carried by the neighboring pixels. This process is then followed by quantization and encoding of the prediction error using a variable-length coding method covered previously. The encoded image data are then transmitted through a communication channel.

The block diagram of Figure 11.8 represents the entire predictive encoding process with its CODEC and main components. The input to the coder is the image pixel $x(m, n)$ from which the predicted value $\hat{x}^o(m, n)$ is subtracted to produce the residual or prediction error $e(m, n)$. This error is then quantized to produce $e^o(m, n)$ and subsequently encoded using one of the methods, e.g. Huffman encoding, in the previous section. At the decoder, the quantized residual is first decoded to recover $e^o(m, n)$. The decoder subsystem also uses the same predictor with identical parameters to generate the predicted pixel $\hat{x}^o(m.n)$, which is added to the decoded error $e^o(m, n)$ to produce the reconstructed pixel $x^o(m, n)$.

To perform prediction at the coder, 2-D linear AR models (see Chapter 9) are usually employed. Assuming that the image is scanned from left-to-right and top-to-bottom, Figure 11.9 shows the NSHP ROS (red dashed line region) of a first-order 2-D AR model given by:

$$x(m, n) = a_{0,1}x(m, n-1) + a_{1,1}x(m-1, n-1) + a_{1,0}x(m-1, n) + a_{1,-1}x(m-1, n+1) + e(m, n), \tag{11.12}$$

where $a_{i,j}$ are AR model coefficients and $e(m, n)$ is the driving process. It is assumed that the model parameters have been computed using the 2-D extension of the Yule-Walker method in Chapter 9 based on a *training* dataset with similar statistics. Then, the best (in the MMSE sense) linear predictor of $x(m, n)$ is:

$$\hat{x}(m, n) = a_{0,1}x(m, n-1) + a_{1,1}x(m-1, n-1) + a_{1,0}x(m-1, n) + a_{1,-1}x(m-1, n+1). \tag{11.13}$$

Thus, the prediction of each new pixel, $\hat{x}(m, n)$, requires four previously processed pixels in its NSHP ROS. Figure 11.9 also shows all the pixels that need to be stored in a *global state vector* for this first-order 2-D predictor. As indicated in Section 10.2.5, the role of global state is to serve as a storage to keep all the previously processed pixels needed when prediction moves to a new pixel location along the same row, or when starting the next row. The prediction error, $e(m, n) = x(m, n) - \hat{x}(m, n)$, which has a smaller variance and requires shorter codewords, is then computed and used as input to the quantizer. The quantized error is subsequently encoded and transmitted or stored as shown in Figure 11.8. This procedure is also called differential PCM (DPCM) [7].

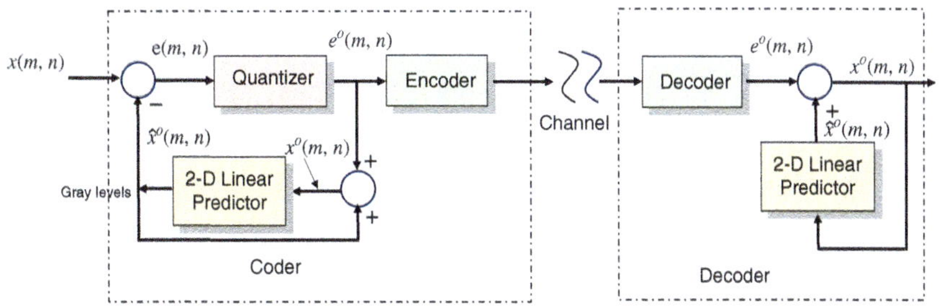

Figure 11.8 Block diagram of 2-D predictive encoding system and its components.

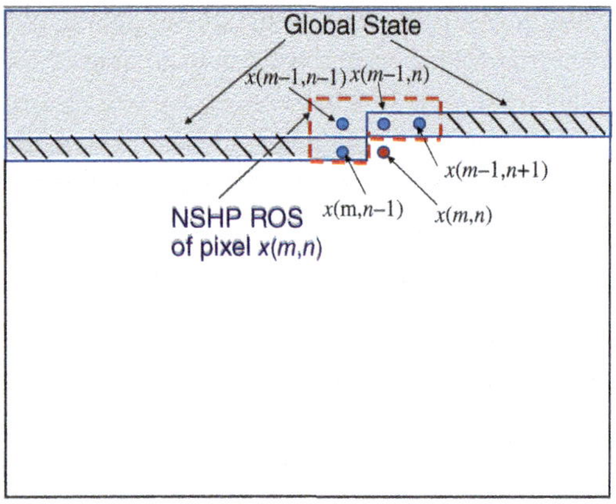

Figure 11.9 A first-order 2-D AR model with NSHP ROS and the global state vector.

Assuming that the reproduced values (quantized) up to $(m, n-1)$, i.e. $x^o(m, n-1)$ are available, the predicted value of the current pixel is computed using:

$$\hat{x}^o(m,n) = a_{1,0}x^o(m,n-1) + a_{1,1}x^o(m-1,n-1) + a_{1,0}x^o(m-1,n) + a_{1,-1}x^o(m-1,n+1). \tag{11.14}$$

Note that since the decoder reconstructs the image based on the decoded and quantized error, $e^0(m, n)$, the reproduced values, $x^o(m, n)$, are also used at the coder (instead of $x(m, n)$) to maintain consistency between the coder and decoder processes. The predicted value is then used to generate the prediction error, which is subsequently applied to the quantizer:

$$e(m,n) = x(m,n) - \hat{x}^o(m,n). \tag{11.15}$$

The quantized version of this prediction error, $e^o(m, n)$, is encoded and transmitted. In the decoder, this error is used along with the predicted value, $\hat{x}^o(m, n)$, to reconstruct the pixel using:

$$x^o(m,n) = e^o(m,n) + \hat{x}^o(m,n), \tag{11.16}$$

where $\hat{x}^o(m, n)$ is generated by the 2-D predictor at the decoder as illustrated in Figure 11.8.

11.3.1.1 Performance Analysis of DPCM

Using (11.15) and (11.16), it follows that the quantization error in both PCM and DCPM is the same:

$$\begin{aligned} q(m,n) &= x(m,n) - x^o(m,n) \quad \text{PCM quantization error} \\ &= e(m,n) - e^o(m,n) \quad \text{DPCM quantization error} \end{aligned}$$

However, for the same quantization error, $q(m, n)$, DPCM requires much fewer bits. This is because, in PCM, the original image $x(m, n)$ is directly quantized, whereas in DPCM, the prediction error, $e(m, n)$, is quantized. For PCM, the rate distortion function gives:

$$R_{\text{PCM}} = \frac{1}{2}\log_2 \sigma_x^2/\sigma_q^2 \quad \text{bits/pixel}$$

where $\sigma_q^2 = E[q^2(m, n)]$ is the quantization noise variance. However, for DPCM, the rate distortion function is:

$$R_{\text{DPCM}} = \frac{1}{2}\log_2 \sigma_e^2/\sigma_q^2 \quad \text{bits/pixel}$$

and since $\sigma_e^2 \ll \sigma_x^2$, it follows that $R_{\text{DPCM}} \ll R_{\text{PCM}}$. The bit reduction offered by DPCM relative to PCM is measured by:

$$R_{PCM} - R_{DPCM} = \frac{1}{2}\log_2 \sigma_x^2/\sigma_e^2 = \frac{1}{0.6}\log_{10}\sigma_x^2/\sigma_e^2 \quad \text{bits/pixel.} \tag{11.17}$$

Thus, the achieved compression depends on the inter-pixel redundancy within the ROS of the model, i.e. for an image with no spatial redundancy (random image) $\sigma_e^2 = \sigma_x^2 \to R_{\text{DPCM}} = R_{\text{PCM}}$. Moreover, model mismatch in certain areas containing edges or textures yields poor compression performance in such areas. This problem arises since the predictor parameters are usually derived based on global image statistics, which may not accurately represent such areas with different local statistics. Thus, in areas where the local statistics deviate from the global ones, a model mismatch will occur. Adaptive predictive encoding techniques have been used [7] to compensate for this and other issues with the standard predictive methods.

To illustrate the predictive encoding process, we used the original "Lena" image in Figure 11.10a as the input to the 2-D linear predictor in (11.13). The coefficients of this first-order 2-D AR model with NSHP ROS were found in Section 9.2.4 using the 2-D Yule-Walker method. These coefficients are: $a_{1,-1} = 0.0959$, $a_{0,1} = 0.4464$, $a_{1,0} = 0.7822$, and $a_{1,1} = -0.3243$. Figure 11.10b shows the predicted image generated at the coder. The error image after using 6-bit uniform quantization is shown in Figure 11.10c. Figure 11.10d, on the other hand, gives the reconstructed image at the decoder. As can be observed, there are no visible artifacts in the reconstructed image. The SNR for this reconstructed image was found to be 21.44 dB, and the bit reduction relative to PCM is 2.866 bits/pixel.

11.3.2 Transform-Based Encoding

The principal idea behind transform-based methods is to reduce redundancy by applying a unitary transformation to image blocks (or to the entire image) prior to data reduction, quantization, and encoding [7, 12]. A subset of the transformed domain coefficients that contain most of the image energy is then retained and subsequently quantized and encoded, while the remaining coefficients that correspond to redundant information are discarded. A typical transform-based encoding process and its components are illustrated in Figure 11.11. The image is first partitioned into nonoverlapping blocks. Each block is then unitary transformed using, e.g. 2-D DCT and the principal coefficients are quantized and encoded. Each coefficient must be quantized and encoded separately using a parallel configuration or a block quantizer and a block encoder. This is because the dynamic range of the coefficients is vastly different, and each coefficient has a different statistical distribution. At the receiver, the coefficients are decoded and subsequently inverse transformed to reconstruct the image on a block-by-block basis.

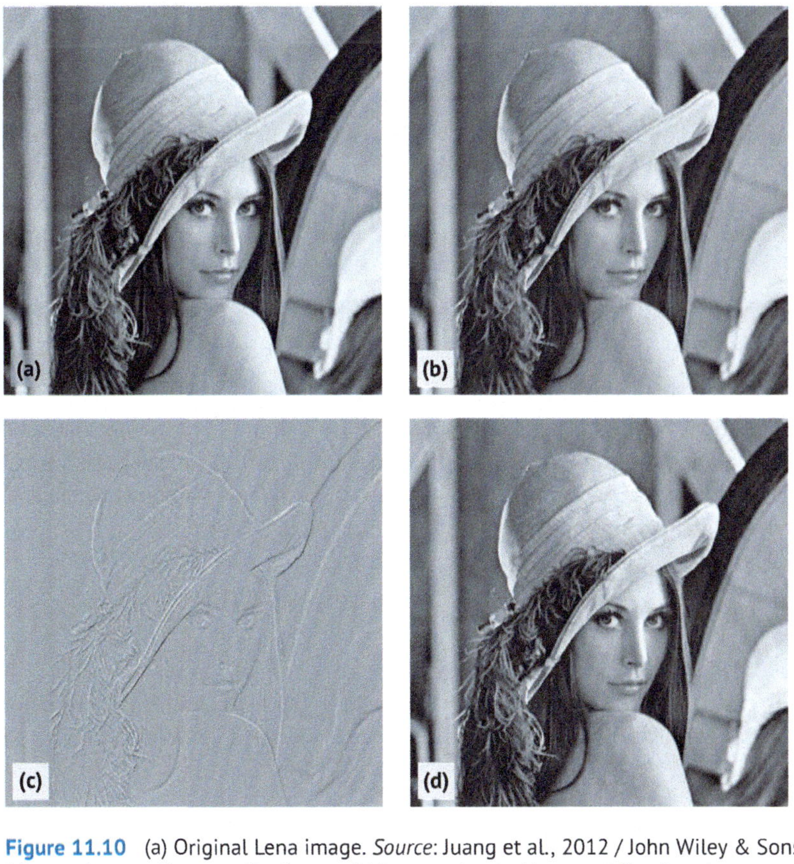

Figure 11.10 (a) Original Lena image. *Source*: Juang et al., 2012 / John Wiley & Sons / CC BY 3.0. (b) predicted image at the coder, (c) error image after quantization, and (d) reconstructed image at the decoder.

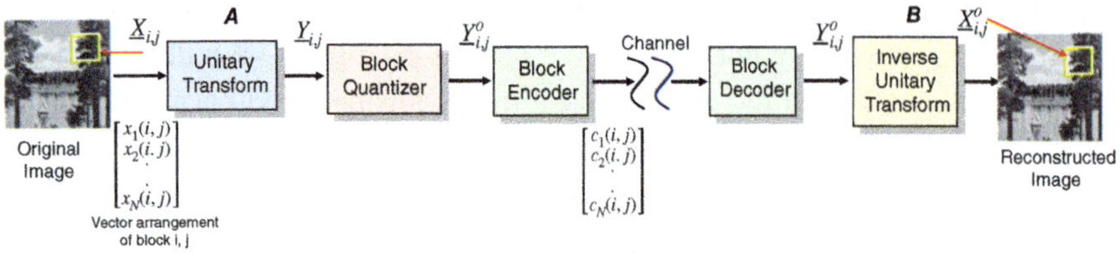

Figure 11.11 Transform-based encoding process and its components. *Source* (boat image): Hu et al., 2014 / John Wiley & Sons / CC BY 3.0.

Now, two fundamental questions are: (a) What are the optimal transform and inverse transform matrices, **A** and **B**, so that maximum redundancy removal is achieved while the distortion due to data reduction is minimized? (b) What is the optimal quantizer that minimizes quantization distortion?

To answer these questions, let us assume that $\underline{X}$ is a random vector representing blocks of an image and $\underline{Y}$ is its unitary transformed, i.e. $\underline{Y} = \mathbf{A}\underline{X}$. A subset of components $Y(k)$ of vector $\underline{Y}$ is retained, quantized, encoded, and transmitted. At the receiver, the coefficients are first decoded and then the image block is reconstructed. If we denote the vector of decoded coefficients by $\underline{Y}^o$, the reconstructed image block is $\underline{X}^o = \mathbf{B}\underline{Y}^o$. The objective is to find optimum transform and inverse transform matrices **A** and **B** and optimum quantizer such that the distortion $D = E[\| \underline{X} - \underline{X}^o \|^2]$ is minimized.

In Chapter 6, we showed that the distortion due to the data reduction operation in the transformed domain is minimized when $\mathbf{A} = \mathbf{\Psi}^H$ and $\mathbf{B} = \mathbf{\Psi}$, i.e. KL transform pair, since for a fixed number of retained coefficients, the KL transform offers the best energy packing property. Moreover, in Chapter 5 we showed that the Lloyd-Max quantizer minimizes quantization distortion for a given number of bits.

Although, in theory the optimum choice for the unitary mapping of image blocks is the KL transform, in practice, however, it is not employed in data compression standards such as JPEG [9–11]. This is because the transform matrix $\mathbf{\Psi}^H$ and its inverse $\mathbf{\Psi}$ are obtained based on the second-order statistics of the image field, and different images will likely result in different transform matrices. Consequently, the transform matrix must also be transmitted or stored, increasing the overall number of bits, hence negatively affecting the achievable compression rate. The energy packing ability of the 2-D DCT is comparable to that of the KL while the basis functions are not image data-dependent. As a result, the older generation of JPEG [9] used 2-D DCT to map each 8×8 image blocks before quantization and encoding.

11.3.2.1 Bit Allocation Strategy

The retained coefficients $Y^o(k)$, after data reduction and quantization, generally have a wide dynamic range, hence requiring different number of bits to encode them. As a result, one cannot uniformly allocate bits to each coefficient. Thus, the problem of bit allocation becomes a constrained optimization problem in which the constraint is the total bit budget M to be allocated to N retained components of $\underline{Y}^o$ (or an average of $\frac{M}{N}$ bits/coefficient), so that the distortion:

$$D = \frac{1}{N}\sum_{k=1}^{N} E[(y(k) - y^o(k))^2] = \frac{1}{N}\sum_{k=1}^{N} \sigma_k^2 f(m_k) \tag{11.18}$$

is minimized subject to $M = \sum_{k=1}^{N} m_k$. In (11.18), $f(m)$ is the distortion function of the m-bit quantizer (see Property 3 of Lloyd-Max quantizer) in block quantization, σ_k^2 is the variance of coefficient $y(k)$, and m_k is the number of bits allocated to the quantized coefficient $y^o(k)$. Thus, this optimal bit allocation involves finding m_k to minimize D subject to a fixed bit budget $M = \sum_{k=1}^{N} m_k$. Note that coefficients with higher variance contain more information than those with lower variance, and hence more bits are assigned to them to improve the performance.

The above constrained optimization problem can be cast as a Lagrangian formulation:

$$\varepsilon(m_1, \dots, m_N) = \frac{1}{N}\left\{\sum_{k=1}^{N} \sigma_k^2 f(m_k) + \lambda\left(M - \sum_{k=1}^{N} m_k\right)\right\} \tag{11.19}$$

where the first term is the distortion D in (11.18) and the second term enforces the constraint with a Lagrangian parameter λ. This constrained problem is minimized by taking the partial derivative of (11.19) with respect to m_j and setting the result to zero:

$$\frac{\partial\, \varepsilon(m_1, \dots, m_N)}{\partial m_j} = \sigma_j^2 f^{(1)}(m_j) - \lambda = 0, \tag{11.20}$$

which yields $f^{(1)}(m_j) = \lambda/\sigma_j^2$ where $f^{(1)}(m_j) = \frac{df(m_j)}{dm_j}$. Now, if $\sigma_j^2 = 1,\ \forall j \in [1, N]$, i.e. unity variance coefficients (normalized), then this result implies that all quantizers must have the same slope λ to the distortion function curve at m_j points [13]. Let g be the inverse of function $f^{(1)}$, i.e. $g(f^{(1)}(x)) = x$, then $m_j = g(\lambda/\sigma_j^2)$. Also, let us define $\lambda = \theta f^{(1)}(0)$ where θ is an unknown parameter which must be found to satisfy the constraint. The followings are different bit allocation strategies for different choices of quantizer distortion function.

i. Shannon's Allocation Strategy

For Shannon's quantizer [7, 13], the distortion function is $f(x) = 2^{-2x}$ for which we have $y = f^{(1)}(x) = -(2\ln 2)2^{-2x}$ and hence $x = g(y) = -\frac{1}{2}\log_2\left(\frac{-y}{2\ln 2}\right)$. Now, since $f^{(1)}(0) = -2\ln 2$, we get:

$$m_j(\theta) = \max\left(0, \frac{1}{2}\log_2\left(\frac{\sigma_j^2}{\theta}\right)\right), \tag{11.21}$$

where θ is chosen to satisfy the average bit constraint $p = \frac{M}{N}$ bpp. The expressions for this average bpp and minimum distortion are:

$$p = \frac{M}{N} = \frac{1}{N}\sum_{k=1}^{N} \max\left(0, \frac{1}{2}\log_2\left(\frac{\sigma_k^2}{\theta}\right)\right) \quad \text{and} \tag{11.22}$$

$$D = \frac{1}{N}\sum_{k=1}^{N} \min(\theta, \sigma_k^2), \tag{11.23}$$

respectively.

ii. Segall Allocation Strategy

As we showed in Chapter 5, the optimal quantizer for sample-by-sample quantization and encoding is the Lloyd-Max quantizer. Although there is no closed-form expression for the distortion function $f(m)$, one can approximate [14] the numerical results for small m and use the asymptotic result for large m. This yields the following distortion function and bit allocation equations [14]:

$$f(m) = \begin{cases} 2^{-1.57m}, & m \le 2.32 \\ \dfrac{2.73\ 2^m}{(2^m + 0.853)^3}, & m > 2.32 \end{cases} \tag{11.24}$$

and

$$m_j(\theta) = \begin{cases} \dfrac{1}{1.57}\log_2\left(\dfrac{\sigma_j^2}{\theta}\right), & \sigma_j^2 \ge \theta > 0.083\,\sigma_j^2 \\ \dfrac{1}{1.78}\log_2\left(\dfrac{1.46\sigma_j^2}{\theta}\right), & 0.083\,\sigma_j^2 \ge \theta > 0 \\ 0, & \theta > \sigma_j^2 \end{cases} \tag{11.25}$$

where θ is found to meet $\sum_{k=0}^{N} m_k(\theta) = M$.

iii. Huang-Schultheiss Allocation Strategy

This bit allocation approximates the optimal nonuniform allocation for Gaussian coefficients [15]. The bit allocation equation is:

$$m_j = \mathrm{Int}\left[p + 2\log_2\sigma_j^2 - \frac{2}{N}\sum_{i=1}^{N}\log_2\sigma_i^2\right] \tag{11.26}$$

where $\mathrm{Int}[x]$ stands for the integer part of x. These bit allocations must satisfy the constraint $M = \sum_{k=1}^{N} m_k$. That is, the real-valued Huang-Schultheiss bit allocation [15] is rounded to the nearest integer, in a manner that does not violate the overall bit budget.

Figure 11.12 (a) and (c) Reconstructed. *Source:* Juang et al., 2012 / John Wiley & Sons / CC BY 3.0. (b) and (d) error images of Lena and Barbara, respectively, using Shannon bit allocation (1.5 bpp).

Figures 11.12 and 11.13 show the reconstructed and error images of the Lena and Barbara using Shannon and Segall bit allocation methods at a fixed average of $p = 1.5$ bpp. The reconstructed images using Shannon's bit allocation exhibit $SNR_{Lena} = 20.5$ dB and $SNR_{Barb} = 17.2$ dB, respectively, while those of Segall method give $SNR_{Lena} = 21.2$ dB and $SNR_{Barb} = 16.9$ dB, respectively. Visual evaluation of the reconstructed and error images reveal the fact that they both had some difficulties in the textured areas though Shannon allocation provided slightly better results, especially on the Barbara image, which has more textured regions such as the scarf, tablecloth, and pants.

11.3.3 Wavelet-Based Encoding and New JPEG Standards

In contrast to the transform-based encoding methods that partition the image into nonoverlapping blocks and apply 2-D DCT to each block followed by coefficient quantization and encoding, wavelet-based approaches, map the entire image into several subband images as described in Chapter 7. The elimination of block-by-block transformation, naturally circumvents the blocking artifacts that are noticeable in the reconstructed images, particularly at high compression ratio. While the lowest-order approximation in the 2-D DWT subband decomposition can be encoded using the standard PCM or DPCM covered earlier, the added details that contain significant redundancy can be separately quantized and encoded. The added details subband images typically contain a large number of pixels with negligible visual significance, which can be discarded via thresholding and quantization operations. As a result, wavelet-based encoding methods provide excellent compression rates and reconstructed images that appear visually lossless [9, 10]. An important consideration is that the histograms of added details are highly skewed due to a large number of very small coefficients. Therefore, uniform quantization does not perform well for quantizing added details, and it is recommended to use nonuniform quantization for each added details in which case the decision and reconstruction levels must be provided to the decoder as well.

Let us compare the quality of the reconstructed images and compression ratios for image compression using 2-D DCT and DWT methods. In Chapter 6, we illustrated how the 2-D DCT is applied to nonoverlapping blocks of images to reduce

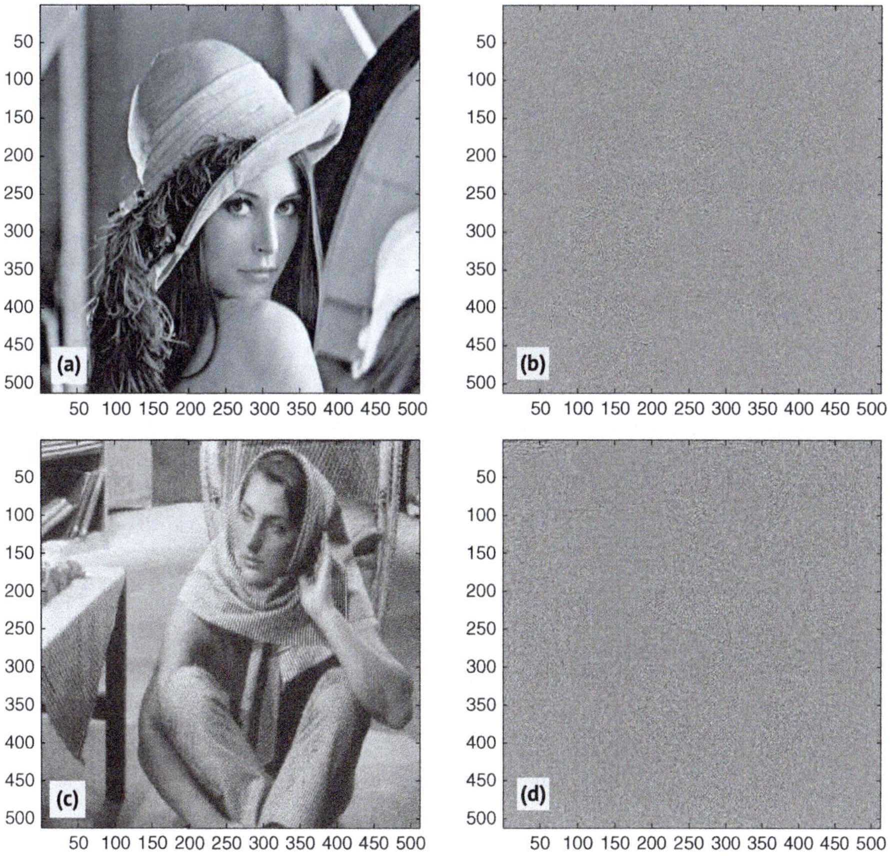

Figure 11.13 (a) and (c) Reconstructed and (b) and (d) error images of Lena and Barbara, respectively, using Segall bit allocation (1.5 bpp). *Source*: Juang et al., 2012 / John Wiley & Sons / CC BY 3.0.

Figure 11.14 Original Peppers image.

the data in the transform domain. Here, we repeat the same experiment on the "Peppers" image in Figure 11.14 using two different block sizes, namely 8×8 and 16×16. The histogram-based thresholding is then applied to zero out low-magnitude coefficients in each block for both block sizes. The threshold is decided such that approximately 98% of the transformed coefficients are set to zero. The image is then reconstructed block-by-block using inverse 2-D DCT. Figure 11.15a–d shows the reconstructed Peppers and error images for 8×8 and 16×16 block sizes, respectively. As can be seen, the reconstructed images exhibit some ringing artifacts due to coefficient truncation in the transform domain. The blocking effects are also noticeable though they are much more pronounced for the smaller block size. The SNR values of the reconstructed images are measured to be SNR = 13.25 dB and SNR = 17.04 dB for 8×8 and 16×16 block sizes, respectively. Figure 11.15e,f, on

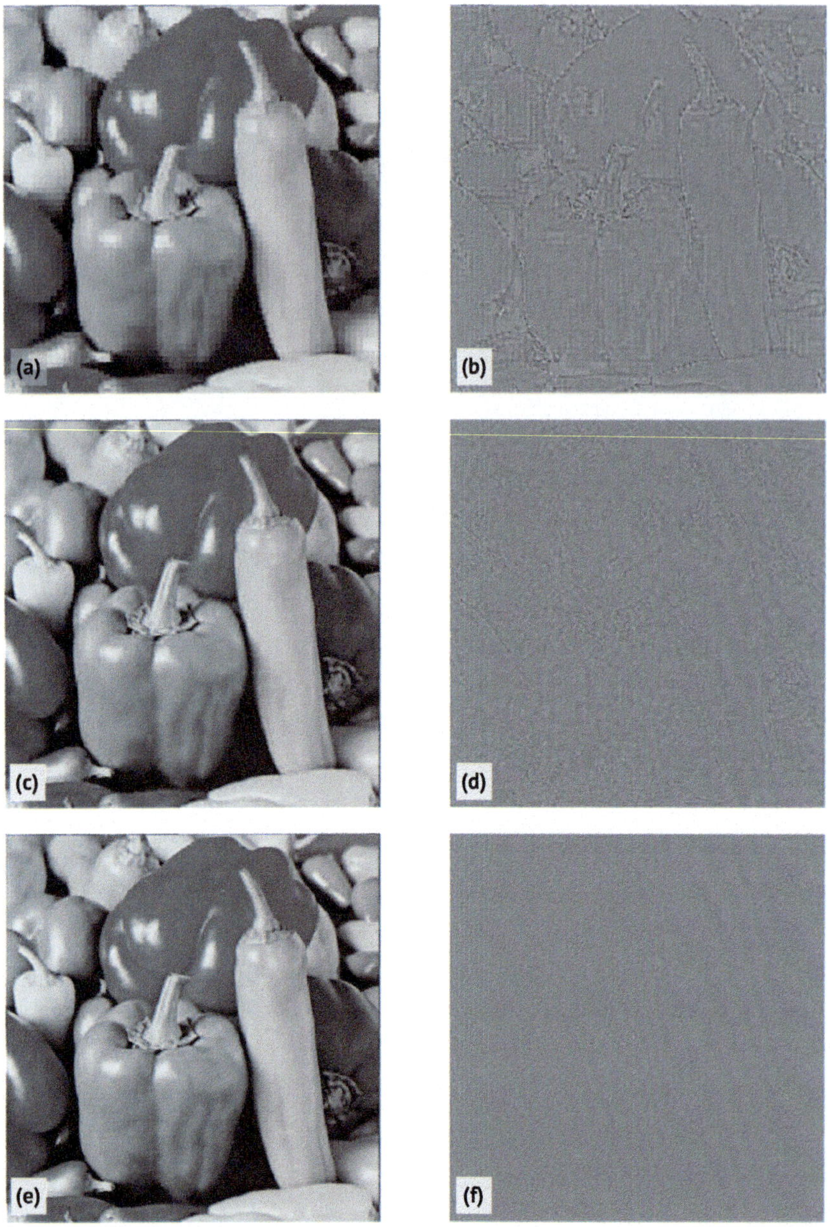

Figure 11.15 (a–d) Reconstructed and error Peppers images using block-based 2-D DCT with 8×8 and 16×16 block sizes, and (e,f) reconstructed and error images using two-level 2-D DWT with 5/3 biorthogonal wavelet.

the other hand, shows the reconstructed and error images using a two-level 2-D DWT decomposition with 5/3 biorthogonal wavelets (see Chapter 7). For approximately the same percent zeros, the DWT-based approach provides a much higher quality reconstructed image with SNR = 20.52 dB, i.e. about 2.5 dB improvements over the 2-D DCT method using 16×16 blocks. Additionally, the error image does not show any blocking artifacts that are noticeable in the DCT-based results.

The filter bank structure of the DWT enables efficient hardware implementation using various platforms such as FPGA and ASIC. Owing to this advantage coupled with its superior compression rate, DWT-based image encoding was adopted in the JPEG-2000 standard [9]. JPEG-2000 supports a high dynamic range of up to 38 bits per color channel and several color systems. The encoding process involves color space transformation (e.g. RGB to YC_bC_r), tiling, 2-D wavelet decomposition, nonuniform quantization, context modeling, and bit-plane arithmetic encoding. The superiority of JPEG-2000 is particularly evident at low bit rates, for example, for compression ratios greater than 10 : 1. Unlike prior DCT-based version, JPEG-2000 can be implemented in lossless or lossy modes, depending on the choice of wavelet and quantizers. The lossless implementation uses 5/3 biorthogonal wavelets with integer coefficients which do not require quantization. However, the lossy mode uses 9/7 biorthogonal wavelets and the fast wavelet transform (FWT) [16] implementation.

A more recent variant of wavelet-based JPEG is JPEG XS [11], which is designed to support specific applications that require low latency, light weight, and high quality CODEC, including virtual reality, self-driving cars, autonomous systems, and broadcast television. The encoder part of the CODEC in JPEG XS consists of a preprocessor, wavelet transformation, quantization, and entropy encoding. The preprocessor carries out the following operations: (a) expanding the bit precision of the image data to 16–20 bits; (b) removing DC level; and (c) converting RGB format into the YC_bC_r color space. The latter operation is performed in lieu of the fact that the human eye is more sensitive to luminance changes of color than to chromatic changes. Thus, in the YC_bC_r color space luminance is captured in the Y component, whereas the chromatic information is separated in the C_b and C_r components, which can be further downsampled to improve efficiency without sacrificing the visual quality. The wavelet transformation employs a 5-level asymmetric 5/3 biorthogonal DWT subband decomposition that leads to one vertical detail and five horizontal details for each component. The reason for this asymmetric filter bank is to minimize latency. The quantizer also uses a nonuniform quantization of the wavelet coefficients.

11.3.4 Vector Quantization

Vector quantization (VQ) is another lossy image compression algorithm [2–6] that is also applied to image blocks as in the DCT-based compression. However, instead of using unitary transforms to reduce redundancy, VQ represents each image block by a *codebook* vector which is closest, in some distance measure, to the image block. The redundancy is reduced since this process is many-to-one, i.e. many image blocks are represented by one codebook vector. In many ways, VQ can be viewed as a pattern matching process where a specific pattern within a block is matched to one of K possible codebook vectors, each representing a typical pattern that could exist in an image block, e.g. background, different vertical, horizontal, and diagonal edge patterns, and various coarse and fine texture patterns. These codebook vectors form a *dictionary* of possible patterns. This dictionary is designed and made available at both the transmitter and receiver ends. Once the codebook vector with the best match to a particular image block is found, its corresponding address or label can be quantized, encoded and transmitted. At the receiver, the codebook address is decoded and the associated codebook vector is reshaped into an image block to represent the reconstructed version of the original image block at the same location. Thus, the decoder operates like a lookup table.

To design the codebook vectors, a clustering algorithm is typically applied to training set consisting of some randomly chosen image blocks. Cluster discovery can be done using many different algorithms, including the *competitive learning* [17] and *K-means* clustering [18]. The clustering process is a memoryless operation, since encoding a particular image block does not depend on the previously encoded blocks. The number of clusters, K, can be determined based on prior knowledge of the image characteristics and desired compression rate. The choice of the block size is another important factor which impacts not only the achievable compression rate but also the quality of the reconstructed images and the computational and memory requirements of the VQ algorithms. Typically, small sizes, e.g. 4×4, blocks are preferred since larger blocks require more codebook vectors to capture many possible pattern combinations that may exist in such blocks, thus increasing the computational and memory requirements. Furthermore, larger block lead to more noticeable blocking artifacts, especially for smaller size dictionaries. In what follows, we present the K-means clustering method to form codebook vectors. The competitive learning algorithm and its application in self-organized feature maps [17] will be presented in Chapters 17 and 18.

11.3.4.1 K-Means Clustering Algorithm

Let $\mathcal{X} = \{\underline{x}(p)\}_{p=1}^{P}$ be a set of P (unlabeled) training image blocks arranged (e.g. row-stacked) into column vectors of dimension N. The goal is to partition these data vectors into $K < P$ clusters using an iterative algorithm with the following steps:.

1. Randomly order the samples in $\mathcal{X}$.
2. *Cluster Initialization:* Select the first K samples as initial centroids $\underline{w}_k$, $k \in [1, K]$ of the clusters, i.e. $\underline{w}_l = \underline{x}(l)$, $l \in [1, K]$.
3. *Cluster Selection:* Assign each sample in $\mathcal{X}$ to the cluster with the nearest centroid, i.e.

$$\text{if } l = \underset{j}{\operatorname{argmin}} ||\underline{x}(m) - \underline{w}_j|| \text{ , } j \in [1, K], \qquad \text{then} \quad \underline{x}(m) \in C_l.$$

   Increment the number of samples Q_l in cluster l, i.e. $Q_l^{\text{new}} = Q_l^{\text{old}} + 1$.

4. *Cluster Updating:* Compute the new centroid for the chosen cluster (only) using:

$$\underline{w}_l^{\text{new}} = \underline{w}_l^{\text{old}} + \alpha\left(\underline{x}(m) - w_l^{\text{old}}\right),$$

or equivalently

$$\underline{w}_l^{\text{new}} = \frac{Q_l^{\text{old}}}{Q_l^{\text{new}}}\,\underline{w}_l^{\text{old}} + \frac{1}{Q_l^{\text{new}}}\,\underline{x}(m),$$

where $\underline{w}_l^{\text{old}}$ is the weight vector of cluster l before updating and $\underline{w}_l^{\text{new}}$ is the updated version. The step size in the first updating equation is chosen to be $\alpha = 0.95$.

5. Repeat Steps 3 and 4 until the weight vectors no longer change. Upon convergence, the weight vector $\underline{w}_l = \underline{c}_l$, i.e. the centroid (or codebook) of cluster $l \in [1, K]$.

Once the training phase is completed, the codebook vectors form dictionary $\mathcal{C} = \{\underline{c}_1, \underline{c}_2 \cdots, \underline{c}_K\}$ used in the matching process and codebook selection for any new image block. This matching process is identical to that in Step 3. Once the cluster address (or label) of the winning codebook l is decided, it will be quantized and encoded for transmission.

11.4 Hybrid Methods

Hybrid image encoding methods [7, 19–21] combine different lossy and lossless methods to achieve better quality reconstructed images and higher compression ratios. These systems may integrate transform-based, predictive, and vector quantization methods, or incorporate modern machine learning approaches.

One classical hybrid approach involved combining transform-based and predictive encoding systems [7, 22]. A 1-D unitary transform, e.g. KL transform, is first applied to each column (or row) of the image to reduce intra-column redundancy. A 1-D scalar predictive encoding is then applied sequentially to each transformed domain coefficient inter-column-wise along the horizontal (or orthogonal) direction. Typically, a 1-D AR(1) model is employed for predictive encoding. The decoder reconstructs the coefficients using the same prediction model followed by the inverse KL transform. To reduce computational complexity, the image may be partitioned into strips, and the hybrid processing is applied independently within each strip.

A hybrid of 2-D block-wise DCT and VQ was proposed in [19] for color image compression. The process involves partitioning each RGB channel into nonoverlapping blocks (e.g. 8×8) and mapping each block using the 2-D DCT transform. The 2-D DCT coefficients are then arranged in a vector using zigzag scanning [19]. The coefficients with small magnitudes are truncated to reduce dimensionality. These vectors are used to establish the codebook vectors using VQ. During encoding phase, each transformed block is matched to the closest codebook vector, and the corresponding index is quantized, encoded, and transmitted. A review of several hybrid encoding methods is presented in [19, 20].

11.5 Computer Projects

The computer projects in this chapter use Python libraries OpenCV, NumPy, matplotlib, scikit-image, and SciPy.

Computer Project 11.1: Using the same steps in Computer Project 6.2, generate the performance plots of SNR (in dB) versus bits/pixel using two bit allocation strategies in Section 11.3.2 for the Baboon image. Provide a detailed discussion on the effectiveness of each bit allocation method; comment on their advantages and disadvantages.

Computer Project 11.2: Using the same setup in Computer Project 7.2, generate the performance plots of SNR (in dB) versus bits/pixel using the Huffman encoding method in Section 11.2.2 for the Baboon image. Provide a detailed discussion on the effectiveness of this encoding method, and comment on its advantages and disadvantages.

Computer Project 11.3: Repeat Computer Project 7.2 using the color Finn image. Generate the performance plots of SNR (in dB) versus bits/pixel using the Huffman encoding method in Section 11.2.2. Provide a detailed discussion on the effectiveness of this encoding method and comment on its strengths and weaknesses.

Computer Project 11.4: The purpose of this computer project is to investigate the application of predictive encoding in Section 11.3.1 for the "Baboon" image data compression. First, use the procedure in Computer Project 9.2 to fit a first-order 2-D AR model with causal NSHP ROS to this image. Using this model, compute the predicted and residual images similar to the experimental results presented in Section 11.3.1. The residual image should then be quantized using a B-bit uniform quantizer and subsequently encoded using Huffman encoding. The codewords are subsequently decoded and dequantize to reproduce the residual image at the decoder side. The predicted version of the reproduced image pixel (see Figure 11.8) is then recursively generated using the same 2-D AR model and added to the corresponding residual image pixel to reconstruct the image. Using the difference image (between original and reconstructed), compute the SNR (in dB) for different number of quantizer's bits $B = 2 - 6$. Generate the performance plot of this 2-D predictive encoding in terms of SNR versus R_{DPCM} and identify an optimal operating point that provides a good balance of reconstruction quality and compression rate. Provide a detailed discussion on the performance of this encoding method and comment on its strengths and weaknesses.

Computer Project 11.5: The purpose of this computer project is to design a vector quantizer for image compression using the K-means clustering method in Section 11.3.4. This computer project has three parts. First, partition the original "Baboon" image into nonoverlapping 4×4 blocks and arrange each block into a 16-dimensional vector (row-stacked). Randomly choose 33% of these vectors as training dataset for K-means clustering method to establish $K = 32$ codebook vectors or the cluster centroids. Second, using the trained system, vector quantize every block in the image using the procedure described in Section 11.3.4 and devise a scheme to encode the index or label of the winning cluster centroid. Then, decode and reconstruct the image block-by-block by reshaping the winning codebook vector for each block. Repeat the first and second parts by changing the number of clusters to $K = 64$, 128, and 256. Third, for each case, generate the performance plot of this VQ method in terms of SNR versus average bits/pixel. Experimentally determine the optimum number of codebook vectors considering both the visual appearance of the reconstructed image and the compression rate versus SNR trade-off. Provide a detailed discussion on your results and point out the advantages and limitations of this method in relation to those of the others presented in this chapter.

11.6 Problems

Problem 11.1: Consider an image represented by a discrete random variable X with values $\{x_1, x_2, \cdots, x_7\}$ and associated probabilities $P_1 = 0.49$, $P_2 = 0.26$, $P_3 = 0.12$, $P_4 = 0.04$, $P_5 = 0.04$, $P_6 = 0.03$, $P_7 = 0.02$.

a. Determine a binary Huffman code for X.
b. Find the expected code length for this encoding.

Problem 11.2: Is it possible to use arithmetic encoding in Problem 11.1? If yes, compare the results with those of Huffman encoding.

Problem 11.3: Consider a discrete r.v. $X \sim \mathcal{N}(0, \sigma^2)$. If a one-bit quantizer is used, find: (a) the optimum reconstruction levels, (b) the quantizer's distortion, and (c) compare the result in part (b) with $D(R) = \sigma^2 2^{-2R}$.

Problem 11.4: Consider an image with pixel intensity represented by a continuous r.v. X with the pdf shown in Figure 11.16. This image is sampled and then quantized using a uniform quantizer.

Figure 11.16 PDF of r.v. X in Problem 11.4.

a. For a $L = 4$-level uniform quantizer, find the reconstruction and decision levels.
b. Determine the Huffman codewords for each reconstruction level.
c. Find the average codeword length using this encoder and comment on its optimality.

Problem 11.5: Consider the 8×8 image shown in Figure 11.17 which has three different gray levels.

a. Find the pmfs associated with the three gray levels and the entropy of this image.
b. Develop Huffman code for this image and find the average codeword length.
c. Develop the arithmetic code for non-overlapping blocks of size 2×2 and find the average codeword length.
d. Comment on the efficiency of the two encoding methods for this image.

Problem 11.6: A binary source generates a block of N iid symbols with probabilities $P_0 = 0.9$ and $P_1 = 0.1$. Find Huffman code for $N = 1, 2, 3$ and compute the average code length. Determine the efficiency of the coding in each case.

Problem 11.7: Repeat Problem 11.6 using arithmetic ending for $N = 2, 3$.

Problem 11.8: Repeat Problem 11.5 using run-length coding and determine the average codeword length and efficiency.

Problem 11.9: Consider an image X that can be modeled as an iid r.v. with pdf shown in Figure 11.18.

If $Y = X_1 - X_2$, where X_1 and X_2 are independent copies of X, determine if applying the encoding to the residual image Y is more efficient than applying the encoding to the original image X. Hint: Use the property that the pdf of the result of two iid r.v.s added/subtracted together is the convolution of their pdfs.

Problem 11.10: Consider an image $x(m, n)$ with four gray levels $\{x_1, x_2, x_3, x_4\}$ and the associated *actual* pmfs $\{P_1, P_2, P_3, P_4\}$. Using some data, these probabilities were measured as $\{\hat{P}_1, \hat{P}_2, \hat{P}_3, \hat{P}_4\}$ where $P_i = \hat{P}_i + \delta_i, \ i \in [1, 4]$ and δ_i is the error in estimating P_i. Huffman codewords were then designed using these measured pmfs.

a. Determine analytically the effects of these probability mismatches on the average codeword length.
b. If the actual pmfs are $\{1/2, 1/4, 1/8, 1/8\}$ and the measured ones are $\{1/3, 1/3, 1/6, 1/6\}$ comment on the robustness of Huffman encoding in terms of average codeword length as a function of estimation error.

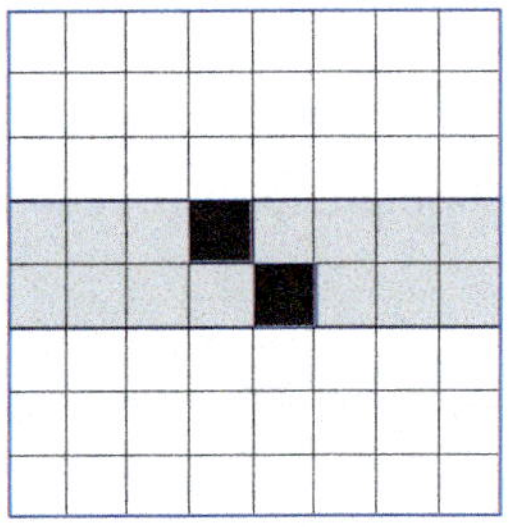

Figure 11.17 Image of Problem 11.5.

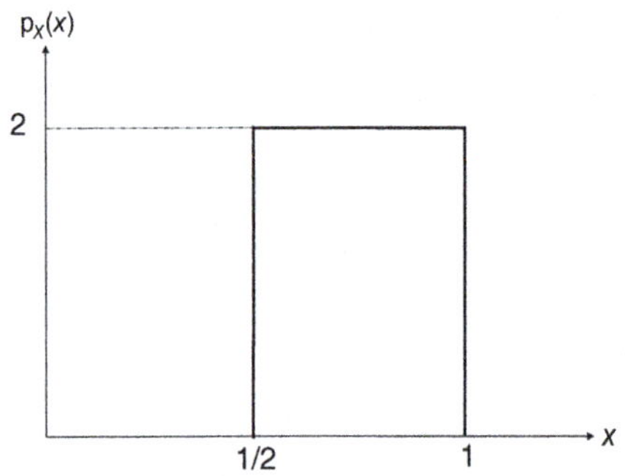

Figure 11.18 PDF of image of Problem 11.9.

Problem 11.11: Consider a zero-mean image $x(m.n)$ with an ACF $R_{xx}(k, l) = \sigma_x^2 \rho^{|k|+|l|}$. This image is applied to the predictive CODEC system in Figure 11.8. The quantizer adds distortion with variance σ_q^2 uncorrelated with the error image. Find expression for variance of the:

a. error image before quantization,
b. error image after quantization,
c. decoded error image, and
d. output image. Hint: Use the results in Example 9.3 to find the transfer function of the predictor.

Problem 11.12: Repeat Problem 11.11, assuming that the degradation process adds zero-mean white noise $\eta(m, n)$ with variance σ_η^2 uncorrelated with the image. Also, determine if the channel noise impacts the rate distortion function for this system.

Problem 11.13: Consider a zero-mean image $x(m.n)$ with variance σ_x^2 represented by an AR model:

$$x(m, n) = \rho x(m - 1, n) + e(m, n)$$

driven by the white process $e(m, n)$. This image is predictive-encoded using the system in Figure 11.8. A B-bit Lloyd-Max quantizer is used with distortion function $f(B)$. Find the expression for the variance of:

a. error image before quantization,
b. error image after quantization,
c. decoded error image, and
d. output image.

Problem 11.14: In Problem 11.13, assume that the communication channel adds zero-mean white noise $\eta(m, n)$ with variance σ_η^2 uncorrelated with the image. Repeat (a–d).

Problem 11.15: A zero-mean image $x(m, n)$ with ACF $R_{xx}(k, l) = \sigma_x^2 \rho^{|k|+|l|}$ is applied to a block transform coder for 2×2 blocks. The transform basis images $U(k, l)$ are:

$$\mathbf{U}(1,1) = \frac{1}{2}\begin{bmatrix} 1 & 1 \\ 1 & 1 \end{bmatrix},\ \mathbf{U}(1,2) = \frac{1}{2}\begin{bmatrix} 1 & 1 \\ -1 & -1 \end{bmatrix},\ \mathbf{U}(2,1) = \frac{1}{2}\begin{bmatrix} 1 & -1 \\ 1 & -1 \end{bmatrix},\ \mathbf{U}(2,2) = \frac{1}{2}\begin{bmatrix} 1 & -1 \\ -1 & 1 \end{bmatrix}.$$

Each block (i, j) is arranged in a vector 4×1, $\underline{x}_{i,j}$ using row-wise arrangement. Assuming that the basis images are also arranged in 4×1 vectors using the same row-wise arrangement to produce the 4×4 mapping matrix $\mathcal{U}$, we can write the transformation as $\underline{y}_{i,j} = \mathcal{U}\underline{x}_{i,j}$ where $\underline{y}_{i,j}$ is the transformed block.

a. Determine the covariance matrix of $\underline{x}$.
b. Determine the matrix $\mathcal{U}$ and then the covariance matrix of the transformed blocks $\underline{y}$.
c. For an average bit rate of p bits/pixel, determine the optimal bit allocation for the coefficients.
d. Determine the quantization distortion to directly quantize each pixel (without transform coding) using p bits/pixel.

Problem 11.16: Repeat Problem 11.15 for 2-D DCT basis images \$\mathbf{C}(k, l)$ instead of \$\mathbf{U}(k, l)$s.

Problem 11.17: A zero-mean $N \times N$ image $x(m.n)$ with ACF $R_{xx}(k, l) = \rho^{|k-l|}$ is transformed column-wise using $\underline{y}_n = \mathbf{U}\underline{x}_n$ where $\underline{x}_n$ represents n^{th} column of image $x(m, n)$, $\underline{y}_n$ is the transformed version and $\mathbf{U}$ is a lower triangular matrix with elements:

$$u(i, j) = \begin{cases} 1, & i = j \\ -\rho, & i = j + 1 \\ 0, & \text{otherwise} \end{cases}$$

a. Show that the elements of $\underline{y}_n$ are uncorrelated.
b. Find correlation matrix of $\underline{y}_n$.
c. If k^{th} element of $\underline{y}_n$, i.e. $y(k, n)$ is quantized using m_k bits, show that the reproduced vector $\underline{x}_n^o = \mathbf{U}^{-1}\underline{y}_n^o$ has distortion:

$$D = \frac{1}{N}\sum_{k=0}^{N-1}\gamma_k f(m_k)$$

where $\gamma_k = 1 - \rho^{2(N-k)}$, $\gamma_0 = \frac{1-\rho^{2N}}{1-1-\rho^2}$, and $f(.)$ is the distortion function of the quantizer.

Problem 11.18: The pixel-by-pixel quantized image may be buffered and then block-encoded by Huffman encoding in which case the required number of bits per symbol will be the entropy of the quantized image. This quantizer has the advantage of being easily implementable. The characteristics of the optimal equally spaced quantizer can be approximated using the distortion function:

$$f(m) = \begin{cases} 2^{-1.6m}, & m \le 2 \\ 1.66\ 2^{-2m}, & m > 2. \end{cases}$$

Derive the optimal bit allocation strategy for this quantizer distortion function.

References

1 T. J. Schmit, J. Li, S. J. Lee, Z. Li, R. Dworak, Y. K. Lee, M. Bowlan, J. Gerth, G. D. Martin, W. Straka, and K. C. Baggett, "Legacy atmospheric profiles and derived products from GOES-16: Validation and applications," *Earth and Space- Wiley Online Library*, vol. 6, pp. 1730–1748, 2019.
2 R. Gonzalez and R. Woods, "*Digital Image Processing*," Pearson, 4th Edition, 2017.
3 A. J. Hussain, A. Al-Fayadh, and N. Radi, "Image compression techniques: A survey in lossless and lossy algorithms," *Journal of Neurocomputing*, pp. 44–69, 2018.
4 R. Kaur and P. Choudhary, "A review of image compression techniques," *International Journal of Computer Applications*, vol. 142, pp. 8–11, 2016.
5 A. K. Jain, "Image data compression: A review," *Proceedings of the IEEE*, vol. 69, pp. 349–389, 1981.
6 T. M. Cover and J. A. Thomas, "*Elements of Information Theory*," Wiley Series in Telecommunications, 1991.
7 A. K. Jain, "*Fundamentals of Digital Image Processing*," Prentice-Hall, 1989.
8 J. P. Nebot, M. Morbee, and E. J. Delp, "Generalized PCM coding of images," *IEEE Transactions on Image Processing*, vol. 21, pp. 3801–3806, 2012.
9 M. Unser and T. Blu, "Mathematical properties of the JPEG2000 wavelet filters," *IEEE Transactions on Image Processing*, vol. 12, pp. 1080–1090, 2003.
10 W. Pennebaker and J. Mitchell, "*JPEG: Still Image Data Compression Standard*," New York: Van Nostrand-Reinhold, 1993.
11 S. Tian, Q. Song, J. He, Y. Wang, K. Nie, G. Du, and L. Bu, "Implementation of JPEG XS entropy encoding and decoding on FPGA," *Journal of Real-Time Image Processing*, vol. 21, p. 34, 2024. https://doi.org/10.1007/s11554-023-01410-8.
12 V. Goyal, "Theoretical foundations of transform coding," *IEEE Signal Processing Magazine*, pp. 9–21, 2001.
13 Y. Shoham and A. Gersho, "Efficient bit allocation for an arbitrary set of quantizers," *IEEE Transactions on Acoustics, Speech and Signal Processing*, vol. pp. 1445–1453, 1988.
14 A. Segall, "Bit allocation and encoding for vector sources," *IEEE Transactions on Information Theory*, vol. 22, pp. 162–169,1976.
15 J. Huang and P. Schultheiss, "Block quantization of correlated Gaussian random variables," *IEEE Transactions on Communications Systems*, vol. 11, pp. 289–296, 1963.
16 S. G. Mallat, "A theory for multiresolution signal decomposition: The wavelet representation," *IEEE Transactions on Pattern Analysis and Machine Intelligence*, vol. 2, pp. 674–693, 1989.
17 S. Hayken, "*Neural Networks and Learning Machines*," Prentice Hall, 3rd Edition, 2007.

18 J. Wu, "*Advances in K-means Clustering-A Data Mining Thinking*," Springer, 2012.

19 A. K. Pal, G. P. Biswas, and S. Mukhopadhyay, "A hybrid DCT-VQ based approach for efficient compression of color images," 2010 International Conference on Computer and Communication Technology (ICCCT), pp. 177–181, 2010.

20 V. J. Rehna and M. K. Jeya Kumar, "Hybrid approaches to image coding: A review," *International Journal of Advanced Computer Science and Applications*, vol. 2, pp. 108–115, 2011.

21 J. Jiang, "Image compression with neural networks-A survey," *Signal Processing: Image Communication*, vol. 14, pp. 737–760, 1999.

22 J. S. Lim, "*Two-dimensional Signal and Image Processing*," Prentice-Hall, 1990.

12

Image Segmentation

Image segmentation is a process by which the image is partitioned into a set of disjoint regions with similar characteristics. The similarity criterion can be based on pixel intensity, color, texture, or other meaningful features. The outcome is an image with isolated regions, objects, or structures (e.g. lines or contours) with identifiable boundaries. Image segmentation is typically the first and one of the most critical step in traditional image analysis and pattern recognition systems. Subsequent steps involve feature extraction and pattern classification as shown in the general block diagram in Figure 12.1. Clearly, the overall performance of the image analysis is highly dependent on the quality of segmentation, since inaccurate or incomplete region delineation can adversely affect all downstream processing stages. The role of the preprocessor in Figure 12.1 is to prepare the image for segmentation and feature extraction by improving its contrast, removing noise, detecting and sharpening edges, or any other low-level operations. The role of the feedback path will be described in greater detail in the introduction section of Chapter 13.

According to Fu and Mui [1, 2], classical image segmentation approaches can be broadly categorized into three general categories: (i) edge-based methods, (ii) feature-based methods, and (iii) region-based methods. However, in a broader sense, one can view image segmentation as a detection/classification or hypothesis testing problem. For example, image segmentation by clustering based on textural or any other features can be represented as an unsupervised classification task, where the goal is to partition the image into disjoint regions (or *classes*) that exhibit similar features.

There are many classical image segmentation approaches whose applicability and effectiveness depend on the image characteristics and the homogeneity criteria for region partitioning. In many scenarios, the choice of an appropriate class of methods is influenced by the type of imaging system that captured the images. For example, image segmentation algorithms that are specifically designed for magnetic resonance imaging (MRI) imaging may not be effective on thermal (infrared) images. In this chapter, our aim is to cover general-purpose segmentation algorithms that may be applied with reasonable success across many different imaging systems.

Most classical image segmentation methods are heuristic in nature, and rely primarily on low-level image features. Thus, they often lack the ability to incorporate the high-level semantic concepts utilized by human visual perception. Consequently, there is a gap between the psycho-physical aspects of human visual perception and the low-level features that the classical segmentation methods use. In recent years, semantic image segmentation [3] has attracted considerable attention, mainly due to the wide-spread emergence of deep neural networks (DNNs) in image processing. Using DNN models, high-level contextual and semantic information can be incorporated into image segmentation rules via the learning. Comprehensive surveys of DNN-based methods that outperform traditional approaches can be found in [3, 4].

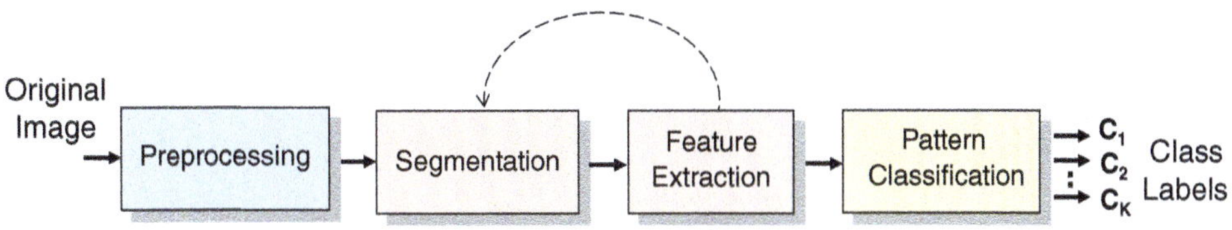

Figure 12.1 Block diagram of a traditional pattern recognition system.

The goal of this chapter is to present several classical and special-purpose image segmentation algorithms. These include: (i) histogram-based, (ii) region-based, (iii) Markov Random Fields (MRF), and (iv) active contours and gradient vector flow (GVF). A morphological-based image segmentation approach will also be introduced in Chapter 14.

12.1 Similarity-Based Image Segmentation

Classical image segmentation algorithms are based on either *similarity* characteristics within disjoint regions or *discontinuities* or boundaries between disjoint regions. Methods in the first category group pixels into disjoint regions based on homogeneous characteristics, whereas methods in the second category identify boundaries by detecting abrupt changes in image properties. An example of the similarity-based methods is clustering, such as the K-means covered in Chapter 11 or the competitive learning [5]. In these methods, similarity criteria are applied to image pixel intensity or to the mapped version in the feature domain. In contrast, boundary- and contour-based methods exploit the *discontinuity* principle, i.e. abrupt changes in intensity, to delineate regions within an image. A common characteristic of methods in both categories is that they are heuristic in nature and, as a result, do not take into account any *prior* information about the specific class of image being segmented. Consequently, assessing the quality of segmentation results is not straightforward. Typically, performance is determined using simple criteria, such as the degree of disjointness of the regions or objects boundaries for boundary-based methods, and classification accuracy for similarity-based methods. Moreover, as stated before, the effectiveness of each method depends on the type of image used for evaluation. This section presents several similarity-based image segmentation methods and their results on a variety of representative images.

12.1.1 Histogram Thresholding

Histogram-based segmentation methods partition an image into homogeneous regions according to some predefined similarity criteria. As discussed in Chapter 8, the modes in the histogram of an image, in a crude way, correspond to distinct regions within the image. A particular region can be segmented by applying a contrast mapping operation (see Figure 8.6a) and selecting appropriate thresholds. The choice of the thresholds depends on how well-separated the modes are. That is, if the modes are well-separated with distinct valleys between them, then appropriate thresholds can be selected within these valleys to successfully segment the image into disjoint regions. However, if the modes are not separated, optimal threshold selection becomes more challenging in order to avoid merging (or splitting) different regions and loss of important structures.

The threshold selection may be done manually based on visual inspection of the histogram or decided automatically using adaptive threshold selection methods [6, 7]. Figure 12.2a,b shows an image of red blood cells and its bimodal histogram with a wide valley between the two modes. Despite this clear separation of the modes, applying manual thresholding can lead to vastly different segmentation results for different choices of the threshold. Figure 12.3a,b shows segmentation results after edge extraction using the LoG operator (see Chapter 8) for thresholds $t = 60$ and $t = 130$, respectively. As can be seen, for the lower threshold ($t = 60$), portions of cells in the mid-intensity range are removed, although the shapes of the *central pallor* are preserved. For the higher threshold $t = 130$, Figure 12.3b also illustrates poor results since some cells merged and

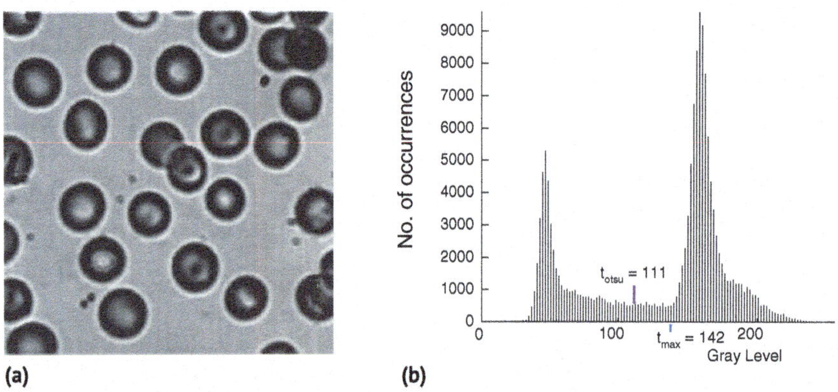

Figure 12.2 (a) Original red blood cell image and (b) its bimodal histogram. *Source*: (a) MDPI / CC BY 4.0 / https://www.mdpi.com/2073-8994/18/6/961 / last accessed June.19, 2026.

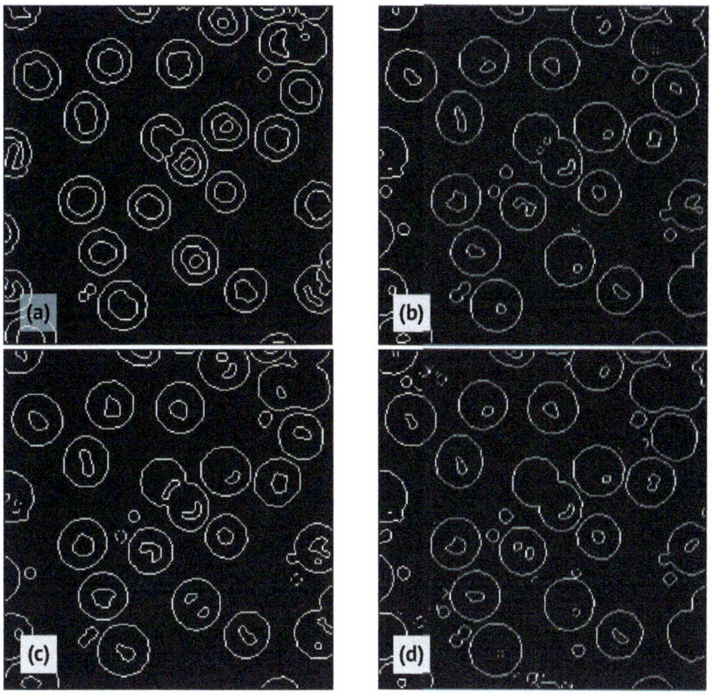

Figure 12.3 Boundaries of segmented regions for three different choices of thresholds: (a) $t = 60$, (b) $t = 130$, (c) $t_{\text{opt}} = 111$ obtained using Otsu's method, and (d) $t_{\text{opt}} = 142$ obtained using maximum entropy method.

the structures of their *central pallor* are distorted. These results are indicative of the fact that even in this case where the two modes are well-separated performance of histogram-based segmentation is highly dependent on threshold selection.

12.1.2 Adaptive Threshold Selection

The main drawback of fixed histogram thresholding is the lack of a reliable mechanism to choose optimally the threshold values. This is particularly true when: (a) there are no distinguishable boundaries or valleys between adjacent modes of the histogram, or (b) the valley region is wide and flat, as in the histogram of the previous example. Additionally, noise (e.g. quantization effects) can further perturb histogram structure and complicate appropriate threshold selection. To this end, there are several adaptive thresholding methods that treat threshold selection as an optimization problem by optimizing a discriminant function with respect to the unknown threshold. Among such methods include Otsu's method [6, 7], which maximizes between-class variance, and also the maximum relative entropy method [8–10]. These methods are presented next.

A. Otsu's Method

Otsu [6, 7] formulates the problem of optimal threshold selection as a binary classification problem based on the image histogram. A discriminant function is used as a measure of separation between two classes, i.e. background and the region of interest. The goal is to find the threshold that maximizes this measure of class separability. The original derivation of Otsu's method [7, 11] is based on the Bayes rule. In this section, we present an alternative and simpler approach.

Let us represent the pixel intensities of an image $x(m, n)$ as realizations of a discrete random variable (r.v.) X in the range $\{x = i,\ i \in [0, L-1]\}$. The pixel values X are drawn from a probability mass function (pmf) $P(X = i),\ i \in [0, L-1]$ that is estimated from the image histogram, i.e. $P(X = i) = \frac{n_i}{N}$ where n_i is the frequency of occurrences of pixel with intensity $X = i$ and N is the total number of pixels, and we have $\sum_{i=0}^{L-1} P(X = i) = 1$. Now, the problem is to find the threshold t to partition the pixels into two classes C_1 and C_2 according to the rule $X \in C_1$ if $X = i \leq t$ or $X \in C_2$ if $X = i > t$ (see Figure 12.4). The pmf of r.v. X may be written as a mixture of two pmfs, i.e.

$$\begin{aligned} P(X = i) &= \frac{N_1}{N}\frac{n_i}{N_1}\mathbb{I}_{i \leq t} + \frac{N_2}{N}\frac{n_i}{N_2}\mathbb{I}_{i > t} \\ &= P(C_1)P(X = i|C_1) + P(C_2)P(X = i|C_2), \quad (12.1) \\ &= P(C_1)P(X_1 = i) + P(C_2)P(X_2 = i) \quad (12.2) \end{aligned}$$

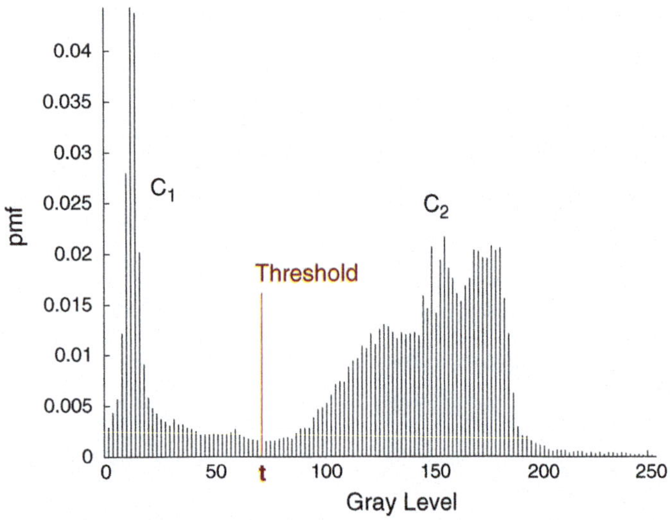

Figure 12.4 Distribution under the two threshold-dependent classes.

where $\mathbb{I}_A$ stands for the *indicator function* which satisfies $\mathbb{I}_A = 1$ if the event A is true, otherwise $\mathbb{I}_A = 0$; $N_1 = \sum_{i=0}^{t} n_i$ and $N_2 = \sum_{i=t+1}^{L-1} n_i$ are the number of pixels assigned to classes C_1 and C_2 based on threshold t, respectively; $P(C_1) = \frac{N_1}{N}$ and $P(C_2) = \frac{N_2}{N}$ with $P(C_1) + P(C_2) = 1$ are the prior probabilities of classes C_1 and C_2, respectively; and $P(X = i|C_1) = P(X_1 = i) = \frac{n_i}{N_1}\mathbb{I}_{i\le t}$ and $P(X = i|C_2) = P(X_2 = i) = \frac{n_i}{N_2}\mathbb{I}_{i>t}$ are pmf of r.v.s X_1 and X_2, respectively. That is, for $i \in [0, t]$, we have $P(X = i) = P(C_1)P(X_1 = i)$, while for $i \in [t+1, L-1]$, we have $P(X = i) = P(C_2)P(X_2 = i)$. Using these pmfs, equations for the mean of X, X_1, and X_2:

$$\mu_X = \sum_{i=0}^{L-1} iP(X = i), \quad \mu_1(t) = \sum_{i=0}^{t} iP(X_1 = i), \quad \text{and} \quad \mu_2(t) = \sum_{i=t+1}^{L-1} iP(X_2 = i) \tag{12.3}$$

can be rewritten as:

$$\begin{aligned} \mu_X &= P(C_1)\mu_1(t) + P(C_2)\mu_2(t) \\ \mu_1(t) &= \frac{1}{P(C_1)} \sum_{i=0}^{t} iP(X = i) \\ \mu_2(t) &= \frac{1}{P(C_2)} \sum_{i=t+1}^{L-1} iP(X = i), \end{aligned} \tag{12.4}$$

Similarly, the variances of r.v.s X_1 and X_2 are:

$$\begin{aligned} \sigma_X^2 &= \sum_{i=0}^{L-1} (i - \mu_X)^2 P(X = i) \\ \sigma_1^2(t) &= \frac{1}{P(C_1)} \sum_{i=0}^{t} (i - \mu_1(t))^2 P(X = i) \\ \sigma_2^2(t) &= \frac{1}{P(C_2)} \sum_{i=t+1}^{L-1} (i - \mu_2(t))^2 P(X = i). \end{aligned} \tag{12.5}$$

A commonly used measure of separation for two classes is given by the Fisher discriminant function [11]:

$$D(t) = \frac{|\mu_1(t) - \mu_2(t)|^2}{\sigma_1^2(t) + \sigma_2^2(t)}. \tag{12.6}$$

This measure is maximized with respect to the threshold. However, as can be noted, these first- and second-order statistics are all threshold-dependent. To simplify computations, Otsu [6] used the discriminant function:

$$\eta(t) = \frac{\sigma_B^2(t)}{\sigma_X^2}, \tag{12.7}$$

where $\sigma_B^2(t)$ is referred to as the *between-class* variance [6, 7, 11] given by:

$$\sigma_B^2(t) = P(C_1)(\mu_1(t) - \mu_X)^2 + P(C_2)(\mu_2(t) - \mu_X)^2. \tag{12.8}$$

Using (12.4), the between-class variance can be simplified to:

$$\sigma_B^2(t) = P(C_1)P(C_2)(\mu_2(t) - \mu_1(t))^2 = \frac{P(C_1)(\mu_X - \mu_1(t))^2}{1 - P(C_1)}. \tag{12.9}$$

The threshold-dependent terms in (12.9), are the prior probability $P(C_1)$ and the mean $\mu_1(t)$ that are computed based on r.v. X_1. Now, since σ_X^2 in (12.7) is independent of the threshold, Otsu's optimal threshold is obtained using

$$t_{\text{opt}} = \underset{t}{\text{argmax}}\ \sigma_B^2(t). \tag{12.10}$$

To implement Otsu's method, the threshold is swept over all admissible gray levels, e.g. $t \in [0, L-2]$. For each threshold, we compute $\sigma_B^2(t)$ and choose the value of t that maximizes this measure. First, let us examine the usefulness of this method on the red blood cell image and its histogram in Figure 12.2a,b, respectively. Figure 12.3c shows the edge map of the segmented image obtained using the Otsu's method. The optimal threshold for this case was found to be $t_{\text{opt}} = 111$, resulting in an improved segmentation result compared to those in Figure 12.3a,b for the manual thresholding.

Figure 12.5a,b, on the other hand, shows a section of a bone marrow image and its histogram, respectively. In this case, the histogram modes are not well-separated, making the optimal threshold selection more difficult. The optimal threshold obtained using Otsu's method was found to be $t_{\text{opt}} = 185$. The corresponding edge map of the segmented results for this optimal threshold is shown in Figure 12.5c. As can be observed, this method overestimated the threshold, resulting in the removal of some lower-intensity regions (missed detections).

An alternative and simpler algorithm uses only the mean values of the two regions associated with r.v.s X_1 and X_2. The procedure can be implemented as follows: (i) start with a low value initial threshold t; (ii) compute the mean values $\mu_1(t)$ and $\mu_2(t)$ of the resulting two regions using (12.4); (iii) set the new threshold to $t = (\mu_1(t) + \mu_2(t))/2$; and (iv) repeat steps (ii) and (iii) until t no longer changes.

The edge map of the segmented image using this algorithm for the bone marrow image is shown in Figure 12.5d. The resulting threshold using this method was $t = 153.9$. As can be seen, this method underestimated the threshold, which led to increased false detections and merger of adjacent regions. The latter effect is even more problematic since the shapes of these regions have changed significantly, thus adversely impacting the subsequent feature extraction and pattern recognition.

B. Maximum Entropy Method

Image segmentation based on entropy and relative entropy measures was proposed in [8–10] as an alternative and more general approach for optimal threshold selection. In [8], the authors showed that under the assumption that the density functions are Gaussian, the relative (or cross) entropy criterion reduces to Otsu's method, i.e. the latter is a special case of the former.

Again, assume that the r.v.s associated with the segmented regions for threshold t are X_1 and X_2 with PMFs $P(X_1 = i) = \frac{P(X=i)}{P(C_1)}$, $i \in [0, t]$ and $P(X_2 = i) = \frac{P(X=i)}{P(C_2)}$, $i \in [t+1, L-1]$, respectively, where $P(C_1)$ and $P(C_2)$ were defined before. Now, the entropies of these r.v.s for threshold t are:

$$\begin{aligned} H(X_1) &= -\sum_{i=0}^{t} \frac{P(X=i)}{P(C_1)} \log_2 \frac{P(X=i)}{P(C_1)}, \\ H(X_2) &= -\sum_{i=t+1}^{L-1} \frac{P(X=i)}{P(C_2)} \log_2 \frac{P(X=i)}{P(C_2)}. \end{aligned} \tag{12.11}$$

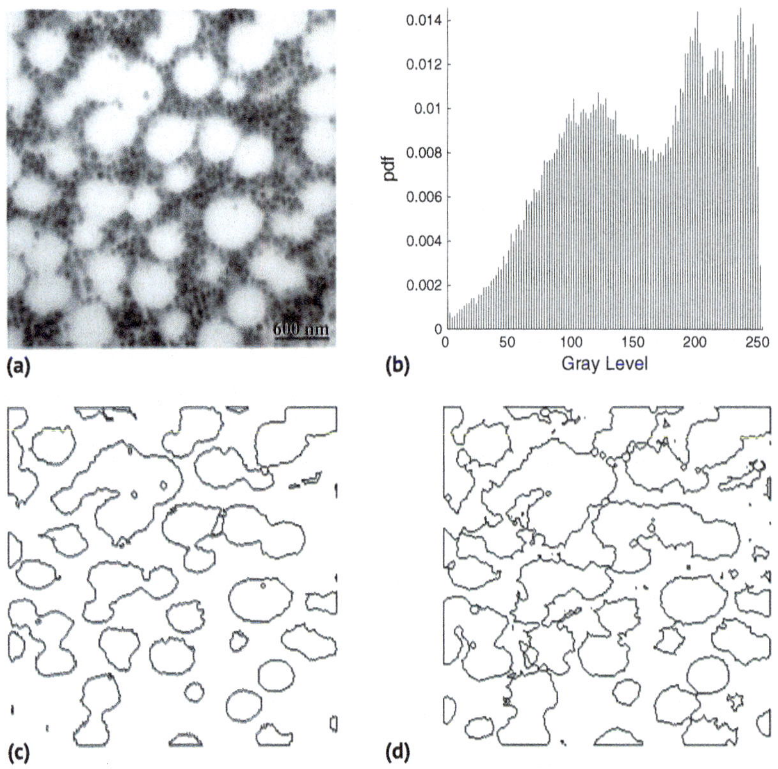

Figure 12.5 (a) Original bone marrow image, (b) histogram of the original image, (c) edge map of segmented image using Otsu's method $t_{opt} = 185$, and (d) edge map of segmented image using an alternative method $t = 153.9$. *Source*: (a) Rejeeth et al. 2012 / Springer Nature / CC BY 2.0.

These equations can be simplified [9] as:

$$\begin{aligned} H(X_1) &= \frac{H_t}{P(C_1)} + \log_2 P(C_1), \\ H(X_2) &= \frac{H(X) - H_t}{P(C_2)} + \log_2 P(C_2) = \frac{H(X) - H_t}{1 - P(C_1)} + \log_2(1 - P(C_1)) \end{aligned} \tag{12.12}$$

where $H(X) = -\sum_{i=0}^{L-1} P(X = i) \log_2 P(X = i)$ and $H_t = -\sum_{i=0}^{t} P(X = i) \log_2 P(X = i)$. Thus, we have $H(X) - H_t = -\sum_{i=t+1}^{L-1} P(X = i) \log_2 P(X = i)$. The sum of these two entropies:

$$\mathcal{H}(t) = H(X_1) + H(X_2) = \log_2(P(C_1)(1 - P(C_1))) + \frac{H_t}{P(C_1)} + \frac{H(X) - H_t}{1 - P(C_1)} \tag{12.13}$$

is maximized with respect to t to find the optimal threshold, i.e. $t_{opt} = \arg\max_t \mathcal{H}(t)$.

To compare the performance of this method with Otsu's, we applied this method to the same image of red blood cells in Figure 12.2a. The optimal threshold obtained using this method was found to be $t_{opt} = 142$. The corresponding edge map is shown in Figure 12.3d together with those of other methods to facilitate comparison. The maximum entropy criterion overestimated the threshold in this case, producing inferior segmentation result when compared to that of the Otsu's method.

Example 12.1 *Find the expression for $\mathcal{H}(t)$ in (12.12) and the optimal threshold assuming that the image intensity represented by r.v. X is uniformly distributed over $L = 2^B$ gray levels.*

Solution: For uniform distributions, the terms in (12.12) become: $P(X = i) = 1/L$, $P(C_1) = \sum_{i=0}^{t} P(X = i) = (t + 1)/L$, $P(C_2) = 1 - (t + 1)/L$, $H_t = -\sum_{i=0}^{t} P(X = i) \log_2 P(X = i) = (t + 1)B/L$, $H(X) = -\sum_{i=0}^{L-1} P(X = i) \log_2 P(X = i) = B$. Therefore, substituting these into (12.13) gives:

$$\mathcal{H}(t) = \log_2[(t + 1)(L - t - 1)].$$

Maximizing $\mathcal{H}(t)$ with respect to t yields $t = \frac{L}{2} - 1$, i.e. the midpoint of the gray levels.

12.1.2.1 Summary Remarks

- Both Otsu's and maximum entropy methods can be extended to multi-threshold segmentation. As an example, in color image segmentation, the histogram of each color channel can be thresholded independently, and the results combined in an appropriate color space (see Chapter 4).
- It is important to note that the binary image segmentation methods are incapable of segmenting regions that are highly textured. An example of such a case is the MRI image in Figure 12.6a with histogram shown in Figure 12.6b. The edge maps resulting from Otsu's optimal threshold selection and maximum entropy methods are shown in Figure 12.6c,d, respectively. The thresholds selected using these methods are $t_{otsu} = 118$ and $t_{maxe} = 100$, respectively. As can be observed from these results, texture regions that are essential for the discrimination of benign and malignant tumors are lost.

C. Binary Hypothesis Testing – Bayesian Framework

As discussed previously, image segmentation can be regarded as a hypothesis testing problem, i.e. detection or classification. Detection or binary hypothesis testing involves deciding between the true (or alternative) hypothesis (i.e. object present) and the null hypothesis (i.e. background only). In contrast, classification or M-ary hypothesis testing corresponds to deciding the class membership of a region as one of M possible classes. The general framework for solving such problems is based on Bayesian decision theory [11], which seeks to minimize the expected risk (or probability of error) in making decisions by finding an optimal threshold. In binary hypothesis testing, two types of errors may occur: false positives (false alarm) and false negatives (missed detections). A false alarm corresponds to making wrong decisions in favor of C_1 when hypothesis C_0 is actually true, whereas a missed detection corresponds to deciding in favor of C_0 when C_1 is true. The false alarm probability, P_{FA}, (or false positive rate P_{FP}) is given by the area under class-conditional density $p(x|C_0)$ for values of $x > t$. Similarly, the probability of missed detection, P_M, (or false negative rate P_{FN}) is given by the area under $p(x|C_1)$ for

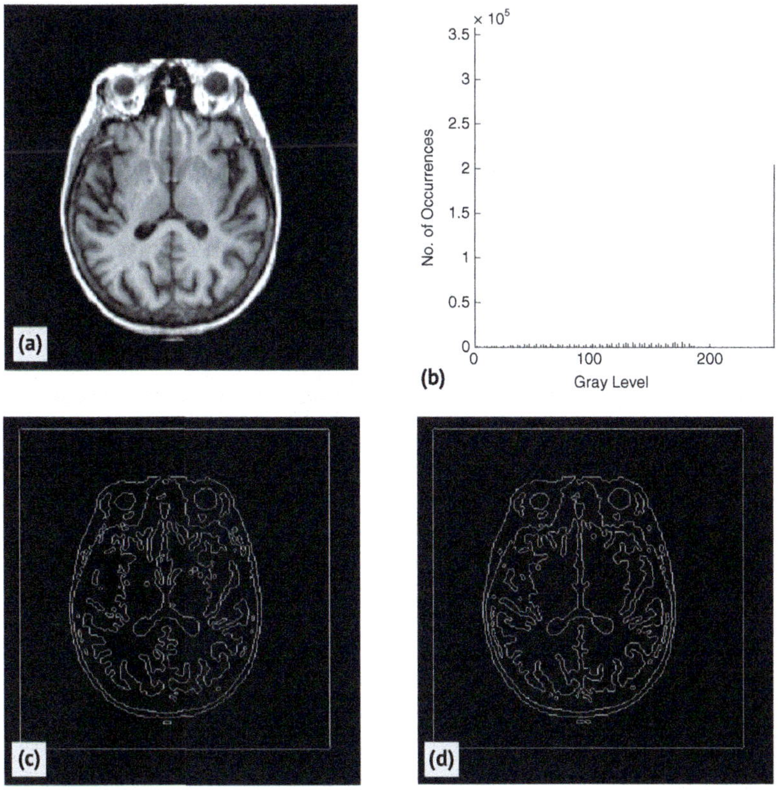

Figure 12.6 (a) Original brain MRI image, (b) histogram of MRI image, (c) edge map of segmented regions using Otsu's method $t_{otsu} = 118$, and (d) edge map of segmented regions using maximum entropy method $t_{maxe} = 100$. *Source*: (a) Photo by Keith A. Johnson and J. Alex Becker.

values of $x \leq t$. In Section 16.2.2, we discuss these types of error in both binary and M-ary hypothesis testing and provide measures for evaluating the performance of the detector or classifier.

12.1.3 Region-Based Segmentation

In region-based segmentation, an image is partitioned into regions with connected set of pixels, that are homogeneous with respect to intensity, color, texture, or other selected features. Thus, similar to thresholding methods, region-based segmentation methods also belong to the category of similarity-based approaches. However, unlike thresholding, these methods take into account the spatial relationship among neighboring pixels.

Assuming that $\mathcal{R}_i$ is the ith segmented region and $\mathcal{R}$ denotes the entire image, the conditions that must be satisfied when performing region-based segmentation are: (i) *completeness:* $\bigcup_{i=1}^{n} \mathcal{R}_i = \mathcal{R}$, i.e. the union of all segmented regions must cover the entire image, (ii) *disjointness:* $\mathcal{R}_i \bigcap \mathcal{R}_j = \emptyset, i \neq j$, i.e. the regions do not overlap, (iii) *connectivity:* Each region $\mathcal{R}_i$ consists of spatially connected pixels, forming a meaningful and compact structure, and (iv) *maximal homogeneity:* $HC(\mathcal{R}_i \bigcup \mathcal{R}_j) =$ False, where region $\mathcal{R}_i$ is adjacent to region $\mathcal{R}_j$ and HC is a set function that measures homogeneity. This condition ensures that the two adjacent regions are not merged unless they satisfy the homogeneity requirement.

There are two general categories of region-based image segmentation methods, namely region growing and region splitting and merging. These are discussed next.

A. Region Growing

Pixel aggregation is a region growth method that starts with a *seed* consisting of one pixel or a set of pixels of a potential region and iteratively grows the region by incorporating neighboring pixels that satisfy a predefined similarity criterion. Typically, an 8-point adjacency connectivity is used. The similarity or homogeneity criterion may be based on many image attributes, including intensity, color, texture, or other features. These steps are repeated until there are no more neighboring pixels that satisfy the similarity criterion for inclusion in a region. The process then moves to a different seed for another region. The choice of similarity or homogeneity measure and stopping criterion has a significant impact on the segmentation results.

Several similarity tests have been proposed, most of which are application-specific or image-dependent. One approach is based on statistical testing using the distance between the local mean of a segmented region and the candidate neighboring pixels, and deciding if a test statistic is satisfied.

Let $\mathcal{R}_n$ be an N_n-pixel region at iteration n for a particular seed. Also, define a discrete r.v. X_n whose values are the pixel intensities within $\mathcal{R}_n$. Then, the local sample mean and sample variance at iteration n are: $\bar{X}_n = \frac{1}{N_n} \sum_{i,j \in \mathcal{R}_n} x(i,j)$ and $\sigma^2_{X_n} = \frac{1}{N_n} \sum_{i,j \in \mathcal{R}_n} (x(i,j) - \bar{X}_n)^2$, respectively. A neighboring pixel $x(p,q)$ is tested to determine whether it belongs to the same region. A simple approach to perform this decision-making or hypothesis testing is to use the Student's-t test [12], which is used for small sample size populations to determine if a particular new sample belongs to the same population (the null hypothesis) or not (the alternative hypothesis). The test statistic is given by:

$$T = \frac{\bar{X}_n - x(p,q)}{\sigma_{X_n}/\sqrt{N_n}}.$$

Once the test statistic T is calculated for a given degrees of freedom $N_n - 1$ and a risk level (risk=1-confidence), e.g. $\alpha = 0.05$ or 5% (confidence level 95%), the statistical tables (for normal distributions) give us critical t-value or the threshold for decision-making. Now, if the test statistics $|T| > t$ then the pixel doesn't belong to the same region; otherwise the pixel is associated with the region, i.e. the null hypothesis is accepted, and we add $x(p,q)$ to this region. In this case, $\mathcal{R}_{n+1} = \mathcal{R}_n \bigcup x(p,q)$ and we update $\bar{X}_n \rightarrow \bar{X}_{n+1}$ and $\sigma^2_{X_n} \rightarrow \sigma^2_{X_{n+1}}$, and the process is repeated. This algorithm is computationally intensive, and its effectiveness depends on the statistical properties of the image. Figure 12.7 shows the segmented bacteria image using the pixel aggregation approach. As can be observed, comparing to the thresholding methods, this method produces well-isolated regions, while meeting the conditions for successful region-based segmentation.

B. Region Splitting and Merging

Region splitting operates in a manner opposite to region growing in the sense that it starts with the entire image and iteratively splits it into smaller subregions that satisfy a specified homogeneity criterion. In this top-down approach, there is no need for seed points. Instead, the image is divided into a set of disconnected regions. The splitting process follows a

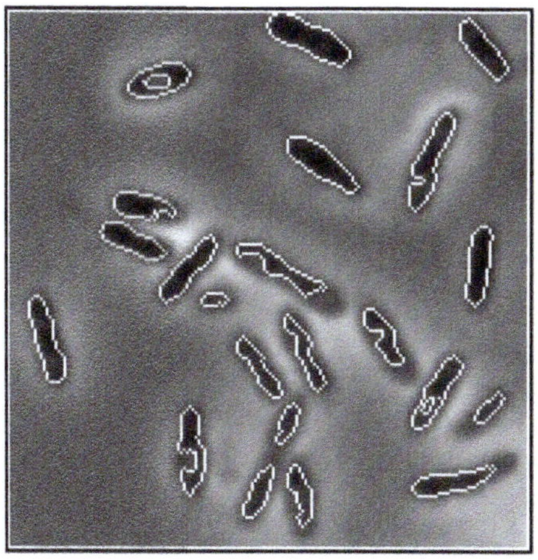

Figure 12.7 Segmented regions using pixel aggregation method. *Source*: Claudio Rosito Jung (2002)/Federal University of Rio Grande do Sul.

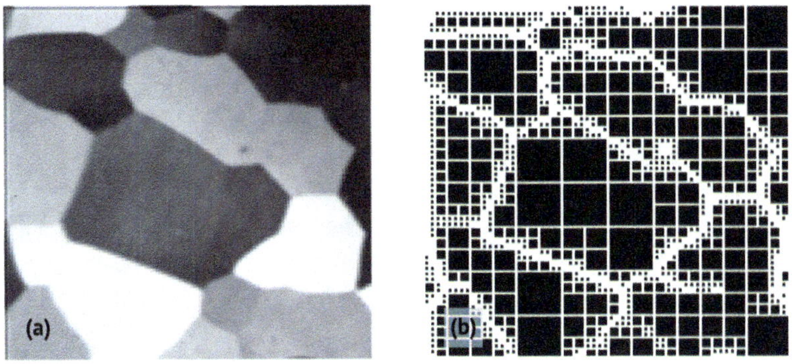

Figure 12.8 (a) Original image and (b) region splitting result using quad-tree decomposition. *Source*: (a) Avrithis and Rapantzikos 2011 / IEEE.

quad-tree decomposition [7, 13], where the root of the tree corresponds to the entire image and each node corresponds to a subregion. A typical splitting criterion is based on the local variance within a region. If the local variance exceeds a chosen threshold, then that region does not satisfy the homogeneity requirement and hence is subdivided into four equal-sized subregions. This process is repeated for each subregion until the splitting condition is no longer met, i.e. the homogeneity condition is satisfied. The result of this process is an image partitioned into many squares of varying sizes. Figure 12.8a,b shows the original image and segmented subregions of various sizes using quad-tree decomposition algorithm. Smaller subregions (or square blocks) define the boundaries of different homogeneous regions in the image. For proper implementation, the dimensions of the original image must be powers of 2. Otherwise, the image can be zero-padded to satisfy this condition.

After the splitting process, adjacent subregions or squares with similar properties are merged to form larger regions with well-defined boundaries. The merging process typically involves two criteria: a homogeneity test and a connectivity test. If variance is used for the splitting criterion, two or more adjacent subregions are merged if the variance of their union does not exceed a chosen threshold (homogeneity test). The selection or connectivity test ensures that the merged regions form a valid connected structure. If a connectivity test is met, the subregions are merged together, and the resulting variance is computed. These steps are repeated iteratively until no further changes occur. Post-processing may also be applied to eliminate small regions or merge them with the closest neighboring regions. Figure 12.9a shows the results of applying region splitting to the original brain MRI in Figure 12.6a, where the image is partitioned into blocks of varying sizes. Figure 12.9b, on the other hand, shows the result of region merging applied to the image in Figure 12.9a. As can be observed, the region split-merge method performed much better than the histogram-based thresholding methods. Different regions in the MRI image are well-separated without loss of valid information, which are prevalent with the thresholding schemes. In some cases, it is prudent to apply an edge-based detection method (Chapter 8) to improve the boundary definition in the results of region merging [14].

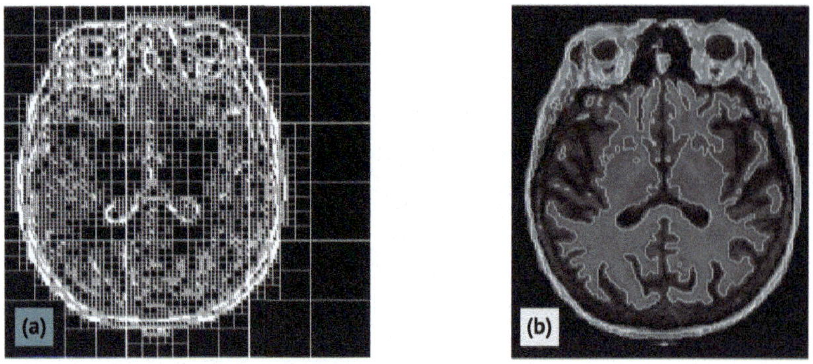

Figure 12.9 (a) Brain MRI image after region splitting using quad-tree algorithm and (b) final segmented image obtained by applying region merging to the image in (a). (The color rendition of this image is available to view at www.wiley.com/go/Azimi449).

12.1.3.1 Summary Remarks

- Region-based methods generally produce superior segmentation results compared to histogram-based thresholding, particularly when the distributions of the object and background overlap. However, this improved performance comes at the cost of increased computational complexity.
- The split-and-merge approach has an advantage over region growing methods in that it does not require the selection of seed points, which can significantly influence the final segmentation.
- Feature-based similarity criteria [14, 15] may also be adopted depending on the specific objectives of image segmentation.

12.1.4 Clustering Methods

The idea behind clustering is to partition data into various clusters based on a chosen similarity criterion. The similarity criterion may be as simple as pixel intensity, which is typically employed for region-based segmentation of grayscale images, or may involve multi-dimensional data such as spectral information in multispectral image segmentation. In Section 11.3.4, we used clustering as a tool to perform vector quantization (VQ) for lossy image compression. In that context, similarity was defined in terms of local edge and texture characteristics in small-size image blocks. In general, clustering algorithms are iterative in nature. They learn to discover data partitions or clusters in an unsupervised manner directly from the training data. The partitioning is based on the underlying data distributions. Thus, clustering can also be viewed as a data classification, where cluster centroids serve as prototypes representing classes or subclasses. Once the clusters are formed using unlabeled data, they may be associated with specific classes or subclasses using a small set of labeled data samples. Note that labeled and unlabeled data refer to data samples with known- and unknown-class labels, respectively.

There are many different iterative algorithms for cluster learning, among which is the K-means method introduced in Section 11.3.4. In general, clustering methods can be categorized into two groups: hierarchical and partition-based [16]. Hierarchical methods cluster the data following a tree-like structure using either a top-down or a bottom-up approach corresponding to *Divisive* and *Agglomerative* clustering, respectively. Divisive clustering works in a manner similar to region splitting in that it starts with a single cluster containing all the data samples and successively splits it into smaller clusters according to a similarity criterion. This process is repeated until a stopping condition is met. On the other hand, the agglomerative methods work similar to region merging in the sense that small clusters are successively merged to form larger clusters satisfying a similarity criterion until a stopping condition is met. Although hierarchical methods are effective for general data clustering, their high computational requirements limit their applicability in image segmentation.

Partition-based methods are more popular and preferred for image segmentation mainly due to their simplicity and lower computational complexity. There are two types of partition-based clustering methods: hard and soft clustering methods, which are described next.

A. Partition-Based Methods: Hard Clustering

Hard clustering methods include K-means and other centroid-based methods. They partition data into clusters with maximal intra-cluster similarity and minimal inter-cluster similarity. They use iterative learning algorithms that find the cluster centroid that has the shortest Euclidean distance to a data sample and then update it by including that sample in the cluster. The number of clusters, K, must be determined based on some prior knowledge of the number of regions to be segmented. The K-means clustering algorithm was described in Section 11.3.4 for vector quantization applications. Here, we summarize its key principles.

Let $\mathcal{S} = \{\underline{x}(p)\}_{p=1}^{P}$ be a set of P unlabeled data vectors of dimension N. The goal is to partition these *training* vectors into K clusters by minimizing the sum of squared error (SSE):

$$J(\underline{\mu}_1, \ldots \underline{\mu}_K) = \sum_{k=1}^{K} \sum_{\underline{x}(p) \in C_k} ||\underline{x}(p) - \underline{\mu}_k||^2 \tag{12.14}$$

where C_k represents the kth cluster and $\underline{\mu}_k$ is its centroid or local mean vector of the samples inside the cluster. Each sample $x(p)$ is assigned to the nearest centroid using:

$$l = \underset{k}{\text{argmin}} ||\underline{x}(p) - \underline{\mu}_l||^2 \quad \Rightarrow \quad \underline{x}(p) \in C_l \tag{12.15}$$

and the suboptimal estimate of the associated cluster centroid is:

$$\underline{\mu}_l = \frac{\sum_{\underline{x}(p) \in C_l} \underline{x}(p)}{|C_l|} \tag{12.16}$$

where $|C_l|$ is the number of data samples in cluster l. This solution is suboptimal because the objective function in (12.14) is nonconvex and hence there are many local minima that can be reached as possible solutions depending on the initialization of the centroids. Therefore, the optimality of the solution is not guaranteed, and many random initializations should be tried, which in turn increase the computational burden of the algorithm.

When using clustering-based algorithms, the choice of data representation is directly related to the goal of segmentation problem. For grayscale image segmentation, the goal is to delineate areas based on intensity, in which case clustering is performed using scalar pixel intensity values. For color image segmentation, the data vectors are typically 3-D, corresponding to the specific color channels, e.g. RGB or other color spaces such as HSI. For multispectral or hyperspectral image segmentation, the data vectors may consist of pixel intensities at a specific location across all spectral bands or their reduced-dimensional versions. Finally, if the goal is to perform image segmentation based on textural features, then the data vectors may correspond to small image blocks or patches arranged in vectors similar to the VQ process discussed in Chapter 11. The next example illustrates one of these applications.

Figure 12.10a–c shows the original MRI image and the segmented images using K-means clustering for $K = 4$ and $K = 5$ number of clusters, respectively. In this application, the clustering is performed based on individual pixel intensity. As can be seen, this algorithm performed well in segmenting different areas of the MRI image compared to those of the thresholding methods in Figure 12.6c,d. In addition, performance is comparable to that of region splitting and merging in Figure 12.9b with much reduced computational complexity. Increasing the number of clusters from $K = 4$ to $K = 5$ indeed improved the

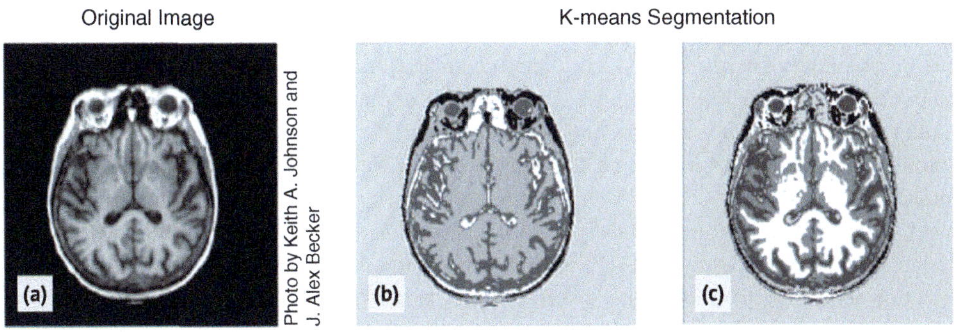

Figure 12.10 (a) Original MRI image, (b) and (c) K-means clustering result for $K = 4$ and $K = 5$, respectively. (The color rendition of this image is available to view at www.wiley.com/go/Azimi449).

performance by revealing finer textural and structural details. Finally, clusters may be labeled to represent different classes, e.g. benign and malignant regions in the MRI images, a capability that was not shared by the previously covered methods.

B. Partition-Based Methods: Soft Clustering

In soft clustering, each data sample is assigned a degree of membership to each cluster rather than belonging exclusively to a single cluster. These membership degrees are continuous values in the range [0, 1]. A widely used soft clustering method is fuzzy c-means [16], which can be viewed as the fuzzy extension of K-means clustering by incorporating membership degrees in the objective function in (12.14) as:

$$J(\{\underline{\mu}_k, \beta_{kp}\}_{k=1}^{K}) = \sum_{k=1}^{K}\sum_{p=1}^{P} \beta_{kp}^{m} \, ||\underline{x}(p) - \underline{\mu}_k||^2, \quad m > 1 \tag{12.17}$$

where $\beta_{kp} \in [0, 1]$ represents membership degree of the pth data sample in the kth cluster, and m is a parameter. The sub-optimal solution to this optimization problem is obtained iteratively by updating β_{kp} and $\underline{\mu}_k$ using:

$$\begin{aligned} \underline{\mu}_k &= \frac{\sum_{p=1}^{P} \beta_{kp}^{m} \, \underline{x}(p)}{\sum_{p=1}^{P} \beta_{kp}^{m}}, \\ \beta_{kp} &= \frac{1}{\sum_{j=1}^{K} \left\{ \frac{||\underline{x}(p) - \underline{\mu}_k||}{||\underline{x}(p) - \underline{\mu}_j||} \right\}^{\frac{2}{m-1}}}, \end{aligned} \tag{12.18}$$

The parameter m controls the degree of fuzziness: larger values of m reduce the influence of outliers or other less reliable samples and yield smoother cluster boundaries. The coupled updating equations in (12.18) are iterated until the change in cluster centers between two successive iterations falls below a specified threshold value [16].

12.1.5 Markov Random Field Segmentation

As stated earlier, image segmentation can be viewed as a classification problem in which a class label is assigned to each pixel or a group of pixels. Markov random fields (MRF) [17–21] provides a probabilistic framework for modeling such problems by incorporating spatial dependencies or regularities among neighboring pixels. More specific applications of MRF in image processing include remote sensing, biomedical imaging, change detection, color- and motion-based segmentation, and texture discrimination.

MRF provides a well-suited method for modeling spatial homogeneity and other local features, and hence is a natural choice for image segmentation. As in other region-based methods, the underlying assumption is that the image consists of many disjoint regions in which pixel intensity varies smoothly, while transition from one region to another is abrupt. The goal is to infer the hidden class labels associated with different regions using a stochastic framework that accounts for the local spatial interactions.

Since MRF is related to undirected graphs, here we first introduce some key concepts in graph theory [17, 18]. Let $\mathcal{G} = (\mathcal{V}, \mathcal{E})$ be a graph, where $\mathcal{V} = \{v_1, \cdots v_N\}$ is the set of *vertices or sites* and $\mathcal{E}$ is the set of edges. Two vertices v_i and v_j are *neighbors* if $\exists\, e_{ij} \in \mathcal{E}$ that connects them. The set of vertices that are neighbors of a site v is called the neighborhood of v and is denoted by $\mathcal{G}_v$. The set $\mathcal{G} = \{\mathcal{G}_v | v \in \mathcal{V}\}$ is a *neighborhood system* for $\mathcal{V}$ meaning any collection of subsets of $\mathcal{V}$ for which (a) $v \notin \mathcal{G}_v$ and (b) $w \in \mathcal{G}_v \leftrightarrow v \in \mathcal{G}_w$. A subset of vertices $C \subseteq \mathcal{V}$ is called a *clique* if every pair of vertices in C are neighbors. We denote $\mathcal{C}$ as the set of such cliques. Figure 12.11 illustrates first- and second-order neighborhood systems, together with the set of their cliques. The cliques for the first-order neighborhood lattice are $\{(i, j)\}, \{(i, j), (i, j+1)\}, \{(i, j), (i+1, j)\}$ as shown in Figure 12.11a. To simplify the notation, we used i and j for the vertices indices. Obviously, the number of clique types grows rapidly with the order.

Now, let us assume that $\underline{X} = [X_1, \cdots X_N]$ is a random vector representing the *hidden* class labels associated with N sites and X_j is a discrete random variable taking possible values in $\{1, \cdots, L\}$, with L being the number of possible classes. Also, let $\mathcal{Y} = \{\underline{y}_1, \cdots \underline{y}_N\}$ be the observation set, where $\underline{y}_i \in \mathbb{R}^M$ is the observed vector at site i, e.g. texture-based feature vector.

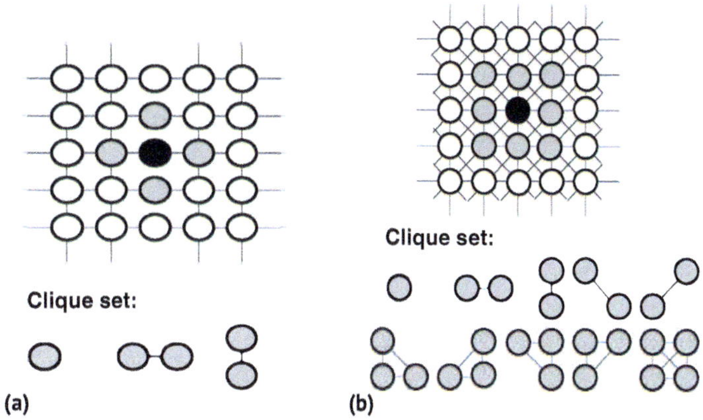

Figure 12.11 (a) First- and (b) second-order neighborhood systems and associated clique sets.

MRF casts the segmentation problem as estimating the hidden labels given the site observations by maximizing the *a posteriori* (MAP) [22] conditional pdf:

$$\hat{\underline{X}} = \underset{\forall X_j}{\operatorname{argmax}}\ p(\underline{X} = \underline{x} | \mathcal{Y}) \tag{12.19}$$

Using Bayes rule $p(\underline{X} = \underline{x} | \mathcal{Y}) = \frac{p(\mathcal{Y}|\underline{X}=\underline{x})p(\underline{X}=\underline{x})}{p(\mathcal{Y})}$, this MAP estimation [22] problem can be converted to:

$$\hat{\underline{X}} = \underset{\forall X_j}{\operatorname{argmax}}\ p(\mathcal{Y}|\underline{X} = \underline{x})p(\underline{X} = \underline{x}), \tag{12.20}$$

since the denominator $p(\mathcal{Y})$ is independent of X_j. Alternatively, we can maximize the ln of $p(\mathcal{Y}|\underline{X} = \underline{x})p(\underline{X} = \underline{x})$:

$$\hat{\underline{X}} = \underset{\forall X_j}{\operatorname{argmax}}\ \{\ln p(\mathcal{Y}|\underline{X} = \underline{x}) + \ln p(\underline{X} = \underline{x})\}, \tag{12.21}$$

where the first term is the *ln* of the likelihood function, maximizing which corresponds to the Maximum Likelihood (ML) solution [22]. If the observations are statistically independent, then $p(\mathcal{Y}|\underline{X} = \underline{x}) = \prod_{i=1}^{N} p(\underline{y}_i|x_i)$ where $p(\underline{y}_i|x_i)$ is the conditional pdf for observation $\underline{y}_i$ given its label x_i.

It must be pointed out that the conditional pdf $p(\mathcal{Y}|\underline{X} = \underline{x})$ is a function of a set of parameters $\boldsymbol{\Theta}$. For example, if Gaussian mixture models (GMMs) [17] are used to model this pdf, the mean vectors, covariance matrices, and mixture weights form the parameter set of this pdf. One could use a training dataset with labels to estimate these unknown parameters using a supervised method. Alternatively, if both parameters and hidden labels are to be estimated, then the MAP estimation problem becomes a joint parameter estimation and classification problem, i.e.

$$\{\hat{\underline{X}}, \hat{\boldsymbol{\Theta}}\} = \underset{\forall X_j, \boldsymbol{\Theta}}{\operatorname{argmax}}\ p(\underline{X}, \boldsymbol{\Theta} | \mathcal{Y}). \tag{12.22}$$

Among commonly used methods to find the suboptimal solutions of this unsupervised (or incomplete data) problem are the Expectation Maximization (EM) [17] or Iterated Conditional Modes (ICM) [20, 23], which will be briefly reviewed later in this section.

For the prior $p(\underline{X} = \underline{x})$ in (12.20) or (12.21), we use the Gibbs distribution [19] to capture contextual information considering the class (label) dependencies in neighboring sites, i.e.

$$p(\underline{X} = \underline{x}) = \frac{1}{Z} \exp\{-U(\underline{x})\} \tag{12.23}$$

where $U(\underline{x})$ is an appropriate *Gibbs energy function* and Z is called the *partition* function defined by $Z = \sum_{\forall X_j \in \Lambda} \exp\{-U(\underline{x})\}$, i.e. sum over all possible labeling. The energy function $U(\underline{x})$ in (12.23) enforces local homogeneity in labeling and is defined as:

$$U(\underline{x}) = \sum_{C \in \mathcal{C}} V_C(\underline{x}_C) \tag{12.24}$$

where $V_C(\underline{x}_C)$ is the *clique potential* of clique $C \in \mathcal{C}$ with the label configuration $\underline{x}_C$. It is customary to use pairwise cliques corresponding to a pair of neighboring pixels to enforce class homogeneity. Thus, the clique potential is:

$$V_C(\underline{x}_C) = \theta(x_i, x_j) = \begin{cases} 1, & \text{if } x_i \neq x_j \\ -1, & \text{otherwise.} \end{cases} \tag{12.25}$$

The local potential $\theta(x_i, x_j)$, known as the Potts model [18], is the general potential between a pair of pixels. It takes its minimum for neighboring pixels of the same class. Substituting (12.24) and (12.25) in (12.23) gives:

$$p(\underline{X} = \underline{x}) = \frac{1}{Z} \exp\{-\sum_{i,j \in \mathcal{C}} \theta(x_i, x_j)\}, \tag{12.26}$$

which takes large values for homogeneous regions.

Multivariate Gaussian distribution [17, 23] is typically used in (12.20) or (12.21) to model the conditional distribution (or likelihood function) $p(\mathcal{Y}|\underline{X} = \underline{x})$. The practical implication is that the feature vectors in the observation set $\mathcal{Y}$ tend to be close to the mean vectors of different segmented homogeneous regions with some variations represented by the covariance matrix. If we further assume that individual features are statistically independent then the conditional distribution can be written as:

$$p(\mathcal{Y}|\underline{X} = \underline{x}) = \prod_{i=1}^{N} p(\underline{y}_i|x_i) = \prod_{i=1}^{N} \frac{1}{\sqrt{(2\pi)^M \det \mathbf{C}_i}} \exp\{-\frac{1}{2}(\underline{y}_i - \underline{\mu}_i)^T \mathbf{C}_i^{-1}(\underline{y}_i - \underline{\mu}_i)\} \tag{12.27}$$

where $\underline{\mu}_i$ and $\mathbf{C}_i$ are the mean vector and covariance matrix of data with class label x_i and $\det \mathbf{C}_i$ represents the determinant of matrix $\mathbf{C}_i$. Now, combining (12.26) and (12.27), the log-likelihood function in (12.21) becomes:

$$\ln\; p(\mathcal{Y}|\underline{X} = \underline{x})p(\underline{X} = \underline{x}) \propto -U(\mathcal{Y}, \underline{X}) \tag{12.28}$$

where $U(\mathcal{Y}, \underline{X})$ is given by:

$$U(\mathcal{Y}, \underline{X}) = -\sum_{i=1}^{N} \left[\frac{1}{2}\{M \ln(2\pi) + \ln(\det \mathbf{C}_i)\} + \frac{1}{2}(\underline{y}_i - \underline{\mu}_i)^T \mathbf{C}_i^{-1}(\underline{y}_i - \underline{\mu}_i)\right] + \beta \frac{1}{2} \sum_{i,j \in \mathcal{C}} \theta(x_i, x_j), \tag{12.29}$$

where β is the proportionality constant which controls homogeneity of the segmented region, i.e. if $\beta = 0$, then no homogeneity is imposed by the Gibbs prior. In [18], $\beta = 2$ was recommended to satisfy the homogeneity condition. Thus, the MAP solution corresponds to:

$$\hat{\underline{X}} = \underset{\forall X_j \in \Lambda}{\operatorname{argmin}}\; U(\mathcal{Y}, \underline{X}). \tag{12.30}$$

Since the parameter β and the number of classes L are assumed to be known, the likelihood function in (12.28) is dependent on the unknown parameters $\underline{\mu}_i$ and $\mathbf{C}_i$ and class labels, all of which need to be identified. In this case, we have a joint parameter and label (classification) estimation problem. If a labeled training dataset is available, then estimates of these parameters can be made using sample mean vectors and covariance matrices, leading to the ML estimates of these parameters (Problem 12.12). However, this requires manual labeling of different regions in images, which may not be practical in realistic applications, e.g. medical imaging. In these situations, unsupervised methods that do not require labeling are the preferred choices. As mentioned earlier, among the popular methods used for incomplete unlabeled data are EM [17] and ICM [20, 23]. The latter method is introduced next.

12.1.5.1 Iterated Conditional Modes (ICM)

Solving this MAP estimation problem for MRF-based image segmentation is computationally costly. Geman and Geman [19] proposed Gibbs sampler and Simulated Annealing (SA) to reduce the computational requirements of this optimization problem. In [17], EM was used to estimate the parameters of the Gaussian likelihood function. In this section, we introduce an alternative and computationally efficient method proposed by Besag [20] known as ICM in which labeling of each site is iteratively updated by maximizing local conditional probabilities. That is, the goal is to find a class label which maximizes conditional probability, given the observation at that site and the current label estimates in its neighborhood. The local conditional probabilities are defined by $P(X_i = x_i | \underline{X}_{\diamond i} = \underline{\hat{x}}_{\diamond i}, \underline{\mathcal{Y}} = \underline{y}_i)$ where $\underline{X}_{\diamond i}$ is a vector of class labels of all immediate neighbors of x_i for the site i and $\underline{\hat{x}}_{\diamond i}$ is the current estimate vector of these labels.

Let us begin with the joint probability $P(x_i, \underline{\hat{x}}_{\diamond i}, \underline{y}_i)$ and use the Bayes rule to relate it to $P(X_i = x_i | \underline{X}_{\diamond i} = \underline{\hat{x}}_{\diamond i}, \underline{\mathcal{Y}} = \underline{y}_i)$ or in simple form:

$$P(x_i | \underline{\hat{x}}_{\diamond i}, \underline{y}_i) = \frac{P(x_i, \underline{\hat{x}}_{\diamond i}, \underline{y}_i)}{P(\underline{y}_i | \underline{\hat{x}}_{\diamond i}) P(\underline{\hat{x}}_{\diamond i})}$$

Applying the chain rule, the joint probability can be expressed as:

$$P(x_i, \underline{\hat{x}}_{\diamond i}, \underline{y}_i) = P(\underline{y}_i | x_i, \underline{\hat{x}}_{\diamond i}) P(x_i | \underline{\hat{x}}_{\diamond i}) P(\underline{\hat{x}}_{\diamond i})$$

If we further assume $P(\underline{y}_i | x_i, \underline{\hat{x}}_{\diamond i}) = P(\underline{y}_i | x_i)$, which implies that only the label x_i, and not the labels of neighboring pixels, determines the observation $\underline{y}_i$, then the local conditional probabilities can be related to:

$$P(X_i = x_i | \underline{X}_{\diamond i} = \underline{\hat{x}}_{\diamond i}, \underline{\mathcal{Y}} = \underline{y}_i) \propto p(\underline{y}_i | x_i) P(x_i | \underline{\hat{x}}_{\diamond i}). \tag{12.31}$$

When applied to each pixel, this procedure defines a single cycle of the iterative algorithm for estimating x_i. The algorithm starts with an ML-based initial estimate of the class label, which ignores the neighborhood lattice structure and chooses x_i to maximize $p(\underline{y}_i | x_i)$ at each site i separately. We then apply the algorithm for a fixed number of cycles or until convergence to produce the final estimate. Thus, the label at site i is updated by selecting the class that maximizes the product of the local likelihood and the contextual prior induced by the neighboring labels.

As shown before, the local conditional probability in (12.31) can be written in terms of the energy function as:

$$P(X_i = x_i | \underline{X}_{\diamond i} = \underline{\hat{x}}_{\diamond i}, \underline{\mathcal{Y}} = \underline{y}_i) = \frac{1}{Z_i} \exp\{-U_i(x_i, \underline{\hat{x}}_{\diamond i}, \underline{y}_i)\}, \tag{12.32}$$

where Z_i is the partition function for site i and

$$U_i(x_i, \underline{x}_{\diamond i}, \underline{y}_i) = -\ln\ p(\underline{\mathcal{Y}} = \underline{y}_i | X_i = x_i) + \beta \sum_{c \in \mathcal{C}_i} V_C(\underline{\hat{x}}_{\diamond i}), \tag{12.33}$$

where the summation is for all cliques to which the site i belongs. Now, maximizing the conditional probability in (12.31) is equivalent to minimizing the energy function in (12.33). The steps of the ICM algorithm are given below.

1. Initialize $\hat{x}_i$ at each site by maximizing the local log-likelihood $\hat{x}_i^{(0)} = \arg\max_{x_i} p(\underline{y}_i \mid x_i)$, i.e. the conventional ML estimate.
2. For each site i, update the class label according to $\hat{x}_i = \arg\min_{x_i} U_i(x_i, \underline{\hat{x}}_{\diamond i}, \underline{y}_i)$, where $U_i(x_i, \underline{x}_{\diamond i}, \underline{y}_i)$ is given in (12.33).
3. Repeat Step 2 until convergence or until a prescribed maximum number of iterations is reached.

To assess the effectiveness of MRF-based image segmentation utilizing the ICM algorithm, we apply it to the brain MRI image in Figure 12.12a. Initial estimates of class labels are generated from the K-means algorithm for both $K = 4$ and $K = 5$ classes as shown in Figure 12.10b,c, respectively. The energy function specified in (12.33) is constructed using a first-order neighborhood system and its corresponding clique set, as depicted in Figure 12.11. A Gaussian Mixture Model (GMM) is employed to model the conditional probability $p(\mathcal{Y} = y_i | X_i = x_i)$, assuming statistical independence of observations. The GMM parameters are estimated using the ML approach described earlier. The value of β was set to $\beta = 2$ as suggested in [18]. The ICM algorithm demonstrated rapid convergence, reaching a solution after 5 and 7 iterations for the $K = 4$ and $K = 5$ cases, respectively. The resulting segmented images are displayed in Figure 12.12b,c. Comparison of these results with those of K-means clustering reveals better spatial consistency segmentation of different regions. This improved performance is primarily attributed to the incorporation of the neighborhood system within the MRF framework, which enforces spatial continuity and smoothness in the final segmentation.

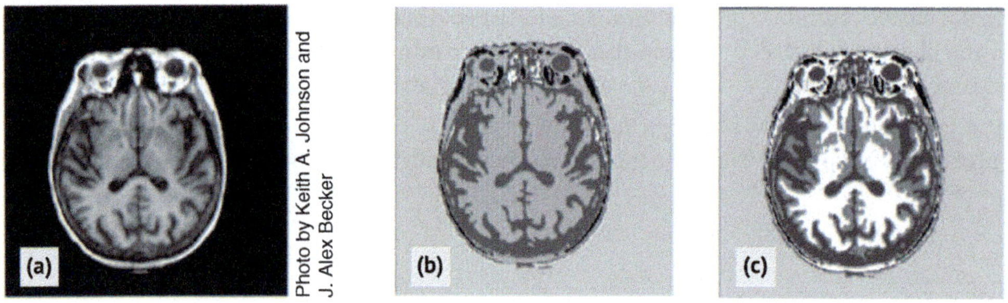

Figure 12.12 (a) Original MRI image, (b) MRF-ICM results for $K = 4$, and (c) MRF-ICM results for $K = 5$. (The color rendition of this image is available to view at www.wiley.com/go/Azimi449.)

12.1.5.2 Summary Remarks

- ICM offers several important advantages for implementing MRF models. First, it is computationally less demanding compared to SA and EM. Second, it circumvents the undesirable large-scale properties of the MRF prior models. More specifically, MRF models have been shown to exhibit a phenomenon known as *phase transition*, in which the prior favors overly large homogeneous regions or nearly uniform label fields. That is, the resulting images generated by MRF process tend to be unicolor within large regions. The ICM method was effectively designed [20, 21] to overcome this problem by making the algorithm insensitive to these large-scale properties.
- Comparative studies [24] involving synthetic and real images corrupted by additive white Gaussian noise indicated that ICM robust segmentation performance, even under low signal-to-noise ratio (SNR) conditions. Its robustness was also demonstrated on images corrupted by correlated noise and for images containing textured regions.

12.2 Edge and Contour-Based Segmentation

Edge- and contour-based detection methods can be applied separately to segment regions of interest in digital images. These methods are typically used in image analysis systems that rely on shape-dependent discrimination, regardless of the features inside the segmented regions. Alternatively, they can be applied jointly with similarity-based methods in the previous section to enhance overall segmentation performance in applications that require both accurate boundary localization and feature homogeneity within segmented regions. These edge- and contour-based methods work on the premise that intensity discontinuities occur when transitioning from one region to another or from the background. Thus, if these transition boundaries or edges/contours can be extracted reliably and without missing parts, then the objects of interest can be effectively isolated. Edge detection methods and operators were extensively treated and examined in Chapter 8. In this section, we will focus on contour-based segmentation methods using active contour models and their variants [25–27].

12.2.1 Active Contour Segmentation

Traditional active contour or snake models [25–27] have been extensively used in computer vision and image processing. In general, there are two types of active contour models: the parametric active contours and geometric active contours [28]. The parametric active contours synthesize parametric curves within an image domain and allows them to move toward the boundaries of the objects. The curves evolve as a result of *external forces* consisting of potential and pressure forces, as well as *internal forces* which are designed to hold the curve together (elastic forces) and keep it from bending too much. Parametric models suffer from two major limitations, including sensitivity to initialization and inability to progress into boundary concavities. Moreover, they lack the ability to handle topological changes (e.g. splitting and merging of contours), though they are popular due to their simplicity and the ease with which contour smoothness constraints can be incorporated. Geometric active contours, on the other hand, are introduced to circumvent the problem of required topological changes during the curve evolution. Geometric active contours are implicitly represented as level sets of 2-D distance functions that evolve according to partial differential equations in a Eulerian formulation rather than the parametric approaches represented in a Lagrangian formulation framework [28].

An extension of the parametric active model is the gradient vector flow (GVF) snake [26, 27], which alleviates the sensitivity to initialization and allows the contour to conform to boundary concavities, hence improving the capability in delineating object boundaries for image segmentation. The GVF snake employs a force field derived from the image gradients, which guides the contour toward the edges of objects within an image. Unlike traditional gradient-based forces, this vector field is not confined to the gradient directions but is diffused over the entire image domain, allowing the contour to capture object boundaries even when the initial contour is placed far away from them. This capability makes GVF particularly useful in applications where precise contour detection is of utmost importance.

12.2.1.1 Active Contour or Snake Formulation

The traditional snake [25] is a parameterized curve, $\underline{v}(s) = [u(s)\ v(s)]^T$ of normalized parameter $s \in [0, 1]$, which evolves in the 2-D spatial coordinates (u, v) to minimize the energy functional:

$$E^* = \int_0^1 \Big(E_{\text{int}}(\underline{v}(s)) + E_{\text{ext}}(\underline{v}(s))\Big)ds. \tag{12.34}$$

where E_{int} and E_{ext} are the *internal* and *external* energy functions [7, 25], respectively. The internal energy corresponds to elastic and bending behavior of the snake. While the elastic energy is proportional to the integral of the square of first derivative of $\underline{v}(s)$, the bending/stiffness is proportional to the integral of the square of second derivative of $\underline{v}(s)$. Thus, we can express the internal energy function as:

$$E_{\text{int}}(\underline{v}(s)) = \frac{1}{2}[\alpha||\underline{v}^{(1)}(s)||^2 + \beta||\underline{v}^{(2)}(s)||^2] \tag{12.35}$$

where $\underline{v}^{(k)}(s) = \frac{\partial^k \underline{v}(s)}{\partial s^k}$ is the kth derivative of $\underline{v}(s)$ with respect to s and α and β control the tension (elasticity) and rigidity (bending), respectively. The external energy term is intended to move the snake to the boundaries of segmented objects when it is minimized, and hence it is derived from the image being segmented. For a continuous gray level image, $x(u, v)$, typical choices for the external energies are [26]:

$$\begin{aligned} E_{\text{ext}}(u, v) &= -||\nabla x(u, v)||^2 = -\left(\left(\frac{\partial x(u, v)}{\partial u}\right)^2 + \left(\frac{\partial x(u, v)}{\partial v}\right)^2\right) \\ E_{\text{ext}}(u, v) &= -||\nabla[G_\sigma(u, v) ** x(u, v)]||^2 \end{aligned} \tag{12.36}$$

where ∇ is the gradient operator and $G_\sigma(u, v)$ is a Gaussian smoothing function with smoothing parameter (standard deviation) σ. If instead of gradient we use Laplacian, then $E_{\text{ext}}(u, v)$ in the second equation in (12.36) corresponds to the LoG operator covered in Chapter 8 for edge detection. Minimizing the energy function in (12.34) leads to the following Euler–Lagrange equation [7] that must be satisfied:

$$\alpha\underline{v}^{(2)}(s) - \beta\underline{v}^{(4)}(s) - \nabla E_{\text{ext}} = \underline{0}. \tag{12.37}$$

Now, since the spatial derivative of an energy function is a force, the above equation can be viewed as a force balance equation:

$$\underline{f}_{\text{int}} + \underline{f}_{\text{ext}} = \underline{0}, \tag{12.38}$$

where $\underline{f}_{\text{ext}} = -\nabla E_{\text{ext}} = -[\frac{\partial E_{\text{ext}}}{\partial u}\ \frac{\partial E_{\text{ext}}}{\partial v}]^T$ and $\underline{f}_{\text{int}} = \alpha\underline{v}^{(2)}(s) - \beta\underline{v}^{(4)}(s)$. The internal force $\underline{f}_{\text{int}}$ discourages stretching and bending, while the external force $\underline{f}_{\text{ext}}$ pulls the snake toward the boundary of the segmented object.

The force balance equation in (12.37) is usually solved iteratively by setting the partial derivative of $\underline{v}(s)$ with respect to the iteration index t equal to the left-hand side of (12.37) as follows:

$$\frac{\partial \underline{v}(s, t)}{\partial t} = \alpha\underline{v}^{(2)}(s, t) - \beta\underline{v}^{(4)}(s, t) + \underline{f}_{\text{ext}}(s, t). \tag{12.39}$$

At equilibrium, where $\frac{\partial \underline{v}(s,t)}{\partial t} = 0$, (12.39) reduces to (12.38) and the force balance condition is satisfied. The derivative terms in (12.39) can be approximated using a forward-difference method [7, 27] leading to linear recursive equations that can be solved recursively.

Let us assume that the parameter s and the iteration index t are discretized using steps Δs and Δt, respectively, i.e. $s = m\Delta s$ and $t = n\Delta t$. Then, using the forward difference method, we obtain the discrete version of (12.39) as:

$$\begin{aligned}\frac{\underline{v}(m,n)-\underline{v}(m,n-1)}{\Delta t} = {} & \alpha\frac{\underline{v}(m+1,n)-2\underline{v}(m,n)+\underline{v}(m-1,n)}{\Delta^2 s} \\ & -\beta\frac{\underline{v}(m+2,n)-4\underline{v}(m+1,n)+6\underline{v}(m,n)-4\underline{v}(m-1,n)+\underline{v}(m-2,n)}{\Delta^4 s} \\ & +\underline{f}_{ext}(\underline{v}(m,n-1)).\end{aligned} \tag{12.40}$$

Defining $\gamma_1 = \frac{\alpha\Delta t}{\Delta^2 s}$ and $\gamma_2 = \frac{\beta\Delta t}{\Delta^4 s}$, and rearranging (12.40) yields:

$$\begin{aligned}&(1+2\gamma_1+6\gamma_2)\underline{v}(m,n)-(\gamma_1+4\gamma_2)\underline{v}(m+1,n)+\gamma_2\underline{v}(m+2,n) \\ &-(\gamma_1+4\gamma_2)\underline{v}(m-1,n)+\gamma_2\underline{v}(m-2,n)=\underline{v}(m,n-1)+\underline{f}_{ext}(\underline{v}(m,n-1)).\end{aligned} \tag{12.41}$$

Now, let us define $\underline{\vartheta}(n) = \left[\underline{v}(0,n)\underline{v}(1,n)\cdots\underline{v}(P-1,n)\right]^T$. Therefore, (12.41) becomes:

$$(\mathbf{I}+\boldsymbol{\Gamma})\underline{\vartheta}(n) = \underline{\vartheta}(n-1)+\underline{\boldsymbol{\phi}}(\underline{\vartheta}(n-1)) \tag{12.42}$$

or

$$\underline{\vartheta}(n) = (\mathbf{I}+\boldsymbol{\Gamma})^{-1}\left[\underline{\vartheta}(n-1)+\underline{\boldsymbol{\phi}}(\underline{\vartheta}(n-1))\right] \tag{12.43}$$

where $\underline{\boldsymbol{\phi}}(\underline{\vartheta}(n-1)) = \left[\underline{f}_{ext}(\underline{v}(0,n-1)),\cdots\underline{f}_{ext}(\underline{v}(P-1,n-1))\right]^T$ and the matrix $\boldsymbol{\Gamma}$ is defined as:

$$\boldsymbol{\Gamma} = \begin{bmatrix} (2\gamma_1+6\gamma_2) & -(\gamma_1+4\gamma_2) & \gamma_2 & 0 & 0 & \dots & 0 \\ -(\gamma_1+4\gamma_2) & (2\gamma_1+6\gamma_2) & -(\gamma_1+4\gamma_2) & \gamma_2 & 0 & \ddots & 0 \\ \gamma_2 & (\gamma_1+4\gamma_2) & (2\gamma_1+6\gamma_2) & \ddots & \ddots & \ddots & \vdots \\ 0 & \gamma_2 & (\gamma_1+4\gamma_2) & \ddots & \ddots & \ddots & \dots \\ \vdots & \ddots & \ddots & \ddots & \ddots & \dots & \dots \\ 0 & \dots & 0 & \gamma_2 & -(\gamma_1+4\gamma_2) & (2\gamma_1+6\gamma_2) & -(\gamma_1+4\gamma_2) \\ 0 & \dots & \dots & 0 & \gamma_2 & -(\gamma_1+4\gamma_2) & (2\gamma_1+6\gamma_2) \end{bmatrix} \tag{12.44}$$

Since the matrix $(\mathbf{I}+\boldsymbol{\Gamma})$ is a Toeplitz matrix, the linear equation in (12.42) can be solved efficiently.

12.2.2 Gradient Vector Flow (GVF) Snake

The traditional snake model [25], which relies on edge-based external forces, suffers from several limitations. The initial contour must be close to the true boundary; otherwise it may converge to the wrong result. In addition, it has poor convergence to boundary concavities. The GVF snake [26] solves these problems by extending the capture range of the external force field and guiding the contour into concave regions. In this case, the Euler–Lagrange iterative equation in (12.39) is modified to:

$$\frac{\partial\underline{v}(s,t)}{\partial t} = \alpha\underline{v}^{(2)}(s,t)-\beta\underline{v}^{(4)}(s,t)+\underline{\zeta}(u(s,t),v(s,t)) \tag{12.45}$$

where $\underline{\zeta}(u,v) = [z_1(u,v)\; z_2(u,v)]^T$ known as a GVF vector is used instead of the external force based on the edge map in the original snake formulation. GVF is defined as the minimizer of the energy functional:

$$\mathcal{E}(\underline{\zeta}) = \iint\left[\mu\left(\|\nabla z_1(u,v)\|^2+\|\nabla z_2(u,v)\|^2\right)+\|\nabla g(u,v)\|^2\|\underline{\zeta}(u,v)-\nabla g(u,v)\|^2\right]du\,dv \tag{12.46}$$

where $\nabla z_1(u,v) = \left[\frac{\partial z_1(u,v)}{\partial u} \; \frac{\partial z_1(u,v)}{\partial v}\right]^T$ and similarly for $\nabla z_2(u,v)$ and $g(u,v) = -E_{\text{ext}}(u,v)$ takes one of the two functions in (12.36). The vector $\nabla g(u,v)$ is normal to the edges of segmented objects. When $\nabla g(u,v)$ is small (i.e. homogeneous regions), the first term in (12.46), which enforces smoothness of the vector field, dominates, causing the field to vary slowly. In contrast, when $\nabla g(u,v)$ is large, the second term dominates. This term is minimized by setting $\underline{\zeta}(u,v) = \nabla g(u,v)$, hence producing the desired effect of keeping $\underline{\zeta}(u,v)$ nearly equal to the gradient of the edge map. The choice of the regularization parameter μ presents a trade-off between the two effects. Also, low SNR images require larger values of μ to encourage smoothing.

Using the calculus of variations [26], the GVF field satisfies the following Euler–Lagrange equations:

$$\begin{aligned} \mu\nabla^2 z_1(u,v) - (z_1(u,v) - g_u(u,v))(g_u^2 + g_v^2) &= 0 \\ \mu\nabla^2 z_2(u,v) - (z_2(u,v) - g_v(u,v))(g_u^2 + g_v^2) &= 0 \end{aligned} \tag{12.47}$$

where ∇^2 is the Laplacian operator and $g_u(u,v) = \frac{\partial g(u,v)}{\partial u}$ and $g_v(u,v) = \frac{\partial g(u,v)}{\partial v}$. These equations provide insight into how GVF works. In homogeneous areas of the image $x(u,v)$ regions where g_u and g_v are small, these equations reduce approximately to Laplace equations, solutions of which give the GVF field from the region boundary. This explains why GVF yields vectors that point to boundary concavities. In [26], it has been shown that the equations in (12.47) can be discretized using a forward difference method and solved iteratively using the following recursive equations:

$$\begin{aligned} z_1(i,j,n+1) &= (1 - b_{i,j}\triangle t)z_1(i,j,n) + r\big[z_1(i+1,j,n) + z_1(i,j+1,n) + z_1(i-1,j,n) \\ &\quad + z_1(i,j-1,n) - 4z_1(i,j,n)\big] + c_{i,j}\triangle t \\ z_2(i,j,n+1) &= (1 - b_{i,j}\triangle t)z_2(i,j,n) + r\big[z_2(i+1,j,n) + z_2(i,j+1,n) + z_2(i-1,j,n) \\ &\quad + z_2(i,j-1,n) - 4z_2(i,j,n)\big] + d_{i,j}\triangle t \end{aligned} \tag{12.48}$$

where i, j, and n correspond to the discretized spatial coordinates and iteration index (time), respectively, and $\triangle t$ is the iteration step size. The variables $b_{i,j}$, $c_{i,j}$, $d_{i,j}$, and r are defined as: $b_{i,j} = g_u^2(i,j) + g_v^2(i,j)$, $c_{i,j} = b_{i,j}g_u(i,j)$, $d_{i,j} = b_{i,j}g_v(i,j)$, and $r = \frac{\mu\triangle t}{\triangle u \triangle v}$ with $\triangle u$ and $\triangle v$ being the spatial sampling intervals. To have convergent solutions, $b_{i,j}$, $c_{i,j}$, and $d_{i,j}$ must be bounded and moreover $r \leq 1/4$ or $\triangle t \leq \frac{\triangle u \triangle v}{4\mu}$.

Algorithm for GVF Snake:

i. Apply Canny edge detection operator (see Chapter 8) to obtain the edge map. We can use any edge linking method to prevent disconnected edges.
ii. Based on the edge map, use (12.48) to compute GVF field $\underline{\zeta}(u,v)$ as an external force.
iii. Initialize the snake. The initial snake can be placed automatically using the first-order geometric moments of the image. The $(k+l)^{\text{th}}$ order geometric moments are given by $m_{k,l} = \sum_{\forall i,j} i^k j^l x(i,j)$ where $x(i,j)$ is the pixel intensity at (i,j). The first-order normalized moments, $i_{\text{center}} = m_{10}/m_{00}$, $j_{\text{center}} = m_{01}/m_{00}$, are used as the coordinates of the center of the initial snake. Then, for all possible edge points on the binary edge map, the distances between the edge points and the snake center are calculated, and the median value of the distance is used as the radius of the initial snake.
iv. Use (12.43) to update the snake until convergence.

Figure 12.13a illustrates an example of the GVF ability to conform to a U-shaped object with concavity. The evolution and convergence of the GVF snake and its close conformance to the boundaries of the object is clearly evident from this figure. The process was initialized with an outer circle with a large radius. As can be observed, the GVF snake has a large capture range, i.e. it can converge to the boundaries of the object even when the initial snake is far from the actual object boundaries. This characteristic attests to independence of this method on the initial choice of snake. Figure 12.13b–d shows the edge map, the external energy, and GVF field, respectively. Figure 12.13d illustrates that the gradient vectors around the boundary of the concavity have downward trends, indicating a better conformance to the object boundaries compared with those of the regular active contour [26]. In this experiment, we used control parameters $\alpha = 0.6$, $\beta = 0$ and regularization parameter $\mu = 0.2$. The edge maps used in GVF were normalized to the range [0, 1].

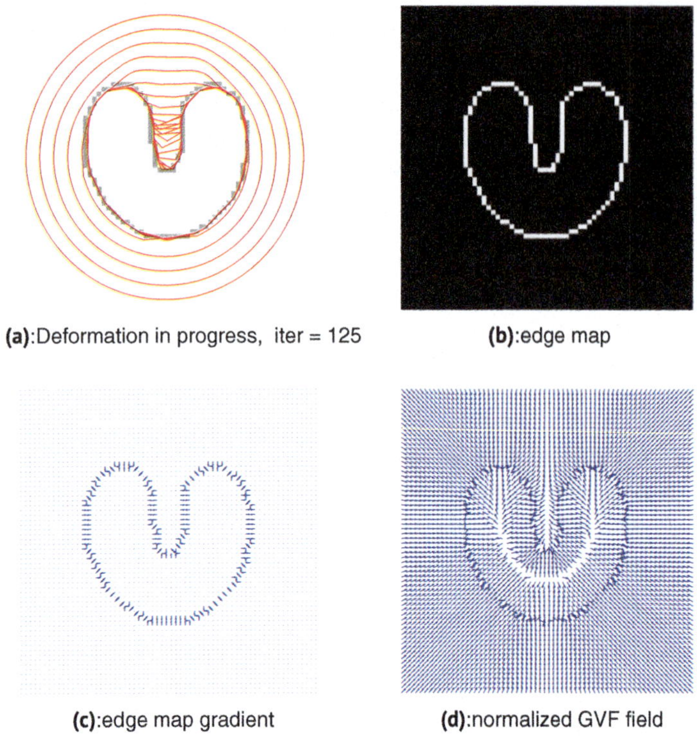

Figure 12.13 (a) Convergence of GVF snake to object boundaries, (b) edge map, (c) external energy, and (d) external force field.

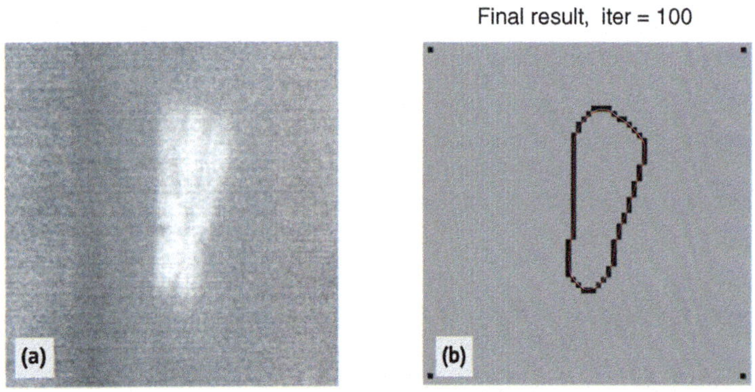

Figure 12.14 (a) Original range image and (b) GVF snake segmented region. *Source*: Naval Surface Warfare Center (NSWC)/Public domain.

Next, we show an application of the GVF snake for object segmentation in range imagery. Laser-based optical devices [29] provide good optical contrast images as well as range maps with depth information, which are essential for underwater object identification. A critical step in such applications is reliable image segmentation prior to feature extraction and recognition. Typically, range images are challenging to process due to noise and other sensor imperfections. Figure 12.14a shows a 256×256-pixel range image of a trapezoidal-shaped underwater object. To reduce sensitivity to noise and potential missing edge segments, and also to reduce computational burden, a lower-resolution version of size 64×64 pixels is used. The corresponding converged snake segmentation result for this image is shown in Figure 12.14b. Figure 12.15a shows the evolution of the GVF snake and its convergence to the object boundaries, with the initial snake being the outer circle. Figure 12.15b–d shows the edge map, external energy, and force fields (or GVF) field, respectively.

We conclude this chapter by providing another example of the application of the GVF snake for segmenting regions in medical imagery. In particular, we consider one slice of a lung CT scan in Figure 12.16a that shows a relatively large tumor located in the upper part of the right lung. The objective is to assess the capability of the GVF snake to accurately isolate the tumor from surrounding lung tissue and the chest cavity. Figure 12.16b shows the edge map obtained using the LoG operator

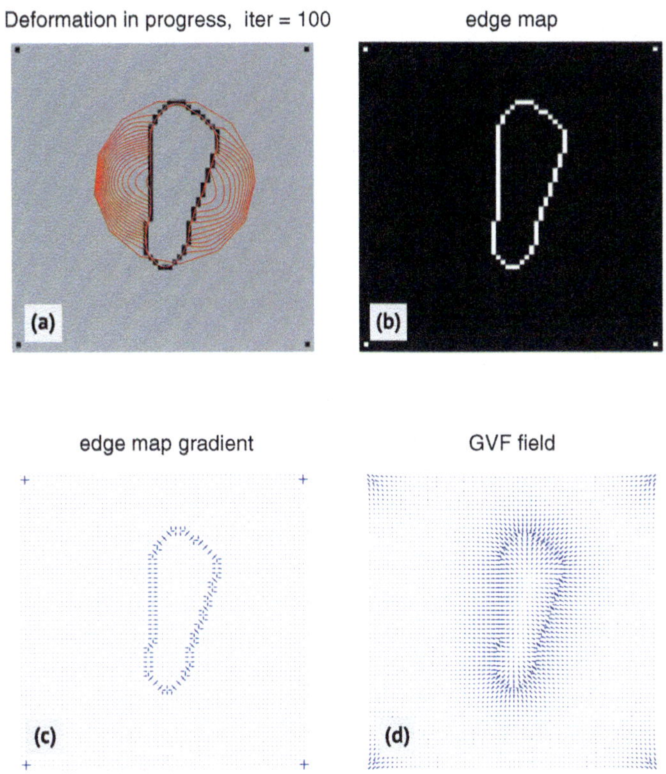

Figure 12.15 (a) GVF snake convergence process for a range image in Figure 12.14a, (b) edge map, (c) external energy, and (d) external force or GVF field.

(see Chapter 8). The edge map gradient is then used to generate the external force field, as shown in Figure 12.16c. Using this external force field, we iterated the GVF snake process for 1500 iterations with control parameters $\alpha = 0.3$, $\beta = 0.3$ and regularization parameter $\mu = 0.2$. Figure 12.16d shows the initial starting area in blue, the evolution process in green, and the final converged segmented region of the tumor in red. Despite the low contrast and ambiguity of the tumor boundary within the surrounding right lung tissue, the GVF snake was able to successfully isolate the region in which the tumor occupies.

12.3 Computer Projects

The computer projects in this chapter use Python libraries OpenCV, NumPy, Matplotlib, ImageIO, scikit-image, and SciPy.

Computer Project 12.1: This computer project focuses on the implementation and testing of histogram thresholding algorithms covered in Section 12.1.1 for image segmentation. The image to be considered is the lung CT image used in the last experiment of the previous section. The main goal is to isolate the region in which the tumor occupies in the chest cavity. Based on the histogram of this image, develop Python-based scripts to implement both Otsu's and the maximum entropy methods and test their performance for separating the tumor from the surrounding lung tissues. You may use histogram specification before adaptive threshold selection to better isolate the histogram mode associated with the tumor. Provide a detailed comparison of your results with those presented in Section 12.2.2 using the GVF snake and clearly describe the strengths and limitations of each method.

Computer Project 12.2: The main goal of this computer project is similar to that of Computer Project 12.1, using the same lung image. However, region-based segmentation methods covered in Section 12.1.2 are to be used instead of histogram thresholding. Develop Python scripts to implement and test the region growing and split-and-merge methods for segmenting the tumor from other tissues. Follow the procedure outlined in the relevant sections for these algorithms. Provide a detailed comparison of your results with those provided in Section 12.2.2 using the GVF snake method and clearly describe the advantages and disadvantages of each method for this particular application.

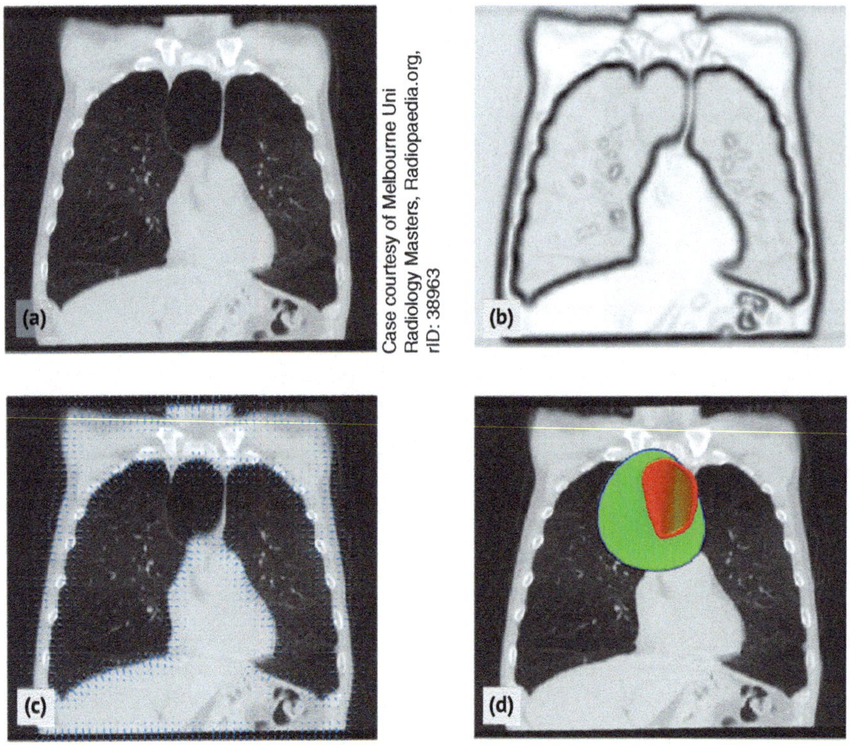

Figure 12.16 (a) CT-scan image of right lung with a tumor, (b) edge map, (c) external force field, (d) snake evolution (green) from initial condition (blue) to final localization of the tumor (red). (The color rendition of this image is available to view at www.wiley.com/go/Azimi449.)

Computer Project 12.3: The main goal of this computer project is identical to that of Computer Project 12.1 using the same lung image. However, clustering-based segmentation methods covered in Section 12.1.3 must be employed instead of histogram thresholding. Develop Python scripts to implement and test the hard and soft (fuzzy) clustering methods described in Section 12.1.3 for segmenting the tumor from other tissues. Provide a detailed comparison of your results with those presented in Section 12.2.2 using the GVF snake method and clearly describe the advantages and disadvantages of each method for this particular application.

Computer Project 12.4: In this computer project, we study the effectiveness of the MRF method in Section 12.1.4 for image segmentation. The goal is the same as that in the previous computer projects, namely to separate the tumor in the chest cavity from the rest of the lung image. To simplify the formation of the Gibbs energy function in (12.24), use a first-order neighborhood lattice and the corresponding cliques. Develop a Python-based script using the ICM algorithm in Section 12.1.4 to estimate the class labels of different regions within the lung image. Initial estimates of the class labels can be generated using K-means clustering or the ML method. Use maximum of 10 iterations for the ICM algorithm. Provide a detailed analysis of your algorithm in terms of convergence and segmentation accuracy. Use the K-means clustering result as a benchmark for the correct classification rate comparison. Compare your results with those obtained using the GVF snake. Describe the advantages and disadvantages of each method for this particular application.

Computer Project 12.5: This computer project involves using the GVF snake described in Section 12.2.2 for image segmentation. The test image is the color image of Finn and his toy. The goal is to isolate his toy from the rest of the image using this contour-based segmentation method. First, apply Canny or LoG operator to produce the edge map, which is subsequently used to generate the external force field. For the internal force field, the control and regularization parameters α, β, and μ should be chosen empirically to achieve accurate segmentation. Demonstrate the impact of using different initialization by following the steps in the algorithm in Section 12.2.2. The number of iterations should be at least 1500. In order to compare the performance of your GVF-based segmentation method with a baseline method, implement a color-based (3-D RGB components) segmentation using K-means clustering algorithm. A subset of pixels in the Finn image can be used for training and the remaining pixels for testing the performance of the trained system. The clusters can be assigned class labels using a small set of data samples with known labels. Investigate the number of clusters required to achieve good segmentation accuracy. Provide a detailed analysis of the GVF algorithm in terms of segmentation quality, as well as conformance to the toy boundaries. How do the results compare to those obtained using K-means benchmark? Describe the advantages and limitations of the GVF snake method.

12.4 Problems

Problem 12.1: Consider the image X in Example 12.1 and use Otsu's method to find the optimum threshold. How does this result differ from that in Example 12.1?

Problem 12.2: A digital image X has a bimodal pmf consisting of two uniformly distributed modes corresponding to the background and target, respectively. The pmf of X in these regions is defined by $P(X=i)=\frac{1}{2L},\ i\in[0,L_1-1]$ and $P(X=i)=\frac{A}{L},\ i\in[L_1,L-1]$ where $L=\frac{3L_1}{2}$.

a. Find amplitude A for a valid bimodal pmf.
b. Use Otsu's method to find the optimum threshold.

Problem 12.3: Repeat Problem 12.2 using the Maximum Entropy method.

Problem 12.4: In Problem 12.2, assume that the threshold $t<L_1$.

a. Determine $P(C_1)$ and $P(C_2)$ and the probability of error. What is the error type in this case?
b. Repeat part (a) for $t>L_1$.
c. What is the optimum threshold for this bimodal pmf that gives the best segmentation performance? Use your intuition.

Problem 12.5: Consider an image X with a bimodal pdf whose modes correspond to a target (true hypothesis C_1) and background (null hypothesis C_0) regions. Assume this bimodal pdf can be expressed as a Gaussian mixture,

$$p(x)=P(C_0)p(x|C_0)+P(C_1)p(x|C_1)$$

where $P(C_0)+P(C_1)=1$ and $p(x|C_i), i=0,1$ are Gaussian pdfs with known mean μ_i and variance σ_i^2. The prior probabilities $P(C_i)$ are also assumed to be known. If t is the threshold to successfully segment target from the background, find:

a. Expression for the probability of error as a function of t.
b. Expression for the optimal threshold that gives minimum probability of error.
c. Repeat part (b) when $\sigma_1^2=\sigma_2^2$.

Problem 12.6: In Problem 12.5, we assume that the pdfs are Laplacian with $p(x|C_i)=\frac{1}{2\alpha_i}exp\{-\frac{|x-\mu_i|}{\alpha_i}\}$ where μ_i is the mean (assume $\mu_0<\mu_1$) and α_i is the scale parameter. Assuming that the prior probabilities $P(C_i),\ i=0,1$ are known and t is the threshold to successfully segment target from the background, find:

a. Expression for the probability of error as a function of t.
b. Expression for the optimal threshold that gives minimum probability of error.
c. Repeat part (b) when $\alpha_1=\alpha_2$.

Problem 12.7: The background and target in an image are distributed using the following pdfs:

$$p_{X_0}(x)=\begin{cases}x, & \text{when } 0\le x\le 1\\ -x+2, & \text{when } 1\le x\le 2\end{cases},$$

$$p_{X_1}(x) = \begin{cases} x - \frac{3}{2}, & \text{when } \frac{3}{2} \le x \le \frac{5}{2} \\ -x + \frac{7}{2}, & \text{when } \frac{5}{2} \le x \le \frac{7}{2} \end{cases},$$

respectively.

a. Determine the range of possible thresholds for object segmentation.
b. If the background occupies 3/4 of the total image, find the threshold that minimizes the probability of error.

Problem 12.8: The background and target in an image are distributed using the following pdfs:

$$p_{X_0}(x) = \begin{cases} \frac{\pi}{4} \cos \frac{(x-1)\pi}{2}, & \text{for } 0 \le x \le 2 \\ 0, & \text{else} \end{cases}$$

$$p_{X_1}(x) = \begin{cases} \frac{\pi}{8} \cos \frac{(x-3)\pi}{4}, & \text{for } 1 \le x \le 5 \\ 0, & \text{else} \end{cases}$$

respectively.

a. Determine the range of possible thresholds for object segmentation.
b. If the background occupies 2/3 of the total image, find the threshold that minimizes the probability of error.

Problem 12.9: A discrete r.v. X with two regions has a geometric PMF $P(X = i) = (1 - p)^{i-1}p, \; i = 1, 2, \ldots,$ where p is the probability in each trial. Find the expression for the optimal threshold t for separating these regions using the maximum entropy method.

Problem 12.10: Consider a discrete r.v. X with Poisson PMF $P(X = i) = \frac{\lambda^i e^{-\lambda}}{i!}, i = 0, 1, \cdots$ where the parameter $\lambda > 0$ determines the shape of the distribution. If this r.v. consists of two regions that must be segmented, find the expression for the optimal threshold t using Otsu's method.

Problem 12.11: Derive Eq. (12.29) for MRF and explain the assumptions made to obtain this equation.

Problem 12.12: As stated in Section 12.1.4, the MRF-based image segmentation model requires finding the parameters $\underline{\mu}_i$ and $\mathbf{C}_i$ for every Gaussian (or class). Assuming that labeled training data are available for each class, devise a scheme to estimate these parameters.

Problem 12.13: Explain all the assumptions behind Eqs. (12.31) and (12.32) for the ICM algorithm. Based on these assumptions, derive the equations.

Problem 12.14: An alternative application of MRF is noise filtering. Consider a noisy 1-D signal $z(n) = x(n) + \eta(n)$ where $z(n)$ is the observation at time n and $\eta(n)$ is a white Gaussian noise. If N observations are received, estimate $x(n), n = 1, \ldots, N$ by minimizing the energy function:

$$E(\underline{x}) = \sum_{n=1}^{N} (z(n) - x(n))^2 + \lambda \sum_{n=1}^{N} (x(n+1) - x(n))^2$$

where the first term is the filtering term and the second term is the data related, which enforces that the adjacent samples on the grid to have similar values. The constant λ controls the trade-off between smoothness and data fit. Develop a system of N linear equations to find $\underline{x} = [x(1) \cdots x(N)]^T$.

Problem 12.15: In Problem 12.14, assume that the boundary conditions are neglected (e.g. signal is much longer than the filter support).

a. Show that solution in Problem 12.14 can be approximated by the convolution equation $x(n) * (\delta(n) + \lambda g(n)) = z(n)$ where $g(n) = \{-1, 2, -1\}$ is an FIR filter impulse response.
b. Find the inverse filter $h(n)$ that restores $x(n)$ from $z(n)$. What type of filter is $h(n)$?

Problem 12.16: Show that the discretization of the derivative of $\underline{v}(s, t)$ with respect to the iteration index t in (12.39) results in (12.41) and subsequently (12.42).

Problem 12.17: Show that the discretization of the Euler–Lagrange equations in (12.47) leads to those in (12.48). What are the assumptions in these derivations?

Problem 12.18: In equation (12.43), assume $P = 2$ and $\Delta t = \Delta s = 1$. Find:

a. Matrix $\mathbf{\Gamma}$ and $(\mathbf{I} + \mathbf{\Gamma})^{-1}$,
b. The conditions on α and β parameters so that the final converged result does not depend on the initial choice of snake.

References

1 K. S. Fu and J. K. Mui, "A survey on image segmentation," *Pattern Recognition*, vol. 13, pp. 3–16, 1981.
2 N. R. Pal and S. K. Pal, "A review of image segmentation techniques," *Pattern Recognition*, vol. 26, pp. 1277–1294, 1993.
3 X. Liu, Z. Deng, and Y. Yang, "Recent progress in semantic image segmentation," *Artificial Intelligence Review*, vol. 52, pp. 1089–1106, 2019.
4 S. Minaee, Y. Boykov, F. Porikli, A. Plaza, N. Kehtarnavaz, and D. Terzopoulos, "Image segmentation using deep learning: A survey," *IEEE Transactions on Pattern Analysis and Machine Intelligence (PAMI)*, vol. 44, pp. 3523–3542, 2022.
5 S. Hayken, "*Neural Networks and Learning Machines*," Prentice Hall, 3rd Edition, 2007.
6 N. Otsu, "A threshold selection method from gray-level histograms," *IEEE Transactions on Systems, Man, and Cybernetics*, vol. 9, pp. 62–66, 1979.
7 R. C. Gonzalez and R. E. Woods, "*Digital Image Processing*," Pearson, 2018.
8 A. D. Brink and N. E. Pendcock, " Minimum cross-entropy threshold selection," *Pattern Recognition*, vol. 29, pp. 179–188, 1996.
9 J. N. Kapur, P. K. Sahoo and A. K. C. Wong, "A new method for gray-level picture thresholding using the entropy of the histogram," *Computer Vision, Graphics, and Image Processing*, 29, pp. 273–285, 1985.
10 T. Acharya and A. K. Ray, "*Image Processing Principles and Applications*," Wiley-Interscience, 2005.
11 R. O. Duda, P. E. Hart, and D. G. Stark, "*Pattern Classification*," Wiley, 2nd Edition, 2000.
12 E. L. Lehmann and J. P. Romano, "*Testing Statistical Hypotheses*," Springer, 4th Edition, 2005.
13 S. L. Horowitz and T. Pavlidis, "Picture segmentation by a tree traversal algorithm," *Journal of the ACM*, vol. 23, pp. 368–388, 1976.
14 Y. L. Chang and X. Li, "Adaptive-region growing," *IEEE Transactions on Image Processing*, vol. 3, pp. 868–872, 1994.
15 K. Haris, S. N. Efstratiadis, N. Maglaveras, and A. K. Katsaggelos, "Hybrid image segmentation using watersheds and fast region merging," *IEEE Transactions on Image Processing*, vol. 7, pp. 1684–1699, 1998.
16 H. Mittal, A. C. Pandey, M. Saraswat, S. Kumar, R. Pal, and G. Modwel, "A comprehensive survey of image segmentation: Clustering methods, performance parameters, and benchmark datasets," *Multimedia Tools and Applications*, vol. 81, pp. 35001–35026, 2022.
17 Z. Kato and T-C. Pong, "Markov random field image segmentation model for color textured images," *Image and Vision Computing*, vol. 24, pp. 1103–1114, 2006.
18 M. Berthod, Z. Kato, S. Yu, and J. Zerubia, "Bayesian image classification using Markov random fields," *Image and Vision Computing*, vol. 14, pp. 285–295, 1996.
19 S. Geman and D. Geman, "Stochastic relaxation, Gibbs distributions and the Bayesian restoration," *IEEE Transactions on Pattern Analysis and Machine Intelligence*, vol. 6, pp. 721–741, 1984.
20 J. Besag, "On the statistical analysis of dirty pictures," *Journal of the Royal Statistical Society. Series B (Methodological)*, vol. 48, pp. 259–302, 1986.

21 R. Chellappa and A. Jain, "*Markov Random Fields- Theory and Application*," Academic Press, 1993.

22 S. M. Kay, "*Fundamentals of Statistical Signal Processing-Estimation Theory*," Prentice-Hall, vol. I, 2009.

23 K. Held, E. R. Kops, B. J. Krause, W. M. Wells, R. Kikinis, and H.-W. Muller-Gartner, "Markov Random Field Segmentation of Brain MR Images," *IEEE Transactions on Medical Imaging*, vol. 16, pp. 878–886, 1997.

24 R. C. Dubes, A. K. Jain, S. G. Nadabar, and C. C. Chen, "MRF model-based algorithms for image segmentation," *Proceedings. 10th International Conference on Pattern Recognition,* vol. 2, pp. 808–814, 1990.

25 M. Kass, A. Witkin, and D. Terzopoulos, "Snakes: Active models," *International Journal of Computer Vision*, vol. 1, pp. 321–331, 1988.

26 C. Xu and J. L. Prince, "Snakes, shapes, and gradient vector flow," *IEEE Transactions on Image Processing*, vol. 7, pp. 359–369, 1998.

27 L. A. Vese and C. L. Guyader, "*Variational Methods in Image Processing*," CRC Press, 2016.

28 C. Xu, A. Yezzi, and J. L. Prince, "On the relationship between parametric and geometric active contours," Proceedings of the Thirty-Fourth Asilomar Conference on Signals, Systems and Computers, pp. 483–489, 2000.

29 G. Tao, M. R. Azimi-Sadjadi, and A. Nevis, "Underwater target identification using GVF snake and Zernike moments," Proceedings of MTS/IEEE Oceans, 2002.

13

Feature Extraction

Feature extraction is a cornerstone of data analysis, extending far beyond the realm of image analysis to underpin a diverse array of important applications. Its utility is integral to fields such as robotic vision, content-based image retrieval (CBIR), and data mining, and in 1-D signal processing across various fields, such as speech recognition, acoustic analysis, seismic data interpretation, and radar and sonar signal processing.

In Chapter 12, we illustrated a traditional image analysis system and its components in Figure 12.1 and introduced different methods for image segmentation. Once the objects or regions within an image have been successfully segmented, the subsequent step is the feature extraction process. Figure 12.1 also indicated a feedback path from the feature extraction to the image segmentation block, raising the question about the role of this feedback mechanism. Consider a scenario where the image segmentation subsystem erroneously merges two or more disjoint regions or incorrectly splits one region into two or more subregions, due to subtle or less distinguishable differences in their homogeneity properties. Now, if the feature extraction process can produce discriminative features (e.g. textural) that capture these subtle differences, then the feedback from the feature extraction block enables the segmentation subsystem to identify these errors and initiate corrective actions. Therefore, this feedback mechanism plays a critical role in improving the overall performance of the image analysis system by creating a self-correcting loop.

There are many ways to categorize various image-based feature extraction methods. In this chapter, we categorize them according to their specific function. These categories are: (a) color, (b) energy, (c) spectral, (d) textural, (e) contour, and (f) shape features. Each category will detail commonly employed methods and their respective characteristics. Prior to exploring these methods, the fundamental objectives and essential requirements for effective feature extraction in a traditional image analysis system, as depicted in Figure 12.1, will be outlined.

13.1 Objectives and Requirements

13.1.1 Objectives

The main objective of any feature extraction system is to extract a set of representative and salient features with a high discriminatory ability. That is, when designing an image analysis system for a particular class of images, one should search for features that closely capture distinguishing characteristics of different object classes to facilitate their successful discrimination. As an example, consider the task of discriminating between benign and malignant masses in breast mammograms. Radiologists typically rely on shape and texture characteristics of detected masses to determine their likelihood of malignancy. Thus, an image analysis system designed for this task should also utilize feature extraction methods that capture these shape and texture-based attributes of the segmented areas within mammogram images. However, there are many challenges in selecting or designing appropriate feature extraction methods for a specific image analysis task. First, the choice of feature extraction is heavily dependent on prior knowledge of the visually significant attributes of the objects that enable effective discrimination. This is a nontrivial task, particularly in situations involving a large number of object classes with feature similarities. Moreover, a thorough understanding of the image formation system and its underlying properties is often required. These considerations rely heavily on the expertise of the system designer. Even when an appropriate feature extraction method is identified that closely represents the characteristics of the objects to be discriminated, a fundamental gap remains between the high-level concepts utilized by human perception and the low-level features extracted by traditional systems. Human perception is inherently inferential and remarkably good at learning novel concepts even from limited data. Traditional feature extraction methods, which are not learning-based, lack this capability and therefore struggle to capture semantic attributes that humans use to make decisions. Recent advances in artificial intelligence (AI), in particular in deep learning, have significantly narrowed this gap between human- and computer-based perception by automatic learning of hierarchical and semantically meaningful features. Chapters 18 and 19 explore some of the applications of deep learning in this area.

Another key objective of feature extraction in traditional image analysis systems is to reduce the dimension of the data. *Bellman's Curse of Dimensionality* [1, 2] states that the performance of a classifier does not improve with increasing the dimension of the input space. This is because large dimensional inputs require more parameters to be estimated in the classification system, which can lead to increased estimation error when limited training data are available. In that case, the trained classifier will be so fine-tuned (or over-fitted) to the training data that will lack *generalization* ability [3] on unseen data. Therefore, the dimensionality reduction is another essential aspect of feature extraction.

Furthermore, statistical dependency among features can negatively impact classifier performance, and may lead to bias in decision-making. Feature decorrelation has been found to be an effective strategy for mitigating these effects and improving robustness. Although there are many ways by which we can achieve feature decorrelation, Principal Component Analysis (PCA), covered in Chapter 6, is a well-established method for both feature decorrelation and dimensionality reduction. However, problem arises when PCA is to be used for region-based feature extraction, since different object classes tend to have significantly different statistical characteristics and distributions. As a consequence, one PCA transformation matrix cannot be used to map data of different classes. Nevertheless, PCA is more effective for pixel-based feature extraction tasks, such as multispectral or hyperspectral imaging where the feature vectors consist of pixels at the same spatial location across multiple spectral bands. In such cases, the data share common statistical properties, making PCA an effective tool for dimensionality reduction, decorrelation, and feature representation.

13.1.2 Invariance and Equivariance Requirements

An important requirement in any feature extraction method is the *invariance* property, by which we mean that the extracted features must remain unchanged when the segmented objects undergo geometric and affine transformations including rotation, translation, scaling, and even ideally distortion/deformation (e.g. due to shearing). Such effects are typically caused by perspective variations of the imaging sensor relative to the scene being imaged, e.g. in satellite imaging. The invariance property is critical in order to design pattern classification systems that remain robust under these variations. However, achieving feature invariance with respect to a broad class of transformations is highly challenging if not unlikely. For example, if the segmented object has a circular shape, the extracted features may exhibit invariance to rotation and translation, though scaling can destroy the invariance properties. Nevertheless, certain *equivariance or covariance* can still exist between features that are not invariant, under that specific class of transformations. In general, equivariance refers to a structured relationship between two sets of data-dependent variables, in this case image features, when the data undergo transformations. Hence, it is a critical concept in data analysis to identify patterns and trends in data sets. To illustrate distinction between invariance and equivariance, consider a scenario where the imaging sensor, e.g. a radar or sonar platform, is collecting a sequence of observations from a ground or an underwater target at different aspect angles. For an arbitrary-shaped target, it is unrealistic to expect feature invariance with respect to aspect angle variations, especially when additional changes in range and elevation are present. However, it may be possible to look for features that are equivariant (i.e. related) with respect to changes in aspect angle. In addition to geometric transformations, captured images may also be subject to photometric changes that lead to global or local intensity variations depending on specific conditions. Global intensity change can be caused by such effects as illumination changes, e.g. in visible imaging, which can be mitigated by normalization or histogram manipulation methods. However, local intensity changes are more difficult to remedy.

To further clarify these concepts, assume that we have several different object classes where $S = \{s_1, s_2, \ldots, s_M\}$ represents the set of all these M objects. Consider an object $s \in S$ subjected to a geometric transformation $g \in G$ where G is a transformation group. The transformed object is denoted by $t = g \circ s$. The set of all such transformed instances form the *orbit* of s: $T = \{t = g \circ s | g \in G\}$ as shown in Figure 13.1. All elements of this set correspond to different views (or poses) of the same object.

When imaged by a particular imaging system, transformed object $t = g \circ s$ forms an observation $G \times S \rightarrow \{x(\underline{v}; t)\}$ where $\underline{v} = [u\ v]^T$ is the spatial coordinate vector, and $x(\underline{v}; t)$ is the observed image. Let us denote the feature vector extracted from this image by $\underline{y}(t)$. If the feature vectors corresponding to nearby poses on the orbit remain unchanged, i.e.

$$\underline{y}(t(p_i)) = \underline{y}(g \circ s(p_0)) = \underline{y}(s(p_0)), \quad \forall\, i \in \mathcal{N} \tag{13.1}$$

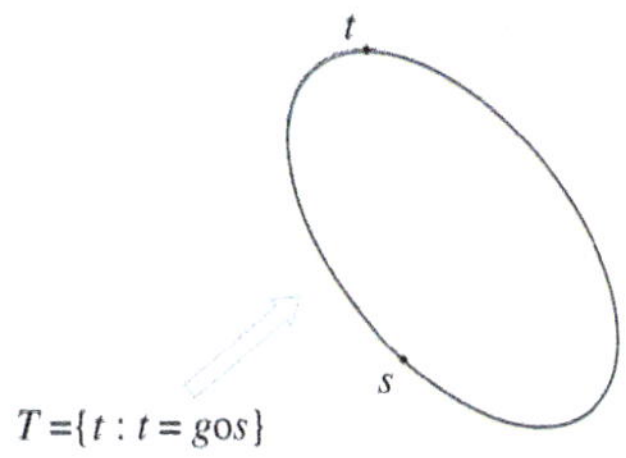

Figure 13.1 Orbit of object s.

where p_i is a nearby point to p_0 on the orbit within a neighborhood $\mathcal{N}$, and $s(p_0)$ is the *template*, then the feature vector is said to be *invariant* under geometric transformation g. Now, if the feature vectors corresponding to poses that are not within the neighborhood $\mathcal{N}$ satisfy:

$$\underline{y}(t(p_j)) = \underline{f}(\underline{y}(s(p_0))), \quad \forall j \notin \mathcal{N}, \tag{13.2}$$

where $\underline{f}$ is an action that belongs to a transformation group $\underline{F}$, then the feature vectors are said to be *equivariant* (or covariant) with respect to this transformation group.

An important example of transformation group G is the *affine group*, whose actions on the image coordinates is given by:

$$x(\underline{v}; t) = x(\mathbf{A}\underline{v} + \underline{b}; s) \tag{13.3}$$

where $x(\underline{v}; t) = x(\underline{v}; g \circ s)$ is the observed image at a pose close to that of the template $x(\underline{v}; s)$ and $x(\mathbf{A}\underline{v} + \underline{b}; s)$ is the image of object s after coordinate transformation. The matrix $\mathbf{A}$ is of size 2×2 representing linear transformations (e.g. rotation, scaling, shear) and $\underline{b}$ is a 2×1 translation vector. An example of a transformation group $\underline{F}$ is also a linear group with action $\underline{z} = \mathbf{B}\underline{y}$ where $\mathbf{B}$ represents a feature mapping matrix.

13.2 Color Features

Color-based features are among the most widely used features for object discrimination and related applications, including, e.g. CBIR [4]. These features are defined depending on the choice of color space such as RGB, HSI, or YC_bC_r, which were discussed in Chapter 4. Color representations are inherently invariant to rotation and translation of segmented regions. Scale invariance can also be achieved by a suitable normalization of color histograms, which can be represented either as a single 3-D histogram or three separate 1-D histograms. Once color space is specified, color features can be extracted from the entire image or from segmented regions.

There are several color features that can be used for both color-based discrimination and image retrieval applications. Among the widely used are the color histogram, color correlogram, color coherence vector, and color moments [4–9]. A color histogram describes the global color distribution in an image or in a segmented region. Although color histogram is simple and easy to compute, it does not include any spatial information, making it prone to misclassification, especially in scenarios where spatial relationships are important.

The color correlogram, proposed in [6], provides statistical features that are dependent on spatial distance and represent the distribution of local spatial dependencies of colors. Here, we describe the principle of this method. For more efficient implementation, the reader is referred to [6].

Consider a color image $x(m, n)$ of size $N \times N$, where the colors are quantized into c discrete levels indexed by $l \in [1, c]$. Let us define the pixel coordinate vector $\underline{p} = [m\ n]^T$ and let $X_l = \{\underline{p} | x(\underline{p}) = l\}$ be a set of all pixels of image $x(\underline{p})$ with color index l. Then, the *correlogram* of image $x(\underline{p})$ is defined as:

$$\gamma_{k,l}^d = \mathrm{Prob}_{\underline{p}_1 \in X_k}\left\{\underline{p}_2 \in X_l \,\middle|\, ||\underline{p}_1 - \underline{p}_2||_\infty = d\right\}, \tag{13.4}$$

where $d \in [1, N]$ is a distance parameter and $||.||_\infty$ represents l_∞-norm, i.e. $||\underline{p}_1 - \underline{p}_2||_\infty = \max\left\{|m_1 - m_2|, |n_1 - n_2|\right\}$. Thus, given a pixel of color index k, the correlogram $\gamma_{k,l}^d$ in (13.4) gives the probability that a pixel located at distance d away from the given pixel has color l. The computational requirement of this method depends on c and d and is on the order $\mathcal{O}(c^2 d)$. In [6], two different algorithms for small and large values of d were proposed. Note that when $k = l$, γ_k^d, $k \in [1, c]$, becomes *autocorrelogram*, which captures spatial color dependencies among identical colors, and is a subset of the correlogram. In general, compared to other color feature extraction methods, color correlograms are computationally demanding, and also sensitive to noise. Moreover, they are not inherently invariant to rotation and scale transformations.

Figure 13.2a,b shows the original Finn image and its vector-quantized (VQ) version using the K-means clustering algorithm. The number of clusters (or colors) was chosen to be $K = 8$ in order to reduce the computational requirements of the correlogram method. Figure 13.2c,d illustrates the correlogram matrices for $d = 1$ and $d = 2$, respectively. The diagonal elements of these matrices are much larger than the off-diagonal ones, due to the fact that pixels adjacent to or very near each other are highly likely to share the same color, especially in a VQ image with a limited color palette. Clearly, a wider search distance d or a more diverse color palette would reduce the probabilities along the diagonal while increasing the off-diagonal ones.

A simpler and computationally more efficient method for extracting color-based features is based on color moments [7]. Color moments provide statistical measures that characterize the color distribution in an image and are analogous to the central moments of a probability distribution. Commonly used color moments [7, 8] are the mean, standard deviation, and skewness.

Let us define $x_l(m, n)$, $l = 1, 2, 3$ as the image corresponding to the color channel l in the RGB or HSI color space. The color moments for channel l are defined as:

$$
\begin{aligned}
\mu_l &= \frac{1}{N^2}\sum_{m=0}^{N-1}\sum_{n=0}^{N-1} x_l(m, n), \\
\sigma_l &= \sqrt{\frac{1}{N^2}\sum_{m=0}^{N-1}\sum_{n=0}^{N-1} (x_l(m, n) - \mu_l)^2}, \\
\xi_l &= \sqrt[3]{\frac{1}{N^2}\sum_{m=0}^{N-1}\sum_{n=0}^{N-1} (x_l(m, n) - \mu_l)^3},
\end{aligned}
\tag{13.5}
$$

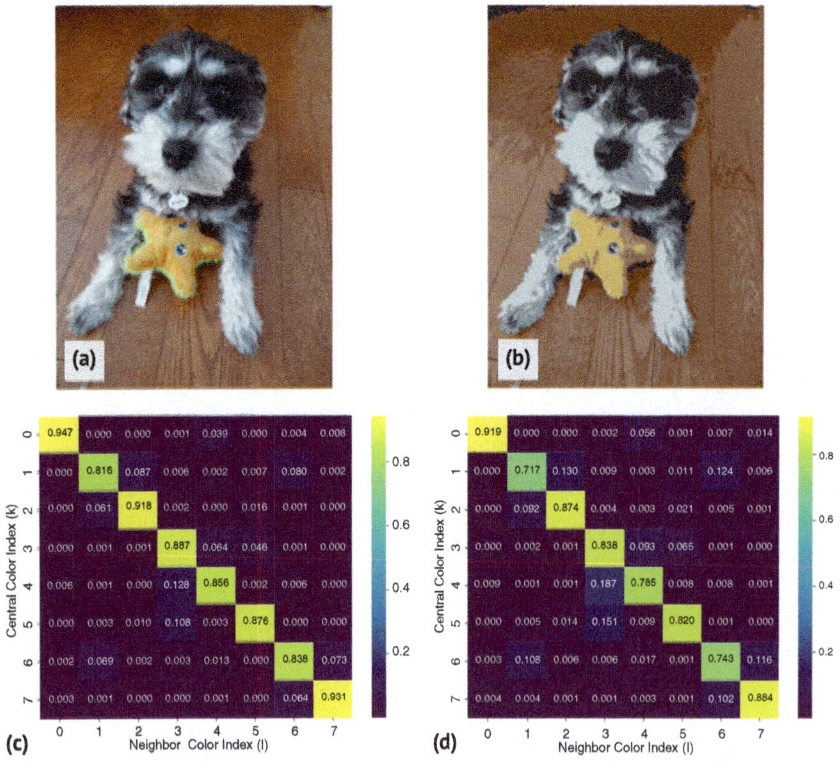

Figure 13.2 (a) Original Finn image, (b) vector-quantized Finn to 8 colors using K-means algorithm, (c) correlogram matrix for $d = 1$, and (d) correlogram matrix for $d = 2$.

where μ_l, σ_l, and ξ_l are the mean, standard deviation, and skewness, respectively. Although these expressions are given for the entire image, they can also be applied to segmented regions. The resulting feature vector across all three color channels is therefore nine-dimensional. Compared to color correlogram, color moments have several advantages: they are simpler and more efficient to compute, and inherently scale- and rotation-invariant. Nevertheless, unlike correlograms, they don't convey any spatial dependency information.

13.3 Energy Features

In Section 6.4.3, an efficient algorithm was presented for computing the eigenvalues and corresponding eigenvectors of the sample covariance matrix for a facial image dataset. As stated, the principal eigenvalues represent the energy (or variance) contributions of the image data when projected along the corresponding principal eigenvectors. The eigenvectors, when reshaped into image forms, give the eigen-images, which capture certain low- to high-level characteristics of the images in the database. For example, in the facial image database of Section 6.4.4, eigenfaces associated with large eigenvalues tend to capture certain low-level facial characteristics such as head, lips, nose, eyes, and mouth. Eigenfaces associated with smaller eigenvalues tend to represent finer, high-level characteristics such as facial hair, glasses, or other subject-specific attributes.

Given a new image $y(m, n)$, it can be represented by a linear combination of only a few principal eigenimages weighted by the corresponding principal components (PC). That is, if we denote the eigen-images by $\mathbf{K}(k, l)$ and PCs of the image matrix $\mathbf{y}$ with elements $y(m, n)$, $m, n \in [0, N-1]$ by $Y(k, l)$, then the best reconstruction, in the minimum mean squared error (MMSE) sense, of this image matrix using a subset of $P \times Q$ eigen-images and PCs is given by:

$$\hat{\mathbf{y}} = \sum_{k=1}^{P} \sum_{l=1}^{Q} Y(k, l)\mathbf{K}(k, l) + \boldsymbol{\mu} \tag{13.6}$$

where $\boldsymbol{\mu}$ is the sample mean image computed from the training dataset, i.e. $\boldsymbol{\mu} = \frac{1}{M} \sum_{i=1}^{M} \mathbf{x}_i$ with $\mathbf{x}_i$, $i \in [1, M]$ being the ith image matrix in the training dataset containing M images. If all the eigen-images and PCs are used, then the reconstruction will be exact, i.e. $\hat{\mathbf{y}} = \mathbf{y}$. The PCs, Y(k,l), of the image $\mathbf{y}$ are obtained by projecting the mean-subtracted image onto the eigen-images, $\mathbf{K}(k, l)$, using:

$$Y(k, l) = <\mathbf{y} - \boldsymbol{\mu}, \mathbf{K}(k, l)> /\gamma_i, \quad i = (k-1)Q + l, \quad k \in [1, P], \ l \in [1, Q] \tag{13.7}$$

where γ_i are the eigenvalues of the sample covariance matrix and as before $< \mathbf{A}, \mathbf{B} >$ stands for the inner product operation of matrices $\mathbf{A}$ and $\mathbf{B}$. Note that equation $Y(k) = \underline{\xi}_k^H \underline{\bar{y}}/\gamma_k$, in the paragraph after (6.45) is the vector-arranged version of (13.7).

From (13.6), it is evident that each PC, $Y(k, l)$, determines the contribution of the corresponding eigen-image $\mathbf{K}(k, l)$ to the representation of image $\mathbf{y}$. The importance of each PC in this representation is determined by its variance (or energy) computed based on the training dataset. Although the eigenvalues and eigenvectors (or eigen-images) are obtained based on the entire training dataset, the PCs are image-dependent, and hence can be viewed as *energy-based features* of the image. These features, when arranged into a *feature vector,* can be used for discrimination or other tasks.

Several important remarks are in order. First, if all training images are subjected to the same rotation and translation, the eigenvectors of the covariance matrix of the transformed data will be rotated so that the projection of the data on the transformed coordinates and their variances remain unchanged. In other words, PCs and their variances are invariant under rotation and translation operations. These invariance properties will be shown in the next subsection. Second, the PCA-based method presented in this section is inherently a *global* feature extraction process that is applied to the entire image and not to specific segmented regions. As a result, it is not suited for capturing local statistical characteristics from different segmented regions within an image. In such cases, one may apply *local* PCA to image blocks taken from various segmented regions or alternatively use statistical, textural, or shape-dependent feature extraction methods.

13.3.1 Invariance Properties

To show the invariance properties under rotation and translation, let $\underline{x}$ be a N-D random vector with zero mean and covariance matrix $\mathbf{C}_x$. We assume that all the realizations of this random vector (e.g. in the training data) are rotated and translated to yield new random vectors $\underline{y} = \mathbf{R}_\theta \underline{x} + \underline{\upsilon}$ where $\underline{\upsilon}$ is a deterministic translation vector and $\mathbf{R}_\theta$ is an orthogonal rotation matrix by angle θ. The covariance matrix of the transformed data is:

$$\mathbf{C}_y = E[(\underline{y} - \underline{\upsilon})(\underline{y} - \underline{\upsilon})^T] = \mathbf{R}_\theta E[\underline{x}\,\underline{x}^T]\mathbf{R}_\theta^T = \mathbf{R}_\theta \mathbf{C}_x \mathbf{R}_\theta^T.$$

As shown in Section 6.4, the PCA transformation $\mathbf{\Psi}_x^H$ diagonalizes the covariance matrix $\mathbf{C}_x$ as, $\mathbf{\Psi}_x\mathbf{\Lambda}\mathbf{\Psi}_x^H = \mathbf{C}_x$ where $\mathbf{\Psi}_x$ contains all orthonormal eigenvectors as its columns and $\mathbf{\Lambda}$ is the diagonal matrix that contains the corresponding eigenvalues $\mathbf{\Lambda} = \text{Diag}[\lambda_0, \cdots, \lambda_{N-1}]$ with $\lambda_0 > \lambda_1 \cdots \lambda_{N-1}$. Using this result in the expression for $\mathbf{C}_y$ we get:

$$\mathbf{C}_y = \mathbf{R}_\theta \mathbf{\Psi}_x \mathbf{\Lambda} \mathbf{\Psi}_x^H \mathbf{R}_\theta^T = \mathbf{\Psi}_y \mathbf{\Lambda} \mathbf{\Psi}_y^H \tag{13.8}$$

or $\mathbf{\Psi}_y = \mathbf{R}_\theta \mathbf{\Psi}_x$. Thus, each eigenvector is rotated by the same transformation, i.e. $\underline{\xi}_y(k) = \mathbf{R}_\theta\, \underline{\xi}_x(k)$. The eigenvalues or the variances of PCs λ_i, however, remain invariant under rotation and translation operations.

We now examine whether the PCs of any realization remain invariant to these transformations. For a particular realization $\underline{x}_m$, the PC vector is $\underline{X}_m = \mathbf{\Psi}_x^H \underline{x}_m$ and similarly for the rotated and translated version $\underline{Y}_m = \mathbf{\Psi}_y^H(\underline{y}_m - \underline{v})$. Using the relation between the two transformation matrices, $\mathbf{\Psi}_y = \mathbf{R}_\theta \mathbf{\Psi}_x$, and the orthogonality of rotation matrix, i.e. $\mathbf{R}_\theta^T \mathbf{R}_\theta = \mathbf{I}$ or $\mathbf{R}_\theta^T = \mathbf{R}_\theta^{-1}$, we get:

$$\underline{Y}_m = \mathbf{\Psi}_y^H(\underline{y}_m - \underline{v}) = \mathbf{\Psi}_y^H \mathbf{R}_\theta \underline{x}_m = \mathbf{\Psi}_x^H \mathbf{R}_\theta^T \mathbf{R}_\theta \underline{x}_m = \mathbf{\Psi}_x^H \underline{x}_m = \underline{X}_m \tag{13.9}$$

i.e. the PCs also remain invariant to these transformations. It is important to mention that these invariance properties are valid for the entire data population and not for an individual sample. That is, if the training data samples are not rotated, but for a new test sample PCs are computed before and after rotation operation, they will indeed be different.

13.4 Spectral Features

There are many different ways spectral feature extraction methods can be utilized, depending on the specific application, some of which are described in the following subsections.

13.4.1 Multi-Channel Spectral Features

Multispectral or hyperspectral image data consist of pixels at the same spatial coordinates across many spectral bands. These spectral bands typically correspond to visible, near-infrared, and infrared sensor channels. For example, GOES-16 satellite has 16 spectral bands, 2 visible channels, 4 near-infrared channels, and 10 infrared channels, designed to capture important characteristics of clouds, oceans, and land areas of certain geographical regions on Earth. Data from these channels are used to generate meteorological and atmospheric products, such as precipitation estimation [10]. In such cases, data vectors are typically arranged in spectral bands for every pixel. That is, if $x_i(m, n)$ represents pixel (m, n) at spectral channel $i \in [1, L]$ then the spectral data vector is:

$$\underline{x}(m,n) = \begin{bmatrix} x_1(m,n) \\ x_2(m,n) \\ \vdots \\ x_L(m,n) \end{bmatrix}$$

where each image $x_i(m, n)$ is assumed to be of the same size $N \times N$.

These spectral vectors, or their mapped versions, are then applied to classifiers for discrimination purposes [10]. One effective method to map the multichannel spectral data is the PCA, which finds the principal spectral components that contribute the most to the total energy of the data. It is important to note that PCA is applied differently here than in the previous section. In this case, the data vector at each pixel is formed across L spectral bands, rather than across the entire image. Consequently, the PCs, $X_i(m, n)$, of spectral band $i \in [1, L]$, form an image capturing certain spectral-based features of the multichannel data. In contrast, the previous application resulted in scalar PCs that convey the contributions of different eigen-images toward representing the original image. Moreover, the displayable images are the eigen-images (e.g. eigenfaces), as opposed to this case where displayable images are the PC images. Finally, unlike the global-based method in the previous section, it is possible to apply the method in this section to segmented regions of the multichannel data.

To apply PCA method to multichannel data, we compute the sample mean vector and covariance matrix using:

$$\underline{\mu}_x = \frac{1}{N^2}\sum_{m=0}^{N-1}\sum_{n=0}^{N-1}\underline{x}(m,n),$$

$$\mathbf{C}_x = \frac{1}{N^2}\sum_{m=0}^{N-1}\sum_{n=0}^{N-1}(\underline{x}(m,n)-\underline{\mu}_x)(\underline{x}(m,n)-\underline{\mu}_x)^T$$

where the covariance matrix $\mathbf{C}_x$ can be diagonalized as $\boldsymbol{\Psi}^H\mathbf{C}_x\boldsymbol{\Psi} = \boldsymbol{\Lambda}$ and the transformation matrix $\boldsymbol{\Psi}^H$ is used to map each spectral data vector $\underline{x}(m,n)$ to its PC vector using $\underline{X}(m,n) = \boldsymbol{\Psi}^H(\underline{x}(m,n) - \underline{\mu}_x)$. Each $\underline{X}(m,n) \in \mathbb{R}^{L\times 1}$ contains elements $X_i(m,n)$, $i \in [1,L]$ that are the PCs of $\underline{x}(m,n)$. The importance of each PC image $X_i(m,n)$ is determined by its corresponding variance, λ_i, or the ith eigenvalue of the covariance matrix $\mathbf{C}_x$. In practice, only a subset $K < L$ of PC images $X_i(m,n)$, $i \in [1,K]$, which capture important information of the original data, is used. If needed, we can reconstruct the original multichannel data using the inverse mapping $\hat{\underline{x}}(m,n) = \boldsymbol{\Psi}_{red}\underline{X}_{red}(m,n) + \underline{\mu}_x$, where $\underline{X}_{\text{red}}(m,n) = [X_1(m,n) \cdots X_K(m,n)]^T$ contains the first K PCs and $\boldsymbol{\Psi}_{\text{red}}$ consists of the first K columns of matrix $\boldsymbol{\Psi}$. Clearly, when $K = L$, this reconstruction is exact, i.e. we have $\hat{\underline{x}}(m,n) = \underline{x}(m,n)$.

To show how this method can be applied to multispectral data, six (out of 16) infrared channels of GOES-16 data are chosen in this experiment. The selected channels correspond to channel 8 (upper-level water vapor), channel 9 (middle-level water vapor), channel 10 (lower-level water vapor), channel 11 (cloud-top phase), channel 14 (long-wave window), and channel 15 (dirty long-wave window). Note that channel 15 is referred to as "dirty" as it is more sensitive to water vapor than the other infrared window channels. These channels are typically used [10] for precipitation estimation from satellite imagery. The other near-infrared and visible channels are less sensitive to water vapor and precipitations, and hence are not used here. Figure 13.3 shows these six IR images for July 18, 2024 at 15:00 UTC each of size 1500 × 3000 pixels that cover mainly the south-eastern part of the United States.

The data vectors, arranged across these six spectral bands at every pixel, are used to generate the sample covariance matrix using all pixels. This matrix is then diagonalized to produce the PCA transformation matrix $\boldsymbol{\Psi}^H$. Based on the observation of the eigenvalues of the sample covariance matrix, we used only four (out of six) PC images which are obtained by pre-multiplying the multispectral data vector at each pixel by the reduced PCA transformation matrix,

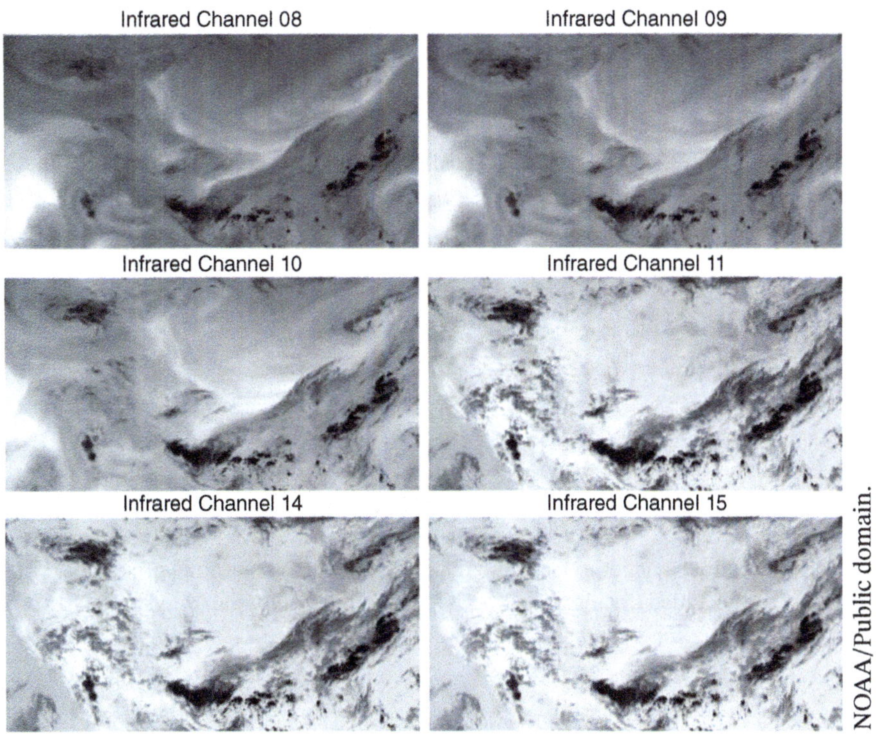

Figure 13.3 Six GOES-16 IR images commonly used for precipitation estimation.

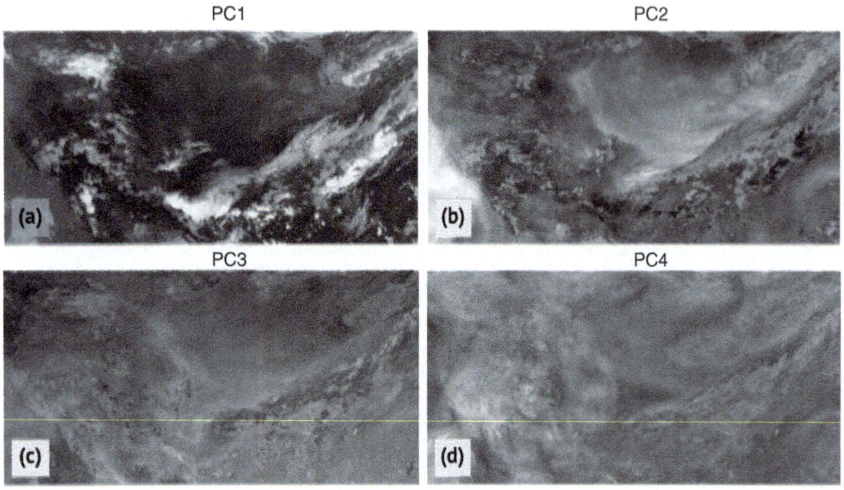

Figure 13.4 (a–d) The first, second, third, and fourth PC images corresponding to the six spectral images in Figure 13.3.

i.e. $\underline{X}_{\text{red}}(m, n) = \mathbf{\Psi}^H_{\text{red}}(\underline{x}(m, n) - \underline{\mu}_x)$, where $\underline{x}(m, n)$ is a six-dimensional data vector. Figure 13.4 shows these four PC images that capture most of the relevant information of the original multispectral data and, hence, could be used for precipitation estimation and retrieval purposes. Notably, the ratio of the first to the second eigenvalues was around 184, which indicates that the first PC image captures most of the information of the six spectral channels.

13.4.2 Single-Channel Spectral Features

The goal here is to extract spectral-based features from a single image, acquired by an imaging sensor, or from its segmented regions. There are many candidate methods that can be applied to this problem, many of which were covered in Chapters 6 and 7, including 2-D DFT, DCT, and DWT. In particular, 2-D DWT has been extensively used [11–14] in many subband-based feature extraction applications. The multi-resolution property of the DWT can be utilized to decompose the image into subimages that capture spatial and scale (or frequency) characteristics. More specifically, the added details subimages resulted from cascaded combinations of several low-pass and high-pass half-band filters (see Chapter 7) contain the edge and texture information, making them well-suited as features for image classification and retrieval.

Traditional approaches assume that the energy distribution in the subband domain is a good descriptor of texture characteristics. Under this assumption, the l_1 and l_2 norms of the 2-D DWT subband coefficients corresponding to the added details are calculated and used as *texture-based* features. Specifically, given the ith subband image, $X_i(k, l)$, $k \in [1, K_i]$, $l \in [1, L_i]$, these features are given by:

$$\begin{aligned} f_{i,1} &= \frac{1}{K_i L_i} \sum_{k=1}^{K_i} \sum_{l=1}^{L_i} |X_i(k, l)|, \\ f_{i,2} &= \sqrt{\frac{1}{K_i L_i} \sum_{k=1}^{K_i} \sum_{l=1}^{L_i} X_i^2(k, l)}, \quad i \in [1, D] \end{aligned} \tag{13.10}$$

where D is the number of added details used. Higher order statistical moments have also been suggested to augment these features. In Section 13.5.1, we introduce an alternative statistical texture feature extraction method that can be applied to these subimages.

A natural extension of the energy-based feature extraction method is to model texture using the marginal probability density functions of wavelet subband coefficients for the added details. In [11, 12], a generalized Gaussian density (GGD) was proposed to represent these DWT coefficients in particular for image retrieval applications. The GGD is given by:

$$p(X_i; \alpha_i, \beta_i) = \frac{\beta_i}{2\alpha_i \Gamma(1/\beta_i)} e^{-\left(\frac{|X_i|}{\alpha_i}\right)^{\beta_i}} \tag{13.11}$$

where $\Gamma(.)$ is the Gamma function given by $\Gamma(y) = \int_0^\infty e^{-t} t^{y-1} dt, \quad y > 0$, α_i is the scale parameter that controls the width of the PDF peak (e.g. standard deviation for Gaussian) for subband i, and β_i is the shape parameter that is inversely proportional to the decreasing rate of the peak. As special cases, when $\beta_i = 2$, GGD becomes Gaussian while for $\beta_i = 1$ we get a Laplacian PDF.

Given the coefficients at each subband, the corresponding parameters α_i and β_i of GGD can be estimated using the Maximum Likelihood (ML) method [11]. Assuming that the data in each subimage are i.i.d., the likelihood function for the ML estimation is

$$\mathcal{L}(\alpha_i, \beta_i) = \ln \prod_k \prod_l p(X_i(k,l); \alpha_i, \beta_i),$$

where $\mathbf{X}_i$ represents the ith subimage matrix and $X_i(k,l)$ is its $(k,l)^{\text{th}}$ element. It was shown [11] that the ML estimate for α_i is dependent on β_i and is given by:

$$\hat{\alpha}_i = \left(\frac{\beta_i}{K_i L_i} \sum_{k=1}^{K_i} \sum_{l=1}^{L_i} |X_i(k,l)|^{\beta_i} \right)^{\frac{1}{\beta_i}} \tag{13.12}$$

To find the estimate of β_i, an iterative root finding procedure was proposed in [11] using the Newton–Raphson method, which converges rapidly. The estimated parameters for each added detail subimage are then used as features for retrieval or classification. Experimental results in [11] on a database of 640 natural texture images of sizes 128×128 revealed that β_i typically lies in the range $[0.5, \ 1]$. In these experiments, they used 3-level 2-D DWT using db4 orthogonal wavelet resulting in $D = 9$ added detail subimages. Comparison with the results of energy-based features in (13.10), indicated substantially better image retrieval performance of the GGD-based features.

13.5 Textural Features

Textural features are another class of image attributes that humans use to perceive and interpret visual scenes. Texture is an innate characteristic of images acquired from many different surface types, facilitating their discrimination. As a result, texture-based features are widely used in many image analysis applications including biomedical, remote sensing, meteorology, and military. Texture types may be qualitatively classified in terms of being fine, coarse, smooth, rippled, irregular, or lineated. Unlike pixel-wise image features, e.g. color or spectral features, textural features depend on the local spatial dependencies among groups of pixels. Methods used for texture feature extraction can be categorized into two general categories: statistical and transformed domain.

Among popular statistical methods is the Gray Level Co-occurrence Matrix (GLCM) [15, 16]. It computes frequency of co-occurrence of pairs of pixels' gray levels separated by a certain distance and angle to produce the co-occurrence matrix, from which a set of 14 textural features are derived. An alternative statistical-based feature extraction approach is to represent texture images using coefficients of 2-D linear image models. Assuming local wide-sense stationarity (WSS), one can fit a 2-D AR model with preferably noncausal ROS to the image data within the segmented region. The model coefficients computed by solving the corresponding Yule-Walker equation in Section 9.2.3, characterize local spatial dependencies among a group of pixels. Among the commonly used transformed domain methods are the 2-D wavelet transform [11–14] and the 2-D Gabor filter [17–20] which will be covered in Section 13.5.2.

13.5.1 Gray Level Co-Occurrence Matrix

The GLCM is based on the conjecture that the human visual system uses second-order statistics for texture discrimination in images. As stated earlier, co-occurrences matrices count how often pairs of pixel gray levels that are separated by a certain distance and lie along a certain direction occur in a textured digital image. The features are then derived from these co-occurrence matrices.

Let us assume that we are given a textured image $x(m,n)$ of size $N \times N$ with L gray levels. Also, let us define matrix $\mathbf{G}$ with elements $g(i,j)$ that represent frequency of co-occurrence of pixel intensity levels $i, j \in [0, L-1]$ at a certain distance d and an angle θ. For example, $d = 1$ corresponds to immediate neighboring pixels while $d = 2$ means pixels that are separated by one pixel. Also, typical choices for θ are $0°$, $45°$, $90°$, and $135°$ along which the pairs of pixels lie. These angles are measured with respect to the horizontal axis. For $d = 1$ and these values of θ we get four co-occurrence matrices each of

Table 13.1 A 4×4 image with $L = 4$ gray levels.

0	0	1	1
0	0	1	1
0	2	2	2
2	2	3	3

size $L \times L$. The $(i, j)^{\text{th}}$ elements of these matrices for these particular choices of θ and $i, j \in [0, L-1]$ are defined as:

$$\begin{aligned} g_{0^\circ}(i,j) &= \left|\left\{x(m,n) = i,\ x(m, n \pm 1) = j\right\}\right| \\ g_{45^\circ}(i,j) &= \left|\left\{x(m,n) = i,\ \left(x(m-1, n+1) = j \vee x(m+1, n-1) = j\right)\right\}\right| \\ g_{90^\circ}(i,j) &= \left|\left\{x(m,n) = i,\ x(m \pm 1, n) = j\right\}\right| \\ g_{135^\circ}(i,j) &= \left|\left\{x(m,n) = i,\ \left(x(m+1, n+1) = j \vee x(m-1, n-1) = j\right)\right\}\right| \end{aligned} \tag{13.13}$$

where the symbol $|S|$ stands for the cardinality of the set S and $\vee$ means OR operation.

As an example, let us consider the image in Table 13.1 with $L = 4$ gray levels [15]. The corresponding four co-occurrence matrices for $d = 1$ are:

$$\mathbf{G}_{0^\circ} = \begin{bmatrix} 4 & 2 & 1 & 0 \\ 2 & 4 & 0 & 0 \\ 1 & 0 & 6 & 1 \\ 0 & 0 & 1 & 2 \end{bmatrix} \quad \mathbf{G}_{45^\circ} = \begin{bmatrix} 4 & 1 & 0 & 0 \\ 1 & 2 & 2 & 1 \\ 0 & 2 & 4 & 1 \\ 0 & 0 & 1 & 0 \end{bmatrix} \quad \mathbf{G}_{90^\circ} = \begin{bmatrix} 6 & 0 & 2 & 0 \\ 0 & 4 & 2 & 0 \\ 2 & 2 & 2 & 2 \\ 0 & 0 & 2 & 0 \end{bmatrix} \quad \mathbf{G}_{135^\circ} = \begin{bmatrix} 2 & 1 & 3 & 0 \\ 1 & 2 & 1 & 0 \\ 3 & 1 & 0 & 2 \\ 0 & 0 & 2 & 0 \end{bmatrix}$$

For example, the element $g_{45^\circ}(2,3) = 1$ in matrix $\mathbf{G}_{45^\circ}$ is the number of times gray levels 2 and 3 occur diagonally adjacent at 45° angle.

Once these matrices are generated, the elements are normalized by the sum of all elements $\sum_i \sum_j g(i,j)$ to convert the frequency of co-occurrences to probabilities $p_{ij} = g(i,j) / \sum_i \sum_j g(i,j)$ from which many features are computed. Haralick et al. [15] proposed 14 features that measure statistical-based characteristics of textured images. In many texture discrimination applications, only a subset of these features have been found to be adequate. These are briefly described next.

1. **Contrast:** This feature is given by:

$$f_1 = \sum_{i=0}^{L-1} \sum_{j=0}^{L-1} (i-j)^2 p_{ij}. \tag{13.14}$$

   This feature measures local spatial variations present in an image. Highly textured images produce higher values of f_1, while homogeneous images or regions lead to a lower contrast measure.

2. **Correlation:** This feature is defined by:

$$f_2 = \sum_{i=0}^{L-1} \sum_{j=0}^{L-1} \frac{(i-\mu_1)(j-\mu_2)p_{ij}}{\sigma_1 \sigma_2} \tag{13.15}$$

   where $\mu_1, \mu_2, \sigma_1,$ and σ_2 are the mean values and standard deviations of the marginal distributions, $p_i = \sum_{j=0}^{L-1} p_{ij}$ and $p_j = \sum_{i=0}^{L-1} p_{ij}$, respectively. That is, we have:

$$\mu_1 = \sum_{i=0}^{L-1} i p_i, \quad \mu_2 = \sum_{j=0}^{L-1} j p_j$$

$$\sigma_1^2 = \sum_{i=0}^{L-1} (i - \mu_1)^2 p_i, \quad \sigma_2^2 = \sum_{j=0}^{L-1} (j - \mu_2)^2 p_j. \tag{13.16}$$

The correlation feature, f_2, is a measure of the linear dependence of gray levels of neighboring pixels in an image. Large values of this feature indicate a more structured texture.

3. **Homogeneity:** This feature is given by:

$$f_3 = \sum_{i=0}^{L-1}\sum_{j=0}^{L-1} \frac{p_{ij}}{1 + (i - j)^2} \tag{13.17}$$

represents local similarity in the image. That is, for locally uniform or smooth-textured images where neighboring pixels have similar intensities, this measure takes large values.

4. **Angular Second Moment:** This feature is given by:

$$f_4 = \sum_{i=0}^{L-1}\sum_{j=0}^{L-1} p_{ij}^2. \tag{13.18}$$

It measures energy or uniformity of the distribution. It takes high values when structured or repetitive textures are encountered. For a constant image (maximum homogeneity), this measure becomes 1.

5. **Entropy:** This feature is defined by:

$$f_5 = -\sum_{i=0}^{L-1}\sum_{j=0}^{L-1} p_{ij} \log_2 p_{ij}. \tag{13.19}$$

This feature is a measure of local randomness or disorder in the image. That is, for a more complex texture, it takes high values, whereas when there is local redundancy (homogeneity), it takes small values. It takes its maximum when $p_{ij} = \frac{1}{L^2}$, i.e. when all possible gray-level co-occurrences are equally likely.

6. **Dissimilarity:** Measures the average difference in intensity between neighboring pixels and is given by:

$$f_6 = \sum_{i=0}^{L-1}\sum_{j=0}^{L-1} p_{ij}|i - j|. \tag{13.20}$$

High values of this feature indicate greater heterogeneity or "roughness" in the texture, while low values indicate smoother, more homogeneous textures.

The major drawback of the GLCM method is that the size of the co-occurrence matrices becomes large for large number of gray levels, leading to increased computational and storage requirements. Reducing the number of gray levels using re-quantization can partly ease this problem. However, this method is still more time-consuming than most of the other textural feature extraction methods. Additionally, the choice of distance parameter d is image-dependent. While $d = 1$ is commonly used, for some textured images, one may need to increase d, which in turn significantly increases the complexity of this process.

To examine the effectiveness of the GLCM method for textural image discrimination, we used the three images in Figure 13.5a–c. Each image is composed of 6 subimages where each subimage of size 300×400 pixels is labeled 1 to 6 from left to right and top to bottom. These images illustrate a variety of fabric textures with different structures. To capture localized textural features, the GLCM was applied to $k \times k$ size windows for various values of $k \in [8, 16, 21]$. Windows are swept in both dimensions one pixel at a time. We also chose a combination of distances $d \in [1, 2, 4, 6, 8]$, angles $\theta_k = \frac{k\pi}{8}$, $k \in [0, 7]$ and five GLCM features, namely contrast, correlation, energy (or angular second moment), homogeneity, and dissimilarity. These combinations lead to a 600-dimensional textural feature vector for each window.

Using this window-based GLCM method, image segmentation (or discrimination) can be carried out by assigning each pixel in the image to a specific class (or region) while taking into account the local spatial properties of the surrounding pixels within the sliding window. This can be done using either supervised classification or unsupervised clustering methods, e.g. K-means covered in Section 12.1.3. In this experiment, we used multinomial or categorical logic regression [17] for

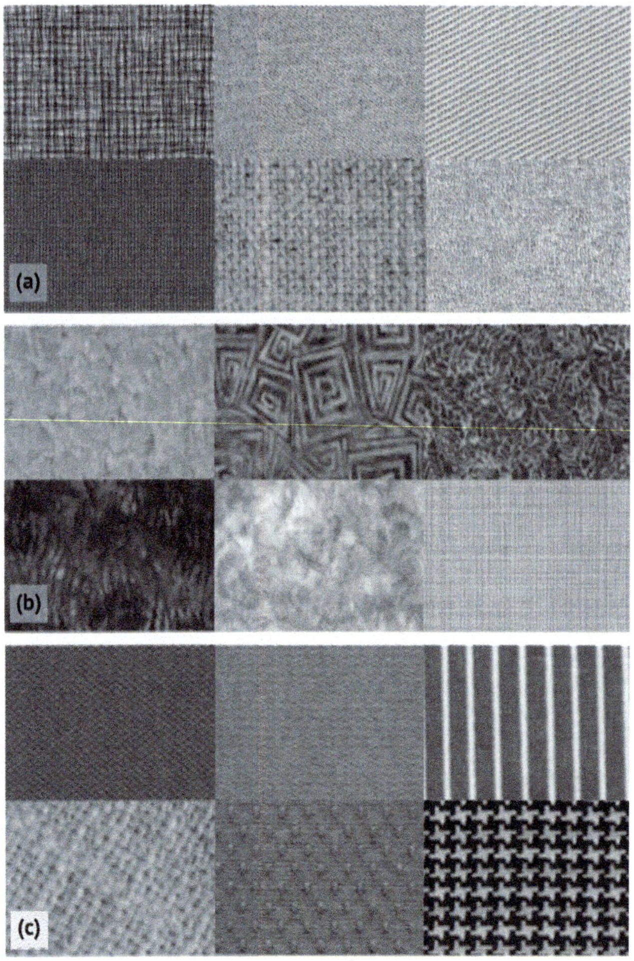

Figure 13.5 (a–c) Fabrics 1–3 and their subimages. *Source*: AITEX Fabric Image Database(www.kaggle.com).

multi-class pattern classification with a sparsity constraint, as discussed in Section 17.3.2. The sparsity constraint added to the loss function is based on the l_1-norm of the weight vectors, which encourages the model to ignore GLCM features that don't contribute to texture discrimination. The regularization parameter was chosen using cross-validation while monitoring the average number of nonzero weights. It must be pointed out that increasing the regularization parameter for the sparsity penalty, causes more weights to be driven to zero.

The Python scikit-learn library was used to implement the categorical logistic regression model [17]. The correct classification rates for each texture subimage in Fabrics 1–3 are presented in Table 13.2. As can be seen, the classifier performed very well on Fabric 1 while requiring only 5 (out of 600) nonzero weights or GLCM features (see the last row of Table 13.2). The largest number of selected weights (or features) was 84 for Fabric 2, indicating increased complexity in its texture structure. The reported correct classification rates in Table 13.2 are computed for the entire region using the best model selected via cross-validation.

Categorical logistic regression outputs posterior probability of class membership for each pixel. The color overlays of these probability maps on the original images are shown in Figure 13.6a–c, where each subimage is assigned a specific color depending on the class. As can be observed, most subimages in some of the fabrics are uniformly classified as belonging to one region. These color overlay maps also reveal misclassification regions particularly near boundaries. This limitation is more noticeable for Fabric 3 mainly due to the regular structures of their texture and lack of support around the boundaries. Further investigation of the selected features revealed that a window size of 32 was chosen consistently for all subimages and texture types irrespective of their specific characteristics. Although, the GLCM-based approach achieved high accuracy of texture representation using a relatively small subset of features, it remains computationally intensive.

Table 13.2 GLCM categorical logistic regression accuracy and weight count.

	Fabric 1	Fabric 2	Fabric 3
Subimage 1 accuracy	99.93%	56.82%	82.00%
Subimage 2 accuracy	98.01%	75.62%	83.74%
Subimage 3 accuracy	97.12%	87.37%	89.82%
Subimage 4 accuracy	96.04%	77.62%	91.37%
Subimage 5 accuracy	99.05%	75.06%	95.22%
Subimage 6 accuracy	96.83%	95.79%	99.98%
All subimages accuracy	97.83%	78.05%	90.36%
Weight count (sparsity)	5	84	20

Figure 13.6 (a–c) Categorical logistic regression performance color map for Fabrics 1–3 using GLCM features, respectively.

13.5.2 Gabor Filter

In Chapter 7, we introduced the Gabor transform to overcome some of the time-frequency localization limitations of the standard Fourier transform for 1-D signals. The Gabor transform basis functions are the translated and modulated versions of a Gaussian window function in the time domain. Alternatively, in the frequency domain, the 1-D Gabor transform can be interpreted as a bank of constant-bandwidth filters with center frequencies determined by the modulating frequency. In a similar manner, we can define 2-D Gabor functions [18–20, 22] that represent localized information in both spatial and spatial frequency domains. These 2-D Gabor functions have extensively been used in image texture feature extraction and analysis among many other important image processing applications including image retrieval, segmentation, registration, and motion tracking. In practice, an image is filtered using a bank of 2-D Gabor filters that are the translated and rotated versions of the 2-D Gabor prototype function. Various choices for the orientation angle and spatial frequency result in a bank of filters, where each filter is tuned to extract certain types of localized image feature. The Gaussian envelope defines the elliptical region of support and its elongation. A feature vector can then be constructed from this set of localized responses and used for discrimination purposes.

The 2-D Gabor elementary function (filter PSF), $g(u, v)$ is given by:

$$g(u,v) = \frac{1}{2\pi\sigma_u\sigma_v} e^{-\left(\frac{u^2}{\sigma_u^2}+\frac{v^2}{\sigma_v^2}\right)} e^{2\pi j\xi u+\varphi} \tag{13.21}$$

where σ_u^2 and σ_v^2 are the variances of the 2-D Gaussian envelope along the u and v directions, respectively, ξ is the modulating spatial frequency, and φ is the phase offset. Studies have indicated [20] that the real part of 2-D Gabor functions is a good representative of the functionality of the visual cortex receptive field in various species of mammalian. Therefore, the real part of the prototype function in (13.21) can be used to generate a complete 2-D Gabor filter dictionary by a spatial rotation angle θ, i.e.

$$g_r(u',v') = \frac{1}{2\pi\sigma_u\sigma_v} e^{-\left(\frac{u'^2}{\sigma_u^2}+\frac{v'^2}{\sigma_v^2}\right)} \cos\left(\frac{2\pi}{\lambda}u' + \varphi\right) \tag{13.22}$$

where $u' = u\cos\theta + v\sin\theta$ and $v' = -u\sin\theta + v\cos\theta$ are the rotated spatial variables, and $\lambda = \frac{1}{\xi}$ is the wavelength. Varying the orientation θ in the Gabor filter allows capturing directional textures, while σ_u and σ_v control the support of the Gabor basis functions and their ellipticity or elongation. Now, given an image $x(u, v)$, the 2-D Gabor-filtered image is:

$$y_{\theta,\lambda,\varphi}(u,v) = \iint x(\alpha,\beta) g_r(u-\alpha, v-\beta) d\alpha\, d\beta \tag{13.23}$$

In [19], two choices for the phase offset were suggested, $\varphi = 0$ and $\varphi = -\frac{\pi}{2}$ leading to symmetric and antisymmetric filter banks, respectively. Additionally, they used eight orientations $\theta = \frac{k\pi}{8}$, $k \in [0, 7]$ and three spatial frequencies ξ =23, 31, and 47 cycles per image, hence generating a total of 24 Gabor filters for each filter bank. That is, for each point in the image $y_{\theta,\lambda,\varphi}(u, v)$ a 24-dimensional feature vector is generated for each phase offset. These two feature vectors can then be combined into one vector in which each element is $e_{\theta,\lambda}(u,v) = \sqrt{y^2_{\theta,\lambda,\varphi=0}(u,v) + y^2_{\theta,\lambda,\varphi=-\frac{\pi}{2}}(u,v)}$. This 24-dimensional vector is closely related to the 2-D spectrogram of the 2-D Gabor-transformed image. Figure 13.7 shows all 40 2-D Gabor basis images generated using (13.22) for orientations $\theta = \frac{k\pi}{8}$, $k \in [0, 7]$ and spatial frequencies $\xi = \frac{0.25}{2^{\frac{l}{2}}}$, $l \in [0, 4]$. The combinations of these basis images can represent textures of various types and fineness/coarseness. There are many other ways to produce texture-based features from the output of the 2-D Gabor filter in (13.23). In [22], the mean and standard deviation of the output image of each filter are computed for the two filter banks. These statistical features provide a low-dimensional and computationally efficient representation for image classification and retrieval.

For digital images, a simplified 2-D Gabor filter pair (real and imaginary parts) is given by:

$$\begin{aligned} g_r[m,n] &= A_r e^{-\frac{(m^2+n^2)}{2\sigma^2}} \cos(2\pi\xi(m\cos\theta + n\sin\theta)) \\ g_i[m,n] &= A_i e^{-\frac{(m^2+n^2)}{2\sigma^2}} \sin(2\pi\xi(m\cos\theta + n\sin\theta)) \end{aligned} \tag{13.24}$$

where A_r and A_i are normalization factors that depend on the choices of σ and θ.

It must be mentioned that Gabor filters are closely related to Gabor wavelets [23], which are generated through dilation and rotation of the mother Gabor wavelet $g(u, v)$ in (13.21). Thus, the filter dictionary for Gabor wavelets can be obtained using:

$$g_{k,l}(u,v) = a^{-k} g(u',v'), \quad a > 1 \quad k \in [1, K], l \in [0, L-1] \tag{13.25}$$

Figure 13.7 2-D Gabor basis images for eight orientations (left-to-right) and five spatial frequencies (top-to-bottom).

where $u' = a^{-k}(u \cos\theta_l + v \sin\theta_l)$ and $v' = a^{-k}(-u \sin\theta_l + v \cos\theta_l)$ and $\theta_l = \frac{l\pi}{L}$, $l \in [0, L-1]$ with L being the number of orientations. Then, Gabor wavelet filter output for an input image $x(u, v)$ is:

$$y_{k,l}(u, v) = \iint x(\alpha, \beta) g^*_{k,l}(u - \alpha, v - \beta) d\alpha \, d\beta. \tag{13.26}$$

Assuming that local texture regions are spatially homogeneous, mean and standard deviation of the magnitude of the Gabor wavelet filter coefficients can be used as attributes to represent the segmented region [23]. These features are:

$$\begin{aligned} \mu_{k,l} &= \iint |y_{k,l}(u, v)| du \, dv, \\ \sigma_{k,l} &= \sqrt{\iint (|y_{k,l}(u, v)| - \mu_{k,l})^2 du \, dv} \end{aligned} \tag{13.27}$$

The feature vector constructed using these components for specific choices of L and K is then used for texture-based image classification or retrieval.

To compare the effectiveness of 2-D Gabor-based features with GLCM features for texture-based discrimination, we applied the 2-D Gabor filter bank to the same fabric images in Figure 13.5a–c. The hyper-parameters of the Gabor filter that are varied include: orientation angle $\theta_k = \frac{\pi k}{8}$, $k \in [0, 7]$; phase offset of $\varphi = 0$; variance $\sigma_u \in \{4, 8, 12, 16, 24\}$ with aspect ratio of $\gamma = \frac{\sigma_v}{\sigma_u} \in \{0, 0.5, 1\}$ allowing for a focus on nearby or far away features along one or both axis; wavelengths $\lambda \in \{2, 4, 8, 12, 24\}$ which allow for periodicity in the image texture to be as low as adjacent pixels or as large as 24 pixels; and the filter lengths $M \in \{32, 64\}$. This combination leads to 1200 Gabor filter kernels, i.e. for each pixel we have a 1200-dimensional Gabor feature vector. The categorical logistic regression with sparsity constraint was also applied to these 1200 dimensional feature vectors using the same training, validation, and testing datasets as in the GLCM experiment. Table 13.3 gives the correct classification rates for each texture subimage in Fabrics 1–3. Comparison of these results with those of the GLCM in Table 13.2 clearly shows the superiority of GLCM features for texture-based classification or

Table 13.3 Gabor categorical logistic regression accuracy and weight count.

	Fabric 1	Fabric 2	Fabric 3
Subimage 1 accuracy	77.24%	55.98%	74.45%
Subimage 2 accuracy	77.24%	62.64%	69.26%
Subimage 3 accuracy	73.46%	51.80%	78.43%
Subimage 4 accuracy	80.62%	41.56%	60.52%
Subimage 5 accuracy	84.38%	40.17%	67.63%
Subimage 6 accuracy	78.06%	97.28%	91.91%
All subimages accuracy	78.50%	58.24%	73.70%
Weight count (sparsity)	536	473	508

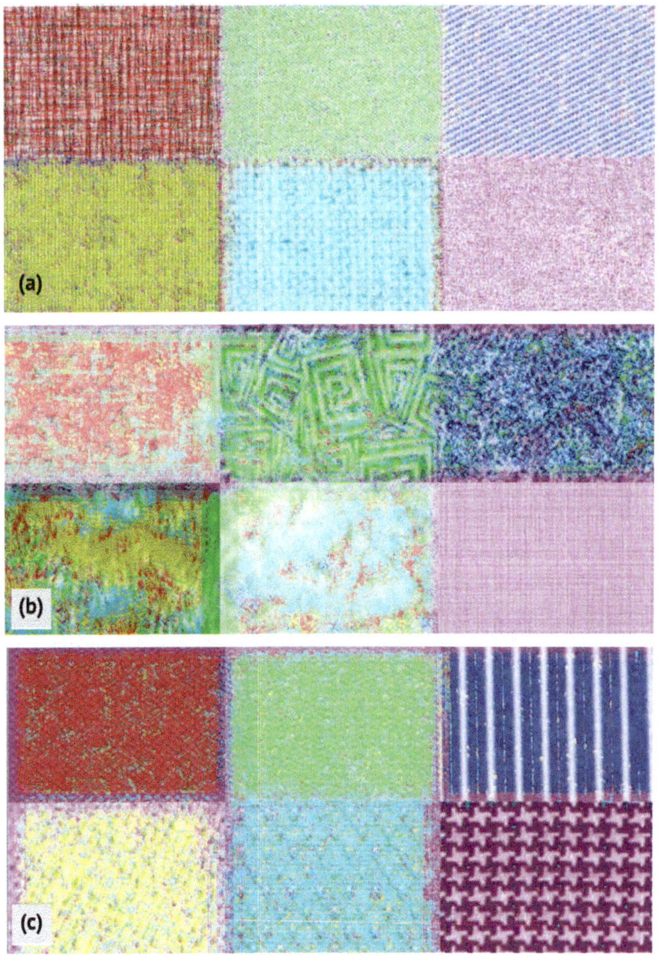

Figure 13.8 (a–c) Categorical logistic regression performance color map for fabrics 1–3 using Gabor features.

segmentation. Moreover, the Gabor filter requires significantly more features compared to those of the GLCM in Table 13.2 as evident from the last rows of these tables. Figure 13.8 shows the color overlays of the posterior probability maps on the original images. These results demonstrate substantially larger misclassified regions in each subimage when compared to those of the GLCM method.

13.6 Line and Contour Features

The methods in this category exploit the shape of the detected boundary for object discrimination. These methods can be categorized into two general categories: global and structural [24]. Global methods extract a feature vector that describes the entire contour shape, whereas structural methods decompose the shape boundary into segments called *primitives* using specific criteria. The final representation is usually a string, or graph, from which a higher-level description of the object is derived. In the following subsections, we present two global contour-based feature extraction methods, namely the Fourier Descriptor (FD) [24–26] and the Wavelet Descriptor (WD) [26–28]. Among the structural type methods, are the Hough transform [29, 30], chain code [31, 32], and B-spline representation [21].

13.6.1 Fourier Descriptor

The Fourier Descriptor (FD) [24] is a global method that extracts shape-dependent features of contours (or curves). The contour must satisfy a few conditions for the FD to work properly including: (i) the initial and final points are mapped onto the same point, i.e. *closed contour*; (ii) it does not intersect itself, that is, a *simple contour*; and (iii) it must contain a sufficient number of points to adequately represent the shape. If the first condition is not satisfied, due to the connectivity issues of the edge detection process, an edge tracing/linking method [31] can be adopted to generate a closed contour.

A common approach involves representing a closed contour in the complex plane, forming a periodic sequence of complex-valued data points. The features used for contour-based shape discrimination are derived from the magnitudes of the Discrete Fourier Series (DFS) coefficients of this periodic sequence. These features are known as Fourier Descriptors (FDs). The underlying principle is that smooth, regular-shaped (e.g. circular) contours yield FDs concentrated primarily in the low-order coefficients, while objects with more complex irregular shapes produce FDs that span a broader range of high-order coefficients. As a consequence, the feature vector formed from these FD coefficients can serve as a good representation of the contour shape for discrimination purposes. An adequate sampling density (number of points) along the contour is required to ensure a representative and robust set of FDs. The periodicity of the closed contour data ensures that the resulting FD features are inherently invariant to translation and rotation.

The steps to compute the FD are:

1. Consider a simple closed contour with M points as shown in Figure 13.9.
2. Map each contour point $(x(n), y(n))$ to real and imaginary parts of a complex-valued finite-extent periodic sequence, $p(n) = x(n) + jy(n),\ n \in [0, M-1]$ with period M, i.e. $p(n) = p(n + lM),\ l \in \mathbb{Z}$.
3. Compute the 1-D DFT (or DFS for periodic case) of this complex-valued sequence, $p(n)$, to generate FD coefficients, i.e.

$$P(k) = \mathrm{DFT}\{p(n)\} = \sum_{n=0}^{M-1} p(n)e^{-j\frac{2\pi nk}{M}}, \quad k \in [0, M-1] \tag{13.28}$$

The first FD coefficient $P(0)$ represents the centroid of the contour in the complex domain, while the second FD coefficient $P(1)$ corresponds to the best-fit circular-shaped approximation of the contour. Therefore, a perfect circle can closely be described by the first two FD coefficients. As k increases, higher-order FD components reveal finer details of the contour shape. However, they are also affected by noise and other perturbations along the object boundary. Consequently, in practical situations, a truncated set of low-order FD coefficients is often used to achieve a balance between robustness and discrimination ability.

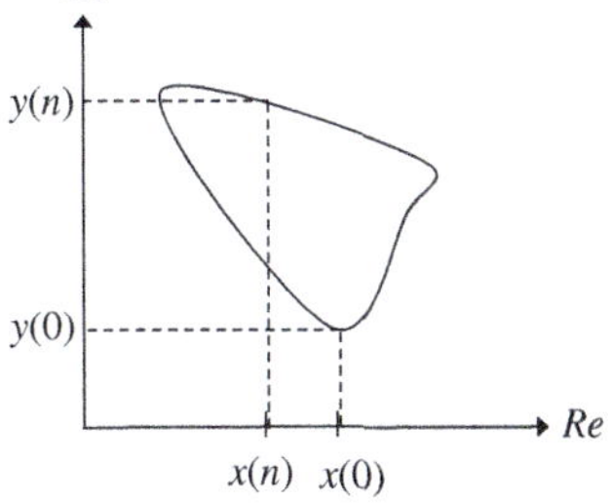

Figure 13.9 A simple closed contour with M points.

There are important considerations when using FD for contour-based discrimination. First, when applying the FD method to discriminate among different object shapes, it is necessary to ensure that all contours have the same number of points M. This can be done by resampling the contours that contain a larger number of points. This process also has the effect of smoothing the noise and small irregularities along the contour. There are many different ways we can achieve this resampling process. Comparative studies have shown [25] that the method of *equal arc length sampling* provides superior performance. Using this nonuniform sampling method, pixels that are equally spaced in arc length along the contour are selected given the constraint of M points. The spacing between consecutive points is given by ρ/M, where ρ is the perimeter of the contour. Alternatively, we can use the *equal-point resampling* method, which selects candidate pixels spaced at equal number of pixels along the contour, i.e. the spacing between two consecutive candidate pixels is given by L/M, where L is the total number of original contour points. Although, this method is simpler, it does not usually preserve geometric properties of the contours as effective as the arc-length sampling. Additionally, since we do not know the maximum frequency associated with the contour sequence, one cannot accurately determine the sampling rate, even though the accuracy of FD representation is directly related to the number of points M along the contour, i.e. larger values of M leads to a more faithful description of the contour. Second, as mentioned earlier, the lower-order FD coefficients capture information about the regularity of the shape, while the higher FDs convey information about the finer details of the shape or deviations from the regularity. Typically, a small subset of the low-order FDs is enough to represent the essential shape characteristics, implying that high-order FDs have limited discriminatory power and more susceptible to noise and perturbations and hence can be discarded. This leads to a low-dimensional FD feature vector, which is subsequently applied to the classifier for contour-based discrimination. Third, there are many applications in which object contours are neither closed nor simple, e.g. alphanumeric patterns. In such cases, in order to apply the FD method, we will trace the boundary once and then retrace it in the opposite direction to form a closed contour [24]. It can be shown (Problem 13.9) that in this case, the FD coefficients satisfy symmetry property, $P(K) = P(-K)$, when the path begins traversing at one of the endpoints.

13.6.1.1 Invariance Properties

1. Contour translation only adds a fixed bias to $P(0)$, while other FD coefficients remain invariant (Problem 13.10). Therefore, this component is usually excluded from the feature set.
2. Rotation of the contour does not change the magnitude, $|P(k)|$, of FD coefficients (Problem 13.11).
3. Size scaling of the object changes the number of sampled points and hence changes the FD coefficients. Thus, FD does not provide invariance or even covariance with respect to scaling.

Figure 13.10a–h shows binary silhouettes of eight different animals with variables contour lengths. The corresponding resampled contours with $M = 200$ points are shown in Figure 13.11a–h. The equal point resampling method was used for this example due to its simplicity. The magnitudes of the FD coefficients were then computed for each resampled contour and the distance matrix in Figure 13.12a was generated to show the Euclidean distance of each FD feature vector to those of the other animals. Each element $d_{i,j} = d_{j,i}$ in this matrix is the Euclidean distance of the FD feature vectors of animals i and j. These distances provide a measure of similarity between contour shapes and hence can represent separability of FD coefficients for these classes of animals. The largest distance was observed between the shark and the dog because of significant contour shape differences. The smallest distance measure was noted between the cow and the cat. This is

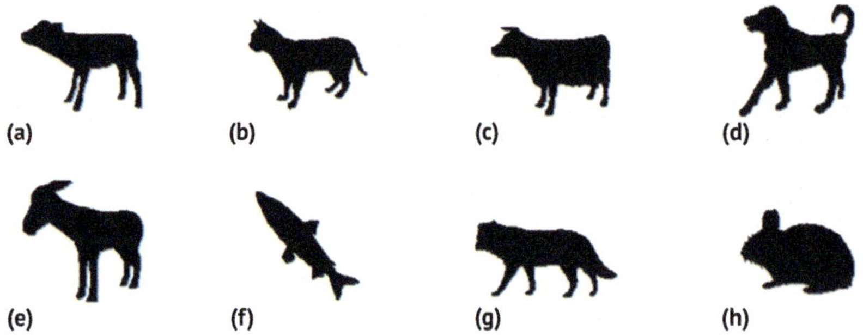

Figure 13.10 Binary silhouettes of eight different animals. (a) Calf, (b) cat, (c) cow, (d) dog, (e) donkey, (f) shark, (g) fox, and (h) rabbit. *Source*: Ishit Bansal / GitHub / https://github.com/ishitbansal/Shape-Alignment-using-Template-based-Orientation-Matching/tree/main/OneDrive_2023-09-27/KIMIA%20dataset / last accessed Aug. 5, 2026.

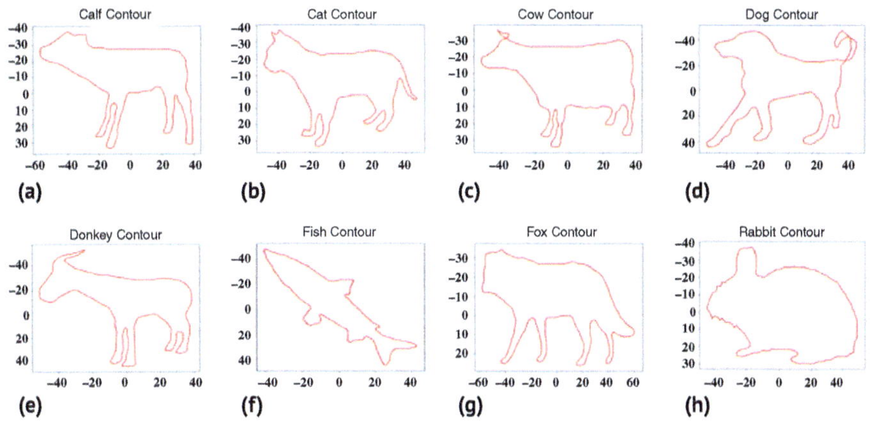

Figure 13.11 Resampled contours of binary images in Figure 13.10a–h. (a) Calf, (b) cat, (c) cow, (d) dog, (e) donkey, (f) shark, (g) fox, and (h) rabbit.

expected, since size differences are not taken into account, and both cows and cats have shorter legs compared to their torsos.

Figure 13.12b gives the distance matrix for the FD feature vectors of contours rotated by 90°. The equivalence of this matrix with that of the original silhouettes in Figure 13.12a proves the invariance to rotation of FD features. Figure 13.12c shows the distance matrix when the silhouettes were scaled down (in size) by 25%. The minor changes in the elements of this matrix when compared with those in Figure 13.12a shows the robustness of the FD features to nominal size scaling. To study the robustness to noise, the binary images were corrupted by additive Gaussian noise with PSNR $\approx$ 10 dB. The noisy images along with their corresponding contours are illustrated in Figures 13.13a–h and 13.14a–h, respectively. The distance matrix in Figure 13.12d was generated between the FD feature vectors of the original (noise-free) contours and those of the noisy ones. Unlike the previous distance matrices, this matrix is neither symmetrical nor has zero diagonal elements. However, for each noisy case (each column), the smallest distance measure occurs for the same animal type, i.e. along the diagonal. Furthermore, the distances between different classes remain consistent compared to the noise-free case in Figure 13.12a, pointing the fact that the FD feature vectors are well-suited for invariance pattern recognition based on contour shapes. However, the contours must be connected and properly sampled.

Example 13.1 *An alternative representation of the points* $\{x(n), y(n)\}_{n=0}^{M-1}$ *along a closed curve is to use vector form* $\underline{p}(n) = [x(n)\ y(n)]^T$. *Derive the FD expression using this representation.*

Solution: Let us start by rewriting the FD equation in (13.28) as:

$$P(k) = \sum_{n=0}^{M-1} (x(n) + jy(n))\left(\cos\frac{2\pi nk}{M} - j\sin\frac{2\pi nk}{M}\right)$$

If we denote the real and imaginary parts of $P(k)$ by $X(k)$ and $Y(k)$, respectively, using the above equation, we get:

$$X(k) = \sum_{n=0}^{M-1}\left(x(n)\cos\frac{2\pi nk}{M} + y(n)\sin\frac{2\pi nk}{M}\right), \quad \text{and}$$

$$Y(k) = \sum_{n=0}^{M-1}\left(-x(n)\sin\frac{2\pi nk}{M} + y(n)\cos\frac{2\pi nk}{M}\right)$$

Now, if we define FD vector, $\underline{P}(k) = [X(k)\ Y(k)]^T$ then FD representation in vector form becomes:

$$\underline{P}(k) = \sum_{n=0}^{M-1}\begin{bmatrix} \cos\frac{2\pi nk}{M} & \sin\frac{2\pi nk}{M} \\ -\sin\frac{2\pi nk}{M} & \cos\frac{2\pi nk}{M} \end{bmatrix}\underline{p}(n).$$

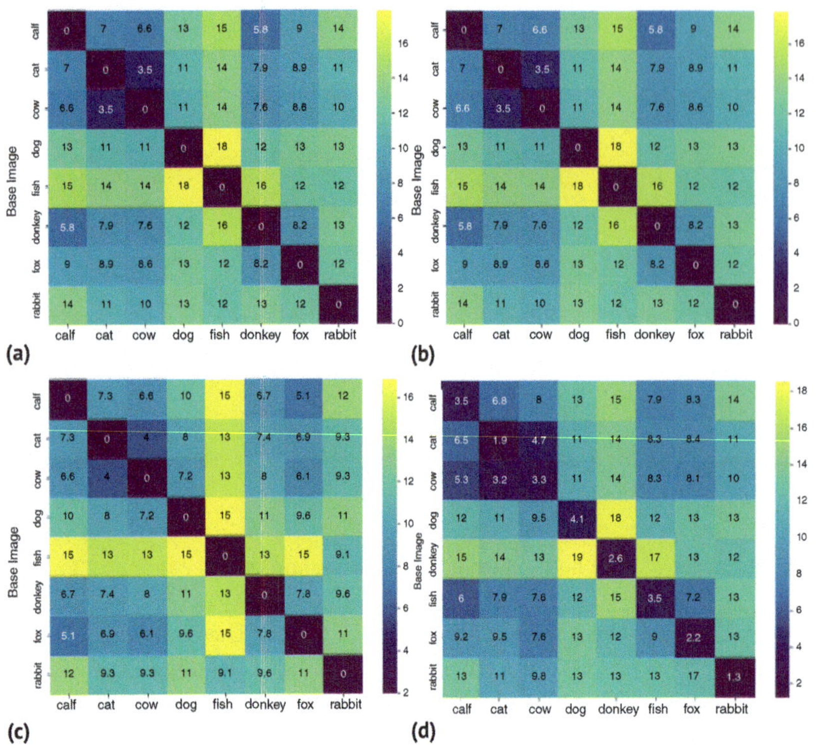

Figure 13.12 Distance matrices of FD features, of (a) original image, (b) rotated version, (c) scaled by 25%, and (d) original versus noisy images.

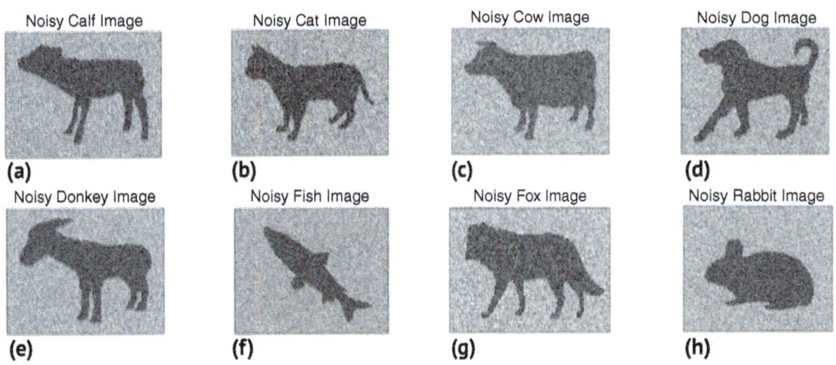

Figure 13.13 Binary silhouettes of eight different animals corrupted with additive noise. (a) Calf, (b) cat, (c) cow, (d) dog, (e) donkey, (f) shark, (g) fox, and (h) rabbit. (The color rendition of this image is available to view at www.wiley.com/go/Azimi449.)

13.6.2 Wavelet Descriptor

Despite the advantages and simplicity of the FD method, the extracted features are obtained based on the global structure of the contour. Experimental studies have shown [26] that low-frequency FDs provide good noise immunity and robustness to small details of the shape. Although this insensitivity is beneficial in applications such as hand-printed character recognition [28] where the robustness of features to variations in writing style is required, it limits the ability of FDs to distinguish shapes that are similar in structure but have subtle differences.

As shown in Chapter 7, DWT provides a localized multi-resolution representation for extracting features at varying coarse to fine scales, hence enabling discrimination of contours with subtle differences. These features are referred to as WDs. The DWT representation is applied to points $\{x(n), y(n)\}_{n=0}^{M-1}$ along a closed contour by considering the variations of $x(n)$ and

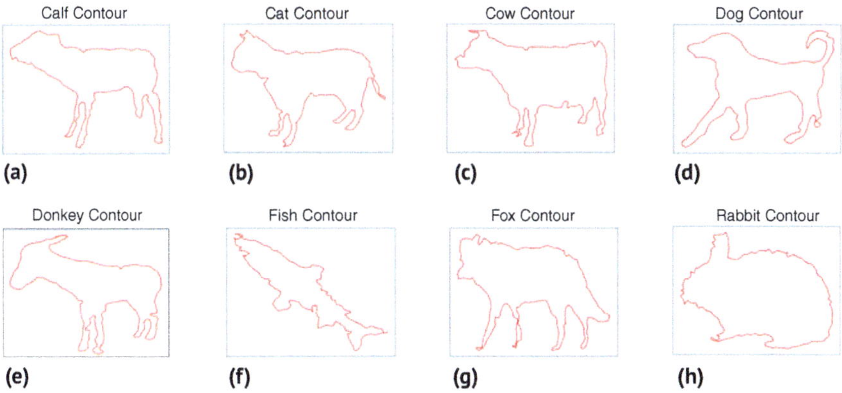

Figure 13.14 Contours of noisy binary images in Figure 13.13a–h. (a) Calf, (b) cat, (c) cow, (d) dog, (e) donkey, (f) shark, (g) fox, and (h) rabbit.

$y(n)$ as 1-D finite-extent sequences of n (surrogate for time). Now, using the same notation used in Chapter 7 for an L-level DWT, the synthesis equations can be written as:

$$\begin{bmatrix} x(n) \\ y(n) \end{bmatrix} = \begin{bmatrix} x_a(n) \\ y_a(n) \end{bmatrix} + \sum_{j=1}^{L} \begin{bmatrix} x_d^{(j)}(n) \\ y_d^{(j)}(n) \end{bmatrix}, \tag{13.29}$$

where $x_a(n)$ and $y_a(n)$ are the synthesized versions of $x(n)$ and $y(n)$ based on the lowest-order approximation sequences, and $x_d^{(j)}(n)$ and $y_d^{(j)}(n)$ are the synthesized versions of $x(n)$ and $y(n)$ based on the added details at the jth decomposition level. These signals can be related to the impulse responses (see Section 7.4.3) of the corresponding synthesis filters, namely $g_L(n-2^L p)$ for the lowest-order approximation and $h_j(n-2^j i),\ j \in [1, L]$ for the added details at j^{th} decomposition level. More specifically, we can write:

$$\begin{aligned} x_a(n) &= \sum_p X_0^{(L)}(p) g_L(n-2^L p) \\ y_a(n) &= \sum_p Y_0^{(L)}(p) g_L(n-2^L p) \\ x_d^{(j)}(n) &= \sum_i X_1^{(j)}(i) h_j(n-2^j i) \\ y_d^{(j)}(n) &= \sum_i Y_1^{(j)}(i) h_j(n-2^j i) \end{aligned} \tag{13.30}$$

where $X_0^{(L)}(p)$ and $Y_0^{(L)}(p)$ are the lowest-order approximations of $x(n)$ and $y(n)$ at the Lth decomposition level and $X_1^{(j)}(i)$ and $Y_1^{(j)}(i),\ j \in [1, L]$ are the added details at the jth decomposition level. These wavelet subband coefficients can, in turn, be expressed in terms of the impulse responses of the analysis filter. Using the results in Section 7.4.3, we have the following equations for these DWT coefficients:

$$\begin{aligned} X_0^{(L)}(p) &= \sum_n x(n) g_L(2^L p - n) \\ Y_0^{(L)}(p) &= \sum_n y(n) g_L(2^L p - n) \\ X_1^{(j)}(i) &= \sum_n x(n) h_j(2^j i - n) \\ Y_1^{(j)}(i) &= \sum_n y(n) h_j(2^j i - n), \end{aligned} \tag{13.31}$$

where $g_L(2^L p - n)$ is the impulse response of the analysis low-pass filter at the Lth decomposition level and $h_j(2^j i - n),\ j \in [1, L]$ is the impulse response of the analysis filter (band-pass or high-pass) associated with the added detail at j^{th} decomposition level.

A subset of the added details coefficients, $X_1^{(j)}(i)$ and $Y_1^{(j)}(i)$, can be selected based on their ability to accurately reproduce the contour details. To provide invariance to rotation, a normalization procedure [28] must be implemented before computing the wavelet transform.

To describe this normalization procedure, let us use the complex variable $p(n)$ representation for the contour as shown in Figure 13.9 and let $\tilde{p}(n) = p((n - n_0))_M$ be the rotated (or circularly shifted) version of $p(n)$ by known value n_0. We are looking for a normalization procedure that yields invariance to the choice of initial point. Using the circular shifting property of DFT of $\tilde{p}(n)$, we have $\tilde{P}(k) = e^{\frac{2\pi j n_0 k}{M}} P(k) = w^k P(k)$ where $w = e^{\frac{2\pi j n_0}{M}}$. Then, the normalization steps [28] are:

1. For $k = 1$, $\tilde{P}(1) = wP(1) = w|P(1)|e^{j\phi_1}$ where ϕ_1 is the known phase of $P(1)$.
2. Normalize each DFT coefficient $\tilde{P}(k)$ of $\tilde{p}(n)$ by removing the phase dependence, i.e. the DFT of $\tilde{p}(n)$ becomes $\frac{\tilde{P}(k)}{1\angle \tilde{P}^k(1)} = \tilde{P}(k) w^{-k} e^{-j\phi_1 k}$. But using $\tilde{P}(k) = w^k P(k)$, we get the normalized coefficients $\tilde{P}_N(k) = P(k)e^{-j\phi_1 k}$ which is invariance to any circular shift of the contour.
3. Compute these IDFT of the normalized coefficients. The wavelet coefficients of this sequence form the WDs.

Example 13.2 *Show that the added details wavelet coefficients are invariant to contour translation.*

Solution: Let us define the wavelet coefficients vector $\underline{Z}_1^{(j)}(i) = [X_1^{(j)}(i)\ Y_1^{(j)}(i)]^T$. Then from (13.31), we have:

$$\underline{Z}_1^{(j)}(i) = \sum_n h_j(2^j i - n)\underline{p}(n)$$

where $\underline{p}(n) = [x(n)\ y(n)]^T$. Also, let us define the translated version of the contour by $\underline{p}'(n) = \underline{p}(n) + \underline{\alpha}$ where $\underline{\alpha}$ is a fixed translation vector. Now, using this translated contour $\underline{p}'(n)$, we get:

$$\underline{Z}_1^{\prime(j)}(i) = \sum_n h_j(2^j i - n)(\underline{p}(n) + \underline{\alpha}) = \underline{Z}_1^{(j)}(i) + \Big(\sum_n h_j(2^j i - n)\Big)\underline{\alpha},$$

where $\underline{Z}_1^{\prime(j)}(i)$ is the added details wavelet coefficients vector after the translation. However, the second term is zero since $\sum_n h_j(2^j i - n) = 0$. To see the reason, let us write this term as $\sum_n h_j(2^j i - n)z^{-n}|_{z=1}$, or using the transfer function of the corresponding analysis filter:

$$\sum_n h_j(2^j i - n)z^{-n} = z^{-2^j i}\sum_l h_j(l)z^l = z^{2^j i}H_j(z^{-1}).$$

Using the recursive relationship $h_j(n) = \sum_i h(i)g_{j-1}(n - 2i)$ and taking z-transform we get $H_j(z) = G_{j-1}(z)H(z^2)$. Now, since $H(1) = 0$ for high-pass filters, then $H_j(1) = 0$ and consequently $\sum_n h_j(2^j i - n) = 0$. This proves $\underline{Z}_1^{\prime(j)}(i) = \underline{Z}_1^{(j)}(i)$, i.e. invariant to translation.

13.6.3 Chain Code

Chain code belongs to the category of structural contour feature extraction methods. The basic idea is to describe the points on a simple closed contour using a sequence of codes that represent the transitions between successive points along the contour.

Given a selected starting point, a chain code can be generated using either 4- or 8-directional pixel transitions. For each boundary pixel, the direction of the adjacent pixel is determined. In the 4-directional case, the possible directions are east, north, west, and south. Each possible direction is coded using integers 0–3 or 0–7 for 4-directional or 8-directional transitions, respectively, as shown in Figure 13.15. A chain code is formed by concatenating the numbers associated with the successive pixel transition directions along the contour until the starting point is reached and the closed contour is completely analyzed.

Figure 13.16 shows a 9×9 binary image with a simple closed contour. Assuming that we start from the red pixel, the chain codes for the 4-directional and 8-directional pixel connectivity are $A_1 = \{0, 0, 0, 0, 3, 0, 3, 0, 3, 3, 3, 2, 2, 2, 1, 1, 2, 1, 2, 1, 2\}$ and $A_2 = \{0, 0, 0, 7, 7, 7, 6, 5, 4, 4, 2, 3, 3, 3\}$, respectively. Since 8-directional connectivity allows 4 diagonal transitions,

Figure 13.15 Connectivity diagram for 4-directional and 8-directional codes.

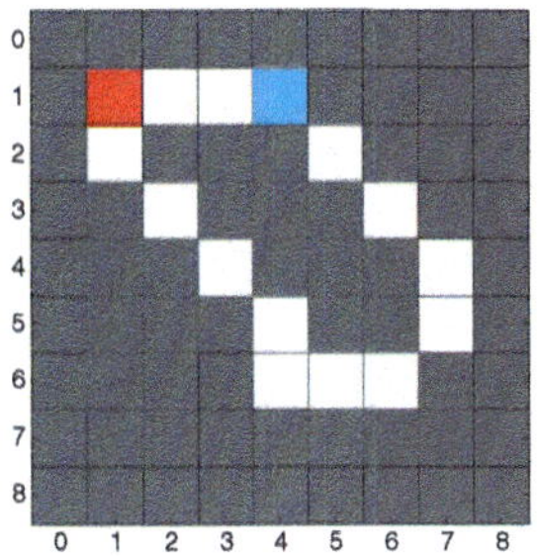

Figure 13.16 A 9 × 9 binary image with a closed contour. Staring pixel is indicated in red.

the resulting code is considerably shorter. Another important observation is that the choice of the starting point changes the chain code. For example, starting from the blue pixel, the 4-directional connectivity yields chain code $B_1 = \{0, 3, 0, 3, 0, 3, 3, 3, 2, 2, 2, 1, 1, 2, 1, 2, 1, 2, 1, 0, 0\}$. To achieve invariance to the choice of starting point and rotation, one can normalize the code using differences between successive directions instead of representing the contour by relative directions. More specifically, the difference between an element c_i of the chain code and the previous element c_{i-1} is calculated and the result is converted to base $N = 4$ or $M = 8$ for the 4-directional and 8-directional transitions, respectively. Thus, the difference code consists of code, $c_i' = ((c_i - c_{i-1}))_N$. Then, a rotation-invariant chain code is obtained by a cyclic permutation that produces the smallest sequence. Applying this difference method to A_1 gives $A_1' = \{0, 0, 0, 3, 1, 3, 1, 3, 0, 0, 3, 0, 0, 3, 0, 1, 3, 1, 3, 1\}$. However, this difference code is still variable with respect to scaling and is also sensitive to noise near the boundary points. Moreover, the dimension of the code becomes large for contours with a large number of pixels.

To address these limitations, a normalized chain code histogram (CCH) was proposed in [25]. The CCH corresponds to the probability distribution of direction occurrences along the contour given by $P(c_i = k) = n_k/n$ where n_k is the number of occurrences of direction k and n is the number of chain code elements. The normalization method [25] produces a short feature set with translation, rotation, and scale invariance. If l_k is the length of direction k, and l is the length of contour, the weighted CCH features are $f(k) = l_k n_k / l$. Alternatively, a set of statistical features can be generated from the CCH and used for discrimination purposes.

Compared to FD and WD methods, chain codes offer a simple structural representation of contours. Nonetheless, without normalizations, they don't provide invariance properties, and further, they are more sensitive to noise. Despite these limitations, histogram-based methods such as CCH significantly improve robustness of this approach.

13.6.4 Hough Transform

The Hough transform also belongs to the structural feature extraction category. It is widely used for detecting shapes that are represented parametrically such as lines, circles, and ellipses. Its fundamental idea is to map points from the image domain to the underlying parameter space. The polar representation of a line is given by:

$$\rho(\theta) = u \cos\theta + v \sin\theta, \tag{13.32}$$

where (u, v) denotes a pixel location in the image domain, θ is the angle of the normal to the line, and ρ is the perpendicular distance from the origin to the line (see Figure 13.17a). Each pixel on a line segment of an edge map in the (u, v)-plane is mapped to the parameter space (θ, ρ) as a sinusoidal function (see Figure 13.17b). This can be seen by writing

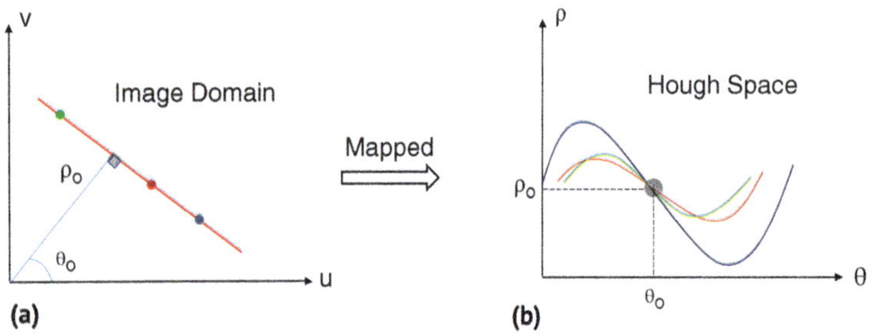

Figure 13.17 (a) A line segment in image space and (b) mapped sinusoids in Hough space.

the above equation as $\rho'(\theta) = \sin(\alpha + \theta)$ where $\rho'(\theta) = \rho(\theta)/\sqrt{(u^2 + v^2)}$, $u/\sqrt{(u^2 + v^2)} = \sin\alpha$, and $v/\sqrt{(u^2 + v^2)} = \cos\alpha$. This parameter space is also called the Hough space. If several points (u_i, v_i) lie on the same line, the intersections of their associated sinusoids in the Hough space give the parameters (θ_o, ρ_o) that represent the line segment. By accumulating all these sinusoids, peaks emerge in the Hough space, the intensity of which determines the confidence in the parameter estimates and the presence of a corresponding line. This process is similar to the Radon transform in Chapter 15, although the applications are different.

Before applying the Hough transform, the edge map of the segmented area must be extracted using one of the methods described in Chapter 8. The Hough space must be discretized in $\rho \in [-\rho_{\max}, \rho_{\max}]$ and $\theta \in [0, \pi]$ into grids where $\rho_{\max}$ is the maximum distance from the origin. The choice of grid size or resolution presents a trade-off between accuracy and computational complexity. We then initialize an accumulator array $A(\rho_m, \theta_n)$ to zero. For each pixel (u_i, v_i) on the edge map that satisfies the parametric line (13.32) for possible pairs of (ρ_k, θ_l) in the accumulator array, the corresponding array elements are incremented. As more edge pixels are associated with the same pair of parameters, local maxima emerge in the array indicating the presence of a line segment or collinear points on the edge map (see Figure 13.17b). The higher the intensity of these local maxima, the stronger the evidence of such local structures [29, 30].

Using the classical Hough transform, closed contours can be approximated by piecewise linear segments. The parameters (ρ, θ) that pass a certain confidence level (or threshold), determined by the corresponding values of the accumulator array A, can be combined into a feature vector that collectively represents the contour for image analysis applications.

The method can be extended to more complicated shape detection and feature extraction at the price of increasing the dimension of the Hough space and hence increasing the computational and storage requirements. For example, for a general ellipsoidal shape, we have:

$$\alpha\, u^2 + \beta\, v^2 + 2\gamma\, uv + 2\eta\, u + 2\xi\, v + \rho = 0, \tag{13.33}$$

which leads to a six-dimensional Hough space. However, by normalizing all the parameters based on α, we can reduce the number of parameters to five. One possible way to further reduce the dimension of the Hough space is to incorporate the gradient local direction $\frac{dv}{du}$ into the process [29].

There are many variants of the Hough transform for different applications. A good survey of these methods can be found in [30]. The generalized Hough transform (GHT) [29] has been developed for more complicated contour representation and shape detection. The GHT can be viewed as a matched filter in the Hough space with much better performance than the standard matched filter in the image space. The key benefits of the GHT are its robustness to disconnected edges and occlusion, as well as its insensitivity to noise and edge map irregularities.

13.6.5 B-Spline

The B-spline (basis spline) method is also a structural contour-based feature extraction that uses piecewise polynomial curves to locally approximate an arbitrary contour or curve [21]. The parameters of these polynomial curves are representative of shape information and can be used for various applications including computer graphics, contour shape recognition, and reconstruction. Spline functions are constructed as linear combinations of B-splines with a set of *control points* that correspond to vertices of the contour polygonal approximation. The B-spline can easily be manipulated by modifying the corresponding polygons. However, due to the local support property of B-splines, modifying the position of a vertex (or a control point) only affects a local portion of a contour. Clearly, this is a major advantage over global methods such as FD.

As before, denote a contour or a curve by the vector $\underline{p}(s) = [x(s)\ y(s)]^T$ where s is the curve parameter. This curve can be approximated using a linear combination of piecewise polynomial functions $B_{i,k}(s)$ using:

$$\underline{p}(s) = \sum_{i=0}^{L-1} \underline{c}_i B_{i,k}(s) \tag{13.34}$$

where $\underline{c}_i$ are called *control points* and $B_{i,k}(s)$ represents the normalized B-spline basis function of degree $k-1$ (order k). The control points (vertices of a polygon) reveal how the splines conform to the curve, and can also be used to reproduce the curve, if needed. Thus, they are good candidates for feature-based discrimination of objects based on their contour shapes. The points s_i (non-descending order $s_i \le s_{i+1}$) where the spline functions meet are called the *knots*. The B-spline functions can be generated recursively using the following formula:

$$B_{i,k}(s) = \frac{(s-s_i)B_{i,k-1}(s)}{s_{i+k-1}-s_i} + \frac{(s_{i+k}-s)B_{i+1,k-1}(s)}{s_{i+k}-s_{i+1}}, \quad k = 2, 3, \ldots \tag{13.35}$$

with the initial function $B_{i,1}(s) = \begin{cases} 1, & s_i \le s < s_{i+1} \\ 0, & \text{else} \end{cases}$.

Some B-spline functions for $i = 0$ and $k = 1, 2, 3$ are given [21] below.

$$B_{0,1}(s) = \begin{cases} 1, & 0 \le s < 1 \\ 0, & \text{else,} \end{cases} \qquad B_{0,2}(s) = \begin{cases} s, & 0 \le s < 1 \\ 2-s, & 1 \le s < 2 \\ 0, & \text{else} \end{cases}$$

and

$$B_{0,3}(s) = \begin{cases} s^2/2, & 0 \le s < 1 \\ -s^2+3s-1.5, & 1 \le s < 2 \\ (3-s)^2/2, & 2 \le s < 3 \\ 0, & \text{else} \end{cases} \tag{13.36}$$

respectively.

If the knots are uniformly spaced, i.e. $s_i - s_{i-1} = \Delta s, \forall i$ then it is easy to show (Problem 13.16) that for $\Delta s = 1$ we get:

$$B_{i,k}(s) = B_{0,k}(s-i), \quad i = k-1, k, \ldots, L-k. \tag{13.37}$$

That is, the spline functions become shifted version of $B_{0,k}(s)$ and (13.34) converts to a linear convolution-like sum for a shift-invariant system, i.e.

$$\underline{p}(s) = \sum_{i=0}^{L-1} \underline{c}_i B_{0,k}(s-i). \tag{13.38}$$

For closed curves (or periodic) the knots can be chosen as $s_i = ((i))_L$ where L is the number of distinct knots. In this case, we get:

$$B_{i,k}(s) = B_{0,k}((s-i))_L \tag{13.39}$$

and (13.34) becomes similar to a circular convolution [21].

Typically, B-spline of order $k = 3$ is sufficient to capture most types of curvature changes.

Now, given a set of discrete contour points t_j, $j \in [0, L-1]$, the goal is to find the control points $\underline{c}_i$ using (13.34). Plugging the discrete contour points, t_j, in (13.34) gives a system of linear equations to solve for the control points, i.e.

$$\underline{p}(t_j) = \sum_{i=0}^{L-1} \underline{c}_i B_{i,k}(t_j), \quad j \in [0, L-1] \tag{13.40}$$

or in matrix form:

$$\mathbf{P} = \mathbf{B}_k \mathbf{C} \tag{13.41}$$

where $\mathbf{P} = [\underline{p}(t_0)\underline{p}(t_1) \cdots \underline{p}(t_{L-1})]^T \in \mathbb{R}^{L\times 2}$, $\mathbf{C} = [\underline{c}_0\underline{c}_1 \cdots \underline{c}_{L-1}]^T \in \mathbb{R}^{L\times 2}$ and $\mathbf{B}_k = \left\{B_{i,k}(t_j)\right\}_{i,j} \in \mathbb{R}^{L\times L}$. However, in most situations, the number of control points is much smaller than the number of points on the curve and these points are usually nonuniformly spaced even if the image is sampled on a uniform grid. This leads to an overdetermined system of equations that can be solved using constrained least squares method [33]. The problem can be stated as follows: given matrices $\mathbf{P}$ and $\mathbf{B}_k$, find the estimate of the control point matrix $\mathbf{C}$ such that:

$$\hat{\mathbf{C}} = \underset{\mathbf{C}}{\operatorname{argmin}} \ ||\mathbf{P} - \mathbf{B}_k\mathbf{C}||_F^2 \quad s.t. \ ||\mathbf{C}||_F^2 = \alpha \tag{13.42}$$

where α is a constant and $||\mathbf{A}||_F = \sqrt{\operatorname{tr}(\mathbf{A}^T\mathbf{A})}$ is the Frobenius norm of matrix $\mathbf{A}$. This problem can alternatively be posed as:

$$\hat{\mathbf{C}} = \underset{\mathbf{C}}{\operatorname{argmin}} \ \left\{\operatorname{tr}\{(\mathbf{P} - \mathbf{B}_k\mathbf{C})^T(\mathbf{P} - \mathbf{B}_k\mathbf{C})\} + \lambda(\operatorname{tr}(\mathbf{C}^T\mathbf{C}) - \alpha)\right\} \tag{13.43}$$

where λ is the Lagrange parameter. Taking the derivative with respect to the unknown matrix $\mathbf{C}$ and setting the results to zero gives the following LS solution:

$$\hat{\mathbf{C}} = (\mathbf{B}_k^T\mathbf{B}_k + \lambda\mathbf{I})^{-1}\mathbf{B}_k^T\mathbf{P}. \tag{13.44}$$

The constraint term also ensures that the matrix in (13.44) is always invertible. When $\lambda = 0$, we obtain the standard LS solution $\hat{\mathbf{C}} = (\mathbf{B}_k^T\mathbf{B}_k)^{-1}\mathbf{B}_k^T\mathbf{P} = \mathbf{B}_k^{\dagger}\mathbf{P}$, where $\mathbf{B}_k^{\dagger} = (\mathbf{B}_k^T\mathbf{B}_k)^{-1}\mathbf{B}_k^T$ is the left pseudo-inverse of $\mathbf{B}_k$ [21].

Figure 13.18a,b illustrates two contours that are approximated using the B-spline method with $L = 11$ and $L = 8$ control points, respectively. The first case shows the versatility of the B-spline method for representing contours with concavity, while the second example illustrates the applicability of this method for approximating contours with crossing points. In both cases, the control points are good representatives of the contour properties and thus can be used as features for contour-based discrimination.

Although the B-spline methods offer compact representation for contours with concavities and intersections, the representation is not inherently invariant to geometric transformations.

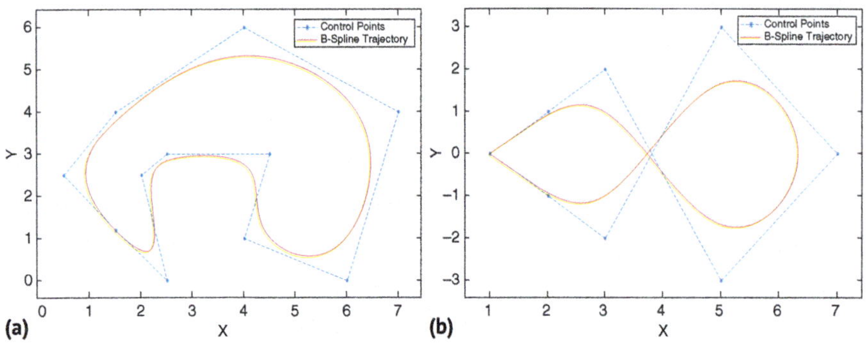

Figure 13.18 (a) Contour with concavity and $L = 11$ control points, (b) contour with crossing pattern and $L = 8$ control points.

13.7 Shape Features

Shape-based (or region-based) feature extraction methods differ fundamentally from contour-based methods in several aspects. First, they take into account all pixels within a segmented region to arrive at a feature set that represents the entire shape of the object rather than only its boundary. Second, they do not require extracting the boundary of a segmented object and converting it into a binary closed contour. However, similar to contour-based methods, the shape-based feature extraction methods can be categorized into global versus structural methods [25]. In this section, we will focus on widely used global shape descriptors, namely moment invariant (MI) [21, 31, 34] and Zernike moments [35–37].

13.7.1 Moments Invariant

The method of moment invariants (MI) method was originally proposed by Hu [34] to extract shape-dependent features for image analysis and pattern recognition. This method extracts features from the entire segmented region and generates a set of seven features that are nonlinear functions of the normalized second and third order central moments of the image. These features are invariant to rotation, scaling, and translation hence the name MI [34]. As a consequence of these invariance properties, simple shape discrimination applications can easily be performed by using a minimum distance classifier (see Section 16.1.2), which assigns an object to the class whose prototype feature vector is closest in distance to the object's MI feature vector.

Recall that the relationship between two discrete random variables (r.v.s) X and Y is described by their joint PMF $p_{X,Y}(x_m, y_n)$ where x_m and y_n are the possible values of these r.v.s. In many cases, the joint PMF can be characterized by a limited set of moments, e.g. jointly Gaussian r.v.s can be described by their first-and second-order moments. Thus, the joint moments are good descriptors of the shape of the joint distribution of two or more r.v.s. In analogy, moments of a segmented image $x(m, n)$ capture essential shape-dependent features of the object. For a discrete image $x(m, n)$, the $(p+q)^{\text{th}}$ order regular moments can be defined as:

$$\mu_{p,q} = \sum_m \sum_n m^p n^q x(m, n). \tag{13.45}$$

The zeroth order regular moment $\mu_{0,0}$ relates to the total image intensity or area for binary images. The first-order regular moments, $\mu_{1,0}$ and $\mu_{0,1}$, when normalized by $\mu_{0,0}$, correspond to the centroid coordinates, i.e. $\overline{m} = \frac{\mu_{10}}{\mu_{00}}$ and $\overline{n} = \frac{\mu_{01}}{\mu_{00}}$.
Now, the central moments of $(p+q)^{\text{th}}$ order, which are translation invariant, are defined as:

$$\xi_{p,q} = \sum_m \sum_n (m - \overline{m})^p (n - \overline{n})^q x(m, n) \tag{13.46}$$

where $\xi_{0,0} = \mu_{0,0}$. The MI method uses second and third order $(p + q = 2, 3)$ central moments namely:

$$\xi_{1,1} = \mu_{1,1} - \overline{n}\mu_{1,0}$$

$$\xi_{2,0} = \mu_{2,0} - \overline{m}\mu_{1,0}$$

$$\xi_{0,2} = \mu_{0,2} - \overline{n}\mu_{0,1}$$

and

$$\xi_{3,0} = \mu_{3,0} - 3\overline{m}\mu_{2,0} + 2\mu_{1,0}\overline{m}^2$$

$$\xi_{0,3} = \mu_{0,3} - 3\overline{n}\mu_{0,2} + 2\mu_{0,1}\overline{n}^2$$

$$\xi_{1,2} = \mu_{1,2} - 2\overline{n}\mu_{1,1} - \overline{m}\mu_{0,2} + 2\overline{n}^2\mu_{1,0}$$

$$\xi_{2,1} = \mu_{2,1} - 2\overline{m}\mu_{1,1} - \overline{n}\mu_{2,0} + 2\overline{m}^2\mu_{0,1}$$

to generate the relevant MI features. To achieve scale invariance, these central moments are normalized to produce:

$$\eta_{p,q} = \frac{\xi_{p,q}}{\xi_{0,0}^{\gamma}}, \quad p + q = 2, 3 \tag{13.47}$$

where $\gamma = \frac{p+q}{2} + 1$. Hu [34] derived a set of seven features that are obtained by nonlinear combinations of these normalized moments and offer the desirable properties of being invariant under image translation, scaling, and rotation. These seven invariant moments are:

$$\begin{aligned}
\phi_1 &= \eta_{2,0} + \eta_{0,2} \\
\phi_2 &= (\eta_{2,0} - \eta_{0,2})^2 + 4\eta_{1,1}^2 \\
\phi_3 &= (\eta_{3,0} - 3\eta_{1,2})^2 + (3\eta_{2,1} - \eta_{0,3})^2 \\
\phi_4 &= (\eta_{3,0} + \eta_{1,2})^2 + (\eta_{2,1} + \eta_{0,3})^2 \\
\phi_5 &= (\eta_{3,0} - 3\eta_{1,2})(\eta_{3,0} + \eta_{1,2})\left[(\eta_{3,0} + \eta_{1,2})^2 - 3(\eta_{2,1} + \eta_{0,3})^2\right] \\
&\quad + (3\eta_{2,1} - \eta_{0,3})(\eta_{2,1} + \eta_{0,3})\left[3(\eta_{3,0} + \eta_{1,2})^2 - (\eta_{2,1} + \eta_{0,3})^2\right] \\
\phi_6 &= (\eta_{20} - \eta_{02})\left[(\eta_{30} + \eta_{12})^2 - (\eta_{21} + \eta_{03})^2\right] \\
&\quad + 4\eta_{1,1}(\eta_{3,0} + \eta_{1,2})(\eta_{2,1} + \eta_{0,3}) \\
\phi_7 &= (3\eta_{2,1} - \eta_{3,0})(\eta_{3,0} + \eta_{1,2})[(\eta_{3,0} + \eta_{1,2})^2 - 3(\eta_{2,1} + \eta_{0,3})^2] \\
&\quad + (3\eta_{1,2} - \eta_{3,0})(\eta_{2,1} + \eta_{0,3})[3(\eta_{3,0} + \eta_{1,2})^2 - (\eta_{2,1} + \eta_{0,3})^2].
\end{aligned} \tag{13.48}$$

As shown in [34] the symmetric form of these features ensures invariance to rotation and translation. Scaling invariance is also achieved via the normalization procedure in (13.47). For many simple shape discrimination problems, one may use only two features known as *Spread* and *Slenderness*, which are defined in terms of the second-order central moments as $S_P = \mu_{20} + \mu_{02}$ and $S_L = \sqrt{(\mu_{20} - \mu_{02})^2 + 4\mu_{11}^2}$, respectively.

Regular moments in (13.45) can be viewed as an expansion of the image $x(m, n)$ onto the nonorthogonal basis functions $m^p n^q$. These moments, in their original forms, lack rotational invariance and do not support image reconstruction. To address these limitations, Teague [35] introduced a set of orthogonal moments that are known as Zernike moments. These moments are constructed using orthogonal polynomials and possess desirable properties, including rotation invariance and the ability to reconstruct the original image.

Comparative studies in [36] revealed the fact that Zernike moments outperform regular and other types of moments (e.g. Legendre moments), in terms of image representation ability, noise robustness, and reduced redundancy. More specifically, experiments in several character recognition tasks indicated that the classification results of the Zernike moments were less sensitive to additive noise effects in the images compared to the other moments. In [37], a similar study was carried out in which regular moments and Zernike moments were used to extract features that are applied to a backpropagation neural network (BPNN) classifier to classify 26 uppercase characters (A to Z) in English alphabets. The silhouettes were allowed to have varying scale, translation, and orientation forming 24 sets of images. In addition, random noise with varying SNR from 5 to 50 dB was added to these images. The results showed superior classification accuracy and resilience to noise of Zernike moments. In what follows, we present a brief coverage of the Zernike moments method and its important properties.

13.7.2 Zernike Moments

Zernike basis functions form a complete orthogonal set of complex polynomials defined in the interior of the unit circle $(u^2 + v^2) < 1$. For an image, $x(u, v)$, the complex Zernike moments of order p with repetition q are given [35, 36] as:

$$A_{p,q} = \frac{p+1}{\pi} \int_0^{2\pi} \int_0^1 x(\rho\cos\theta, \rho\sin\theta)\, V_{p,q}^*(\rho, \theta)\rho\, d\rho\, d\theta, \tag{13.49}$$

where $p = 0, 1, \ldots, \infty$ and q is an integer such that $|q| \leq p$ and $p - |q|$ is even. Note that (13.49) is expressed in polar coordinates where $u = \rho\cos\theta$, $v = \rho\sin\theta$, ρ is the length of a vector from the origin to the (u, v) point, and θ is the corresponding phase angle. The basis functions satisfy the orthogonality condition:

$$\int_0^{2\pi} \int_0^1 V_{p,q}^*(\rho, \theta)\, V_{r,s}(\rho, \theta)\, \rho\, d\rho\, d\theta = \frac{\pi}{p+1}\delta(p - r)\delta(q - s). \tag{13.50}$$

The Zernike polynomials, $V_{p,q}(\rho, \theta)$, are given by:

$$V_{p,q}(\rho, \theta) = R_{p,q}(\rho)e^{jq\theta}, \tag{13.51}$$

where the radial polynomial $R_{p,q}(\rho)$ is defined as:

$$R_{p,q}(\rho) = \sum_{s=0}^{(p-|q|)/2} \frac{(-1)^s(p-s)!}{s!\left(\frac{p+|q|}{2}-s\right)!\left(\frac{p-|q|}{2}-s\right)!}\rho^{p-2s}, \tag{13.52}$$

where we have symmetry in q, i.e. $R_{p,-q}(\rho) = R_{p,q}(\rho)$, and also $A^*_{p,q} = A_{p,-q}$ due to $x(u, v)$ being real-valued. Unlike MI method, the orthogonality allows approximate reconstruction of the image using the synthesis equation:

$$x(u, v) \approx \sum_{p=0}^{M-1}\sum_{q} A_{p,q}V_{p,q}(u, v), \tag{13.53}$$

where the summation over q is for all $|q| \leq p$ and $p - |q|$ is even.

13.7.2.1 Invariance Properties

The magnitude of the Zernike moments, i.e. $|A_{p,q}|$, $p \geq 0$, are widely used as *rotation invariant* features in many pattern analysis applications, since image rotation only affects the phase of the Zernike moments (see Example 13.3) while leaving the magnitude unchanged. To make the Zernike moments translation invariant, the image is transformed to a new coordinate system by moving the origin to the centroid prior to the moment computation. In this coordinate system, the first-order moments of the image become zero. To achieve scale invariance, the image is scaled in size by mapping its zeroth-order moment to a predetermined value [37]. A general mapping of type $x(u, v) = y(\bar{u} + \frac{u}{a}, \bar{v} + \frac{v}{a})$ is applied before computing the Zernike moments, where $(\bar{u}, \bar{v})$ are coordinates of the centroid of the original image $x(u, v)$ and a is a scaling parameter. The scale and translation normalization process affects only two Zernike features, namely $|A_{0,0}|$ and $|A_{1,1}|$. However, $|A_{0,0}|$ will be the same for all the images and $|A_{1,1}|$ will be zero. As a result, they are excluded from the feature set, and the selected features start from the second-order moments. In addition, as mentioned before, previous studies have shown that Zernike moments offer better robustness to noise when compared with the regular moments.

Example 13.3 *Show that the magnitude of the Zernike moments* $|A_{pr}|$, $p \geq 0$ is invariant under image rotation.

Solution: In the polar coordinate system, the rotated version of the original image $x(\rho, \theta)$ by angle ϕ is given by $x'(\rho, \theta) = x(\rho, \theta - \phi)$, since rotation only impacts the angle and not the location. Then, using (13.49) and substituting for $V_{p,q}(\rho, \theta) = R_{p,q}(\rho)e^{jq\theta}$, gives the Zernike moments of image $x'(\rho, \theta)$:

$$A'_{p,q} = \frac{p+1}{\pi}\int_0^{2\pi}\int_0^1 x(\rho, \theta - \phi)R_{p,q}(\rho)e^{-j\,q\theta}\rho\;d\rho\;d\theta.$$

Changing variable $\theta - \phi = \theta'$, it is easy to show that $A'_{p,q} = A_{p,q}e^{-jq\phi}$, i.e. rotation only introduces a phase shift in the Zernike moments while the magnitude remains unaffected.

Both MI and Zernike moments are applied to the same binary silhouettes of eight different animals shown in Figure 13.10a–h. Although these methods can be applied to grayscale images, this particular dataset was chosen here to enable a direct performance comparison with the contour-based feature extraction method, namely the FD in Section 13.6.1. Performance comparisons are made under three conditions: (i) 50% scaling (zooming out), (ii) additive noise corruption, and (iii) occlusion. The effects of rotation and translation were not considered since these invariance properties have already been proven theoretically for these feature extraction methods.

To implement scaling by approximately 50% size reduction, the original silhouettes were downsampled. Additionally, as in Section 13.6.1, additive noise was added to the images to produce PSNR $\approx$ 10 dB. Finally, to implement the effects of common partial occlusion, approximately 5% of each silhouette was removed, primarily from the lower-right region for consistency.

For the MI method, since the images are binary, the magnitudes of higher features are very small. Thus, normalization was required prior to computing the distance matrices. Figure 13.19 shows the silhouettes of all the animals with approximately 5% occlusion. Figure 13.20a–d shows the distance matrices whose elements $d_{i,j}$ represent the Euclidean distances between

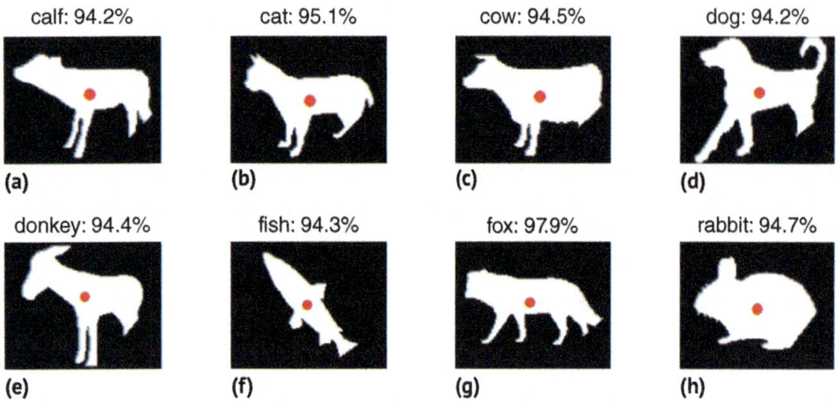

Figure 13.19 Binary silhouettes of eight animals in Figure 13.10a–h with partial occlusion.

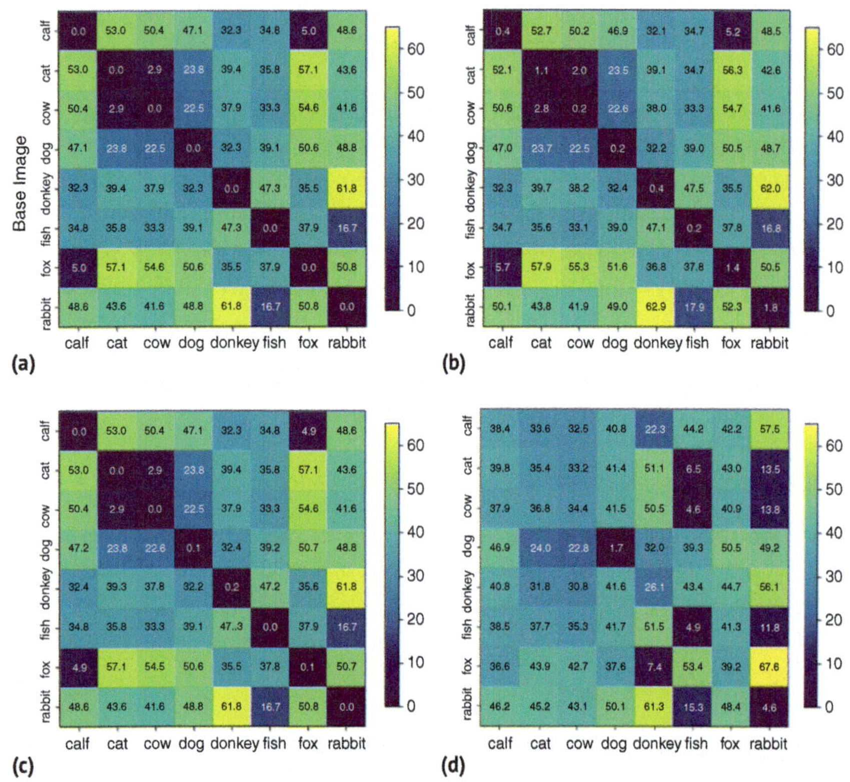

Figure 13.20 Distance matrices of MI features of (a) original image, (b) scaled image by 50%, (c) noisy image, and (d) occluded image.

the MI feature vector of the original animal i silhouette versus those of (a) animals j, (b) scaled version of animals j, (c) noisy version of animals j, and (d) occluded version of animals j, respectively. Several observations can be made from these results. First, as expected, the distance matrix in Figure 13.20a is symmetric with diagonal elements that are zero; while those in Figure 13.20b–d do not have the same properties. Second, even scaling by 50% and additive noise with PSNR of approximately 10 dB had minimal impact on the MI features, demonstrating their scale invariance and robustness to noise. Moreover, compared to the FD distance matrices in Figure 13.12, the elements of these distance matrices exhibit improved class separability, although the cow and cat classes remain less distinguishable. In contrast, occlusion significantly affects the MI features, as evident from the elements of the distance matrix in Figure 13.20d, with dog, shark, and rabbit being the only classes that are less impacted. More specifically, the increased intra-class distances (diagonal elements) result in reduced discrimination performance.

For the Zernike moments, a circle with a specific radius should be used to encompass each silhouette. A reasonable choice for this radius presents a trade-off between good class separability and maintaining consistent nominal Zernike feature magnitude. Figure 13.21 shows the radii for different animals that were selected experimentally. In this study, magnitudes of the first 25 Zernike moments were used to represent each animal silhouette. Figure 13.22a–d shows the corresponding distance matrices under the same experimental conditions as those in Figure 13.20a–d for the MI features. Several important observations can be made from these results. First, unlike MI features, which exhibited poor separability for the cow and cat classes, the Zernike moments provided strong feature separability across all classes. Second, scaling by 50% led to larger change in the diagonal elements compared to those of the MI features, though the inter-class separability remains strong. Third, the Zernike moments features demonstrate much better robustness to additive noise as evident from the distance matrices in Figures 13.22c and 13.20c. Finally, as can be observed from the distance matrix in Figure 13.22d, Zernike moments features remained robust to 5% occlusion as the intra-class distances remain smaller than inter-class distances, maintaining discrimination performance even with missing shape information.

Overall, these results indicate that Zernike moments provide much better performance for shape-based feature extraction compared to standard MI method. Although their implementation is more involved, they offer improved feature separability, enhanced robustness to noise, and greater resilience to occlusion and scaling, making them a powerful tool for many practical pattern classification applications.

13.8 Computer Projects

The computer projects in this chapter use Python libraries OpenCV, gc, os, NumPy, pyefd, Pandas, itertools, Mahotas, matplotlib, skimage, and Seaborn.

Computer Project 13.1: The goal of this computer project is to investigate the application of the color-based feature extraction methods covered in Section 13.2. First, capture a full-color image containing five different objects with distinct shapes and colors. Record the locations and orientations of the objects. Develop Python scripts to implement both the correlogram and color moments methods. For the color correlogram method, use $d = 1, 2$. For the color moments, covered in Section 13.2, compute the mean, standard deviation, and skewness. For each object in the image, extract features that represent the entire object pixels with one feature vector. To evaluate feature separability, produce a distance matrix similar to those in Figure 13.12 for each feature set. Comment on feature separability and other advantages and disadvantages of each method. Capture new images of the same objects after rotating and translating them within the scene.

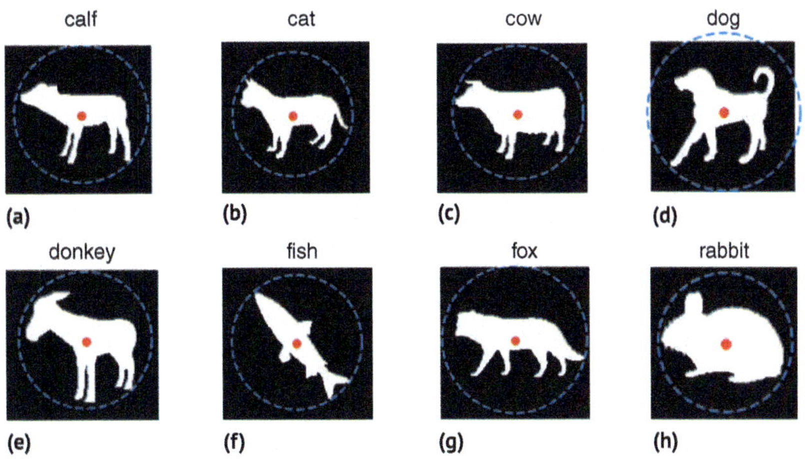

Figure 13.21 Binary silhouettes of eight animals in Figure 13.10a–h where each silhouette is enclosed within its circular support region.

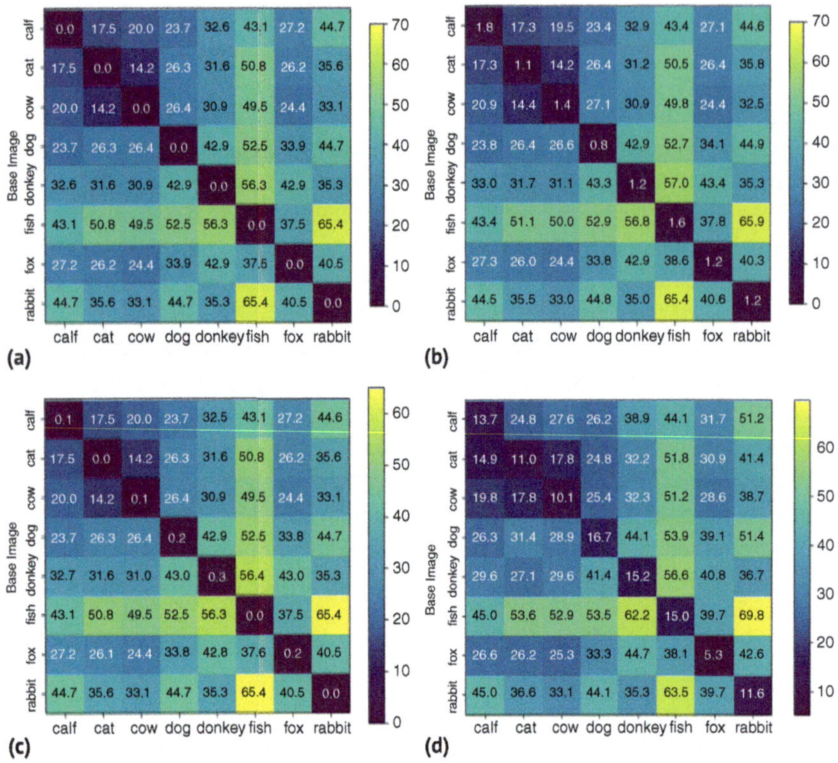

Figure 13.22 Distance matrices of the Zernike moments features of (a) original image, (b) scaled image by 50%, (c) noisy image, and (d) occluded image.

Keep the camera distance and viewing angle approximately unchanged. Repeat feature extraction for these images. Next, return the objects to their original locations and orientations, but increase the camera distance so that the objects undergo 25–30% size reduction. Repeat the feature extraction process for this image. Determine the invariance properties of these methods with respect to rotation, translation, and size scaling by comparing the distance matrices for these cases with those of the original image. Provide your detailed observations and conclusions on the results.

Computer Project 13.2: This computer project involves spectral feature extraction using 2-D DWT (see Section 13.4.2). The image used in this project is Fabric 1 from Section 13.5.1, which contains six subimages with different texture properties. Apply a 2 level 2-D DWT using the db4 wavelet to each subimages. Then, compute the local statistical features in Section 13.4.2 using 25×25 pixel blocks within each added details subimage. The overlap between adjacent blocks is 10 pixels in each direction. The estimated local statistics for each added detail subimage are then averaged to produce two statistics-based features. When combined for all 9 subimages of a 2-level 2-D DWT, this produces an 18-dimensional feature vector for each fabric subimage. Using these feature vectors, generate a Euclidean distance matrix similar to that in Figure 13.12. Comment on the separability of these DWT-based statistical features for these six textural images. Explain if these features offer any invariance property, and, if so, demonstrate at least one such property experimentally. Discuss whether these features can be used with a simple classifier, e.g. unsupervised clustering, to distinguish different texture types.

Computer Project 13.3: In this computer project, we investigate the usefulness of the GLCM method for medical image analysis. The CT scan image in Figure 12.16 is used, which contains a tumor in the upper region of the right lung. The goal is to determine whether the textural features extracted from the tumor region and other parts of the lungs are sufficiently different to aid the discrimination process. Apply the GLCM method to the segmented region of the tumor and to the segmented regions of the left and right lungs using the edge map image. Use the window sizes of 32×32 pixels, $d = 2$, and angles $\theta_k = \frac{k\pi}{8}$, $k \in [0, 7]$ to produce the GLCM features. The windows are swept with 50% overlap in both directions. Then, compute the six GLCM features (see Section 13.5.1) in each window of the tumor and lung areas and average the estimates over all windows for each segmented region. This process results in three GLCM feature vectors each of dimension 48×1 for the tumor, left, and right lungs. Using these feature vectors, generate the Euclidean distance matrix similar to that in Figure 13.12. Comment on the effectiveness of the GLCM features in discriminating the tumor from normal lung tissues. Explain whether these features offer any invariance properties and, if so, show at least one invariance property experimentally. Provide a detailed discussion on your results and advantages and limitations of the GLCM method.

Computer Project 13.4: The goal of this computer project is to study and compare two contour-based feature extraction methods: the Fourier descriptor (FD) and B-spline contour representation. First, capture several images containing three objects with different shapes and geometries. From one image to another, vary the positions and orientations of the objects, and record their locations and orientations. Then, apply one of the edge extraction methods, e.g. LoG or Canny, in Chapter 8 to produce the objects' contours. Before using the feature extraction methods, apply one of the resampling schemes discussed in Section 13.6.1 to get contours with the same number of points. Note that the contours must be connected for the FD method to work properly. Develop Python scripts to implement both the FD and B-spline methods for each object contour. The number of points on each contour, the B-spline order k in (13.36), and number of control points are design parameters that must be chosen depending on the complexity of the object contours. The magnitudes of the FD coefficients arranged in a vector represent the FD feature vector of each object. For the B-spline method, the control points augmented into a vector represent the feature vector. To evaluate feature separability, form the distance matrices of the feature vectors of the three objects (see Figure 13.12) for each method and under each location and orientation condition. Compare the performance of the two methods in terms of feature separability and invariance properties and describe the advantages and limitations of each method for this particular application.

Computer Project 13.5: This computer project involves using shape-based feature extraction methods, namely the MI and Zernike moments covered in Section 13.7. Use the same type of images described in the first part of Computer Project 13.4. For each image, segment each object within a sufficiently large window that contains the object and portions of the surrounding background. This process is necessary because both of these methods are applied to individual objects and not to the whole image. Develop Python scripts using OpenCV library to implement the MI method and use Python library Mahotas, to compute Zernike moments. For each object at a particular location and orientation, the MI produces a seven-dimensional feature vector, while the number of Zernike features depends on the design parameters. The experiments in Section 13.7.2 showed that 25 Zernike features provided good object representation. When using this algorithm, the radius of the circle that encompasses an object should be carefully chosen to allow good class separability and maintain balanced feature magnitudes. To compare the separability of the features for each method, generate the distance matrices of the feature vectors for all objects (see Figure 13.12) and under each location and orientation condition. Study the invariance properties of these methods. Thoroughly compare the results of these methods and describe the advantages and limitations of each.

13.9 Problems

Problem 13.1 In Section 13.3.1, we showed the invariance property of PCA to data rotation and translation, when the training data samples are assumed to be rotated and/or translated. Now, assuming that this assumption is not satisfied and considering a test sample $\underline{y}_m$ which is a rotated version of the training sample $\underline{x}_m$, i.e. $\underline{y}_m = \mathbf{R}_\theta \underline{x}_m$. Explain the relationship between PCs of $\underline{x}_m$ and $\underline{y}_m$ samples. How does this relationship change if the test sample is also translated?

Problem 13.2 In Section 13.4.2, generalized Gaussian density (GGD) was used to model the marginal density of DWT subband coefficients for texture-based feature extraction. The GGD pdf is given in (13.11). Assuming that $\beta_i = 1, \forall i$, find the ML estimate for parameter α_i using the expression for the likelihood function $\mathcal{L}(\mathbf{X}_i, \alpha_i)$.

Problem 13.3 Repeat Problem 13.2 when $\beta_i = 2, \forall i$.

Problem 13.4 Consider the image shown below. Find the GLCM matrices for $d = 1$ and vertical directions 90 and 270 degrees and horizontal directions 0 and 180 degrees. Then normalize the results. What are your conclusions?

Problem 13.5 Consider image $x(u, v)$ and its 2-D Gabor filtered version $y(u, v)$. Determine effects of:

a. translating the original image by (u_f, v_f) on the filter output and comment on the invariance property of this method to translation.
b. scaling the spatial coordinates of the original image by α and β and comment on the invariance property of this method to scaling.

1	1	0	0
2	3	2	3
1	2	3	3
3	3	3	0

Problem 13.6 In Problem 13.5, if the coordinates are rotated about the origin by angle ϕ, the rotated image is $x(u', v')$ where $u' = u \cos \phi + v \sin \phi$ and $v' = -u \sin \phi + v \cos \phi$. Determine the effect of this rotation on the 2-D Gabor filtered image and comment on the invariance property of this method to rotation.

Problem 13.7 Consider image $x(u, v)$ and its filtered image $y_{k,l}(u, v)$ using 2-D Gabor wavelet filter in (13.26). Determine effects of:

a. translating the original image by (u_f, v_f) on the filter output and comment on the invariance property of this method to translation.
b. scaling the spatial coordinates of the original image by α and β and comment on the invariance property of this method to scaling.

Problem 13.8 In Problem 13.7, if the coordinates are rotated about the origin by angle ϕ, the rotated image is $x(u', v')$ where $u' = u \cos \phi + v \sin \phi$ and $v' = -u \sin \phi + v \cos \phi$. Determine the effect of this rotation on the 2-D Gabor wavelet filtered image and comment on its invariance property to rotation.

Problem 13.9 Assume that we have a curve that is not closed, and start from one of the endpoints and retrace the curve in the opposite direction to make it a closed contour. Show that the FD coefficients of this closed curve have the symmetry property, i.e. $P(k) = P(-k)$ if a certain condition is satisfied. What is that condition?

Problem 13.10 Prove the translation invariance of the FD coefficient in Section 13.6.1 using DFT properties in Chapter 2.

Problem 13.11 Prove the rotation invariance property of the FD coefficient in Section 13.6.1 using the DFT properties in Chapter 2. What is the effect on the phase of the FD coefficients of the rotated object?

Problem 13.12 Using the DFT properties in Chapter 2, show the effects of upsampling (scaling up) and downsampling (scaling down) the contour points/samples on the FD coefficients.

Problem 13.13 Repeat Problem 13.12 assuming that the contour is a continuous function of spatial coordinates u and v and we use the magnitude of the Fourier transform as FD feature.

Problem 13.14 Consider the closed contour in Figure 13.23. Develop a chain code for this contour using both the 4- and 8-directional connectively in Figure 13.15. Consider the red pixel as the starting point. Comment on the effectiveness of each chain code sequence as a feature set.

Problem 13.15 Using the contour in Figure 13.16, show that the normalized chain code histogram features $f(k) = l_k n_k / l$ where n_k is the number of occurrence of the value k in a chain code sequence, l_k is the length of the direction k, and l is the length of the contour, are invariant to translation, scaling and rotation. Use four-directional connectivity.

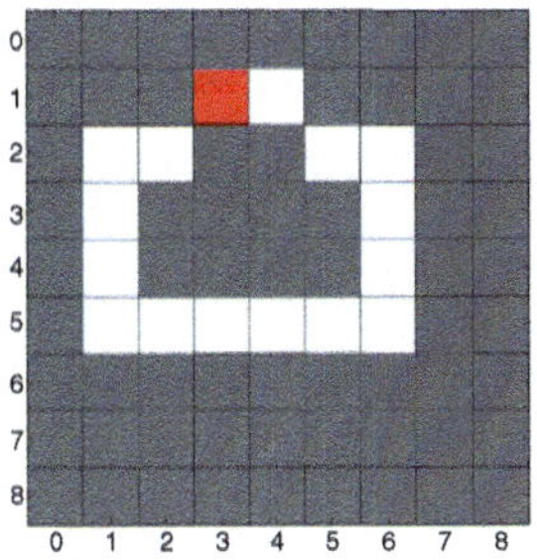

Figure 13.23 A 9×9 binary image with a closed contour. Staring pixel is indicated in red.

Problem 13.16 Prove that the shift property in (13.37) holds, if the *knots* for the B-splines (Section 13.6.5) are uniformly spaced, i.e. $s_i - s_{i-1} = 1, \forall i$.

Problem 13.17 Show that circular shift in (13.38) holds for a closed contour (i.e. periodic), if the knots are chosen as $s_i = ((i))_L$.

Problem 13.18 The $(p+q)^{\text{th}}$ order central moments for a continuous image $x(u, v)$ are given by:

$$\xi_{p,q} = \iint (u - \bar{u})^p (v - \bar{v})^q x(u, v) du dv$$

where $\bar{u} = \frac{\mu_{1,0}}{\mu_{0,0}}$ and $\bar{v} = \frac{\mu_{0,1}}{\mu_{0,0}}$ with $\mu_{0,0}$, $\mu_{1,0}$ and $\mu_{0,1}$ being the zeroth and first-order regular moments. Prove the scaling invariance property of the normalized moments using the normalization in (13.47).

Problem 13.19 In Problem 13.18, determine the effect of translation by (u_f, v_f) on the $(p+q)^{\text{th}}$-order central moments and comment on the invariance.

Problem 13.20 Consider the scenario of reflection across a line through the origin, rotated through a positive angle ϕ with respect to the v axis. The effect of reflection (mirror image of an object) can be described by mapping $x'(\rho, \theta) = x(\rho, 2\phi - \theta)$. Show that this is valid and then develop the expression for the Zernike moments of the reflected image. Is there any useful invariance in this case? Explain.

References

1 R. E. Bellman, "*Adaptive Control Processes*," Princeton, NJ: Princeton University Press, 1961.
2 R. O. Duda, P. E. Hart, and D. G. Stork, "*Pattern Classification*," Wiley, 2nd Edition, 2000.
3 S. Haykin, "*Neural Networks and Learning Machines*," Prentice-Hall, 3rd Edition, 2008.
4 A. O. Salau and S. Jain, "Feature extraction: A survey of the types, techniques, applications," Proceedings of the IEEE 2019 International Conference on Signal Processing and Communication (ICSC), 2019.
5 G. Kumar and P. K. Bhatia, "A detailed review of feature extraction in image processing systems," Proceedings of the 2014 Fourth International Conference on Advanced Computing and Communication Technologies, 2014.
6 J. Huang, S. R. Kumar, M. Mitra, W. Zhu, and R. Zabih, "Image indexing using color correlograms," Proceedings of the IEEE Computer Society Conference on Computer Vision and Pattern Recognition, pp. 762–768, 1997.
7 H. Yu, M. Li, H. Zhang, and J. Feng, "Color texture moments for content-based image retrieval," Proceedings of the IEEE Interference Conference on Image Processing, pp. 929–932, 2002.
8 D. P. Tian, "A review on image feature extraction and representation techniques," *International Journal of Multimedia and Ubiquitous Engineering*, vol. 8, pp. 385–395, 2013.
9 A. K. Jain and A. Vailaya, "Image retrieval using color and shape," *Pattern Recognition*, vol. 29, pp. 1233–1244, 1996.
10 Y. Yang, H. Chen, K. A. Hilburn, R. J. Kuligowski, and R. Cifelli, "Deep learning for precipitation retrievals using ABI and GLM measurements on the GOES-R series," *IEEE Transactions on Geoscience and Remote Sensing*, vol. 61, pp. 1–14, 2023.

11 M. N. Do and M. Vetterli, "Wavelet-based texture retrieval using generalized Gaussian density and Kullback-Leibler distance," *IEEE Transactions on Image Processing*, vol. 11, pp. 146–158, 2002.

12 G. V. Wouwer, P. Scheunders, and D. V. Dyck, "Statistical texture characterization from discrete wavelet representations," *IEEE Transactions on Image Processing*, vol. 8, pp. 592–598, 1999.

13 A. Laine and J. Fan, "Texture classification by wavelet packet signature," *IEEE Transactions on Pattern Analysis and Machine Intelligence*, vol. 15, pp. 1186–1191, 1993.

14 X. Guo, X. Huang, and L. Zhang, "Three-dimensional wavelet texture feature extraction and classification for multi/hyperspectral imagery," *IEEE Geoscience and Remote Sensing Letters*, vol. 11, pp. 2183–2187, 2014.

15 R. M. Haralick, K. Shanmugam, and I. H. Dinstein, "Textural features for image classification," *IEEE Transactions on System, Man, and Cybernetics*, vol. 3, pp. 610–621, 1973.

16 C. C. Gotlieb and H. E. Kreyszig, "Texture descriptors based on co-occurrence matrices," *Computer Vision, Graphics, and Image Processing*, vol. 51, pp. 70–86, 1990.

17 G. C. Cawley, N. L. Talbot, and M. Girolami, "Sparse multinomial logistic regression via bayesian L1 regularization," Proceedings of the 20th International Conference on Neural Information Processing Systems, 2006.

18 I. Fogel and D. Sagi, "Gabor filters as texture discriminator," *Biological Cybernetics*, vol. 61, pp. 103–113, 1989.

19 S. E. Grigorescu, N. Petkov, and P. Kruizinga, "Comparison of texture features based on Gabor filters," *IEEE Transactions on Image Processing*, vol. 11, pp. 1160–1167, 2002.

20 J. P. Jones and L. A. Palmer, "An evaluation of the two-dimensional Gabor filter model of simple receptive fields in cat striate cortex," *Journal of Neurophysiology*, vol. 58, pp. 1233–1258, 1987.

21 A. K Jain, "*Fundamentals of Digital Image Processing*," Prentice-Hall, 1989.

22 F. Bianconi and A. Fernández, "Evaluation of the effects of Gabor filter parameters on texture classification," *Pattern Recognition*, vol. 40, pp. 3325–3335, 2007.

23 B. S. Manjunathi and W. Y. Ma, "Texture features for browsing and retrieval of image data," *IEEE Transactions on Pattern Analysis and Machine Intelligence*, vol. 18, pp. 837–842, 1996.

24 E. Persoon and K. S. Fu, "Shape discrimination using Fourier descriptors," *IEEE Transactions On Systems, Man and Cybernetics*, vol. 7, pp. 170–179, 1977.

25 D. Zhang and G. Lu, "Review of shape representation and description techniques," *Pattern Recognition*, vol. 37, pp. 1–19, 2004.

26 S. Osowski and D. D. Nghia, "Fourier and wavelet descriptors for shape recognition using neural networks- a comparative study," *Pattern Recognition*, vol. 35, pp. 1949–1957, 2002.

27 G. Chuang and C. J. Kuo, "Wavelet descriptor of planar curves: Theory and application," *IEEE Transactions on Image Processing*, Vol. 5, pp. 56–70, 1996.

28 P. Wunsch and A. F. Laine, "Wavelet descriptors for multiresolution recognition of handwritten characters," *Pattern Recognition*, vol. 28, pp. 1237–1249, 1995.

29 D. H. Ballard and C. M. Brown, "*Computer Vision*," Prentice-Hall, 1982.

30 P. Mukhopadhyaya and B. B. Chaudhuri, "A survey of Hough transform," *Pattern Recognition*, pp. 993–1010, 2015.

31 R. C. Gonzalez and R. E. Woods, "*Digital Image Processing*," Pearson, 2018.

32 M. Sonka, V. Hlavac, and R. Boyle, "*Image Processing, Analysis, and Machine Vision*," Thomson Pub, 3rd Edition, 2008.

33 L. L. Scharf, "*Statistical Signal Processing: Detection, Estimation, and Time Series Analysis*," Addison-Wesley Pub, 1991.

34 M. K. Hu, "Visual pattern recognition by moment invariants," *IEEE Transactions on Information Theory*, vol. 8, pp. 179–187, 1962.

35 M. R. Teague, "Image analysis via the general theory of moments," *Journal of the Optical Society of America*, vol. 70, pp. 920–930, 1980.

36 C. H. Teh and R. T. Chin, "On image analysis by the methods of moments," *IEEE Transactions on Pattern Analysis and Machine Intelligence*, vol. 10, pp. 496–513, 1988.

37 A. Khotanzad and J. H. Lu, "Classification of invariant image representation using a neural network," *IEEE Transactions on Acoustics, Speech and Signal Processing*, vol. 38, pp. 1028–1038, 1990.

14

Morphological Operations

Mathematical morphology has its origin in studies related to biological forms and structures. However, the field evolved in the 1960s as a powerful tool for image analysis in quantitative electron microscopy. In this context, the primary objective is to analyze the topological or structural properties of objects or regions in digital images for extracting information useful to represent shapes, boundaries, skeletons, and convex hull.

The theoretical foundations of mathematical morphology, along with its wide range of applications, were formally introduced in several seminal works, including [1]. Since then, morphological operations have found numerous applications in areas such as biomedical image processing, remote sensing, astronomy, and computer vision and robotics. Recent emerging applications are related to nonlinear image filtering, edge detection, image segmentation, shape representation, image recognition, and image encoding.

The mathematical morphology is based on set-theoretic concepts and nonlinear operations. A wide variety of morphological algorithms can be constructed from a set of primitive operations using a well-established morphological algebra. These operations are commonly applied to binary digital images, although they can also be extended to grayscale and color images [2–6].

In this chapter, we present morphological operations and applications for various digital image processing tasks. These operations involve nonlinear mappings that locally modify the geometric structure of images within a neighborhood by applying a set of operations via some compact sets of simple shapes and sizes known as *structuring elements*. The choice of the structuring element and its shape and size plays a crucial role on the outcome of these operations and hence is highly application dependent.

We begin by defining some key concepts from set theory specifically for morphological operations. Digital grayscale images are represented by 2-D *functions* whose domains are subsets of $\mathbb{Z}^2$ (i.e. discrete spatial coordinates), and whose ranges (pixel intensity) are subsets of $\mathbb{R}$ or $\mathbb{Z}$, depending on the absence or presence of pixel intensity quantization. Here, $\mathbb{Z}$ and $\mathbb{R}$ refer to sets of integers and real numbers, respectively. In contrast, binary images are defined by 2-D *subsets* corresponding to the foreground pixels with values that are usually one. Consequently, in the sequel we use the term *subsets* for binary images and *functions* for grayscale images [3]. Note that a grayscale image can be represented as a *family* of binary images (or subsets) obtained via thresholding at all possible gray levels. Binary images are represented by an ensemble set $B \subseteq \mathbb{Z}^2$. Thus, a binary image X is a subset of B, i.e. $X \subseteq B$. Additionally, we denote set $S \subset B$ to stand for a compact set of *structuring elements* consisting of small size and simple shapes, e.g. 3×3 cross-shaped. The selection of different structuring elements results in different morphological operations when applied to an image. A pixel in the foreground region (nonzero) of a binary image X, or a structuring element S, is usually represented by a vector of its spatial coordinates, i.e. $\underline{z}$ or $\underline{s}$, respectively.

Our treatment in this chapter begins with morphological operations for binary images consisting of foreground and background pixels. We will then extend these concepts to grayscale and color images, followed by applications of morphological operations in some specific digital image processing tasks.

14.1 Basic Morphological Operations – Binary Images

14.1.1 Dilation and Erosion

We begin by introducing several fundamental operations that form the basis for constructing more advanced morphological operators. Although these operations differ from those studied in the previous chapters, they still follow the general paradigm of mapping an input image to an output image.

The first two operations are *translation* and *reflection*. Translation of a binary image X by a vector $\underline{s} \in \mathbb{Z}^2$ is denoted by $X \pm \underline{s} = \{\underline{z} \pm \underline{s} : \underline{z} \in X\}$, which translates every pixel location $\underline{z} \in X$ by $\pm\underline{s}$. Reflection is another important basic operation defined as, $\check{X} = -X = \{-\underline{z} : \underline{z} \in X\}$, which represents mirror reflection of X about the origin. Equivalently, $\check{X}$ can be viewed as the rotated version of X about the origin by 180°.

Next, we define Minkowski Addition and Subtraction [2], both of which can be represented in terms of translation operation. Given a binary image X and structuring element S, the Minkowski addition is given by:

$$X \oplus S = \bigcup_{\underline{s} \in S} (X + \underline{s}) = \{\underline{z} + \underline{s} : \underline{z} \in X,\ \underline{s} \in S\}, \tag{14.1}$$

which gives an image obtained by translating X by each element of S and then forming the union of the translated images. The Minkowski subtraction of binary image X and structuring element S also uses translation and is given by:

$$X \ominus S = \bigcap_{\underline{s} \in S} (X + \underline{s}) = \{\underline{z} + \underline{s} \in X,\ \forall \underline{s} \in S\}, \tag{14.2}$$

where in this case, every pixel in X is translated by each vector in S and the intersection of the translated images is formed.

More general versions of these operations are known as *Dilation* and *Erosion,* which are the cornerstone of many morphological operations [1–5]. The dilation operation results in the growth of pixels along the boundaries of regions within an image, while erosion removes pixels around the boundaries. Dilation is defined as:

$$\mathcal{D}(X, S) = X \oplus S = \{\underline{z} : (\check{S} + \underline{z}) \cap X \neq \emptyset\}. \tag{14.3}$$

That is, dilation can be viewed as the non-empty set of intersections of the original binary image with the reflected and translated version of the structure element $\check{S}$. The process is graphically similar to the 2-D linear convolution process in the sense that the structuring element (similar to the PSF of a 2-D system) is rotated about the origin by 180° and then translated by $\underline{z}$ with respect to the origin of $\check{S}$ and overlayed on the image X. Then, the result intersected with the original image as opposed to forming a sum of the products of the overlapped pixels as in 2-D linear convolution. Figure 14.1a–c shows a simple structural element with three pixels, a binary image of the letter “F” and the dilated version. Note that in the next several figures, white pixels represent the foreground, dark gray pixels correspond to the background, and light gray pixels in S are “don’t care” and should be ignored. The origin of the structuring element is marked by “X.” As can be seen, the dilated image contains larger number of foreground pixels, i.e. dilation has the effect of *filling* an image.

Erosion of X by S is defined as:

$$\mathcal{E}(X, S) = X \ominus S = \{\underline{z} : (S + \underline{z}) \subset X\}, \tag{14.4}$$

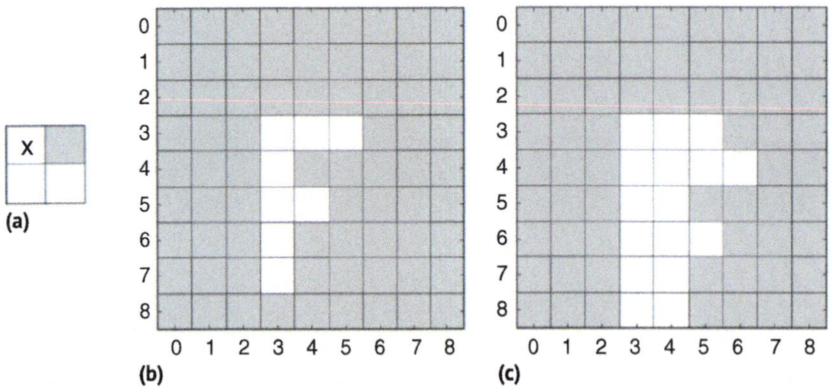

Figure 14.1 (a) Structuring element, (b) a binary image, and (c) dilated version.

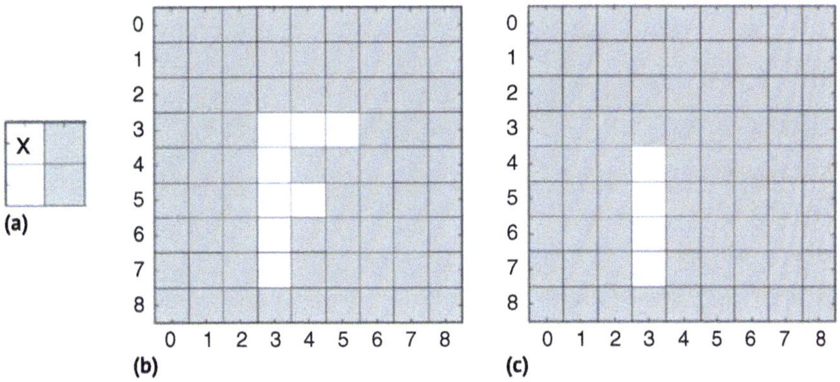

Figure 14.2 (a) Structuring element, (b) a binary image, and (c) eroded version.

that is, the set of translated points of S contained in the image set X. Note that the same symbols are used for Minkowski addition and subtraction. The reason being when $S = \check{S}$, i.e. invariant with respect to reflection, the definitions in (14.3) and (14.4) reduce to those in (14.1) and (14.2). This symmetry property is usually satisfied for most structuring elements considered here. Figure 14.2a–c shows the structuring element with two pixels, a binary image of the letter "F" and the eroded result, respectively. In this case, the eroded image contains fewer foreground pixels, i.e. erosion has the effect of *shrinking* an image.

14.1.1.1 Algebraic Properties

The following are the key algebraic properties of the dilation and erosion operations [2].

1. ***Commutative:***

$$X \oplus S = S \oplus X = \bigcup_{\underline{s} \in S} (X + \underline{s}) = \bigcup_{\underline{x} \in X} (S + \underline{x}), \qquad \text{or} \qquad \mathcal{D}(X, S) = \mathcal{D}(S, X).$$

2. ***Associative:***

$$X \oplus (S_1 \oplus S_2) = (X \oplus S_1) \oplus S_2, \qquad \text{or} \qquad \mathcal{D}\big(X, \mathcal{D}(S_1, S_2)\big) = \mathcal{D}\big(\mathcal{D}(X, S_1), S_2\big).$$

3. ***Distributive:***
Dilation is distributive over the union operation, i.e.

$$X \oplus (S_1 \cup S_2) = (X \oplus S_1) \cup (X \oplus S_2), \qquad \text{or} \qquad \mathcal{D}\big(X, (S_1 \cup S_2)\big) = \mathcal{D}(X, S_1) \cup \mathcal{D}(X, S_2).$$

However, erosion of union satisfies

$$X \ominus (S_1 \cup S_2) = (X \ominus S_1) \cap (X \ominus S_2), \qquad \text{or} \qquad \mathcal{E}\big(X, (S_1 \cup S_2)\big) = \mathcal{E}(X, S_1) \cap \mathcal{E}(X, S_2).$$

Note that the distributive property is not valid for intersection operation.

4. ***Translation Invariance:***

$$X \oplus (S + \underline{s}) = (X \oplus S) + \underline{s}$$

that is, dilation of an image with a translated structuring element is equivalent to translated version of their dilation result. This property is a special case of associative property when $S_2 = \{\underline{s}\}$. Also, we have:

$$X \ominus (S + \underline{s}) = (X \ominus S) + \underline{s}.$$

5. ***Duality:***

$$X \oplus S = (X^c \ominus S)^c, \quad \text{and} \quad X \ominus S = (X^c \oplus S)^c$$

where ^{c} represents *complement* operation, i.e. $X^c = \{\underline{z} : \underline{z} \notin X\}$.

Table 14.1 Morphological dilation and erosion properties.

Property	Formulation
Commutative	$X \oplus S = S \oplus X$
Associative	$X \oplus (S_1 \oplus S_2) = (X \oplus S_1) \oplus S_2$
Distributive	$X \oplus (S_1 \cup S_2) = (X \oplus S_1) \cup (X \oplus S_2)$
Translation Invariance	$X \oplus (S + \underline{s}) = (X \oplus S) + \underline{s}$
Duality	$X \oplus S = (X^c \ominus S)^c$ and $X \ominus S = (X^c \oplus S)^c$
Erosion of Union	$X \ominus (S_1 \cup S_2) = (X \ominus S_1) \cap (X \ominus S_2)$
Erosion of Dilation	$X \ominus (S_1 \oplus S_2) = (X \ominus S_1) \ominus S_2$
Anti-distributive	$X \oplus (S_1 \cap S_2) \subset (X \oplus S_1) \cap (X \oplus S_2)$
Increasing 1	If S is fixed and $X_1 \subset X$, then $\mathcal{D}(X_1, S) \subset \mathcal{D}(X, S)$
Increasing 2	If S is fixed and $X_1 \subset X$, then $\mathcal{E}(X_1, S) \subset \mathcal{E}(X, S)$
Decreasing	If X is fixed and $S_1 \subset S_2$, then $\mathcal{E}(X, S_1) \supset \mathcal{E}(X, S_2)$

These and other important algebraic properties of morphological dilation and erosion operations are given in Table 14.1. These operations are used in the following subsections to arrive at other key operators.

Example 14.1 *Prove the decreasing property in Table 14.1.*

Solution: Since $S_1 \subset S_2$, according to the definition of erosion $S_2 + \underline{z} \subset X$ directly implies $S_1 + \underline{z} \subset X$. Thus, $\underline{z} \in \mathcal{E}(X, S_2)$ $\Rightarrow \underline{z} \in \mathcal{E}(X, S_1)$ and the proof follows.

14.1.2 Opening and Closing

In addition to the basic morphological operators discussed in the previous section, in this section, we introduce two additional operators that play a significant role in image analysis. These are *opening* and *closing* operations that are constructed using dilation and erosion. The opening operation is defined as:

$$\mathcal{O}(X, S) = X \circ S = (X \ominus S) \oplus S. \tag{14.5}$$

Alternatively, the opening of the foreground image X by structuring element S can be generated by taking the union of all subsets of translated versions of S that fit inside X, that is:

$$\mathcal{O}(X, S) = \bigcup \{(S + \underline{z}) : (S + \underline{z}) \subset X\}. \tag{14.6}$$

Thus, this operator is also called *fitting* operator. To see the effects of this operator on an image, consider the image together with the structuring element in Figure 14.3. The result of opening operation is shown in Figure 14.3c. As can be seen, this operator removed the transition pixel (or noise around the boundary) and also the narrow segment that connected the two parts. In general, depending on the choice of the structuring element, opening operator smooths sharp transitions and corners and may result in disjoining regions that are narrowly connected.

The closing operation is defined as:

$$\mathcal{C}(X, S) = X \bullet S = (X \oplus S) \ominus S. \tag{14.7}$$

Alternatively, closing can be represented as a union of all subsets of translated versions of S that roll around the outside of X and not the inside as opposed to opening, i.e.

$$\mathcal{C}(X, S) = \left[\bigcup \{(S + \underline{z}) : (S + \underline{z}) \cap X = \emptyset\}\right]^c. \tag{14.8}$$

Figure 14.4 illustrates the effect of closing operation on the letter "H" image using the same structuring element in Figure 14.3a. In contrast to the opening operation that removes narrow links, the result of closing operation produces an

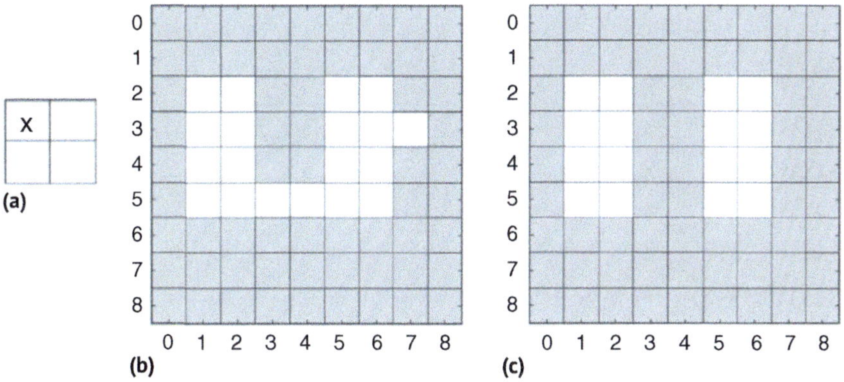

Figure 14.3 (a) Structuring element, (b) a binary image, and (c) processed image by *opening* operation.

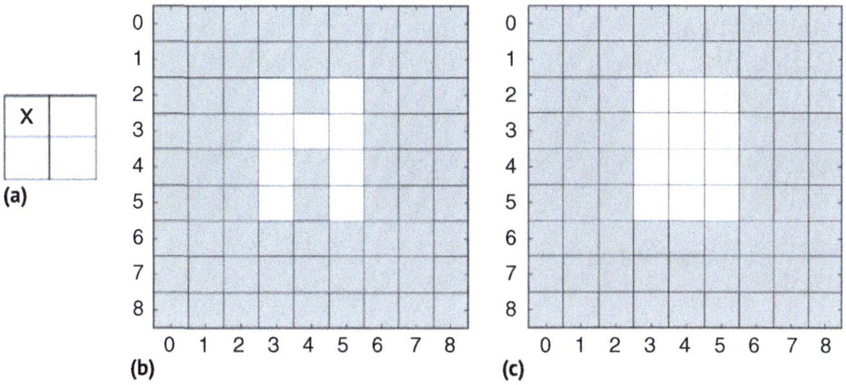

Figure 14.4 (a) Structuring element, (b) a binary image, and (c) processed image by *closing* operation.

image in which the gaps and holes have been filled. Both opening and closing can be viewed as nonlinear filters with smoothing effects of the contours. Clearly, the shape and size of the structuring element determines the nature and degree of smoothing.

As a final note, if $\mathcal{O}(X, S) = X$ the image is called *S-open*. Similarly, if $\mathcal{C}(X, S) = X$ the image is called *S-close*. Table 14.2 lists important properties [2] of opening and closing operations. While many of these properties are intuitively clear some have important practical implications. In particular, the last property $\mathcal{O}(X, Y) \subset \mathcal{O}(X, S)$, when Y is S-open, implies that Y removes more features from image X than S. Similarly, its dual, i.e. $\mathcal{C}(X, Y) \subset \mathcal{C}(X, S)$ states that the structuring element S generates more smoothing (i.e. filling gaps) than Y. These properties are fundamental in developing morphological filters.

Example 14.2 *Prove the S-open idempotence 1 property in Table 14.2.*

Solution: We want to show if image Y is S-open, i.e. $\mathcal{O}(Y, S) = Y$, then (i) $\mathcal{O}\big(\mathcal{O}(X, Y), S\big) = \mathcal{O}(X, Y)$ and (ii) $\mathcal{O}\big(\mathcal{O}(X, S), Y\big) = \mathcal{O}(X, Y)$.

For part (i), let $\underline{z} \in \mathcal{O}(X, Y)$, then according to the definition of opening operation, we have $\underline{z} \in Y + \underline{y} \subset X$ or $\underline{z} \in \mathcal{O}(Y, S) + \underline{y} \subset X$ since Y is S-open. Again, we have $\underline{z} \in S + \underline{s} + \underline{y} \subset Y + \underline{y} \subset X$, which implies that $\underline{z} \in S + (\underline{s} + \underline{y}) \subset \mathcal{O}(X, Y)$. Thus, $\underline{z} \in \mathcal{O}\big(\mathcal{O}(X, Y), S\big)$ or $\mathcal{O}(X, Y) \subset \mathcal{O}\big(\mathcal{O}(X, Y), S\big)$. Finally, using anti-extensivity property the reverse inclusion follows, i.e. $\mathcal{O}(X, Y) = \mathcal{O}\big(\mathcal{O}(X, Y), S\big)$.

For part (ii), we first use the anti-extensivity property $\mathcal{O}(X, S) \subset X$. Using increasing, i.e. $\mathcal{O}\big(\mathcal{O}(X, S), Y\big) \subset \mathcal{O}(X, Y)$. Now, since Y is S-open $\mathcal{O}(X, Y) \subset \mathcal{O}(X, S)$ by the S-open decreasing property. Next, we invoke the idempotence property, i.e. $\mathcal{O}(X, Y) = \mathcal{O}\big(\mathcal{O}(X, Y), X\big) \subset \mathcal{O}\big(\mathcal{O}(X, S), Y\big)$, which implies (ii).

Table 14.2 Morphological opening and closing properties.

Property	Formulation
Complement 1	$\mathcal{O}(X, S) = \mathcal{C}(X^c, S)^c$
Complement 2	$\mathcal{C}(X, S) = \mathcal{O}(X^c, S)^c$
Anti-extensivity	$\mathcal{O}(X, S) \subset X$
Increasing 1	If $Y \subset X$ then $\mathcal{O}(Y, S) \subset \mathcal{O}(X, S)$
Idempotence 1	$\mathcal{O}\big(\mathcal{O}(X, S), S\big) = \mathcal{O}(X, S)$
Extensivity	$X \subset \mathcal{C}(X, S)$
Increasing 2	If $Y \subset X$ then $\mathcal{C}(Y, S) \subset \mathcal{C}(X, S)$
Idempotence 2	$\mathcal{C}\big(\mathcal{C}(X, S), S\big) = \mathcal{C}(X, S)$
Translation Invariance 1	$\mathcal{O}(X + \underline{z}, S) = \mathcal{O}(X, S) + \underline{z}$
Translation Invariance 2	$\mathcal{C}(X + \underline{z}, S) = \mathcal{C}(X, S) + \underline{z}$
***S*-open Idempotence 1**	If Y is S-open then $\mathcal{O}\big(\mathcal{O}(X, Y), S\big) = \mathcal{O}(X, Y)$ and $\mathcal{O}\big(\mathcal{O}(X, S), Y\big) = \mathcal{O}(X, Y)$
***S*-open Idempotence 2**	If Y is S-open then $\mathcal{C}\big(\mathcal{C}(X, Y), S\big) = \mathcal{C}(X, Y)$ and $\mathcal{C}\big(\mathcal{C}(X, S), Y\big) = \mathcal{C}(X, Y)$
***S*-open Decreasing**	If Y is S-open then $\mathcal{O}(X, Y) \subset \mathcal{O}(X, S)$ and $\mathcal{C}(X, S) \subset \mathcal{C}(X, Y)$

14.2 Morphological Transforms – Binary Images

14.2.1 Hit-or-Miss

The hit-or-miss transformation is a morphological operation that is typically used to detect specific local patterns in binary images. These local patterns could simply correspond to, e.g. corners or other small structures of interest. Unlike the operations introduced earlier, the hit-or-miss transform simultaneously looks for such local patterns in both foreground and background. As a consequence, there are two disjoint structuring elements, S_1, which looks for a match in the foreground, and S_2, which searches for the relevant patterns in the background. We collectively represent them as a *composite structuring element* $\mathrm{S} = (S_1, S_2)$ that contains both foreground and background pixels. Now, the hit-or-miss transform expressed in terms of erosion operation is:

$$X \odot \mathrm{S} = (X \ominus S_1) \cap (X^c \ominus S_2) \tag{14.9}$$

or

$$X \odot \mathrm{S} = \{\underline{z} : (S_1 + \underline{z}) \subseteq X \text{ and } (S_2 + \underline{z}) \subseteq X^c\}. \tag{14.10}$$

Thus, according to (14.10), the result of the hit-or-miss transform is the set of translations where the first structuring element, S_1, *hits* a match in the foreground of image X, while the second structuring element S_2 *misses* it (or not finding it) in the background. Note that *hitting* a match in X^c is equivalent to missing it in X. Figure 14.5a shows a composite structuring element S consisting of foreground white pixels in S_1 and background dark gray pixels in S_2. This structuring element is designed to extract corner pixels with −90° orientation. Figure 14.5b shows the input image with foreground and background pixels, and Figure 14.5c shows the detected corner pixels by the hit-or-miss transformation. This operation can be considered as a form of template matching discussed in Chapter 8, where the template is the composite structuring element.

14.2.2 Thinning

This morphological operator is defined in terms of the hit-or-miss transform and is primarily used to reduce the thickness of foreground regions in binary images. The basic *thinning* operation is defined as:

$$X \otimes S = X - (X \odot S) \tag{14.11}$$

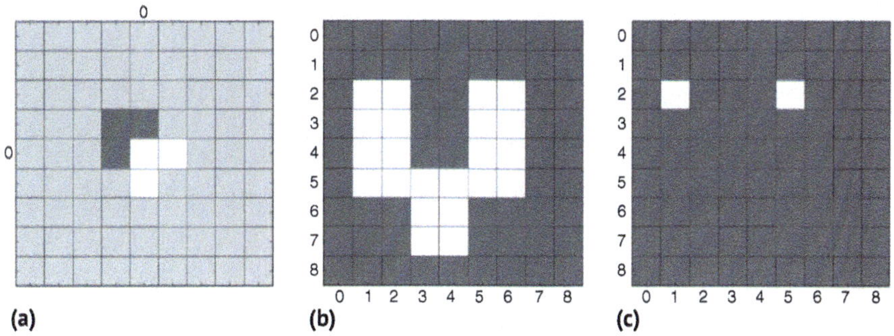

Figure 14.5 (a) Structuring element S consisting of foreground (white) and background (dark gray), (b) a binary image, and (c) hit-or-miss corner detection result.

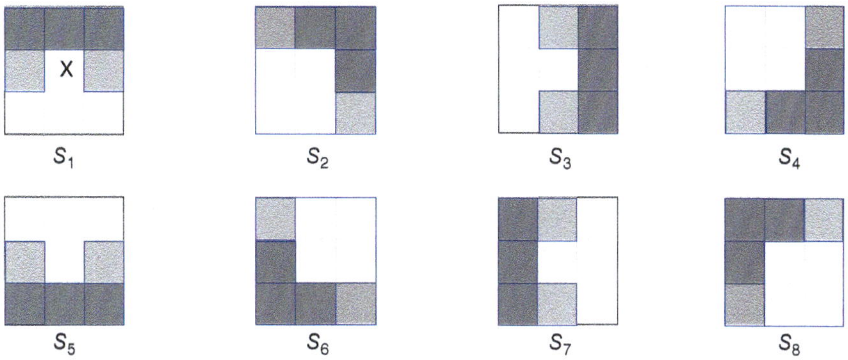

Figure 14.6 Eight structuring elements used for thinning operation. The origin is marked with X.

where the *difference* operation is with respect to the set. However, the thinning operation is usually implemented in a repeated and sequential manner by using two structuring elements S_1 and S_2 and their rotated versions by 90°, 180°, and 270°, i.e. a total of eight structuring elements that are illustrated in Figure 14.6. Again, white pixels represent the foreground, dark gray pixels represent background, and light gray pixels are don't care. The process of thinning involves translating the origin of a selected structuring element (indicated by X) to each foreground pixel, and if the foreground and background pixels of the template structuring element match the overlayed foreground and background pixels of the image, then the corresponding pixel is set to background; otherwise, it remains unchanged.

If we denote the set of the structuring elements by $\mathcal{S} = \{S_1, \cdots S_8\}$, then the sequential thinning operation using these structuring elements yields:

$$X \otimes \mathcal{S} = (((X \otimes S_1) \otimes S_2) \otimes S_3) \ldots \otimes S_8 \tag{14.12}$$

that is, the image X is first processed using S_1 and the result is processed using S_2 and so on. This process is repeated until the resultant image no longer changes.

Figure 14.7a,b shows the original binary image of the letter "Y" image and its thinned version, respectively. This thinned image is obtained by consecutive application of the structuring elements S_2, S_3, S_4, and S_6 in the first iteration, followed by one iteration of the structuring element S_4, i.e. $Y = ((((X \otimes S_2) \otimes S_3) \otimes S_4) \otimes S_6) \otimes S_4$. Note that the other structuring elements, i.e. S_1, S_5, S_7, and S_8, don't change the image. As can be seen, the thinning operation left a *spur* pixel on the right of the letter "Y". These common unwanted spurious pixels must be removed prior to any automated image analysis. In Section 14.2.6, a *pruning* algorithm is introduced for removing these spurious pixels.

14.2.3 Thickening

This morphological operator is also defined in terms of hit-and-miss transform using:

$$X \circledast \mathcal{S}^c = X \cup (X \odot \mathcal{S}^c) \tag{14.13}$$

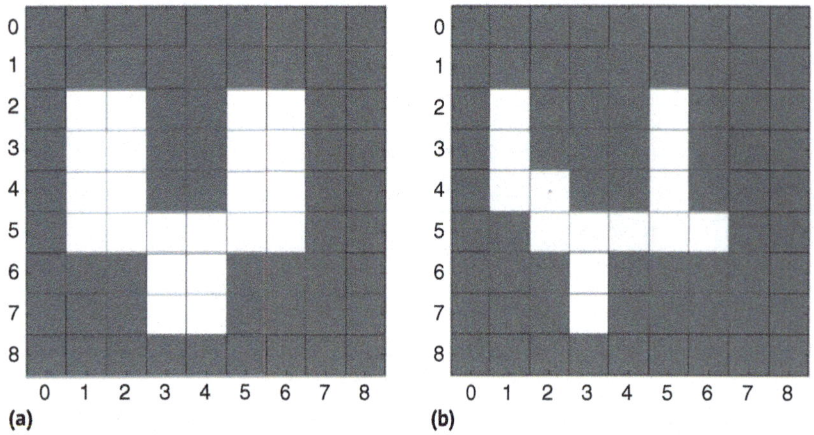

Figure 14.7 (a) Original thick Y image, and (b) thinned image using all structuring elements.

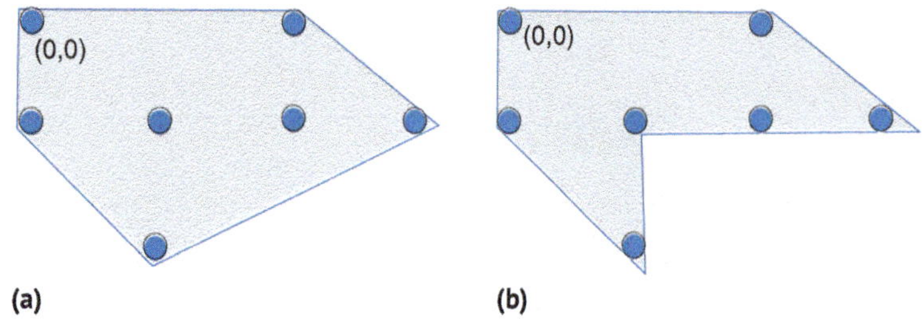

Figure 14.8 (a) Convex hull of the pixel set and (b) a non-convex region.

where $\mathcal{S}^c$ is a set of structuring elements $\mathcal{S}^c = \{S_1^c, \cdots S_8^c\}$ that are complements of those of thinning operator in Figure 14.6. Similar to thinning, the thickening operation in (14.13) is also implemented by repeated iterative applications of the structuring elements in the set S^c until convergence.

The thickening operation is the dual of the thinning operator and is generally used to thicken or grow selected regions of foreground pixels in binary images depending on the choice of the structuring elements. There are many morphological algorithms that use thickening operation including: convex hull construction and skeleton by zone of interest, which will be covered in subsequent sections. The procedure of thickening is analogous to that of thinning, with the key difference that when a match with the translated structuring element is found, the image pixel at the position of the translated origin of the structuring element is set to a foreground pixel, thereby expanding the foreground areas.

14.2.4 Convex Hull

Convex sets play an important role in many morphological operations. Generally speaking, an image set X is said to be convex if, for any pair of points in X, the straight line connecting them lies entirely inside X. For an arbitrary image set X, its *convex hull* denoted by $\mathbb{CH}(X)$ is the smallest (or minimal) convex set containing X. The set difference $\mathbb{CH}(X) - X$ is called the *convex deficiency* of X, which is often used for structural characterization. It is worth noting that the convex hull of the set X is the intersection of all convex sets that contain X. For a digital image, the convex hull corresponds to the smallest convex polygon enclosing all foreground pixels. Formally, a convex polygon is one that does not intersect itself and in which any line segment connecting every pair of points on the edges never lies outside the polygon. Figure 14.8a shows a convex polygon, while Figure 14.8b is a non-convex region.

There are several methods to construct convex polygons for a set of foreground pixels [10]. Here, we present a convex hull approximation algorithm based on the hit-or-miss transform. This algorithm uses four 3×3 structuring elements, S_1, and its three rotated versions S_2, S_3, and S_4 rotated by 90°, 180°, and 270°, respectively. These structuring elements are shown in Figure 14.9. The iterative algorithm for building a convex polygon for a binary image X is given below.

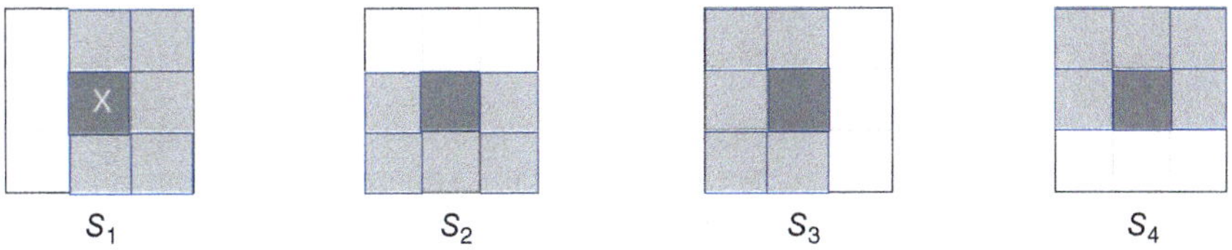

Figure 14.9 Structuring elements for convex hull construction.

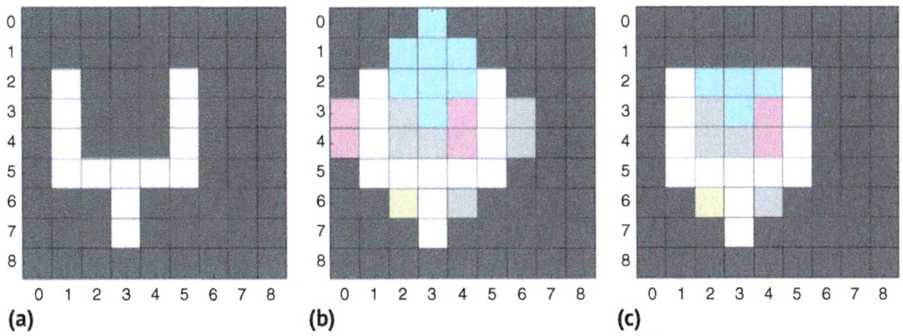

Figure 14.10 (a) Original image, (b) convex polygon result, and (c) convex hull result after trimming.

1. Initialize $k = 0$, $i = 1$, and $Y_1(0) = X$.
2. For fixed i, iterate $Y_i(k) = (Y_i(k-1) \odot S_i) \cup Y_i(k-1)$ with respect to k until convergence, i.e. until $Y_i(k) = Y_i(k-1)$.
3. If $i = 4$, go to Step 4. Otherwise, set $i = i + 1$, $Y_i(0) = X$ and $k = 0$ and go to Step 2.
4. The converged convex hull of X is $\mathbb{CH}(X) = \bigcup_{i=1}^{4} Y_i$.
5. To guarantee that the minimal convex set is achieved, one may need to trim $\mathbb{CH}(X)$ to the bounding box of the original pixel set.

Figure 14.10a shows the original image "Y" in background (dark gray pixels). The structuring element S_1 is first applied to the image using the algorithm described above. The converged result Y_1 after two iterations with the structuring element S_1 added six pixels indicated in light gray color in Figure 14.10b to the original image. The structuring element S_2 is applied only once to produce the converged result Y_2 added one new pixel (light yellow) as shown in Figure 14.10b. The structuring element S_3 is also applied only once to add four new pixels (light pink). Finally, the converged result Y_4 of the structuring element S_4 was obtained after five iterations leading to those newly added pixels that are shown in light blue in Figure 14.10b. The union of all these results yields the convex polygon in Figure 14.10b. Using the trimming scheme in Step 5 of the above algorithm, the final minimal set (or convex hull) is shown in Figure 14.10c.

14.2.5 Skeleton

As the name suggests, the *skeleton* operation removes foreground pixels from a binary image to leave behind only the skeleton, which preserves the essential structure and connectivity of the object's shape. This makes it particularly useful for applications in object recognition based on shape, structure, and topology. Alternatively, one can view the skeleton as the loci of the centers of *maximal disks* that fit within the image X. A maximal disk is the largest circular subset of X that touches the foreground boundary in at least two points. To form a complete skeleton, the size of the maximal disks is varied in order to reach every corner of the foreground image.

Morphological skeleton uses erosion and opening operations [5, 7, 8, 17] and can be viewed as a structural thinning process that produces a representation with only one pixel wide. A key advantage of this morphological operation is that it allows the reconstruction of the original image from its skeleton subsets, if needed.

Before we describe the morphological skeleton algorithm, let us first denote the symbol kS to represent a sequence of k successive operations on an image using structuring element S. For example, $X \ominus kS$ implies $X \ominus kS = (((X \ominus S) \ominus S) \ldots \ominus)S$, i.e. k successive erosion on image X. For $k = 0$, by definition $0S = (0, 0)$, i.e. the origin. Then, the skeleton algorithm is given as follows:

1. Initialize $k = 0$.
2. Compute skeleton subsets $L_k(X) = (X \ominus kS) - [(X \ominus kS) \circ S]$.

3. If $(X \ominus kS) = \emptyset$ set $K = k - 1$ and proceed to Step 4. Otherwise, increment $k \leftarrow k + 1$ and repeat Step 2.
4. Form the final skeleton of X using the union of all the skeleton subsets, i.e. $L(X) = \bigcup_{k=0}^{K} L_k(X)$.

As pointed out earlier, the skeleton algorithm has a nice property that allows reconstructing the original image from the skeleton subsets or *shape components* $L_k(X)$ which are disjoint. That is, we can recover the original image using:

$$X = \bigcup_{k=0}^{K} (L_k(X) \oplus kS) \tag{14.14}$$

where $L_k(X) \oplus kS = (((L_k(X) \oplus S) \oplus S) \ldots \oplus)S$. This reconstruction is multiscale for different values of k. More specifically, small values of k lead to reconstruction that lack smoothness of the boundary, while the shape components corresponding to large k relate to the interior parts of X that are shaped similar to kS. It should be noted that one cannot reconstruct X from $L(X)$. The morphological skeleton has many important properties. First, the skeleton subsets $L_k(X)$, $k \in [0, K]$ are mutually disjoint, and the overall skeleton set transformation $L(X)$ is translation invariant, anti-extensive and idempotent (Problem 14.16).

Figure 14.11a shows the original image X used for the skeleton operation. A 3×3 cross-shaped structuring element is used here. Figure 14.11b,c gives the corresponding eroded images $(X \ominus kS)$ for $k = 1, 2$, respectively. Note that for $k = 0$, $X \ominus 0S = X$. In Figure 14.11d, the light blue, gray, and pink pixels correspond to skeleton subsets $L_0(X)$, $L_1(X)$, and $L_2(X)$, respectively. The union of these subsets produces the complete skeleton in Figure 14.11e. This example highlights several important observations. First, depending on the image and the choice of the structuring element, the resulting skeleton may be disconnected as in this example. Second, due to the discrete nature of digital images, one cannot closely approximate a maximal disk. Third, as can be noted, the very first subset, i.e. $L_0(X)$, typically captures the corners, tips, or other sharp transition points of the shape that are in contact with the outside or background, while $L_1(X)$ and $L_2(X)$ progressively capture more information about the internal structural characteristics of the object. Finally, the skeleton process is sensitive to noise and small boundary perturbations, which may lead to undesirable artifacts.

A closely related operation to the morphological skeleton is the Medial Axis Transform (MAT) [11, 12], which is usually generated by applying the *distance transform* to the foreground image. The distance transform maps a binary foreground image to a grayscale image in which the pixel intensities are determined based on a distant measure of each pixel to the closest boundary point. That is, the MAT is defined as the loci of local maxima on the distance map. Consequently, MAT

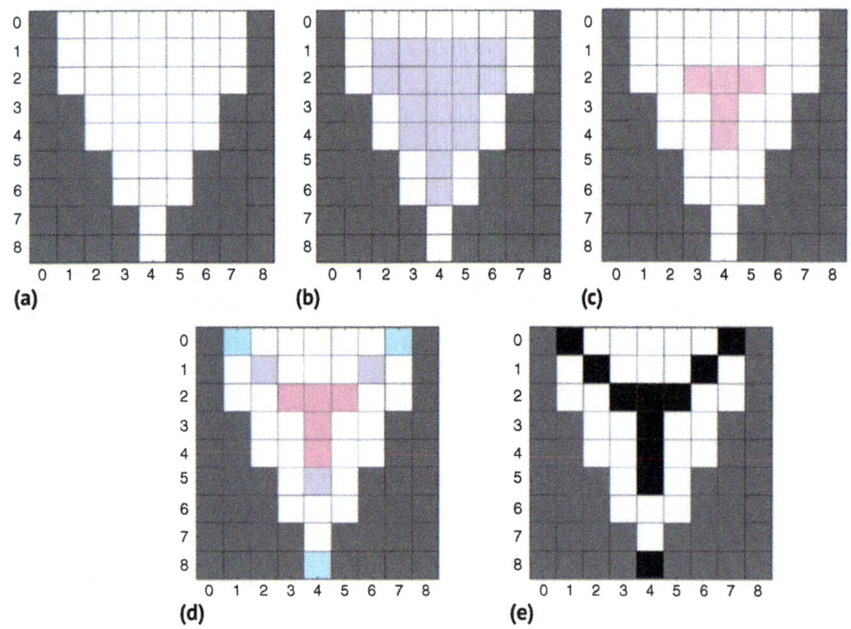

Figure 14.11 (a) Original binary image, (b) and (c) results of erosion operations for $(X \ominus kS)$ for $k = 1, 2$, respectively, (d) skeleton subsets, and (e) final skeleton result.

can be viewed as the skeleton with the maximal disk radius retained at each point. Owing to the latter property, the original image can be reconstructed from the MAT result, although this was not possible from the skeleton $L(X)$.

One way to implement MAT using the morphological operations is to apply successive erosion operations with a 3×3 square or cross-shaped structuring element that represents the distance measure. The process is repeated until no more foreground regions are eroded away. Then, a gray level value is assigned to each pixel according to the number of iterations it took to erode that pixel. The result is the distance transform of the image. As with the skeleton algorithm, one major issue with most of the standard algorithms for generating MAT, is the sensitivity to noise or small perturbations along the boundary, which can produce *spikes* in the results. These spiky imperfections may, in fact, negatively impact the accuracy of structural and geometrical definition of the MAT or skeleton. The other issue, of course, is the computational demand of the algorithm. There are many published methods to improve the geometrical stability and computational efficiency of MAT. As an example, in [13] the authors proposed an efficient method, referred to as Q-MAT, that uses quadratic error minimization to compute a compact, geometrically accurate and structurally simple representation.

14.2.6 Pruning

The morphological *pruning* operation is typically used to remove or prune spurious pixels or small offshoot branches that are often generated when using skeleton or thinning processes [8, 9]. These spurious structures do not represent the core structure of the original object, and their presence may negatively affect subsequent image analysis tasks, e.g. in hand-printed character recognition or in automatic part recognition in assembly line. Therefore, they must be removed using the pruning operation before any further high-level processing. An example was presented in Figure 14.7c where a spur pixel appeared on the right side of the letter "Y" after the thinning process.

The pruning operation uses two structuring elements S_1 and S_2 and their rotated versions by $90°$, $180°$, and $270°$, resulting in a total of eight structuring elements, as shown in Figure 14.12. These specific structuring elements are designed to detect and remove end-pixels of the spurious off-shoot branches with different orientations one at a time. Once again, white pixels represent the foreground, dark gray pixels represent background, and light gray pixels are "don't care." The pruning procedure uses a thinning, hit-or-miss transform, and dilation operation and consists of four steps, which are given below.

1. Apply successive thinning operation using the set of structuring elements $\mathcal{S} = \{S_1, S_2, \ldots, S_8\}$ using $X_k = X_{k-1} \otimes \mathcal{S}$, where $X_0 = X$ until convergence, i.e. until $X_k = X_{k-1}$. Then set $X_k = P_1$ and go to the nest step. Note that the number of thinning iterations k is determined by the maximum length of the spurious offshoot branches that must be pruned.
2. Apply the hit-or-miss transform to the image P_1 using structuring elements S_i and form the union of the results, i.e. $P_2 = \bigcup_{i=1}^{8} P_1 \odot S_i$. The goal of this step is to get the end-points of the thinned image P_1.
3. Perform dilation using a 33 square structuring element D of all ones, followed by intersection with the original image to restore the original image with the end-pixels added but without the spurious offshoots, i.e. $P_3 = (P_2 \oplus D) \cap X$.
4. Form the union of P_1 and P_3 to get the final filtered image, i.e. $P = P_1 \cup P_3$.

Here, we consider the image in Figure 14.7b as an input to the pruning operation. This image is reproduced in Figure 14.13a for convenience. The goal of pruning is to remove the spur pixel on the right side of the letter "Y." Since the length of this offshoot branch is one pixel, only a single iteration of thinning operation with the structuring elements in $\mathcal{S}$ is

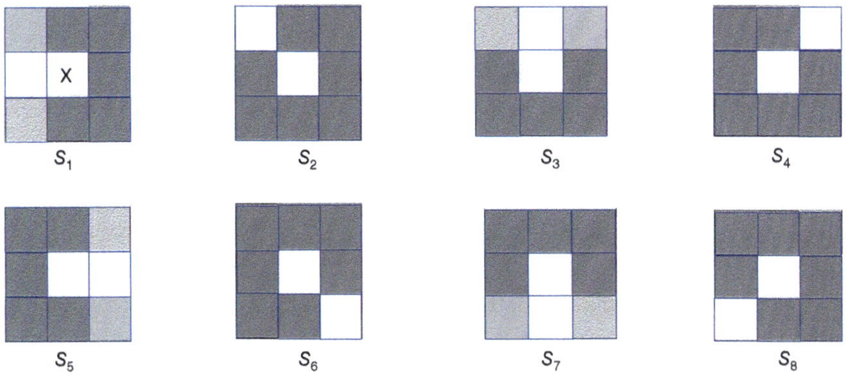

Figure 14.12 Eight structural elements used for pruning operation. The origin is marked with X.

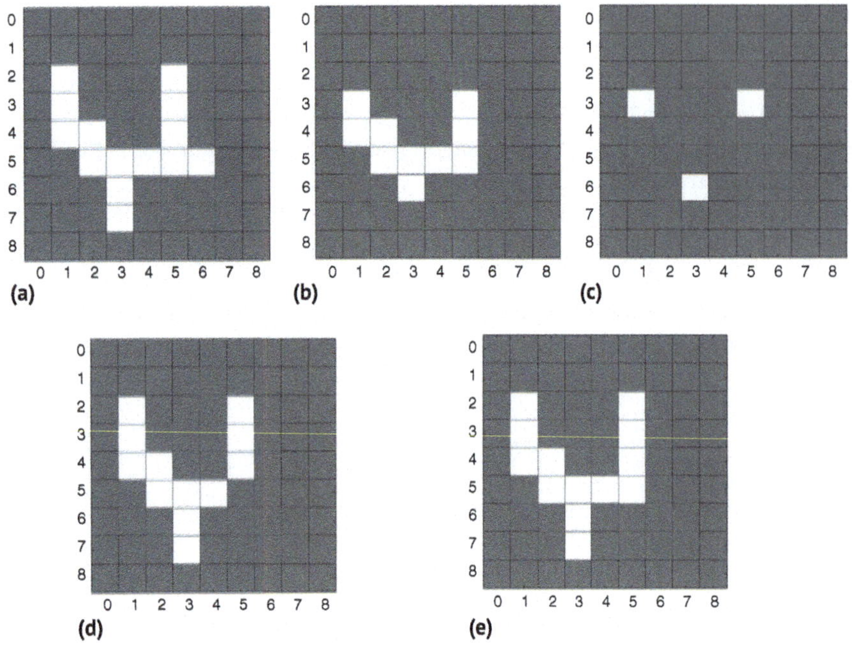

Figure 14.13 (a) Original binary image, (b) thinned image P_1, (c) end-pixels image P_2 (d) dilation and intersection result P_3, and (e) final image $P = P_1 \cup P_3$.

required. The thinning result P_1 is shown in Figure 14.13b. Next, the end-pixels of image P_1 are extracted using the union of hit-or-miss operation with the same structuring elements, producing image P_2 illustrated in Figure 14.13c. The end-points are then dilated using a 3×3 square structuring element consisting of all ones and the resultant image is intersected with the original image to avoid overgrowth of foreground pixels. This process yields image P_3, shown in Figure 14.13d. Finally, the union of P_3 and P_1 is generated to provide the final filtered image P illustrated in Figure 14.13e. As can be seen, the undesirable offshoot branch on the right side of the original image is successfully removed.

14.3 Morphological Operations – Grayscale Images

In this section, we extend binary morphological image processing to general grayscale images. Morphological operations for grayscale images have numerous applications including contrast enhancement, edge detection, noise removal and smoothing (e.g. speckle noise), and image segmentation.

The basic binary morphological operations in the previous sections can be extended to grayscale images by using simple modifications. Here, we consider the grayscale image $x(\underline{z})$, where $\underline{z} \in \mathbb{Z}^2$ represents the spatial coordinates vector, and $x \in \mathbb{Z}$ or $x \in \mathbb{R}$ depending on whether the image is quantized. We further assume that the structuring element S is constant over its region of support and therefore can be treated as a binary set. Grayscale structuring elements are also used in some specific cases [8].

There are various ways to extend morphological operations to grayscale images. One classical approach introduced by Serra [1, 5] is to represent the image in terms of its threshold set. That is, define $\mathbb{T}_t(x) = \{\underline{z} : x(\underline{z}) \geq t, \ \forall t\}$ for all threshold levels t. The threshold set satisfies: (i) $t_2 > t_1 \Rightarrow \mathbb{T}_{t_2}(x) \subseteq \mathbb{T}_{t_1}(x)$ and (ii) reconstruction of image $x(\underline{z})$ from its threshold set using $x(\underline{z}) = \max\{t : \underline{z} \in \mathbb{T}_t(x), \ \forall \underline{z}\}$.

The above equation implies that the threshold set of the image $x(\underline{z})$ has all the necessary information about the image to allow its reconstruction. Based on this idea, one can perform grayscale morphological operations between the threshold set of the image and the set of structuring element $\mathcal{S}$. This process gives the threshold set of the output image from which the grayscale output image can be reconstructed. For example, grayscale dilation of image $x(\underline{z})$ by a structuring element set S yields the threshold set $\mathbb{T}_t(x) \oplus S$, from which the grayscale dialed output image can be generated using:

$$[x \oplus S](\underline{z}) = \max\{t : \underline{z} \in \mathbb{T}_t(x) \oplus S\}.$$

An alternative and perhaps more intuitive approach is to view the pixel intensity variations of a grayscale image as a height map with hills and valleys. In this context, morphological operation of an image by a translated structuring element of finite support correspond to performing local operations within the region of overlap. Two important local operations that are used are the min and max filters, which assign to each image pixel a new value equal to the minimum or maximum value in a local neighborhood around that pixel. The shape and size of the neighborhood are determined by the structuring element used.

14.3.1 Grayscale Dilation and Erosion

Let $\underline{z} \in \mathbb{Z}^2$ be the spatial coordinate vector corresponding to the origin of a flat structuring element S. The grayscale dilation operation involves translating the reflected structuring element by vector $\underline{z}$, i.e. $S(\underline{z} - \underline{s})$, over the image and finding the maximum pixel value of the image within the overlap region. Then, at location $\underline{z}$ the dilation result of the image x by the structuring element S is:

$$[x \oplus S](\underline{z}) = \max_{\underline{s} \in S}\{x(\underline{z} - \underline{s})\} \tag{14.15}$$

or

$$[x \oplus \check{S}](\underline{z}) = \max_{\underline{s} \in S}\{x(\underline{z} + \underline{s})\} \tag{14.16}$$

where $\check{S}$ is the reflected version of S about the origin.

This process is also analogous to a 2-D linear convolution with a finite region of support filter, except here we find the maximum value of the pixels instead of the sum of products of the image pixels with the filter PSF. The process is also similar to median filtering, where instead of the median of the pixels within the mask, we choose the maximum. Therefore, we can consider these types of operation as special cases of order statistics [3, 4].

Grayscale erosion operation is also defined in a similar manner with two key differences. First, there is no reflection of the structuring element. Second, it uses local minimum instead of maximum. Thus, erosion of image x by a flat structuring element S at location $\underline{z}$ is given by:

$$[x \ominus S](\underline{z}) = \min_{\underline{s} \in S}\{x(\underline{z} + \underline{s})\} \tag{14.17}$$

where $\underline{z}$ is the coordinate vector of the origin of S. As in the binary case, these basic operations are used to develop other grayscale morphological operations.

Figure 14.14a shows the original image of the letter "i" with four gray levels. The result of the grayscale dilation operation using the structuring element S_1 shown to the left of the original image is illustrated in Figure 14.14b. As expected, grayscale dilation expands regions with high pixel intensities, while suppressing low intensity regions. The extent and direction of the growth depends on the shape of the structuring element used as in the binary case. Figure 14.14c, on the other hand, illustrates the result of grayscale erosion using the structuring element S_2. In contrast to the grayscale dilation, this operation resulted in shrinking the regions with high pixel intensities and hence growing those with low intensities.

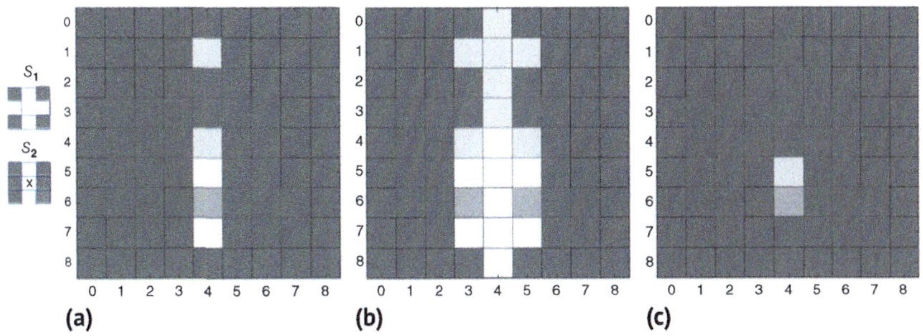

Figure 14.14 (a) Original grayscale image of the letter "i", (b) dilation result with structuring element S_1, and (c) erosion result with structuring element S_2.

14.3.2 Grayscale Opening and Closing

The grayscale dilation and erosion operators can be used to construct grayscale opening and closing operators as in the binary case. The grayscale opening of an image $x(\underline{z})$ with structuring element S involves performing a grayscale erosion followed by dilation:

$$[x \circ S] = (x \ominus S) \oplus S, \tag{14.18}$$

which corresponds to removing small bright features and smoothing boundaries by a sequence of a local minimum followed by a local maximum operation within the support of the structuring element. The grayscale closing operation, on the other hand, involves performing grayscale dilation followed by erosion, i.e.

$$[x \bullet S] = (x \oplus S) \ominus S, \tag{14.19}$$

which corresponds to removing small dark features by a sequence of a local maximum followed by a local minimum operation within the support of the structuring element. The overall effect is to fill small holes within foreground objects.

To show examples of grayscale opening and closing operations, we use the results of the grayscale dilation and erosion operations in Figure 14.14. To obtain the grayscale open version of the image in Figure 14.14a, we apply grayscale dilation to the image in Figure 14.14c using the same structural element S_2. The result of this operation is illustrated in Figure 14.15a, which shows partial recovery of the letter "i." To show an example of the grayscale closing operation, we apply grayscale erosion with the structuring element S_1 to the dilated image in Figure 14.14b. The result of this operation is shown in Figure 14.15b. In this particular example, it happened that the closed image coincided with the original image. However, in general, dilation and erosion operations are not reversible, and exact recovery is not guaranteed.

If we treat a grayscale image as a 3-D height or topographical map with hills and valleys where the third dimension represents pixel intensities, opening removes peaks smaller than the structuring element, whereas the grayscale closing fills valleys smaller than the structuring element. The residuals of these operations known as *top-hat* and *bottom-hat* transforms:

$$\begin{aligned} \delta_T(x) &= x(\underline{z}) - [x \circ S](\underline{z}), \\ \delta_B(x) &= [x \bullet S](\underline{z}) - x(\underline{z}), \end{aligned} \tag{14.20}$$

capture these small scale structures, where $\delta_T(x)$ and $\delta_B(x)$ denote the top-hat and bottom-hat, respectively. The size and shape of the structuring element in these transforms depend on the morphology of the structures to be extracted. For example, to detect bright features of width smaller than l, a top-hat transform with a structuring element slightly larger than l should be considered. These transforms play an important role in grayscale segmentation of localized structures in slow varying background.

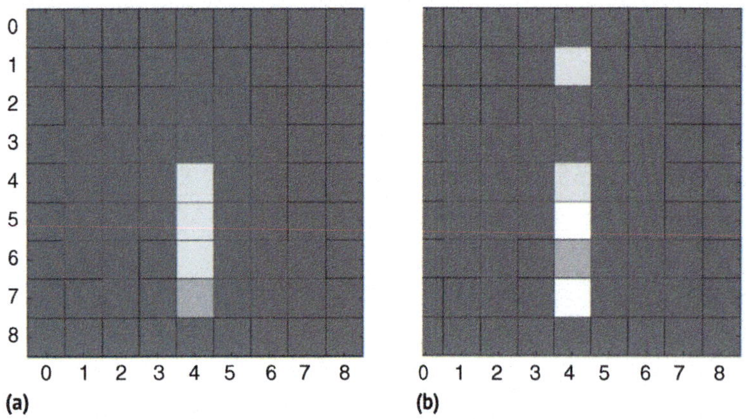

Figure 14.15 (a) Grayscale opened version of image in Figure 14.14a and (b) grayscale closed version of the same image.

14.3.3 Grayscale Skeleton and Shape Representation

In Section 14.2.5, we introduced two morphological operations, namely skeleton and MAT, which can be used to represent the shapes of the foreground regions in binary images. In this section, we extend the skeleton operation to grayscale digital images [3, 16].

Let us consider grayscale digital image $x(\underline{z})$ with $x \in \mathbb{Z}$ and $\underline{z} \in \mathbb{Z}^2$. The structuring element S is assumed to be binary with compact support. The grayscale skeleton procedure is similar to that of the binary in Section 14.2.5 with the exception of a few changes. The skeleton subsets or shape components for grayscale image $x(\underline{z})$ are given by:

$$l_k(x(\underline{z})) = (x \ominus kS)(\underline{z}) - [(x \ominus kS) \circ S](\underline{z}), \quad k \in [0, K], \tag{14.21}$$

where $K = \max\,\{k \geq 0, x \ominus kS \neq \emptyset\}$. Then, the grayscale skeleton can be obtained by point-wise summation of these components,

$$l(x) = \sum_{k=0}^{K} l_k(x). \tag{14.22}$$

Thus, unlike the binary case, the grayscale skeleton is not formed by simple union or summation, but rather by retaining all scale-dependent components. The set $\{l_k(x)\}_{k=0}^{K}$ also conveys the shape information of the grayscale image $x(\underline{z})$ and hence can be used to reconstruct the image using:

$$x \circ kS = \bigcup_{k \leq m \leq K} l_m(x) \oplus mS, \quad k \in [0, K]. \tag{14.23}$$

If all the skeleton subsets are used in (14.23), i.e. $k = 0$, we get $x \circ kS = x$, hence exact reconstruction. This reconstruction process is multiscale, in the sense that components corresponding to small values of k capture fine details and sharp transitions, while larger values of k describe broader, smoother structures.

If only components with $k \geq k_0$ are used, the reconstruction yields a progressively smoother approximation of the original image. Thus, the grayscale skeleton provides a natural framework for multiscale shape representation.

14.3.4 Extension of Morphological Operations to Functions

An extension of morphological operators to functions was first proposed by Sternberg [14] who considered a grayscale image as a 3-D surface with the first two coordinates spanning the domain of the function and the third coordinate corresponding to the range of values of the function (i.e. the pixel intensities). In this topographical map with hills and valleys, the *top surface* refers to the maximum of the set $x(\underline{z})$, $\forall \underline{z}$, that is, $\tau(x) = \max\{t, (\underline{z}, t) \in \mathbb{Z}^3\}$. For example, the top view of a 3-D surface with hills and valleys is the top surface. The *umbra* (shallow in Latin) of the surface includes the set of all the points on or below this surface. Figure 14.16a illustrates both the top surface (purple-shaded area) and its umbra, which is the entire set below the top (blue dotted lines). It is essentially the complete shaded region beneath the 3-D surface. Figure 14.16b shows the example of the letter "Y" top surface (purple) and umbra (light blue). More specifically, we can define the umbra of function $x(\underline{z})$ as:

$$U(x) = \{(\underline{z}, t) \in \mathbb{Z}^3, t \leq x(\underline{z})\} \tag{14.24}$$

Given the umbra, one can reconstruct [3] the function using:

$$x(\underline{z}) = \max\{t : (\underline{z}, t) \in U(x)\}, \quad \forall \underline{z}. \tag{14.25}$$

Now, suppose that we have two grayscale images $x(\underline{z})$ and $y(\underline{z})$ and we would like to perform their dilation and erosion. In other words, we treat one image, for example, y, as a grayscale structuring element.This operation can be carried out by forming the top surface of the dilation or erosion of the umbrae of images y and x. That is, dilating or eroding the umbra of image x by the umbra of image y gives the umbra of a new function $U(x \oplus y) = U(x) \oplus U(y)$ or $U(x \ominus y) = U(x) \ominus U(y)$, from which we can obtain the dilation result, $x \oplus y$, by taking top surface of the corresponding umbra:

$$x \oplus y = \tau(U(x) \oplus U(y)). \tag{14.26}$$

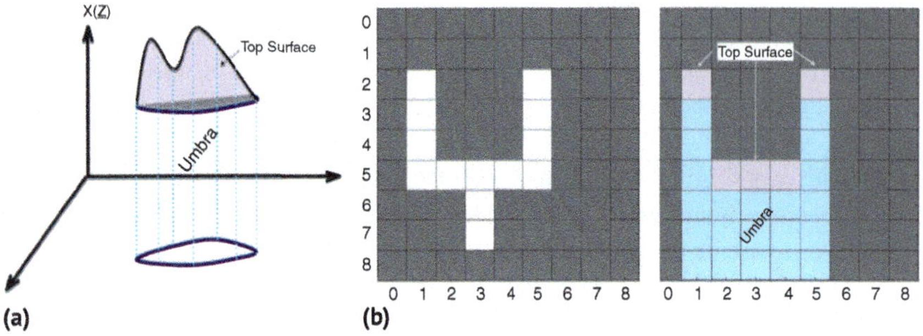

Figure 14.16 (a) Top surface and umbra indicated by purple shaded region and blue dotted lines, respectively, and (b) 1-D example of the letter "Y," its top surface and umbra.

and similarly for the erosion $x \ominus y = \tau(U(x) \ominus U(y))$. Alternatively, we can perform grayscale dilation and erosion of the two images directly using the following equations [3]:

$$\begin{aligned}
[x \oplus y](\underline{z}) &= \max_{\underline{w}}\{x(\underline{z} - \underline{w}) + y(\underline{w})\}, \\
[x \ominus y](\underline{z}) &= \min_{\underline{w}}\{x(\underline{z} + \underline{w}) - y(\underline{w})\},
\end{aligned} \tag{14.27}$$

where for each $\underline{z}$, $\underline{w}$ ranges over the intersection of the support of y with the translated support of x. The function y is assumed to have a compact support and plays a similar role as a grayscale structuring element.

As in the binary case, grayscale dilation and erosion are fundamental building blocks for the design of many other useful operations. In particular, grayscale opening and closing with a grayscale structuring element y are given by:

$$\begin{aligned}
x \circ y &= (x \ominus y) \oplus y \\
x \bullet y &= (x \oplus y) \ominus y
\end{aligned} \tag{14.28}$$

14.3.5 Geodesic Transforms and Reconstruction Filters

The geodesic transforms and morphological reconstruction filters differ from the classical morphological operations covered in the previous sections in many ways that listed below:

1. Classical operations generally apply to a single binary or grayscale image using fixed structure elements. In contrast, morphological geodesic transforms are applied to two images and a structuring element that may be adaptive to the local properties of the image. The two images are referred to as Marker (image X) and Mask (image K). The goal is to reconstruct image K from possibly the distorted image X by iterative application of dilation or erosion.
2. In contrast to the classical morphological operations that use Euclidean metric distance measure, the geodesic transforms use a non-Euclidean *geodesic* distance metric. The geodesic distance between two points on a 3-D surface (or in a set) corresponds to the shortest path connecting the two points and remains entirely on the surface (or in the set). This corresponds to the minimal number of edges of the grid to cross from one pixel to another and is highly dependent on the type of structuring element and its connectivity. These geodesic operations may be implemented in a manner that affects only part of an object or a scene.
3. Morphological reconstruction filters have been shown to offer many advantages over the traditional methods in various image processing tasks owing to their robustness to noise and effective feature extraction and image segmentation with no undesirable artifacts.

14.3.5.1 Binary Geodesic Dilation and Erosion

Let us consider discrete sets X and K for binary marker and mask images, respectively, where $X \subseteq K$. The geodesic dilation [15] of size $n \geq 0$ of image X within mask K is the set of the pixels of K whose geodesic distance to X is smaller or equal to n, i.e.

$$\mathcal{D}_K^{(n)}(X) = \{\underline{w} \in K;\ d_g(\underline{w}, X) \leq n\} \tag{14.29}$$

where $d_g(\underline{w}, X)$ is the geodesic distance between pixel $\underline{w} \in K$ and pixels in the marker image X. If there is no path connecting two points, then $d_g(\underline{w}, \underline{z}) = \infty$. This geodesic dilation of *size n* can be obtained by iterating n elementary (or size one) geodesic dilation operations:

$$\mathcal{D}_K^{(n)}(X) = \mathcal{D}_K^{(1)}(\mathcal{D}_K^{(n-1)}(X)) = \mathcal{D}_K^{(1)}(\mathcal{D}_K^{(1)}(\mathcal{D}_K^{(1)} \ldots \mathcal{D}_K^{(1)}(X))) \tag{14.30}$$

Geodesic transformations (both dilation and erosion) should be iterated until convergence is reached after a finite number of iterations, at which point idempotence is achieved, i.e. until $\mathcal{D}_K^{(n+1)}(X) = \mathcal{D}_K^{(n)}(X)$. We will show shortly how the results of this operation are used in morphological reconstruction of K from X by dilation.

Note that the elementary geodesic dilation of size one of the marker image with the mask image can be obtained via a standard dilation of size one followed by an intersection operation, i.e.

$$\mathcal{D}_K^{(1)}(X) = (X \oplus S) \cap K \tag{14.31}$$

where $(X \oplus S)$ is the binary Euclidean dilation of the set X with the basic structuring element S and the intersection confines (or masks) the result to only inside the set K. Iterating this process n times using (14.30) gives the geodesic dilation of size n.

To see an example of this operation, consider the marker and mask images in Figure 14.17a,b, respectively. A 3×3 structuring element of all ones was used. The result of geodesic dilation of *size one* after the intersection operation is illustrated in Figure 14.17c by white pixels. The gray pixels in Figure 14.17c correspond to those of $X \oplus S$ that were outside of the set K. For this example, size one geodesic dilation suffices since idempotence is achieved after one iteration.

Similarly, the geodesic erosion of the size $n \geq 0$ of image X within the mask K is the set of pixels of K whose geodesic distance from X is greater than n, i.e.

$$\mathcal{E}_K^{(n)}(X) = \{\underline{w} \in K;\ d_g(\underline{w}, X) > n\}. \tag{14.32}$$

The elementary geodesic erosion of size one of the marker image X with the mask image K can be obtained via a standard erosion of size one followed by the union operation:

$$\mathcal{E}_K^{(1)}(X) = (X \ominus S) \cup K \tag{14.33}$$

where $(X \ominus S)$ is the binary Euclidean erosion of the set X with the basic structuring element S and the union ensures it grows over the mask K. Iterating this process n times gives the geodesic erosion of size n, i.e.

$$\mathcal{E}_K^{(n)}(X) = \mathcal{E}_K^{(1)}(\mathcal{E}_K^{(n-1)}(X)). \tag{14.34}$$

This process is repeated until idempotence is reached, i.e. until $\mathcal{E}_K^{(n+1)}(X) = \mathcal{E}_K^{(n)}(X)$.

14.3.5.2 Binary Geodesic Reconstruction

In Figure 14.17, we illustrated an example of geodesic reconstruction of a mask image K, from the marker image X, which may be considered as an incomplete version of the mask. This is precisely the goal of the reconstruction filters. However, in general, the mask image K may consist of multiple disjoint sets of connected components. In this case, when performing successive geodesic dilation, the set of markers that reside in some components and have nonempty intersection will

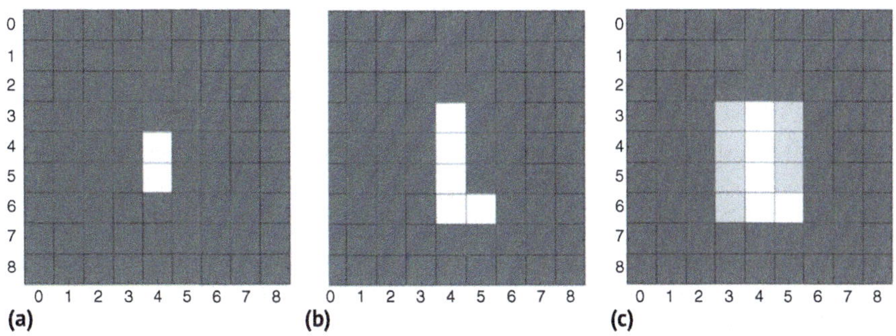

Figure 14.17 (a) Marker image, (b) mask image of the letter "L," and (c) geodesic dilation result (white pixels).

get progressively enlarged within those components until convergence (idempotence) is reached. That is, the reconstruction of K from the marker image X where $X \subseteq K$ is obtained by iterating elementary geodesic dilation of X inside K until idempotence. Then, the reconstruction is done using:

$$\rho_K(X) = \bigcup_{n \geq 1} \mathcal{D}_K^{(n)}(X) \tag{14.35}$$

where n is the iteration at which idempotence is reached. The algorithm for geodesic reconstruction [15] for binary images follows the following steps.

1. Label (unique identifiers) the connected components of the mask image K.
2. Determine the labels of the connected components which contain at least one pixel of the marker image X, and ignore the rest of the components. Reconstruction is determined directly based on the retained components.
3. Repeat steps in (14.30) and (14.31) until stability is reached.

14.3.5.3 Grayscale Geodesic Reconstruction

For grayscale marker image X and grayscale mask image K with gray levels in the range $[0, L-1]$, we employ the threshold decomposition principle [15] introduced earlier. In this case, the union operator in (14.35) is replaced by a supremum operation (the smallest upper bound denoted by $\vee$). Then, grayscale reconstruction of the mask image from the marker image using a flat structuring element S is obtained by iterative application of elementary grayscale geodesic dilation of X *under* K until idempotence is reached. That is:

$$\rho_K(X) = \bigvee_{n \geq 1} \mathcal{D}_K^{(n)}(X) \tag{14.36}$$

where the elementary geodesic dilation of image $X \leq K$ *under* K (i.e. for all $\underline{z}$ in the domain $X(\underline{z}) \leq K(\underline{z})$) is:

$$\mathcal{D}_K^{(1)}(X) = (X \oplus S) \wedge K. \tag{14.37}$$

Here, $\wedge$ stands for the point-wise minimum, and $(X \oplus S)$ is the dilation by a flat structuring element S. The same recursive equation as in (14.30) is then applied repeatedly until convergence is reached.

Reconstruction by grayscale erosion is also similar to (14.36). Let X and K be two grayscale images defined in the same domain such that $X \geq K$, i.e. X lies *above* K. Then, the dual reconstruction equation of the mask K from the marker X is obtained by:

$$\rho'_K(X) = \bigwedge_{n \geq 1} \mathcal{E}_K^{(n)}(X), \tag{14.38}$$

where $\bigwedge$ denotes the infimum (greatest lower bound) of a set. This operation is also implemented iteratively until idempotence is achieved. The elementary geodesic erosion of the image $X \geq K$ *above* K is:

$$\mathcal{E}_K^{(1)}(X) = (X \ominus S) \vee K \tag{14.39}$$

where $\vee$ stands for the point-wise maximum and $(X \ominus S)$ is the erosion of X by a flat structuring element S. The same recursive equation as in (14.34) is then applied until a stable result is reached.

14.4 Applications of Morphological Operations

Morphological filters play an essential role in many different aspects of an image analysis system, including noise removal, edge detection, feature extraction, and segmentation. These techniques have been widely applied in diverse areas such as biomedical imaging, remote sensing, computer vision, and radar/range imaging. In this section, we first explore how basic morphological operations, specifically dilation and erosion, can be utilized for edge extraction. We then present a morphological texture segmentation algorithm and compare its performance with that of the Gray Level Co-occurrence Matrix (GLCM) method discussed in Chapter 13.

14.4.1 Morphological Edge Detection

A simple and yet effective way to perform edge detection using morphological operations is based on the difference between an image and its eroded or dilated version using a small symmetric structure element S. A commonly used choice is the 3×3 cross-shaped structuring element. For binary images, the internal and external edges can be extracted using set difference of the image X with its eroded or dilated versions:

$$\begin{aligned} X_i &= X - (X \ominus S) \\ X_e &= (X \oplus S) - X \end{aligned} \tag{14.40}$$

where X_i and X_e are the internal and external edge sets, respectively.

Figure 14.18a–c illustrates the original binary diamond image and its internal and external extracted edge maps obtained using operations in (14.40). A 3×3 cross-shaped structuring element with the origin at the center was used to obtain these edge maps.

For grayscale images, edge extraction can be defined using the algebraic difference of the image $x(\underline{z})$ with its eroded or dilated versions:

$$\begin{aligned} x_i(\underline{z}) &= x(\underline{z}) - [x \ominus S](\underline{z}) \\ x_e(\underline{z}) &= [x \oplus S](\underline{z}) - x(\underline{z}). \end{aligned} \tag{14.41}$$

These edge operators can also be combined [3] to provide specific edge operators for different tasks, e.g. distinguishing smooth edges versus rippled edges. Examples include: (i) morphological gradient $G(x) = x_i(\underline{z}) + x_e(\underline{z})$, (ii) morphological edge-strength operators $\min\{x_i(\underline{z}), x_e(\underline{z})\}$ or $\max\{x_i(\underline{z}), x_e(\underline{z})\}$, or (iii) nonlinear Laplacian operator $\Delta(x) = x_e(\underline{z}) - x_i(\underline{z})$. In practice, the performance of these edge operators improves when the input images are first low-pass filtered to suppress noise and small-scale variations. In general, they are more effective and computationally less demanding compared to the gradient-based operators in Chapter 8.

As an example, consider the original lung CT image in Figure 14.19a. The grayscale dilated and eroded versions of this image obtained using a disk-shaped structuring element of radius 2 are shown in Figure 14.19b,c, respectively. Figure 14.19d shows the final edge map generated using the morphological gradient operator $G(x) = [x \oplus S](\underline{z}) - [x \ominus S](\underline{z})$, which combines the two images in Figure 14.19b,c. Unlike the spatial gradient operators of Chapter 8, this nonlinear morphological filter produces an edge map with well-connected edges while offering robustness against noise and other intensity variations.

As discussed in Section 14.3.2, the top-hat and bottom-hat transforms are commonly used to detect localized structures such as bright and dark blobs [3]. A bright blob corresponds to a local peak in the 3-D intensity surface, whose spatial support is determined by the shape of the structuring element S, while its scale is controlled by the size of S. Conversely, dark blobs correspond to local valleys in the 3-D intensity surface with a spatial support shaped like S. The bottom-hat transform can be used to extract these dark blobs.

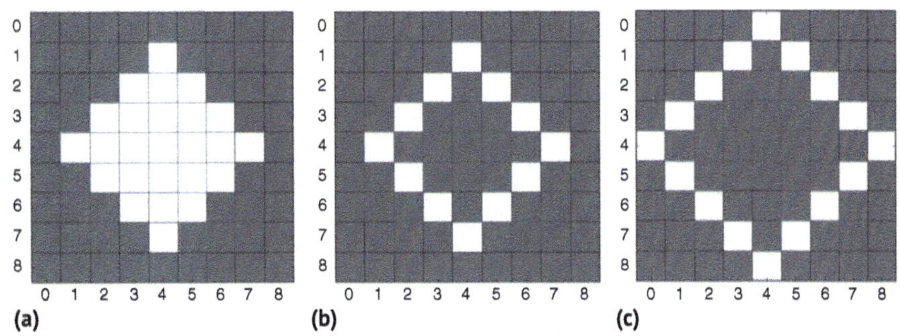

Figure 14.18 (a) Original binary image, (b) extracted internal edge map, and (c) extracted external edge map.

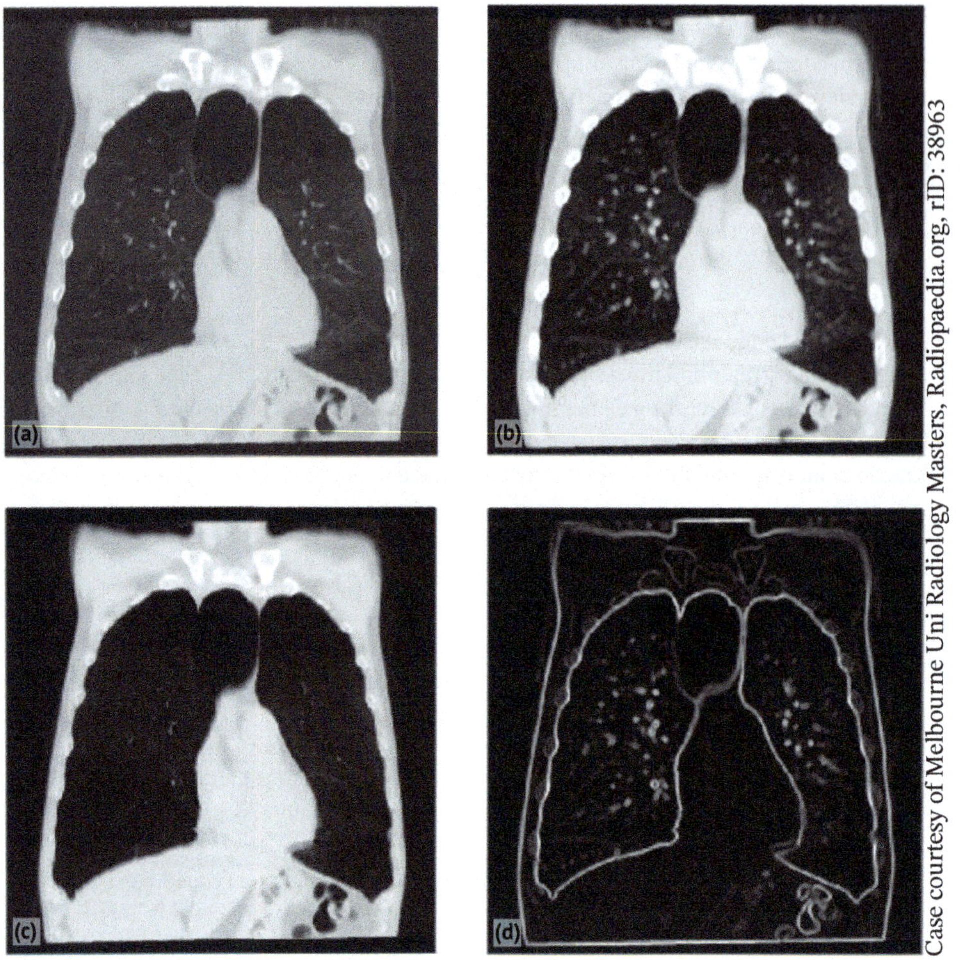

Figure 14.19 (a) Original lung CT scan image, (b) dilated version, (c) eroded version, and (d) gradient image. (The color rendition of this image is available to view at www.wiley.com/go/Azimi449.)

14.4.2 Morphological Image Segmentation

Many existing morphological segmentation approaches rely on boundary detection. One commonly used method in this category is the *watershed algorithm* [3, 8, 19], which treats an image as a topographical surface of different regions that can be separated based on their catchment basins. Although there are many different variants of watershed transformations that are effective for some boundary-based image segmentation problems mainly for separating large regions, they often struggle with segmentation of textured regions or those containing complex structures.

Hierarchical Texture Decomposition (HTD) [18] is a systematic multi-resolution approach for segmenting textured regions in grayscale images. The HTD-based segmentation accounts for the spatial structure of different textured patterns and sizes by their fundamental building blocks, known as *primitives*. A textured image is then decomposed into a number of texture component images, each containing texture primitives of certain sizes and structures. Since natural texture primitives are irregular and often dissimilar, non-flat structuring elements are not suited for matching their shapes. Thus, typically flat structuring elements are used for this morphological-based textural filtering.

A family of structuring elements S_i of different sizes is constructed to segment texture primitives at multiple scales, where $S_{i+1} = S_i \oplus S,\ \ i \in [0, n-1]$ with $S_0 = \{0, 0\}$ is the origin and S is a basic structuring element. Now, using the morphological opening [18], one can perform an iterative multi-scale texture decomposition by keeping textural primitives larger than the structuring element, while eliminating sharp peaks and ridges in an image that are smaller than the structuring element.

More specifically, let S_n be the largest structuring element, then the opening of grayscale image $x(\underline{z})$ by S_n gives the coarsest decomposition image:

$$y_0(\underline{z}) = [x \circ S_n](\underline{z}),$$

which only retains texture primitives larger than the structuring element S_n. To extract finer primitives that are smaller, we form the residual image $x(\underline{z}) - y_o(\underline{z})$ and apply the opening operation using the next smaller structuring element S_{n-1}. This gives:

$$y_1(\underline{z}) = [(x - y_0) \circ S_{n-1}](\underline{z}).$$

This process can be repeated to find image decompositions for smaller size texture primitives. This multi-scale iterative image decomposition process [18] can be summarized in the following steps:

$$\begin{cases} x_0(\underline{z}) = x(\underline{z}), \\ y_i(\underline{z}) = [x_i \circ S_{n-i}](\underline{z}), \quad i = 0, 1, \dots, n \\ x_{i+1}(\underline{z}) = x_i(\underline{z}) - y_i(\underline{z}), \end{cases} \tag{14.42}$$

where $y_i(\underline{z})$ are the decomposed image components corresponding to various textural primitive sizes and shapes. These component images that result from this multi-scale morphological decomposition carry all the information about the original image $x(\underline{z})$. Consequently, the original image can be reconstructed from these decomposed component images using:

$$x(\underline{z}) = \sum_{i=0}^{n} y_i(\underline{z}). \tag{14.43}$$

Let $\mathcal{M}_i\{\}$ denote this multi-scale morphological decomposition operator such that $y_i(\underline{z}) = \mathcal{M}_i\{x(\underline{z}), S_j, \ j \in [0, n]\}$, for all scales $i \in [0, n]$. This operator has several important properties [18]. First, it is complete and idempotent, i.e.

$$\mathcal{M}_k\{y_i(\underline{z}), S_j, \ j \in [0, n]\} = \begin{cases} y_i(\underline{z}), & \text{for } k = i \\ 0, & \text{for } k \neq i. \end{cases} \tag{14.44}$$

The operator is complete since, for any $k \neq i$, the component image $y_k(\underline{z})$ does not contain any primitive texture corresponding to $y_i(\underline{z})$. In addition, repeated application of this algorithm does not change the decomposition result.

The second property is given by:

$$\mathcal{M}_i\{\alpha x(\underline{z}) + \beta, S_j, \ j \in [0, n]\} = \begin{cases} \alpha y_0(\underline{z}) + \beta, & \text{for } i = 0 \\ \alpha y_i(\underline{z}), & \text{for } i \neq 0 \end{cases} \tag{14.45}$$

which shows that the bias term β only affects the first (coarsest) component, while the scaling α affects all the decomposed components.

It should be pointed out that the decomposition generated using (14.42) depends not only on the primitive textural sizes but also on their intensity variations. This is due to the fact that opening with a flat structuring element can cut down the sharp peaks of texture primitives, thereby splitting one primitive into multiple components across different scales. From these component images $y_i(\underline{z})$, various features such as local mean, variance, gradient mean, and area ratios [18], can be extracted and used for segmentation and classification purposes.

The same textured fabric images in Figure 13.5a–c of Section 13.5 are used here to test the performance of the HTD method. To apply HTD, three basic structuring element shapes are used, namely rectangle, diamond, and ellipse. For each shape, four base structuring element sizes 3×3, 3×7, 3×11, and 3×15 are initialized. For every combination of shape and size, eight orientations of the basic structuring element are generated using rotation angles, $\theta_k = \frac{k\pi}{8}$, $k \in [0, 7]$. Furthermore, for all combinations of shapes, sizes, and rotations, three maximum decomposition levels $n = \{8, 16, 32\}$ are used. Each maximum decomposition level produces a corresponding set of texture primitives $y_i(\underline{z})$, leading to 6144 total texture primitives for the HTD method. As a result, each pixel is represented by a 6144-dimensional feature vector, where the ith

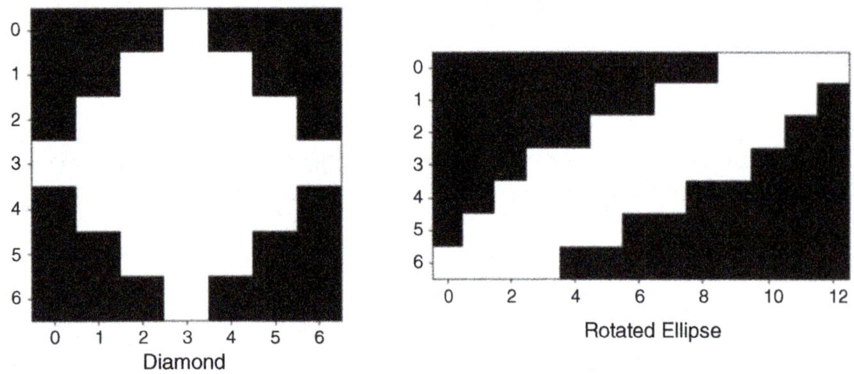

Figure 14.20 Two structuring elements at $i = 3$.

Table 14.3 HTD logistic regression accuracy and coefficient count.

	Fabric 1	Fabric 2	Fabric 3
Subimage 1 accuracy	97.30%	78.83%	97.39%
Subimage 2 accuracy	97.38%	88.47%	93.87%
Subimage 3 accuracy	96.72%	83.35%	93.32%
Subimage 4 accuracy	99.89%	96.21%	82.68%
Subimage 5 accuracy	95.86%	89.21%	92.17%
Subimage 6 accuracy	96.79%	94.50%	98.75%
All subimages accuracy	97.32%	88.43%	93.03%
Coefficient count (sparsity)	276	350	236

element of this vector corresponds to pixel value in the component image $y_i(\underline{z})$. For visual reference, two structuring elements at iteration $i = 3$ (i.e. S_3) are shown in Figure 14.20: a 3×3 diamond (left) with zero rotation and a 3×5 ellipse (right) rotated by $\frac{\pi}{4}$ radians.

The correct classification results of the multinomial logistic regression (see Chapter 17) with l_1 norm sparsity using HTD features are given in Table 14.3. These results show substantially better performance of the HTD-based texture segmentation method for Fabric 2 and moderate improvements for Fabrics 1 and 3 compared to those in Table 13.2 for the GLCM method. The sparse multinomial logic regression picked at most 350 features out of 6144 to achieve 100% cross-validation performance. Figure 14.21 shows color overlays of the probability of correct classification on the original images for all the fabrics. As can be observed, near perfect segmentation performance is achieved in all patches of Fabric 1 and across several patches of Fabrics 2 and 3.

Performance comparison with those of other texture feature extraction methods, namely the GLCM and Gabor filter approaches discussed in Section 13.5, indicates that the HTD achieved superior overall segmentation results. Moreover, the GLCM methods incurs significantly higher computational cost compared to HTD, making it less efficient for large-scale practical applications. By all accounts, the Gabor filters performed the worst. It is likely that using more advanced nonlinear classifiers, such as convolutional neural networks (CNNs), can potentially enhance overall classification performance, especially for more challenging cases, e.g. Fabric 2.

14.5 Computer Projects

The computer projects in this chapter use Python libraries matplotlib and scikit-image.

Computer Project 14.1: The goal of this computer project is to investigate the application of various morphological operations for speckle noise removal from a synthetic aperture radar (SAR) image of a farm. As stated in Section 10.1, speckle

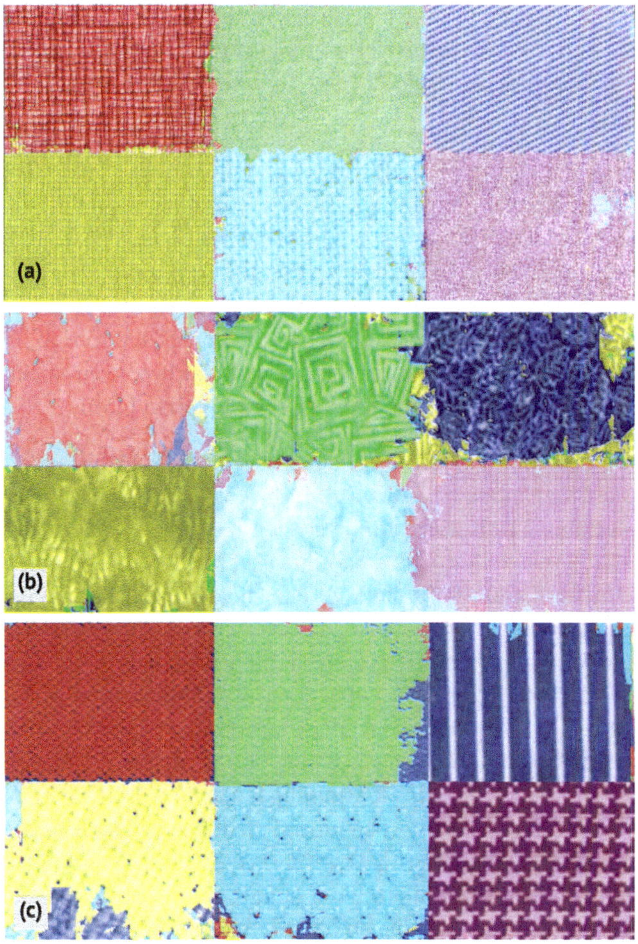

Figure 14.21 (a)–(c) Logistic regression performance color map for Fabrics 1-3 using HTD features, respectively.

noise has the characteristics of a multiplicative noise and appears in the image as tiny spots with varying intensity. Using grayscale dilation, erosion, opening, and closing operators, devise a Python-based algorithm that removes the effects of speckle noise while preserving the essential characteristics of the original image. Reference [20] presents a review of traditional speckle removal methods and also introduces a new morphological filter that adapts to the local properties of the image. Implement this algorithm and compare the results with those of your own approach. Describe the strengths and weaknesses of each method for speckle noise removal from SAR images.

Computer Project 14.2: This computer project focuses on developing specific morphological operations to enhance the contrast in medical images. Contrast enhancement is a critical preprocessing step in medical imaging pipelines to improve the accuracy of subsequent diagnosis by human experts or AI-based systems. The core objective is to improve the definition and visibility of abnormalities relative to the surrounding background tissue. Reference [21] provides examples of how morphological operators have been previously applied to improve the contrast of X-ray images. For implementation and testing, you can use the lung image in Figure 14.19a, or any similar medical image that contains abnormalities. Develop a Python-based algorithm using grayscale top-hat and bottom-hat operators. These operators are designed to isolate bright and dark features of a specific size from a relatively uniform background. Provide a detailed analysis of your algorithms' effectiveness when compared with those of the traditional contrast enhancement methods such as histogram modeling in Chapter 8. This analysis should include visual assessment as well as objective image quality metrics.

Computer Project 14.3: In this computer project, we use the grayscale skeleton introduced in Section 14.3.3, or the medial axis transform (MAT) [16] for shape representation. First, generate images of at least three objects with distinct shapes, such as bananas, pears, and oranges, or various mechanical parts. Develop a Python-based code to find all the skeleton subsets of the grayscale images. Use $k = 3$ and choose appropriate binary structuring elements for the selected objects. Experimentally demonstrate the ability to partially and fully reconstruct the images from their skeleton subsets (see

Section 14.3.3). Then, use the skeleton subsets to discriminate among the three objects based on their shape features. Devise an algorithm to represent each object using attributes derived from the skeleton subsets to allow discrimination using a simple distance-based classifier (see Section 16.1.2). Describe the strengths and weaknesses of this grayscale skeleton-based shape representation method.

Computer Project 14.4: In this computer project, we investigate the application of the reconstruction filters discussed in Section 14.3.5 in electron microscopy imaging. Use the red blood cell image in Figure 12.2a, or any similar image containing overlapping or partially occluded cells. Develop a Python-based algorithm using scikit-image morphology module to implement morphological reconstruction for separating overlapping or occluded cells so that the total number of cells can be counted more easily and automatically. Compare the performance of different grayscale masks and structuring elements, e.g. disks of various radii. Describe the advantages and limitations of your algorithm for this application. In addition, describe how the same goal can be achieved using a traditional image processing method introduced in earlier chapters. Further details on grayscale reconstruction filters and their implementation can be found in [15].

Computer Project 14.5: This computer project involves the implementation, testing, and analysis of the multiscale HTD method in Section 14.4.2. Use the same medical image selected for Computer Project 14.2. The goal is to investigate the applicability of the HTD method to provide texture-based features for discriminating between tumors and normal tissues. Before applying HTD method, improve the contrast of the image using a histogram-based method in Chapter 8. Then, develop a Python-based texture segmentation using the multiscale HTD method. The choices of the base structuring elements are critical to the success of this approach. These choices should be made based on visual evaluation of the textural differences between benign and malignant tissues. Use a maximum decomposition level of $n = 10$, producing 10 component images $y_i(\underline{z})$ for each structuring element. As in the experiment of Section 14.4.2, the feature vector for each pixel will be of $10 \times m$-dimension, where m is the number of combinations of the structuring elements. Next, use K-means clustering method (see Section 11.3.4) to partition a training subset, consisting of 40% of the total feature vectors extracted from benign and malignant regions, into at most four clusters. After training is completed, clusters can be labeled by using a limited number of feature vectors with known labels, namely benign and malignant. Determine the average true and false positive and negative rates on all the remaining (60% of the total) feature vectors extracted from the tumor and normal tissue areas. Note that the number of cases to average over must be the same for both classes to avoid biased estimates of these rates. Provide a thorough analysis of the results, including observations about the effectiveness and limitations of the HTD-based texture feature extraction method.

14.6 Problems

Problem 14.1 Consider the binary image for the letter "F" in Figure 14.1b. Using a 3×3 cross-shaped structuring element (origin in the center), find:

a. dilated version of the image,
b. eroded version of the image. State your observations for each result.

Problem 14.2 Using set algebra, prove commutative, associative, and distributive properties in Table 14.1 for the dilation and erosion operations.

Problem 14.3 Using set algebra, prove translation invariance, duality, erosion of union, and erosion of dilation properties in Table 14.1 for the dilation and erosion operations.

Problem 14.4 Using set algebra, prove anti-distributive and increasing 1 and 2 properties in Table 14.1 for the dilation and erosion operations.

Problem 14.5 Consider the binary image in Figure 14.22b. Using the three structuring elements (origin in the center) shown in Figure 14.22a, find the opened versions of the image and comment on the practical usefulness of each structuring element.

Problem 14.6 Consider the binary image in Figure 14.22c. Using the three structuring elements (origin in the center) shown in Figure 14.22a, find the closed versions of the image and comment on the practical usefulness of each structuring element.

Problem 14.7 Using set algebra, prove complement 1 and 2 and increasing 1 and 2 properties in Table 14.2 for the opening and closing operations.

Problem 14.8 Using set algebra, prove idempotence 1 and 2, anti-extensive, and extensive properties in Table 14.2 for the opening and closing operations.

Problem 14.9 Using set algebra, prove translation invariance 1 and 2 and S-open decreasing properties in Table 14.2 for the opening and closing operations.

Problem 14.10 The multi-scale opening and closing operations at scale k are defined by $X \circ kS = (X \ominus kS) \oplus kS$ and $X \bullet kS = (X \oplus kS) \ominus kS$, respectively. Using the binary image in Figure 14.22b and the first structuring element in Figure 14.22a, find:

a. the multi-scale opened set of images for $k = 3$.
b. the multi-scale closed set of images for $k = 3$.

Problem 14.11 Apply the hit-or-miss transform to the binary images "H" and "J" in Figure 14.23a,b using the first two structuring elements in Figure 14.12 and comment on the practical usefulness of each structuring element.

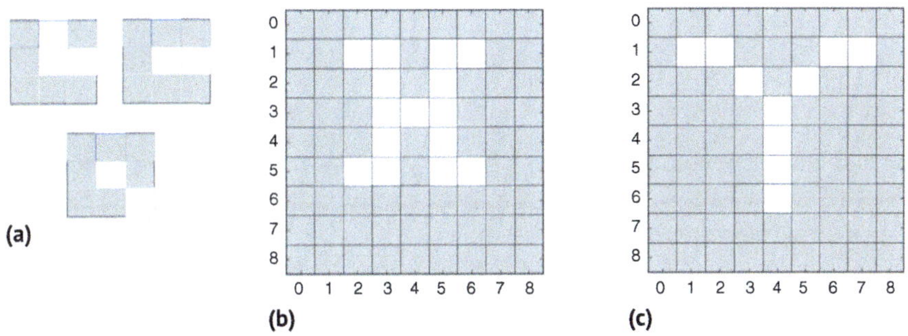

Figure 14.22 (a) Three structuring elements (origin in the center), (b) binary image "H", and (c) binary image "Y."

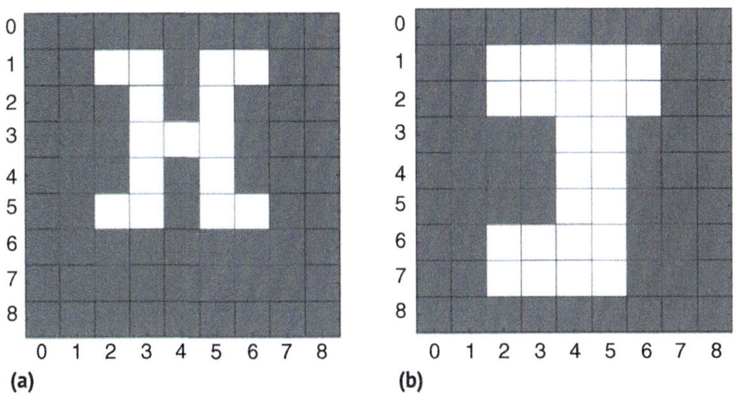

Figure 14.23 (a) Binary image "H" and (b) binary image "J." Foreground: white pixels, background: black pixels, and don't care: gray pixels.

Problem 14.12 Apply the thickening and thinning transforms to the binary images "H" and "J" in Figure 14.23a,b using the structuring elements in Figure 14.6.

Problem 14.13 Find the convex hull of the binary images "H" and "J" in Figure 14.23a,b using the structuring elements in Figure 14.9.

Problem 14.14 Show the following properties of the skeleton operation for a binary image set X and structuring element S.

a. If $S = \{0\}$, then $L(X) = \emptyset$.
b. If $X \ominus \check{S} = \emptyset$ where $\check{S}$ is the reflected version of S then $L(X) = X$.

Problem 14.15 Show the following properties of the skeleton set transformation $L(X)$.

a. Translation invariant, i.e. if $X \to X_{\underline{w}} = X + \underline{w}$, then $L(X_{\underline{w}}) = L(X) + \underline{w}$.
b. Anti-extensive, i.e. $L(X) \subseteq X$.
c. and idempotent, i.e. $L[L(X)] = L(X)$.

Problem 14.16 Apply the pruning transform to the binary images "H" in Figure 14.23a.

Problem 14.17 Prove the following properties of the grayscale morphological operations for the image $x(\underline{z})$ and the structure element set S where ^{c} is the complement operation, i.e. $x^c(\underline{z}) = -x(\underline{z})$ and $\check{S}$ is the reflected version of S.

a. $[x \ominus S](\underline{z}) = [x^c \oplus \check{S}]^c(\underline{z})$,
b. $[x \oplus S]^c(\underline{z}) = [x^c \ominus \check{S}](\underline{z})$,
c. $[x \circ S]^c(\underline{z}) = [x^c \bullet \check{S}](\underline{z})$,
d. $[x \bullet S]^c(\underline{z}) = [x^c \circ \check{S}](\underline{z})$.

Problem 14.18 The multi-scale opening and closing operations at scale k for the grayscale image $x(\underline{z})$ and structuring element S are defined by $x \circ kS = (x \ominus kS) \oplus kS$ and $x \bullet kS = (x \oplus kS) \ominus kS$, respectively. Using the grayscale image for the letter "i" in Figure 14.14 and the first structuring element (cross-shaped), find:

a. the multi-scale opened set of images for $k = 3$.
b. the multi-scale closed set of images for $k = 3$.

Problem 14.19 Consider the mask and marker images in Figure 14.24a,b, respectively. Using a 3×3 cross-shaped structuring element (origin at center) find:

a. reconstruction by dilation of the marker image with respect to the mask,
b. reconstruction by erosion of the marker image with respect to the mask.

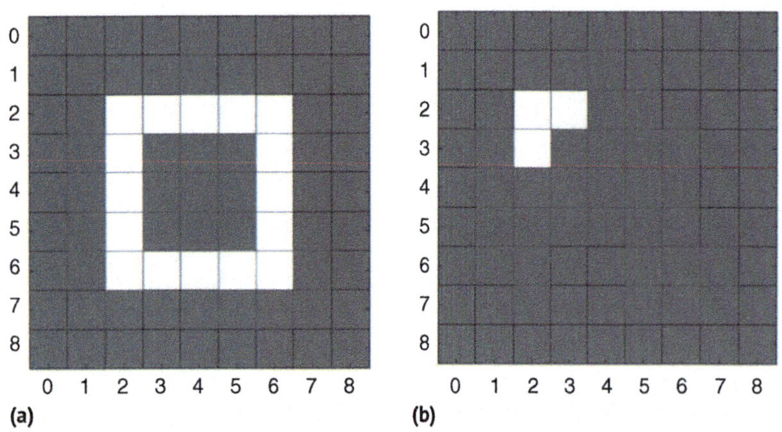

Figure 14.24 (a) Binary mask image "O" and (b) binary marker image.

Problem 14.20 Prove the validity of the reconstruction equation (14.43) for the multi-scale morphological decomposition using (14.42).

References

1 J. Serra, "*Image Analysis and Mathematical Morphology*," Academic Press, 1982.
2 C. R. Giardina and E. R. Dougherty, "*Morphological Methods in Images and Signal Processing*," Prentice-Hall, 1988.
3 P. Maragos and R. W. Schafer, "Morphological systems for multidimensional signal processing," *Proceedings of the IEEE*, vol. 78, pp. 690–710, 1990.
4 P. Maragos, "Tutorial on advances in morphological image processing and analysis," *Optical Engineering*, vol. 26, pp. 623–632, 1987.
5 J. Serra and L. Vincent, "An overview of morphological filtering," *Circuits, Systems, and Signal Processing*, vol. 11, pp. 47–108, 1992.
6 M. L. Comer and E. J. Delp, "Morphological operations for color image processing," *Journal of Electronic Imaging*, vol. 8, pp. 279–289, 1999.
7 A. K. Jain, "*Fundamentals of Digital Image Processing*," Prentice-Hall, 1989.
8 R. C. Gonzalez and R. E. Woods, "*Digital Image Processing*," Pearson, 2018.
9 M. Sonka, V. Hlavac, and R. Boyle, "*Image Processing, Analysis, and Machine Vision*," Thomson Pub, 3rd Edition, 2008.
10 M. A. Jayaram and H. Fleyeh, "Convex hulls in image processing: A scoping review," *American Journal of Intelligent Systems*, vol. 6, pp. 48–58, 2016.
11 D. Ballard and C. Brown, "*Computer Vision*," Prentice-Hall, 1982.
12 R. Haralick and L. Shapiro, "*Computer and Robot Vision*," Addison-Wesley Publishing Company, vol. 1, 1992.
13 P. Li, b. Wang, F. Sun, C. Zhang, and W. Wang, "Q-MAT: Computing medial axis transform By quadratic error minimization," *ACM Transactions on Graphics*, vol. 35, no. 8, pp. 1–16, 2015.
14 S.R. Sternberg, "Grayscale morphology," *Computer Vision, Graphics, and Image Processing*, vol. 35, pp. 333–355, 1986.
15 L. Vincent, "Morphological grayscale reconstruction in image analysis: Applications and efficient algorithms," *IEEE Transactions on Image Processing*, vol. 2, pp. 176–201, 1993.
16 S. Peleg and A. Rosenfeld, "A min-max medial axis transformation," *IEEE Transactions on Pattern Analysis and Machine Intelligence*, vol. 3, pp. 208–210, 1981.
17 P. A. Maragos and R. W. Schafer, "Morphological skeleton representation and coding of binary images," *IEEE Transactions on Acoustics, Speech, and Signal Processing*, vol. 34, pp. 1228–1244, 1986.
18 D. Wang, V. Haese-Coat, A. Bruno, and J. Rosin, "Texture classification and segmentation based on iterative morphological decomposition," *Journal of Visual Communication and Image Representation*, vol. 4, pp. 197–214, 1993.
19 M. Pesaresi and J. A. Benediktsson, "A new approach for the morphological segmentation of high-resolution satellite imagery," *IEEE Transactions on Geoscience and Remote Sensing*, vol. 39, pp. 309–320, 2001.
20 A. S. Mashaly, E. E. AbdElkawy, and T. A. Mahmoud, "Speckle noise reduction in SAR images using adaptive morphological filter," Proceedings of the IEEE Interference Conference on Intelligent Systems Design and Applications, 2010.
21 R. Firoz, S. Ali, N. U. Khan, K. Hossain, K. Islam, and Shahinuzzaman, "Medical image enhancement using morphological transformation," *Journal of Data Analysis and Information Processing*, vol. 4, pp. 1–12, 2016.

15

Image Reconstruction

The problem of image reconstruction from projections can be framed as an inverse problem, similar to image restoration from noisy and blurred observations discussed in Chapter 10. However, in this case, the inverse problem involves reconstructing a 2-D distribution or image from its 1-D projections or line integrals taken along a finite number of lines of known locations. The problem can also be extended to the reconstruction of 3-D surfaces from their 2-D projections. One can view projections as shadow-like measurements obtained by illuminating an object with penetrating radiation, e.g. X-ray, from different perspectives. These types of problems arise in numerous applications, including computerized tomography (CT) for diagnostic medicine, astrophysics to reconstruct radio emission maps from celestial sources using radio telescopes, reconstruction of molecular structures from electron microscopy data, and nondestructive testing and evaluation using a variety of imaging systems [1].

In CT imaging, X-ray-type radiation is used to produce cross-sectional *slices* of a body part in various planes. A series of slices is then used to create 3-D representations of anatomical structures, such as tumors or other abnormalities, for diagnostic purposes. A CT image reflects variations in density of the tissue being imaged. As bones, muscles, organs, and blood vessels have different densities, they appear with different gray-level intensities on a CT scan. Various types of CT scans can be adapted depending on the anatomical region being imaged. In some cases, contrast agents are administered to enhance the visibility of the small blood vessels inside the organs to check for abnormalities.

From a broad viewpoint, the methods used in image reconstruction from projections can be categorized into two general categories: *analytical* and *iterative methods*. Analytical approaches are generally noniterative and use mathematical inversion methods, such as the filtered back-projection (FBP) [1–3], to perform fast reconstruction. The FBP is commonly adopted in some commercial CT scanners. The iterative approaches [1, 4, 5], on the other hand, perform reconstruction by repeated refinement of the image from a finite number of measurements. They tend to produce more accurate reconstruction at the expense of higher computational cost. Among the commonly used iterative methods are the algebraic reconstruction technique (ART), statistical reconstruction algorithms based on maximum likelihood (ML), weighted least squares, and modern iterative methods based on machine learning [6]. In this chapter, we describe one method in each category described above, including FBP, ART, and ML, using a conjugate gradient-based algorithm.

15.1 Basics of Image Reconstruction

In most X-ray imaging systems, the measured projections of an object at a set of angles produce a collection of line integrals, which are known as sinograms [7]. The term sinogram is adopted since the line integral of a point source resembles a blurred sine wave. These line integrals are dependent on the absorption characteristics of the object when illuminated by a penetrating X-ray beam traveling along straight lines through the object. Figure 15.1 illustrates the geometry of projections for a 2-D object $x(u, v)$ illuminated by parallel beams at a specific incident angle θ and radial position r. Each projection leads to a set of line integrals computed along lines perpendicular to the direction of propagation.

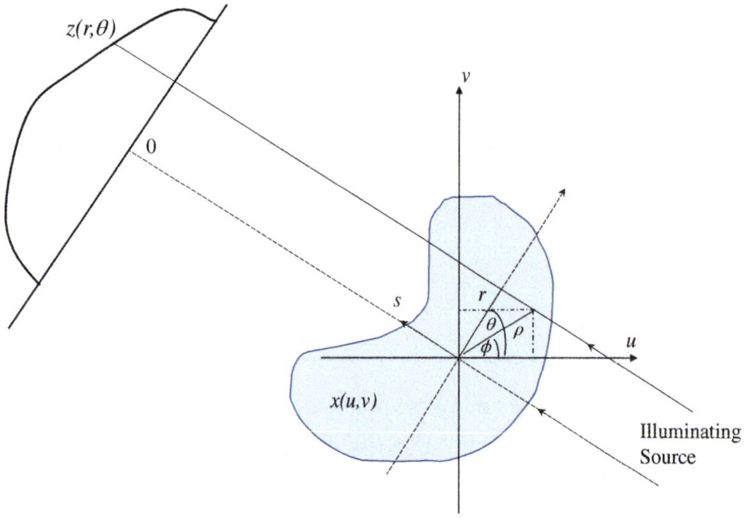

Figure 15.1 Geometry of projections in image reconstruction.

Let $x(u, v)$ be the unknown density function of the object being irradiated at the continuous-valued spatial location (u, v). As a consequence, $x(u, v)$ can be viewed as an image whose gray level at any point (u, v) is proportional to the relative linear attenuation at that point. This image usually corresponds to an infinitesimally thin slice of a 3-D object. The detected/recorded intensity of the beam can be expressed [8] as:

$$I = I_0 \exp\left\{-\int_L x(u, v)ds\right\}, \tag{15.1}$$

where I_0 is the incident beam intensity and s is the distance along the ray path L, which represents the path of all the X-ray photons for a particular source–detector pair. Thus, the observed data at position r and incident angle θ can be expressed as the line integral of the image [1, 8]

$$z(r, \theta) = \ln \frac{I_0}{I} = \int_L x(u, v)ds, \quad r \in (-\infty, \infty), \ \ \theta \in [0, \pi]. \tag{15.2}$$

That is, for a fixed (r, θ), (15.2) gives an integral of $x(u, v)$ along a line L perpendicular to the line through the origin at angle θ and radial distance r.

Now, the problem of image reconstruction in tomographic imaging is to recover the original unknown image $x(u, v)$ from its projections $z(r, \theta)$. The mathematical foundation of this problem was established by Johann Radon in 1917 [9] and is known as the *Radon transform*. This framework provides a mapping between the spatial domain (u, v) and the projection domain (r, θ).

15.1.1 Radon Transform

As stated before, the relationship between the relative linear attenuation function $x(u, v)$ and the family of all its line integrals $z(r, \theta)$, is given by (15.2). Now, using the coordinate rotation,

$$\begin{bmatrix} r \\ s \end{bmatrix} = \begin{bmatrix} \cos\theta & \sin\theta \\ -\sin\theta & \cos\theta \end{bmatrix} \begin{bmatrix} u \\ v \end{bmatrix} \tag{15.3}$$

relationship in (15.2) can be expressed as:

$$z(r, \theta) = [\mathcal{R}x](r, \theta) = \int_{-\infty}^{\infty}\int_{-\infty}^{\infty} x(u, v)\delta(u\cos\theta + v\sin\theta - r)du\, dv. \quad r \in (-\infty, \infty), \ \ \theta \in [0, \pi]. \tag{15.4}$$

This is known as the Radon transform or ray sum of $x(u, v)$ with the symbol $[\mathcal{R}x]$. The role of $\delta(u \cos\theta + v \sin\theta - r)$ is to confine the integration along the ray path $u \cos\theta + v \sin\theta = r$. Alternatively, if we express (u, v) in terms of (r, s) we can write (15.4) as:

$$z(r,\theta) = [\mathcal{R}x](r,\theta) = \int_{-\infty}^{\infty} x(r\cos\theta - s\sin\theta, r\sin\theta + s\cos\theta)ds, \quad r \in (-\infty, \infty), \quad \theta \in [0, \pi]. \tag{15.5}$$

It is important to distinguish the domain of the function $[\mathcal{R}x]$, or the projection coordinates, from the polar coordinates (ρ, ϕ), where $u = \rho \cos\phi$ and $v = \rho \sin\phi$. In fact, when $r = 0$, the Radon transform $[\mathcal{R}x](0, \theta)$ is not zero and corresponds to the line integral of $x(u, v)$ along a line passing through the origin, making angle θ with the positive u-axis. In contrast, when $\rho = 0$, $x(0, \phi)$ in polar form maps to the origin regardless of ϕ. Nevertheless, if the original image is represented in polar coordinates, i.e. $x(\rho, \phi) = x(\rho\cos\phi, \rho\sin\phi)$, then (15.4) can be expressed as:

$$z(r,\theta) = \int_{-\infty}^{\infty} \int_{-\pi}^{\pi} x(\rho,\phi)\delta(\rho\cos(\theta-\phi) - r)\,\rho\, d\rho\, d\phi, \quad r \in (-\infty, \infty) \quad \theta \in [0, \pi]. \tag{15.6}$$

Some of the key properties of the Radon transform operator $[\mathcal{R}x]$ are given [8] in Table 15.1. The notation $*^r$ denotes convolution with respect to the variable r.

Example 15.1 *Find the impulse response of the operator Radon transform* $[\mathcal{R}]$.

Solution: Using (15.4), the impulse response for input image $x(u, v) = \delta(u - u_0, v - v_0)$ is

$$[\mathcal{R}\delta](r,\theta) = h(r,\theta; u_0, v_0) = \delta(u_0\cos\theta + v_0\sin\theta - r) = \begin{cases} 1, & \text{when} \quad r = u_0\cos\theta + v_0\sin\theta \\ 0, & \text{otherwise,} \end{cases}$$

or $[\mathcal{R}\delta](r,\theta) = 1$ when $r = \sqrt{u_0^2 + v_0^2}\,\sin(\theta + \alpha)$ and $\alpha = \tan^{-1}\left(\frac{u_0}{v_0}\right)$ otherwise $[\mathcal{R}\delta](r,\theta) = 0$. This corresponds to a sinusoidal trajectory in the (r, θ) domain, hence the name *sinogram*. The amplitude and phase of this sinogram are dependent on u_0 and v_0. This result also indicates that the operator $[\mathcal{R}]$ is shift-varying.

15.1.2 Inverse Radon Transform

Now, given the continuum of projections $z(r, \theta)$ for all angles $\theta \in [0, \pi]$, the goal of the inversion is to recover the unknown image $x(u, v)$. The *central slice or projection-slice theorem* [10] provides the theoretical solution to this problem through the relationship between the Fourier transform of the image and the Fourier transforms of its parallel projections. This is described next.

Table 15.1 Radon transform properties.

Property	2-D function	Radon transform
Linearity	$\alpha_1 x_1(u,v) + \alpha_2 x_2(u,v)$	$\alpha_1 z_1(r,\theta) + \alpha_2 z_2(r,\theta)$
Finite extent	If $x(u,v) = 0$ for $\lvert u\rvert, \lvert v\rvert > \frac{W}{2}$	then $z(r,\theta) = 0$ for $\lvert r\rvert > \frac{\sqrt{2}W}{2}$
Symmetry	For $x(u,v)$	$z(r,\theta) = z(-r, \theta \pm \pi)$
Periodic	For $x(u,v)$	$z(r,\theta) = z(r, \theta + 2k\pi)$, k integer
Shifting	If $y(u,v) = x(u - u_0, v - v_0)$	$[\mathcal{R}y](r,\theta) = z(r - u_0\cos\theta - v_0\sin\theta, \theta)$
Rotation	Rotate by ψ: $x(\rho, \phi + \psi)$	$z(r, \theta + \psi)$
Scaling	If $y(u,v) = x\left(\frac{u}{a}, \frac{v}{b}\right)$	$[\mathcal{R}y](r,\theta) = z\left(\frac{r}{\sqrt{a^2\cos^2\theta + b^2\sin^2\theta}}, \tan^{-1}\left\{\frac{b}{a}\tan\theta\right\}\right)$
Mass conservation	If $M = \int_{-\infty}^{\infty}\int_{-\infty}^{\infty} x(u,v)du\,dv$	$M = \int_{-\infty}^{\infty} z(r,\theta)dr, \ \forall\theta$
Convolution	$y(u,v) = x(u,v) ** h(u,v)$	$[\mathcal{R}y] = [\mathcal{R}x]*^r[\mathcal{R}h] = \int_{-\infty}^{\infty}[\mathcal{R}x](\eta,\theta)\,[\mathcal{R}h](r-\eta,\theta)d\eta$

15.1.2.1 Projection-Slice Theorem

For a fixed angle θ, the projection function $z(r, \theta)$ is a 1-D function of $r \in \mathbb{R}$. Taking its 1-D Fourier transform in r and using (15.4) yields

$$\begin{aligned} Z(\omega, \theta) = \mathcal{F}_r\{z(r, \theta)\} &= \int_{-\infty}^{\infty} z(r, \theta)e^{-j\omega r}\, dr \\ &= \int_{-\infty}^{\infty}\int_{-\infty}^{\infty} x(u, v)e^{-j\omega(u\cos\theta + v\sin\theta)}\, du\, dv \\ &= X(\omega\cos\theta, \omega\sin\theta), \end{aligned} \tag{15.7}$$

where $\mathcal{F}_r\{.\}$ is the 1-D FT operator in r. This result implies that the 1-D FT (in variable r) of projections at a specific angle θ corresponds to the 2-D FT of the unknown original image $x(u, v)$ evaluated along a line through the origin at angle θ, defined by $\omega_1 = \omega\cos\theta$ and $\omega_2 = \omega\sin\theta$ in the (ω_1, ω_2)-plane. That is, the 2-D FT is sliced through the origin at an angle θ. This remarkable result allows the recovery of $x(u, v)$ using the inverse 2-D FT of $X(\omega_1, \omega_2)$ or the inverse Radon transform.

The inverse 2-D FT of $X(\omega_1, \omega_2)$ is

$$x(u, v) = \frac{1}{4\pi^2}\int_{-\infty}^{\infty}\int_{-\infty}^{\infty} X(\omega_1, \omega_2)e^{j(u\omega_1 + v\omega_2)}d\omega_1 d\omega_2. \tag{15.8}$$

However, using the relationship $\omega_1 = \omega\cos\theta$ and $\omega_2 = \omega\sin\theta$, the Jacobian of the transformation yields $d\omega_1 d\omega_2 = \det(\mathbf{J})d\omega\, d\theta = |\omega| d\omega\, d\theta$, where $\mathbf{J}$ is the Jacobian matrix $\mathbf{J} = \begin{bmatrix} \cos\theta & -\omega\sin\theta \\ \sin\theta & \omega\cos\theta \end{bmatrix}$. As a result, the inverse 2-D FT in (15.8) can be rewritten as:

$$\begin{aligned} x(u, v) &= \frac{1}{4\pi^2}\int_0^{\pi}\int_{-\infty}^{\infty} X(\omega\cos\theta, \omega\sin\theta)e^{j\omega(u\cos\theta + v\sin\theta)}|\omega|\, d\omega\, d\theta \\ &= \frac{1}{4\pi^2}\int_0^{\pi}\left\{\int_{-\infty}^{\infty} |\omega| Z(\omega, \theta)e^{j\omega(u\cos\theta + v\sin\theta)}\, d\omega\right\} d\theta \\ &= \frac{1}{2\pi}\int_0^{\pi} g(u\cos\theta + v\sin\theta, \theta)\, d\theta, \end{aligned} \tag{15.9}$$

where the inner integral $g(u\cos\theta + v\sin\theta, \theta) = \frac{1}{2\pi}\int_{-\infty}^{\infty} |\omega| Z(\omega, \theta)e^{j\omega(u\cos\theta + v\sin\theta)}d\omega$ is the 1-D inverse FT of $|\omega| Z(\omega, \theta)$ or

$$g(r, \theta) = \frac{1}{2\pi}\int_{-\infty}^{\infty} |\omega| Z(\omega, \theta)e^{j\omega r}d\omega = \frac{1}{2\pi}\int_{-\infty}^{\infty} \omega\, \mathrm{sgn}(\omega)Z(\omega, \theta)e^{j\omega r}d\omega, \tag{15.10}$$

where $\mathrm{sgn}(x)$ is the signum function. Now, applying the convolution property yields

$$g(r, \theta) = \mathcal{F}^{-1}\{\mathrm{sgn}(\omega)\, \omega Z(\omega, \theta)\} = \mathcal{F}^{-1}\{\mathrm{sgn}(\omega)\} * \mathcal{F}^{-1}\{\omega Z(\omega, \theta)\}. \tag{15.11}$$

Alternatively, we can write (15.11) as $g(r, \theta) = z(r, \theta) * k(r)$ where $k(r)$ is referred to as the Radon kernel function [9], which is the inverse FT of $|\omega|$. Thus, the process involves taking a 1-D FT of the sinogram, i.e. $\mathcal{F}_r\{z(r, \theta)\} = Z(\omega, \theta)$, filtering $Z(\omega, \theta)$ with a filter with frequency response $|\omega|$, and taking inverse 1-D FT to produce $g(r, \theta)$, from which $x(u, v)$ can be reconstructed using (15.9). The latter step corresponds to the back-projection operation, which is discussed in the next section.

The first term on the right side is $\mathcal{F}^{-1}\{\mathrm{sgn}(\omega)\} = -\frac{1}{j\pi r}$ and for the second term, we use the differentiation property of FT $\mathcal{F}^{-1}\{\omega Z(\omega, \theta)\} = \frac{1}{j}\frac{\partial z(r,\theta)}{\partial r}$. Therefore, convolution in (15.11) becomes

$$\begin{aligned} g(r, \theta) &= \frac{1}{\pi}\int_{-\infty}^{\infty} \frac{\left[\partial z(\xi, \theta)/\partial \xi\right]}{r - \xi}d\xi \\ &= \mathcal{H}\left\{\frac{\partial z(r, \theta)}{\partial r}\right\}, \quad \forall\theta, \end{aligned} \tag{15.12}$$

where $\mathcal{H}\{f(t)\}$ represents the Hilbert transform [10] of a function $f(t)$, which is the convolution of $f(t)$ with $h(t) = \frac{1}{\pi t}$ or $\mathcal{H}\{f(t)\} = \frac{1}{\pi}\int_{-\infty}^{\infty} \frac{f(\tau)}{t-\tau} d\tau$. In the FT domain $\mathcal{F}\{\mathcal{H}\{f(t)\}\} = -j\text{sgn}(\omega)F(\omega)$.

Now, combining (15.12) and (15.9) gives the inverse Radon transform or reconstruction formula,

$$x(u, v) = [\mathcal{R}^{-1}z](u, v) = \frac{1}{2\pi^2}\int_0^{\pi}\int_{-\infty}^{\infty} \frac{\left[\partial z(r,\theta)/\partial r\right]}{u\cos\theta + v\sin\theta - r} dr\, d\theta \tag{15.13}$$

or in polar form,

$$x(\rho, \phi) = \frac{1}{2\pi^2}\int_0^{\pi}\int_{-\infty}^{\infty} \frac{[\partial z(r,\theta)/\partial r]}{\rho\cos(\theta - \phi) - r} dr\, d\theta. \tag{15.14}$$

The integral in (15.12) is improper due to singularity at $r = \xi$ and must be treated using Cauchy's principal value method [11]. Even if this integral exists, its numerical evaluation may be far from straightforward. Thus, the inversion formulae in (15.13) or (15.14) are not of practical use in their original forms. To address these practical issues, a more computationally efficient formulation based on *filtered back-projection* is typically used. This approach relies on the back-projection operator, which will be discussed in the next section.

15.1.3 The Back-Projection Operator

The *adjoint operator* (see Example 15.2) to the Radon transform is the *back-projection operator* [7, 12]. This operator maps the measured projections $z(r, \theta)$ back into the image domain by accumulating all the measurements or projections that pass through the point (u, v) or (ρ, ϕ) for all θ. The back-projection operator is defined by

$$y(u, v) = [\mathcal{B}z](u, v) = \int_0^{\pi} z(u\cos\theta + v\sin\theta, \theta)\, d\theta, \tag{15.15}$$

where $y(u, v)$ is the back-projected version of $z(r, \theta)$ and $r = u\cos\theta + v\sin\theta$. It must be emphasized that the back-projection operation is not the inverse Radon transform, i.e. $y(u, v) \neq x(u, v)$. In fact, it can be shown that projection followed by back-projection leads to a blurred version of the original image and not the image itself. Specifically, $y(u, v)$ and $x(u, v)$ are related via the following 2-D convolution,

$$y(u, v) = [\mathcal{B}z](u, v) = [\mathcal{B}\mathcal{R}x](u, v) = x(u, v) ** h(u, v), \tag{15.16}$$

where the PSF of the blur $h(u, v) = \frac{1}{\sqrt{u^2+v^2}}$ or in polar coordinate $h(\rho, \phi) = \frac{1}{|\rho|}$ [13]. It can be shown that the frequency response of this filter is $H(\omega_1, \omega_2) = \frac{4\pi^2}{\sqrt{\omega_1^2+\omega_2^2}}$. This result may suggest that we can simply use back-projection on projections or observations $z(r, \theta)$ followed by an inverse filter $\frac{1}{H(\omega_1,\omega_2)}$. However, this filter is not physically realizable, though discrete approximations are possible. Moreover, the back-projected image $y(u, v)$ has an unbounded region of support (ROS) [8], whereas image reconstruction should give image $x(u, v)$ with a finite ROS.

Comparing the inverse Radon transform in (15.9) with the back-projection operator formula in (15.15) indicates that we can reconstruct $x(u, v)$ from the filtered projections $g(r, \theta)$ via back-projection, i.e.

$$x(u, v) = \frac{1}{2\pi}[\mathcal{B}g](u, v). \tag{15.17}$$

However, in (15.12) we showed that $g(r, \theta)$ can be written as:

$$g(r, \theta) = \{\mathcal{H}[\mathcal{D}z]\}(r, \theta) \tag{15.18}$$

where $\mathcal{D}$ denotes differentiation with respect to r, and $\mathcal{H}$ denotes the Hilbert transform. Combining (15.17) and (15.18) and simplifying the notation gives the desired analytical image reconstruction equation,

$$x(u, v) = \frac{1}{2\pi}[\mathcal{B}\mathcal{H}\mathcal{D}z](u, v). \tag{15.19}$$

This result implies $\mathcal{R}^{-1} \equiv \mathcal{B}\mathcal{H}\mathcal{D}$. In other words, the inverse Radon transform $[\mathcal{R}^{-1}z]$ of a function $z(r, \theta)$ can be generated by the following steps: (i) compute derivative of $z(r, \theta)$ with respect to r; (ii) take the Hilbert transform of the result of step

(i) with respect to its first variable to obtain the function $g(r, \theta)$; and (iii) back-projection $g(r, \theta)$ and multiply by $\frac{1}{2\pi}$. This approach is referred to as *convolution back-projection* due to the convolution with $\frac{1}{\pi r}$ inherent in the Hilbert transform. Again, the issue here is the singularity problem stated earlier. Nevertheless, filtering may be applied to observations before back-projection, leading to the widely used filtered back-projection (FBP) method discussed in Section 15.2.1.

Example 15.2 *Show that the Radon transform and back-projection operators are adjoint. That is, for the spatial domain function $p(u, v)$ and the projection domain function $q(r, \theta)$, we have $\langle \mathcal{R}p, q\rangle = \langle p, \mathcal{B}q\rangle$.*

Solution: The Radon transform of $p(u, v)$ is

$$[\mathcal{R}p](r, \theta) = \int_{-\infty}^{\infty} p(r\cos\theta - s\sin\theta, r\sin\theta + s\cos\theta)ds.$$

Thus, the inner product in (r, θ)-space can be expressed as:

$$\begin{aligned}\langle [\mathcal{R}p], q\rangle &= \int_{-\infty}^{\infty}\int_{0}^{\pi} [\mathcal{R}p](r, \theta)\, q(r, \theta)\, dr\, d\theta \\ &= \int_{-\infty}^{\infty}\int_{-\infty}^{\infty}\int_{0}^{\pi} p(r\cos\theta - s\sin\theta,\ r\sin\theta + s\cos\theta)\, q(r, \theta)\, dr\, d\theta\, ds.\end{aligned}$$

Now, using the coordinate transformation in (15.3) we have $drds = \det(\mathbf{J})du\,dv = du\,dv$ since the determinant of the Jacobian matrix is $\det(\mathbf{J}) = 1$. Thus, we get

$$\begin{aligned}\langle [\mathcal{R}p], q\rangle &= \int_{-\infty}^{\infty}\int_{-\infty}^{\infty}\int_{0}^{\pi} p(u, v)\, q(u\cos\theta + v\sin\theta, \theta)\, du\, dv\, d\theta \\ &= \int_{-\infty}^{\infty}\int_{-\infty}^{\infty}\left\{\int_{0}^{\pi} q(u\cos\theta + v\sin\theta, \theta)\, d\theta\right\} p(u, v)\, du\, dv \\ &= \int_{-\infty}^{\infty}\int_{-\infty}^{\infty} [\mathcal{B}q](u, v)\, p(u, v)\, du\, dv = \langle p, [\mathcal{B}q]\rangle.\end{aligned}$$

This is an important result.

15.2 Analytical Image Reconstruction Methods

Analytical image reconstruction methods provide direct mathematical solutions for inverting the Radon transform. Among multitude of approaches, the most widely used in practical CT scanners is the filtered back-projection (FBP) method. The popularity of FBP method stems from its computational efficiency and numerical stability.

15.2.1 Filtered Back-Projection

The principal idea of the FBP method is to use the projection-slice theorem and apply an interpolation filter to the 1-D FT of projections (sinograms) before back-projection operation. More specifically, we combine the results in (15.9) and (15.17) to recover $x(u, v)$ as:

$$x(u, v) = \frac{1}{2\pi}[\mathcal{B}g](u, v) = \frac{1}{2\pi}\int_{0}^{\pi} g(u\cos\theta + v\sin\theta, \theta)d\theta, \tag{15.20}$$

or in polar coordinates,

$$x(\rho, \phi) = \frac{1}{2\pi}\int_{0}^{\pi} g(\rho\cos(\theta - \phi), \theta)d\theta. \tag{15.21}$$

But $g(r,\theta) = \mathcal{F}^{-1}\{|\omega| Z(\omega,\theta)\}$, where $Z(\omega,\theta)$ is the 1-D FT of the projections $z(r,\theta)$ in variable r. Furthermore, the product of $Z(\omega,\theta)$ by $|\omega|$ amounts to filtering the observation using a filter with a ramp frequency response. Using these relations in (15.20) gives

$$x(u,v) = [\mathcal{B}\mathcal{P}z](u,v), \tag{15.22}$$

where the filtering operator $[\mathcal{P}z](r,\theta)$ is defined as

$$[\mathcal{P}z](r,\theta) = \frac{1}{2\pi}\mathcal{F}^{-1}\{|\omega| Z(\omega,\theta)\} = h(r) * z(r,\theta). \tag{15.23}$$

This result implies that to find $g(r,\theta)$ for the back-projection operation in (15.20), the observations $z(r,\theta)$ are filtered with a 1-D filter that has an impulse response $h(r) = \frac{1}{2\pi}\mathcal{F}^{-1}\{|\omega|\}$, hence the name filtered back-projection or FBP. The steps in the FBP methods can be summarized as: (i) compute 1-D FT of each projection $z(r,\theta)$ to obtain $Z(\omega,\theta)$; (ii) multiply $Z(\omega,\theta)$ by the ramp filter $H(\omega) = \frac{1}{2\pi}|\omega|$; (iii) compute the 1-D inverse FT to obtain the filtered projections $g(r,\theta)$; and (iv) back-project $g(r,\theta)$ over all angles to recover $x(u,v)$. To obtain an accurate representation of the original image, projections over the full angular range $[0,\pi]$ (i.e. 180°) are required. The entire process is depicted in the block diagram in Figure 15.2.

However, the ramp filter introduces several undesirable effects. First, this filter sets the mean of the reconstructed image to zero, since its frequency response $H(0) = 0$. Second, the ramp filter tends to amplify the effects of high-frequency noise in the observations, which is particularly problematic in medical imaging systems. Finally, the ramp filter is not physically realizable and must be approximated.

In practice, these issues are mitigated by applying a window function to limit the response of the ramp filter to the frequency range $\pm\frac{\omega_s}{2} = \pm\frac{\pi}{\beta}$, where ω_s and β denote the sampling frequency and the sampling interval, respectively. There are a few different choices for the window function. The simplest choice is to use a rectangular window function with frequency response $W(\omega) = \mathrm{rect}[\beta\omega]$, which leads to a filter known as Ram-Lak [8]. The frequency response of the windowed filter will be $H(\omega) = \frac{1}{2\pi}|\omega| W(\omega)$. A more commonly used choice is the generalized Hamming window function [1, 8] with frequency response,

$$W(\omega) = \begin{cases} \alpha + (1-\alpha)\cos\beta\omega, & \text{for } \omega \in \left[0, \frac{\pi}{\beta}\right] \\ 0, & \text{otherwise.} \end{cases}$$

For $\alpha = 0.54$ or $\alpha = 0.5$, the generalized Hamming window becomes the Hamming and Hanning windows, respectively. Another widely used window function is the Shepp–Logan filter [8] whose frequency response is

$$W(\omega) = \begin{cases} \mathrm{sinc}(\beta\omega), & \text{for } \omega \in \left[0, \frac{\pi}{\beta}\right] \\ 0, & \text{otherwise.} \end{cases}$$

Although the Radon transform and FBP provide a solid theoretical framework for image reconstruction in tomography, many practical issues remain. These include: (i) input data to a reconstruction algorithm usually consist of only a finite number of measurements as opposed to idealized line integrals, making the reconstruction problem inherently ill-posed; (ii) measurements are often imperfect due to many factors such as the limited beamwidth, scattering effects, photon statistics, and detector noise; and (iii) the Radon and back-projection assume continuous spatial variables, whereas for efficient implementation, discretization of the formulations is required, thereby introducing additional approximation errors. In the next section, we address the implications of finite and discrete measurements in the context of FBP-based reconstruction.

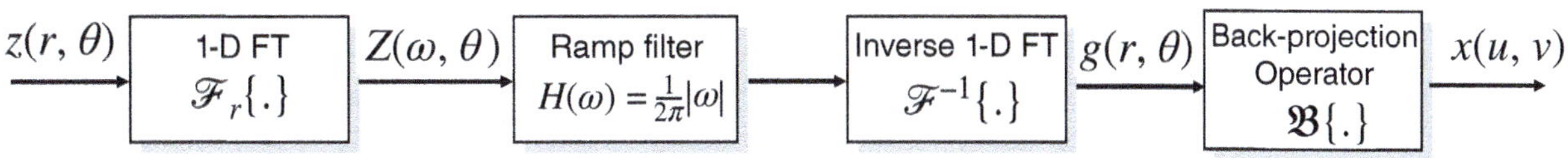

Figure 15.2 Block diagram of the FBP process for image reconstruction from projections.

15.2.2 Discrete Implementation of Filtered Back-Projection

The goal here is to implement the FBP operation $x(u, v) = [\mathcal{BP}z](u, v)$ using sampled projection data $z(m, n) = z(r_m, \theta_n)$ and for a limited number of observations. To this end, we consider the parallel data collection scenario with N equally spaced angular views, i.e. $\theta_n = n\Delta\theta,\ \ n \in [0, N-1]$, where $\Delta\theta = \frac{\pi}{N}$, and $2M + 1$ equally spaced parallel lines with distance Δr between the parallel lines in each view, i.e. $r_m = m\Delta r,\ \ m \in [-M, M]$. There are two equations for the FBP that must be discretized: convolution and back-projection operations in (15.23) and (15.20), respectively. The discrete version of the convolution in (15.23) is a convolution summation of the form,

$$g(m\Delta r, n\Delta\theta) = \Delta r \sum_{l=-M}^{M} z(l\Delta r, n\Delta\theta)h((m-l)\Delta r), \quad n \in [0, N-1]. \tag{15.24}$$

The discrete version of the back-projection equation in (15.20) is

$$x(i, j) = \Delta\theta \sum_{n=0}^{N-1} g(i\cos(n\Delta\theta) + j\sin(n\Delta\theta), n\Delta\theta), \tag{15.25}$$

where v and v are assumed to be sampled, i.e. $u = i\triangle u$ and $v = j\triangle v$, with typical sampling intervals $\triangle u = \triangle v = 1$. To make sure that this is a good approximation to the integral in (15.20), we need to perform an interpolation to compute $g(i\cos(n\Delta\theta) + j\sin(n\Delta\theta), n\Delta\theta)$ for each n from $g(m\Delta r, n\Delta\theta)$ in (15.24) for $m \in [-M, M]$. This is due to the fact that the argument $i\cos(n\Delta\theta) + j\sin(n\Delta\theta)$ does not generally lie on the discrete grid r_m. Among the commonly used interpolation methods are nearest-neighbor and linear interpolation [1]. In the nearest-neighbor method, we approximate $g(i\cos(n\Delta\theta) + j\sin(n\Delta\theta), n\Delta\theta)$ by $g(m\Delta r, n\Delta\theta)$, where m is chosen such that $[m\Delta r - i\cos(n\Delta\theta) + j\sin(n\Delta\theta)]$ is as small as possible. In this case, the back-projection in (15.25) is done by adding together the ray sums of the lines at each view and multiplying the result by $\Delta\theta$. Alternatively, we can use a linear interpolator in which case we select m so that $m\Delta r \leq i\cos(n\Delta\theta) + j\sin(n\Delta\theta) < (m+1)\Delta r$ and approximate,

$$g(i\cos(n\Delta\theta) + j\sin(n\Delta\theta), n\Delta\theta) \approx g(m\Delta r, n\Delta\theta) + \left[\frac{i\cos(n\Delta\theta) + j\sin(n\Delta\theta) - m\Delta r}{\Delta r}\right] \Big(g((m+1)\Delta r, n\Delta\theta) - g(m\Delta r, n\Delta\theta)\Big).$$

More accurate interpolation methods such as spline or cubic interpolation may also be used, at the expense of increased computational complexity.

Discrete FBP Algorithm:

1. The discrete convolution in (15.24) is usually carried out in the frequency domain using the 1-D DFT. To ensure that circular convolution yields the desired linear convolution results, the sampled projections, $z(m, n)$, must be zero-padded to size $K = 2(2M + 1)$. The data samples for $k \in [0, K-1]$ of the periodically extended projections are then used in the 1-D FFT of size K to produce $Z(k, n)$.
2. The windowed frequency response of the ramp filter $H(\omega)$ is then sampled uniformly $H(k) = H(k\Delta\omega)$ to the same size K so that the conjugate symmetry is satisfied, i.e. $H(K-k) = H^*(k),\ \ k \in [0, K/2]$.
3. Multiply (pointwise) $H(k)$ by $Z(k, n)$ and take 1-D IDFT (using FFT) of size K. Discard $K/2$ samples that do not correspond to the desired linear convolution result and take samples for $m \in [-M, M]$ of $g(m\Delta r, n\Delta\theta)$.
4. Perform interpolation (nearest-neighbor, linear, or higher-order methods) to evaluate $g(\cdot, n\Delta\theta)$ at non-grid locations. Alternatively, a zero-order hold (see Section 5.1.2) can be used for more practical implementation [8].
5. Perform discrete back-projection using (15.25).

Figure 15.3a shows the phantom head generated to mimic a 2-D slice of the human head with different regions and small and large structures. Figure 15.3b shows the corresponding sinogram or the Radon transform of this image obtained over projection angles from 0° to 180° with 3° spacing. The FBP reconstruction was then applied using a cubic interpolation and a Hamming-windowed ramp filter. To produce the desired linear convolution result using a circular convolution, the projection image was zero-padded before taking the DFT. The procedure used in this experiment was intended to show the actual operations adopted in CT image reconstruction. Figure 15.3c illustrates the reconstructed image, which exhibits

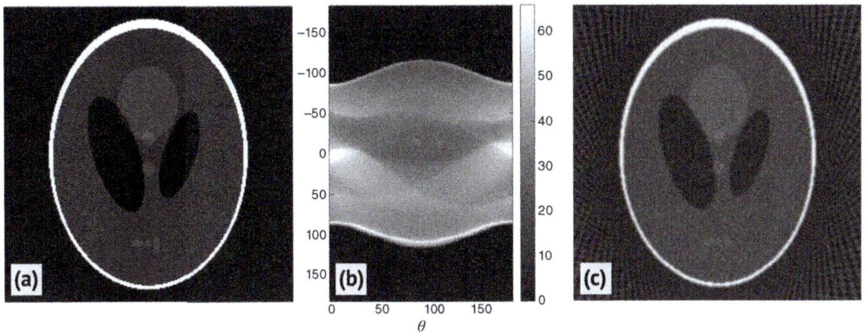

Figure 15.3 (a) Original phantom head image; (b) Radon transform or sinogram image; and (c) reconstructed image using FBP. (The color rendition of this image is available to view at www.wiley.com/go/Azimi449.)

some blurring artifacts due to windowing or truncation in the frequency domain. Furthermore, Moiré-type patterns are also observed in the reconstructed image due to angular sampling of the parallel beams. This particular artifact cannot be eliminated, even with increased angular resolution. The reconstruction error, measured by the root-mean-square (RMS) difference between the reconstructed and original images, was found to be 0.05.

15.3 Iterative Image Reconstruction Methods

There are several fundamental differences between analytical and iterative image reconstruction methods. As observed from Figure 15.3, the presence of artifacts is inevitable in analytical methods such as FBP due to filter approximations, discretization, and using a limited number of measurements. Iterative methods address these limitations by refining the reconstruction through repeated updates, often yielding improved reconstruction at the expense of a higher computational cost. Additionally, in contrast to analytical methods, iterative methods allow incorporation of prior knowledge about the image and noise properties into the process, resulting in a better representation of the imaging system. Iterative approaches generally use a linear observation model that relates the measured projection data to the unknown image through a projection matrix determined by the image acquisition system. These methods operate directly on the discretized representation of the inversion problem and produce reconstructed images from a finite set of equations that are solved numerically. As such, they are closely related to the image restoration methods discussed in Chapter 10.

In this section, we cover two different classes of iterative-based image reconstruction methods, namely ART [1, 14, 15] and statistical reconstruction methods [16–20].

15.3.1 Algebraic Reconstruction Technique

ART is based on a series-expansion approach to represent the image to be reconstructed. Starting from a discretized model, one is then concerned with estimating the unknown image by finding a finite set of unknown parameters. Let us assume that there exists a set of known basis functions $\{b_1, \dots, b_L\}$, each of which has a specified ROS and captures different attributes of the image. We will discuss a few possible choices of these basis functions later. Now, the unknown image $x(u, v)$ can be represented in polar coordinates as a linear combination of these basis functions:

$$x(\rho, \phi) = \sum_{l=1}^{L} \lambda_l b_l(\rho, \phi), \tag{15.26}$$

where $u = \rho \cos \phi$ and $v = \rho \sin \phi$ and as before $x(\rho, \phi) = x(\rho \cos \phi, \rho \sin \phi)$. The coefficients λ_l represent the contribution of each basis function $b_l(\rho, \phi)$ to the representation of the image $x(\rho, \phi)$. Thus, (15.26) defines a linear model for the image $x(\rho, \theta)$, where the parameters λ_l are unknown.

Let us assume that projection (sinograms) measurements $[\mathcal{R}x](r_k, \theta_k)$ are available for $k \in [1, K]$. Then, define the functional $[\mathcal{R}_k]$ that represents the sampled value of projections $[\mathcal{R}]$ at sample k. That is, we have $[\mathcal{R}_k x] = [\mathcal{R}x](r_k, \theta_k)$, $\forall k \in [1, K]$. Now, applying this functional to (15.26) and using the linearity property of $[\mathcal{R}]$ or $[\mathcal{R}_k]$ (see Table 15.1) yields,

$$z_k = \sum_{l=1}^{L} \lambda_l h_{k,l}, \quad k \in [1, K] \tag{15.27}$$

where $z_k = z(r_k, \theta_k) = [\mathcal{R}_k x]$ and $h_{k,l} = [\mathcal{R}_k b_l]$. Since b_l are user-defined functions, $h_{k,l}$ can be found by analytical means. Arranging (15.27) in matrix-vector form gives

$$\underline{z} = \mathbf{H}\underline{\lambda} + \underline{n}, \tag{15.28}$$

where $\mathbf{H} \in \mathbb{R}^{K \times L}$ is a matrix with elements $h_{k,l}$, $z = [z_1, ..., z_K]^T$, and $\underline{\lambda} = [\lambda_1, ..., \lambda_L]^T$ are the observation and parameter vectors, respectively. The vector $\underline{n} = [n_1, ..., n_K]^T$ represents the effects of modeling errors or random noise due to imperfections of the imaging system.

Now, our discrete image reconstruction problem is: given the data vector $\underline{z}$ and the knowledge of the matrix $\mathbf{H}$, estimate the image parameter vector $\underline{\lambda}$. The observation equation in (15.28) is identical to that in Problem 10.3 for image restoration, whose deterministic solution using the method of least squares (LS) [21, 22] gives $\underline{\lambda}_{LS} = (\mathbf{H}^T\mathbf{H})^{-1}\mathbf{H}^T\underline{z} = \mathbf{H}^{\dagger}\underline{z}$ where $\mathbf{H}^{\dagger} = (\mathbf{H}^T\mathbf{H})^{-1}\mathbf{H}^T$ is known as the left pseudo-inverse or Moore-Penrose inverse. Thus, ART can be viewed as an iterative way to find an approximate solution to this problem efficiently without performing any matrix inversion. Clearly, other iterative approaches, such as Krylov subspaces [23], exist for finding efficient solutions of similar problems.

Once the estimate of the parameter vector $\hat{\underline{\lambda}}$ is generated, the image can be reconstructed using

$$\hat{x}(\rho, \phi) = \sum_{l=1}^{L} \hat{\lambda}_l b_l(\rho, \phi). \tag{15.29}$$

or in Cartesian coordinates,

$$\hat{x}(u, v) = \sum_{l=1}^{L} \hat{\lambda}_l b_l(u, v). \tag{15.30}$$

ART updates the estimate of the parameter vector iteratively using only one measurement at a time. That is, at iteration $j + 1$, the parameter vector $\underline{\lambda}^{(j+1)}$ is updated from $\underline{\lambda}^{(j)}$ using one of the K measurements. This iterative process is repeated for multiple passes or epochs over the measurement set until convergence is reached. All ART methods have an update equation of the following general form:

$$\underline{\lambda}^{(j+1)} = f_j\left(\underline{\lambda}^{(j)}, \underline{h}_p, z_p\right), \tag{15.31}$$

where $\underline{h}_p^T$ is the pth row of $\mathbf{H}$ and z_p is the pth element of the measurement vector $\underline{z}$. The sequence of nonlinear functions f_j defines how the iterative process is performed using the previous estimate $\underline{\lambda}^{(j)}$ at iteration j, a single measurement z_p, and the corresponding row $\underline{h}_p^T$, of matrix $\mathbf{H}$. This single measurement is chosen using a cyclic schedule $p = ((j))_K + 1$ where $((.))$ stands for modulo K operation, e.g. for $j = nK$, $n = 0, 1, ...,$ we get $p = 1$.

Different ART variants differ in the way they define the f_j functionals. One popular ART method uses the following update equation,

$$\underline{\lambda}^{(j+1)} = \underline{\lambda}^{(j)} + \gamma_j \frac{\left(z_p - \underline{h}_p^T\underline{\lambda}^{(j)}\right)}{||\underline{h}_p||^2}\underline{h}_p^T, \tag{15.32}$$

where $0 < \gamma_j \leq 1$ is called *relaxation* parameter. Now, if we define $\hat{z}_p = \underline{h}_p^T\underline{\lambda}^{(j)}$, which is basically the estimate of the measurement z_p based on the previous estimate of the parameter vector $\underline{\lambda}^{(j)}$, then the updating equation in (15.32) becomes

$$\underline{\lambda}^{(j+1)} = \underline{\lambda}^{(j)} + \gamma_j \frac{(z_p - \hat{z}_p)\underline{h}_p^T}{||\underline{h}_p||^2}, \tag{15.33}$$

or simply

$$\underline{\lambda}^{(j+1)} = \underline{\lambda}^{(j)} + \gamma_j \varepsilon_p \underline{e}_{p,} \tag{15.34}$$

where $\varepsilon_p = z_p - \hat{z}_p$ is the error between the actual measurement and its estimated value and $\underline{e}_p = \frac{\underline{h}_p^T}{||\underline{h}_p||^2}$ is the unit norm vector along the direction of $\underline{h}_p^T$. This vector is normal to the hyperplane defined by $\hat{z}_p = \underline{h}_p^T \underline{\lambda}^{(j)}$. From the geometrical perspective, this implies that the updating of the parameter vector takes place along the same direction as the unit-norm vector $\underline{e}_p$, while its magnitude is controlled by the error ε_p and the relaxation parameter γ_j. The relaxation parameter plays the same role as the step size in the updating equation, and choosing its value presents a trade-off between accuracy and speed of convergence. That is, a larger value of γ_j speeds up the convergence while sacrificing the accuracy, and vice versa. In some cases, this parameter is initially chosen to be relatively large to promote faster convergence at the early stages of the updating and then reduced, either linearly or exponentially, to provide more accurate estimates during the later stages of the updating process.

In [13] a reasonable choice was suggested to initialize the iterative process in (15.34). Using this method, all the elements of the parameter vector are initialized to the same value related to the mean of the measurements, i.e.

$$\underline{\lambda}^{(0)} = \frac{\sum_{k=1}^{K} zk}{\sum_{k=1}^{K} \sum_{l=1}^{L} h_{k,l}} \underline{1}, \tag{15.35}$$

where $\underline{1}$ is a vector of all ones.

Figure 15.4a,c shows the reconstructed phantom head images using one and three iterations of a variant of the ART method, respectively. The iradon_sart function in Python skimage.transform was used for this implementation. This function is based on the Simultaneous Algebraic Reconstruction Technique (SART) in [14]. As with the other variants of the ART algorithm, SART uses pixel grids as the basis functions for the linear approximation of the image. Using simple pixel/voxel basis functions is computationally straightforward and effective for these algebraic methods. The same sinogram in Figure 15.3b with 3° angle spacing was used as the measurements for this image reconstruction. The corresponding error images in Figure 15.4b,d illustrates the quality of image reconstruction of the ART method. The root-mean-square (RMS) values of the error were found to be 0.06 and 0.03 for reconstruction using one and three iterations of SART, respectively. Comparing these results with that of the FBP method in Figure 15.3c reveals several important observations. The reconstruction quality of the ART method using one iteration over the measurements is inferior to that of the FBP method, though without any Moiré-type artifacts. ART performed significantly better when the number of iterations increased to three with no visible artifacts. In general, increasing the number of iterations improves the recovery of the high-frequency attributes but can also lead to more noise in the reconstructed image. Finally, the computational cost of the ART algorithm is much higher than that of the FBP method.

15.3.1.1 Choice of Basis Functions

As mentioned earlier, a simple way to define basis functions is to partition the 2-D (ρ, ϕ) or (u, v) plane into pixel grids or the 3-D space into voxels, and then define each basis function to be unity inside a specific pixel grid (or voxel) and 0 elsewhere, i.e. $b_l(u, v) = \text{rect}\left[\frac{u-u_l}{\Delta}\right]\text{rect}\left[\frac{v-v_l}{\Delta}\right]$ where rect[.] represents rectangular function of unit width, Δ is the grid size, and (u_l, v_l) are the spatial coordinates of the center of grid l. Using this pixel-based representation, $h_{k,l}$ corresponds to the length of intersection of the projection line associated with the kth measurement for the lth pixel grid. Although this choice of basis function leads to a sparse matrix **H** and is computationally efficient, it typically produces blocking artifacts. Improved reconstruction quality can be achieved by using circularly symmetric basis functions with smooth transitions. The smoothness of the basis functions eliminates some of the undesirable ringing artifacts in the reconstructed images. One particular approach proposed in [1, 14] is to use the generalized Kaiser-Bessel window functions that give a family of functions characterized by parameters α, d, and a. A member of this family is given [1] by,

$$b(\rho, \phi; \alpha, d, a) = \begin{cases} \frac{\sqrt{3}d^2\alpha}{4\pi a^2 I_3(\alpha)}\left[1 - \left(\frac{\rho}{a}\right)^2\right] I_2\left(\alpha\sqrt{1 - \left(\frac{\rho}{a}\right)^2}\right), & \text{for } 0 \le \rho \le a. \\ 0, & \text{otherwise,} \end{cases} \tag{15.36}$$

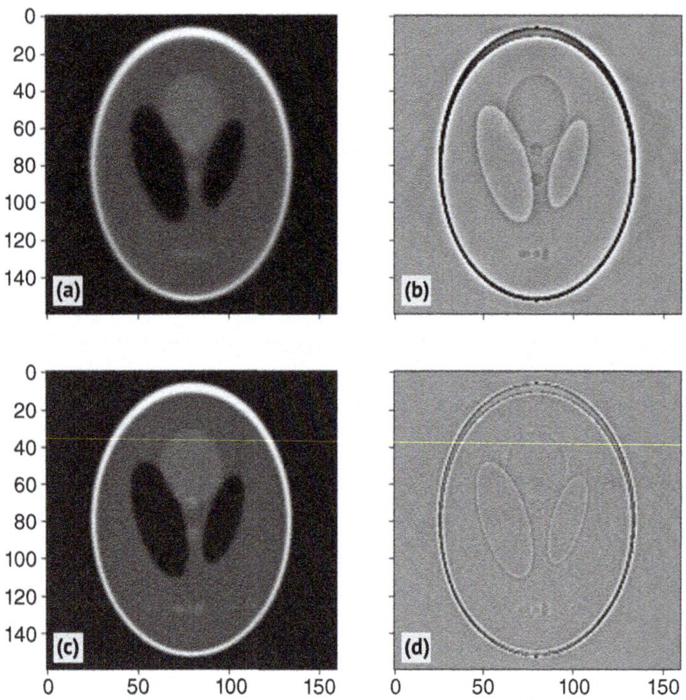

Figure 15.4 (a) Reconstructed phantom head image using ART method (one iteration); (b) error image for reconstructed image of (a); (c) Reconstructed phantom head image using ART method (three iterations); and (d) error image for reconstructed image of (c). (The color rendition of this image is available to view at www.wiley.com/go/Azimi449.)

where $I_m(.)$ is the modified Bessel function of the first kind of order m and a is the radius of the circularly symmetric function. It is important to note from (15.36) that these basis functions are radially symmetric and do not depend on ϕ. Unlike pixel-based basis functions, these basis functions have a bell-shaped profiles that decay smoothly with radial distance ρ from the center. The parameter α controls the degree of tapering amount, while d is the sampling interval in a hexagonal sampling grid to obtain b_l, $l \in [1, L]$ from $b(\rho, \phi; \alpha, d, a)$. More specifically, if we denote the grid points by δ_l, $l \in [1, L]$, then b_l are obtained by shifting the centers of $b(\rho, \phi; \alpha, d, a)$ from the origin to these grid points. These grid points in hexagonal sampling are

$$\delta_l = \left(\frac{md}{2}, \frac{\sqrt{3}nd}{2} \right), \quad (m, n) \in \mathbb{Z} \text{ with } m + n \text{ even.} \tag{15.37}$$

However, the challenge in this approach is how to choose appropriate values for the parameters α and d to achieve good reconstruction performance. A simpler alternative that offers smooth, bell-shaped profile, is given by the polynomial basis functions:

$$b(\rho; a) = \begin{cases} \left[1 - \left(\frac{\rho}{a}\right)^2\right]^2, & \text{for } 0 \leq \rho \leq a, \\ 0, & \text{otherwise.} \end{cases} \tag{15.38}$$

An important advantage of these radially symmetric basis functions is that their projections can be expressed in closed-form [14]. That is, we have,

$$\begin{aligned} h(r, \theta) &= [\mathcal{R}b](r, \theta) = 2\int_0^{\sqrt{(a^2-r^2)}} b\left(\sqrt{(r^2+s^2)}\right) ds, \quad |r| \leq a \\ &= \frac{16a}{15}\left[1 - \left(\frac{r}{a}\right)^2\right]^{3/2}. \end{aligned} \tag{15.39}$$

These basis functions are then shifted so that the lth shifted version b_l has circular symmetry about the pixel grid δ_l at the (u_l, v_l) location.

15.3.1.2 Stochastic Gradient Descent and ART

Stochastic gradient descent (SGD) is a widely used method for iterative solution of optimization problems [24]. The term stochastic stems from the use of an estimate of the gradient of an objective function instead of the actual gradient. This method will be covered in more depth when we discuss performance learning methods in Chapter 17. However, here our goal is to shed some light on: (i) the objective function of the ART and (ii) the connection between the ART update rule in (15.34) and the SGD algorithm.

In general, optimization problems involve minimizing or maximizing an objective function, $J(\underline{\lambda})$, with respect to the unknown parameters $\underline{\lambda}$ to be estimated. Using gradient descent for minimization, the iterative update rule [24] is given by

$$\underline{\lambda}^{(j+1)} = \underline{\lambda}^{(j)} - \frac{1}{2}\gamma_j \nabla J\left(\underline{\lambda}^{(j)}\right), \tag{15.40}$$

where $\nabla J\left(\underline{\lambda}^{(j)}\right) = \frac{\partial J\left(\underline{\lambda}^{(j)}\right)}{\partial \underline{\lambda}^{(j)}}$ is the gradient of the objective function, and γ_j is the step size. In the ART framework, the update at each iteration is based on a single measurement z_p. Thus, given the previous estimate of the parameter vector at iteration j, an instantaneous squared error

$$J\left(\underline{\lambda}^{(j)}\right) = \left(z_p - \hat{z}_p\right)^2 = \left(z_p - \underline{h}_p^T \underline{\lambda}^{(j)}\right)^2,$$

can be used as the objective function for this minimization. Taking the gradient of this objective function with respect to $\underline{\lambda}^{(j)}$ yields

$$\nabla J\left(\underline{\lambda}^{(j)}\right) = 2\left(z_p - \underline{h}_p^T \underline{\lambda}^{(j)}\right)\underline{h}_p^T = 2\varepsilon_p \underline{h}_p^T.$$

Substituting this result into (15.40) gives

$$\underline{\lambda}^{(j+1)} = \underline{\lambda}^{(j)} + \gamma_j \varepsilon_p \underline{h}_p.$$

Finally, normalizing the direction vector $\underline{h}_p$ by $\|\underline{h}_p\|^2$ yields the ART update rule in (15.34).

This derivation establishes that the ART algorithm can be interpreted as a SGD method applied to the minimization of an instantaneous squared error objective function.

15.3.2 Statistical Image Reconstruction

So far, we have ignored the effects of noise and interference in the observation or measurement equation (15.28). As stated earlier, the ART algorithm can be viewed as an iterative method that provides a deterministic solution to the image reconstruction problem. This is analogous to the inverse filtering and LS methods presented in Chapter 10 for image restoration from blurred observations. In the absence of noise, our inverse problem in (15.28) closely resembles the deblurring problem, where basis functions play the same role as the PSF of the blurring process.

Statistical image reconstruction methods, however, account for the effects of random noise in the observation equation of (15.28). In this case, the inverse problem can be stated as: given the observation vector $\underline{z}$ or a sequence of scalar measurements $z_k,\ k \in [1, K]$, where z_k is the kth element of the vector $\underline{z}$, the system matrix $\mathbf{H}$, and the pdf of noise $\underline{n}$, the goal is to generate estimates of the parameter vector $\underline{\lambda}$ iteratively. Note that although in some statistical-based image reconstruction methods [8] the parameter vector is also assumed to be random, here we restrict our attention to the case where this vector is unknown but deterministic, that is to say, no prior probability distribution is assigned to $\underline{\lambda}$.

Our approach begins with the pdf of $\underline{Z}$, parameterized by $\underline{\lambda}$. We then construct the corresponding likelihood function and maximize it with respect to $\underline{\lambda}$ [22]. For a sequence of independent and identically distributed (iid) measurements $z_k,\ k \in [1, K]$, this likelihood function is

$$\ell\left(\underline{\lambda}; \{z_k\}_{k=1}^K\right) = \prod_{k=1}^{K} p_Z(Z = z_k; \underline{\lambda}), \tag{15.41}$$

which we will maximize with respect to $\underline{\lambda}$,

$$\hat{\underline{\lambda}} = \underset{\underline{\lambda}}{\operatorname{argmax}} \ \ell\left(\underline{\lambda}; \{z_k\}_{k=1}^{K}\right). \tag{15.42}$$

The result of (15.42) gives the ML solution, which requires knowledge of the likelihood function or, equivalently, the noise distribution $p(\underline{n})$. It is well known [19] that under the Gaussian assumption for the noise distribution, the ML solution reduces to that of the LS, which may be solved using variants of ART described in the previous section. However, in many imaging systems that produce sinogram data, additive noise is typically modeled as an iid Poisson random variable [1]. This is because noise in these systems corresponds to the number of photons recorded by the detector over a period of time. Now, using the Poisson assumption for noise with PMF $p(N = n) = \frac{\mu^n e^{-\mu}}{n!}$, where μ is the ensemble mean, measurements $z_k,\ k \in [1, K]$ can be considered as realizations of independent Poisson random variables $Z_k \sim \text{Poisson}\{\bar{z}_k\}$ with mean

$$\bar{z}_k = E[Z_k] = E\left[\underline{h}_k^T\underline{\lambda} + N_k\right] = \underline{h}_k^T\underline{\lambda} + \mu, \qquad k \in [1, K]$$

where μ represents the mean of the additive noise component and $\underline{h}_k^T$ represents the kth row of the matrix $\mathbf{H}$. Therefore, the likelihood function in (15.41) can be expressed as:

$$\ell\left(\underline{\lambda}; \{z_k\}_{k=1}^{K}\right) = \prod_{k=1}^{K} e^{-\bar{z}_k} \frac{(\bar{z}_k)^{z_k}}{z_k!}. \tag{15.43}$$

Taking the natural logarithm of (15.43) yields the log-likelihood function,

$$\begin{aligned} \ln \ell\left(\underline{\lambda}; \{z_k\}_{k=1}^{K}\right) &= \sum_{k=1}^{K} \left\{z_k \ln \bar{z}_k - \bar{z}_k - \ln(z_k!)\right\} \\ &= \sum_{k=1}^{K} \left\{z_k \ln\left(\underline{h}_k^T\underline{\lambda} + \mu\right) - \left(\underline{h}_k^T\underline{\lambda} + \mu\right) - \ln(z_k!)\right\}. \end{aligned} \tag{15.44}$$

Substituting this result into (15.42), our ML problem becomes

$$\hat{\underline{\lambda}} = \underset{\underline{\lambda}}{\operatorname{argmax}} \left\{\sum_{k=1}^{K} z_k \ln\left(\underline{h}_k^T\underline{\lambda} + \mu\right) - \left(\underline{h}_k^T\underline{\lambda} + \mu\right)\right\}. \tag{15.45}$$

The last term $\ln(z_k!)$ in (15.44) is omitted because it does not depend on $\underline{\lambda}$. As can be observed, the resulting objective function for this maximization problem is a nonlinear function of $\underline{\lambda}$, and hence the solution must be obtained numerically as a closed-form solution does not exist. One possible approach to solve this maximization problem iteratively is to use the gradient ascent algorithm. However, the gradient ascent algorithm suffers from accuracy-speed trade-offs and converges to suboptimal solutions, particularly for non-quadratic objective functions. Several alternative iterative reconstruction algorithms have been proposed to solve this optimization problem. In what follows, we describe one method using the conjugate gradient method [20].

15.3.2.1 Conjugate Gradient Method

The conjugate gradient algorithm [18] is designed to overcome some of the limitations of the gradient descent/ascent algorithms. While it was primarily developed for linear optimization problems with quadratic objective functions, it can also be extended to nonlinear problems by locally approximating the objective function, up to a second order, using Taylor series expansion.

If we denote the log-likelihood objective function in (15.44) by $L(\underline{\lambda})$,

$$L(\underline{\lambda}) = \sum_{k=1}^{K}\left\{z_k \ln\left(\underline{h}_k^T\underline{\lambda} + \mu\right) - \left(\underline{h}_k^T\underline{\lambda} + \mu\right)\right\}, \tag{15.46}$$

then our goal is to maximize this function iteratively with respect to the unknown parameter vector $\underline{\lambda}$. More specifically, given an initial estimate $\underline{\lambda}^{(0)}$, the objective is to find a search direction vector $\underline{d}^{(j)}$ such that

$$\underline{\lambda}^{(j+1)} = \underline{\lambda}^{(j)} + \gamma_j \underline{d}^{(j)} \quad j = 0, 1, \dots, \tag{15.47}$$

converges to a maximizer of the objective function $L(\underline{\lambda})$ as $j \to \infty$, where γ_j is a scalar step size. To achieve good estimates, it is ideally desired that the update at each iteration j, leads to vanishing gradient, i.e. $\nabla L\left(\underline{\lambda}^{(j+1)}\right) = 0$. Approximating the objective function using a first-order Taylor expansion, this gradient becomes

$$\nabla L\left(\underline{\lambda}^{(j)} + \gamma_j \underline{d}^{(j)}\right) \approx \nabla L\left(\underline{\lambda}^{(j)}\right) + \gamma_j \mathbf{S}\underline{d}^{(j)} = 0, \tag{15.48}$$

where $\mathbf{S} = \nabla^2 L\left(\underline{\lambda}^{(j)}\right)$ is the Hessian matrix of the objective function $L\left(\underline{\lambda}^{(j)}\right)$ with $(p, q)^{\text{th}}$ element given by

$$s_{p,q} = -\sum_{k=1}^{K} \frac{h_{k,p} h_{k,q} z_k}{(\bar{z}_k)^2}. \tag{15.49}$$

Solving (15.48) for the search direction vector gives $\underline{d}^{(j)} \approx -\frac{1}{\gamma_j}\mathbf{S}^{-1}\nabla L\left(\underline{\lambda}^{(j)}\right)$. However, computing $\mathbf{S}^{-1}$ at each iteration is computationally inefficient.

The conjugate gradient method circumvents this difficulty by constructing a sequence of mutually conjugate search directions. The first iteration coincides with the gradient ascent direction. For subsequent iterations, the search direction $\underline{d}^{(n)}$ at iteration n must be chosen such that the gradient $\nabla L\left(\underline{\lambda}^{(j)}\right)$ is orthogonal to $\underline{d}^{(j)},\ j < n$. That is, the new search direction $\underline{d}^{(n)}$ should satisfy the mutual conjugate condition [24] given by $\underline{d}^{(n)^T}\mathbf{S}\underline{d}^{(j)} = 0,\ \ n \neq j$.

The conjugate gradient without preconditioning is based on the Polak–Ribiere algorithm [20], which does not make use of the Hessian matrix. At each iteration, it updates the gradient, constructs a new search direction, and applies a line search along the new direction. The steps in this algorithm are given below. For detailed treatment of this algorithm, the reader is referred to [18, 20].

1. *Initialization*: Select $\underline{\lambda}^{(0)}$ and $\underline{g}^{(0)} = \nabla L\left(\underline{\lambda}^{(0)}\right)$. Set $\underline{d}^{(0)} = -\underline{g}^{(0)}$.
2. *Optimal Step Size:* $\gamma_j = \underset{\gamma>0}{\arg\max}\ L\left(\underline{\lambda}^{(j)} + \gamma \underline{d}^{(j)}\right)$.
3. *Updating:* $\underline{\lambda}^{(j+1)} = \underline{\lambda}^{(j)} + \gamma_j \underline{d}^{(j)}$.
4. *Gradient:* $\underline{g}^{(j+1)} = \nabla L\left(\underline{\lambda}^{(j+1)}\right)$.
5. *Polak–Ribière Parameter:* $\eta_j = \dfrac{\underline{g}^{(j+1)^T}\left(\underline{g}^{(j+1)} - \underline{g}^{(j)}\right)}{\|\underline{g}^{(j)}\|^2}$.
6. *New Search Direction:* $\underline{d}^{(j+1)} = \underline{g}^{(j+1)} + \eta_j \underline{d}^{(j)}$.
7. Increment j and go to step 2.

A diagonal preconditioning is typically employed in the gradient ascent algorithm to speed up convergence. A similar approach can be employed in the conjugate gradient method for faster convergence. A preconditioning matrix $\mathbf{C}$ modifies the gradient direction to account for scaling differences in the parameter space. The preconditioned conjugate gradient algorithm can be summarized as follows.

1. Steps 1–4 are identical to the original conjugate gradient given before.
5. *Preconditioned Gradient:* $\underline{p}^{(j+1)} = \mathbf{C}^{(j+1)}\underline{g}^{(j+1)}$.
6. *Search Step Size:* $\eta_j = \dfrac{\left(\underline{g}^{(j+1)} - \underline{g}^{(j)}\right)^T \underline{p}^{(j+1)}}{\underline{g}^{(j)^T}\underline{p}^{(j)}}$.
7. *Search Direction:* $\underline{d}^{(j+1)} = \underline{p}^{(j+1)} + \eta_j \underline{d}^{(j)}$.
8. Increment j and go to step 2.

The preconditioning matrix $\mathbf{C}$ must be symmetric positive definite (PD) in order to guarantee convergence. The ideal preconditioning matrix is $\mathbf{S}^{-1}$, which is impractical to compute at each iteration. However, the diagonal elements of this matrix can be used to form the diagonal preconditioning matrix $\mathbf{C} = \text{diag}[1/s_{1,1} \cdots 1/s_{L,L}]$. An alternative method was proposed in [17], which does not require the computation of diagonal elements of the Hessian matrix. Using this method, the diagonal preconditioning matrix is given by

$$\mathbf{C}^{(j)} = \text{diag}\left\{\frac{\lambda_n^{(j)} + \delta}{\sum_m h_{m,n}}\right\}, \tag{15.50}$$

where δ is a small positive number to ensure that $\mathbf{C}^{(j)}$ is positive definite and $h_{m,n}$ is the $(m, n)^{\text{th}}$ element of matrix $\mathbf{H}$ in (15.28). In [18], the authors approximated the Hessian matrix using a circulant matrix, the inverse of which can be efficiently computed using the FFT [21].

Finally, for non-quadratic objective functions, there is no guarantee that the search direction $\underline{d}^{(j)}$ is an ascent direction. This can be verified by checking the condition $\underline{g}^{(j)^T}\underline{d}^{(j)} > 0$ at each iteration. Otherwise, we reset $\underline{g}^{(j)}$ and $\underline{p}^{(j)}$, if this condition is violated.

15.4 Computer Projects

The computer projects in this chapter use the Python libraries SciPy, scikit-image, NumPy, and Matplotlib.

Computer Project 15.1: The purpose of this computer project is to investigate the Radon transform and back-projection for image reconstruction. Use the lung CT image in Figure 14.19a, or any similar medical image. First, the contrast of the image must be enhanced using a histogram-based method discussed in Chapter 8. Apply the Radon transform to this contrast-enhanced image to generate the sinograms over the angular range 0 and 180° with 1°, 2°, 5°, and 10° spacing. Experimentally show that the Radon transform followed by the back-projection operation leads to a blurred reconstructed image as illustrated in (15.16). Next, design an inverse filter to remove the effects of this blur using the methods in Section 10.2.1? Then, apply the FBP in Section 15.2.1 using at least two different interpolation methods and two different filters to all the sinograms you generated. Plot the frequency responses of the filters used for FBP. Provide a comprehensive evaluation of all the results and state important observations about the performance of different methods that you used in this computer project. What are the strengths and weaknesses of the FBP method?

Computer Project 15.2: The first part of this computer project is essentially the same as Computer Project 15.1, in order to generate a set of sinograms for image reconstruction. However, in this project, these sinograms are used to reconstruct the lung image using the ART presented in Section 15.3.1 instead of the FBP method. Develop a Python-based code to reconstruct the lung image from these sinograms using ART or its variants. The number of iterations plays an important role in the accuracy of reconstruction, as illustrated in the example of Figure 15.4. Test several numbers of iterations and compare the quality of the reconstruction using both visual inspection and the RMS error, where the error image is the difference between the reconstructed image and the original image. Provide a comprehensive analysis of the ART results in terms of reconstruction quality, computational time, and undesirable artifacts in the reconstructed images. Describe the advantages and limitations of the ART method.

Computer Project 15.3: The first part of this computer project is also the same as Computer Project 15.1, in order to generate a set of sinograms with the various angle spacings. However, in this project, our goal is to implement the statistical approach discussed in Section 15.3.2 to reconstruct the lung image from the generated sinograms. Develop a Python-based code to implement the conjugate gradient iterative method for solving the ML-based image reconstruction. Implement and test both versions of the conjugate gradient, i.e. with and without a diagonal preconditioning matrix. Provide a comprehensive analysis and comparison of the results of both methods in terms of image reconstruction quality, computational time, and undesirable artifacts. Describe the benefits and shortcomings of these statistical-based methods.

15.5 Problems

Problem 15.1: Derive the Radon transform equation in (15.6) in polar coordinates.

Problem 15.2: Prove the following properties of Radon transform in Table 15.1:
(a) Linearity (b) Symmetry (c) Periodic.

Problem 15.3: Prove the following properties of Radon transform in Table 15.1:
(a) Shifting (b) Rotation (c) Scaling.

Problem 15.4: Prove the following properties of Radon transform in Table 15.1:
(a) Mass Conservation (b) Convolution.

Problem 15.5: Find the Radon transform of the following 2-D functions.

a. Gaussian $x(u, v) = e^{-\pi\alpha(u^2+v^2)}$,

b. $x(u, v) = \begin{cases} A, & (u^2 + v^2) \leq \sigma^2 \\ 0, & \text{otherwise} \end{cases}$,

c. $x(u, v) = e^{-2\pi j(ku+lv)}, \quad |u|, |v| \leq 1/2$.

Problem 15.6: Repeat Example 15.1 for the back-projection operator $[\mathcal{B}]$. What is your conclusion?

Problem 15.7: Find the impulse responses of the systems in Figure 15.5.

Problem 15.8: Consider an unknown separable image $x(u, v) = x_1(u)x_2(v)$. Show that this image can be reconstructed from two projections and comment on the projection angles.

Problem 15.9: Does the following property hold for Radon transform?

$$\int_{-\infty}^{\infty}\int_{-\infty}^{\infty} |x(u, v)|^2 du\, dv = \int_{0}^{\pi}\int_{-\infty}^{\infty} |z(r, \theta)|^2 dr\, d\theta.$$

Problem 15.10: In Section 15.1.3, it was stated that projection followed by back-projection leads to a blurred version of the original image, i.e.

$$y(u, v) = [\mathcal{B}\mathcal{R}x](u, v) = x(u, v) ** h(u, v).$$

Show that the PSF of the blur $h(u, v) = \frac{1}{\sqrt{u^2+v^2}}$ or in polar coordinates $h(\rho, \phi) = \frac{1}{|\rho|}$.

Problem 15.11: In Problem 15.10, show that the blur frequency response is $H(\omega_1, \omega_2) = \frac{4\pi^2}{\sqrt{\omega_1^2+\omega_2^2}}$.

Problem 15.12: Another windowed filter that can be employed in filtered back-projection (FBP) method is called Ram-Lak filter, which has frequency response $H(\omega) = |\omega|\text{rect}\left[\frac{\omega}{\alpha}\right]$. Find its impulse response $h(r)$. Comment on the performance of this filter for image reconstruction.

Problem 15.13: Find the inverse Radon transform of $z(r, \theta) = e^{2\pi jr}\delta(\theta - \varphi)$ using the direct convolution back-projection method.

Problem 15.14: Repeat Problem 15.13 using the filtered back-projection method with the standard ramp filter.

Problem 15.15: Assume that in Section 15.3.2, the objective function for maximization is changed to

$$J(\underline{\lambda}) = -\sum_{k=1}^{K} \frac{(z_k - \bar{z}_k)^2}{\sigma_k^2}$$

where $\bar{z}_k = \underline{h}_k^T\underline{\lambda}$ is the ensemble mean of the kth measurement. Find the LS estimate for $\underline{\lambda}$ using this objective function.

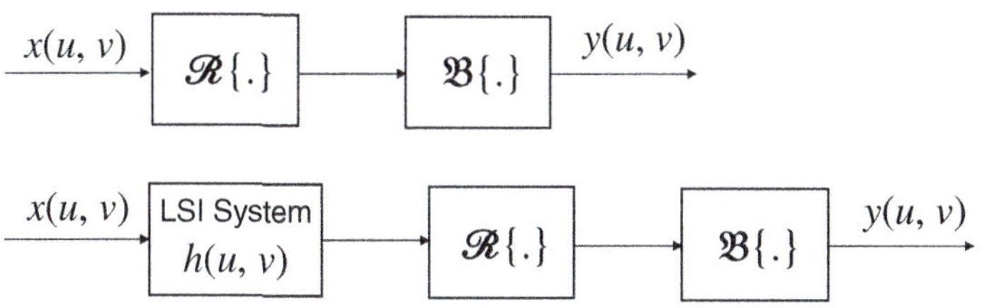

Figure 15.5 Problem 15.7.

Problem 15.16: Repeat Problem 15.15 using the following objective function:

$$J(\underline{\lambda}) = -\sum_{k=1}^{K} \frac{(z_k - \bar{z}_k)^2}{\sigma_k^2} + \nu||\underline{\lambda}||^2.$$

Find the LS estimate for $\underline{\lambda}$ using this objective function and comment on the result.

Problem 15.17: Assume that in Section 15.3.1, the objective function for maximization is changed to

$$J(\underline{\lambda}) = -(\underline{z} - \bar{\underline{z}})^T \mathbf{W}(\underline{z} - \bar{\underline{z}}),$$

where $\bar{\underline{z}} = \mathbf{H}\underline{\lambda}$ is the ensemble mean of the measurement vector $\underline{z}$. Find the LS estimate for $\underline{\lambda}$ using this objective function.

Problem 15.18: In the ML image reconstruction method in Section 15.3.2, assume that additive noise samples N_k are independent Gaussian distributed with mean zero and variance σ_k^2. Derive the ML optimization problem to find the parameter vector $\underline{\lambda}$. How does this assumption change the iterative approach to this problem?

Problem 15.19: In the ML image reconstruction method in Section 15.3.2, assume that the additive noise vector is multivariate Gaussian distributed with mean zero and covariance matrix $\mathbf{R}_n$. Derive the ML optimization problem for finding the parameter vector $\underline{\lambda}$. Comment on the results.

Problem 15.20: Assume that we use the Hessian matrix $\mathbf{S}$ as a preconditioning matrix in the preconditioned conjugate gradient method in Section 15.3.2. Is the mutual conjugate condition $\underline{d}^{(0)^T}\mathbf{S}\underline{d}^{(1)}$ for $n = 0, j = 1$ satisfied for this choice of preconditioning matrix ?

References

1 G. T. Herman, "*Fundamentals of Computerized Tomography-image Reconstruction from Projections*," Springer, 2009.

2 M. J. Willemink and P. B. Noel, "The evolution of image reconstruction for CT-from filtered back projection to artificial intelligence," *European Radiology*, vol. 29, pp. 2185–2195, 2019.

3 R. Schofield, L. King, U. Tayal, I. Castellano, J. Stirrup, F. Pontana, J. Earls, and E. Nicol, "Image reconstruction: Part 1-understanding filtered back projection, noise and image acquisition," *Journal of Cardiovascular Computed Tomography*, vol. 14, pp. 219–225, 2020.

4 M. Bertero, H. Lanteri, and L. Zanni, "Iterative image reconstruction: A point of view," Proceedings of the Interdisciplinary Workshop on Mathematical Methods in Biomedical Imaging and Intensity-Modulated Radiation Therapy (IMRT), Pisa, Italy, 2007.

5 W. Stiller, "Basics of iterative reconstruction methods in computed tomography: A vendor-independent overview," *European Journal of Radiology*, vol. 109, pp. 147–154, 2018.

6 G. Wang, J. C. Ye, K. Mueller, and J. A. Fessler, "Image Reconstruction Is a New Frontier of Machine Learning, " *IEEE Transactions on Medical Imaging*, vol. 37, pp. 1289–1296, 2018.

7 J. Nuyts and S. Matej, "*Chapter 13-Image Reconstruction*," 2014.

8 A. K. Jain, "*Fundamentals of Digital Image Processing*," Prentice-Hall, 1989.

9 D. E. Dudgeon and R. M. Mersereau, "*Multidimensional Digital Signal Processing*," Prentice-Hall, vol. 7, 1984.

10 J. Proakis and D. Manolakis, "*Digital Signal Processing: Principles, Algorithms and Applications*," Pearson, 4th Edition, 2006.

11 R. P. Kanwal, "*Linear Integral Equations: Theory and Technique*," Boston, MA, 2nd Edition, 1996.

12 S. Webb, A. Hilger, "*From the Watching of Shadows: The Origins of Radiological Tomography*," Bristol, CRC Press, 1990.

13 M. Noori-Asl and A. Sadremomtaz, "Analytical image reconstruction methods in emission tomography," *Journal of Biomedical Science and Engineering*, vol. 6, pp. 100–107, 2013.

14 A. H. Andersen and A. C. Kak "Simultaneous algebraic reconstruction technique (SART): A superior implementation of ART algorithm," *Ultrasonic Imaging*, vol. 6, pp. 81–94, 1984.

15 G. T. Herman and L. B. Meyer, "Algebraic reconstruction techniques can be made computationally efficient," *IEEE Transactions on Medical Imaging*, vol. 12, pp. 600–609, 1993.

16 R. M. Lewitt, "Alternatives to voxels for image representation in iterative reconstruction algorithms," *Physics in Medicine & Biology*, vol. 37, pp. 705–716, 1992.

17 R. M. Lewitt and S. Matej, "Overview of methods for image reconstruction from projections in emission computed tomography," *Proceedings of the IEEE*, vol. 91, pp. 1588–1611, 2003.

18 J. Qi and R. M Leahy, "Iterative reconstruction techniques in emission computed tomography," *Physics in Medicine & Biology*, vol. 51, pp. R541–578, 2006.

19 L. Kaufman, "Maximum likelihood, least squares, and penalized least squares for PET," *IEEE Transactions on Medical Imaging*, vol. 12, pp. 200–214, 1993.

20 J. A. Fessler and S. D. Booth, "Conjugate-gradient preconditioning methods for shift-variant PET image reconstruction," *IEEE Transactions on Medical Imaging*, vol. 8, pp. 688–699, 1999.

21 S. Kay, "*Fundamentals of Statistical Signal Processing, Volume I: Estimation Theory*," Prentice-Hall, 1st Edition, 1993.

22 L. L. Scharf, "*Statistical signal processing : Detection, Estimation, and Time Series Analysis*," Addison-Wesley, 1991.

23 C. G. Broyden and T. V. Maria, "*Krylov Solvers for Linear Algebraic Systems: Krylov Solvers*," Elsevier, vol. 11. 2004.

24 E. K. P. Chong, W-S. Lu, and S. H. Zak, "*An Introduction to Optimization with Applications to Machine Learning*," Wiley, 5th Edition, 2024.

16

Traditional Image Classification

Pattern recognition is a fundamental cognitive function that humans and many animals possess, enabling them to interpret and respond to stimuli encountered in their environments. These stimuli may originate from diverse sources and are perceived via different sensory channels, including vision, audition, olfaction, and tactile sensation. These sensory stimuli are processed in the cerebral cortex, where decision-making tasks take place based on one or all of such stimuli and using rules that are inferred through *learning* and contact with the environment. The outcome of the decision-making process is manifested in the form of *actions*, which are executed via the motor organs and their associated end-effectors.

In Chapter 12, we presented the block diagram of a traditional image analysis system with its components in Figure 12.1. The final stage of this process is pattern classification, whose function is to assign an unknown pattern to one of several known classes or categories. In a general context, a *pattern* may correspond to features extracted from an image, a fingerprint, handwritten characters, acoustic signals, ECG signals, radar signals, or sonar data. However, in this chapter, we are primarily concerned with image patterns, and in particular feature vectors extracted from segmented objects or regions. Nevertheless, the methods are broadly applicable to any type of pattern classification problem. The feature vector represents different attributes such as shape, color, texture, or other discriminatory information of the segmented object as discussed in Chapter 14.

The classifier then maps the feature or pattern space of dimension N to the decision (label) space of dimension $M \ll N$. This mapping is governed by a decision rule, which is constructed during a training process using labeled patterns from different classes. Each class or category corresponds to a set of patterns representing objects with similar characteristics, such as color, size, shape, or texture.

The methods used to design a pattern classification system [1, 2] can be categorized into many different categories such as linear versus nonlinear, statistical versus structural, supervised versus unsupervised, parametric vs nonparametric, and generative vs discriminative. Despite this diversity, we can partition these methods into two broad categories: *traditional* and *modern*. This chapter focuses mainly on the fundamental principles of pattern classification, emphasizing classical approaches of both deterministic and statistical types. Subsequent chapters will discuss modern pattern classification methods that leverage machine learning models.

16.1 Deterministic Pattern Classification

Let us assume that $\underline{x} \in \mathbb{R}^N$ represents an unknown pattern or feature vector containing features $x_1, \dots, x_N$, that needs to be classified (or labeled) into one of M possible classes. A classifier performs the mapping $\underline{x} \to \ell \in [1, M]$, where $\ell = f(\underline{x})$ and $\ell = m$ represents the mth class. Figure 16.1a illustrates this mapping between the feature or pattern space and the decision space. The function $f(\cdot)$ specifies the relationship between the inputs and outputs of the classifier, and is referred to as the *decision rule*. The decision rule partitions the feature space into M disjoint regions, $\mathcal{R}_k$, $k \in [1, M]$, known as *decision regions*. These regions are partitioned by decision *boundaries*, $S_{ij} = S_{j1}$, which separate regions $\mathcal{R}_i$ and $\mathcal{R}_j$. Figure 16.1b shows a two-dimensional feature space partitioned into three decision regions $\mathcal{R}_1$, $\mathcal{R}_2$, and $\mathcal{R}_3$ associated with three classes, and with decision boundaries S_{12}, S_{23}, and S_{31}. The data samples corresponding to these classes are indicated by black $*$, red $\circ$, and green $\diamond$. Ideally, as depicted in this figure, the decision rule should be designed to yield disjoint decision regions, thereby guaranteeing perfect classification with no misclassifications.

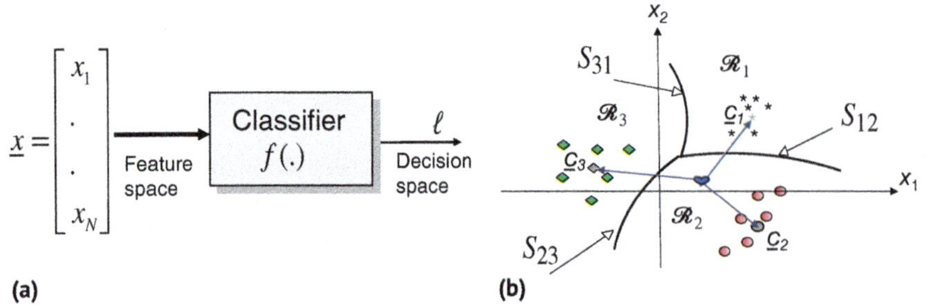

Figure 16.1 (a) A classifier maps feature vectors to decisions; (b) decision rule partitions the feature space into three decision regions/clusters separated by decision boundaries. The centroid of each cluster is denoted by $\underline{c}_i$.

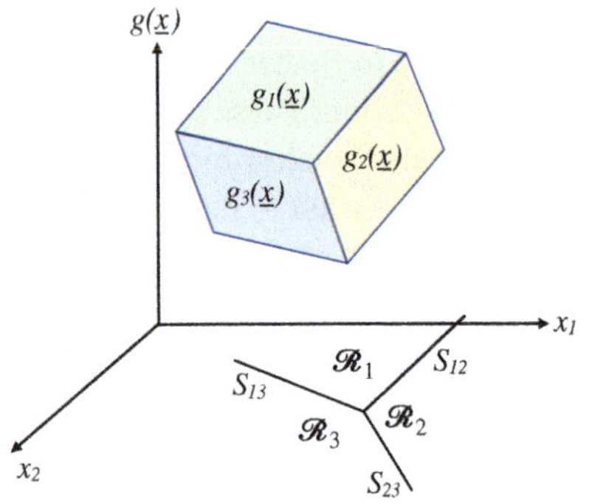

Figure 16.2 Decision boundaries are projections of intersection of discriminant functions onto the feature space.

16.1.1 Discriminant Functions and Linear Classifiers

From the above discussion, it is evident that the construction of the decision rule is the main goal in the design of any classification system, regardless of its type. Two important questions naturally arise: (1) how do we construct a decision rule for a specific pattern classification problem so that the incidence of misclassifications is minimized? and (2) assuming that the classifier is already designed, how do we make decisions about the class membership of an unknown input pattern?

Let us first address the second question. Assuming that the classifier is already designed, the classification decision for an unknown pattern $\underline{x}$ is made by comparing M scalar functions $g_1(\underline{x}), g_2(\underline{x}), \dots g_M(\underline{x})$, known as *discriminant functions*. More specifically, an unknown pattern $\underline{x}$ is assigned to class j (or C_j) if and only if (iff)

$$g_j(\underline{x}) > g_k(\underline{x}), \quad \forall k \in [1, M], k \neq j \iff \underline{x} \in C_j, \tag{16.1}$$

or equivalently

$$j = \underset{k}{\operatorname{argmax}}\, g_k(\underline{x}), \quad \forall k \in [1, M] \iff \underline{x} \in C_j, \tag{16.2}$$

that is, the class corresponding to the largest discriminant function value is selected. The special case of two-class pattern classification (binary hypothesis testing) can be carried out using a single function $g(\underline{x}) = g_1(\underline{x}) - g_2(\underline{x})$. In this case, the decision reduces to the evaluation of the sign of $g(\underline{x})$: if $g(\underline{x}) > 0 \Rightarrow \underline{x} \in C_1$; otherwise if $g(\underline{x}) < 0 \Rightarrow \underline{x} \in C_2$.

Clearly, the decision boundary, S_{kl}, separating two contiguous decision regions $\mathcal{R}_k$ and $\mathcal{R}_l$ satisfies $g_k(\underline{x}) - g_l(\underline{x}) = 0 \Rightarrow S_{kl}$. That is, if the pattern $\underline{x}$ lies on the decision boundary between two classes, then their discriminant functions are equal, implying that the pattern cannot be uniquely assigned to either class. From a geometric point of view, the decision boundary is the projection of the intersection of two discriminant functions onto the feature space $\underline{x} \in \mathbb{R}^N$. Figure 16.2 illustrates three 2-D planes corresponding to three different discriminant functions $g_1(\underline{x})$, $g_2(\underline{x})$, and $g_3(\underline{x})$ for a three-class classification problem. As shown in this figure, the projections of the intersections of these planes onto the 2-D feature space $\underline{x} = [x_1\ x_2]^T$ define the decision boundaries between different classes, which in this case are linear due to the planar form of the discriminant functions. In general, for an N-dimensional feature space, these decision boundaries will be $N - 1$

dimensional hypersurfaces. For many practical pattern classification problems, linear decision boundaries do not yield the disjoint decision regions required for good classification performance.

Now, let us address the first question posed earlier. The type of classifier and the complexity of its decision boundaries are directly dependent on the type of discriminant functions. Once the type of discriminant function is selected, the learning process determines the unknown parameters of these functions. Among typical discriminant functions or classifiers are: linear, Bayesian, radial basis (kernel machines), and artificial neural networks. In this chapter, we will show how the first two classifiers can be designed and parameterized using training data.

Figure 16.2 illustrates an example of 3-D hyperplane discriminant functions leading to *linear* classifiers with linear decision boundaries. In general, a linear classifier in an N-dimensional feature space produces $(N-1)$-dimensional hyperplanes as decision boundaries. In this case, the ith discriminant function can be written as:

$$g_i(\underline{x}) = \underline{w}_i^T \underline{x} + w_{i,N+1} = \sum_{j=1}^{N} w_{i,j} x_j + w_{i,N+1}, \quad \forall i \in [1, M], \tag{16.3}$$

where $\underline{w}_i = [w_{i,1} \cdots w_{i,N}]^T$ is the weight vector with weights $w_{i,j}$, $j \in [1, N]$ that is normal to the hyperplane, and $w_{i,N+1}$ determines the intercept. These parameters are computed using a learning algorithm. A pattern classification problem that can successfully be solved using such hyperplane discriminant functions is referred to as a *linearly separable* problem.

Example 16.1 *Six patterns in 2-D feature space need to be separated into two classes. Class 1 (C_1) patterns are: $\{\underline{x}_1 = [0, 0]^T, \underline{x}_2 = [-.5, -1]^T, \underline{x}_3 = [-1, -2]^T\}$ and class 2 (C_2) patterns are: $\{\underline{x}_4 = [2, 0]^T, \underline{x}_5 = [1.5, -1]^T, \underline{x}_6 = [1, -2]^T\}$. Design a two-class classifier to separate these patterns [3].*

Solution: Based on the locations of these six patterns shown in Figure 16.3, it is clear that this two-class classification problem is linearly separable, and there exists an infinite number of possible lines that can separate these patterns in the 2-D feature space without any misclassifications. One possible decision boundary is the line $g(\underline{x}) = [-2\ 1]\underline{x} + 2 = 0$ for which the decision rule is $g(\underline{x}) > 0 \Rightarrow \underline{x} \in C_1$ and $g(\underline{x}) < 0 \Rightarrow \underline{x} \in C_2$.

The corresponding discriminant functions $g_1(\underline{x})$ and $g_2(\underline{x})$ can be constructed such that their difference yields the decision function $g(\underline{x})$. One possible choice is $2x_1 - x_2 + 2g_1(\underline{x}) - 4 = 0$ and $-2x_1 + x_2 + 2g_2(\underline{x}) = 0$, which leads to the discriminant functions,

$$g_1(\underline{x}) = \underline{w}_1^T \underline{x} + b_1 \ \text{ and } \ g_2(\underline{x}) = \underline{w}_2^T \underline{x} + b_2,$$

where $\underline{w}_1 = [-1, \frac{1}{2}]^T$, $\underline{w}_2 = [1, -\frac{1}{2}]^T$, $b_1 = 2$, and $b_2 = 0$. To validate these choices for $g_1(\underline{x})$ and $g_2(\underline{x})$, we apply two patterns $\underline{x}_1 = [0, 0]^T$ and $\underline{x}_4 = [2, 0]^T$ with known labels and determine the results. For $\underline{x}_1$, we obtain $g_1(\underline{x}_1) = 2 > g_2(\underline{x}_1) = 0$ hence $\underline{x}_1 \in C_1$; while for $\underline{x}_4$ we get $g_2(\underline{x}_4) = 2 > g_1(\underline{x}_4) = 0$ hence $\underline{x}_4 \in C_2$.

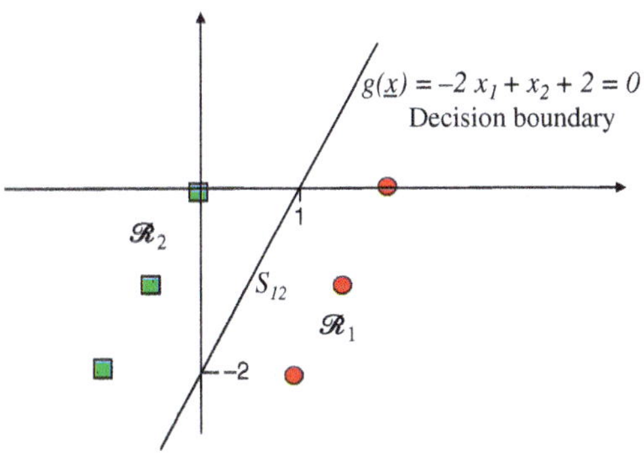

Figure 16.3 A linear decision line separating six patterns of two classes.

16.1.2 Minimum Distance Classifier

In Figure 16.1, let us assume that each class is represented by its *prototype* pattern $\underline{c}_i$, $i \in [1, M]$ as shown for the three data clusters. Typically, the prototype pattern is chosen to be the centroid (or local mean) of the cluster of patterns in each class. The minimum distance classifier assigns an unknown pattern $\underline{x}$ (shown by a blue heart) to the class whose prototype has the shortest distance:

$$d(\underline{x}, \underline{c}_j) < d(\underline{x}, \underline{c}_k), \quad \forall k \in [1, M], k \neq j \iff \underline{x} \in C_j, \tag{16.4}$$

where $d(.,.)$ is usually the Euclidean distance $d(\underline{x}, \underline{c}_j) = \| \underline{x} - \underline{c}_j \|^2$ although other distance measures, e.g. the Mahalanobis distance [1, 2], may also be used. The Euclidean distance can be expressed as:

$$\| \underline{x} - \underline{c}_j \|^2 = (\underline{x} - \underline{c}_j)^T (\underline{x} - \underline{c}_j) = \underline{x}^T \underline{x} - 2\underline{c}_j^T \underline{x} + \underline{c}_j^T \underline{c}_j.$$

Since the term $\underline{x}^T \underline{x}$ is common to all $j \in [1, M]$, it suffices to examine the class-dependent terms $\underline{c}_j^T \underline{x} - \frac{1}{2} \underline{c}_j^T \underline{c}_j$. Thus, the decision rule in (16.3) can alternatively be expressed as:

$$\underline{c}_j^T \underline{x} - \frac{1}{2} \underline{c}_j^T \underline{c}_j > \underline{c}_k^T \underline{x} - \frac{1}{2} \underline{c}_k^T \underline{c}_k, \quad \forall k \in [1, M], k \neq j \iff \underline{x} \in C_j. \tag{16.5}$$

Comparing (16.4) with the general decision rule in (16.1), several important observations can be made. First, it is clear that the discriminant functions of the minimum distance classifier are $g_j(\underline{x}) = \underline{c}_j^T \underline{x} - \frac{1}{2} \underline{c}_j^T \underline{c}_j$, implying that this classifier is a linear. Furthermore, comparing these discriminant functions with the general linear discriminant functions in (16.3), the weight vector and the bias are $\underline{w}_j = \underline{c}_j$ and $w_{j,N+1} = -\frac{1}{2} \underline{c}_j^T \underline{c}_j$, respectively. This result provides a mechanism for training linear classifiers using clustering algorithms that estimate the centroids $\underline{c}_i$ of the clusters. One such method is the K-means algorithm discussed in Section 11.3.4. Other approaches include various unsupervised clustering methods that will be covered in Chapters 17 and 18.

We now present the results of an experiment using principal component (PC) features extracted from each image in the Yalefaces dataset (see Section 6.4.4). The classification task is to distinguish between happy (Class 1) and sad (Class 2) facial expressions using a minimum distance classifier trained on PC features. Figure 16.4 shows examples of four subjects from each class in this dataset.

The training dataset consists of PC feature vectors extracted from 10 randomly selected facial images per class, while the test dataset contains the remaining samples, i.e. five samples per class. The number of retained PC features is chosen such that at least 95% of the total variance (sum of eigenvalues of retained components) is preserved. The centroid of each cluster

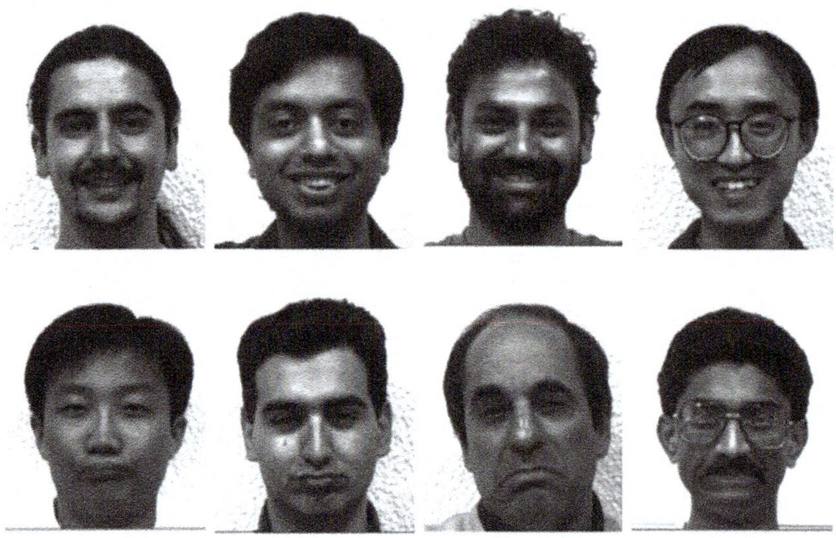

Figure 16.4 Examples of subjects with happy and sad facial expressions in the Yale facial dataset. *Source:* [4]/ Yale Center for Computational Vision and Control.

of training feature vectors is computed using the K-means algorithm. Classification decisions are then made by assigning each test sample to the class associated with the nearest centroid. The mean correct classification rates calculated based on 200 random trials, in which the training and testing data samples are shuffled, were found to be approximately 65 and 71% for classes 1 and 2, respectively, with corresponding standard deviations of 0.85 and 0.94. Although these results are obtained using small training and testing datasets, they indicate a reasonable degree of separability in the PC feature space. Improved performance can be achieved using modern classifiers, such as neural networks covered in Chapters 17 and 18.

16.2 Statistical Pattern Classification

The minimum distance classifier described in the previous section is simple and easy to implement; however, it ignores the probability distribution of data in each class. Additionally, its applicability is limited to linearly separable problems, rendering it inappropriate for most practical pattern recognition scenarios. The distribution of the data in each class is generally multivariate, since the feature vectors are inherently multidimensional. Statistical methods that explicitly account for these probability distributions are therefore more appropriate for pattern classification tasks, although they are more challenging to implement.

Bayesian decision theory [5–7] provides a solid theoretical framework for constructing optimal classifiers that minimize loss or risk in decision-making. A crucial assumption in this framework is that the underlying distributions of the data are known. In practice, however, this assumption rarely holds, and these distributions must be estimated from training data. Simplifying assumptions about these distributions are often made to render the problem tractable; however, such assumptions may lead to suboptimal or even inappropriate classifiers. These issues will be discussed later in this section.

16.2.1 Bayes Classifier

Assume that we are given an unknown pattern $\underline{x} \in \mathbb{R}^N$ that belongs to class C_k, and we make a decision, D_j in favor of class C_j. If $j = k$, the decision is correct, while for $j \neq k$ an error occurs in decision-making. Let $L(D_j|C_k)$ denote the loss incurred when deciding D_j while the true class in C_k. Bayesian decision theory is based on minimizing the expected loss. For an M-class classification (or M-ary hypothesis testing), the expected loss (or *conditional risk*) given the feature vector $\underline{x}$ is

$$r(D_j|\underline{x}) = \sum_{k=1}^{M} L(D_j|C_k)p(C_k|\underline{x}), \tag{16.6}$$

where $p(C_k|\underline{x})$ is the posterior distribution of class C_k. This posterior distribution is generally not known a priori, but it can be related to the class-conditional distribution (or the likelihood function), which is measurable from the training data using the Bayes rule:

$$p(C_k|\underline{x}) = \frac{p(\underline{x}|C_k)P(C_k)}{p(\underline{x})}, \tag{16.7}$$

where $p(\underline{x}|C_k)$ is the class-conditional distribution, $P(C_k)$ is the *prior* probability of class k, and $p(\underline{x}) = \sum_{k=1}^{M} p(\underline{x}|C_k)P(C_k)$ is the density of all the data samples. The conditional risk $r(D_j|\underline{x})$ in (16.5) is averaged across all classes for a specific decision D_j. Now, the overall risk or the expected value of this conditional risk over all possible input vectors $\underline{x}$ is

$$\mathcal{R} = E_{\underline{x}}[r(D_j|\underline{x})] = \int r(D_j|\underline{x})p(\underline{x})d\underline{x}. \tag{16.8}$$

To minimize $\mathcal{R}$, it suffices to minimize the conditional risk $r(D_j|\underline{x})$, $\forall j \in [1, M]$, since $p(\underline{x})$ is not dependent on the decision rule. Also, it is intuitive and convenient to choose 0–1 loss, i.e. $L(D_j|C_k) = 1 - \delta(j-k)$, which implies that we assign no loss for making a correct decision and 1 for making a wrong decision. Substituting this into (16.5) yields

$$r(D_j|\underline{x}) = \sum_{k=1}^{M}(1 - \delta(j-k))p(C_k|\underline{x}) = \sum_{k=1}^{M} p(C_k|\underline{x}) - p(C_j|\underline{x}). \tag{16.9}$$

But the first term in (16.9) is unity since $\sum_{k=1}^{M} p(C_k|\underline{x}) = \frac{\sum_{k=1}^{M} p(\underline{x}|C_k)P(C_k)}{p(\underline{x})} = \frac{p(\underline{x})}{p(\underline{x})} = 1$ and we get

$$r(D_j|\underline{x}) = 1 - p(C_j|\underline{x}). \tag{16.10}$$

Thus, minimizing the conditional risk is equivalent to maximizing the posterior probability, and the Bayes decision rule becomes

$$\text{If} \quad p(C_j|\underline{x}) > p(C_k|\underline{x}), \ \forall k \in [1, M], k \neq j \iff \underline{x} \in C_j. \tag{16.11}$$

That is, the pattern $\underline{x}$ is assigned to the class with the largest posterior probability. This classifier is known as the *Maximum A Posteriori (MAP)* classifier [5–7]. Using Bayes' rule in (16.7), since the denominator $p(\underline{x})$ is common to all classes, the decision rule in (6.11) can alternatively be expressed as

$$\text{If} \quad p(\underline{x}|C_j)P(C_j) > p(\underline{x}|C_k)P(C_k), \ \forall k \in [1, M], k \neq j \iff \underline{x} \in C_j. \tag{16.12}$$

Comparing this result with the general form of discriminant functions, it is clear that the discriminant function for the Bayes classifier is

$$g_i(\underline{x}) = p(\underline{x}|C_i)P(C_i) = p(\underline{x}, C_i), \tag{16.13}$$

or using the natural log operation,

$$g_i(\underline{x}) = \ln[p(\underline{x}|C_i)P(C_i)] = \ln p(\underline{x}, C_i). \tag{16.14}$$

Thus, unlike the minimum distance classifier whose discriminant functions were deterministic functions of $\underline{x}$, the discriminant functions of the Bayes classifier are probabilistic.

For two-class or binary hypothesis testing (or dichotomizer), the condition in (16.9) can be converted to the likelihood ratio test,

$$\text{If} \quad \Lambda(\underline{x}) = \frac{p(\underline{x}|C_1)}{p(\underline{x}|C_2)} > \frac{P(C_2)}{P(C_1)} = \lambda \iff \underline{x} \in C_1 \quad \text{else} \quad \underline{x} \in C_2, \tag{16.15}$$

where $\Lambda(\underline{x})$ is the likelihood ratio and λ is the decision threshold, which is chosen to provide a desired classification performance.

If $P(C_j) = P(C_k), \ \forall k, j \in [1, M], \ k \neq j$, i.e. equally likely classes, the condition in (16.9) reduces to

$$\text{If} \quad p(\underline{x}|C_j) > p(\underline{x}|C_k), \ \forall k \in [1, M], k \neq j \iff \underline{x} \in C_j, \tag{16.16}$$

which corresponds to the maximum likelihood (ML) test for class membership.

16.2.1.1 Special Cases

1. **Bayes Classifier for Gaussian Data with Identical Covariance Matrix**

   The Bayes classifier requires knowledge of the likelihood functions $p(\underline{x}|C_j), \ \forall j$ in order to implement the decision rule. However, in many practical situations, the exact form of these likelihood functions is unknown and must be approximated. Suppose that we can approximate the likelihood function in each class C_i using a multivariate Gaussian distribution:

$$p(\underline{x}|C_i) = \frac{1}{(2\pi)^{N/2}(\det(\mathbf{R}_i))^{1/2}} \exp\left[-\frac{1}{2}(\underline{x} - \underline{\mu}_i)^T \mathbf{R}_i^{-1}(\underline{x} - \underline{\mu}_i)\right], \tag{16.17}$$

   where $\underline{\mu}_i$ and $\mathbf{R}_i$ represent the mean vector and covariance matrix of class C_i, respectively, estimated from the training data and det(.) denotes the determinant.

   Let us consider the special case in which all classes have the same covariance matrix, i.e. $\mathbf{R}_i = \mathbf{R}, \forall i$, while they have different mean vectors. Under this assumption, the discriminant function can be given by

$$\begin{aligned} g_i(\underline{x}) &= \ln \ p(\underline{x}|C_i) + \ln \ P(C_i) \\ &= -\frac{N}{2} \ln 2\pi - \frac{1}{2} \ln \det(\mathbf{R}) - \frac{1}{2}(\underline{x} - \underline{\mu}_i)^T \mathbf{R}^{-1}(\underline{x} - \underline{\mu}_i) + \ln \ P(C_i). \end{aligned} \tag{16.18}$$

   Ignoring all class-independent terms in (16.18), including $\underline{x}^T\mathbf{R}^{-1}\underline{x}$, which is common to all discriminant functions, (6.18) simplifies to

$$g_i(\underline{x}) = \underline{x}^T\mathbf{R}^{-1}\underline{\mu}_i - \frac{1}{2}\underline{\mu}_i^T\mathbf{R}^{-1}\underline{\mu}_i + \ln \ P(C_i). \tag{16.19}$$

Comparing (16.18) with the general form of a linear discriminant function reveals the fact that when the class-conditional distributions are assumed to be Gaussian with identical covariance matrix but different mean vectors, the Bayes classifier reduces to a linear classifier with parameters $\underline{w}_i = \mathbf{R}^{-1}\underline{\mu}_i$ and $w_{i,N+1} = -\frac{1}{2}\underline{\mu}_i^T\mathbf{R}^{-1}\underline{\mu}_i + \ln P(C_i)$.

Using this result, the decision boundary between two contiguous decision regions $\mathcal{R}_k$ and $\mathcal{R}_l$, which is the difference of their discriminant functions, becomes

$$S_{kl} : g_k(\underline{x}) - g_l(\underline{x}) = \underline{v}_{kl}^T\underline{x} + v_{kl} = 0, \tag{16.20}$$

where $\underline{v}_{kl} = \mathbf{R}^{-1}(\underline{\mu}_k - \underline{\mu}_l)$ and $v_{kl} = (w_{k,N+1} - w_{l,N+1})$. This result implies that the decision boundary between each pair of contiguous decision regions is a hyperplane. The orientation of the hyperplane is determined by the covariance matrix $\mathbf{R}$ and the mean vectors $\underline{\mu}_k$ and $\underline{\mu}_l$. The intersection location of the hyperplane that separates the decision regions is determined by the prior probabilities.

2. **Bayes Classifier for Gaussian Data with Distinct Covariance Matrix**
Now, consider the general case [1] where the Gaussian distribution of each class C_i has its own covariance matrix $\mathbf{R}_i$ and mean vector $\underline{\mu}_i$. In this case, the expression of the discriminant function in (16.14) becomes

$$\begin{aligned} g_i(\underline{x}) &= -\frac{N}{2}\ln 2\pi - \frac{1}{2}\ln\det(\mathbf{R}_i) - \frac{1}{2}(\underline{x} - \underline{\mu}_i)^T\mathbf{R}_i^{-1}(\underline{x} - \underline{\mu}_i) + \ln P(C_i) \\ &\equiv -\frac{1}{2}\underline{x}^T\mathbf{R}_i^{-1}\underline{x} + \underline{x}^T\mathbf{R}_i^{-1}\underline{\mu}_i - \frac{1}{2}\underline{\mu}_i^T\mathbf{R}_i^{-1}\underline{\mu}_i - \frac{1}{2}\ln\det(\mathbf{R}_i) + \ln P(C_i). \end{aligned}$$

Note that since the term $\left(-\frac{N}{2}\ln(2\pi)\right)$ is independent of the class index i, it does not affect the classification decision and is omitted. The discriminant function is then expressed in the quadratic form as

$$g_i(\underline{x}) = \underline{x}^T\mathbf{W}_i\underline{x} + \underline{w}_i^T\underline{x} + w_{i0}, \tag{16.21}$$

where $\mathbf{W}_i = -\frac{1}{2}\mathbf{R}_i^{-1}$, $\underline{w}_i = \mathbf{R}_i^{-1}\underline{\mu}_i$, $w_{i0} = -\frac{1}{2}\underline{\mu}_i^T\mathbf{R}_i^{-1}\underline{\mu}_i - \frac{1}{2}\ln\det(\mathbf{R}_i) + \ln P(C_i)$. Unlike special case 1 with identical covariance matrix, in this case, the discriminant functions is a quadratic function of the feature vector $\underline{x}$. Consequently, the decision boundaries between classes are generally quadratic surfaces (hyperquadrics) rather than hyperplanes. This classifier is known as *Quadratic Discriminant Analysis (QDA)*, which is suited for solving certain nonlinearly separable pattern classification problems.

16.2.2 Error Analysis and Performance Metric

As discussed before, the objective of Bayes decision theory is to minimize the expected risk in making decisions. In this section, we quantify the sources of error and their associated probabilities for the two-class or binary hypothesis testing. Again, let D_i be the decision in favor of class C_i, $i = 0, 1$. Then, four conditional probabilities characterize the performance of a binary classifier. The *true negative rate*, $P(D_0|C_0)$, corresponds to the probability of correctly deciding D_0 when the true class is C_0. Similarly, the *true positive rate*, $P(D_1|C_1)$, corresponds to the probability of correctly deciding D_1 when the true class is C_1. The *false positive or false alarm rate*, $P_{\text{FP}} = P(D_1|C_0)$, is the probability of making an erroneous decision D_1 when C_0 is true; while the *false negative rate* or *Missed Detection Probability*, $P_{\text{FN}} = P(D_0|C_1)$, is the probability of making an erroneous decision D_0 when C_1 is true. Thus, the overall probability of error is

$$P_E = P(D_1, C_0) + P(D_0, C_1) = P(D_1|C_0)P(C_0) + P(D_0|C_1)P(C_1). \tag{16.22}$$

For continuous-valued features, these error probabilities correspond to the integrals of the class-conditional distributions in the overlap regions, as shown in Figure 16.5. That is, if $\mathcal{R}_0$ and $\mathcal{R}_1$ denote the decision regions associated with classes C_0 and C_1, respectively, then these error probabilities are:

$$\begin{aligned} P_{FP} &= P(D_1|C_0) = \int_{\mathcal{R}_1} p(\underline{x}|C_0)d\underline{x}, \\ P_{FN} &= P(D_0|C_1) = \int_{\mathcal{R}_0} p(\underline{x}|C_1)d\underline{x}. \end{aligned} \tag{16.23}$$

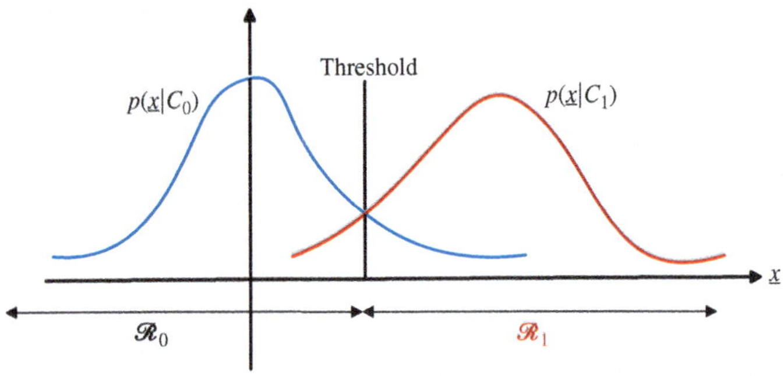

Figure 16.5 Distributions of data in two classes and Bayes decision threshold.

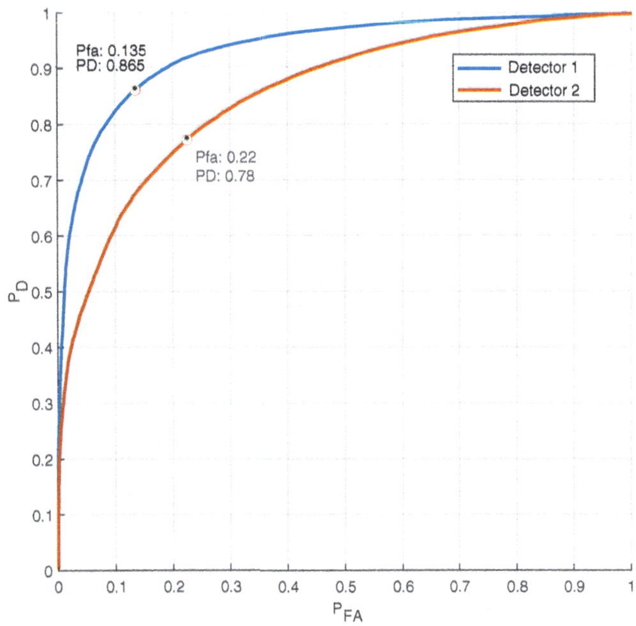

Figure 16.6 ROC curves of two detectors.

The Bayes classifier chooses the decision boundary that minimizes the total probability of error, i.e. Bayes classifier is the minimum probability of error classifier. This optimality result extends naturally to the general M-ary classification problem.

For detection or binary hypothesis testing, the decision threshold is typically chosen to provide an acceptable false alarm or false positive probability $P_{\text{FA}} = P_{\text{FP}}$. At that threshold, the detector provides probability of detection or true positive rate, $P_D = P_{\text{TP}}$. The plot of P_D versus P_{FA} over a wide range of threshold values is called the *Receiver Operating Characteristic* (ROC) curve [5], which is an important metric for evaluating the performance of a detector. Since the area under the distributions is unity, the probability of missed detection (or false negative rate) $P_M = P_{\text{FN}}$ is related to the true positive rate (or detection probability) $P_D = 1 - P_M$, and the false alarm rate $P_{\text{FA}} = P_{\text{FN}}$ is also related to the true negative rate, $P_{\text{TN}} = 1 - P_{\text{FN}}$. At one particular threshold value on the ROC curve, we have $P_D + P_{\text{FA}} = 1$. This point is called the knee-point of the ROC curve. An ideal detector achieves a high P_D while maintaining a low P_{FA}. Consequently, ROC curves that approach the upper-left corner of the plot indicate superior detector performance. Figure 16.6 shows examples of two ROC curves for two different detectors. Detector 1 exhibits a much better ROC performance with a knee-point of $P_D = 86.5\%$ and $P_{FA} = 13.5\%$.

For M-ary hypothesis testing or M-class classification problems, a commonly adopted measure of performance is the *confusion matrix* of the classifier. This is a matrix of dimension $M \times M$ whose off-diagonal entry (i, j), $i, j \in [1, M]$ $i \neq j$ represents the probability of incorrectly classifying samples from class i as class j, while the diagonal elements for

$i = j$, correspond to the correct classification rates of individual classes. Since each row represents all possible classification outcomes for a given class, the entries along every row sum to one.

Table 16.1 shows an example confusion matrix for a three-class classifier where the classes correspond to Cats, Dogs, and Racoons. The correct classification rates of these three classes along the diagonal are 72, 81, and 95%, respectively. The off-diagonal elements indicate the misclassification probabilities. For example, among Dog samples, most misclassifications (16%) occurred for Cats, while only 3% were misclassified as Racoons.

16.2.3 Naive Bayes Classifiers

The Naive Bayes classifiers [8, 9] are probabilistic machine learning models widely employed for classification tasks, particularly with discrete-valued data. Such applications include: text and document classification, spam filtering, health monitoring, and diagnostic systems. In these applications, one often deals with very high-dimensional feature vectors whose elements are binary-valued, indicating the presence or absence of certain words, tokens, or attributes. Alternatively, the features may represent the frequency of occurrence of such events. This classifier can be applied to both binary or M-ary classification problems, as we will see later in this section.

The reason this classifier is called Naive Bayes is due to the simplistic, or *naive*, assumption that the features are *class-conditionally independent*. This means that, for a given class, the individual features are assumed to be statistically independent of one another. For example, in spam filtering, the presence of words like "Bank," "Credit," or "Social Security" might each independently increase the likelihood of an email being spam, without influencing the dependencies among those words. While this assumption is often not strictly true in real-world data, Naive Bayes classifiers have proven remarkably effective in various difficult classification tasks. In particular, the independence assumption enables the model to handle feature vectors with very large and potentially variable dimensions. Because of this class conditional independence assumption of the features and using the chain rule of probability, we can write

$$\begin{aligned} p(\underline{x}|C_i) &= p(x_1, \dots, x_N|C_i) \\ &= p(x_1|C_i)p(x_2|C_i, x_1)p(x_3|C_i, x_1, x_2) \dots p(x_N|C_i, x_1, x_2, \dots, x_{N-1}) \\ &= \prod_{j=1}^{N} p(x_j|C_i) \end{aligned} \tag{16.24}$$

where each $p(x_j|C_i)$ may be modeled by a Bernoulli distribution for binary features (Case 1) or by a multinomial distribution for general discrete-valued features (Case 2). Note that class conditional independent doesn't mean x_j are independent, i.e. $p(x_j, x_i) \neq p(x_j)p(x_i), \forall i, j$. As in the original Bayes classifier, the discriminant function for decision-making is

$$g_i(\underline{x}) = p(\underline{x}|C_i)P(C_i) = \prod_{j=1}^{N} p(x_j|C_i)P(C_i) = \prod_{j=1}^{N} p(x_j, C_i). \tag{16.25}$$

To construct a Naive Bayes classifier, we need to find the class-conditional and prior probabilities $\alpha(j|i) = p(x_j|C_i)$ and $P(C_i)$, respectively. Next, we consider three widely used versions of the Naive Bayes classifier for different applications and show how these probabilities can be estimated from training data with class labels.

Table 16.1 A confusion matrix for a three-class classifier.

Actual class	Cat	Dog	Racoon
Cat	72%	23%	5%
Dog	16%	81%	3%
Racoon	2%	3%	95%

Case 1: Bernoulli Features

Binary features occur in many applications, including text and document classification based on the occurrences of certain words. In this case, the unknown parameters are $\alpha(j|0)$ and $\alpha(j|1)$ and the class priors $P_0 = P(C_0)$ and $P_1 = P(C_1)$ with, where $P_0 = 1 - P_1$ for binary hypothesis testing.

Given a labeled training dataset $\mathcal{X} = \{\underline{x}_k, y_k\}_{k=1}^{K}$ where $y_k = \{0, 1\}$ represents the class label (C_0 or C_1) of the feature vector $\underline{x}_k \in \mathbb{R}^N$ that are assumed to be independent and identically distributed (iid), we form the joint likelihood function

$$\mathcal{L}\left(\{\alpha(j|0), \alpha(j|1)\}_{j=1}^{N}, P_1\right) = \prod_{k=1}^{K} p(\underline{x}_k, y_k). \tag{16.26}$$

Using the class conditional independence assumption inherent in the Naive Bayes classifier, we have

$$\mathcal{L}\left(\{\alpha(j|0), \alpha(j|1)\}_{j=1}^{N}, P_1\right) = \prod_{k=1}^{K}\prod_{j=1}^{N} p(x_{k,j}|y_k)p(y_k), \tag{16.27}$$

where $x_{k,j}$ is the jth element of $\underline{x}_k$. Using Bernoulli distribution for the features, $p(x_{k,j}|y_k) = \alpha(j|y_k = 1)^{x_{k,j}}\alpha(j|y_k = 0)^{1-x_{k,j}}$, (16.27) becomes

$$\mathcal{L}(\{\alpha(j|0), \alpha(j|1)\}_{j=1}^{N}, P_1) = \prod_{k=1}^{K}\prod_{j=1}^{N} \alpha(j|y_k = 1)^{x_{k,j}}\alpha(j|y_k = 0)^{1-x_{k,j}}\, p(y_k). \tag{16.28}$$

Maximizing the log-likelihood function with respect to the unknown parameters yields

$$\begin{aligned}
\hat{\alpha}(j|1) &= \frac{\sum_{k=1}^{K} \mathbb{I}\{x_{k,j} = 1, y_k = 1\}}{\sum_{k=1}^{K} \mathbb{I}\{y_k = 1\}} \\
\hat{\alpha}(j|0) &= \frac{\sum_{k=1}^{K} \mathbb{I}\{x_{k,j} = 1, y_k = 0\}}{\sum_{k=1}^{K} \mathbb{I}\{y_k = 0\}}, \quad j \in [1, N] \\
\hat{P}_1 &= \frac{\sum_{k=1}^{K} \mathbb{I}\{y_k = 1\}}{K}
\end{aligned} \tag{16.29}$$

where $\mathbb{I}\{e\}$ is the *indicator* function defined as $\mathbb{I}\{e\} = \begin{cases} 1, & \text{if } e \text{ is true} \\ 0, & \text{if } e \text{ is false} \end{cases}$.

Thus, $\alpha(j|i)$ is the percentage of class $y_k = i,\ i = 0, 1$ samples in which the jth feature $x_{k,j}$ is present. To avoid ambiguous results when the denominator becomes very small, equations for $\alpha(j|1)$ and $\alpha(j|0)$ in (16.29) can be modified using *Laplace Smoothing* [8, 9] as follows:

$$\begin{aligned}
\hat{\alpha}(j|1) &= \frac{\sum_{k=1}^{K} \mathbb{I}\{x_{k,j} = 1, y_k = 1\} + 1}{\sum_{k=1}^{K} \mathbb{I}\{y_k = 1\} + 2} \\
\hat{\alpha}(j|0) &= \frac{\sum_{k=1}^{K} \mathbb{I}\{x_{k,j} = 1, y_k = 0\} + 1}{\sum_{k=1}^{K} \mathbb{I}\{y_k = 0\} + 2}, \quad j \in [1, N].
\end{aligned} \tag{16.30}$$

Once the training phase is completed and parameters have been estimated, a new pattern is assigned to the class associated with the largest discriminant function value, i.e. if $l = \arg\max_i \ln g_i(\underline{x})$, then $\underline{x} \in C_l$. Now, using the Bernoulli distribution for $p(x_j|C_i) = \hat{\alpha}(j|1)^{x_j}\hat{\alpha}(j|0)^{1-x_j}$ in the ln of $g_i(\underline{x})$ in (16.25) yields

$$\ln g_i(\underline{x}) = \sum_{j=1}^{N} x_j \ln \hat{\alpha}(j|i) + \ln \hat{P}(C_i) = \underline{w}_i^T \underline{x} + w_{i,N+1} \tag{16.31}$$

where $\underline{w}_i = [\hat{\alpha}(1|i), \cdots, \hat{\alpha}(N|i)]^T$ and $w_{i,N+1} = \ln \hat{P}(C_i)$. This result shows that the Bernoulli Naive Bayes classifier is a linear classifier in the log domain.

Case 2: Multinomial Features

Multinomial distribution for modeling $p(\underline{x}|C_i)$ is widely used in a number of applications, including text classification in Natural Language Processing (NLP) [8, 9], where each document is represented by a vector of frequency counts of the words or tokens selected for a specific text classification task. In essence, the feature vector $\underline{x}$ is a histogram of a document and each feature x_j represents the frequency of occurrence of token j in a document. This differs from the Bernoulli model, where the features are either 0 or 1 indicating the presence or absence of a particular token. As an example, consider a document containing M words selected from a vocabulary of $N \ll M$ distinct tokens. If $x_i = l$, then token i appears l times in the document. Clearly, $M = \sum_{i=1}^{N} x_i$ is the total number of words in the document. The multinomial distribution is therefore a natural model to represent the likelihood of observing a histogram $\underline{x}$ given class C_i,

$$p(\underline{x}|C_i) = \frac{M!}{x_1!x_2!\cdots x_N!}\prod_{j=1}^{N}\alpha(j|i)^{x_j}, \tag{16.32}$$

where $\alpha(j|i)$ denotes the probability of observing token j in class C_i, satisfying $\sum_{j=1}^{N} \alpha(j|i) = 1$. Assuming that we use the labeled training dataset $\mathcal{X} = \{\underline{x}_k, y_k\}_{k=1}^{K}$, where as before $y_k \in \{0, 1\}$ represents the class label (C_0 or C_1) of the feature vector (or document) $\underline{x}_k \in \mathbb{R}^N$, the likelihood function in (16.26) becomes

$$\begin{aligned}\mathcal{L}\left(\{\alpha(j|0), \alpha(j|1)\}_{j=1}^{N}, P_1\right) &= \prod_{k=1}^{K} p(\underline{x}_k, y_k) = \prod_{k=1}^{K} p(\underline{x}_k|y_k)p(y_k)\\ &= \prod_{k=1}^{K}\left[\frac{M_k!}{x_{k,1}!x_{k,2}!\cdots x_{k,N}!}\prod_{j=1}^{N}\alpha(j|y_k)^{x_{k,j}}\right]p(y_k).\end{aligned} \tag{16.33}$$

Maximizing the log-likelihood function with respect to the unknown parameters gives

$$\begin{aligned}\hat{\alpha}(j|1) &= \frac{\sum_{k=1}^{K}\mathbb{I}\{y_k = 1\}x_{k,j} + \beta}{\sum_{k=1}^{K}\mathbb{I}\{y_k = 1\}\gamma_k + \beta N}\\ \hat{\alpha}(j|0) &= \frac{\sum_{k=1}^{K}\mathbb{I}\{y_k = 0\}x_{k,j} + \beta}{\sum_{k=1}^{K}\mathbb{I}\{y_k = 0\}\gamma_k + \beta N}\\ \hat{P}_1 &= \frac{\sum_{k=1}^{K}\mathbb{I}\{y_k = 1\}}{K},\end{aligned} \tag{16.34}$$

where $\gamma_k = \sum_{j=1}^{N} x_{k,j}$ is the total number of tokens in document $\underline{x}_k$, and β is a smoothing parameter added to avoid numerical instability in the ML estimates of $\alpha(j|i)$. The choice $\beta = 1$ corresponds to Laplace smoothing [8, 9]. The quantities $\hat{\alpha}(j|i)$ represent the relative frequency of token j among all tokens observed in class C_i.

The discriminant functions of the multinomial Naive Bayes classifier can be obtained similarly to the previous case leading to a linear classifier in the log-domain.

Case 3: Multi-Class Features

For the multi-class (M-ary) categorical features, each element x_j, $j \in [1, N]$ of the feature vector $\underline{x} \in \mathbb{R}^N$ can take one of L possible discrete values (or categories), i.e. $x_j \in \{1, \ldots, L\}$. Note that binary features are a special case of categorical features where $L = 2$.

Let us assume that we want to use a multi-class Naive Bayes classifier for discriminating among species of Cats, Dogs, and Racoons ($M = 3$ classes) based on categorical features such as ear size, height, length, and fur color. Each feature is quantized (encoded) into $L = 4$ possible discrete values. For example, for the fur color feature may be encoded as white = 1, gray = 2, brown = 3, and black = 4. Then, $x_4 = 2$ implies the fourth feature in data vector $\underline{x}$ corresponds to gray fur, and $p(x_4 = 2|C_3)$ gives the probability that Racoons exhibit gray color fur. In general, $p(x_j = l|C_i) = \alpha_l(j|i)$ is the probability of feature j having the value l given that the class label is i, where $\sum_{l=1}^{L} \alpha_l(j|i) = 1$ for each i.

For categorical features, the appropriate probabilistic model is the Multinoulli (or categorical) distribution [9] $p(x_{k,j} = l|y_k = i) = \alpha_l(j|i)^{\mathbb{I}\{x_{k,j}=l\}}$, where k represents the index of the kth training sample with class label $y_k = i$. This leads to the likelihood function

$$\begin{aligned}\mathcal{L}(\{\alpha_l(j|i)\}_{l=1}^{L}, P(C_i)) &= \prod_{k=1}^{K}\prod_{j=1}^{N} p(x_{k,j} = l|y_k = i)p(y_k = i) \\ &= \prod_{k=1}^{K}\prod_{j=1}^{N} \alpha_l(j|i)^{\mathbb{I}\{x_{k,j}=l\}} p(y_k = i).\end{aligned} \tag{16.35}$$

Maximizing the ln of this function with respect to the unknown parameters gives

$$\begin{aligned}\hat{\alpha}_l(j|i) &= \frac{\sum_{k=1}^{K} \mathbb{I}\{y_k = i\}\mathbb{I}\{x_{k,j} = l\} + \beta}{\sum_{k=1}^{K} \mathbb{I}\{y_k = i\} + \beta L} \quad i \in [1, M], \;\; j \in [1, N], \;\; l \in [1, L] \\ \hat{P}(C_i) &= \frac{\sum_{k=1}^{K} \mathbb{I}\{y_k = i\}}{K},\end{aligned} \tag{16.36}$$

where β is a smoothing parameter, with $\beta = 1$ corresponding to the Laplace smoothing as before. As can be observed, $\hat{\alpha}_l(j|i)$ is the ratio of number of samples of class C_i that have feature $x_{k,j} = l$ to the total number of class C_i samples. Using these estimated parameters, the classification decision for a new pattern $\underline{x}$ can be made as in the previous cases. More specifically, the class membership of a new pattern $\underline{x}$ can be determined using

$$\underline{x} \in C_p \Longleftrightarrow p = \arg\max_i \left\{\hat{P}(C_i)\prod_{j=1}^{N} \hat{\alpha}_l(j|i)\right\}. \tag{16.37}$$

Taking the logarithm of the discriminant function again yields a linear classifier in the transformed feature space, similar to the Bernoulli and multinomial Naive Bayes classifiers.

To show the applicability of the categorical Naive Bayes for image segmentation, we use the original Finn color image in Figure 16.7a. The goal is to partition the image into five disjoint regions or classes ($M = 5$) corresponding to the floor areas (regions 1 and 2), the star toy (region 3), the areas of white and gray fur (region 4), and the black fur areas (region 5). The image is first binned or vector quantized into $l = 8$ codebook vectors using the K-means clustering applied to the RGB color attributes for each pixel. This step is necessary for applying categorical Naive Bayes classifier. Small representative regions from each class are then selected and used to train the classifier. The trained classifier is subsequently applied to the entire image. Figure 16.7b shows the segmented regions and their color codes. As can be seen, the categorical Naive Bayes provided excellent segmentation results by identifying even subtle color differences, especially in the star toy region.

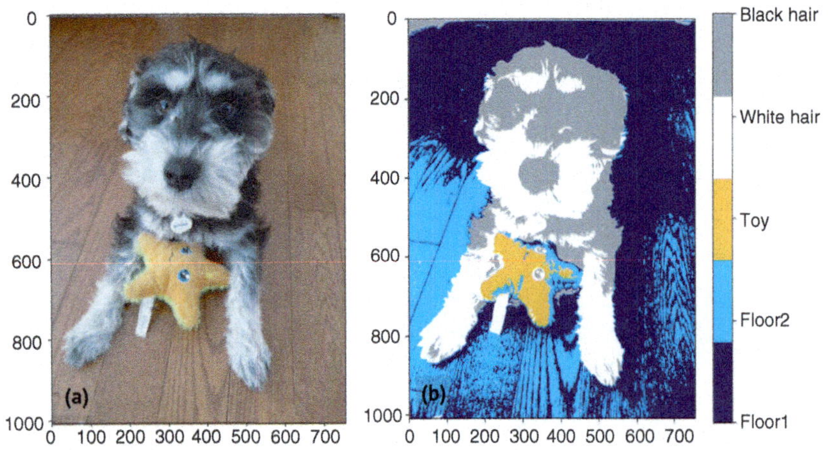

Figure 16.7 (a) Original Finn image and (b) categorical Naive Bayes segmentation result with five distinct regions.

16.2.3.1 Important Remark

At this juncture, it is appropriate to categorize the classifiers covered in this section as *generative* learning models [8, 9]. Generative models construct a parameterized joint distribution, $p(\underline{x}_k, y_k; \Theta)$, of the input data $\underline{x}_k$ and class labels y_k using the priors $p(y_k = i)$ and class conditional distributions $p(\underline{x}_k | y_k = i; \Theta) = p(\underline{x}_k | C_i; \Theta)$. The model parameters Θ are estimated directly from training data and then used in Bayes decision rule for classification. Fundamentally, generative models attempt to characterize the statistical distribution of the data within each class. The Bayes classifier and the various Naive Bayes classifiers discussed in this section all fall into this generative learning category. These models employ supervised learning, a methodology that inherently requires labeled training data to successfully estimate the unknown parameters of the model. Later in this book, we will study *discriminative learning models*, which directly model the posterior probabilities $p(y|\underline{x})$ or the decision boundaries between classes without explicitly modeling the class-conditional distributions.

16.3 Computer Projects

The computer projects in this chapter use Python libraries sklearn, PIL, NumPy, OS, random, and Matplotlib.

Computer Project 16.1: The purpose of this computer project is to design and evaluate a minimum distance classifier to classify different handwritten digits. Use a subset (12%) of the MNIST database of handwritten digits, which contains 60,000 training images and 10,000 testing images each of size 28×28. The digit images have been size-normalized and centered within a fixed-size frame. The MNIST dataset is readily available and can be downloaded from a variety of websites. Randomly divide this subset of the MNIST database into training and testing datasets while maintaining an equal number of samples from each digit class in the training set. Arrange each image into a vector using either a row- or column-stacking scheme. Using the training data, find the centroid (or mean vector) of each cluster and use it as the *prototype* to represent that class. Then, apply the testing data vectors to the trained classifier and record the classification results for each class. Repeat the experiment by choosing a different random selection of the training and testing datasets. Try at least 50 random trials to get relatively reliable mean classification rates for each class. Using the mean classification rates, produce the confusion matrix (10×10) of the classifier. Provide a detailed discussion on your results and state important observations as well as the strengths and limitations of this classifier.

Computer Project 16.2: This computer project investigates the application of the multi-class or categorical Naive Bayes classifier for color-based image segmentation. Choose five different color images from the training subset of the Berkeley Segmentation Dataset (BSDS300) https://www2.eecs.berkeley.edu/Research/Projects/CS/vision/bsds/. To apply the Naive Bayes classifier, the images must first be quantized (or binned) into a fixed number of levels depending on the number of categories. Use the K-means clustering algorithm discussed in Section 11.3.4 to perform vector quantization based on 3-channel RGB color vectors of image pixels. Then, choose various regions with known class labels such as sky, water, forest, mountains, buildings, or any other meaningful categories to train the categorical Naive Bayes classifier. Choose five additional images from the testing subset of the BSDS300 database that have similar theme (e.g. nature scenes, animals, people, urban scenes, etc.) to those used for training. Apply the trained classifier to all the binned testing images and determine the overall segmentation performance by comparing your segmentation results with those of the benchmark methods in the same datasets. These benchmark results are available for the BSDS300 database on the above website. Provide a detailed discussion on your results and state important observations as well as the strengths and weaknesses of this classifier for image segmentation applications.

16.4 Problems

Problem 16.1: Consider the six 2-D patterns in Example 16.1.

a. Design a minimum distance classifier to separate the two classes.
b. Find the expression for decision boundary. How does that differ from the result in Example 16.1.
c. Determine the class membership of unknown patterns $[2\ \ 5]^T$ and $[-1\ \ -1]^T$ using the classifier designed in part (a).

Problem 16.2: Consider the following six 3-D patterns that belong to two classes.

Class 1 Patterns: $\underline{x}_1 = [0.8\ 0.5\ 0]^T$, $\underline{x}_2 = [0.9\ 0.7\ 0.3]^T$, $\underline{x}_3 = [1\ 0.8\ 0.5]^T$
Class 2 Patterns: $\underline{x}_4 = [0\ 0.2\ 0.3]^T$, $\underline{x}_5 = [0.2\ 0.1\ 1.3]^T$, $\underline{x}_6 = [0.2\ 0.7\ 0.8]^T$

a. Find the centroid vector of each class and then design a minimum distance classifier to separate the patterns.
b. Find the expression for decision boundary.

Problem 16.3: The parameters w_1 and w_2 of a classifier must be determined by minimizing the following objective function with respect to these parameters.

$$J(\underline{w}) = \frac{1}{2}\left[(w_2 - w_1)^2 + (1 - w_1)^2\right]$$

a. Find the gradient vector $\nabla J(\underline{w})$ and its orientation.
b. Find the optimal parameter vector $\underline{w} = [w_1\ w_2]^T$ that minimizes this function.
c. Find the minimum of $J(\underline{w})$.
d. Determine the type of classifier, i.e. linear vs. nonlinear.

Problem 16.4: A minimum distance classifier must be designed by minimizing an appropriate objective function that relates the parameters to the centroid of different classes. Find this objective function and show that when minimized, the parameter vectors of hyperplanes correspond to the centroid of different clusters.

Problem 16.5: An M-class classifier must be designed to minimize the following objective function with respect to the unknown parameter vectors $\underline{w}_i$ of each class i.

$$J(\{\underline{w}_i\}_{i=1}^{M}) = \sum_{k=1}^{K}\sum_{i=1}^{M} ||\underline{x}_{i,k} - \underline{w}_i||^2$$

where $\underline{x}_{i,k}$ represents the kth training sample of class i and K is the number of training samples in each class.

a. Minimize this objective function with respect to each weight vector $\underline{w}_i$.
b. Find the expressions of the optimal weight vectors.
c. Does the result in part (b) coincide with that of the minimum distance classifier?

Problem 16.6: The likelihood functions $P(x|C_i)$, $i = 0, 1$ for a two-class classifier are univariant Gaussian with means μ_0 and μ_1 and equal variance σ^2.

a. Find the expression for P_E or probability of error. Assume equally likely classes.
b. Find the expression for decision boundary. Is this a linear classifier?

Problem 16.7: The likelihood functions $P(\underline{x}|C_i)$, $i = 0, 1$ for a two-class classifier are multivariate Gaussian with mean vectors $\underline{\mu}_0$ and $\underline{\mu}_1$ and equal covariance matrix $\sigma^2\mathbf{I}$.

a. Find the expression for P_E or probability of error. Assume equally likely classes.
b. Find the expression for decision boundary. Is this a linear classifier?

Problem 16.8: In Problem 16.7,

a. Find the expression for the log-likelihood ratio for equally likely classes.
b. Repeat part (a) when the classes are not equally likely and the ratio of prior probabilities is $\lambda = \frac{P(C_0)}{P(C_1)}$.
c. Find the expression for the decision boundary of the classifier in part (b).

Problem 16.9: Repeat Problem 16.8 when covariance matrices of the two classes are $\mathbf{R}_0 = \text{diag}[\alpha\ 3\alpha]$ and $\mathbf{R}_1 = \text{diag}[2\alpha\ \alpha]$, where α is assumed to be known.

Problem 16.10: Consider a signal detection problem (or a binary hypothesis testing) where under the null hypothesis $\mathcal{H}_0$ the measurements consist of noise alone and under the alternative hypothesis $\mathcal{H}_1$ the measurements contain a known signal in noise. That is, we have the following equations for the measurements under these hypotheses:

$$\mathcal{H}_0 : \ \underline{z} = \underline{n}$$

$$\mathcal{H}_1 : \ \underline{z} = \underline{s} + \underline{n}$$

where $\underline{n}$ is white Gaussian noise with zero-mean vector and covariance matrix $\mathbf{R} = \sigma_n^2 \mathbf{I}$ and $\underline{s}$ is the signal vector. Design a minimum probability error detector for this problem.

Problem 16.11: In Problem 16.7, assume that the multivariate Gaussian distribution parameters $\underline{\mu}_i$, $i = 0, 1$ and σ^2 are unknown, and we would like to estimate them before any decision-making. Given some independent and identically distributed (iid) training data $\{\underline{x}_k, y_k\}_{k=1}^K$ where y_k is the class label of the data sample $\underline{x}_k$, use ML method to estimate these unknown parameters. The appropriate log-likelihood function for this problem is

$$\mathcal{L}(\underline{\mu}_0, \underline{\mu}_1, \sigma^2) = ln \prod_{k=1}^{K} p(\underline{x}_k, y_k).$$

a. Find the expressions of the ML estimates of the unknown parameters assuming that $p(y_k)$ are known. What is your observation about these results?

b. Repeat part (a) assuming that each distribution has a different variance σ_i^2, $i = 0, 1$.

Problem 16.12: The class conditional distributions of scalar data for a binary classifier are Laplacian with the location parameter μ_i and scaling parameter α_i for class $i = 0, 1$, i.e.

$$p(x|C_i) = \frac{1}{2\alpha_i} e^{\frac{-|x-\mu_i|}{\alpha_i}}.$$

a. Find the expression for the log-likelihood ratio when the classes are equally likely.

b. Find the expression for the decision boundary for this classifier and comment on the result.

Problem 16.13: In Problem 16.12, assume that the distribution parameters μ_i and α_i are unknown, and we would like to estimate them before any decision-making. Given some independent and identically distributed (iid) training dataset $\{x_k, y_k\}_{k=1}^K$ where y_k is the class label of the data sample x_k, use ML method to estimate the unknown parameters. The appropriate log-likelihood function for this problem is

$$\mathcal{L}(\mu_0, \mu_1, \alpha_0, \alpha_1) = ln \prod_{k=1}^{K} p(x_k, y_k).$$

Find the expressions of the ML estimates of the unknown parameters assuming that $p(y_k)$ are known. What is your observation about these results?

Problem 16.14: The class conditional distributions of scalar data for a two-class classifier are Rayleigh with scaling parameter α_i for class $i = 0, 1$, i.e.

$$p(x|C_i) = \frac{x}{\alpha_i^2} \exp \frac{-x^2}{2\alpha_i^2}, \quad x \geq 0.$$

a. Find the expression for the log-likelihood ratio assuming equally likely classes.

b. Find the expression for the decision boundary for this classifier and comment on the result. Is this a linear classifier?

Problem 16.15: Two-photon detectors for two different medical imaging systems have Poisson distributions $p(x|\lambda_i) = e^{\lambda_i} \frac{\lambda_i^x}{x!}$, $i = 1, 2$ where $\lambda_1 > \lambda_2$. We would like to design a classifier to determine which system was used to capture a given recorded image. Assuming that $P(\lambda_1) = P(\lambda_2)$, find the Bayes decision boundary.

Problem 16.16: Consider a three-class classifier with a univariate Gaussian class conditional distribution with mean values $\mu_1 = 0$, $\mu_2 = 0.5$, and $\mu_3 = 1$ and identical unity variance. Assume $P(C_1) = 1/2$ and $P(C_2) = P(C_3) = 1/4$. We sample a sequence of four points $x = 0.6, 0.1, 0.9, 1.1$.

a. Find the probability that the sequence corresponds to C_1, C_3, C_3, C_2.
b. Repeat part (a) for C_1, C_2, C_2, C_3.
c. What is the sequence with maximum probability?

Problem 16.17: It is desired to design a two-class Naive Bayes classifier to determine admission chances of high school students to a university based on factors such as high school GPA (High, Low), SAT/ACT score (High, Low) and extracurricular activity (Yes, No). The output of the classifier will be Accept or Reject.

Problem 16.18: Consider a Naive Bayes classifier when the features are Bernoulli distributed. Derive the expressions in (16.29) for the model parameters.

Problem 16.19: Consider a Naive Bayes classifier when the features are multinomial distributed. Derive the expressions in (16.34) for the model parameters.

Problem 16.20: Consider a Naive Bayes classifier for the multi-class categorical features.

a. Derive the expressions in (16.36).
b. Derive the expression for the discriminant function in the log-domain and comment on the result.

References

1 R.O. Duda, P.E. Hart, and D.G. Stork, "*Pattern Classification*," 2nd Edition, 2001.
2 C.M. Bishop, "*Pattern Recognition and Machine Learning*," Springer, 2006.
3 J. M. Zurada, "*Introduction to Artificial Neural Networks*," West Pub. Co., 1992.
4 A. S. Georghiades, P. N. Belhumeur, P. N., and D. J. Kriegman, "From few to many: Illumination cone models for face recognition under variable lighting and pose." Yale University, Center for Computational Vision and Control, 2001.
5 L. L. Scharf, "*Statistical Signal Processing: Detection, Estimation, and Time Series Analysis*," Addison Wesley, 1991.
6 S. Kay, "*Fundamentals of Statistical Signal Processing, Volume I: Estimation Theory*," Prentice-Hall, 1st Edition, 1993.
7 S. Kay, "*Fundamentals of Statistical Signal Processing, Volume II: Detection Theory*," Prentice-Hall, 1st Edition, 1993.
8 A. Ng, "*Part IV: Generative Learning Algorithms*," Stanford University lecture notes (CS229), 2019.
9 K. P. Murphy, "*Machine Learning- A Probabilistic Perspective*," The MIT Press, 2012.

17

Modern Image Classification: Elements of Machine Learning

Machine learning (ML) models have become ubiquitous in many image processing and more specifically in image pattern classification. This widespread adoption is driven by several critical factors. First, the volume of data that needs to be stored, transmitted, processed, analyzed, and screened is growing exponentially. The capability to process such massive volume of data and make reliable inference in near real-time often exceeds the capabilities of traditional computational models. Traditional pattern classification methods rely on representing the data in a suitable reduced-dimensional form by extracting a set of salient features prior to any image analysis. Designing such a system is a formidable challenge even for an experienced designer, due to the semantic gap between human visual perception mechanisms and the outputs of conventional image segmentation and feature extraction algorithms. *Deep neural network* (DNN) models close this gap by extracting features directly from the raw data, thereby obviating the need for explicit, hand-engineered image segmentation and feature extraction subsystems.

More broadly, ML provides a set of powerful and robust computational tools that facilitate *learning from experience* in terms of a hierarchy of concepts. These models can discover and represent complex relationships in data and have demonstrated remarkable success in applications such as anomaly detection, image understanding, speech recognition, pattern retrieval, and decision-making. Furthermore, ML models allow for handling extremely large language models (LLMs) that are used to process, translate, understand, and respond to human speech and language and generate new texts and speech. The automated analysis of large-scale data and the extraction of meaningful information from it require a diverse set of ML methodologies. Consequently, numerous learning paradigms and model architectures have been developed, each designed to address different classes of problems and data types.

The objective of this and the next three chapters is to introduce the reader to the fundamental concepts of this dynamic and rapidly evolving field and to provide an appreciation of its broad range of applications in image processing and pattern recognition. This chapter begins with a brief review of important aspects of biological neural networks. We then introduce several artificial neuron models together with important supervised and unsupervised learning algorithms and discuss their properties and applications.

17.1 Biological Neural Networks

Biological neural networks (BNNs) [1, 2] possess a complex and elaborate structure with an incredibly large number of interconnections. The elementary processing units of BNNs are neurons, or nerve cells. A neuron's structure consists of (i) cell body or *soma*: a large round central body with a diameter varying from 5 to 100 microns; (ii) *axon*: an elongated electrically active fiber attached to the soma that conducts nerve impulses generated by the neuron; and (iii) *dendrites*: branching extensions that receive inputs from other neurons [1] through specialized junctions called *synapses*. A synaptic connection is the site at which the axon terminal of one neuron communicates with the dendrite of another neuron and facilitates the transfer of signals between neurons [2].

Communication between neurons occurs through the release of chemical substances called *neurotransmitters* from the presynaptic cell. Specialized structures called terminal boutons facilitate the storage and subsequent release of these chemicals, which affect neighboring neurons and cause them to generate new signals. A neuron responds to the aggregate effect of all incoming signals over a short time interval, known as the *period of temporal and spatial summation*. The neuron produces a response and "fires" if the total membrane potential reaches a certain threshold, sending a pulse response down its axon. Incoming impulses are classified as excitatory if they promote firing or inhibitory if they hinder the neuron from firing. For firing to occur, total excitation must exceed total inhibition by the neuron's firing threshold of approximately 55 mV. Because synaptic connections cause these excitatory and inhibitory reactions in the receiving neuron, it is standard practice to assign positive (+) or negative (−) weight values to such connections, respectively. In general, firing requires a specific number of incoming pulses generated by neighboring neurons and potentially the neuron itself via self-feedback. Pulses that are closely spaced in time and arriving synchronously are more likely to trigger an action potential. After conducting a pulse, an axon fiber enters a state of reduced or complete non-excitability for a certain duration called the *refractory period*. This physiological mechanism can be viewed as a form of memory that influences the neuron's state at time $t + 1$ based on its state at time t.

The human cerebral cortex contains approximately 10^{11} neurons, with each neuron forming connections to roughly 10^3 other neurons through dendritic branches and synapses. Consequently, the brain contains on the order of 10^{14} synaptic connections. Although individual biological neurons operate relatively slowly compared with modern electronic devices, the enormous degree of parallelism provided by these interconnections enables the brain to perform highly sophisticated perceptual, cognitive, and decision-making tasks with remarkable efficiency.

17.2 Artificial Neural Networks

Artificial neural networks (ANNs) have been studied for many decades with the aim of achieving human-like performance in speech and pattern recognition. Initial research on developing mathematical models for neurons dates back to the 1940s. However, groundbreaking advances in learning algorithms for layered machines did not emerge until the late 1980s and early 1990s [3, 4]. Subsequent decades have seen extraordinary growth in the development of modern deep neural networks and numerous ML models [5]. Among the important contributing factors are advances in graphical processing units (GPUs) [6] that enabled the implementation of very large neural networks with millions or even billions of parameters; development of new AI models for real-time image data and natural language processing; and finally, the emergence of new and high-impact applications in fields such as medicine, autonomous systems, robotics, and remote sensing has further accelerated research and development in this area.

In the following sections, we first introduce several neuron models and discuss their characteristics and limitations. This is then followed by the development of important supervised and unsupervised learning algorithms for a single neuron and a single-layer neural networks, which form the foundation for understanding more advanced multi-layer and deep learning architectures.

17.2.1 Neuron Models

Inspired by biological neuron structure, artificial neurons (or cells) consist of connections that play the same roles as dendrites and synapses. Each connection has an adaptable weight that determines the contribution of the corresponding input signal to neuron's firing. The weighted sum of all inputs (excitatory and inhibitory) is compared against a threshold associated with each neuron. The threshold acts similar to the membrane potential in BNNs. The neuron fires if the weighted sum of inputs exceeds the threshold, in which case the neuron sends its processed information to other interconnected neurons.

The first neuron model was introduced by McCulloch and Pitts in 1943 [1, 2]. This model produces a binary response depending on the weighted sum of the inputs and its relation to the threshold. Although this neuron model is suitable for a variety of digital Boolean operations, e.g. NAND, NOR, and memory cells [2], it ignores many important aspects of biological neurons. Among its main shortcomings are: (a) works with binary states only; (b) operates under discrete-time assumption; and (c) assumes synchrony of operations of all neurons in a network. A simple generalization of the McCulloch–Pitts model, which circumvents most of these issues and limitations, was proposed by Rosenblatt's Perceptron model in 1958 [7].

A generalized version of the Rosenblatt's Perceptron model uses a nonlinear activation (or squashing) function $f(\cdot)$ for each neuron. This function mimics how a biological neuron's membrane potential leads to firing [1, 2]. Figure 17.1 depicts the structure of the Perceptron neuron i within a neural network. The neuron receives inputs $x_j,\ j \in [1, N]$ from other neurons in its vicinity. These inputs produce an aggregated activation to the neuron, which is then compared against the

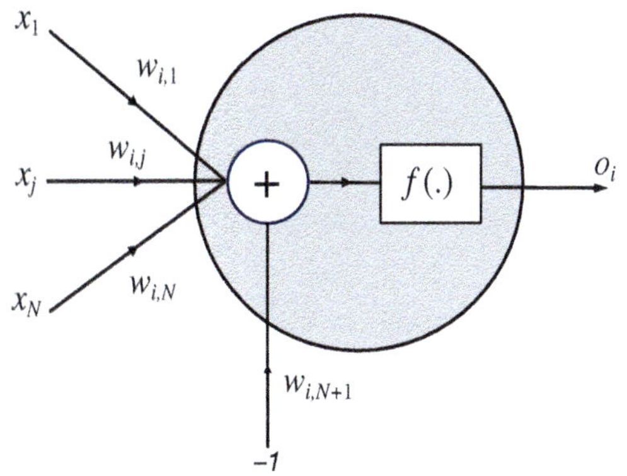

Figure 17.1 A modern perceptron neuron model.

neuron's threshold t_i. This threshold plays a role similar to the firing threshold of biological neurons. Thus, the aggregated input going to the activation function $f(\cdot)$ of the neuron is

$$y_i = \sum_{j=1}^{N} w_{i,j} x_j - t_i = \underline{w}_i^T \underline{x} - t_i \tag{17.1}$$

where $w_{i,j}$ is the synaptic weight of the jth input of neuron i, $\underline{w}_i = [w_{i,1} \cdots w_{i,N}]^T$ is the weight vector of neuron i, and $\underline{x} = [x_1 \cdots x_N]^T$ is the input vector. This equation represents a hyperplane in the N-dimensional space and can serve as a decision boundary for a two-class linear classifier. The final output o_i of this neuron is

$$o_i = f(y_i). \tag{17.2}$$

The synaptic weights $w_{i,j}$ and threshold t_i are the parameters of the neuron that must be estimated during the training process. Thus, the threshold t_i can be incorporated as a bias weight, i.e. $w_{i,N+1} = t_i$, into the weight vector, where the corresponding input is $x_{N+1} = -1$ as shown in Figure 17.1. In this model, the activation function is a continuous and differentiable function of y_i. Typical choices for activation function $f(.)$ are defined below.

1. **Unipolar Sigmoidal or Logistic:**
   One of the most popular activation function,

   $$o_i = f(y_i) = \frac{1}{1 + \exp(-\lambda y_i)}, \tag{17.3}$$

   is used when the output must be in the region $0 \le o_i \le 1$. The parameter λ determines the sharpness or width of the transition region between the inactive, when $o_i = 0$, and active (firing), when $o_i = 1$, states of the neuron. As $\lambda \to \infty$, the sigmoid activation approaches the hard-limiter threshold or the original version of the Perceptron model. A very important property of the unipolar sigmoid is that its first derivative can be expressed in terms of the output of the neuron,

   $$f'(y_i) = \lambda(1 - o_i)o_i. \tag{17.4}$$

   This property facilitates implementation of gradient-based learning in multi-layer neural networks.

2. **Bipolar Sigmoidal:**
   This activation function is defined by,

   $$o_i = f(y_i) = \frac{2}{1 + \exp(-\lambda y_i)} - 1, \tag{17.5}$$

   and is used when the output must be in the interval $-1 \le o_i \le 1$. The unipolar and bipolar sigmoidal activation functions are shown in Figure 17.2a, b, respectively.

   The hyperbolic tangent function, expressed as $o_i = \tan\text{h}(y_i)$, is also a popular alternative to the unipolar sigmoid activation function. Its main advantage over the unipolar sigmoid stems from its derivative, $f'(y_i) = 1 - f^2(y_i) = 1 - o_i^2$,

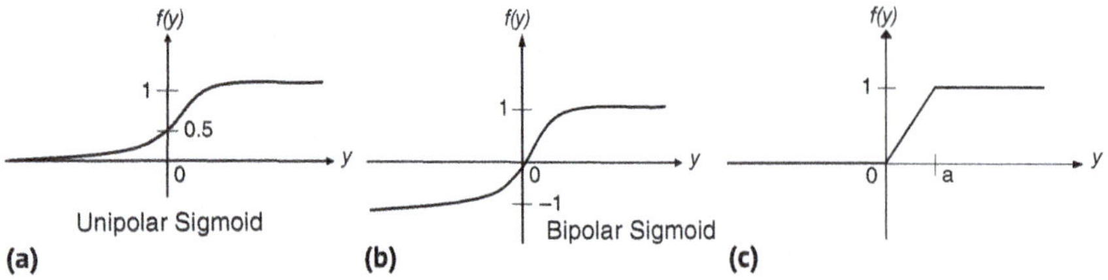

Figure 17.2 (a) Unipolar sigmoid, (b) bipolar sigmoid, and (c) threshold logic activation functions.

which remains non-zero when the neuron output o_i is close to zero. In contrast, the derivative of the unipolar sigmoid approaches zero in regions where the output o_i is small, hence, leading to the vanishing gradients during the training phase.

3. **Threshold Logic:**

This is a piecewise linear activation function given by,

$$o_i = f(y_i) = \begin{cases} 0, & y_i \leq 0 \\ y_i/a, & 0 < y_i < a, \\ 1, & y_i \geq a \end{cases}$$

where a determines the slope of the line in the transition region. This activation function is shown in Figure 17.2c.

4. **Rectified Linear Unit:**

The rectified linear unit (ReLU) activation function is given by

$$o_i = f(y_i) = \begin{cases} 0, & y_i \leq 0 \\ y_{i,} & y_i > 0, \end{cases}$$

which is a unit ramp function. This activation function is also very popular, particularly in deep neural networks and computational neuroscience, mainly due to its computational simplicity, reduced susceptibility to vanishing gradient problem, and faster convergence compared with the unipolar sigmoidal activation.

5. **Hard-limiter:**

This activation function is defined as follows:

$$o_i = f(y_i) = \begin{cases} 1, & y_i > 0 \\ 0, & y_i \leq 0. \end{cases}$$

The bipolar version is called the signum function:

$$o_i = f(y_i) = \begin{cases} 1, & y_i > 0 \\ 0, & y_i = 0 \\ -1, & y_i < 0. \end{cases}$$

6. **Softmax:**

Softmax is not a neuron model; rather, it is used extensively as an activation function, particularly in the final output layer of multi-class classifiers, including convolutional neural networks (CNNs) and multinomial logistic regression. The purpose of a softmax layer is to convert the outputs of the neurons into posterior probabilities, i.e. $P(C_i|\underline{x})$, so that the classification decision can be carried out by selecting the class associated with the largest posterior probability. This corresponds to the MAP classifier, as we showed in the previous chapter. If y_i denotes the aggregated input to neuron i, then the posterior probability is obtained by normalization,

$$P(C_i|\underline{x}) = \frac{e^{y_i}}{\sum_{m=1}^{M} e^{y_m}}, \tag{17.6}$$

where M is the number of neurons in the output layer, or the number of classes, and we have $\sum_{i=1}^{M} P(C_i|\underline{x}) = 1$.

7. **Radial Basis Functions:**

Radial basis functions (RBFs) are neuron models inspired by the way biological neurons in sensory pathways (e.g. retina or cochlea) respond selectively to stimuli occurring within their localized receptive fields. This localized-response mechanism makes RBF neurons a fundamental component in several ML models. They are extensively used in RBF networks, support vector machines (SVMs), support vector regression (SVR), and many other kernel-based machines [1]. Typical choices of the activation functions that provide localized receptive fields are Gaussian and Inverse Multiquadrics,

$$\begin{aligned} k(\underline{x}) &= \exp\left(\frac{-r^2}{2\sigma^2}\right) \quad \text{Gaussian,} \\ k(\underline{x}) &= \frac{1}{\sqrt{r^2+\alpha^2}} \quad \text{Inverse Multiquadrics,} \end{aligned} \tag{17.7}$$

where $r^2 = ||\underline{x} - \underline{c}||^2$ is the radial distance of pattern $\underline{x}$ from the center of the receptive field $\underline{c}$.

As a final note, the choice of activation function has a significant impact on the learning behavior, convergence properties, and representational power of a neural network. In general, differentiable activation functions are preferred in modern neural networks because they facilitate gradient-based learning algorithms. In the following sections, these activation functions will be used in conjunction with supervised and unsupervised learning procedures for estimating the neuron parameters.

17.3 Learning Algorithms

We can begin this section by posing a fundamental question: what is machine learning (ML)? A widely accepted definition due to Mitchell [8] states that an ML algorithm *learns* from *experience E* with respect to a particular *task T* and a *performance measure* ρ, if its performance on T, as measured by ρ, improves with experience E. This is a well-defined framework for formulating learning problems in terms of the triplet (ρ, T, E).

To illustrate this definition, we use two examples. First, consider the problem of classifying benign and malignant tumors from medical images, in which case the task T is to construct a binary hypothesis test that maps medical image data to binary decisions where the null hypothesis ($\mathcal{H}_0$) corresponds to a benign tumor and the alternative hypothesis ($\mathcal{H}_1$) corresponds to a malignant tumor. The performance metric ρ is the probability of error, which includes the overall false positive and false negative decisions and should be minimized (see Section 16.2.2). The experience E consists of samples of benign and malignant tumors from medical images. In this case, learning involves designing a classifier that performs binary hypothesis testing with the minimum probability of error when applied to previously unseen medical images. The second example involves dimensionality reduction for data compression. The task T is to extract a set of low-dimensional attributes from image data by constructing a mapping (linear or nonlinear) that maps image data to a reduced-dimensional subspace while preserving the essential information content. The appropriate performance metric ρ may be the sum squared error (SSE) between the original image and the reconstructed version obtained from the reduced-dimensional representation. Experience E consists of a collection of images with similar statistical properties. In this case, learning involves designing a mapping system that minimizes the reconstruction error for a desired reduction rate.

In general, we define learning as any change in behavior or performance brought about by experience or interaction with the environment. Learning in humans is largely an inferred process, we cannot observe it happening directly, but we can infer its effect through changes in behavior. Learning in ANN, however, is a more explicit process in the sense that we can capture each learning step in a cause-effect relationship. Consequently, designing an ML algorithm involves learning an unknown mapping that transforms inputs to desired outputs from a set of examples drawn from Experience E.

The distinction between learning paradigms is primarily determined by the nature of the available experience E. In some applications, training data are accompanied by desired outputs or labels, whereas in others only unlabeled data are available.

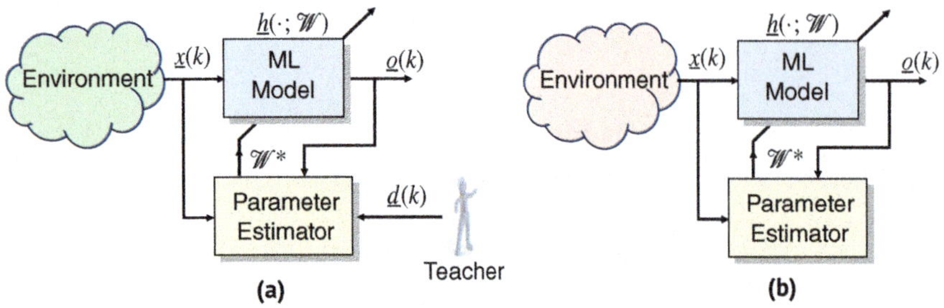

Figure 17.3 Block diagrams of typical (a) supervised and (b) unsupervised systems.

This distinction gives rise to the two major categories of learning considered in this chapter. Although there are several categories of learning, each suited for a specific application, we focus on two general categories with the most widespread application in DIP: (a) supervised and (b) unsupervised learning.

(a) Supervised Learning:

This category of learning algorithms is used in applications such as pattern detection and classification, prediction, regression, and function approximation. Supervised learning methods require an *external teacher* to provide the class labels or desired outputs for every training sample and to monitor the training process. The block diagram of a typical supervised learning system is shown in Figure 17.3a. The training dataset $\mathcal{X} = \{\underline{x}(k), \underline{d}(k)\}_{k=1}^{K}$ consists of pairs of $\underline{x}(k)$ and $\underline{d}(k)$ where $\underline{x}(k) \in \mathbb{R}^N$ is the kth training sample (pattern) drawn from the environment and $\underline{d}(k) \in \mathbb{R}^M$ is the corresponding *desired* output vector provided by the external teacher. The objective of training is to approximate an unknown multivariate mapping function $\underline{d}(k) = \underline{g}(\underline{x}(k))$ by a parameterized function $\underline{o}(k) = \underline{h}(\underline{x}(k); \mathcal{W})$, where $\mathcal{W}$ is the set of unknown parameters of the model. The quality of this approximation is measured by a prespecified performance metric $\rho(\underline{h}(\underline{x}(k); \mathcal{W}), \underline{g}(\underline{x}(k)))$. The optimal parameter set $\mathcal{W}^*$ is the one that minimizes or maximizes the selected performance metric over the training data. That is, for the minimization case, we should have $\rho(\underline{h}(\underline{x}(k); \mathcal{W}^*), \underline{g}(\underline{x}(k))) \leq \rho(\underline{h}(\underline{x}(k); \mathcal{W}^+), \underline{g}(\underline{x}(k))),\ \forall k \in [1, K]$ where $\mathcal{W}^+$ is any other estimate of the parameters. Such learning mechanisms are referred to as *Performance Learning*. Typical choices of the performance metric $\rho(.)$ are MSE, classification error, cross-entropy, and mutual information. Some common challenges associated with supervised learning algorithms are as follows: (i) they require a large number of labeled training data, which may be costly or impractical to collect in many realistic situations; (ii) incremental learning using new data requires access to the prior data and can lead to catastrophic forgetting of prior training; (iii) overtraining based on correlated or repetitive data leads to performance degradation; (iv) imbalanced training data causes bias in decision-making; (v) they suffer from speed-accuracy tradeoff in the sense that faster convergence leads to less accurate solutions of unknown parameters, and (vi) overfitting to the training data leads to poor generalization performance on new unseen data.

(b) Unsupervised Learning:

These learning algorithms use unlabeled data and therefore do not require desired outputs or an external teacher to correct mistakes. They are widely used in applications, including data clustering, vector quantization, data compression, dimensionality reduction, and feature extraction. Figure 17.3b illustrates the block diagram of a typical unsupervised learning system. *Self-organization* principles [9, 10] are used to form topologically meaningful clusters and discover latent features from data based on their underlying statistical distributions. They are unconditional density estimators as opposed to the supervised learning methods, which are class-conditional density estimators. Their structures and learning algorithms are simpler than those of the supervised ones, though their convergence is often slower and more difficult to assess. Furthermore, it is generally not possible to determine *a priori* the optimal number of clusters. Consequently, selecting the appropriate model complexity remains one of the principal challenges in unsupervised learning.

17.3.1 Generalized Learning Rule

Amari [11, 12] proposed a generalized learning rule for updating the weight vector of a neuron that encompasses a broad class of learning mechanisms. According to this generalized learning rule, the weight vector $\underline{w}_i$ of neuron i is updated in proportion to the input $\underline{x}(k)$ and a *learning* signal $r(.)$. The learning process is driven by this learning signal, which provides feedback to the adaptation mechanism on how to make adjustments to the synaptic weights in order to meet the performance metric iteratively. Using this learning rule, the weight vector at iteration $k + 1$ is adjusted using

$$\underline{w}_i(k+1) = \underline{w}_i(k) + \Delta\underline{w}_i(k) = \underline{w}_i(k) + \mu r\left(\underline{w}_i(k), \underline{x}(k), d_i(k)\right)\underline{x}(k), \tag{17.8}$$

where $\underline{w}_i(k)$, $\underline{x}(k)$, and $d_i(k)$ are the weight vector, input vector, and desired signal at learning iteration k, respectively, and μ is the learning rate or step size. The weight vector is initialized randomly. Amari's generalized learning rule provides a unified representation for several learning paradigms, including: (i) performance learning, (ii) coincidence learning, (iii) competitive learning, and many others not covered in this text. Depending on the type learning paradigm, the learning signal may be computed using input vector, weight vector, desired signal, output vector, or combinations thereof. It is important to note that Amari's learning rule can be viewed as a generalization of the Robbins–Monro stochastic approximation method [13], although the former offers faster convergence and better numerical stability. In the subsequent sections, we will define the learning signal for each algorithm corresponding to several important learning paradigms and show how they can be related to the generalized learning rule in (17.8).

17.3.2 Performance Learning

The performance learning methods are based on minimizing or maximizing a prescribed performance measure, objective function or cost function. Many supervised learning algorithms can be formulated as performance learning methods, where the learning signal is derived directly from a measure of error between the desired and actual outputs of the system. In the following subsections, we present several important performance learning algorithms that form the foundation of modern neural networks.

17.3.2.1 ADALINE Learning

ADAptive LINearElement or ADALINE, proposed by Widrow [14], uses a neuron model with a bipolar hard-limiter activation function, although the output is taken from the aggregated input y_i rather than the actual output o_i. The goal of this learning mechanism is to find the optimal weight vector $\underline{w}_i^*$, known as the Wiener-Hopf solution [1, 14], of neuron i by minimizing the mean squared error (MSE) between the aggregated input y_i and the desired output d_i, that is, $E[(d_i - y_i)^2]$ where $E[.]$ is the ensemble average or expectation operation (see Wiener filter in Section 10.2.2). However, in practice, only a limited labeled training data $\mathcal{X} = \{\underline{x}(k), d_i(k)\}_{k=1}^{K}$ is available, where $\underline{x}(k) \in \mathbb{R}^N$ is the input vector and $d_i(k) \in \mathbb{R}$ (or $d_i(k) \in \mathbb{Z}$ for integers) is the corresponding desired output. In this case, the appropriate objective function to minimize is the sum-squared error (SSE) given by

$$J(\underline{w}_i) = \frac{1}{K}\sum_{k=1}^{K}(d_i(k) - y_i(k))^2 = \frac{1}{K}\sum_{k=1}^{K} e_i^2(k), \tag{17.9}$$

where $y_i(k) = \underline{x}^T(k)\underline{w}_i$ and $\underline{w}_i = [w_{i,1} \cdots w_{i,N}]^T$. The summation in (17.9) can be more compactly expressed as

$$\begin{aligned} J(\underline{w}_i) &= \frac{1}{K}\|\underline{e}_i\|^2 = \frac{1}{K}(\underline{d}_i - \underline{y}_i)^T(\underline{d}_i - \underline{y}_i) \\ &= \frac{1}{K}(\underline{d}_i - \mathbf{X}\underline{w}_i)^T(\underline{d}_i - \mathbf{X}\underline{w}_i) \end{aligned} \tag{17.10}$$

where $\underline{e}_i = [e_i(1) \cdots e_i(K)]^T$, $\underline{d}_i = [d_i(1), \dots, d_i(K)]^T$, and $\underline{y}_i = [y_i(1), \dots, y_i(K)]^T$ are the error, desired output, and aggregated input vectors, respectively, and $\mathbf{X} \in \mathbb{R}^{K\times N}$ is the data matrix whose rows contain the training samples, i.e. $\mathbf{X} = [\underline{x}(1) \cdots \underline{x}(K)]^T$. Minimizing (17.10) with respect to the weight vector $\underline{w}_i$ leads to the classical *Least squares* (LS) solution. Taking the gradient of $J(\underline{w}_i)$ with respect to $\underline{w}_i$ and setting it equal to zero yields

$$\nabla J(\underline{w}_i) = \frac{\partial J(\underline{w}_i)}{\partial \underline{w}_i} = \frac{-2}{K}\mathbf{X}^T(\underline{d}_i - \mathbf{X}\underline{w}_i) = 0, \tag{17.11}$$

which upon solving for $\underline{w}_i$, gives the LS solution

$$\underline{\hat{w}}_i = (\mathbf{X}^T\mathbf{X})^{-1}\mathbf{X}^T\underline{d}_i = \mathbf{X}^{\dagger}\underline{d}_i, \tag{17.12}$$

where $\mathbf{X}^{\dagger} = (\mathbf{X}^T\mathbf{X})^{-1}\mathbf{X}^T$ is the left *pseudo inverse* [15] of matrix $\mathbf{X}$. Clearly, this solution requires the matrix $(\mathbf{X}^T\mathbf{X})$ to be non-singular, which happens if the $N \leq K$ columns of $\mathbf{X}$ are linearly independent or rank $(X) = N$.

Now, an important question is how this solution relates to the optimal solution $\underline{w}_i^*$. Using the definitions of $\mathbf{X}$ and $\underline{d}_i$, we get

$$\frac{1}{K}\mathbf{X}^T\mathbf{X} = [\underline{x}(1) \cdots \underline{x}(K)] \begin{bmatrix} \underline{x}^T(1) \\ \vdots \\ \underline{x}^T(K) \end{bmatrix} = \frac{1}{K}\sum_{k=1}^{K} \underline{x}(k)\underline{x}^T(k) = \hat{\mathbf{R}}_{xx}$$
$$\frac{1}{K}\mathbf{X}^T\underline{d}_i = [\underline{x}(1) \cdots \underline{x}(K)] \begin{bmatrix} d_i(1) \\ \vdots \\ d_i(K) \end{bmatrix} = \frac{1}{K}\sum_{k=1}^{K} \underline{x}(k)d_i(k) = \hat{\underline{r}}_{xd}, \tag{17.13}$$

where $\hat{\mathbf{R}}_{xx}$ and $\hat{\underline{r}}_{xd}$ are the *sample* correlation matrix of the training data $\underline{x}(k)$ and the sample cross-correlation vector between the training samples $\underline{x}(k)$ and the desired outputs $d_i(k)$, respectively. This implies that the LS solution in (17.12) can alternatively be expressed as $\hat{\underline{w}}_i = \hat{\mathbf{R}}_{xx}^{-1}\hat{\underline{r}}_{xd}$. Now, if the data samples $\underline{x}(k)$ are drawn from an ergodic process with common second-order statistics [16], then the sample averages can be used in place of ensemble averages, i.e. in this case, the sample correlation matrix and cross-correlation vector $\hat{\mathbf{R}}_{xx}$ and $\hat{\underline{r}}_{xd}$ are good representations of the ensemble correlation matrix $\mathbf{R}_{xx}$ and the ensemble cross-correlation vector $\underline{r}_{xd}$ when K becomes large. In this case, the LS solution converges (in the mean-squared sense) to the minimum MSE (MMSE) solution, i.e. $\hat{\underline{w}}_i \to \underline{w}_i^*$ [17, 18]. In essence, the LS estimator provides a practical approximation of the theoretically optimal Wiener (MMSE) filter when using a finite amount of data from a WSS process, and the approximation improves as more data is used.

It is illuminating to investigate the geometrical interpretation of the LS solution. To do so, we begin by defining the *orthogonal project matrix* $\mathbf{P}_X = \mathbf{X}(\mathbf{X}^T\mathbf{X})^{-1}\mathbf{X}^T$, which projects vectors onto the N-dimensional subspace spanned by the columns of the data matrix $\mathbf{X}$. We denote this subspace by the symbol $< \mathbf{X} >_N$. It is clear that $\mathbf{P}_X\mathbf{X} = \mathbf{X}$. Moreover, this matrix has two important properties (i) $\mathbf{P}_X^T = \mathbf{P}_X$ (symmetry), and (ii) $\mathbf{P}_X^2 = \mathbf{P}_X$ (idempotency), or in general $\mathbf{P}_X^n = \mathbf{P}_X, \forall n \geq 1$. Now, since $\underline{y}_i = \mathbf{X}\underline{w}_i$, then $\hat{\underline{y}}_i = \mathbf{X}\hat{\underline{w}}_i$ is the estimate of $\underline{y}_i$ based on the LS estimate of $\underline{w}_i$. As can be seen in Figure 17.4, both $\underline{y}_i$ and $\hat{\underline{y}}_i$ lie in subspace $< \mathbf{X} >_N$. Now, using the expression for $\hat{\underline{w}}_i$ in (17.12) we get

$$\hat{\underline{y}}_i = \mathbf{X}\hat{\underline{w}}_i = \mathbf{X}(\mathbf{X}^T\mathbf{X})^{-1}\mathbf{X}^T\underline{d}_i = \mathbf{P}_X\underline{d}_i, \tag{17.14}$$

which implies that $\hat{\underline{y}}_i$ is the projection of $\underline{d}_i$ onto the subspace $< \mathbf{X} >_N$ as shown in Figure 17.4. Thus, the corresponding error vector $\hat{\underline{e}}_i$ of the LS solution is

$$\hat{\underline{e}}_i = \underline{d}_i - \hat{\underline{y}}_i = \underline{d}_i - \mathbf{P}_X\underline{d}_i = \mathbf{P}_X^{\perp}\underline{d}_i. \tag{17.15}$$

where $\mathbf{P}_X^{\perp} = \mathbf{I} - \mathbf{P}_X$ is the orthogonal complement of $\mathbf{P}_X$ or the projection matrix of the subspace orthogonal to $< \mathbf{X} >_N$. The actual error vector is $\underline{e}_i = \underline{d}_i - \underline{y}_i$. Using (17.16), this error vector can be expressed as $\underline{e}_i = \hat{\underline{e}}_i + (\hat{\underline{y}}_i - \underline{y}_i)$, which is illustrated geometrically in Figure 17.4.

Example 17.1 Linear Regression *Configure an ADALINE as a linear regression to predict college students' GPAs based on their high school performance in the SAT and ACT tests.*

Solution: Let $x_1(k)$ and $x_2(k)$ denote the SAT and ACT scores, respectively of the kth student. Using a linear regression model, the GPA of this student can be modeled as

$$y(k) = w_0 + w_1x_1(k) + w_2x_2(k) = \underline{w}^T\underline{x}(k),$$

where $\underline{x}(k) = [1\ x_1(k)\ x_2(k)]^T$, $\underline{w} = [w_0\ w_1\ w_2]^T$, and w_0, w_1, and w_2 are the unknown model parameters (w_0 is the bias) that need to be estimated. Using a training dataset $\{d(k), x_1(k), x_2(k)\}_{k=1}^K$ consisting of the GPA, SAT, and ACT scores of K students, we can find the weight vector $\underline{w}$ using the LS solution in (17.12) or more efficiently using the least mean squares (LMS) learning method that will be introduced later in this section. Once the weights are computed, the model can be used to predict the GPA, $y(k)$, of a new student from the corresponding SAT and ACT scores.

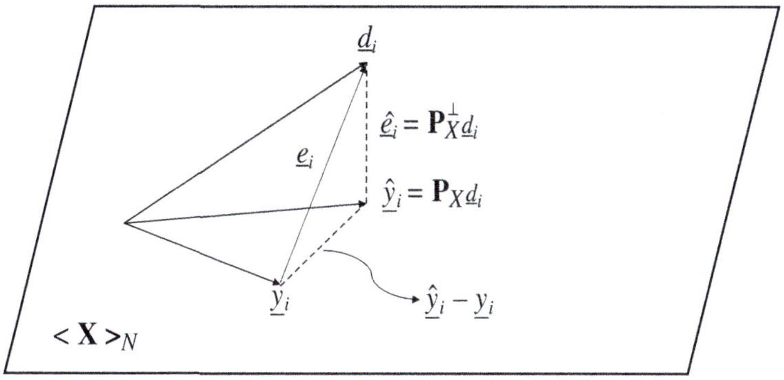

Figure 17.4 Geometrical perspective of the LS solution.

ADALINE and Statistical Linear Regression

So far, no assumptions have been made about the statistical properties of the data, since the LS solution in (17.12) is purely deterministic. Here, we lift this assumption and consider the case where the error $e_i(k)$ is independent identically distributed (iid) random process [14] with Gaussian or normal distribution, i.e. $e_i(k) \sim \mathcal{N}(0, \sigma_e^2)$.

As discussed earlier, the desired signal is related to the error using $d_i(k) = \underline{w}_i^T \underline{x}(k) + e_i(k)$ or in vector form $\underline{d}_i = \mathbf{X}\underline{w}_i + \underline{e}_i$ where $\underline{e}_i \sim \mathcal{N}(\underline{0}, \sigma_e^2\mathbf{I})$. Now, since the training data samples $\underline{x}(k)$, $k \in [1, K]$ are known and $\underline{w}_i$ is unknown but deterministic, it follows that the desired-output vector $\underline{d}_i$ conditioned on X is also Gaussian with mean $\mathbf{X}\underline{w}_i$ and covariance matrix $\sigma_e^2\mathbf{I}$. That is, we have $\underline{d}_i|X \sim \mathcal{N}(\mathbf{X}\underline{w}_i, \sigma_e^2\mathbf{I})$ or $p(\underline{d}_i|\mathbf{X}; \underline{w}_i) \sim \mathcal{N}(\mathbf{X}\underline{w}_i, \sigma_e^2\mathbf{I})$. To estimate the weight vector $\underline{w}_i$ in this statistical formulation of the linear regression, we use the ML method and maximize the log-likelihood function

$$
\begin{aligned}
l(\underline{w}_i) = \ln p(\underline{d}_i|\mathbf{X}; \underline{w}_i) &= \ln\left\{\frac{1}{(2\pi\sigma_e^2)^{K/2}} \exp\left[-\frac{(\underline{d}_i - \mathbf{X}\underline{w}_i)^T(\underline{d}_i - \mathbf{X}\underline{w}_i)}{2\sigma^2}\right]\right\} \\
&= -\frac{K}{2}\ln(2\pi\sigma_e^2) - \frac{1}{2\sigma_e^2}(\underline{d}_i - \mathbf{X}\underline{w}_i)^T(\underline{d}_i - \mathbf{X}\underline{w}_i),
\end{aligned}
\tag{17.16}
$$

with respect to the unknown weight vector $\underline{w}_i$. Ignoring the first term on the right-hand side, which is independent of $\underline{w}_i$, maximizing the second term in (17.16) with respect to $\underline{w}_i$ gives the same result as the LS solution in (17.12). Thus, we can conclude that the ML estimate of the parameters of a linear regression model is the same as the LS solution when error $e_i(k)$ are iid normal.

In the next two sections, we present two alternative approaches to parameterizing the ADALINE model that do not require any matrix inversion operation and are therefore more suitable for large-scale applications.

17.3.2.2 Gradient Descent Algorithm

An alternative procedure based on the gradient descent method presented in Section 15.3.1 can also be used to estimate the weights of the ADALINE model without requiring the inversion of the sample correlation matrix $\hat{\mathbf{R}}_{xx}$. This method involves the following steps:

1. Start with an initial guess for $\underline{w}_i(0)$. Usually, random initialization with uniform distribution is used.
2. Compute the gradient vector of the objective function in (17.10) $\nabla J(\underline{w}_i) \triangleq \frac{\partial J(\underline{w}_i)}{\partial \underline{w}_i}$.
3. Update the weight vector in a direction opposite the gradient to minimize the objective function $J(\underline{w}_i)$.

$$\underline{w}_i(k+1) = \underline{w}_i(k) - \frac{1}{2}\mu\nabla J(\underline{w}_i(k)), \tag{17.17}$$

   where μ is the learning rate or step size that controls the magnitude of the weight corrections.
4. Return to Step 2 and repeat until a convergence criterion is met.

From the result in (17.11), the gradient at iteration k is $\nabla J(\underline{w}_i(k)) = -2\mathbf{X}^T\left(\underline{d}_i(k) - \mathbf{X}\underline{w}_i(k)\right)$. Substituting this result in (17.17) gives

$$\begin{aligned}\underline{w}_i(k+1) &= \underline{w}_i(k) + \frac{\mu}{K}\mathbf{X}^T\left(\underline{d}_i - \mathbf{X}\underline{w}_i(k)\right) = \underline{w}_i(k) + \frac{\mu}{K}\mathbf{X}^T\underline{\hat{e}}_i \\ &= \underline{w}_i(k) + \frac{\mu}{K}\sum_{l=1}^{K}\underline{x}(l)\hat{e}_i(l),\end{aligned} \tag{17.18}$$

where $\underline{\hat{e}}_i = [\hat{e}_i(1)\cdots\hat{e}_i(K)]^T$, $\hat{e}_i(l) = d_i(l) - \hat{y}_i(l)$, and $\hat{y}_i(l) = \mathbf{X}\underline{w}_i(k)$ is the estimate of $y_i(l)$ based on the weight vector $\underline{w}_i(l)$. This result is referred to as *batch gradient descent* algorithm because the gradient is computed using all K training samples at each iteration. Under certain conditions, the algorithm converges to the LS solution (see Problem 17.10). However, unlike the LS solution, the updating equation in (17.18) does not require any matrix inversion. In what follows, we present a computationally efficient iterative update rule for the weight vector of the ADALINE model using the Least Mean-Squared (LMS) or the stochastic gradient descent (SGD) algorithm in which the weight vector is updated using only a single training sample at each iteration.

17.3.2.3 Least Mean-Squared (LMS) Algorithm

The LMS method was developed by Widrow in the 1960s [17, 18] and plays a crucial role in adaptive systems and many ML models. As stated earlier, the gradient descent algorithm requires knowledge of the exact gradient vector $\nabla J(\underline{w}_i(k))$. However, in practice, the gradient is not known and must be estimated based on the instantaneous value of the squared error, $e_i^2(k) = (d_i(k) - y_i(k))^2$ where $y_i(k) = \underline{w}_i^T(k)\underline{x}(k)$. If we consider this instantaneous squared error as our objective function at iteration k, then taking its partial derivative with respect to the unknown weight vector $\underline{w}_i(k)$ gives

$$\frac{\partial J(\underline{w}_i(k))}{\partial \underline{w}_i(k)} = -2\underline{x}(k)e_i(k).$$

Using this instantaneous estimate of the gradient in the gradient descent update equation (17.17) yields the LMS learning rule,

$$\underline{w}_i(k+1) = \underline{w}_i(k) + \mu\underline{x}(k)e_i(k) = \underline{w}_i(k) + \mu\underline{x}(k)\left(d_i(k) - \underline{w}_i^T(k)\underline{x}(k)\right). \tag{17.19}$$

Unlike the batch gradient descent approach, which uses all training samples to estimate the gradient, the LMS algorithm updates the weight vector using only a single training sample at each iteration. As a result, the computational complexity of LMS is substantially lower than the previously discussed methods.

Referring to Amari's generalized learning rule in (17.8), the learning signal for the LMS learning algorithm is the instantaneous error signal $r(\underline{w}_i(k), \underline{x}(k), d_i(k)) = e_i(k)$. The LMS learning rule is also known as the stochastic gradient descent (SGD) algorithm due to the uncertainties in estimating the gradient instead of the actual gradient.

Example 17.1 showed how the ADALINE model can be used to solve linear regression problems. We now present a more realistic example of linear prediction using the LMS learning rule. Suppose that we wish to predict the daily value of a stock index based on its previous three days' values. We employ a third-order auto-regressive (AR) model (see Section 9.1),

$$y(n) = a_1y(n-1) + a_2y(n-2) + a_3y(n-3) + e(n)$$

where $y(n)$ is the actual stock value on day n, $e(n)$ is the driving noise or inaccuracy in modeling, and a_1, a_2, and a_3 are unknown model parameters that must be estimated from historical data. Once these parameters are estimated, the linear predictor,

$$\hat{y}(n) = a_1y(n-1) + a_2y(n-2) + a_3y(n-3),$$

is used to predict future values based on the last three days' values.

To configure an ADALINE for this problem, we use the input vector $\underline{x}(n) = [y(n-1)\ y(n-2)\ y(n-3)]^T$ and the weight vector $\underline{w} = [a_1\ a_2\ a_3]^T$. Thus, the aggregated input of the ADALINE is $\hat{y}(n) = \underline{w}^T\underline{x}(n)$, which is compared to the desired $d(n) = y(n)$ during the training phase. In this case, the LMS update equation for the weight vector becomes

$$\underline{w}(k+1) = \underline{w}(k) + \mu\,\underline{x}(k)[y(k) - \underline{w}^T(k)\underline{x}(k)].$$

Portions of the 2018 stock-price data during low volatility periods were used to estimate the parameters of the AR (3) model. The predictor was then applied to predict the daily values of the stock index using the unseen 2019 data. Figure 17.5 shows the plot of actual (blue) versus the predicted values (red) during this period of time. Clearly, this predictor tracks the overall trend of stock price changes only when the actual testing data match closely those of the training data. That is, the model is incapable of predicting abrupt price changes during high-volatility periods. Nevertheless, this example serves to illustrate how to configure a simple ADALINE as a linear predictor using the LMS learning rule.

Properties of LMS Learning

Several important properties of this learning rule are summarized below. The reader is referred to [17, 18] for more detailed treatment on these properties.

1. The condition for the convergence of the LMS algorithm is that the step size μ should satisfy $0 < \mu < \frac{2}{\lambda_{\max}}$ where $\lambda_{\max}$ is the largest eigenvalue of the ensemble correlation matrix $\mathbf{R}_{xx}$.
2. The plot of objective function $J(\underline{w}_i(k))$ as a function of iteration k is called the *learning curve* of the LMS and is used to monitor the convergence of the weight vector as it moves through the weight space toward a local or global minimum of the objective-function surface. The convergence behavior of LMS learning depends not only on the choice of the step size μ, but also on the eigenvalue spread of the correlation matrix $\mathbf{R}_{xx}$. More specifically, the slowest learning mode is determined by the smallest eigenvalue $\lambda_{\min}$, while the fastest mode is dictated by the largest eigenvalue $\lambda_{\max}$. Thus, if the eigenvalues are widely spread, the overall convergence rate is dominated by the smallest eigenvalue, resulting in a longer settling time. However, in practice, to limit the number of iterations, usually one preselects an error goal, and once $J(\underline{w}_i(k))$ falls at or below this error goal, the learning is terminated.
3. Because the LMS algorithm uses a noisy estimate of the true gradient, the weight vector does not converge exactly to the global minimum even when the objective function is quadratic. Instead, the weight vector fluctuates around the optimal solution ($\underline{w}_i^\star$). This phenomenon is referred to as *misadjustment*, and is measured by

$$\mathcal{M} = \frac{\text{Average Excess MSE}}{\text{Min.MSE}} = \frac{E[J(\mathbf{w}_i(k)] - J_{\min}}{J_{\min}},$$

where $J_{\min} = J(\underline{w}_i^\star)$. It can be shown [17, 18] that $\mathcal{M} = \mu\, tr[\mathbf{R}_{xx}] = \mu \sum_{i=1}^{N} \lambda_i$ where λ_i is the ith eigenvalue of $\mathbf{R}_{xx}$. This result illustrates an important tradeoff in LMS learning. Reducing the step size μ decreases the misadjustment and improves the steady-state accuracy of the solution. However, smaller values of μ also result in slower convergence. Conversely, larger values of μ accelerate convergence but increase the misadjustment.

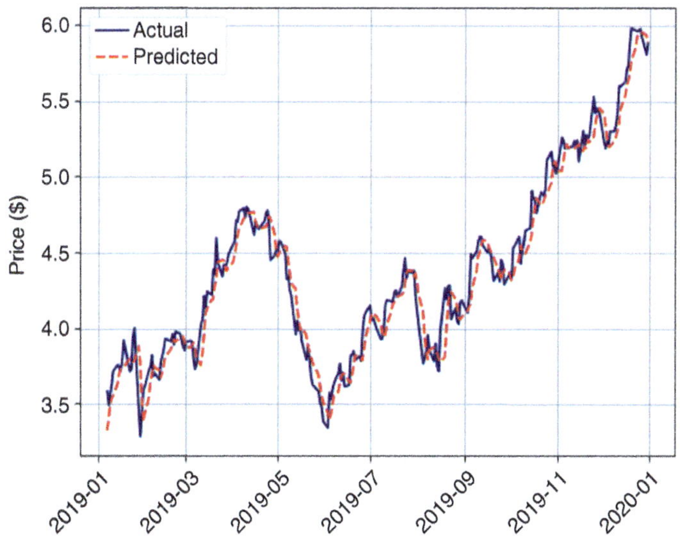

Figure 17.5 Actual vs. predicted values of a stock index on testing dataset.

17.3.2.4 Perceptron Learning Rule

The Perceptron learning rule was originally proposed by Rosenblatt [18] for a single neuron with a bipolar hard-limiter activation function. In contrast to Widrow's ADALINE, which utilizes the aggregated input $y_i(k)$ to generate the learning signal, this supervised learning uses the actual output $o_i(k)$, while the desired signal is binary $d_i = \pm 1$. Making these modifications to the LMS rule gives the Perceptron learning rule as:

$$\underline{w}_i(k+1) = \underline{w}_i(k) + \mu[d_i(k) - o_i(k)]\underline{x}(k) = \underline{w}_i(k) + \mu e_i(k)\underline{x}(k), \tag{17.20}$$

where $o_i(k) = sgn(y_i(k))$ is the output of neuron i, $y_i(k) = \underline{w}_i^T(k)\underline{x}(k)$ is the aggregated input, and $e_i(k) = d_i(k) - o_i(k)$ is the error or learning signal. In this learning, the weights are only updated when o_i is incorrect, that is, if $d_i = 1$ and $o_i = -1$, then $e_i(k) = 2$, or if $d_i = -1$ and $o_i = 1$, then $e_i(k) = -2$. Otherwise, the error will be zero and no updating takes place. For a two-class classification problem, the first case corresponds to misclassification of class 1 (or C_1) as class 2 (or C_2) patterns and vice versa. Thus, the learning rule in (17.20) is simplified to

$$\underline{w}_i(k+1) = \underline{w}_i(k) \pm 2\mu\underline{x}(k), \tag{17.21}$$

If the weight vector is initialized to zero $\underline{w}_i(0) = \underline{0}$, after one complete pass over the entire training dataset, we get

$$\hat{\underline{w}}_i = 2\mu \sum_{k \in \mathcal{S}_1} \underline{x}(k) - 2\mu \sum_{k \in \mathcal{S}_2} \underline{x}(k), \tag{17.22}$$

where $\mathcal{S}_1$ and $\mathcal{S}_2$ are sets of indices of the misclassified training samples from C_1 and C_2 classes, respectively. Thus, $\mathcal{S}_1 \cup \mathcal{S}_2$ is a set of all misclassified data sample indices. In other words, Perceptron learns by its mistakes! Additionally, only a finite number of passes or epochs over the training data is needed to form the decision surface of a linear dichotomizer. This implies that Rosenblatt's Perceptron rule is indeed the proper learning for finding the parameters of a linear machine. This is demonstrated in the following example.

Example 17.2 *Show that a linear machine can be implemented using a single Perceptron neuron for a two-class classification (dichotomizer) problem.*

Solution: The configuration of a single neuron that implements a hyperplane decision surface to separate two classes is the same as that shown in Figure 17.1, where $f(.)$ represents the bipolar hard-limiter activation function of the neuron. To learn the parameters of the separating hyperplane, we use the Perceptron learning rule in (17.21) according to the following procedure.

1. Choose an arbitrary initial weight vector $\underline{w}(0)$.
2. Choose $\underline{x}(k) \in \mathcal{X}$ from the training dataset $\mathcal{X}$.
3. If $\underline{x}(k) \in C_1$ and $\underline{w}^T(k)\underline{x}(k) > 0$, i.e. $o(k) = 1$, then go to Step 2 (no learning).
   If $\underline{x}(k) \in C_1$ and $\underline{w}^T(k)\underline{x}(k) < 0$, $\underline{w}(k+1) = \underline{w}(k) + 2\mu\underline{x}(k)$.
   If $\underline{x}(k) \in C_2$ and $\underline{w}^T(k)\underline{x}(k) < 0$, i.e. $o(k) = -1$, then go to Step 2 (no learning).
   If $\underline{x}(k) \in C_2$ and $\underline{w}^T(k)\underline{x}(k) > 0$, $\underline{w}(k+1) = \underline{w}(k) - 2\mu\underline{x}(k)$.

Figure 17.6 illustrates the process of learning a decision boundary in the 2-D feature space using the Perceptron learning rule. As can be observed, the initial decision boundary is $\underline{w}^T(0)\underline{x} = 0$. When a training sample is misclassified, the weight adjustment to produce $\underline{w}(1)$ occurs in the direction normal to this initial decision boundary hyperplane. Once the weight update is completed, the next decision boundary hyperplane becomes $\underline{w}^T(1)\underline{x} = 0$ and the process continues until all training samples have been applied to the neuron. The Perceptron learning rule updates the weight vector by adding or subtracting a fraction (dependent on μ) of the misclassified input vector. Geometrically, this operation rotates the decision boundary toward a configuration that correctly separates the training patterns. Repeated application of this learning rule gradually improves the classification performance until a satisfactory separating hyperplane is defined.

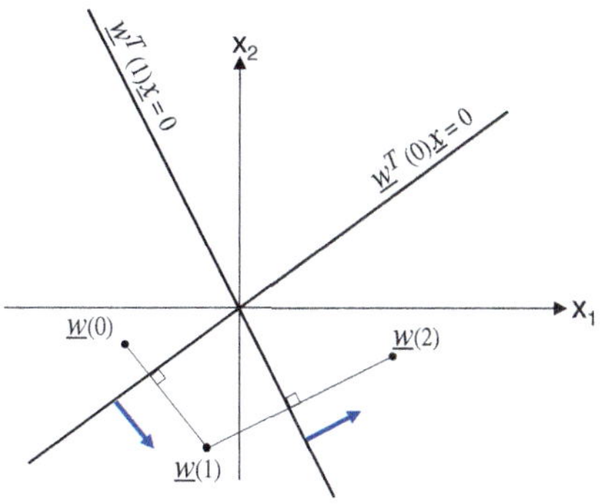

Figure 17.6 Perceptron learns the hyperplane decision boundary iteratively.

Extension to Multi-Class Linear Machines

Example 17.2 showed that the Perceptron learning rule can be applied to train a dichotomizer and find the hyperplane decision boundary that separates patterns of the two classes. In this section, we extend the result in Example 17.2 to a multi-class or M-ary classification problem with the assumption that the classes are *pairwise* linearly separable. A single-layer Perceptron network consisting of M neurons with bipolar hard-limiter activation function $f(.)$ is used, where neuron i with weight vector $\underline{w}_i$ implements the discriminant function $g_i(\underline{x}) = f(\underline{w}_i^T \underline{x})$ for class C_i. The training procedure in this case is given below.

At iteration k, if the pattern $\underline{x}(k)$ is correctly classified, no weight updating takes place. However, if pattern $\underline{x}(k) \in C_i$, but the classifier incorrectly assigns it to class C_j, where $j \neq i$, then both weight vectors $\underline{w}_i$ and $\underline{w}_j$ are updated according to

$$\begin{aligned} \underline{w}_i(k+1) &= \underline{w}_i(k) + 2\mu \underline{x}(k) \\ \underline{w}_j(k+1) &= \underline{w}_j(k) - 2\mu \underline{x}(k). \\ \underline{w}_m(k+1) &= \underline{w}_m(k), \quad \forall m \neq i, j \end{aligned} \tag{17.23}$$

The results in this section have two important implications. First, a single-layer network with M neurons and bipolar hard-limited threshold activation function can be successfully trained using the supervised Perceptron learning rule. Upon convergence, each neuron implements one of M discriminant functions or the hyperplane decision boundaries. Second, a single-layer Perceptron network cannot implement complex nonlinear decision boundaries even if activation functions of the neurons are changed to nonlinear sigmoid. The nonlinearity of the sigmoidal introduces a *shading* effect around the decision boundary by making the transition between classes smooth. This shading particularly affects the patterns near the decision boundaries by assigning real-valued outputs between 0 and 1, which may be interpreted as confidence measures or as estimates of posterior class probabilities (under certain conditions). However, the geometry of the decision boundaries remains linear. Figure 17.7 shows this shading effect of sigmoidal activation for a Perceptron neuron model with two inputs x_1 and x_2 and output $o = \frac{1}{1+\exp -\lambda(x_1+x_2)}$ with $\lambda = 1$.

17.3.2.5 Delta Learning Rule

Rumelhart and McClelland [19] extended the Perceptron learning rule to continuous-valued outputs by employing differentiable continuous activation functions such as the sigmoid. This supervised learning rule overcomes many of the limitations of the previous learning methods and paved the way for most modern neural network models that use error backpropagation learning. One may also view this learning strategy as a generalization of Widrow's LMS or SGD rule, where the instantaneous squared error is formed using the difference between the desired and the actual neuron output rather than the aggregated input $y_i(k)$, i.e., $e_i^2(k) = (d_i(k) - o_i(k))^2 = (d_i(k) - f(y_i(k)))^2$. To derive this learning rule, we take the partial derivative of this instantaneous squared error with respect to the weight vector $\underline{w}_i(k)$,

$$\frac{\partial e_i^2(k)}{\partial \underline{w}_i(k)} = -2 \frac{\partial f(y_i(k))}{\partial \underline{w}_i(k)} e_i(k). \tag{17.24}$$

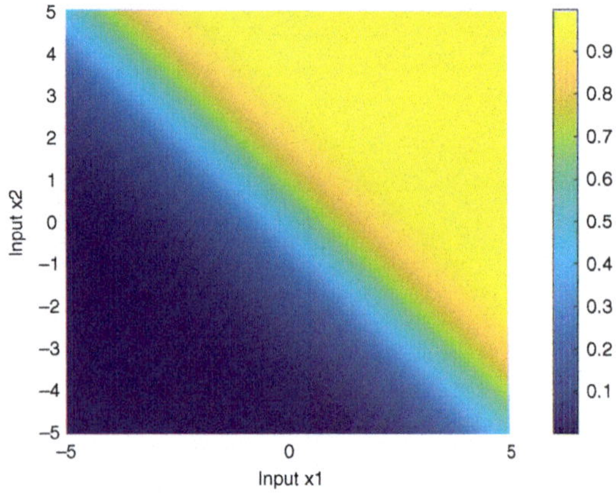

Figure 17.7 Shading effect of the decision boundary of a linear dichotomizer using a sigmoidal activation function.

Now, using the chain rule of differentiation, we have $\frac{\partial f(y_i(k))}{\partial \underline{w}_i(k)} = \frac{\partial f(y_i(k))}{\partial y_i(k)} \frac{\partial y_i(k)}{\partial \underline{w}_i(k)}$. Since $y_i(k) = \underline{w}_i^T(k)\underline{x}(k)$, we get $\frac{\partial y_i(k)}{\partial \underline{w}_i(k)} = \underline{x}(k)$ and (17.24) can be expressed as:

$$\frac{\partial e_i^2(k)}{\partial \underline{w}_i(k)} = -2f'(y_i(k))\,\underline{x}(k)\,e_i(k). \tag{17.25}$$

Using this expression for $\frac{\partial e_i^2(k)}{\partial \underline{w}_i(k)}$ in the gradient descent update equation (17.17) yields the following result, known as delta learning rule

$$\begin{aligned} \underline{w}_i(k+1) &= \underline{w}_i(k) - \frac{1}{2}\mu \nabla e_i^2(k) \\ &= \underline{w}_i(k) + \mu\, f'(y_i(k))\,\underline{x}(k)\,e_i(k). \end{aligned} \tag{17.26}$$

If a unipolar sigmoidal activation function is used, we showed earlier in Section 17.2.1 that $f'(y_i(k)) = \lambda o_i(k)(1 - o_i(k))$ and thus (17.26) becomes

$$\underline{w}_i(k+1) = \underline{w}_i(k) + \mu\, \lambda o_i(k)(1 - o_i(k))\,\underline{x}(k)\,e_i(k). \tag{17.27}$$

Comparing (17.27) with Amari's generalized learning rule in (17.8), we observe that the learning signal is $r(\underline{w}_i(k), \underline{x}(k), d_i(k)) = \lambda o_i(k)(1 - o_i(k))\,e_i(k)$ where $e_i(k) = d_i(k) - o_i(k)$. That is, the learning signal in the delta rule depends on the output as well as the error signal.

17.3.2.6 Logistic Regression and LMS Rule

Logistic regression models are used in many statistical regression problems, including risk analysis and prediction in social sciences, medical, remote sensing, and other areas. In this section, we restrict our attention to the binary case. To this end, we consider the binary hypothesis testing problem in which the desired result takes two possible values $d = \{0, 1\}$ representing the class membership of a pattern. We model this problem using a single neuron with a logistic (or sigmoidal) activation function $o = f(y) = \frac{1}{1+e^{-y}}$, where $y = \underline{w}^T\underline{x}$ and $f(.)$ is the logistic function. Given a training dataset consisting of pairs of $\mathcal{X} = \{\underline{x}(k), d(k)\}_{k=1}^K$, the objective is to estimate the weight vector $\underline{w}$ of the neuron. As noted earlier, even though the logistic (sigmoid) function is nonlinear, logistic regression is nevertheless a linear classifier. Thus, the nonlinearity affects the posterior probability estimates, but does not alter the geometry of the decision surface, which remains a hyperplane in the feature space.

Given an input sample $\underline{x}$ and the weight vector $\underline{w}$, the outputs of the model can be interpreted as the posterior probabilities of class membership, i.e. $o = P(d = 1|\underline{x}; \underline{w})$ and $1 - o = P(d = 0|\underline{x}; \underline{w})$ where $o = f(\underline{w}^T\underline{x})$. Thus, we can represent the posterior distribution as a Bernoulli of the form

$$p(d|\underline{x};\underline{w}) = f(\underline{w}^T\underline{x})^d(1 - f(\underline{w}^T\underline{x}))^{1-d}. \tag{17.28}$$

Now, assuming that the training dataset $\mathcal{X}$ contains iid samples, the likelihood function for estimating the model parameters can be written as

$$\mathcal{L}(\underline{w}) = \prod_{k=1}^{K} p(d(k)|\underline{x}(k);\underline{w}) \tag{17.29}$$

Using (17.28) in the likelihood function of (17.29) and taking the logarithm gives the log-likelihood function that we maximize to find the maximum likelihood (ML) estimates of the model parameters,

$$l(\underline{w}) = \ln \mathcal{L}(\underline{w}) = \sum_{k=1}^{K}[d(k)\ln f(\underline{w}^T\underline{x}(k)) + (1 - d(k))\ln(1 - f(\underline{w}^T\underline{x}(k)))]. \tag{17.30}$$

Alternatively, we can minimize the negative of the log-likelihood function $\zeta(\underline{w}) = -l(\underline{w})$, which is commonly known as the *cross-entropy* loss function. The iterative gradient descent algorithm in (17.17) is then used to compute the ML estimate of $\underline{w}$. To do so, we first determine an instantaneous estimate of the gradient of the cross-entropy loss using (17.30). This gradient is

$$\nabla\zeta(\underline{w}(k)) = -d(k)\underline{x}(k)\frac{f'(y(k))}{f(y(k))} + (1 - d(k))\underline{x}(k)\frac{f'(y(k))}{1 - f(y(k))}. \tag{17.31}$$

Using the derivative property of the logistic function $f'(y) = f(y)(1 - f(y))$ and plugging (17.31) in (17.17) gives the updating rule

$$\underline{w}(k+1) = \underline{w}(k) + \mu(d(k) - f(y(k)))\underline{x}(k) = \underline{w}(k) + \mu e(k)\underline{x}(k), \tag{17.32}$$

where $e(k) = d(k) - f(y(k))$. This learning rule is the same as the LMS or SGD learning rule with the difference that logistic regression uses a nonlinear sigmoidal activation function instead of a linear neuron.

Important Remarks:

1. In contrast to the classifiers covered in the previous chapter, namely the Bayes and Naive Bayes classifiers, which belong to the generative category, logic regression is a *discriminative* model. These algorithms construct the posterior distribution $p(d(k) = i|\underline{x}(k))$ directly from the training data without explicitly modeling the class-conditional probability densities. In essence, they learn the decision boundary that separate the classes. Unlike the generative classifiers, logistic regression does not make any assumption about the distribution of the input data. Consequently, in cases of data model mismatch, discriminative models outperform the generative models.
2. Recall that for linear regression we used a normal distribution for the output, while for logistic regression (binary classifier) we used a Bernoulli distribution. These are special cases of the so-called Generalized Linear Models (GLMs) [20, 21], which use an *exponential family of distributions*, including normal, exponential, gamma, Poisson, and Bernoulli distributions of the general form,

$$p(d;\underline{w}) = \alpha(d)\exp\{\underline{\pi}^T(\underline{w})\underline{t}(d) - \beta(\underline{\pi}(\underline{w}))\}, \tag{17.33}$$

   where $\underline{t}(d) = [t_1(d), \cdots t_M(d)]^T$ consists of M *sufficient statistics* [22], and $\underline{\pi}(\underline{w})$ is called the *natural parameter vector*. For fixed $\underline{t}$ (usually $t(d) = d$), $\alpha(d)$, and $\beta(.)$, varying $\underline{\pi}$ generates a family of probability distributions. For example, for the Bernoulli $p(d;a) = a^d(1-a)^{1-d}$, we can easily show that $t(d) = d$, $\pi = \ln\left(\frac{a}{1-a}\right)$, $\beta(a) = \ln(1 + e^a)$ and $\alpha(d) = 1$.

17.3.2.7 Multi-Class Logistic Regression

Multi-class or categorical logistic regression [23] can be regarded as a GLM discussed in the previous remark. The network in this case consists of M output neurons with softmax activation functions to generate posterior probabilities of class membership. The training dataset $\mathcal{X} = \{\underline{x}(k), \underline{d}(k)\}_{k=1}^{K}$ consists of pairs of data samples $\underline{x}(k)$ and desired outputs $\underline{d}(k)$.

However, for multi-class logistic regression the desired output vector for patterns in class i is $\underline{d}(k) = \vec{e}_i,\ i \in [1, M]$, where $\vec{e}_i = [0 \cdots \underset{i^{th}}{1} \cdots 0]^T$ is a unit vector whose i^{th} element is one and all other elements are zero. The outputs are given by

$$o_i(k) = p(\underline{d}(k) = \vec{e}_i | \underline{x}(k); \underline{w}_i) = \frac{\exp\{\underline{w}_i^T \underline{x}(k)\}}{\sum_{m=1}^{M} \exp\{\underline{w}_m^T \underline{x}(k)\}}, \quad i \in [1, M] \tag{17.34}$$

where $\underline{w}_i$ is the weight vector of the output neuron i. Assuming that the training samples are iid, the negative log-likelihood function is

$$\zeta(\mathbf{W}) = -\sum_{k=1}^{K}\sum_{i=1}^{M} d_i(k) \ln p(\underline{d}(k) = \vec{e}_i|, \underline{x}(k), \underline{w}_i) = -\sum_{k=1}^{K}\sum_{i=1}^{M} d_i(k)\ \ln o_i(k), \tag{17.35}$$

where $\mathbf{W} = [\underline{w}_1, \cdots \underline{w}_M] \in \mathbb{R}^{N\times M}$ is the weight matrix of the network. To develop an iterative learning rule, we compute the gradient of the instantaneous loss function at iteration k with respect to the unknown weight vector $\underline{w}_j(k),\ j \in [1, M]$. It can be shown that

$$\begin{aligned} \nabla_j \zeta(\mathbf{W}(k)) = \frac{\partial \zeta(\mathbf{W}(k))}{\partial \underline{w}_j(k)} &= -d_j(k)(1 - o_j(k))\underline{x}(k) + \sum_{\substack{i=1\\ i\neq j}}^{M} d_i(k) o_j(k)\underline{x}(k) \\ &= -d_j(k)\underline{x}(k) + o_j(k)\underline{x}(k) \sum_{i=1}^{M} d_i(k) \\ &= (o_j(k) - d_j(k))\underline{x}(k) = e_j(k)\underline{x}(k), \end{aligned} \tag{17.36}$$

since $\sum_{i=1}^{M} d_i(k) = 1$. As can be seen, this learning rule is identical to the SGD learning rule for the binary logic regression case.

In many practical applications, the dimension of the input data is very large, and it is desirable to automatically identify and retain only the most discriminatory features of the data. This also produces a model with fewer parameters and often better performance. This can be accomplished by adding a penalty or regularization term to the loss function of the logistic regression to enforce a sparse solution for the parameter vectors $\underline{w}_i\ i \in [1, M]$. In this case, the loss function in (17.35) is modified to

$$\breve{\zeta}(\mathbf{W}) = \zeta(\mathbf{W}) + \lambda \sum_{i=1}^{M} \rho(\underline{w}_i), \tag{17.37}$$

where λ is the regularization parameter and $\rho(\underline{w}_i)$ is an appropriate sparsity constraint. There are many possible choices [24, 25] for this penalty term that directly determine the complexity of the optimization process. Among popular choices are: (i) Lasso or ℓ_1 norm $\rho(\underline{w}_i) = ||\underline{w}_i||_1$, (ii) ridge regression or ℓ_2 norm squared $\rho(\underline{w}_i) = ||\underline{w}_i||_2^2$, and (iii) elastic net which combines (i) and (ii) i.e. $\rho(\underline{w}_i) = \alpha||\underline{w}_i||_2^2 + (1-\alpha)||\underline{w}_i||_1$ with $0 \le \alpha \le 1$. The ℓ_1 norm constraint promotes sparse solutions by driving many coefficients exactly to zero. However, it requires specialized optimization techniques, since the objective function becomes nondifferentiable at zero, which precludes the use of gradient-based minimization methods.

In Sections 13.5.1 and 13.5.2, we used this classifier together with the l_1 norm sparsity constraint to discriminate different fabrics based on their textural features extracted using the GLCM and Gabor filter methods. Moreover, in Section 14.4.2, the same classifier was employed in conjunction with the morphological Hierarchical Texture Decomposition (HTD) method.

17.3.3 Coincidence Learning

In 1949, Donald Hebb, a neuropsychologist, postulated a mechanism for learning at the cellular level based on the principle of coincidence neural activity [1, 2]. This *coincidence learning* is now a widely accepted paradigm for many neural network and ML applications, including associative memories, autoencoders, feature extraction (e.g. PCA) and many other unsupervised learning applications.

According to Hebb's mechanism [1], *"When an axon of a cell A is near enough to excite a cell B and repeatedly or persistently takes part in its firing, some growth process or metabolic change (engrams) takes place in one or both cells such that A's efficiency as one of the cells firing B, is increased."* In short, cells that fire together wire together.

Hebb's mechanism can be implemented in neural networks using two key principles [1]: (i) if two cells (i, j) on either side of a synaptic weight, $w_{i,j}$, are activated simultaneously or synchronously, then the strength of that synapse is selectively

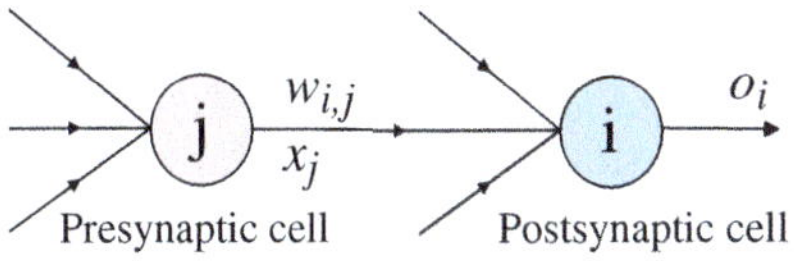

Figure 17.8 Presynaptic and postsynaptic cells and their interaction.

increased, whereas if they are activated asynchronously, then the synapse is selectively weakened or even eliminated; and (ii) causality in firing must also be taken into account. Figure 17.8 depicts presynaptic and postsynaptic cells and their connection $w_{i,j}$. The Hebbian synapse uses a time-dependent, highly localized, and strongly interactive mechanism to increase synaptic efficiency as a function of the instantaneous correlation between *presynaptic* and *postsynaptic* activities. More specifically, the change in the weight connecting presynaptic cell j to postsynaptic cell i is

$$\Delta w_{i,j} = \mu x_j o_i, \tag{17.38}$$

where x_j and o_i are the outputs of presynaptic and postsynaptic cells, respectively. As can be seen from (17.38), repeated coincidence of presynaptic and postsynaptic activities can result in overgrowth of the synaptic weight leading to saturation and numerical instability problems. In Section 19.1, we will show how these issues can be prevented by developing a modified version of this learning mechanism that uses a normalization procedure at every iteration. This modified Hebbian learning rule is then employed in linear autoencoders for feature extraction and dimensionality reduction tasks.

A similar equation can be written for all $j \in [1, N]$ presynaptic cells that connect to cell i, in which case the weight updating equation for neuron i becomes

$$\underline{w}_i(k+1) = \underline{w}_i(k) + \mu \underline{x}(k) o_i(k), \tag{17.39}$$

where $\underline{x}(k) = [x_1 \cdots x_N]^T$ is a vector of all presynaptic inputs. As can be observed from this learning rule, there is no desired signal or external teacher. Consequently, Hebbian learning is an unsupervised learning algorithm whose objective is to discover latent features directly from the input data. Compared to Amari's generalized learning rule, one can see that the learning signal in this case is $r(\underline{w}_i(k), \underline{x}(k)) = o_i(k)$.

Now, if we extend the learning rule in (17.39) to M postsynaptic cells with weight vectors $\underline{w}_i(k),\ i \in [1, M]$ and a common input vector $\underline{x}(k)$ from the presynaptic cells, we obtain

$$\mathbf{W}(k+1) = \mathbf{W}(k) + \mu \underline{o}(k) \underline{x}^T(k), \tag{17.40}$$

where $\mathbf{W}(k) = [\underline{w}_1(k) \cdots \underline{w}_M(k)]$ is the weight matrix and $\underline{o}(k) = [o_1(k), \cdots, o_M(k)]^T$ is the output vector of the network.

The choice of $\underline{o}(k)$ is application-dependent. For example, to construct a *hetero-associative memory* using the Hebbian learning rule in (17.40), the training dataset should consist of pairs of patterns $\{\underline{x}(k), \underline{o}(k)\}_{k=0}^{K-1}$ where $\underline{o}(k)$ is the pattern associated with $\underline{x}(k)$, e.g. "monkey" to "banana." The training or the storage phase of this associative memory involves recursive application of the training data pairs in one epoch. Assuming zero initial weight matrix $\mathbf{W}(0) = \mathbf{0}$ and $\mu = 1$, the final weight matrix after one training epoch is

$$\mathbf{W}(K) = \mathbf{W}^* = \sum_{k=0}^{K-1} \underline{o}(k) \underline{x}^T(k), \tag{17.41}$$

which is the sample cross-correlation matrix between $\underline{o}(k)$ and $\underline{x}(k)$.

The retrieval of a pattern associated with one of the stored patterns, e.g. $x(j)$, is perfect if the stored patterns are mutually orthonormal, i.e. $\underline{x}^T(i)\underline{x}(j) = \delta(i-j)$. To see this, let us apply the stored pattern $\underline{x}(l)$ and find the corresponding output by post-multiplying $\mathbf{W}^*$ by $\underline{x}(l)$ and using the orthonormality property

$$\underline{z}(l) = \mathbf{W}^* \underline{x}(l) = \left(\sum_{k=0}^{K-1} \underline{o}(k) \underline{x}^T(k) \right) \underline{x}(l) = \underline{o}(l), \tag{17.42}$$

which implies that this associative memory is able to perfectly recall the pattern $\underline{o}(l)$ associated with the input pattern $\underline{x}(l)$.

If the applied pattern is a noisy or perturbed version of one of the stored patterns, it is still possible to associate it with its corresponding pattern using a minimum distance criterion. That is, if the retrieved pattern is closest, in the Euclidean distance, to the stored pattern $o(j)$ then we associate the input pattern to $o(j)$ (see Example 17.3). Furthermore, even if the patterns are not orthonormal, accurate recall can often be achieved using recurrent networks [1, 2] that employ feedback mechanisms to iteratively refine the retrieved pattern via state propagation. An associative memory network for retrieval of grayscale or color images is provided in Chapter 20.

Example 17.3 *Consider three pairs of patterns:* $\underline{x}(1) = [1\,0\,0\,0]^T$, $\underline{o}(1) = [5\;\;1\;\;0]^T$, $\underline{x}(2) = [0\,1\,0\,0]^T$, $\underline{o}(2) = [-2\;\;1\;\;6]^T$, *and* $\underline{x}(3) = [0\,0\,1\,0]^T$, $\underline{o}(3) = [-2\;\;4\;\;3]^T$.

a. *Store these patterns using (17.41). Assume* $\mu = 1$.
b. *Apply a new pattern* $\underline{x}' = [.8\;-.15\;.15\;-.2]^T$ *to the trained network and find its associated output* $\underline{z}$. *Which stored pattern is closest to* $\underline{z}$ *in the Euclidean distance sense?*

Solution:

a. It can easily be observed that the patterns $\underline{x}(k)$ are orthonormal. Thus, the network is a perfect associator of the stored pairs $\underline{x}(k)$ and $\underline{o}(k)$. The weight matrix after the storage phase using (17.41) is

$$\mathbf{W}^* = \sum_{k=1}^{3} \underline{o}(k)\underline{x}^T(k) = \begin{bmatrix} 5 & -2 & -2 & 0 \\ 1 & 1 & 4 & 0 \\ 0 & 6 & 3 & 0 \end{bmatrix}$$

b. We note that pattern $\underline{x}' = [.8\;-.15\;.15\;-.2]^T$ is a perturbed version of $\underline{x}(1)$. Applying this pattern to the trained network gives $\underline{z} = \mathbf{W}^*\underline{x}' = [4\;\;1.25\;\;-.45]^T$. The Euclidean distance of this vector to the stored patterns $\underline{o}(k)$, $k = 1, 2, 3$ are $d_1 = 1.12$, $d_2 = 8.81$, and $d_3 = 7.45$. Thus, the memory generated a pattern that is closest to the stored pattern $\underline{o}(1)$.

17.3.4 Competitive Learning

Competitive learning developed by Kohonen [1, 9, 10], is another unsupervised learning algorithm that is typically used for cluster discovery and feature representation applications. The principal idea of this learning mechanism is to allow competition among a group of neurons during the training phase. At each iteration, a winner is neuron declared based on the largest response to an input pattern. The weight vector of the winning neuron is allowed to change, while those of the others (losers) remain unchanged. As a consequence, this learning mechanism is also called *winner-takes-all learning* [1]. Competitive learning extracts the underlying statistical properties of data by forming clusters directly from unlabeled data. Among other important applications of this learning rule include data compression using vector quantization, feature mapping, and also pattern classification when used in a semi-supervised setting.

Consider a single-layer network with M neurons as shown in Figure 17.9. Given unlabeled dataset $\{\underline{x}(k)\}_{k=1}^{K}$, each neuron computes a distance measure between its weight vector, $\underline{w}_i$, and the input vector, $\underline{x}(k)$. The neuron with the smallest distance is chosen as the winner. The winner selection using the Euclidean distance measure involves

$$j = \arg\min_i \|\underline{x}(k) - \underline{w}_i\|, \quad j \in [1, M]. \tag{17.43}$$

However, since $\|\underline{x}(k) - \underline{w}_i\|^2 = (\underline{x}(k) - \underline{w}_i)^T(\underline{x}(k) - \underline{w}_i) = \underline{x}^T(k)\underline{x}(k) + \underline{w}_i^T\underline{w}_i - 2\underline{w}_i^T\underline{x}(k)$, the winner selection can alternatively be made by maximizing $\underline{w}_i^T\underline{x}(k)$, provided that $\|\underline{w}_i\| = \|\underline{w}_j\|$, $\forall i \neq j$. Under this assumption, we can use

$$j = \arg\max_i \underline{w}_i^T\underline{x}(k), \quad i \in [1, M], \tag{17.44}$$

or finding the neuron with the largest response $y_i(k) = \underline{w}_i^T\underline{x}(k)$, $i \in [1, M]$. Once the winner is declared, then $\underline{x}(k)$ is assigned to cluster j whose weight vector is updated using

$$\underline{w}_j(k+1) = \underline{w}_j(k) + \alpha(k)(\underline{x}(k) - \underline{w}_j(k)), \tag{17.45}$$

where $\alpha(k)$ is the learning rate. For all other neurons $i \neq j$, $\underline{w}_i(k+1) = \underline{w}_i(k)$. As can be observed from (17.45) every time a new pattern $\underline{x}(k)$ is assigned to cluster j, the weight vector $\underline{w}_j$ is updated by moving it slightly toward the assigned pattern.

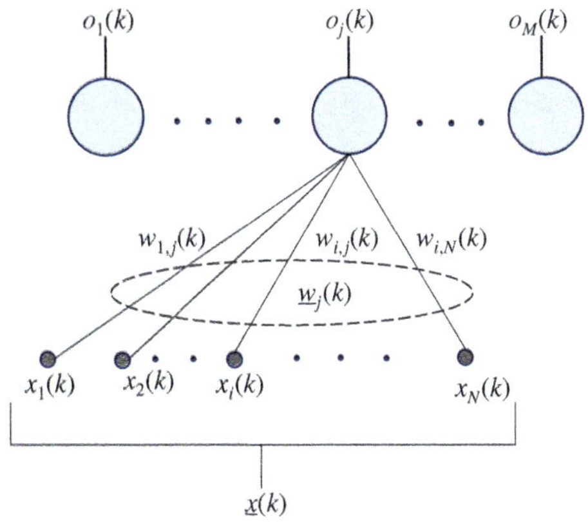

Figure 17.9 A single-layer Kohonen network.

Thus, after learning has been completed, each $\underline{w}_j$ converges to the centroid or local mean vector of all patterns assigned to cluster j. Now, given a new pattern $\underline{x}'$, the trained network assigns this pattern to cluster k if $k = \underset{j}{\text{argmax}}\, o_j$ where $o_j = f(\underline{w}_j^T \underline{x}')$ and $f(.)$ is the activation function of the neurons.

Practical Considerations and Remarks:

1. The parameter $\alpha(k)$ in the competitive learning rule (17.45) plays a role similar to the step size in the SGD method. Depending on the size of the training dataset, $\alpha(k)$ is kept close to 1 (e.g. 0.99) for several thousands iterations and is then reduced monotonically. Examples of such monotonically decreasing functions are: (i) *linear* $\alpha(k) = \alpha_0\left(1 - \frac{k-k_p}{k_q}\right)$ where α_0 is the initial learning rate, k_p is the epoch at which decay starts and k_q is the time constant of the decay; or (ii) *exponential* $\alpha(k) = \alpha_0 \exp\left(-\frac{k-k_p}{k_q}\right)$ with the same parameters as defined for the linear case.
2. The competitive learning rule in (17.45) can also be derived using the gradient descent method by using the instantaneous squared error $\xi(\underline{w}_j) = ||\underline{x}(k) - \underline{w}_j(k)||^2 = (\underline{x}(k) - \underline{w}_j(k))^T(\underline{x}(k) - \underline{w}_j(k))$ as the objective function. Then, using its gradient $\nabla\xi(\underline{w}_j) = -2(\underline{x}(k) - \underline{w}_j(k))$ in the gradient descent rule in (17.17) leads to (17.45).
3. The winner selection criterion in (17.44) assumed equal magnitudes for all the weight vectors $\underline{w}_i$. Thus, if the weight vectors are not normalized, a better criterion for winner selection is

$$j = \arg\max_i \frac{\underline{w}_i^T \underline{x}(k)}{||\underline{w}_i||\ ||\underline{x}(k)||} = \cos \angle(\underline{w}_i, \underline{x}(k)), \quad i \in [1, M], \tag{17.46}$$

   or the cosine of the angle between the two non-zero vectors $\underline{w}_i$ and $\underline{x}(k)$.
4. The convex combination strategy was suggested in [26] as an alternative approach to random weight initialization in the standard Kohonen network. Using this approach, the initial weight vectors are initialized on the unit hypersphere in N-D Euclidean space, i.e. $\underline{w}_i(0) = \frac{1}{\sqrt{N}}[1\,1 \cdots\ 1]^T,\ \ i \in [1, M]$. This approach has several advantages over the random initialization, including improved convergence, facilitates hardware implementation, and helps prevent the *stuck vector* problem, which occurs when the weight vector of a neuron does not move significantly or participates in the competition process.
5. For data compression applications, cluster centroid or weight vectors of the trained network can be used as codebook vectors to perform vector quantization of image blocks (see Section 11.3.4).
6. A trained Kohonen network can also be used as a classifier if a limited labeled data is available to assign class labels to the established clusters. This can be accomplished using a majority vote rule based on the labeled training samples of

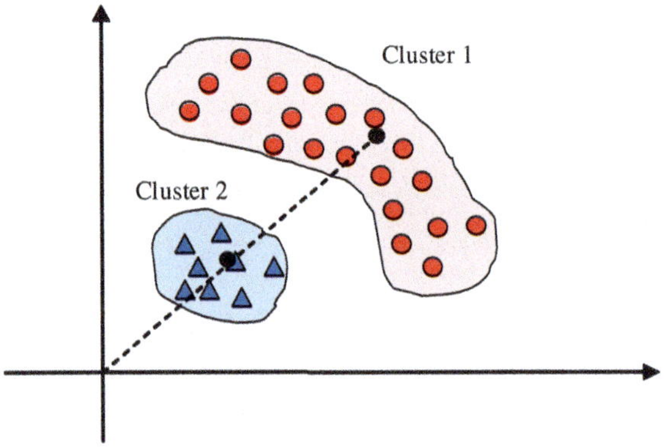

Figure 17.10 Overlapping cluster centroids.

the same class assigned to each cluster. Subsequently, the weight vectors may be refined using various learning vector quantization (LVQ) algorithms (see Section 18.3.4) based on the available labeled training data.

7. In some scenarios, clustering fails when the centroid vectors of two or more clusters overlap, as depicted in an example shown in Figure 17.10. One possible remedy is to increase the number of neurons, thereby partitioning the data into a larger number of smaller clusters.

Here, we demonstrate how competitive learning can be used to perform vector quantization (VQ) for image compression. The Lena image is used as the *training* image to establish the codebook vectors. Four different cases are considered for the number of neurons or codebook vectors, namely K = 32, 64, 128, and 256. The goal is to study the impact of varying the number of codebook vectors on the quality of the reconstructed images and the compression performance. Each network was trained using randomly selected non-overlapping 4×4 blocks taken from the image. The step size was kept fixed during the entire training at $\alpha = 0.9$. Once training is completed and the codebook vectors converge to the centroids of the corresponding clusters, the trained system can be used to determine the winning cluster for any new image block of the same size. In this case, it is only necessary to encode the index associated with the winning cluster for each block.

To reconstruct the image block-by-block, the code associated with the address of the winning cluster is first decoded at the receiver, and the corresponding codebook is reshaped into a 4×4 block and used to reconstruct the image block at that position. Repeating this process for every image block yields the entire reconstructed image. Figure 17.11a–e shows the original and reconstructed Lena images obtained using K=32, 64, 128, and 256 codebook vectors, respectively. As can be seen, increasing the number of codebook vectors significantly improves the reconstruction fidelity, while at the same time increasing the compression rate and the computational and memory requirements.

The optimum number of codebook vectors can be determined experimentally by considering the visual appearance of the reconstructed image, the compression rate versus distortion performance, and the computational and memory requirements of the VQ for K=32, 64, 128, and 256 clusters. From the reconstructed images in Figure 17.11, the corresponding SNR values are measured to be 13.8, 14.8, 15.2, and 16 dB, respectively. For the compression rate, assuming that the image is processed block-by-block from left to right and top to bottom, one only needs to encode the index or address of the winning codebook vector using $k = \log_2 K$ bits/block. Thus, the compression rate for 4×4 size blocks will be $C = k/16$, which yields $C =$ 0.3125, 0.375, 0.4375, and 0.5 bpp for K=32, 64, 128, and 256, respectively. Clearly, better compression rates can be achieved using Huffman encoding of the winning neuron indices. Based on these performance metrics and also on the visual appearance of the reconstructed images, $K = 128$ is the optimum number of codebook vectors for the cases considered.

To evaluate the generalization capability of the learned codebook vectors, the same encoding, decoding, and reconstruction procedures were applied to the Baboon image in Figure 17.12a. The reconstructed image obtained using the learned $K = 128$ codebook vectors is shown in Figure 17.12b. As can be seen, some smearing artifacts are noticeable, especially around the Baboon's nose, whiskers, and hair. These artifacts are expected because the Lena image used for training codebook vectors lacked image blocks with similar texture characteristics. Nonetheless, overall the result shows a very good generalization ability of the learned codebook vectors. The SNR for the reconstructed image is 15.1 dB.

Figure 17.11 (a) Original Lena image and (b)–(e) reconstructed Lena images for K = 32, 64, 128, and 256 codebook vectors, respectively. *Source:* [27]/John Wiley & Sons/CC BY 3.0.

17.4 Computer Projects

The computer projects in this chapter use Python libraries pandas, sklearn, datetime, time, NumPy, random, and Matplotlib.

Computer Project 17.1: In Section 9.2, we showed how the parameters of 2-D AR models with causal, semi-causal, and noncausal region of support (ROS) can be estimated using the 2-D extension of the Yule-Walker method. This computer project illustrates how this parameter estimation can be performed directly from the data without explicitly computing the covariance matrix and its inverse, which can be computationally costly, especially for large-order models. More specifically, the goal of this computer project is to use the LMS learning rule to fit a 2-D AR model with different ROS geometries to the Lena image. First, configure an ADALINE model to represent a first-order 2-D linear predictor using an AR model with NSHP ROS (see Example 17.1). Train the model using the LMS algorithm on the entire Lena image and determine the model parameters. Then, apply the 2-D predictor to the same image and compute the statistics

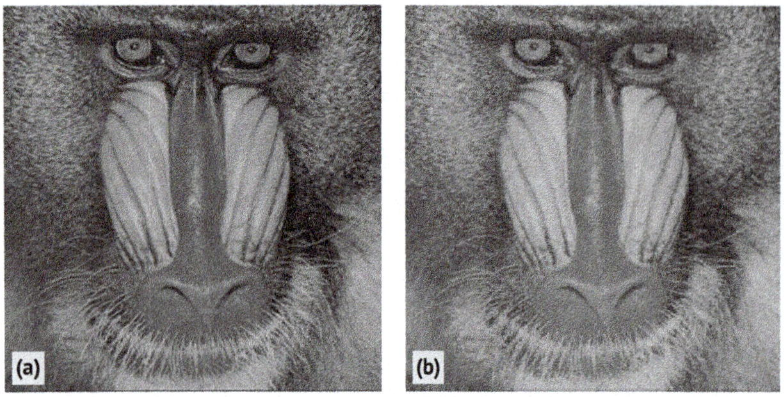

Figure 17.12 (a) Original Baboon image (*Source:* [27]/John Wiley & Sons/CC BY 3.0), and (b) reconstructed Baboon image using $K = 128$ codebook vectors learned from the Lena image. (The color rendition of this image is available to view at www.wiley.com/go/Azimi449.)

(mean and variance) of the prediction error $e(m, n) = y(m, n) - \hat{y}(m, n)$ where $\hat{y}(m, n)$ is the predicted value of the pixel (m, n). Compare your results with those presented in Section 9.2.4 using the Yule-Walker approach and state your observations. Repeat the same experiment using first-order 2-D AR models with semicausal and noncausal ROS. Provide a detailed discussion of your results and findings, as well as the strengths and limitations of this LMS-based parameter estimation method compared to the standard Yule-Walker-based methods in Chapter 9.

Computer Project 17.2: The objective of this computer project is to train a categorical logistic regression classifier to classify ten different handwritten digits in the MNIST dataset. Please refer to Computer Project 16.1 for a description of the training and testing datasets. Train the classifier on the training dataset using the cross-entropy loss function without and with a sparsity constraint. Follow the same experimental setup as in Computer Project 16.1 but perform at least 20 random trials to compute the mean classification rate for each class. Using these rates, construct the confusion matrix of this classifier. Provide a detailed discussion on your experimental results and highlight important observations regarding classification performance, and the effects of sparsity regularization. Discuss the principal strengths and weaknesses of categorical logistic regression for handwritten digit recognition.

Computer Project 17.3: Repeat Computer Project 11.5 on image data compression using competitive learning (see Section 17.3.4) instead of the K-means cluster learning for codebook design and VQ.

Computer Project 17.4: Repeat Computer Project 12.3 for segmentation of the lung image using competitive learning (see Section 17.3.4) instead of hard and soft (fuzzy) clustering methods.

17.5 Problems

Problem 17.1 Configure a McCulloch–Pitts neuron model as NOT, NOR, and NAND gates.

Problem 17.2 Configure a linear neuron to implement a non-recursive or finite impulse response (FIR) filter of order N given by the convolution sum $y(n) = \sum_{k=0}^{N} b_k x(n-k)$ where $x(n)$ is the input signal, $y(n)$ is the output signal, and b_k are the filter coefficients.

Problem 17.3 Configure a linear neuron to implement a 2-D linear predictor $y(m, n) = a_{10}y(m-1, n) + a_{01}y(m, n-1) + a_{11}y(m-1, n-1) + e(m, n)$ where $x(m, n)$ is the input image, $y(m, n)$ is the output image, and a_{ij} are the linear predictor coefficients.

Problem 17.4 Configure a neuron model to fit an nth-order polynomial to data using

$$y = a_0 + a_1 x + a_2 x^2 + a_3 x^3 + \cdots + a_n x^n + e$$

where e is the error in the modeling.

Problem 17.5 Derive the Wiener-Hopf solution for the optimum weight vector, $\underline{w}^*$, of the ADALINE model in Section 17.3.2 by minimizing the mean squared error (MSE) $J(\underline{w}) = E[(d - y)^2]$ between the desired output d and the aggregated input $y = \underline{w}^T \underline{x}$ where $E[.]$ is the ensemble average or expectation operation. Assume the ensemble correlation matrix $\mathbf{R}_{xx} = E[\underline{x}\underline{x}^T]$ and cross-correlation vector $\underline{r}_{xd} = E[\underline{x}d]$ are known.

Problem 17.6

a. In Problem 17.5, show that the estimation error $e_o = d - \underline{w}^{*T}\underline{x}$ is orthogonal to the data, i.e. $E[\underline{x}e_o] = E[\underline{x}(d - \underline{w}^{*^T}\underline{x})] = 0$. Explain the importance of this property.
b. Is the same orthogonality property valid for the LS solution? If yes, derive it. Compare the results in parts (a) and (b) and their interpretations.

Problem 17.7 To guarantee uniqueness and improve the stability of the LS solution, a *regularization term* can be added to the objective function, i.e.

$$J(\underline{w}) = (\underline{d} - \mathbf{X}\underline{w})^T(\underline{d} - \mathbf{X}\underline{w}) + \lambda||\underline{w}||^2.$$

Find the solution of this *regularized (or constrained) LS* and investigate its relation to the LS solution in Section 17.3.2.

Problem 17.8 Consider the problem in Example 17.1 where we want to build a linear model to predict the GPA of college students based on their SAT and ACT scores. Assuming that the correlation and cross-correlation functions of the data are given by $r_{xx}(n) = E[x(k)x(k-n)] = \rho^{|n|}$ and $r_{xd}(n) = E[x(k)d(k-n)] = \alpha^n$, respectively, and further $w_0 = 0$, $\rho = 0.9$ and $\alpha = 0.5$, compute the parameters w_1 and w_2 of the model.

Problem 17.9 Assume that we modify the objective function of the LMS rule from $J(\underline{w}) = e^2(k) = (d(k) - y(k))^2$ where $y(k) = \underline{w}^T(k)\underline{x}(k)$ to $J(\underline{w}) = e^2(k) + \gamma||\underline{w}(k)||^2$ where γ is a proportionality constant. Derive the modified version of the LMS updating equation for this objective function and comment on your result.

Problem 17.10 Determine the conditions under which the batch gradient descent in (17.18) gives the same result as the LS solution.

Problem 17.11 A two-class classifier must be designed to separate four patterns $\underline{x}_1 = [-1\ 2]^T$, $\underline{x}_2 = [-2\ -2]^T$, $\underline{x}_3 = [1\ -1]^T$, $\underline{x}_4 = [-3\ 1]^T$ where $\underline{x}_1, \underline{x}_4 \in C_1$ with desired signal $d = -1$ and $\underline{x}_2, \underline{x}_3 \in C_2$ with desired signal $d = +1$. Use the Perceptron learning rule to design this classifier by applying these patterns in order of their indices. Assume zero initial weights and bias.

Problem 17.12 A single neuron with bipolar sigmoidal activation function is trained using the delta rule for two 3-D patterns $\underline{x}_1 = [2\ 0 - 1]^T$ and $\underline{x}_2 = [1 - 2 - 1]^T$ with desired values of $d_1 = -1$ and $d_2 = 1$, respectively. Assume that the parameter of the bipolar sigmoidal function is $\lambda = 1$, step size is $\mu = 0.2$, and initial weight vector is $\underline{w}(0) = [1\ 0\ 1]^T$.

a. Find the weight vector after the training is completed.
b. Is the trained network a linear or a nonlinear classifier? Explain.
c. Test the trained network using two arbitrary patterns and comment on the results.

Problem 17.13 Repeat Problem 17.12 using the LMS learning rule. Assume linear neurons.

Problem 17.14 Consider a single-layer network with M neurons and input vector $\underline{x} \in \mathbb{R}^N$. Assuming that the delta learning rule is used for each neuron, write the updating equation for the entire network using the weight matrix $\mathbf{W} = [\underline{w}_1, \cdots, \underline{w}_M]$, where $\underline{w}_i$ is the weight vector of the ith neuron using

a. linear activation functions,
b. sigmoidal activation functions.

Problem 17.15 Consider a radial basis function (RBF) neuron $z = k(\underline{x})$ with Gaussian kernel $k(\underline{x}) = \exp\{-||\underline{x} - \underline{c}||^2/2\}$ where $\underline{c}$ is an unknown unit norm vector, i.e. $||\underline{c}|| = 1$ representing the receptive field center (or mean) of the Gaussian kernel. Given a training dataset containing pairs of input-output $\{\underline{x}(k), z(k)\}_{k=1}^{N}$, devise a method to find the unknown center of the kernel. Note that $\underline{c}$ plays a similar role as the weight vector of the RBF neuron.

Problem 17.16

a. Derive the categorical logistic regression weight updating equation in (17.36).
b. Repeat part (a) by adding a ridge regression sparsity constraint term for $\rho(\underline{w}_i) = ||\underline{w}_i||_2^2$ to the cost function.

Problem 17.17 Assuming a linear neuron model $o = \underline{w}^T \underline{x}$, determine the convergence property of the Hebbian learning rule. Is this learning stable?

Problem 17.18 Assume that in the Hebbian learning rule, the stored patterns $\underline{x}(k)$ are not orthogonal. Find the best estimates of the weights for a linear associative memory by minimizing the following cost function with respect to the weight matrix **W**

$$J(W) = \frac{1}{K}\sum_{k=0}^{K-1} ||\underline{o}(k) - \mathbf{W}\underline{x}(k)||^2 = \frac{1}{K}\sum_{k=0}^{K-1} ||\underline{o}(k) - \hat{\underline{o}}(k)||^2.$$

Problem 17.19 Consider the competitive learning rule in (17.45). Prove the convergence to the centroid of each cluster. What are the factors that influence the convergence behavior?

Problem 17.20 Derive a modified competitive learning rule using the Mahalanobis distance measure given by $d_M(\underline{x}, \underline{w}) = (\underline{x} - \underline{w})^T \mathbf{\Sigma}(\underline{x} - \underline{w})$, where $\mathbf{\Sigma}$ is a positive definite (PD) matrix, which can be decomposed as $\mathbf{\Sigma} = \mathbf{S}^T\mathbf{S}$. Comment on the result and differences between this and the original version.

References

1 S. Haykin, "*Neural Networks and Learning Machines*," Prentice-Hall, 3rd Edition, 2009.
2 J. M. Zurada, "*Introduction to Artificial Neural Networks*," West Pub. Co. 1992.
3 D. Rumelhart, G. E. Hinton, and R. J. Williams, "Learning representations by back-propagating errors," *Nature*, vol. 323, pp. 533–536, 1986.
4 Y. LeCun and Y. Bengio, "Convolutional networks for images, speech, and time series," *The Handbook of Brain Theory and Neural Networks*, vol. 3361, p. 10, 1995.
5 I. Goodfellow, Y. Bengio, and A. Courville, "*Deep Learning*," MIT Press, 2016.
6 W. Jeon, G. Ko, J. Lee, H. Lee, D. Ha, and W. W. Ro, "Chapter six - Deep learning with GPUs," *Advances in Computers*, vol. 122, pp. 167–215, 2021.
7 F. Rosenblatt, "The Perceptron: A probabilistic model for information storage and organization in the brain," *Psychological Review*, vol. 65, pp. 386–408, 1958.
8 T. Mitchell, "*Machine Learning*," McGraw Hill, 1997.
9 T. Kohonen, "Self-organized formation of topologically correct feature maps," *Biological Cybernetics*, vol. 43, pp. 59–69, 1982.
10 T. Kohonen, "Essentials of the self-organizing map," *Neural Networks*, vol. 37, pp. 52–65, 2013.
11 S. Amari, "Mathematical foundations of neurocomputing," *Proceedings of the IEEE*, vol. 78, pp. 1443–1463, 1990.
12 S. Amari, "Mathematical theory of neural learning," *New Generation Computing*, vol. 8, pp. 281–294, 1991.
13 H. Robbins and S. Monro, "Stochastic approximation method," Annals of *Mathematical Statistics*, vol. 22, pp. 400–407, 1951.
14 B. Widrow and M. A. Lehr, "30 years of adaptive neural networks: Perceptron, Madaline, and backpropagation," *Proceedings of the IEEE*, vol. 78, pp. 1415–1442, 1990.
15 G. H. Golub and C. F. Van Loan, "*Matrix computation*," John Hopkins University Press, 1996.
16 K. S. Shanmugan and A. M. Breipohl, "*Random Signals-detection, Estimation and Data analysis*," Wiley, 1991.
17 B. Widrow and S. D. Stearns, "*Adaptive Signal Processing*," Prentice-Hall, 1st Edition, 1985.
18 S. Haykin, "*Adaptive Filter Theory*," Prentice-Hall, 4th Edition, 2001.

19 D. E. Rumelhart and J. L. McClelland, "*Parallel Distributed Processing: Explorations in the Microstructure of Cognition Volume 1: Foundation*," MIT Press, 1986.

20 J. Nelder and R. Wedderburn, "Generalized linear models," *Journal of the Royal Statistical Society. Series A*, vol. 135, pp. 370–384, 1972.

21 H. Madsen and P. Thyregod, "*Introduction to General and Generalized Linear Models*," Chapman and Hall, 2011.

22 L. L. Scharf, "*Statistical Signal Processing: Detection, Estimation, and Time Series Analysis*," Addison Wesley, 1991.

23 C. Bishop, "*Pattern Recognition and Machine Learning*," Springer, 2006.

24 C. Fuetterer, M. Nalenz, T. Augustin, and R. M. Pfeiffer, "A powerful penalized multinomial logistic regression approach," *Computational Statistics*, vol. 40, pp. 4565–4587, 2025.

25 G. C. Cawley, N. Talbot, and M. Girolami, "Sparse multinomial logistic regression via Bayesian L1 regularization," *Proceedings of Advances in Neural Information Processing Systems*, vol. 19, 2006.

26 R. Hecht-Nielsen, "*Neurocomputing*," Addison-Wesley, 1990.

27 Y. S. Juang, L. T. Ko, J. E. Chen, Y. S. Shieh, T. Y. Sung, and H. C. Hsin. "Histogram modification and wavelet transform for high performance watermarking," *Mathematical Problems in Engineering*, 164869, 2012. https://doi.org/10.1155/2012/164869.

18

Modern Image Classification-Layered Machines

The learning machines discussed in Chapter 17 consist of a single layer of neurons, which have many limitations in pattern classification applications. More specifically, we showed that a single-layer Perceptron network is suitable only for solving linearly separable problems and is therefore incapable of capturing the complex input–output relationships that characterize many real-world pattern discrimination problems.

However, the human brain's unmatched capability of making difficult contextual-based decisions, creative thinking, and adaptability to changing conditions arise from its complex, deeply layered, and highly interconnected hierarchical organization of neurons with trillions of synaptic connections. Such layered neural structures enable the extraction of pattern attributes, or feature representations, at various levels of complexity. The level of abstraction of these attributes generally increases with the depth of the network. That is, each layer maps its input representation to a higher level of abstraction. The combination of these higher-level feature representations enables complex decision-making to be performed efficiently at the final layers. Consequently, a layered network is capable of generating its own features through learning and contact with the environment without requiring handcrafted feature extraction techniques.

Studies [1] conducted in recent decades have revealed that neurons in the visual cortex respond selectively to localized features corresponding to edges with different orientations. The edge orientation sensitivity is provided by the *simple* cells, whereas the *complex* cells provide a degree of invariance to rotation, translation, skewness, and distortion. Higher cortical layers then combine these primitive responses to form more complex representations corresponding to 2-D and 3-D shapes, surfaces, and objects.

The hierarchical structure of deep convolutional neural networks (CNNs) [2–5] is neurobiologically inspired by the functionalities of *simple* and *complex* cells in the visual cortex. The learned layered representations facilitate decision-making tasks without requiring an external mechanism for feature extraction. Deep-layered networks also provide some level of invariance to translation, rotation, scaling, illumination, and distortion (e.g. warping and elastic deformation), which are essential requirements in many practical image recognition applications. However, increasing the depth of a network also increases the number of weights, or trainable parameters that must be estimated during the training. This, in turn, requires larger labeled (supervised) training dataset, which may be difficult or impossible to collect in certain settings. In such data-scarce situations, the trained network may fail to generalize effectively to previously unseen data.

Building on the neuron models and learning algorithms presented in the previous chapter, this chapter introduces three feedforward neural network architectures and learning paradigms that are widely used in image processing and pattern recognition tasks. These neural network architectures include the multilayer Perceptron (MLP), also known as the back-propagation neural network (BPNN) [6, 7], the CNN [2–5], and the self-organized feature maps (SOFMs) [8, 9]. The first two networks use variants of the delta learning rule, whereas the SOFM uses unsupervised competitive learning.

18.1 Multilayer Perceptron Neural Networks

Recall from Chapter 17 that each neuron in a single-layer network forms a hyperplane discriminant function, which is suitable only for solving linearly separable pattern classification problems. Thus, a single-layer feedforward neural network is incapable of generating complex decision boundaries needed for any practical pattern classification. During the 1960s and 1970s, many researchers attempted to extend the delta learning rule to multilayer Perceptron (MLP) neural networks with nonlinear activation functions, but without major breakthrough. This was mainly due to the lack of a mechanism to provide desired or target values for all the *hidden-layer* neurons that are not exposed to the outside world. However, in the 1980s, several researchers independently developed methods for training MLP neural networks by generalizing the delta rule and using the *error back-propagation* mechanism [6], which enabled training to be implemented on all layers. Thus, MLP networks are also called BPNNs. This development marked the beginning of a new era in machine learning and artificial intelligence (AI), and opened up many important applications of these models in image processing, among other areas.

Using a two-layer MLP, convex polygon-type decision boundaries of arbitrary shape can be created through appropriate choices of the number of neurons in the hidden and output layers. The convex polygons are formed in the output layer as a result of logical combinations of the hyperplanes generated by the hidden-layer neurons. In a three-layer neural network, the output layer neurons receive as input a group of convex polygons; hence, their logical combinations need not be convex. With proper choices of the number of neurons and weights, it is possible to form highly complex decision boundaries. In other words, a three-layer BPNN with an appropriate topology, is sufficient to solve a wide range of pattern classification or function approximation problems. However, the main caveat is that MLP networks still require hand-crafted feature-extraction subsystems to extract features from segmented regions before pattern classification can be performed.

We begin our discussion of MLP neural networks by presenting a simple example that illustrates the generation of decision boundaries of arbitrary shapes by using layered machines.

Example 18.1 *Consider the two-layer feedforward network shown in Figure 18.1a. Determine the decision region formed by this two-layer network when all neurons have bipolar hard-limited (BHL) activation functions. Assume that the network has already been trained and the first-layer weight matrix and second-layer weight vector are*

$$\mathbf{W}^{(1)} = \begin{bmatrix} 1 & 0 & 1 \\ -1 & 0 & -2 \\ 0 & 1 & 0 \\ 0 & -1 & -3 \end{bmatrix} \text{ and } \underline{w}^{(2)} = [1\ 1\ 1\ 1\ 3.5], \text{ respectively.}$$

Solution: Let us define the input vector to the network as $\underline{x} = [x_1\ \ x_2 - 1]^T$, the output vector of the first hidden layer as $\underline{o}^{(1)} = [o_1\ o_2\ o_3\ o_4 - 1]^T$, where the superscript represents layer index, and the single output of the network as o_5. The last elements of the vectors $\underline{x}$ and $\underline{o}^{(1)}$ correspond to the bias terms. For the BHL activation function $f(x) = sgn(x)$, the output of neuron i in the first layer is $o_i^{(1)} = sgn(\underline{w}_i^T \underline{x}),\ i \in [1, 4]$, where $\underline{w}_i^T$ is the ith row of matrix $\mathbf{W}^{(1)}$ matrix or the weight vector for the neuron i in the first layer. Thus, the output vector of the hidden layer is given by $\underline{o}^{(1)} = [sgn(x_1 - 1)\ sgn(-x_1 + 2)\ sgn(x_2)\ sgn(-x_2 + 3)\ \ -1]^T$.

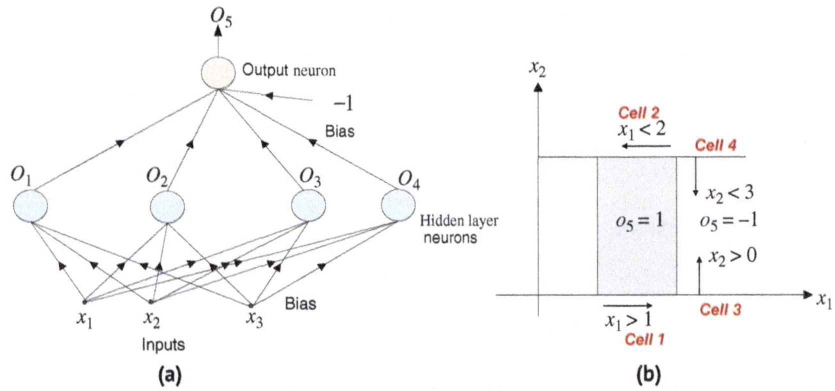

Figure 18.1 (a) Two-layer MLP network of Example 18.1, and (b) decision region generated by the network.

Each element of $\underline{o}^{(1)}$ corresponds to a linear decision boundary. That is, each neuron in the first layer divides the (x_1, x_2)-plane into two half-planes as shown in Figure 18.1b. The half-plane corresponding to an output of +1 (i.e. neuron firing) is indicated by an arrow in this figure. For the second layer, the output is $o_5 = sgn(o_1 + o_2 + o_3 + o_4 - 3.5)$. Consequently, due to the choice of the bias term −3.5, the output neuron fires only when all its inputs are +1. Thus, this network partitions the 2-D feature space into two decision regions, where the positive region (e.g., for class C_2) corresponds to the interior of the rectangle shown in Figure 18.1b, while the negative region (e.g., for class C_1) is everywhere outside of this rectangle. This example illustrates the fact that a two-layer network with an appropriate topology is capable of generating convex polygon decision regions.

18.1.1 Generalized Delta Rule and Error Back-Propagation

The *generalized delta learning rule* is the extension [6] of the original delta rule presented in Section 17.3.2 for the supervised training of multilayer feedforward networks. Although it was originally developed for MLP networks, it provides a powerful framework for training a variety of other layered machines, including CNNs, albeit with some necessary modifications. This section presents the generalized delta learning rule together with the details of error back-propagation, as applied to a general three-layer MLP; since such networks are capable of generating arbitrarily complex decision regions.

Consider the MLP network shown in Figure 18.2, which consists of N inputs, K neurons in the first hidden layer, L neurons in the second hidden layer and M neurons in the output layer for an M-class pattern classification problem. The superscript is used to represent the layer index. In addition, it is assumed that all neurons in the network have the same activation function $f(.)$. Training an MLP using the generalized delta rule involves two passes through the network, namely the *forward pass and the backward pass.* The operations performed in each pass are outlined next.

1. **Forward Pass**
   Given the training dataset $\mathcal{X} = \{\underline{x}(p), \underline{d}(p)\}_{p=1}^{P}$, where $\underline{d}(p)$ is the desired vector corresponding to the input pattern $\underline{x}(p)$, select a training sample $\underline{x}(p)$ and propagate it through the layers of neurons in the forward direction and find the outputs of the final layer. The aggregated activation and output of neuron i in the first hidden layer are

$$y_i^{(1)}(p) = \sum_{j=1}^{N+1} w_{i,j}^{(1)}(p)x_j(p), \quad \text{and} \quad o_i^{(1)}(p) = f(y_i^{(1)}(p)), \ \forall i \in [1, K], \tag{18.1}$$

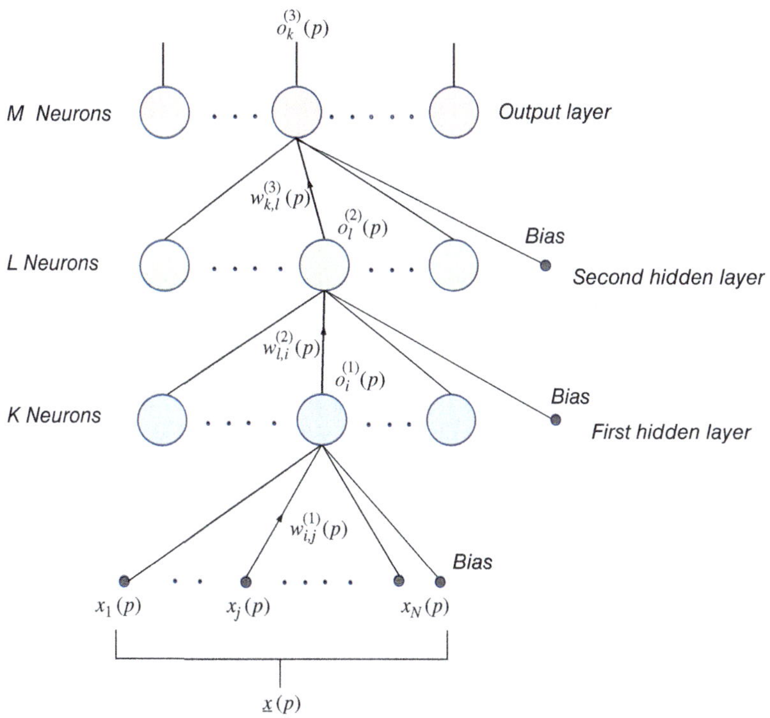

Figure 18.2 A three-layer feedforward MLP.

respectively, where $w_{i,j}^{(1)}$ is the weight connecting input j to neuron i in the first hidden layer, $w_{i,N+1}^{(1)} = b_i^{(1)}$ is the bias with input $x_{N+1}(p) = 1$. Similar equations can be written for neuron l in the second hidden layer and neuron k in the output layer with incoming weights $w_{l,i}^{(2)}$ and $w_{k,l}^{(3)}$, respectively (see Figure 18.2).

The actual outputs $o_k^{(3)}(p)$ of the network are then compared with the corresponding desired outputs $d_k(p)$, producing the prediction error $e_k^{(3)}(p) = d_k(p) - o_k^{(3)}(p), \forall k \in [1, M]$. Now, the objective of learning is to adjust all the weights and biases in the network so that the actual outputs are as close as possible, according to a suitable measure, to the desired outputs for all output neurons and all training samples. The performance measure used here is the sum squared error (SSE) over all the output neurons,

$$\ell(\mathcal{W}, p) = \frac{1}{2} \sum_{k=1}^{M} \left(d_k(p) - o_k^{(3)}(p) \right)^2, \tag{18.2}$$

where $\mathcal{W}$ denotes the set of all network parameters. The performance measure in (18.2) is computed for every training sample $\underline{x}(p)$. In contrast, the aggregated performance measure is the average SSE over all the training samples,

$$\mathcal{L}(\mathcal{W}) = \frac{1}{2P} \sum_{p=1}^{P} \sum_{k=1}^{M} \left[e_k^{(3)}(p) \right]^2 = \frac{1}{P} \sum_{p=1}^{P} \ell(\mathcal{W}, p). \tag{18.3}$$

Minimization of $\ell(\mathcal{W}, p)$ using the delta rule leads to iterative online updates of the network parameters after the presentation of each training sample. In contrast, minimization of $\mathcal{L}(\mathcal{W})$ requires updating the weights only after all training samples have been presented, that is, after one training epoch. The latter batch approach generally incurs higher computational and memory requirements and is therefore less attractive for large datasets. Consequently, online or mini-batch gradient-based learning is often preferred in practice. Using the iterative approach the weights are updated in proportion to the gradient of $\ell(\mathcal{W}, p)$ with respect to the network parameters. Other choices of loss functions will be discussed later in Section 18.2.1 in the context of CNN training.

2. **Backward Pass and Weight Updating**

The weight-updating equation for the output layer weights can be derived by computing the gradient of $\ell(\mathcal{W}, p)$ with respect to the incoming weight $w_{k,l}^{(3)}(p)$, $k \in [1, M]$ of the output layer neurons. Using the chain rule, we get

$$\nabla \ell(\mathcal{W}, p) = \frac{\partial \ell(\mathcal{W}, p)}{\partial w_{k,l}^{(3)}(p)} = -e_k^{(3)}(p) \frac{\partial o_k^{(3)}(p)}{\partial w_{k,l}^{(3)}(p)} = -e_k^{(3)}(p) \frac{\partial f(y_k^{(3)}(p))}{\partial y_k^{(3)}(p)} \frac{\partial y_k^{(3)}(p)}{\partial w_{k,l}^{(3)}(p)}, \quad k \in [1, M]$$

where $\frac{\partial y_k^{(3)}(p)}{\partial w_{k,l}^{(3)}(p)} = o_l^{(2)}(p)$. Thus, the weight adjustment $\Delta w_{k,l}^{(3)}(p)$ for the output layer weights is

$$\Delta w_{k,l}^{(3)}(p) = -\mu \nabla \ell(\mathcal{W}, p) = \mu e_k^{(3)}(p) f'(y_k^{(3)}(p)) o_l^{(2)}(p), \quad \forall k \in [1, M] \text{ and } l \in [1, L].$$

For a unipolar sigmoidal activation with $\lambda = 1$, we have $f'(y_k^{(3)}(p)) = \left(1 - o_k^{(3)}(p)\right) o_k^{(3)}(p)$, in which case, this weight-updating equation becomes

$$w_{k,l}^{(3)}(p+1) = w_{k,l}^{(3)}(p) + \mu e_k^{(3)}(p)\left(1 - o_k^{(3)}(p)\right) o_k^{(3)}(p) o_l^{(2)}(p).$$

Defining the output-layer learning signal, $r_k^{(3)}(p) = f'(y_k^{(3)}(p)) e_k^{(3)}(p) = e_k^{(3)}(p)\left(1 - o_k^{(3)}(p)\right) o_k^{(3)}(p)$, the updating equation for the weights of the output layer can be written compactly as:

$$w_{k,l}^{(3)}(p+1) = w_{k,l}^{(3)}(p) + \mu r_k^{(3)}(p) o_l^{(2)}(p).$$

Clearly, a problem arises when trying to apply the same procedure to the hidden-layer neurons, since desired values are not available for these neurons. The backward pass provides a mechanism for propagating error derivatives from the output layer to the hidden layers, thereby producing surrogates error signals for the hidden-layer neurons. To find the weight update equation for the second hidden-layer neurons, we compute the gradient of $\ell(\mathcal{W}, p)$ with respect to the weights $w_{l,i}^{(2)}(p)$. Using the chain rule, we get

$$\frac{\partial \ell(\mathcal{W}, p)}{\partial w_{l,i}^{(2)}(p)} = \sum_{k=1}^{M} e_k^{(3)}(p) \frac{\partial e_k^{(3)}(p)}{\partial w_{l,i}^{(2)}(p)} = -f'(y_l^{(2)}(p))o_i^{(1)}(p) \sum_{k=1}^{M} w_{k,l}^{(3)}(p) f'(y_k^{(3)}(p)) e_k^{(3)}(p)$$

However, since $r_k^{(3)}(p) = f'(y_k^{(3)}(p))e_k^{(3)}(p)$, the weight adjustment term becomes,

$$\Delta w_{l,i}^{(2)}(p) = \mu f'(y_l^{(2)}(p)) o_i^{(1)}(p) e_l^{(2)}(p), \quad l \in [1, L]$$

where the error signal $e_l^{(2)}(p) = \sum_{k=1}^{M} w_{kl}^{(3)}(p) r_k^{(3)}(p)$ in the output of the second hidden-layer neuron is obtained by back-propagating the learning signals $r_k^{(3)}(p),\ k \in [1, M]$ at the output layer through the weights, $w_{k,l}^{(3)}(p),\ k \in [1, M]$. This enables the updating of the weights of the second layer. Now, if we define the learning signal of the second layer by $r_l^{(2)}(p) = f'(y_l^{(2)}(p)) \sum_{k=1}^{M} w_{k,l}^{(3)}(p) r_k^{(3)}(p) = \left(1 - o_l^{(2)}(p)\right) o_l^{(2)}(p) e_l^{(2)}(p)$, then the updating equation of this layer simplifies to

$$w_{l,i}^{(2)}(p+1) = w_{l,i}^{(2)}(p) + \mu r_l^{(2)}(p) o_i^{(1)}(p).$$

To find the weights of the first layer, we repeat a similar procedure. This yields

$$\begin{aligned}\frac{\partial \ell(\mathcal{W}, p)}{\partial w_{i,j}^{(1)}(p)} &= -\sum_{l=1}^{L} \sum_{k=1}^{M} e_k^{(3)}(p) f'(y_k^{(3)}(p)) w_{k,l}^{(3)}(p) f'(y_l^{(2)}(p)) w_{l,i}^{(2)}(p) f'(y_i^{(1)}(p)) x_j(p) \\ &= -f'(y_i^{(1)}(p)) x_j(p) \sum_{l=1}^{L} r_l^{(2)}(p) w_{l,i}^{(2)}(p) = -f'(y_i^{(1)}(p)) x_j(p) e_i^{(1)}(p),\end{aligned}$$

where the error signal $e_i^{(1)}(p) = \sum_{l=1}^{L} r_l^{(2)}(p) w_{l,i}^{(2)}(p)$ for the first hidden layer is obtained by back-propagating the learning signals $r_l^{(2)}(p),\ l \in [1, L]$ through the weights of the second hidden-layer $w_{l,i}^{(2)}(p)$. Now, if we define the learning signal for the first hidden-layer $r_i^{(1)}(p) = f'(y_i^{(1)}(p)) \sum_l w_{l,i}^{(2)}(p) r_l^{(2)}(p)$. Then, the weight-updating equation for the first layer becomes

$$w_{i,j}^{(1)}(p+1) = w_{i,j}^{(1)}(p) + \mu r_i^{(1)}(p) x_j(p), \quad i \in [1, K].$$

The following step-by-step algorithm summarizes the entire MLP training procedure using the generalized delta learning rule.

1. **Initialization:** Choose the learning rate μ, the maximum allowable error $\varepsilon_{\max}$ for the stopping criterion, and initialize all the weights to small random values.
2. **Forward Pass:** Propagate the training pattern $\underline{x}(p)$ through the network in a layer-by-layer feedforward fashion using the forward pass equations and compute the actual network outputs, $o_k^{(3)}(p), \forall k \in [1, M]$.
3. **Error Computation:** Compute the SSE, $\ell(\mathcal{W}, p) = \frac{1}{2} \sum_{k=1}^{M} \left(d_k(p) - o_k^{(3)}(p)\right)^2$.
4. **Error Back-propagation:** Compute the learning signals for all layers in a top-down fashion, starting from the output layer using,

$$\begin{aligned} r_k^{(3)}(p) &= e_k^{(3)}(p)\left(1 - o_k^{(3)}(p)\right) o_k^{(3)}(p), \quad \forall k \in [1, M], \\ r_l^{(2)}(p) &= \left(1 - o_l^{(2)}(p)\right) o_l^{(2)}(p) \sum_{k=1}^{M} w_{k,l}^{(3)}(p) r_k^{(3)}(p), \quad \forall l \in [1, L], \\ r_i^{(1)}(p) &= \left(1 - o_i^{(1)}(p)\right) o_i^{(1)}(p) \sum_{l=1}^{L} w_{l,i}^{(2)}(p) r_l^{(2)}(p), \quad \forall i \in [1, K]. \end{aligned} \tag{18.4}$$

5. **Weight Updating:** Update the weights of all layers using,

$$\begin{aligned} w_{k,l}^{(3)}(p+1) &= w_{k,l}^{(3)}(p) + \mu r_k^{(3)}(p) o_l^{(2)}(p), \quad \forall k \in [1, M], \\ w_{l,i}^{(2)}(p+1) &= w_{l,i}^{(2)}(p) + \mu r_l^{(2)}(p) o_i^{(1)}(p), \quad \forall l \in [1, L], \\ w_{i,j}^{(1)}(p+1) &= w_{i,j}^{(1)}(p) + \mu r_i^{(1)}(p) x_j(p), \quad \forall i \in [1, K]. \end{aligned} \tag{18.5}$$

6. **Iteration and Error Check (Stopping Criterion):** If $p < P$ increment p and return to Step 2; otherwise, check $\ell(\mathcal{W}, p) < \varepsilon_{max}$; if yes, terminate the training; if not, restart from $p = 1$ and begin a new training epoch from Step 2.

18.1.1.1 MLP and Bayes Classification

Under certain idealized conditions, the performance of an MLP network can approach that of the Bayes classifier. To establish this result, assume that the training data samples $\{\underline{x}(p), \underline{d}(p)\}_{p=1}^{P}$ are realizations of random vectors $\underline{X}$ and $\underline{D}$. Furthermore, assume that the MLP is able to reach the global minimum of the error surface when P is sufficiently large. In this case, the MLP minimizes the mean square error (MSE), $\varepsilon = E[||\underline{D} - \hat{\underline{D}}||^2]$, where $\hat{\underline{D}}$ is the estimate of $\underline{D}$ generated at the output of the MLP, i.e. $\underline{O} = \hat{\underline{D}}$. It is well known [10] that the minimum MSE solution is the conditional mean estimator, i.e. $\underset{\underline{O}}{\text{argmin}}\, E[||\underline{D} - \underline{O}||^2] = E[\underline{D}|\underline{X}]$. Thus, given a pattern $\underline{x}$, this estimator produces

$$\underline{o} = \hat{\underline{d}} = E[\underline{D}|\underline{X} = \underline{x}] = \sum_i \underline{d}(i) p(\underline{D} = \underline{d}(i)|\underline{X} = \underline{x}) \tag{18.6}$$

where $p(\underline{D} = \underline{d}(i)|\underline{X} = \underline{x})$ is the conditional PMF. This result implies that the MLP output is the *conditional mean estimate* of the desired vector for the input vector $\underline{X} = \underline{x}$. Now, for an M-class classification problem, the desired random vector can take one of M possible realizations $\underline{d}(i) = \vec{e}_i$, where $\vec{e}_i = [0 \cdots \underset{i^{th}}{1} \cdots 0]^T$ is the ith canonical basis vector. Using these realizations in (18.6), we can write

$$\begin{aligned} \underline{o} = \hat{\underline{d}} &= \sum_i \underline{d}(i)\, p(\underline{D} = \underline{d}(i)|\underline{X} = \underline{x}) \\ &= p(C_1|\underline{x})\vec{e}_1 + p(C_2|\underline{x})\vec{e}_2 + \cdots + p(C_M|\underline{x})\vec{e}_M = \begin{bmatrix} p(C_1|\underline{x}) \\ \vdots \\ p(C_M|\underline{x}) \end{bmatrix}, \end{aligned} \tag{18.7}$$

That is, the outputs of the MLP are the *posterior* probabilities of class membership $o_i = p(\underline{D} = \underline{d}(i)|\underline{X} = \underline{x}) = p(C_i|\underline{x})$. As a consequence, if $p(C_i|\underline{x}) > p(C_j|\underline{x})$ $\forall j \in [1, M],\ j \neq i$, or the ith output neuron produces the largest response, or $o_i > o_j$ $\forall j \in [1, M],\ j \neq i$, then $\underline{x} \in C_i$, and the MLP implements the Bayes decision rule.

18.1.2 Practical Considerations and Issues

The generalized delta rule, together with error back-propagation, can be computationally intensive, particularly for MLP networks with a large number of hidden-layer neurons. Additionally, the error surface in the feature space is generally non-convex and contains many local minima, saddle points, and flat regions. As a result, there are no rigorous strategies to ensure that the learning algorithm converges to the globally optimal solution or even converge at all. Nonetheless, in what follows, we discuss several important practical factors that impact the learning behavior and convergence speed of the MLP and increase the likelihood of reaching an acceptable solution.

1. **Weight Initialization:** Large values of initial weights can drive the sigmoidal activation function to saturation, resulting in vanishing gradients and consequently slow or failed training. The same problem also occurs for very small initial weights. A commonly recommended [7, 11] scheme is to initialize all the weights randomly using a uniform distribution with zero mean and a standard deviation inversely proportional to the fan-in (i.e. number of incoming connections) of each neuron, i.e. $w_{i,j}^{(l)}(0) \sim \text{Unif}\left[\frac{-\sqrt{3}\eta}{N_l}, \frac{\sqrt{3}\eta}{N_l}\right]$, where N_l is the fan-in of the layer l neurons and $1 \leq \eta \leq 1.5$.

2. **Learning Parameters:** Recall from Chapter 17 that the choice of the learning rate μ presents a trade-off between accuracy and convergence speed in the SGD learning and its variants. A commonly used [11] scheme is to use $\mu_l \propto \frac{1}{\sqrt{N_l}}$. That is, each layer has its own learning rate depending on its fan-in N_l. Many adaptive learning rate schemes have been proposed, which monitor the error characteristic during the training. The reader is referred to [11] and also Section 18.2.3 for additional details.

   To accelerate convergence, a momentum factor $0 < \alpha < 1$ can be incorporated such that the weight adjustment term $\Delta w_{i,j}^{(k)}(p)$ is modified to,

$$\Delta w_{i,j}^{(k)}(p) = \mu r_i^{(k)}(p) o_j^{(k-1)}(p) + \alpha \Delta w_{i,j}^{(k)}(p-1). \tag{18.8}$$

   Clearly, $\alpha = 0$ reduces to the standard generalized delta rule. The effects of momentum become important toward the later stages of the learning. After P iterations,

$$\Delta w_{i,j}^{(k)}(P) = \mu \sum_{p=1}^{P} \alpha^{P-p} r_i^{(k)}(p) o_j^{(k-1)}(p), \tag{18.9}$$

   That is, momentum introduces an exponentially decaying memory of previous gradient updates, where α controls the forgetting of the previous error terms. The effect of the momentum term is similar to the behavior of a ball-bearing rolling down the error surface when the whole system is immersed in a liquid with viscosity α.

   An alternative approach is the *exponential smoothing* method proposed by Sejnowski and Rosenberg [12],

$$\Delta w_{i,j}^{(k)}(p) = \alpha \Delta w_{i,j}^{(k)}(p-1) + (1-\alpha) r_i^{(k)}(p) o_j^{(k-1)}(p),$$

   where $\alpha = 0$ corresponds to no smoothing, whereas $\alpha = 1$ eliminates the weight adjustments.

3. **Local Minima:** MLP learning attempts to solve an N-D nonlinear optimization problem, which can encounter local minima problems. If learning stagnates before reaching the selected error goal, one can use one of the remedies: (i) restart with a different random weight initialization; (ii) modify the learning parameters; (iii) change the network topology, i.e. the number of hidden-layer neurons; (iv) use alternative activation function, e.g. ReLU. Other potential approaches include using more recent optimization algorithms such as Adam or RMSprop (see Section 18.2.3).

4. **Network Topology:** The number of neurons in each hidden layer determines the representational capacity of the network, governing whether an acceptable solution exists and how accurately the model can generate complex non-linear decision boundaries for pattern classification. Unfortunately, no rigorous approach exists for determining the optimum network architecture. However, different empirical guidelines have been proposed. For a two-layer MLP with N inputs, one common rule of thumb is to select $K \approx 2N$ to $2.5N$ hidden-layer neurons. For a three-layer MLP network with N inputs, suggested numbers are $K \approx 2N$ to $2.5N$ for the first hidden layer and $L \approx 1N$ to $1.5N$ for the second hidden layer. Note that these values should be considered only as starting points and not as rigorous design rules.

5. **Training Dataset:** Here are some general guidelines for constructing training datasets: (a) use training samples that produce large learning signals; (b) avoid using highly similar training samples; (c) avoid using samples with questionable labels or outliers; and (d) avoid overtraining or overfitting, where the network learns the training data but fails to generalize on unseen data samples. Overfitting can happen due to many factors, including longer-than-needed training cycle, insufficient number of training samples, or an overly complex network architecture with a large number of system parameters.

   To improve convergence and training performance, data *standardization* and whitening is recommended as a pre-processing step prior to training. One such method involves the following steps:

   - Compute the sample mean vector $\underline{\mu}_x = \frac{1}{P}\sum_{p=1}^{P} \underline{x}(p)$ of all the training samples, irrespective of their class, and subtract it from every training or testing sample, i.e. $\underline{\bar{x}}(p) = \underline{x}(p) - \underline{\mu}_x$.
   - Compute the sample covariance matrix of all the training samples $\mathbf{R}_{xx} = \frac{1}{P}\sum_{p=1}^{P} \underline{\bar{x}}(p)\underline{\bar{x}}^T(p)$. Diagonalize $\mathbf{R}_{xx}$ using, e.g. singular value decomposition (SVD) [13] to $\mathbf{R}_{xx} = \mathbf{U}\mathbf{\Lambda}\mathbf{U}^H$ and then compute $\mathbf{R}_{xx}^{-1/2} = \mathbf{U}\mathbf{\Lambda}^{-1/2}\mathbf{U}^H$.

- Map each mean-subtracted data sample using $\mathbf{R}_{xx}^{-1/2}\underline{\bar{x}} = \underline{\tilde{x}}$. This operation makes the mapped data, $\underline{\tilde{x}}$, decorrelated with the identity covariance matrix,

$$\mathbf{R}_{\tilde{x}\tilde{x}} = \frac{1}{P}\sum_{p=1}^{P} \underline{\tilde{x}}(p)\underline{\tilde{x}}^T(p) = \mathbf{R}_{xx}^{-1/2}\frac{1}{P}\sum_{p=1}^{P} \underline{\bar{x}}(p)\underline{\bar{x}}^T(p)\mathbf{R}_{xx}^{-t/2} = \mathbf{R}_{xx}^{-1/2}\mathbf{R}_{xx}R_{xx}^{-t/2} = \mathbf{I},$$

since $\mathbf{R}_{xx} = \mathbf{R}_{xx}^{1/2}\mathbf{R}_{xx}^{t/2}$.
- Use $\underline{\tilde{x}}$ instead of $\underline{x}$ when training or testing the networks.

Although MLP networks can be used to solve nonlinearly separable classification problems, they suffer from many critical shortcomings that restrict their applicability to complex image pattern classification problems. First, they typically rely on the design of separate image segmentation and feature extraction subsystems, which can become challenging even for a skilled and experienced designer. The features must be designed to provide invariance to translation, scaling, rotation, deformation, distortion, and noise. Traditional handcrafted feature extraction methods often fail to capture the rich local, global, and contextual information that humans naturally exploit in their visual perception for pattern recognition. Second, standard MLP networks are fully connected architectures and therefore suffer from the curse of dimensionality. problems. Processing high-dimensional data requires a network with an overly large number of parameters, increasing both computational complexity, training time, and the amount of training data needed for successful training without overfitting. These limitations motivated the development of convolutional neural networks (CNNs), which will be discussed next.

18.2 Convolutional Neural Networks

Originally proposed by LeCun et al. [2–4], CNNs replace the handcrafted feature extraction in standard MLP networks with a neurobiologically inspired architecture capable of learning subtle discriminative features directly from images. The CNN architecture mimics the locally sensitive and orientation-selective neural information processing mechanisms observed in the visual cortex [1]. Unlike an MLP, which has a fully connected structure, a CNN structure consists of multiple pairs of partially connected *convolutional* and *pooling* layers. Figure 18.3 shows the first pair of convolutional and pooling layers. The convolutional layers perform spatial filtering with a limited support region over local neighborhoods of the input image or outputs of the preceding layers. The weights or filter coefficients are adjusted during the training to extract different local and global image features. Pooling layers perform a form of subsampling operation on the outputs of convolutional layers. Pooling operation achieves a degree of robustness to small to translation, distortion, and deformation, similar to the function of complex cells in the Hubel and Wiesel model [1], which pool inputs from multiple simple cells. A deep CNN consisting of several pairs of convolutional and pooling layers acts like a trainable multiscale filter bank where higher layers learn increasingly abstract feature representations. From this perspective, one may notice similarities between CNN architecture and the filter bank structure of the 2-D wavelet transform covered in Chapter 7. The final layers are typically

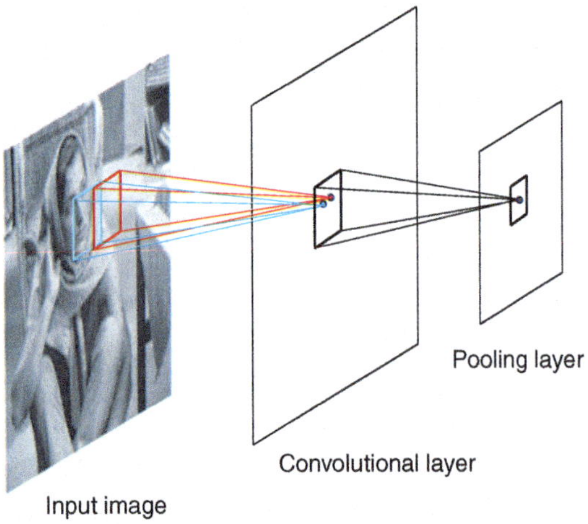

Figure 18.3 A pair of convolutional and pooling layers with local receptive fields.

fully connected with a softmax activation function for pattern classification tasks. The CNN training algorithm is based on a modified error back-propagation that incorporates weight-sharing constraints.

The CNN structure is built around the same principles that govern information processing in the visual cortex. More specifically, the three fundamental principles that are incorporated into the CNN structure are: (1) local receptive fields, (2) weight sharing, and (3) pooling. The principle of local receptive fields in CNN (see Figure 18.3) is inspired by the Cognitron and Neocognitron models proposed by Fukushima [14, 15]. A local receptive field restricts the number of connections to each neuron in the convolutional layer to the size of the convolution kernel or mask. Thus, in contrast to the MLP networks, each neuron extracts local features within its receptive field, while the receptive fields of contiguous neurons overlap. This sparse connectivity has many benefits, including faster training, fewer trainable weights or parameters per layer, hence allowing deeper networks, and improved generalization. The effective kernel or mask size increases with network depth, much like cascading several 2-D finite impulse response (FIR) filters. For example, cascading two 2-D FIR filters with PSF $h_1(m, n)$ and $h_2(m, n)$ of sizes $M_1 \times M_2$ and $N_1 \times N_2$, respectively, is equivalent to a single filter with PSF $h(m, n) = h_1(m, n) ** h_2(m, n)$ of size $(M_1 + N_1 - 1) \times (M_2 + N_2 - 1)$. As the receptive field widens from lower to higher layers, the extracted features become increasingly global and complex in their representation.

The second principle is weight sharing. Convolutional layers consist of neurons with ReLU or sigmoidal activation functions arranged into planes or cubes that share the same set of weights. Synaptic weight sharing allows localized convolution to be performed in small image patches (e.g. 5×5) of pixels with the same kernel or filter coefficients. The output image corresponding to a particular kernel is called a *feature map*. Neurons within a feature map perform the same filtering operation on different blocks of the input image. To extract diverse image features, each convolutional layer typically employs multiple kernels, each with a different set of weights that are trained to produce their corresponding feature maps. Different feature maps at each layer have different sets of filter weights, hence producing a variety of localized features, e.g. edges with different orientations, corners, textures, and increasingly more complex structures at deeper layers. Weight sharing significantly reduces the number of parameters, memory and computational requirements, and introduces the translation *equivariance* property. That is, a translation in the input image results in a corresponding translation of the activations of the feature map by the same amount, hence detecting the presence of an object irrespective of its location within the image.

The third principle is pooling, which performs a local aggregation and subsampling operation on the feature maps. Common pooling operations involve local averaging or weighted averaging followed by subsampling or max pooling. Pooling lowers the spatial resolution, thereby decreasing sensitivity to small shifts, distortions, and noise while reducing the dimensionality of feature representation applied to subsequent layers.

Among the early CNN architectures are LeNet-5 [3], AlexNet, and ImageNet [16]. The first convolutional layer of LeNet-5 consists of six kernels, each with ROS or receptive field of size 5×5. These filters produce six feature maps (see Figure 18.4), each capturing different primitive image features. Each pixel of a feature map is then nonlinearly mapped using an activation function such as ReLU and subsequently applied to a pooling layer. The pooling layers typically use 2×2 nonoverlapping windows and performs either a spatial averaging followed by subsampling or max pooling, which is more common in modern CNNs. The pooled feature map images will be approximately one quarter the size of the original feature map. The second convolutional layer has 16 kernels, each with ROS of size 5×5, producing 16 feature maps. However, unlike the first convolutional layer, each feature map in this layer is generated from several receptive fields at identical locations in a subset of the previous pooling layer outputs [3]. This partial connectivity ensures that the feature maps of the second convolutional layer capture more complex features that are obtained by filtering and combining the features of the

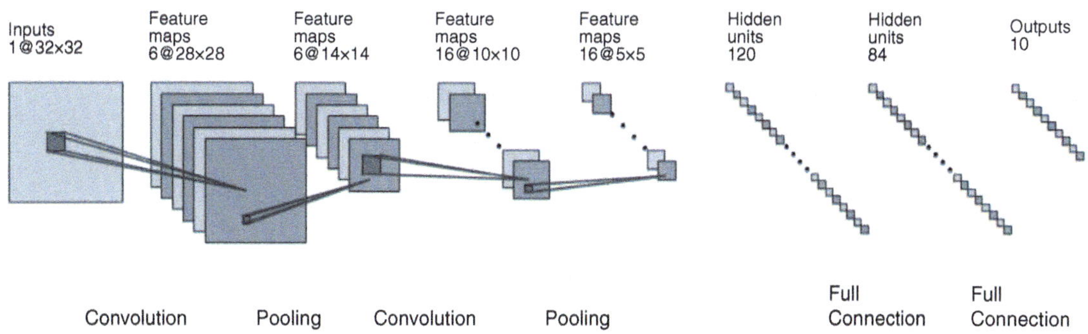

Figure 18.4 LeNet-5 CNN structure [3] consisting of three convolutional layers and two fully connected layer decision-making layers.

preceding layers. These feature maps are subsequently processed by nonlinear activation and pooling operations similar to those used in the previous pooling layer. The third convolutional layer has 120 feature maps, each produced by a kernel with a 5×5 receptive field from all 16 pooled feature maps of the preceding layer. In LeNet-5, there is no need for a pooling layer after the third convolutional layer since each feature map consists of a single pixel [3]. The outputs of the third convolutional layer are fully connected to a hidden layer of 84 neurons followed by the output layer containing 10 neurons. In the original LeNet-5, the activation function of the hidden-layer neurons is a scaled hyperbolic tangent, whereas the output layer uses RBF units whose outputs are dependent on the distance of the hidden layer output vector to the centers of each RBF to generate probabilities of class membership.

Figure 18.5a shows the original 32×32 image of digit 3 from the MNIST handwritten digit dataset [3]. Figure 18.5b,c displays the pooled feature maps produced by the first and second layers of LeNet-5, respectively. The first layer generates six pooled feature maps, each of size 14×14 pixels, whereas the second layer produces 16 pooled feature maps, each of size 4×4 pixels. As can be observed, the learned representations become increasingly abstract as the depth of the network increases.

18.2.1 Learning Rule and Error Back-Propagation for CNN

The same forward and backward passes introduced in Section 18.1.1 for MLP networks also apply to CNNs, albeit with several important modifications due to local connectivity, weigh sharing and pooling operations. We consider a deep CNN consisting of $L+1$ layers, where layers $1, \cdots L$ involve convolution, nonlinear activation, and pooling operations, while the final layer is fully connected and uses a softmax activation function at the output neurons. As before, the training dataset consists of input–output pairs $\mathcal{X} = \{\underline{x}(p), \underline{d}(p)\}_{p=1}^{P}$ where $\underline{x}(p) \in \mathbb{R}^N$ is the input vector and $\underline{d}(p) \in \mathbb{R}^M$ is the desired output vector with $M \ll N$. For an M-ary classification problem, if $\underline{x}(p) \in C_j$ the corresponding desired output vector $\underline{d}(p) = \vec{e}_j$, where as before $\vec{e}_j = [0 \cdots \underset{j\text{th}}{1} \cdots 0]^T$. We define $\underline{w}_{k,m}^{(l)}(p)$ as the augmented weight vector of the $(k, m)^{\text{th}}$ kernel (or filter) of layer l at iteration p. As illustrated in Figure 18.6, this kernel operates on the pooled feature map k of layer $l-1$ and contributes to the activation of $(i, j)^{\text{th}}$ pixel of feature map m of layer l. The aggregated input from all such filters is given by

$$y_{i,j,m}^{(l)}(p) = \sum_{k=1}^{K_{l-1}} \underline{w}_{k,m}^{(l)^T}(p)\underline{x}_{i,j,k}^{(l-1)}(p), \quad m \in [1, K_l] \tag{18.10}$$

where $\underline{x}_{i,j,k}^{(l-1)}$ is the input vector associated with the block centered at $(i, j)^{\text{th}}$ pixel (denoted by $\mathbf{x}_{i,j,k}^{(l-1)}$ block in Figure 18.6) of the pooled feature map $k \in [1, K_{l-1}]$ in layer $l-1$, with K_{l-1} being the number of pooled feature maps. For a specific feature map m, all features $y_{i,j,m}^{(l)}(p)$ share the same kernel parameter vectors $\underline{w}_{k,m}^{(l)}(p)$, $\forall k \in [1, K_{l-1}]$. The resulting feature map is then passed through a nonlinear activation function $f(.)$ (e.g. ReLU) to produce the input image, $v_{i,j,m}^{(l)}(p) = f(y_{i,j,m}^{(l)}(p))$, to the pooling layer, which generates pooled feature map

$$x_{i,j,m}^{(l)}(p) = pool\left(v_{r,s,m}^{(l)}(p)\right), \quad \forall (r, s) \in \mathcal{S}_{i,j} \tag{18.11}$$

where $\mathcal{S}_{i,j}$ is a local neighborhood of $v_{i,j,m}^{(l)}$ and $x_{i,j,m}^{(l)}(p)$ is the $(i, j)^{\text{th}}$ pixel of the pooled feature map or input to the convolutional layer l. The pooling region $\mathcal{S}_{i,j}$ is typically 2×2 (nonoverlapping) block, within which averaging followed by subsampling or max selection is performed.

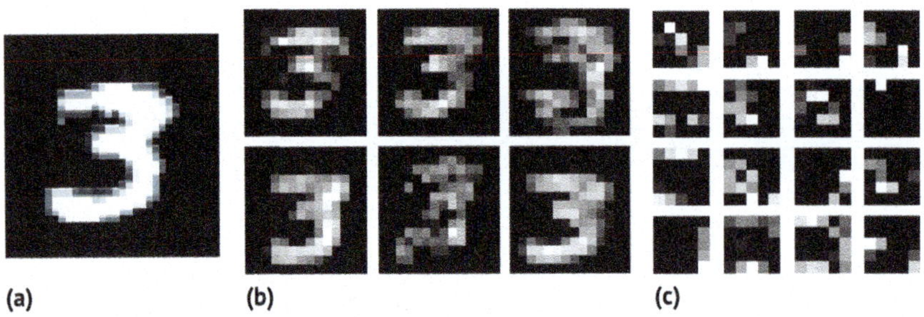

Figure 18.5 Examples of the pooled feature maps of digit "3" produced by the first and second layers of LeNet-5 CNN [3].

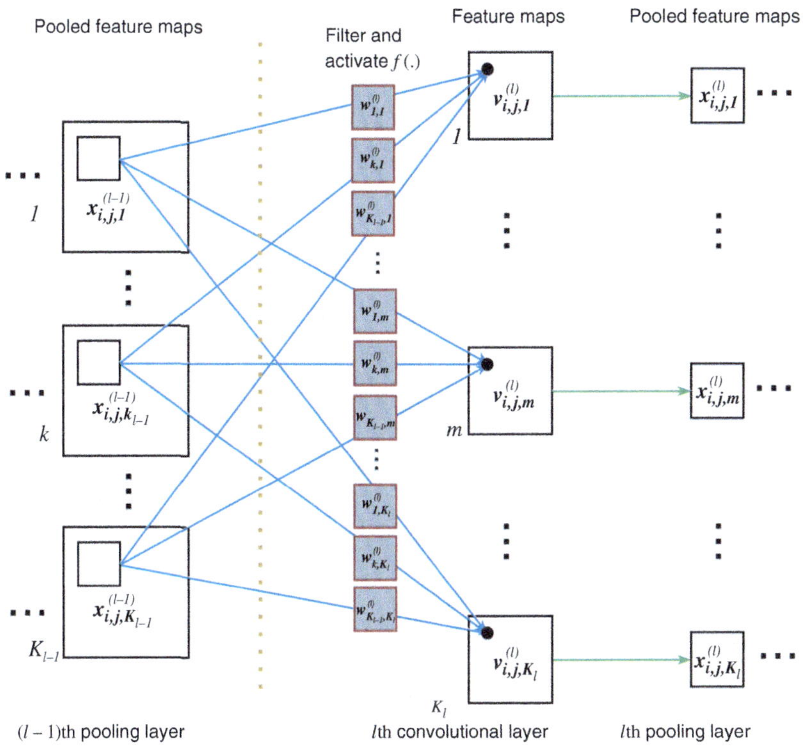

Figure 18.6 Connections between layers $(l-1)$ and l layers in CNN.

After several pairs of convolutional and pooling layers, one or more fully connected layer(s) perform high-level decision-making based on the extracted features produced by the last convolutional and pooling layers. The pooled feature maps of the final convolutional stage are vectorized and concatenated into a single vector that is applied to the fully connected layer. If the size of each pooled feature map in layer L is $Q_L \times Q_L$, the dimension of this concatenated vector is $K' = K_L Q_L^2$. Now, if we denote the concatenation operation by operator $C(.)$ then the kth element of this concatenated vector can be expressed as

$$z_k(p) = C\left(x^{(L)}_{r_1,r_2,q}(p)\right), \quad k \in [1, K'] \tag{18.12}$$

where $x^{(L)}_{r_1,r_2,q}(p)$ is pixel (r_1, r_2) of the pooled feature map image $q \in [1, K_L]$ of layer L and here $k = [(r_1 - 1)Q_L + r_2]q$ is the input index to the fully connected layer. Assuming that *softmax* activation function is used for the output neurons, and we define the net input to the neuron j by $\eta_j(p) = \sum_{k=1}^{K'} w^{L+1}_{k,j}(p) z_k(p)$ where $w^{L+1}_{k,j}$ is the weight connecting the input k to the output neuron j, then the output of this neuron is $o_j = \frac{e^{\eta_j}}{\sum_{m=1}^{M} e^{\eta_m}}$, $\forall j \in [1, M]$. The softmax operation converts CNN outputs to normalized values that can be viewed as posterior probabilities of class membership.

18.2.1.1 Loss Functions

A loss or cost function is a quantitative measure of the performance of a learning machine used during the training phase. For supervised learning, a loss function measures how different the predicted outputs are compared to the desired responses on a set of training data samples. Parameter adjustments are made by driving the predicted outputs as close as possible to the desired values according to the chosen loss function. This is accomplished by minimizing (or maximizing) the loss function with respect to the unknown parameters of the model.

Let $\mathcal{W}$ be the set of all model parameters (weights and biases) of the network. The optimum parameter set is obtained by minimizing (or maximizing) the total loss,

$$\mathcal{L}(\mathcal{W}) = \frac{1}{P} \sum_{p=1}^{P} \ell(\mathcal{W}, \underline{d}(p), \underline{o}(p)), \tag{18.13}$$

where $\ell(W, \underline{d}(p), \underline{o}(p))$ is an appropriate loss function per sample and $\underline{o}(p) \in \mathbb{R}^M$ is the CNN output vector for training sample $\underline{x}(p)$. Some typical choices of loss function for CNN are given below.

1. **Categorical Cross-Entropy Loss:**
   Cross-entropy originates from information theory [17] and measures the discrepancy between a true probability distribution and an estimated probability distribution. If the true and estimated PMFs are denoted by $p(\mathbf{x} = \mathbf{x}_i)$ and $q(\mathbf{x} = \mathbf{x}_i)$, respectively, then their cross-entropy [17] is given by

$$H(p, q) = -\sum_{\mathbf{x}_i \in \mathrm{X}} p(\mathbf{x} = \mathbf{x}_i) \log q(\mathbf{x} = \mathbf{x}_i), \tag{18.14}$$

   where both random variables are assumed to have the same support X.

   The cross-entropy loss function is the most widely used loss function [18] for multi-class classification problems using CNN. However, in this case, the true and predicted distributions correspond to the desired output and the estimated (actual) softmax output of the network, respectively. Thus, for a training sample pair $\{\underline{x}(p), \underline{d}(p)\}$, the per-sample categorical cross-entropy loss function is

$$\ell(W, \underline{d}(p), \underline{o}(p)) = -\sum_{m=1}^{M} d_m(p) \log o_m(p), \tag{18.15}$$

   where $o_m(p)$ is the mth element of the output vector $\underline{o}(p)$. If $\underline{x}(p) \in C_k$, then $\underline{d}(p) = \vec{e}_k$ and the above loss function becomes $\ell\big(W, \underline{d}(p), \underline{o}(p)\big) = -\log o_k(p)$, since $d_k(p) = 1$ while all other elements are zero.

   As discussed earlier, softmax activation is typically used at the output neurons, giving $o_k(p) = \frac{e^{\eta_k(p)}}{\sum_{m=1}^{M} e^{\eta_m(p)}}$ where $\eta_k(p) = \underline{w}_k^{(L+1)^T}(p)\underline{z}(p)$ is the total input to neuron k and $\underline{z}(p)$ and $\underline{w}_k^{(L+1)}(p)$ are the input vector and the weight vector of this neuron. Now, since the softmax outputs can be interpreted as estimates of the posterior probabilities of class membership, i.e. $o_k(p) = p(C_k|\underline{x}(p))$, the cross-entropy loss function attempts to drive the probability associated with the correct class toward 1 while pushing the probabilities of the remaining classes toward zero, hence increasing the confidence in the classification decision. Nevertheless, experimental studies reported in [19] showed that the network trained using this loss function misclassified some of the training samples for the very same reason.

   A variant of this loss function is the weighted categorical cross-entropy loss function [19], which is used when the training dataset contains an unbalanced number of samples. In such cases, the number of samples in different classes differs significantly, leading to poor performance due to bias toward the dominant classes. The weighted categorical cross-entropy loss function modifies the original version by introducing weights that assign a higher penalty to misclassifications of the underrepresented classes. The weighted loss is

$$\ell(W, \underline{d}(p), \underline{o}(p)) = -\sum_{m=1}^{M} \pi_m d_m(p) \log o_m(p), \tag{18.16}$$

   where π_m is the relative weight assigned to class k. A common method for choosing these weights is to make them inversely proportional to the number of samples in each class.

2. **Focal Loss:**
   Another loss function used for imbalanced training data is focal loss [20]. Unlike the weighted cross-entropy loss, which assigns weights based on class frequencies, focal loss assigns weights based on the classification error. More specifically, samples that are less likely to be misclassified are assigned smaller weights compared to those that are likely to be misclassified, hence *focusing* on the problematic classes. For a multi-class classification, the focal loss is

$$\ell(W, \underline{d}(p), \underline{o}(p)) = -\sum_{m=1}^{M} (1 - o_m(p))^{\gamma} d_m(p) \log o_m(p), \tag{18.17}$$

   where $\gamma > 0$ is called the focusing parameter. The term $(1 - o_m(p))^{\gamma}$ in (18.17) acts as a weighting factor to reduce the influence of confidently classified samples and focus on difficult samples. When $\gamma = 0$ focal loss becomes the same as the cross-entropy loss.

3. **Hinge Loss:**
The hinge loss function was originally developed for large margin binary classification using support vector machines (SVM) [7, 21]. Several variants of the hinge loss have been proposed for multiclass problems. However, the per-sample hinge loss function [19] often provides good accuracy and faster convergence than the others and is given by

$$\ell(\mathcal{W}, \underline{d}(p), \underline{o}(p)) = \max\left(0, 1 - \left(\max_{\forall i \neq j} o_i(p) - \underline{d}^T(p)\,\underline{o}(p)\right)\right). \tag{18.18}$$

Now, for a training sample $\underline{x}(p) \in C_j$, three scenarios can happen. If the score of the correct class, $\underline{d}^T(p)\,\underline{o}(p) = o_j(p)$, is greater than or equal to the maximum of all competing scores $o_i(p)$, $\forall i \neq j$ by a margin of at least 1, then the loss is zero. If $o_j(p) > \max_{\forall i \neq j} o_i(p)$ but the margin is less than one, the classification remains correct although the loss is nonzero, indicating lower confidence in the classification decision. In this case, the weight-updating attempts to widen the margin for better confidence in classification. Lastly, if $o_j(p) < \max_{\forall i \neq j} o_i(p)$, i.e. incorrect classification, the per-sample loss measures the extent to which the score of the correct class falls below the largest competing score by a margin of at least 1. As a consequence, hinge loss penalizes both incorrect classifications and correct classifications with low confidence by a factor dependent on the margin. Experimental studies have shown [19] that networks trained with hinge loss offered better generalization performance and faster convergence than those trained using categorical cross-entropy. Although the hinge loss function is convex, it is not differentiable at the margin boundary. Nonetheless, differentiable variants [18] are widely used in practice.

4. **Sum Squared Error Loss:**
This loss function is mostly used in regression problems. The per-sample loss function is

$$\ell(\mathcal{W}, \underline{d}(p), \underline{o}(p)) = \frac{1}{2}||\underline{d}(p) - \underline{o}(p)||^2 = \frac{1}{2}\sum_{m=1}^{M} (d_m(p) - o_m(p))^2. \tag{18.19}$$

The corresponding average loss over the training dataset is $\mathcal{L}(\mathcal{W}) = \frac{1}{P}\sum_{p=1}^{P} \ell(\mathcal{W}, \underline{d}(p), \underline{o}(p))$. This loss function is differentiable and simple to optimize. While SSE is particularly suitable for regression tasks, it is generally less effective than cross-entropy-based losses for multiclass classification problems mainly due to slower convergence.

18.2.1.2 Error Back-Propagation in CNN

The error back-propagation algorithm presented in Section 18.1.1 can also be applied to CNN weight updating. However, there are many factors that make the extension of the original error back-propagation algorithm to CNNs more complicated. Among the key factors are: (a) convolutional layers are not fully connected, (b) there are multiple kernel (filter) types in each convolutional layer the outputs of which merge together, as shown in Figure 18.6 to produce the feature maps of layer l, (c) the weights of a kernel for a feature map are shared, (d) the pooling operation is not reversible, and (e) the connectively of the pooled feature map outputs to the inputs of the next layer varies from one layer to another. In what follows, we presents the modified error back-propagation algorithm for a general CNN structure. We start with the weight-updating equations for the final fully connected output layer and then develop the error back-propagation and updating processes for the pair of convolutional and pooling layers L. The categorical cross-entropy loss function, $\ell(\mathcal{W}, p)$ in (18.15) is used in iterative SGD-based learning.

i. **Fully Connected Layer:** Assume that the training sample $\underline{x}(p)$ is applied to the network and propagated in the forward direction to generate the actual output vector $\underline{o}(p)$, for which the desired output vector is $\underline{d}(p)$. For a deep CNN with L convolutional and pooling layers, the single fully connected layer is the $(L+1)^{\text{th}}$ layer and uses the softmax activation function. The weight connecting input k to output neuron m is denoted by $w_{k,m}^{(L+1)}(p)$. To update these weights using the SGD learning, we need the gradients of the loss function in (18.15) with respect to these weights. For the softmax activation function together with the categorical cross-entropy loss, it can be shown that

$$\frac{\partial \ell(\mathcal{W}, p)}{\partial \eta_m(p)} = o_m(p) - d_m(p), \quad m \in [1, M],$$

where $\eta_m(p) = \sum_{k=1}^{K'} w_{k,m}^{(L+1)}(p) z_k(p)$, is the aggregated input to output neuron j. Now, the chain rule gives

$$\frac{\partial \ell(W,p)}{\partial w_{k,m}^{(L+1)}(p)} = \frac{\partial \ell(W,p)}{\partial \eta_m(p)} \frac{\partial \eta_j(p)}{\partial w_{k,m}^{(L+1)}(p)} = (o_m(p) - d_m(p)) z_k(p), \tag{18.20}$$

where $z_k(p)$ is the kth input to the fully connected layer. Defining the learning signal $r_m(p) = d_m(p) - o_m(p)$, the SGD updating equation becomes

$$w_{k,m}^{(L+1)}(p+1) = w_{k,m}^{(L+1)}(p) + \mu r_m(p) z_k(p). \tag{18.21}$$

That is, the output neurons contribute to the weight updates through their corresponding learning signals. Clearly, if $\underline{x}(p) \in C_j$, then $d_j(p) = 1$, while for all other neurons $d_m(p) = 0 \;\; m \neq j$.

ii. **Convolutional Layer** L: To develop the error back-propagation process for layer L, we first need to back-propagate the output layer errors to the output of the pooling layer L. To do so, we first find the partial derivative of $\ell(W,p)$ with respect to the input $z_k(p)$ of the fully connected layer,

$$\frac{\partial \ell(W,p)}{\partial z_k(p)} = \sum_{m=1}^{M} \frac{\partial \ell(W,p)}{\partial \eta_m(p)} \frac{\partial \eta_m(p)}{\partial z_k(p)} = \sum_{m=1}^{M} (o_m(p) - d_m(p)) w_{k,m}^{(L+1)}(p) = -\sum_{m-1}^{M} r_m(p) w_{k,m}^{(L+1)}(p). \tag{18.22}$$

Thus, the learning signals $r_m(p)$ at the output neurons are back-propagated through the fully connected weights to generate the error signal associated with input $z_k(p)$. Note that for $\underline{x}(p) \in C_j$, the learning signal is $r_j(p) = 1 - o_j(p)$ and $r_m(p) = o_m(p)$, $\forall m \neq j$. However, since the input to the fully connected layer is the vectorized and concatenated version of all K_L pooled feature maps of layer L, the concatenated vector $\underline{z}_p$ must first be reshaped back into its original form. We denote this operation by C^{-1} [22, 23], i.e. the inverse of the concatenation operator $C(.)$. In other words, we can write

$$\frac{\partial \ell(W,p)}{\partial x_{r_1,r_2,q}^{(L)}(p)} = C^{-1}\left(\frac{\partial \ell(W,p)}{\partial z_k(p)}\right), \;\; k \in [1.K'], \;\; q \in [1, K_L], \;\; r_1, r_2 \in [1, Q_L], \tag{18.23}$$

where $\frac{\partial \ell(W,p)}{\partial x_{r_1,r_2,q}^{(L)}(p)}$ represents the error signal associated with pixel (r_1, r_2) of pooled feature map q. Next, the errors must be back-propagated through the pooling layer. For average pooling, this is accomplished by upsampling the error map by a factor of 2, which repeats the pooled error uniformly over the 2×2 pooling block. For max pooling, the error is routed only to the pixel that produced the maximum value during the forward pass, while all other pixels receive zero error. Here, we assume average pooling, which gives,

$$\frac{\partial \ell(W,p)}{\partial v_{s_1,s_2,q}^{(L)}(p)} = \text{upsample}\left(\frac{\partial \ell(W,p)}{\partial x_{r_1,r_2,q}^{(L)}(p)}\right) = \text{upsample}\left\{C^{-1}\left(\frac{\partial \ell(W,p)}{\partial z_k(p)}\right)\right\}, \tag{18.24}$$

where $v_{s_1,s_2,q}^{(L)}(p)$ represents the upsampled version of the pixel (r_1, r_2) of feature map image q or the output of the convolutional layer L. Note that for simplicity, we used the same notation for the upsampled feature maps as the actual feature map before the pooling layer, though they are not typically the same. Now, we are in a position to find the loss function gradient with respect to the convolutional filter weights of layer L. Considering the structure of the CNN, this gradient is given by,

$$\begin{aligned} \frac{\partial \ell(W,p)}{\partial \underline{w}_{l,q}^{(L)}(p)} &= \sum_{s_1=1}^{P_L} \sum_{s_2=1}^{P_L} \frac{\partial \ell(W,p)}{\partial v_{s_1,s_2,q}^{(L)}(p)} \frac{\partial v_{s_1,s_2,q}^{(L)}(p)}{\partial y_{s_1,s_2,q}^{(L)}(p)} \frac{\partial y_{s_1,s_2,q}^{(L)}(p)}{\partial \underline{w}_{l,q}^{(L)}(p)} \\ &= -\sum_{s_1=1}^{P_L} \sum_{s_2=1}^{P_L} upsample\left\{C^{-1}\left(\sum_{m=1}^{M} r_m(p) w_{m,j}(p)\right)\right\} f'(y_{s_1,s_2,q}^{(L)}(p)) \underline{x}_{s_1,s_2,l}^{(L-1)}(p) \end{aligned} \tag{18.25}$$

where $f'(.)$ is the first derivative of the activation function of neurons in the convolutional layers, $\underline{x}_{s_1,s_2,l}^{(L)}(p)$ is the vectorized version of the input pixels within the convolution mask centered around (s_1, s_2)-pixel, and P_L is the dimension of the feature map images at layer L. If ReLU activation is used $f'(x) = u(x)$ or a unit step function. For sigmoidal activation with $\lambda = 1$ we have $f'(y_{s_1,s_2,q}^{(L)}(p)) = v_{s_1,s_2,q}^{(L)}(p)(1 - v_{s_1,s_2,l}^{(L)}(p))$. The summations over s_1 and s_2 reflect the fact that the weight vector $\underline{w}_{l,q}^{(L)}(p)$ is shared for all pixels of the feature map q since $y_{s_1,s_2,q}^{(L)}(p) = \sum_{m'=1}^{K_{L-1}} \underline{w}_{m',q}^{(L)^T}(p)\underline{x}_{s_1,s_2,m'}^{(L-1)}(p)$, $s_1, s_2 \in [1, P_L]$.

Now, if in (18.25) we define the learning signal of this layer by

$$r_{s_1,s_2,q}^{(L)}(p) = \text{upsample}\left\{C^{-1}\left(\sum_{m=1}^{M} r_m(p)w_{m,j}(p)\right)\right\}f'(y_{s_1,s_2,q}^{(L)}(p)) \tag{18.26}$$

the weight-updating equation for this layer becomes

$$\underline{w}_{l,q}^{(L)}(p+1) = \underline{w}_{l,q}^{(L)}(p) + \mu \sum_{s_1=1}^{P_L} \sum_{s_2=1}^{P_L} r_{s_1,s_2,q}^{(L)}(p)\underline{x}_{s_1,s_2,l}^{(L-1)}(p), \tag{18.27}$$

To continue the back-propagation process to layer $L-1$, the learning signals $r_{s_1,s_2,q}^{(L)}(p)$ must be propagated through the convolutional filters. To do so, we need to find the gradient of the cost function with respect to the convolutional layer L inputs,

$$\frac{\partial \ell(\mathcal{W}, p)}{\partial \underline{x}_{s_1,s_2,l}^{(L)}(p)} = \sum_{q=1}^{K_L} \frac{\partial \ell(\mathcal{W}, p)}{\partial v_{s_1,s_2,q}^{(L)}(p)} \frac{\partial v_{s_1,s_2,q}^{(L)}(p)}{\partial y_{s_1,s_2,q}^{(L)}(p)} \frac{\partial y_{s_1,s_2,q}^{(L)}(p)}{\partial \underline{x}_{s_1,s_2,l}^{(L)}(p)}. \tag{18.28}$$

Using the results in (18.25) and (18.26), it is easy to show that (18.28) yields,

$$\frac{\partial \ell(\mathcal{W}, p)}{\partial \underline{x}_{s_1,s_2,l}^{(L)}(p)} = -\sum_{q=1}^{K_L} r_{s_1,s_2,q}^{(L)}(p)\underline{w}_{l,q}^{(L)}(p), \tag{18.29}$$

which is the back-propagated learning signals through the weights of the convolutional layer L. This learning signal is subsequently used to update the weights of layer $L-1$ in a similar manner.

18.2.2 Alternative Learning Algorithms for CNN

As illustrated in the previous section, the generalized delta learning rule in conjunction with error back-propagation discussed in Section 18.1.1 for MLP networks is also applicable to CNN with some modifications. The original SGD learning suffers from accuracy-convergence speed trade-off, especially in the presence of noisy estimates of the gradients. Adding momentum terms to the updating equations improves the convergence speed for both for convex and non-convex cost functions. Nevertheless, the problem of choosing appropriate values for the learning rate and the momentum factor remains critically important for good performance. There are several alternatives to SGD-based learning for situations in which gradient estimates are noisy and/or sparse. The latter scenario happens when gradient estimates contain a large proportion of zero or near-zero values. In what follows, two popular methods, namely root-mean-square propagation (RMSProp) [24] and adaptive moment estimation (Adam) [25], are briefly reviewed.

1. **RMSProp Algorithm:**
   RMSprop can be viewed as a variant of SGD-based learning in which the learning rate for each parameter is adaptively adjusted based on a moving-average estimate of the squared gradients. If we denote the moving-average estimate of the squared gradients of the per sample loss function $\ell(\mathcal{W}, p)$ with respect to weight vector $\underline{w}_{k,m}^{(l)}(p)$ for layer l by $v_{k,m}^{(l)}(p)$, then the stochastic RMSProp updating rule for this weight vector is

$$\underline{w}_{k,m}^{(l)}(p+1) = \underline{w}_{k,m}^{(l)}(p) - \frac{\mu}{\sqrt{v_{k,m}^{(l)}(p)} + \epsilon} \frac{\partial \ell(\mathcal{W}, p)}{\partial \underline{w}_{k,m}^{(l)}(p)}, \tag{18.30}$$

   where ϵ is a small positive constant added for numerical stability. The moving-average estimate is updated according to

$$v_{k,m}^{(l)}(p+1) = \beta v_{k,m}^{(l)}(p) + (1-\beta)\left(\frac{\partial \ell(\mathcal{W}, p)}{\partial \underline{w}_{k,m}^{(l)}(p)}\right)^2, \tag{18.31}$$

where the squared operation is element-wise and $\nu_{k,m}^{(l)}(0) = 0$. The parameter $\beta \in [0, 1)$ determines the relative weighting of the previous moving-average estimate of the squared gradient and the new squared gradient. A typical choice is $\beta = 0.9$. As can be observed from (18.30), the adaptive step size becomes large when the moving-average term is small and vice versa. RMSprop offers a reliable learning particularly for situations with noisy, sparse, or nonstationary gradients.

2. **Adam Algorithm:**
Adam [25] is another efficient stochastic learning method commonly used for training CNNs. It adaptively adjusts the learning rate for different filter parameters using moving average estimates of both the first- and second-order moments of the gradients, hence the name adaptive moment estimation. Adam combines several advantages of RMSProp and momentum-based step size SGD methods with the added benefit of providing a bias correction procedure to compensate for initialization effects in the moving averages. Adam uses moving averages of the gradient and the squared gradient as the estimates of first- and second-order moments of the gradient, together with the bias-correction procedure.

Again, let us denote the biased moving-average estimates of the gradient and squared gradients of the loss function $\ell(\mathcal{W}, p)$ with respect to the weight vector $\underline{w}_{k,m}^{(l)}(p)$ for layer l by $\kappa_{k,m}^{(l)}(p)$ and $\nu_{k,m}^{(l)}(p)$, respectively. These biased moment estimates are updated using

$$\begin{aligned} \kappa_{k,m}^{(l)}(p+1) &= \beta_1 \kappa_{k,m}^{(l)}(p) + (1-\beta_1)\left(\frac{\partial \ell(\mathcal{W}, p)}{\partial \underline{w}_{k,m}^{(l)}(p)}\right), \quad \text{and} \\ \nu_{k,m}^{(l)}(p+1) &= \beta_2 \nu_{k,m}^{(l)}(p) + (1-\beta_2)\left(\frac{\partial \ell(\mathcal{W}, p)}{\partial \underline{w}_{k,m}^{(l)}(p)}\right)^2, \end{aligned} \tag{18.32}$$

where $\beta_1, \beta_2 \in [0, 1)$ control the exponential decay rates of the moving averages. The recommended [25] values for these hyper-parameters are $\beta_1 = 0.9$ and $\beta_2 = 0.99$. The bias-corrected moment estimates are:

$$\hat{\kappa}_{k,m}^{(l)}(p) = \frac{\kappa_{k,m}^{(l)}(p)}{1-\beta_1^p}, \quad \text{and} \quad \hat{\nu}_{k,m}^{(l)}(p) = \frac{\nu_{k,m}^{(l)}(p)}{1-\beta_2^p}. \tag{18.33}$$

Then, the Adam updating rule for this weight vector becomes

$$\underline{w}_{k,m}^{(l)}(p+1) = \underline{w}_{k,m}^{(l)}(p) - \mu \frac{\hat{\kappa}_{k,m}^{(l)}(p)}{\sqrt{\hat{\nu}_{k,m}^{(l)}(p)} + \epsilon}, \tag{18.34}$$

where ϵ is a small positive constant added for numerical stability. As can be observed from (18.33), during the later stages of learning, when p becomes large, very little bias correction is needed.

Here, we demonstrate and compare the performance of MLP and CNN networks for handwritten digit recognition using the entire MNIST dataset. This dataset contains 60,000 training images and 10,000 test images, each of size 28×28 pixels. The digits have been size-normalized and centered in a fixed-size image. For the MLP, a three-layer network with $784 = 28^2$ inputs, 256 neurons in the first hidden layer, 128 neurons in the second hidden layer, and 10 output neurons was configured. For the CNN, we used two convolutional layers with 16 and 32 kernels, respectively. The kernel size was 5×5 with a stride of one pixel to perform local convolution. The activation function of the convolutional layers was ReLU, and max pooling was used in the pooling layers. The classifier was a fully connected two-layer network with $32 \times 7^2 = 1568$ inputs, 128 hidden neurons, and 10 output neurons with softmax activation. The torch.nn package in Python was used to construct and train both networks and subsequently test the trained networks on the MNIST testing dataset. A validation subset was used to select the best trained network from ten random weight initialization trials. For both networks, we used the cross-entropy loss function and the Adam optimizer with an initial learning rate of 0.001.

Figure 18.7a,b shows the confusion matrices of the MLP and CNN networks, respectively, in terms of the actual numbers of correct and incorrect classifications for each class. The reason for reporting the actual numbers instead of percentages is that the number of images in each class is not identical in the MNIST test dataset, and therefore the actual counts are more representative of the classification performance. Examples of misclassified images for each network are also shown in Figure 18.8a,b. As can be seen, some of these cases are difficult to recognize even for humans.

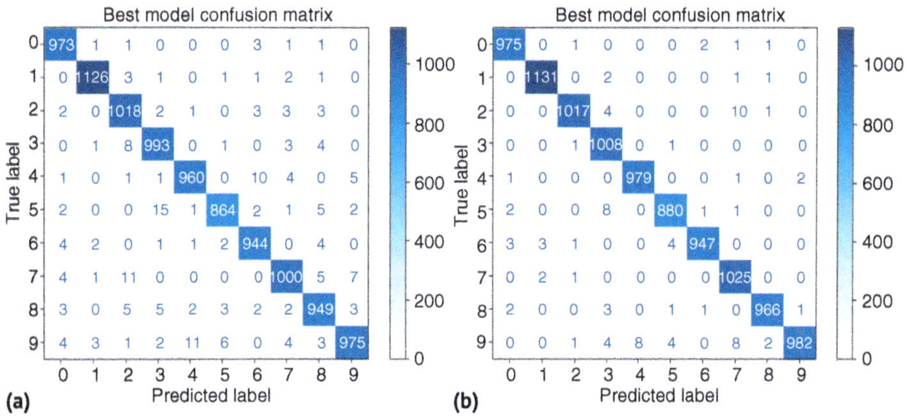

Figure 18.7 Confusion matrices of (a) a three-layer MLP network and (b) a two-layer CNN.

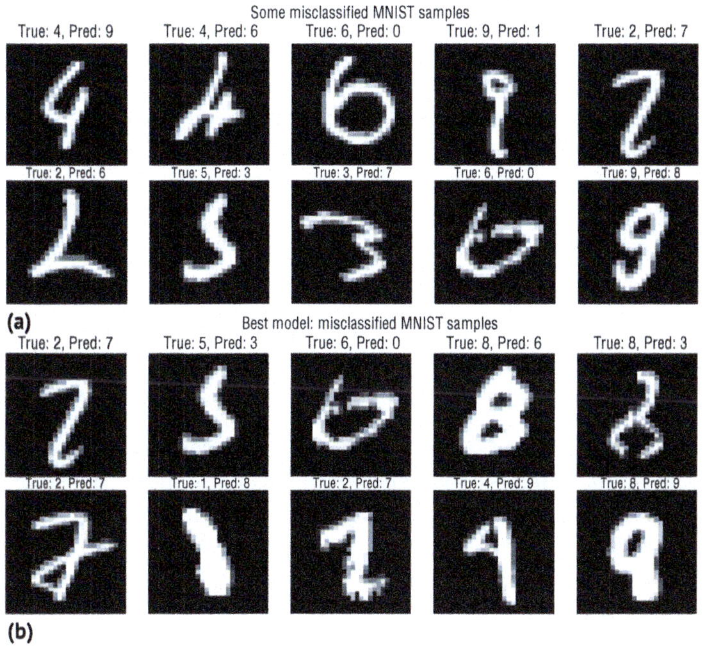

Figure 18.8 Examples of misclassified images in the MNIST test dataset: (a) MLP network and (b) CNN network.
Source: Adapted from Josef Steppan / https://commons.wikimedia.org/wiki/File:MNIST_dataset_example.png / CC BY-SA 4.0 / last accessed Aug. 5, 2026.

It should be mentioned that no attempt was made to optimize the network topologies. The three-layer MLP network performed very well on this dataset, providing an overall correct classification rate of 97.75%. The CNN with two convolutional layers gave an overall classification accuracy of approximately 99% and better overall performance as evident from the confusion matrix in Figure 18.7b.

18.3 Self-Organizing Feature Maps

Internal representations of information in the brain are organized spatially, i.e. individual regions in the cortex exhibit a logical ordering of their functionalities, such as speech, vision, hearing, and motor functions. The self-organizing feature maps (SOFM), also known as *Kohonen map* [7–9] is an unsupervised neural network that attempts to emulate this form of organization. More specifically, the aim of SOFM is to introduce a learning mechanism that can (i) map a high-dimensional pattern space to a lower-dimensional feature space through an unsupervised *self-organization* process, and (ii) capture useful topological properties and structures in the input data. For example, for digital images self-organization into topologically meaningful structures may involve mapping high-dimensional image blocks or regions into lower-dimensional feature representations corresponding to edges with various orientations, texture patterns, or other local image structures. These feature representations tend to be organized topologically based on their statistical similarities. As a result, a high-dimensional pattern space is mapped into a set of low-dimensional *codes*, where each code represents data with similar statistical properties. Patterns with similar characteristics are mapped into adjacent regions of the map much like the biological counterparts. Consequently, SOFM networks have numerous applications in digital image processing, including image segmentation, feature extraction, vector quantization, data compression, and high-dimensional data visualization.

An SOFM consists of a single planar layer of neurons whose associated weight vectors represent the prototype patterns in the input space. The learning rule used to determine these weight vectors is based on a modified version of the competitive learning discussed in the previous chapter. Modifications are incorporated into competitive learning to promote topological ordering of the map during the training phase by allowing *cooperation* [7–9] among neurons within a spatial neighborhood of the winning neuron while enforcing competition as in the original competitive learning. Thus, while neurons still compete to respond to an input pattern, neighboring neurons of the winner are also allowed to participate in the weight adaptation process.

Neuronal cooperation is another biologically inspired mechanism incorporated into SOFM networks [7]. As stated in Chapter 17, biological neural networks consist of layers of neurons that are densely interconnected through lateral synaptic connections. The strengths of these connections vary as a function of lateral distance. The 50–100 micron radius neighborhood is characterized by short-range *excitatory* connections, while 200–500 micron ring corresponds to longer-range (weaker) *inhibitory* connections. The resulting lateral interaction profile resembles a *Mexican hat* function (see Figure 4.2). The response of a planar array of neurons to an external stimuli forms an *activity bubble* whose center corresponds to the most strongly activated neuron (i.e. the winning neuron). Excitatory feedback widens the activity bubble and encourages cooperation among a large number of neighboring neurons, whereas inhibitory feedback narrows the activity bubble to confine cooperation. However, if the net feedback is strongly inhibitory, topological ordering may not form. Thus, cooperation among neighboring neurons of the winning neuron is a fundamental condition for the topological organization of the feature map.

18.3.1 SOFM Unsupervised Training

Consider a 2-D array of neurons connected to the input vector $\underline{x}$ through the weight vector $\underline{w}_i$ for every neuron i, as shown in Figure 18.9. To build topologically ordered maps, self-organization requires both *Cooperation* and *Competition*. Cooperative learning can be implemented through lateral feedback connections and interactions among a topologically related subset of neurons in the spatial neighborhood of the winning neuron. Thus, lateral interaction in the SOFM is imposed by defining *Neighborhood Set* N_m centered around the winning neuron m, as shown in Figure 18.9. The neighborhood N_m can have any shape or geometry, such as square, rectangle, or hexagonal.

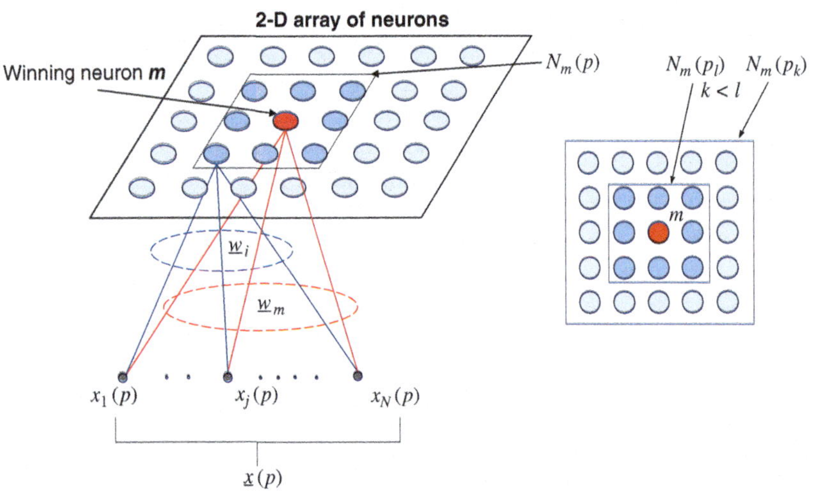

Figure 18.9 2-D lattice of neurons and neighborhood $N_m(p)$ of the winning neuron.

Given an unlabeled training dataset $\{\underline{x}(p)\}_{p=1}^{P}$, a winning neuron m is selected at each learning iteration, and all neurons within the neighborhood $N_m(p)$ of the winner neuron are allowed to update their weights, while the weights of those neurons outside this neighborhood remain unchanged. The radius of $N_m(p)$ is time-varying. That is, during the early stages of learning, $N_m(p)$ is very large to promote higher cooperation and hence more positive feedback to encourage global topological ordering. As learning progresses, the neighborhood radius is gradually reduced, and toward the end of training the neighborhood shrinks to contain only the winning neuron, in which case the SOFM learning reduces to the standard competitive learning discussed in Section 17.3.4. Note that the lateral interaction mechanism of the Mexican Hat model is incorporated implicitly via the localized neighborhood process. The steps in SOFM learning are as follows:

1. **Initialization:** Start with a large number of neurons and a large neighborhood $N_m(0)$. A common recommendation is to select an initial neighborhood $N_m(0)$ radius equal to approximately 50% of the diameter of the network. The weights are initialized using the procedure described for competitive learning in Section 17.3.4. Alternatively, one can use a random initialization procedure similar to that used for MLP in Section 18.1.2.
2. **Winner Selection:** For every training sample $\underline{x}(p)$, find the winning (competition) neuron m using either the minimum distance measure $m = \underset{i}{\text{argmin}}\ ||\underline{x}(p) - \underline{w}_i||$, or the cosine of angle $m = \underset{i}{\text{argmax}}\ \cos \angle(\underline{w}_i, \underline{x}(p)),\ i \in [1, M]$ criterion as in the original competitive learning in Section 17.3.4.
3. **Weight Updating:** Apply the SOFM updating rule to all neurons (cooperation) within the neighborhood $N_m(p)$. This weight-updating rule is given by

$$\begin{aligned} &\underline{w}_i(p+1) = \underline{w}_i(p) + \alpha(p)h_{i,m}(p)[\underline{x}(p) - \underline{w}_i(p)], \quad \text{for} \quad i \in N_m(p), \\ &\underline{w}_j(p+1) = \underline{w}_j(p) \quad \text{for} \quad j \notin N_m(p), \end{aligned} \tag{18.35}$$

where $N_m(p)$ is the neighborhood that shrinks with iteration p, $0 < \alpha(k) < 1$ is the learning rate, and $h_{i,m}(p)$ is a symmetric weighting function that defines the excitatory region within the neighborhood and the degree of participation of neuron i in the learning process. This is a decreasing function of the lateral distance from the winning neuron m (center) and resembles the functionality of the Mexican hat. A typical choice is the Gaussian function [7],

$$h_{i,m}(p) = \exp\left(-\frac{d_{i,m}^2}{2\sigma^2(p)}\right) \tag{18.36}$$

where $d_{i,m}^2 = ||\underline{r}_i - \underline{r}_m||^2$ is the lateral distance between the winning neuron m and neuron $i \in N_m(p)$, $\underline{r}_m$ and $\underline{r}_i$ are the corresponding location vectors of neurons m and i with respect to a reference point, and $\sigma^2(p)$ is the variance of the Gaussian function. The parameter $\sigma(p)$ measures the extent to which excited neurons in the vicinity of the winning neuron participate in the learning process. Since the neighborhood size $N_m(p)$ shrinks with iteration p, $\sigma(p)$ also reduces exponentially using $\sigma(p) = \sigma_0 \exp(-p/\tau)$, where τ is the time constant controlling the rate of this decline and σ_0 is the initial neighborhood width. A commonly used choice [7] for the time constant is $\tau = \frac{1000}{\log \sigma_0}$. As a consequence, $\sigma(p)$ decides the size of the excitatory neighborhood and the number of neurons that participate in the cooperative aspect of this learning.

As can be observed from (18.35), SOFM learning moves the weight vector of the winning neuron, as well as those of the neighboring neurons in $N_m(p)$, toward the input vector $\underline{x}(p)$, although the neighboring neurons get only fractions of the weight updates due to the decreasing effect of $h_{i,m}(p)$ as a function of the lateral distance.

At the end of the training, the SOFM forms a topologically ordered planar neuron map with weight vectors coding the probability distribution of the training data. That is, the neurons map inputs into encoded distributions that represent the computed values of parameters by sites of maximum relative activity within the map. However, the conditions that are essential for successful formation of an organized map are: (i) neurons must be exposed to sufficient number of training data samples; (ii) only weights associated with an excited neighborhood of the map are affected; (iii) the size of the topological neighborhood must decrease monotonically with increasing number of iterations; and (iv) weight adjustments are proportional to the activation received by each neuron within the neighborhood.

To illustrate the process of forming an organized map during the SOFM learning, consider a network with 100 neurons arranged in a 10×10 lattice [7]. The network is trained to organize 2-D input vectors of training data $\underline{x}(p) = [x_1(p)\ x_2(p)]^T$. The network is trained using 1000 2-D input vectors whose elements were generated randomly with uniform distribution

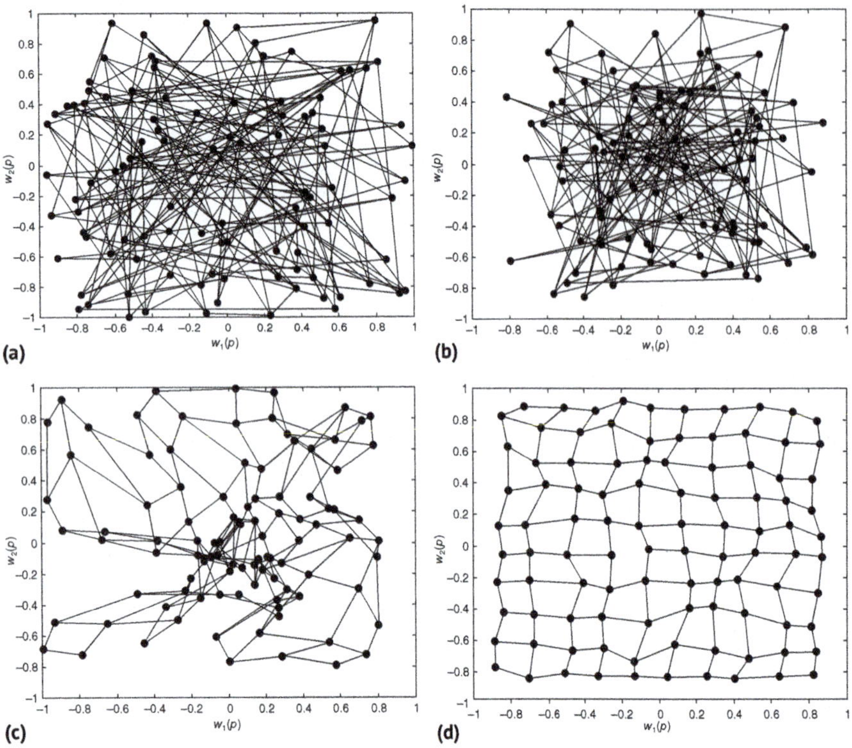

Figure 18.10 Organized map in weight space during SOFM learning after: (a) random initialization, (b) 100 iterations, (c) 1000 iterations, and (d) 10000 iterations.

over the square region $[-1, 1] \times [-1, 1]$, and learning rate parameter was chosen as $\alpha = 0.5$. Figure 18.10 shows different stages of SOFM learning in the weight space (w_1, w_2) until convergence to an ordered map. At convergence, the statistical distribution of the neuron weight vectors closely approximates the underlying distribution of the input data vectors except for some perturbation that will be discussed shortly. Using the results of this example, we can summarize the key features of the SOFM as follows [7]:

1. **Approximation of the Input Space:** The feature map, represented by the set of synaptic weight vectors, provides a good approximation of the input space.
2. **Topological Ordering:** The feature map generated by SOFM is topologically ordered in the sense that the spatial locations of neurons in the lattice correspond to particular domains or features of the input patterns.
3. **Density Matching:** The feature map reflects variations in the statistics of the input distribution. That is, regions in the input space from which sample vectors $\underline{x}$ are drawn with a high probability of occurrence are allocated larger portions of the output space and hence with better representational resolution than regions with lower probability of occurrence. This property explains the slight perturbations observed in the converged map of Figure 18.10d, which occur due to the lack of adequate number of training samples near the boundaries of the square region.
4. **Feature Selection:** Given data drawn from an arbitrary input distribution, SOFM is capable of discovering a set of representative features that approximate the underlying distribution.
5. **Other Applications:** In addition to the aforementioned applications in digital image processing, SOFM has been successfully applied to sensory mapping, robot control, data mining, data analysis and visualization, dimensionality reduction, clustering, and speech processing.

18.3.2 Practical Considerations and Issues

Although SOFM learning rule is simple, its performance depends on the proper selection of learning parameters, neighborhood structure, network size, and training data representation. In what follows, we highlight the important factors that affect the performance.

1. **Choice of Learning Rate $\alpha(p)$:** The learning rate $\alpha(p)$ should decrease monotonically during training. A learning rate of $\alpha(p) \approx 1$, is desirable during the early stages of learning in order to promote rapid global organization of the map. During the later stages, a smaller learning rate is preferred to allow fine local adjustments and convergence. Typically, we use a monotonically decreasing function, for example: (a) *linear:* $\alpha(p) = \alpha_0\left(1 - \frac{p-q}{\tau}\right)$ where $\alpha_0 = 0.9$, q is the epoch at which decay starts, and τ is the time constant; (b) *exponential:* $\alpha(p) = \alpha_0 e^{\left(-\frac{p-q}{\tau}\right)}$; or *Gaussian:* $\alpha(p) = \alpha_0 e^{-\frac{\|\underline{r}_i - \underline{r}_m\|^2}{2\sigma^2(p)}}$, where $\underline{r}_m$, $\underline{r}_i$, and $\sigma^2(p)$ were defined before. Note that if $\alpha(p)$ decreases too rapidly, adequate organization of the map may not take place. Conversely, if $\alpha(p)$ decays too slowly, the map may not converge.
2. **Choice of Neighborhood $N_m(p)$:** As discussed before, the neighborhood $N_m(p)$ determines the excitatory region around the winning neuron m. The size of $N_m(p)$ shrinks as a function of iteration p, i.e. $N_m(p_1) \supset N_m(p_2) \supset \cdots \supset N_m(p_l)$ for $p_1 < p_2 < \cdots < p_l$. This gradual shrinkage is effectively imposed by the weighting function $h_{i,m}(p)$ in the SOFM learning rule of (18.35). Thus, the neighborhood function $h_{i,m}(k)$ controls both the effective neighborhood size and the magnitude of weight updates depending on the lateral distance from the winning neuron. The initial choice of $N_m(0)$ is an important factor for the organization of the map. If $N_m(0)$ is too small to start with, the map will not be ordered globally and the resultant map may exhibit fragmented or mosaic-like parcellations without any specific structure. In contrast, if $N_m(0)$ is too large, important local details are lost leading to an overly smooth map. Toward the end of SOFM learning, the neighborhood will consist of only the winning neuron.
3. **Number of Training Epochs:** The number of training epochs depends on many factors including the complexity of input data distribution, number of neurons, and the convergence criterion. A commonly used rule of thumb is to perform at least several hundred training epochs per neuron, e.g. $500M$, where M is the number of neurons. The number of epochs is not dependent on the dimension of the input data N. The convergence criterion can be decided based on factors such as quantization error or average change in the weight vectors.
4. **Number of Clusters:** One of the main challenges in SOFM and most unsupervised clustering algorithms is the lack of any rigorous method for choosing a suitable number of neurons or clusters M when no little prior information about the data distributions is available. In general, it is preferable to start with a relatively large number of neurons, e.g. $M \approx 5\sqrt{P}$ where P is the number of training samples, rather than a small one. Several variants of SOFM designed to address this problem through adaptive network growth, one of which is the growing SOFM discussed in the next section.

18.3.3 Growing SOFM

A growing self-organizing map (GSOM) was proposed in [26], which dynamically adds neurons during training to automatically produce a network structure with a suitable number of neurons. During the network growth phase, the weight vectors of the neurons continue to self-organize as in the original SOFM. This adaptive growth allows the data analyst to monitor and control the growth process. GSOM learning [26, 27] has three phases that are outlined below.

1. **Phase I – Initialization:** Initialize the network with a minimal topology consisting of four *boundary neurons,* which can generate additional neurons in their neighborhood [26, 27]. A boundary neuron is the one for which at least one of the immediate neighboring positions in the 2-D lattice is unoccupied. The weight vectors of these four boundary neurons are randomly initialized in the interval [0, 1].

   Compute the *growth threshold* (GT) using $GT = -N\ln(SF)$, where N is the dimension of the input data vector and $0 < SF < 1$ is the *spread factor,* which controls the spread and growth of the network. It is recommended to start with a low SF and then gradually increase it for finer cluster formation and a better ordered map.
2. **Phase II – Growing Phase:**
   a. Present the training pattern $\underline{x}(p)$ and identify the winning neuron m using the same procedure as the original SOFM.
   b. Update the weight vectors of the winning neuron and those of its neighbors within $N_m(p)$ using (18.35), where the size of this neighborhood is smaller than that of the original SOFM.
   c. Compute the accumulated quantization error between the input vector and the weight vector of the winner neuron m using $\varepsilon_m(p+1) = \varepsilon_m(p) + \|\underline{x}(p) - \underline{w}_m\|$. It is important to note that this error is accumulated over iterations at which the same neuron m wins the competition.
   d. If $\varepsilon_m(p+1) > GT$, i.e. the neuron is not participating in ordering the map, and that the winning neuron m is a boundary neuron, then create additional neurons in the free positions available around the neuron m in the 2-D

lattice to improve the representation of that region in the input space. The newly created neurons are assigned initial weight values that depend on the place the neuron was added and the weights of its neighboring neurons [26]. One strategy to initialize the weights of the newly grown neuron j for input i, is to use $w_{i,j} = 2w_{i,m} - w_{i,k}$, where neuron k is located opposite the newly created neuron relative to neuron m. If such a neuron does not exist, an alternative estimate is to use $w_{i,j} = w_{i,m} + w_{i,k} - w_{i,l}$, where $w_{i,k}$ and $w_{i,l}$ are weights of the neurons closest to the new neuron. After growth process, reset the learning rate to its initial value and repeat steps in Phase II until all input patterns have been presented and the growth rate of the neurons is reduced to a minimum level.

e. If $\varepsilon_m(p+1) > GT$ but the winning neuron m is not a boundary neuron, the resultant map may not give a proportionate representation of the input data distribution and no growth takes place. In this case, the accumulated error is *distributed* to the neighboring neuron using a simple procedure. First, change the cumulative error of the winning neuron to $\varepsilon_m(p+1) = GT/2$. The accumulated error of the immediate neighboring neurons j of the winner receives a boost using $\varepsilon_j(p+1) = \varepsilon_j(p) + \gamma\varepsilon_m(p)$, where γ is a constant value called the *distribution factor* and controls the increase in the accumulation of distributed errors. Thus, these equations have the net effect of spreading the error outward to the boundary neurons to promote growth.

3. **Phase III – Smoothing Phase:** The goal of this phase is to fine-tune the positions of the weight vectors and improve the continuity and smoothness of the map. Once the growing phase is completed, a conventional SOFM learning is applied using a smaller learning rate and an initial neighborhood size. Unlike the SOFM, each neuron in a GSOM, has a weight vector and 2-D coordinates that identify its position in the lattice. Thus, testing involves identifying the positions of inputs within an already trained network [26].

In Section 17.3.4, we presented an example of the basic competitive learning for vector quantization. Here, we repeated the same experiment using SOFM instead of the original competitive learning to illustrate the formation of a topologically meaningful feature map. The network consists of an 8×16 lattice containing $K = 128$ codebook vectors. The initial size of the neighborhood $N_m(0)$ was chosen to be 4×8 and $\alpha(0) = 0.9$. The other parameters are chosen according to the guidelines provided in Section 18.3.2. The experiment was performed separately for the Lena and Peppers images to illustrate that the organization of the feature map is entirely based on the statistical distribution of local image properties. Figure 18.11a, c shows the reconstructed images obtained using the established maps, providing SNR values of 15.2 dB and 15.8 dB, respectively. Figure 18.11b,d, on the other hand, illustrates the corresponding maps that are ordered according to the distribution of the features in the image blocks. As can be observed, codebook vectors representing similar local image structures within the 4×4 blocks are organized into neighboring regions of the feature map. Furthermore, the spatial resolution of the feature representation in these topologically ordered maps depends on the probability of occurrence of such local features in the training data. Although the local spatial contents within 4×4 of these two images are broadly similar, their statistical distributions differ significantly, resulting in distinctly different topological organizations of their feature maps.

18.3.4 Learning Vector Quantization

To use SOFM as a classifier, the weight vectors of a trained network must be fine-tuned or calibrated using a subsequent supervised learning known as Learning Vector Quantization (LVQ) [28, 29]. LVQ is applied to refine the placement of the weight or *codebook* vectors $\underline{w}_i$ using an iterative learning process, which starts by presenting patterns with known class labels and assigning neurons to different classes by majority voting. Once the codebook vectors have been properly placed and assigned labels, the neuron with the largest response, or the codebook vector closest to the input vector, defines the class label of the input pattern. There are many variants of the LVQ method. Here, we present the original LVQ1 and LVQ2 proposed by Kohonen [28, 29].

1. **LVQ1 Algorithm:**
   It is assumed that codebook vectors, $\underline{w}_i$, have already been established using unsupervised SOFM or classical VQ. The purpose of LVQ1 is to pull the codebook (weight) vectors away from the decision boundary to minimize the occurrences of misclassifications and achieve classification performance close to the Bayes decision boundary. The weight-updating rule for LVQ1 is based on the principle of *Reward and Punishment*, described in the following steps:
   1. Apply SOFM or GSOM to the unlabeled training data $\{\underline{x}(p)\}_{p=1}^{P}$ to form the clusters and associated codebook vectors $\underline{w}_i$, $i \in [1, M]$. The stopping criterion is reached when the codebook vectors have stabilized or a maximum number of epochs has been completed.

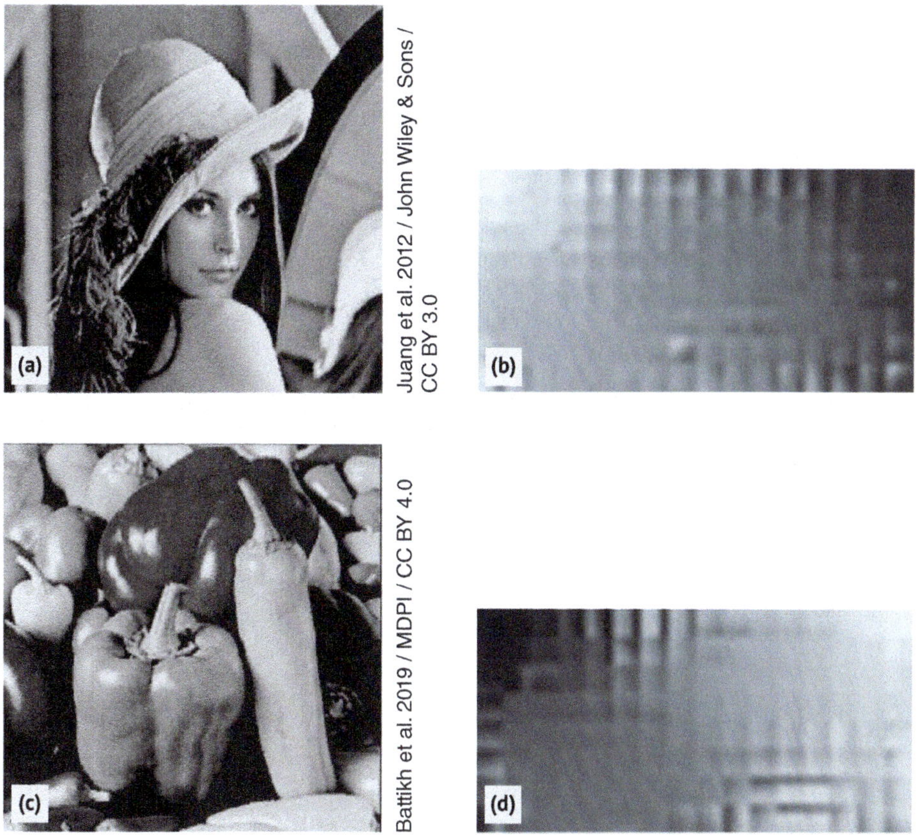

Figure 18.11 (a) and (c) Reconstructed Lena and Peppers images using K=128 codebook vectors; (b) and (d) corresponding topologically ordered feature maps of the trained SOFM.

2. Use a limited set of labeled data $\{\underline{x}(k), d(k)\}_{k=1}^{K}$ with $K \ll P$ to label and fine-tune the clusters in SOFM. This is accomplished by applying $\underline{x}(k) \in C_i$ from the labeled training dataset. If $\underline{x}(k)$ is correctly classified, i.e. the codebook vector $\underline{w}_i$, which belongs to the same class is the nearest codebook vector, then move the codebook vector closer (reward) to this input using,

$$\begin{aligned} \underline{w}_i(k+1) &= \underline{w}_i(k) + \beta(k)\big(\underline{x}(k) - \underline{w}_i(k)\big) \\ \underline{w}_j(k+1) &= \underline{w}_j(k), \quad j \neq i, \end{aligned} \tag{18.37}$$

where the parameter $\beta(k)$ is initialized to a small value, e.g. $\beta(0) \sim 0.01 - 0.03$ and slowly decays to zero over time. However, if $\underline{x}(k)$ is incorrectly classified and $\underline{w}_l$ is the nearest codebook vector, then move the weight vector of this neuron farther (punishment) from the input using,

$$\begin{aligned} \underline{w}_l(k+1) &= \underline{w}_l(k) - \beta(k)\big(\underline{x}(k) - \underline{w}_l(k)\big) \\ \underline{w}_j(k+1) &= \underline{w}_j(k), \quad j \neq l, \end{aligned} \tag{18.38}$$

Figure 18.12a illustrates the LVQ1 codebook vector refinement process. After training, the placement of $\underline{w}_i$ will be such that classification using the nearest-neighbor criterion closely approximates that of the Bayes classifier.

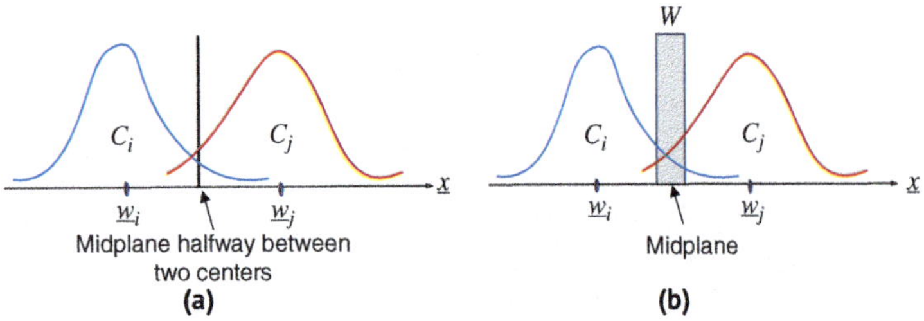

Figure 18.12 Examples of (a) LVQ1 and (b) LVQ2 decision boundary refinement.

2. **LVQ2 Algorithm:**
LVQ2 algorithm provides a better approximation to the Bayes decision boundary than LVQ1. Instead of modifying only the closest codebook vector to an input vector $\underline{x}(k)$, LVQ2 adjusts the two closest codebook vectors, $\underline{w}_i$ and $\underline{w}_j$, i.e. the winner and the runner-up. Assuming that these codebook vectors belong to two different classes, the decision boundary between them is always in the midplane of $\underline{w}_i$ and $\underline{w}_j$. Then, the objective of LVQ2 is to shift this local decision boundary toward the region where the Bayes decision boundary is expected to lie. Define a symmetric window of width W around the midplane, and make corrections to both $\underline{w}_i$ and $\underline{w}_j$ when $\underline{x}(k)$ falls within this window and lies on the wrong side of the midplane. The steps of the LVQ2 algorithm are as follows.

1. Same as LVQ1.
2. Apply $\underline{x}(k) \in C_i$ from the labeled training set and choose two nearest codebook vectors $\underline{w}_i$ and $\underline{w}_j$, i.e. the winner and runner-up.
3. If $\underline{x}(k)$ is closer to $\underline{w}_j$ and falls in W (on the wrong side) then,

$$\begin{aligned} \underline{w}_i(k+1) &= \underline{w}_i(k) + \beta(k)\big(\underline{x}(k) - \underline{w}_i(k)\big) \\ \underline{w}_j(k+1) &= \underline{w}_j(k) - \beta(k)\big(\underline{x}(k) - \underline{w}_j(k)\big) \\ \underline{w}_l(k+1) &= \underline{w}_l(k), \quad \forall l \neq i \text{ or } j \end{aligned} \tag{18.39}$$

   where the parameter $\beta(k)$ is the same as that in LVQ1. Using (18.39), the codebook vector $\underline{w}_i$ is moved closer to the input pattern, whereas the incorrect codebook vector $\underline{w}_j$ is moved farther away. The input vector $\underline{x}(k)$ is considered to be in window W if

$$\min\left\{\frac{d_i}{d_j}, \frac{d_j}{d_i}\right\} > s$$

   where $d_i = \|\underline{w}_i - \underline{x}(k)\|$, $d_j = \|\underline{w}_j - \underline{x}(k)\|$ and $s = \frac{1-W}{1+W}$. In general, the width of W depends on the number of training samples. For a large number of samples, W is a narrower, whereas smaller datasets require a wider window to ensure that a sufficient number of training samples participate in the adaptation process. For good statistical accuracy, a sufficient number of samples should fall into W.
4. If the network correctly classifies $\underline{x}(k) \in C_i$, only the weight vector $\underline{w}_i$ is moved toward the input vector.

Figure 18.12b illustrates how LVQ2 moves the local decision boundary toward the overlap region of the two distributions, hence minimizing the incidence of classification error.

18.4 Computer Projects

The computer projects in this chapter use Python libraries torch, sklearn, NumPy, random, and Matplotlib.

Computer Project 18.1: The purpose of this computer project is to design, train, and test a three-layer MLP network for pattern classification using a subset (20%) of the CIFAR-10 dataset. The CIFAR-10 dataset consists of 60,000 32 × 32 color images of 10 different classes, with 6000 images per class. There are 50,000 training images and 10,000 test images of various animals (birds, cats, deer, dogs, frogs, horses), vehicles (cars, trucks), airplanes, and ships. This dataset can be downloaded from https://www.cs.toronto.edu/~kriz/cifar.html. First, randomly select two subsets of the original training dataset each containing 5000 images for training and a validation. The validation subset is used to select the best performing MLP network. The testing dataset that contains 2000 new images (i.e. no overlap with training or validation sets) should also be selected randomly from the original test dataset. Each subset must contain an equal number of images from every class to avoid class imbalance and ensure unbiased performance evaluation. Design a proper network structure by selecting the number of neurons in the hidden layers. Follow the guidelines suggested in Section 18.1.2 for weight initialization, learning parameters, network architecture, and other implementation issues. Train the network using error back-propagation learning and the generalized delta rule with a momentum factor. Select an appropriate maximum allowable error for the stopping criterion. Monitor the learning process and provide the learning curve of the final trained network. Use the learning curve to determine whether the network is training properly and the selected architecture is appropriate. At least 10 random weight initialization should be tried to select the optimally trained network based on simultaneous evaluation on the validation set. The number of hidden-layer neurons can be adjusted, if necessary, depending on the performance on the validation dataset. Provide a comprehensive performance evaluation of your MLP network based on the overall correct classification rate and the confusion matrix using the testing dataset. Additionally, present several examples of misclassified images and explain possible reasons for the incorrect classifications. Provide a detailed discussion on the strengths and weaknesses of your MLP network.

Computer Project 18.2: This computer project also uses the same CIFAR-10 dataset and experimental setup as Computer Project 18.1. The goal is to compare the generalization performance of different CNN architectures, learning algorithms, and loss functions using the CIFAR-10 testing dataset. Design a three-layer (i.e. three convolutional and three max pooling layers) with ReLu activation functions. The appropriate numbers of convolutional filters in each layer and mask sizes are design choices and should be determined experimentally. The final single-layer neurons should have softmax activations to make classification decisions. At least two different learning algorithms, including standard SGD with a momentum term and an alternative algorithm, e.g. RMSprop or Adam optimizer should be tried. In addition, investigate the effect of two different loss functions on classification performance. At least 5 random weight initializations should be used for each CNN architecture in order to select the best performing network on the validation set. Provide the learning curves and performance plots during the learning for every CNN, learning algorithm, and choice of loss functions for those properly trained networks and determine their generalization ability on the testing dataset. Provide a comprehensive benchmarking of different CNN models based on their overall correct classification rates, the associated confusion matrices, number of filters in each layer, and the total number of parameters or model complexity. Moreover, present several examples of misclassified images, and explain the reasons behind the incorrect decisions. Summarize your findings by discussing the advantages and disadvantages of CNN-based image classification systems and identifying the design choices that had the greatest impact on performance.

Computer Project 18.3: This computer project uses the same CIFAR-10 data subsets as Computer Project 18.1. First, combine the training and validation datasets from Computer Project 18.1 to form a training set containing 10,000 images and use it to train a SOFM. For weight initialization, learning parameters, and neighborhood function, follow the procedures in Section 18.3.2. The network should have 250 neurons arranged in a 2-D lattice of size 25×10. Use at least 10 training epochs to ensure that the cluster weights have properly converged. Investigate the topological ordering of the map by reshaping the codebook vectors into 32×32 images and forming the ordered map image similar to those in Figure 18.11b,d. To use the trained network as a classifier randomly select 200 samples of the testing data subset (20 images per class) and use them to label and fine-tune the cluster centroids using the LVQ2 algorithm. Then, test the performance of the SOFM-based classifier on the rest of the testing dataset. Provide the overall correct classification rates and the confusion matrix of this classifier. In addition, present several examples of misclassified images and explain the possible reasons for the incorrect classifications. Conclude with a detailed discussion of the strengths and weaknesses of the SOFM/LVQ-based classification system.

18.5 Problems

Problem 18.1 Repeat Example 18.1 using the ReLU activation function at all neurons. How does the decision boundary region change comparing to that in Figure 18.1b?

Problem 18.2 Repeat Problem 18.1 using threshold logic (slope = 1) activation function at all neurons.

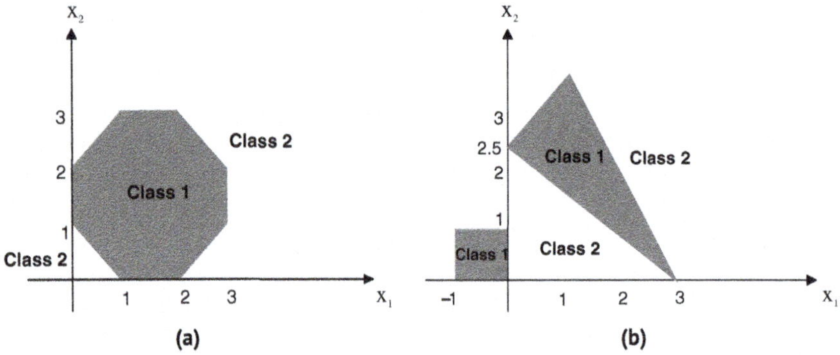

Figure 18.13 (a) Figure for Problem 18.4 and (b) figure for Problem 18.5.

Problem 18.3 Repeat Problem 18.1 using the bipolar sigmoidal activation function at all neurons [30]. You may use an appropriate Python function generate the decision boundaries. Describe your findings.

Problem 18.4 Consider the decision regions in Figure 18.13a [30] for a two-class classification problem. Given the geometry of the decision regions:

a. Design a suitable MLP network to implement this classifier.
b. Determine the weight vectors of neurons in the network of part (a).

Problem 18.5 Repeat Problem 18.4 for the decision regions in Figure 18.13b.

Problem 18.6 Consider the network in Example 18.1. Assume the weight matrix for the hidden layer and the weight vector for the output layer were initialized (bias included) as follows. $\mathbf{W}^{(1)}(0) = \begin{bmatrix} 0 & -1 & 1 \\ 1 & 0 & 0 \\ 0 & -1 & 0 \\ -1 & -1 & 1 \end{bmatrix}$, and $\underline{w}^{(2)}(0) = [0\,0\,-1\,1\,-1]$, we would like to train this two-layer network to classify two classes of patterns. The training samples of Class 1 are $\underline{x}(1) = [1.5\ 1]^T$ and $\underline{x}(2) = [1.75\ 2]^T$ while those of Class 2 are $\underline{x}(3) = [0.5\ 1]^T$ and $\underline{x}(4) = [3\ 2]^T$. The corresponding target values for these samples are $d(1) = d(2) = 1$ and $d(3) = d(4) = 0$, respectively. Assuming that threshold logic nonlinearity (slope $= 1$) is used at all neurons, compute 2 iterations of the error back-propagation algorithm using $\mu = 0.1$, and find the weight matrix and vector after these iterations. Explain your results and observations.

Problem 18.7 A two-layer MLP with two neurons in the hidden layer and two output neurons has the following initial weight matrices (bias included):

$$\mathbf{W}^{(1)}(0) = \begin{bmatrix} 1 & -1 & 0 \\ 1 & 0 & -1 \end{bmatrix} \qquad \mathbf{W}^{(2)}(0) = \begin{bmatrix} 1 & 0 & -1 \\ 1 & -1 & 1 \end{bmatrix}$$

respectively. We would like to train this two-layer network to classify the same patterns in Problem 18.6, though using desired vectors $\underline{d}(1) = \underline{d}(2) = [1\ 0]^T$ and $\underline{d}(3) = \underline{d}(4) = [0\ 1]^T$. Assuming that the threshold logic activation function (slope $= 2$) is used at all neurons, perform 2 iterations of the error back-propagation algorithm using $\mu = 0.1$, and find the final weight matrices. Explain your results and observations.

Problem 18.8 We would like to design a three-class classifier to separate the following nine 2-D data samples. For Class 1, data samples are $\underline{x}(1) = [\frac{1}{4}\ \frac{1}{4}]^T$, $\underline{x}(2) = [\frac{1}{2}\ \frac{3}{4}]^T$, and $\underline{x}(3) = [\frac{3}{4}\ \frac{1}{8}]^T$. For Class 2, data samples are $\underline{x}(4) = [\frac{1}{4}\ \frac{1}{2}]^T$, $\underline{x}(5) = [\frac{1}{2}\ \frac{1}{2}]^T$ and $\underline{x}(6) = [\frac{3}{4}\ \frac{1}{2}]^T$. And for Class 3 data samples are $\underline{x}(7) = [\frac{1}{2}\ \frac{1}{8}]^T$, $\underline{x}(8) = [\frac{3}{4}\ \frac{1}{4}]^T$ and $\underline{x}(9) = [\frac{1}{4}\ \frac{3}{4}]^T$.

a. Is this a linearly separable problem?
b. Design an MLP network with a suitable number of layers and neurons in each layer to separate these patterns.
c. Use the desired vectors $\underline{d}(1) = [1\ 0\ 0]^T$, $\underline{d}(2) = [0\ 1\ 0]^T$ and $\underline{d}(3) = [0\ 0\ 1]^T$ for classes 1–3, respectively, and determine whether a unipolar HL can be used as an activation function to solve this problem.
d. Repeat part (c) with the ReLU activation function.

Problem 18.9 Derive the error back-propagation algorithm for a two-layer MLP with softmax activation function at output layer cells and ReLU activation function at the hidden-layer neurons. How do you compare the capabilities of this network with the one using unipolar sigmoidal activation function at all neurons?

Problem 18.10 Derive the error back-propagation algorithm for a two-layer MLP using the cross-entropy loss instead of the SSE. What are your findings?

Problem 18.11 A single-layer feedforward network with M output neurons uses RBF activation functions with known centers $\underline{\mu}_i$, $i \in [1, M]$ and smoothing parameter $\sigma = 1$ at all neurons.

a. Given a labeled training dataset $\{\underline{x}(p), \underline{d}(p)\}_{p=1}^{P}$, develop a learning scheme to find the unknown weight vectors $\underline{w}_i$, $i \in [1, M]$ of the network.
b. What type of problems is this network capable of solving? Explain with simple examples.

Problem 18.12 Derive equation (18.9) when a momentum term is used and state the practical implications of your result.

Problem 18.13 Consider a simple one-layer CNN structure (one convolutional and one pooling layers) with one fully connected layer with two output neurons for a two-class classification problem. Using the same notations used in Section 18.2.2, derive the forward and error back-propagation equations for the cross-entropy loss function when the activation functions of the convolutional layer neurons and fully connected layer neurons are ReLU and softmax, respectively.

Problem 18.14 Repeat Problem 18.13 for the focal loss function and comment on the results.

Problem 18.15 Consider the Adam learning optimizer in Section 18.2.3, which has three steps, i.e. moving averages updates of gradient and squared gradients, bias correction, and weight updating. If we denote the instantaneous squared gradient by $g^2(p) = \left(\frac{\partial \ell(W,p)}{\partial \underline{w}(p)}\right)^2$, then (18.33) simplifies to

$$\nu(p+1) = \beta_2 \nu(p) + (1 - \beta_2) g^2(p)$$

where $\beta_2 \in [0, 1)$ controls the exponential decay rate of the moving average.

a. Express $\nu(p)$ in terms of the gradients $g(0), g(1), \ldots, g(p)$.
b. Using the result in part (a) and assuming that $g(0), g(1), \ldots g(p)$ are statistically independent and identically distributed (iid), express $E[\nu(p)]$ in terms of $E[g^2(p)]$ where $E[.]$ is the expectation operator.
c. Using the result in part (b), what would happen if we do not perform the bias correction step? Explain the importance of this step.

Problem 18.16 Consider a SOFM consisting of six neurons arranged in a 2-D rectangular lattice. The network has two inputs and the weight vectors connecting these inputs to neurons are $\underline{w}_1 = [-1\ 2]^T$, $\underline{w}_2 = [0\ 4]^T$, $\underline{w}_3 = [3\ -2]^T$, $\underline{w}_4 = [-2\ -3]^T$, $\underline{w}_5 = [3\ 2]^T$, and $\underline{w}_6 = [4\ -1]^T$, respectively. An input pattern $\underline{x} = [2\ -4]^T$ is applied.

a. Determine the winner neuron.
b. Develop the weight-updating equations for part (a) given that $\alpha = 0.5$ and the neighborhood function

$$h = \begin{cases} 1, & \text{for winner neuron} \\ 0.5, & \text{for winner's immediate neighboring neurons} \\ 0, & \text{for neurons outside neighborhood} \end{cases}.$$

Problem 18.17 Repeat Problem 18.16 for a hexagonal array of six neurons.

Problem 18.18 Construct a SOFM architecture using RBF activation functions at neurons, and develop a learning algorithm for its parameters.

Problem 18.19 We would like to form an ordered map of an appropriate size lattice for the following four input vectors $\underline{x}(1) = [0\,0\,0\,1]^T$, $\underline{x}(2) = [0\,0\,1\,0]^T$, $\underline{x}(3) = [0\,1\,0\,0]^T$, and $\underline{x}(4) = [1\,0\,0\,0]^T$. Design a SOFM network for this purpose using $\alpha = 0.5$ and the neighborhood function

$$h = \begin{cases} 1, & \text{for winner neuron and immediate neighbors} \\ 0, & \text{for neurons outside neighborhood.} \end{cases}$$

Comment on the results and your findings.

Problem 18.20 Assuming that the weight vectors are updated at the end of each epoch, derive the batch version of the SOFM and comment on its performance.

References

1 D. H. Hubel and T. N. Wiesel, "Receptive fields, binocular interaction and functional architecture in the cat's visual cortex," *Journal of Physiology*, vol. 160, pp. 106–154, 1962.
2 Y. LeCun and Y. Bengio, "Convolutional networks for images, speech, and time series," *The handbook of brain theory and neural networks*, vol. 3361, p. 10, 1995
3 Y. LeCun, L. Bottou, Y. Bengio, and P. Haffner, "Gradient-based learning applied to document recognition," *Proceedings of the IEEE*, vol. 86, pp. 2278–2324, 1998.
4 Y. LeCun, Y. Bengio, and G. Hinton, "*Deep Learning*," Nature Publishing Group, pp. 436–444, 2015.
5 I. Goodfellow, Y. Bengio, and A. Courville, "*Deep Learning*," MIT Press, 2016.
6 D. E. Rumelhart, G. E. Hinton, and R. J. Williams, "Learning representations by backpropagating errors," *Nature*, 323, pp. 533–536, 1986.
7 S. Haykin, "*Neural Networks and Learning Machines*," Prentice-Hall, 3rd Edition, 2009.
8 T. Kohonen, "Self-organized formation of topologically correct feature maps," *Biological Cybernetics*, vol. 43, pp. 59–69, 1982.
9 T. Kohonen, "Essentials of the self-organizing map," *Neural Networks*, vol. 37, pp. 52–65, 2013.
10 S. Kay, "*Fundamentals of Statistical Signal Processing, Volume I: Estimation Theory*," Prentice-Hall, 1st Edition, 1993.
11 G. Montavon, G. B. Orr, and K. Muller, "*Neural Networks: Tricks of the Trade*," Springer, 2nd Edition, 2012.
12 T. J. Sejnowski and C. R. Rosenberg, "Parallel networks that learn to pronounce English text," *Complex Systems*, vol. 1, pp. 145–168, 1987.
13 G. H. Golub and C. F. Van Loan, "*Matrix Computation*," John Hopkins University Press, 1996.
14 K. Fukushima, "Neocognitron: A self-organizing neural network model for a mechanism of pattern recognition unaffected by shift in position," *Biological Cybernetics*, vol. 36, pp. 193–202, 1980.
15 K. Fukushima, "Neocognitron: A hierarchical neural network capable of visual pattern recognition," *Neural Networks*, vol. 1, pp. 119–130, 1988.
16 A. Krizhevsky, I. Sutskever, and G. E. Hinton, "ImageNet classification with deep convolutional neural networks," *Communications of the ACM*, vol. 60, pp. 84–90, 2017.
17 T. M. Cover and J. A. Thomas, "*Elements of Information Theory*," Wiley, 2nd Edition, 2006.

18 J. Gu, Z. Wang, J. Kuen, L. Ma, A. Shahroudy, B. Shuai, T. Liu, X. Wang, G. Wang, J. Cai, and T. Chen, "Recent advances in convolutional neural networks," *Pattern Recognition*, vol. 77, pp. 354–377, 2018.

19 J. Jin, K. Fu, and C. Zhang, "Traffic sign recognition with hinge loss trained convolutional neural networks," *IEEE Transactions on Intelligence Transportation Systems*, vol. 15. pp. 1991–2000, 2014.

20 Y. Lin, P. Goyal, R. Girshic, K. He, and P. Dollar, "Focal loss for dense object detection," IEEE International Conference on Computer Vision, pp. 299–3007, 2017.

21 T. Mitchell, "*Machine Learning*," McGraw Hill, 1997.

22 Z. Zhang, "*Derivation of Back-propagation in Convolutional Neural Network (CNN)*," University of Tennessee, 2016.

23 Y. Yang, "A recursive least squares training approach for convolutional neural networks," MS Thesis, Colorado State University, 2022.

24 T. Tieleman and G. Hinton, "Lecture 6.5-rmsprop: Divide the gradient by a running average of its recent magnitude," *COURSERA: Neural Networks for Machine Learning*, vol. 4, pp. 26–31, 2012.

25 D. P. Kingma and J. Ba, "Adam: A method for stochastic optimization," *Proceedings of the ICLR*, 2015.

26 D. Alahakoon, S. K. Halgamuge, and B. Srinivasan, "Dynamic self-organizing maps with controlled growth for knowledge discovery," *IEEE Transactions on Neural Networks*, vol. 11, pp. 601–614, 2000.

27 A. L. Hsu, S-L. Tang, and S. K. Halgamuge, "An unsupervised hierarchical dynamic self-organizing approach to cancer class discovery and marker gene identification in microarray data," *Bioinformatics*, vol. 19, pp. 2131–2140, 2003.

28 T. Kohonen, "Improved versions of learning vector quantization," Proceedings of the IJCNN International Joint Conference on Neural Networks, San Diego, 1990.

29 T. Kohonen, "*Learning Vector Quantization*," Springer, 2001.

30 J. M. Zurada, "*Introduction to Artificial Neural Networks*," West Pub. Co., 1992.

19

Dimensionality Reduction Networks and Autoencoders

In Chapters 16–18, we presented several supervised and unsupervised learning machines for pattern classification and data clustering applications. This chapter introduces a new class of unsupervised machines called *autoencoders,* which are widely used for feature extraction and dimensionality reduction among other applications. Autoencoders typically consist of two layers of neurons. The first layer extracts reduced-dimensional features and performs encoding, while the second layer carries out decoding, or reconstruction, of the original data from the reduced-dimensional feature vector. The role of the second layer is critical during the training phase to extract salient and representative features from which the original data can be accurately reconstructed according to an appropriate performance measure. Once training is completed, the network is used mainly in the encoding mode to produce a reduced-dimensional feature set for each input data vector.

In general, many machines that employ unsupervised learning and perform data reduction and feature extraction can be viewed as autoencoders. These machines can be categorized based on whether they are linear or nonlinear, deterministic or stochastic, and so forth. An example of a linear autoencoder is the principal component analysis (PCA), implemented using sequential unsupervised Hebbian-based learning algorithms [1–5]. Examples of nonlinear autoencoders are various manifold learning methods [6–9] and neural network-based approaches [10–12]. Among stochastic autoencoders, the denoising autoencoder [13] learns the conditional distribution of the data from noisy observations. In this chapter, we introduce several conventional and neural network-based autoencoders and examine their applications in several image-processing tasks.

19.1 Linear Autoencoders

A linear autoencoder is a neural network structure, as shown in Figure 19.1, in which the desired vector at the output layer is set equal to the input vector during the training process. In this configuration, the network is trained to reproduce the input at the output layer. At first glance, this might not appear particularly useful. However, when the input data are mapped through a narrower layer of hidden neurons, then the network is expected to discover an efficient representation that compresses the input data at the hidden layer and reconstructs it at the output layer. For a linear autoencoder network with N inputs, N outputs, and M hidden-layer neurons, where $M < N$, it has been shown [14] that the least-squares (LS) solution for the network weights yields a linear combination of the first M principal eigenvectors of the input data covariance matrix. The problem, however, remains on how to couple this network structure with an efficient learning scheme to extract the principal eigenvectors and principal components of the input data directly and without the computationally intensive procedures of the conventional PCA covered in Chapter 6.

Oja [1] introduced a linear neuron model with a constrained Hebbian weight updating that converges to the first principal eigenvector and extracts the first PC of a stationary random vector at its output. Recall that for a pre-synaptic excitation vector $\underline{x}(n) \in \mathbb{R}^N$ and post-synaptic neuron output $y(n)$, the weight vector $\underline{w}(n)$ is updated using the coincident Hebbian learning rule

$$\underline{w}(n+1) = \underline{w}(n) + \mu y(n)\underline{x}(n).$$

It is easy to show that the norm of $\underline{w}(n)$ grows to infinity if there is no constraint imposed on the weight vector. To avoid this instability, Oja [1] modified this Hebbian rule by incorporating a normalization procedure. Let us assume that the updated weight vector $\underline{\tilde{w}}(n+1)$ is normalized at each iteration using,

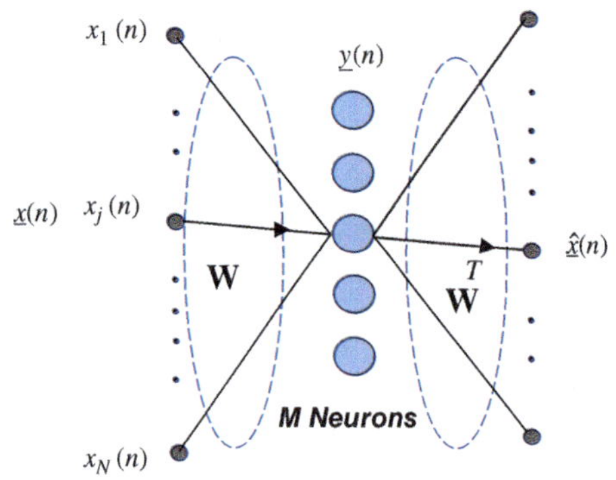

Figure 19.1 A linear autoencoder network for latent feature extraction.

$$\underline{\tilde{w}}(n+1) = \underline{w}(n) + \mu y(n)\underline{x}(n), \tag{19.1}$$

where

$$\underline{w}(n+1) = \frac{\underline{\tilde{w}}(n+1)}{||\underline{\tilde{w}}(n+1)||} \tag{19.2}$$

is the normalized updated weight vector. However, since $\underline{w}^T(n)\underline{x}(n) = y(n)$, the norm of (19.1) can be expressed as

$$||\underline{\tilde{w}}(n+1)||^2 = ||\underline{w}(n)||^2 + 2\mu\underline{w}^T(n)\underline{x}(n)y(n) + O(\mu^2) = 1 + 2\mu y^2(n) + O(\mu^2).$$

Thus, the normalization in (19.2) becomes

$$\underline{w}(n+1) = \underline{\tilde{w}}(n+1)/(1 + 2\mu y^2(n) + O(\mu^2))^{1/2} \approx \underline{\tilde{w}}(n+1)(1 - \mu y^2(n)). \tag{19.3}$$

Now, using (19.1) and ignoring the second- and higher order terms in μ, we get Oja's learning rule [1],

$$\underline{w}(n+1) = \underline{w}(n) + \mu[\underline{x}(n) - y(n)\underline{w}(n)]y(n). \tag{19.4}$$

Observe that the term $y(n)\underline{w}(n) = \hat{\underline{x}}(n)$ corresponds to the reconstructed input vector based on the scalar projection $y(n)$ and the weight vector $\underline{w}^T(n)$ of the second layer (see Figure 19.1), in which case Oja's rule can be simplified to

$$\underline{w}(n+1) = \underline{w}(n) + \mu e(n)y(n), \tag{19.5}$$

where $e(n) = \underline{x}(n) - \hat{\underline{x}}(n)$ is the reconstruction error. As can be seen, this modified Hebbian learning rule is indeed the same as the LMS rule presented in Chapter 17. It can be shown [1] that for a small enough step size using Oja's rule, $\lim_{n\to\infty} \underline{w}(n) \to \underline{\xi}_1$ and $\lim_{n\to\infty} y(n) \to X_1$, i.e. the weight vector converges to the first principal eigenvector while the neuron's output converges to the first PC. Thus, Oja's neuron may be interpreted as a single-neuron linear autoencoder whose objective is to minimize the mean-squared reconstruction error while extracting the dominant principal component.

To extend Oja's rule for computing the first M principal components (PCs), Sanger [2] developed the *generalized Hebbian learning* using a network with multiple neurons and an orthogonalization procedure. Foldiak [3] and Kung [4] subsequently used anti-Hebbian weight updating together with an iterative *deflation process* to obtain orthonormal principal eigenvectors. Assuming that the first $(l-1)$ principal eigenvectors (or weight vectors) $\underline{w}_i^*$, $i \in [1, l-1]$ have already been computed, the weight vector of neuron l can be obtained using the updating rule with the deflation process given by

$$\underline{w}_l(n+1) = \underline{w}_l(n) + \mu y_l(n)\left[\underline{x}(n) - \sum_{i=1}^{l-1} X_i(n)\underline{w}_i^* - y_l(n)\underline{w}_l(n)\right], \tag{19.6}$$

where $X_i(n)$ is the output of the neuron i or the ith PC of the input data vector $\underline{x}(n)$ based on the converged weights $\underline{w}_i^*$. The deflated input corresponds to the residual $\tilde{\underline{x}}_l(n) = \underline{x}(n) - \sum_{i=1}^{l-1} X_i(n)\underline{w}_i^*$, which serves as the desired output for the weight updating of neuron l. Thus, (19.6) can be written as

$$\underline{w}_l(n+1) = \underline{w}_l(n) + \mu y_l(n)[\underline{\tilde{x}}(n) - y_l(n)\underline{w}_l(n)], \tag{19.7}$$

which is identical to Oja's rule in (19.4) except with the deflated input.

As stated in Chapter 17, LMS-based learning rule suffers from accuracy-convergence speed trade-off problems. Moreover, this algorithm does not provide a direct mechanism for estimating the variance (or eigenvalue) associated with each PC. This information is important for determining the number of components needed to achieve an accurate representation of the data. Therefore, there is a need to develop a sequential algorithm that achieves both accuracy and convergence speed, as well as a mechanism to determine the variance of each PC. In [5], a learning algorithm based on recursive least squares (RLS) [15] was developed that offers fast convergence without sacrificing accuracy. In addition, the estimate of the variance of each extracted principal component can be generated as part of the learning process. This method is presented in the following section without going into the details of the derivation, which can be found in [5].

19.1.1 PC Extraction Using RLS-Based Learning

The structure of the network is the same as that shown in Figure 19.1. However, neurons are sequentially trained using the RLS learning algorithm [15] instead of the LMS learning rule. To find the weight vector of the first neuron $\underline{w}_1(n)$, the cost function to be minimized is the running sum of the squared error:

$$J_1(n) = \sum_{k=1}^{n} \left\| \underline{x}(k) - y_1(k)\underline{w}_1(k-1) \right\|^2 = \sum_{k=1}^{n} \left\| \underline{x}(k) - \underline{\hat{x}}(k) \right\|^2 \tag{19.8}$$

where $y_1(k) = \underline{w}_1^T(k-1)\underline{x}(k)$ is the output of this neuron for the input vector $\underline{x}(k)$, n is the present iteration, and $\underline{\hat{x}}(k)$ is the output of the reconstruction (decoder) layer. From this cost function, it is clear that the network seeks to learn an identity mapping, namely, $\underline{\hat{x}}(k) \approx \underline{x}(k)$. Given this cost function, the RLS-based weight-updating equations [5, 15] for the first neuron are

$$\begin{aligned} y_1(n) &= \underline{w}_1^T(n-1)\underline{x}(n) \\ K_1(n) &= \frac{P_1(n-1)y_1(n)}{1 + y_1^2(n)P_1(n-1)} \\ \underline{w}_1(n) &= \underline{w}_1(n-1) + K_1(n)\left[\underline{x}(n) - y_1(n)\underline{w}_1(n-1)\right] \\ P_1(n) &= (1 - K_1(n)y_1(n))P_1(n-1) \end{aligned} \tag{19.9}$$

where $K_1(n)$ is the gain and $P_1(n) = \left(\sum_{l=1}^{n} y_1^2(n)\right)^{-1}$. The initial values for $P_1(0) = \delta$, where δ is a small positive quantity, while the weights are initialized randomly. Note that the above weight update equation is driven by the reconstruction error $\underline{x}(n) - y_1(n)\underline{w}_1(n-1)$.

The gain $K_1(n)$ plays the role of an adaptive learning rate in the RLS algorithm. Unlike the fixed learning rate used in the LMS rule, the gain is adjusted automatically based on the statistical properties of the data. As a result, the RLS learning exhibits superior convergence speed while offering good estimation accuracy. In [5], it was shown that this learning rule converges to the first principal eigenvector, PC, and the corresponding eigenvalue, i.e. $\lim_{n\to\infty} \underline{w}_1(n) \to \underline{\xi}_1$, $\lim_{n\to\infty} y_1(n) \to X_1$ and $\lim_{n\to\infty} \frac{1}{nP_1(n)} \to \lambda_1$. Thus, this algorithm provides an estimate of the variance of each extracted PC. This information can be used to determine the dimensionality required for an accurate data representation.

To find the subsequent weight vectors $\underline{w}_l(n)$, $l \in [2, M]$, the cost function to minimize is the deflated version,

$$J_l(n) = \sum_{k=1}^{n} \| \underline{x}(k) - \sum_{i=1}^{l-1} X_i(k)\underline{w}_i^* - y_l(k)\underline{w}_l(k-1) \|^2 = \sum_{k=1}^{n} \left\| \underline{x}_l(k) - \underline{\hat{x}}_l(k) \right\|^2 \tag{19.10}$$

where $\underline{w}_i^*$ and $X_i(k)$ are the converged weight vector and the output of the neuron i, respectively. Note that $X_i(k)$ is the ith PC of the input data vector $\underline{x}(k)$. The cost function in (19.10) employs the deflation process to ensure that the extracted eigenvectors are orthonormal. The deflated input is $\underline{x}_l(k) = \underline{x}(k) - \sum_{i=1}^{l-1} X_i(k)\underline{w}_i^*$, which acts as the desired vector in the cost function of (19.10), where $\underline{\hat{x}}_l(k) = y_l(k)\underline{w}_l(k-1)$ is the estimate of $\underline{x}_l(k)$. The RLS-based generalized weight-updating equations for the lth neuron are

$$\begin{aligned} y_l(n) &= \underline{w}_l^T(n-1)\underline{x}(n) \\ K_l(n) &= \frac{P_l(n-1)y_l(n)}{1+y_l^2(n)P_l(n-1)} \\ \underline{w}_l(n) &= \underline{w}_l(n-1) + K_l(n)\left[\underline{x}(n) - \sum_{i<l} X_i(n)\underline{w}_i^* - y_l(n)\underline{w}_l(n-1)\right] \\ P_l(n) &= (1 - K_l(n)y_l(n))P_l(n-1), \end{aligned} \tag{19.11}$$

where as before, $P_l(n) = \left(\sum_{k=1}^{n} y_l^2(k)\right)^{-1}$ and $K_l(n)$ is the gain for neuron l. The deflation process in the updating equation of (19.11) uses a desired signal for the neuron l, which is obtained by subtracting the reconstructed version of the input based on the first $l-1$ PCs from the original input. In this learning rule, additional neurons or PCs can be introduced until the ratio $\hat{\lambda}_l/\hat{\lambda}_1$ falls below a prescribed threshold, where $\hat{\lambda}_l$ is the converged estimate of the variance of X_l, i.e. $\lim_{n\to\infty} \frac{1}{nP_l(n)} \to \hat{\lambda}_l$ and similarly for $\hat{\lambda}_1$. This criterion provides a natural mechanism for selecting an appropriate reduced-dimensional representation using linear autoencoders.

19.2 Nonlinear Dimensionality Reduction Using Manifold Learning

In many image processing applications, one encounters a very large volume of high-dimensional data. However, these high-dimensional data are often much simpler than the dimensionality indicates. Specifically, a low-dimensional underlying process exists with a few intrinsic degrees of freedom, relatively low noise, and a nonlinear observation process through which the high-dimensional observations are generated. In other words, low-dimensional processes lend *structures* to the high-dimensional data in the *ambient space*.

In many cases, humans are able to perceive such structural information or perceptually meaningful low-dimensional features directly from images. For example, when we observe a sequence of facial images acquired under different horizontal and vertical poses and varying lighting directions, our brain can naturally infer that the original high-dimensional ambient space associated with the original image data lies in an intrinsically 3-D *manifold space* (or latent space of pose and illumination) that produced it. The goal of *manifold learning* or nonlinear dimensionality reduction methods is to uncover such low-dimensional structures embedded within high-dimensional data. Traditional linear dimensionality reduction methods such as PCA often fail to extract even simple nonlinear structures in the data, e.g. 1-D structure of helix shown in Figure 19.2. The applications of nonlinear dimensionality reduction are numerous, including but not limited to data visualization and exploratory data analysis, image processing, feature extraction, marketing, data mining, bioinformatics, and many other areas.

More formally, given a set of data points $\underline{x}_1 \cdots \underline{x}_K \in \mathbb{R}^N$ that lie on an M-dimensional manifold $\mathcal{M}$, the goal is to determine a low-dimensional representation $\underline{y}_1 \cdots \underline{y}_K \in \mathbb{R}^M$ through a mapping $\underline{f} : \mathbb{R}^N \to \mathbb{R}^M$, such that $\underline{y}_i = \underline{f}(\underline{x}_i)$, $i \in [1, K]$. This mapping must preserve the local geometric structures of the data, including neighborhood relationships, pairwise distances

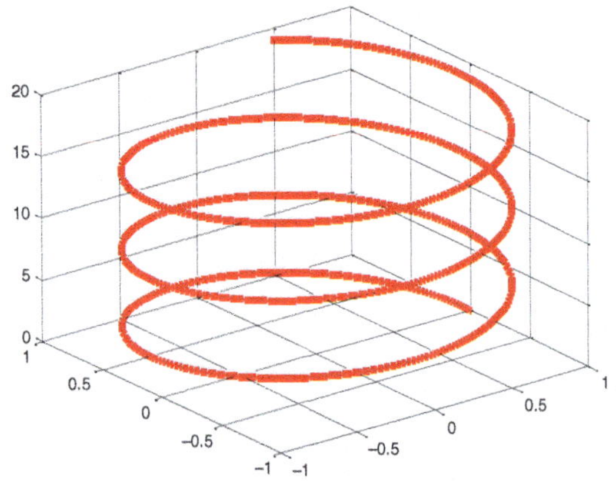

Figure 19.2 1-D helix structure.

between nearby points and their connectivity. By preserving these local properties, the low-dimensional representation captures the intrinsic geometry of the manifold on which the data reside.

There are many different manifold learning methods, including Isometric Feature Mapping (IsoMap) [6], Locally Linear Embedding (LLE) [7], Maximum Variance Unfolding (MVU) [8], and Laplacian Eigenmaps (LE) [9], among others. In this section, however, we focus on LE and show how the method can be extended to embed previously unseen data for continual learning applications.

19.2.1 Laplacian Eigenmaps

This manifold learning method was originally proposed by Belkin and Niyogi [9] to find a low-dimensional embedding that preserves neighborhood relations represented by a graph, where each data point in the neighborhood corresponds to a node in this graph. The graph provides a discrete approximation of the local geometry of the manifold. The connectivity of the graph is represented by a symmetric *edge weight* matrix $\mathbf{W}$, which defines *local similarity*. That is, if nodes i and j are determined to be connected, i.e. i is among the k-nearest neighbors (k-NN) of j or vice versa, then the $(i,j)^{\text{th}}$ element of this matrix $w_{i,j} = 1$; otherwise, $w_{i,j} = 0$ (see Figure 19.3). Alternatively, we can use a Gaussian kernel $w_{i,j} = k(\underline{x}_i, \underline{x}_j) = e^{-\frac{\|\underline{x}_i - \underline{x}_j\|^2}{2\sigma^2}}$ for neighboring nodes identified using k-NN, while $w_{i,j} = 0$ for all other pairs. Now, the key requirement is that if two points $\underline{x}_i$ and $\underline{x}_j$ in the ambient space have a high degree of similarity, as measured by $\mathbf{W}$, i.e. they lie close to one another, the corresponding low-dimensional representations $\underline{y}_i$ and $\underline{y}_j$ should also be close to one another. This condition leads to the following cost function [9] to minimize the distance between similar points,

$$\min_{\mathbf{Y}} \left\{ \sum_i \sum_j w_{i,j} \left\| \underline{y}_i - \underline{y}_j \right\|^2 = tr(\mathbf{Y}^T \mathbf{L} \mathbf{Y}) \right\} \quad s.t. \quad \mathbf{Y}^T \mathbf{D} \mathbf{Y} = \mathbf{I} \tag{19.12}$$

where $\mathbf{Y} \in \mathbb{R}^{M \times K}$ matrix contains the embedding vectors $\underline{y}_i$, $i \in [1, K]$ as its columns, $\mathbf{L} = \mathbf{D} - \mathbf{W}$ is the *graph Laplacian* matrix with $\mathbf{D}$ being a diagonal matrix with elements $D_{i,i} = \sum_j w_{i,j}$. The *graph Laplacian* matrix captures local information about the manifold. The eigenvalues and eigenvectors of the graph Laplacian matrix reveal a wealth of information about the graph, e.g. whether it is complete or connected [9]. Minimizing the cost function in (19.12) ensures that points that are close to each other in the ambient space remain close to each other in the low-dimensional latent space, thereby preserving local connectivity. The constraint $\mathbf{Y}^T \mathbf{D} \mathbf{Y} = \mathbf{I}$ enforces $\mathbf{Y}$ to be of full dimensionality and leads to a closed-form solution of a generalized eigenvalue problem. Using the Lagrange multiplier, the constrained minimization problem [16] in (19.12) can be cast as:

$$\min \left\{ tr(\mathbf{Y}^T \mathbf{L} \mathbf{Y}) - tr(\mathbf{\Lambda}(\mathbf{Y}^T \mathbf{D} \mathbf{Y} - \mathbf{I})) \right\} \tag{19.13}$$

where $\mathbf{\Lambda}$ is a diagonal matrix with diagonal elements that are the Lagrange multipliers λ_i. This minimization leads to the generalized eigenvalue problem [17]:

$$\mathbf{L}\mathbf{Y} = \mathbf{\Lambda}\mathbf{D}\mathbf{Y} \qquad \text{or} \qquad \mathbf{L}\underline{y}_i = \lambda_i \mathbf{D}\underline{y}_i. \tag{19.14}$$

For a connected graph, the graph Laplacian has a trivial eigenvalue equal to zero. Ignoring the eigenvector associated with the trivial case, the eigenvectors corresponding to the smallest M nonzero eigenvalues are used to form matrix $\mathbf{Y}$, i.e.

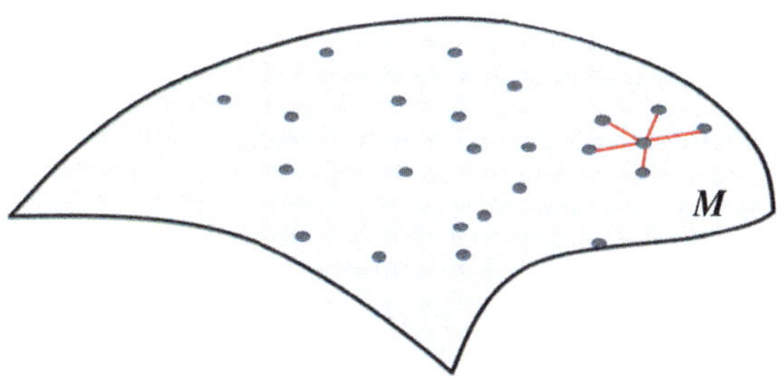

Figure 19.3 A manifold in ambient space and connectivity graph using k-NN with $k = 5$.

$\mathbf{Y} = \left[\underline{y}_1 \underline{y}_2 \cdots \underline{y}_M\right]$. Thus, columns of $\mathbf{Y}$ are the selected M eigenvectors, while the rows correspond to the low-dimensional coordinates of the embedded data points.

The Laplacian eigenmaps method has several shortcomings. First, although it accurately characterizes the local geometry of manifolds, it does not necessarily preserve global geometric relationships. Furthermore, while the original version of this method [9] learns the structure of the discretized manifold from the training data, it does not provide a mechanism to embed previously *unseen data* into the learned manifold.

To embed novel data samples into the manifold, we employ a latent (unseen) variable approach [18, 19]. Let $\mathbf{X}_s$ and $\mathbf{X}_u$ be data matrices corresponding to the previously seen (during training) data and unseen (novel) data to be embedded onto the manifold, respectively. We seek a low-dimensional embedding $\mathbf{Y}_u$ for the unseen data matrix while keeping the embedding matrix $\mathbf{Y}_s$ for the training data unchanged, i.e. no forgetting. To do so, we modify the objective function in (19.13) as:

$$J(\mathbf{Y}_u) = tr\left([\mathbf{Y}_s\ \mathbf{Y}_u]\begin{bmatrix}\mathbf{L}_{ss} & \mathbf{L}_{su}\\ \mathbf{L}_{us} & \mathbf{L}_{uu}\end{bmatrix}\begin{bmatrix}\mathbf{Y}_s^T\\ \mathbf{Y}_u^T\end{bmatrix}\right). \tag{19.15}$$

or

$$J(\mathbf{Y}_u) = tr\left(\mathbf{Y}_s\mathbf{L}_{ss}\mathbf{Y}_s^T\right) + 2tr\left(\mathbf{Y}_s L_{su}\mathbf{Y}_u^T\right) + tr\left(\mathbf{Y}_u L_{uu}\mathbf{Y}_u^T\right). \tag{19.16}$$

where $\mathbf{L}_{ss}$ and $\mathbf{L}_{uu}$ are the graph Laplacians for $\mathbf{X}_s$ and $\mathbf{X}_u$, respectively, and $\mathbf{L}_{su} = \mathbf{L}_{us}^T$ is the graph Laplacian shared between them. Setting $\frac{\partial J(\mathbf{Y}_u)}{\partial \mathbf{Y}_u} = 0$ and solving for $\mathbf{Y}_u$ gives

$$\mathbf{Y}_u = -\mathbf{Y}_s\mathbf{L}_{su}\mathbf{L}_{uu}^{-1}, \tag{19.17}$$

which implements batch embedding of all the novel data samples. This equation reveals an important property of incremental LE: novel samples can be embedded without recomputing the eigen-decomposition of the graph Laplacian.

For incremental manifold updating, consider a single novel sample $\underline{x}$ for $\mathbf{X}_u$ and the associated feature vector $\underline{y}$ for $\mathbf{Y}_u$, then the two graph Laplacians used in embedding this data point onto the manifold are simply given by $\underline{l}_{su} = -\left[k(\underline{x}, \underline{x}_1) \ \cdots \ k(\underline{x}, \underline{x}_K)\right]^T$ and $l_{uu} = \sum_{i=1}^{K} k(\underline{x}, \underline{x}_i)$ where $k(\underline{x}, \underline{x}_i) = \exp\left\{-\frac{\|\underline{x}-\underline{x}_i\|^2}{2\sigma^2}\right\}$ is a Gaussian RBF. Substituting these expressions into (19.17) yields the embedded feature vector,

$$\underline{y} = -\frac{\mathbf{Y}_s\underline{l}_{su}}{l_{uu}} = \sum_{i=1}^{K}\frac{k(\underline{x}, \underline{x}_i)}{\sum_{j=1}^{K} k(\underline{x}, \underline{x}_j)}\underline{y}_i. \tag{19.18}$$

Thus, new data samples can be embedded in the low-dimensional manifold using a weighted combination of the training (seen) feature vectors $\left\{\underline{y}_i\right\}_{i=1}^{K}$ weighted based on the relative connectivity (proximity) of the training data samples $\left\{\underline{x}_i\right\}_{i=1}^{K}$ in the high-dimensional ambient space. The synthesis of the high-dimensional data using the low-dimensional representation $\underline{y}$ can be performed similarly [18] using

$$\hat{\underline{x}} = \sum_{i=1}^{K}\frac{k(\underline{y}, \underline{y}_i)}{\sum_{j=1}^{K} k(\underline{y}, \underline{y}_j)}\underline{x}_i. \tag{19.19}$$

Thus, this algorithm enables (a) the incremental embedding of novel data samples in an unsupervised fashion, hence improving the smoothness of the manifold, and (b) bidirectional mapping between the ambient space and reduced-dimensional manifold space for both data analysis and synthesis.

To illustrate an application of this incremental Laplacian eigenmap manifold learning, a dataset containing 90 images of a mannequin head with varying azimuth angle ranging from 0° to 360° was collected. The illumination conditions were kept constant for all images, and the camera was positioned coplanar with the mannequin head. A subset of these images can be seen in Figure 19.4a. The images were then downsampled to 128 × 128 pixels to reduce computational requirements. A second (novel) dataset containing 26 images was also collected. In this dataset, the tilt of the mannequin head and the orientation of the camera were varied in addition to the azimuth angle. The tilt was set to −15°, 0°, or 15°, while the camera elevation was adjusted to provide a viewing angle of approximately −30°. The illumination conditions were kept constant as before, and the image size was downsampled to be consistent with the training dataset. A subset of these images is shown in Figure 19.4b.

The baseline training of the manifold was performed using (19.14) on the entire training dataset. A Gaussian RBF with $\sigma = 1$ and a $k = 9$-NN were used to form the elements of the matrices **L** and **D** and to compute the low-dimensional manifold features. Figure 19.5a shows the resulting two-dimensional manifold representation of the training data. As expected, the embedded points form a nearly circular trajectory corresponding to the continuous variation of the azimuth angle. The figure also shows the approximate locations that correspond to different mannequin's head orientations when the pictures were taken. A noticeable discontinuity is observed near the 270° region. Moreover, the density of the mapped points is not uniform across the manifold. In particular, the regions near 90° and 270° exhibit higher density of points than those near 0° and 180°, while other regions appear relatively sparse. One possible explanation is that the silhouette of the mannequin's head is wider near the side-view orientations and using only two manifold features, the algorithm found more similarity than dissimilarity. Furthermore, the choices of neighborhood size k and kernel width σ can significantly influence the resulting manifold geometry and point distribution.

Figure 19.5b shows the results of embedding the 26 novel images using the incremental manifold learning in (19.18). As mentioned previously, the novel dataset contained a combination of out-of-manifold images that was generated with variations in azimuth, tilt, and camera elevation. Each category is highlighted using different colors: triangular points represent images with varying azimuth and tilt, hollow circular points denotes image with varying azimuth and camera elevation,

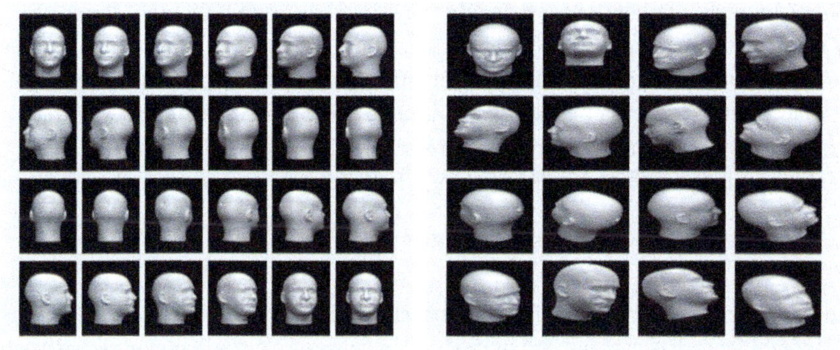

(a) (b)

Figure 19.4 Examples of mannequin head images used in (a) training and (b) novel embedding.

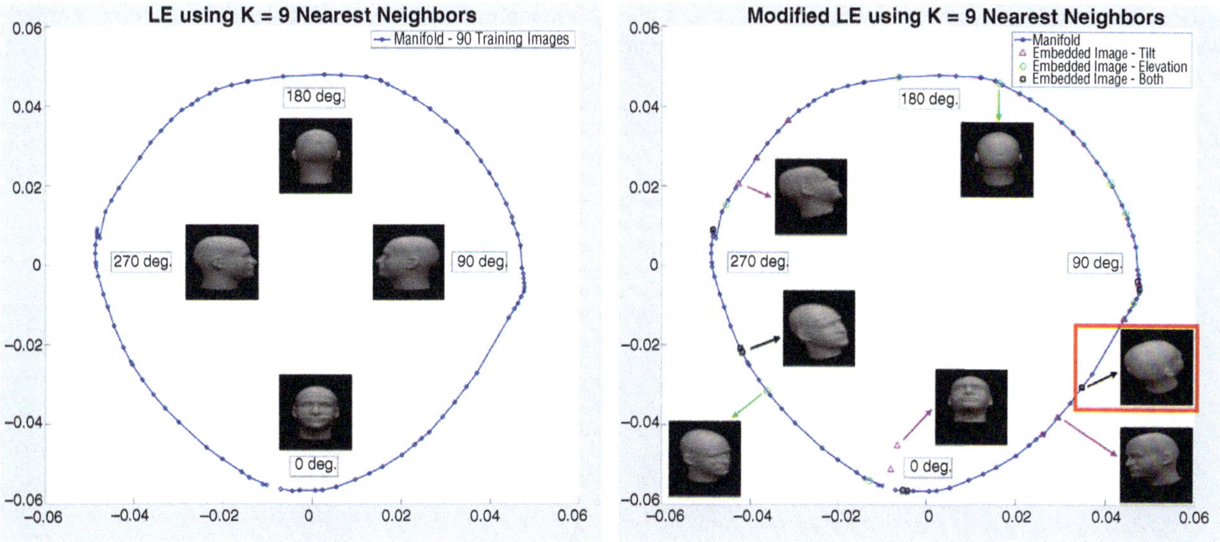

(a) (b)

Figure 19.5 (a) Baseline trained manifold and (b) novel out of manifold embedding results.

and black squares correspond to images containing variations in all three factors. As can be seen, most of the novel images were successfully embedded in the learned manifold. However, two images were not mapped onto the manifold, and eight images were placed at incorrect locations on the manifold; for example, the image highlighted within a box in Figure 19.5b. The embedded point is located near the 45° regions of the manifold, whereas the box shows that the corresponding image actually has an azimuth of ~315°. Nevertheless, the visual similarity between these two poses makes the incorrect association understandable, especially when only two manifold features are employed. Obviously, if the training images had varied degrees of freedom, then such ambiguities would be less likely to happen.

This experiment showed the usefulness of nonlinear dimensionality reduction using manifold learning and, in particular, the Laplacian eigenmaps method. The ability to learn a compact representation, incrementally embed novel samples, and reconstruct data from the manifold makes this approach attractive for numerous applications, including robotic navigation and control, image and signal feature extraction, medical image analysis, and related areas. In the next section, we introduce alternative nonlinear dimensionality reduction techniques using neural network-based autoencoders.

19.3 Nonlinear Autoencoders

Nonlinear autoencoders or auto-associators are feedforward neural networks that use unsupervised learning for feature extraction and dimensionality reduction, similar to their linear counterparts. These networks are capable of learning complex nonlinear relationships that cannot be captured by PCA or other linear dimensionality-reduction techniques. Aside from their traditional use in feature extraction and dimensionality reduction, nonlinear autoencoders can be employed in a multitude of other machine learning and data analysis applications, especially those involving complex high-dimensional data. Among nontraditional examples of nonlinear autoencoders in image processing are image generation using Variational Autoencoders [20], image reconstruction from incomplete or noisy observations [10, 12], and content-based image retrieval (CBIR) [21]. The goal of this section is to introduce several variants of nonlinear autoencoders that are commonly used in these and other related image processing applications. The reader is referred to the relevant resources to gain a deeper understanding of the mathematical foundation of different autoencoders of both deterministic and stochastic types.

19.3.1 Basic Autoencoders

In its simplest form, the basic nonlinear autoencoder has a structure similar to that shown in Figure 19.6. The encoder maps an input data vector $\underline{x} \in \mathbb{R}^N$ to a low-dimensional feature vector $\underline{y} \in \mathbb{R}^M$ at the hidden layer, while the decoder reconstructs an estimate of the input $\hat{\underline{x}}$ from the reduced-dimensional feature representation. However, in contrast to the linear autoencoders, both layers employ nonlinear activation functions, i.e. $\underline{y} = f(\mathbf{W}^{(1)}\underline{x})$ and $\hat{\underline{x}} = g(\mathbf{W}^{(2)}\underline{y})$, where $f(.)$ and $g(.)$ are the activation functions of the first and second layers applied element-wise and $\mathbf{W}^{(1)}$ and $\mathbf{W}^{(2)}$ are the corresponding weight matrices (bias vectors included). Clearly, the objective is to seek a mapping that leads to a substantial reduction in dimensionality with $M < N$ to produce the most representative features similar to the PCA extraction network in Section 19.1.

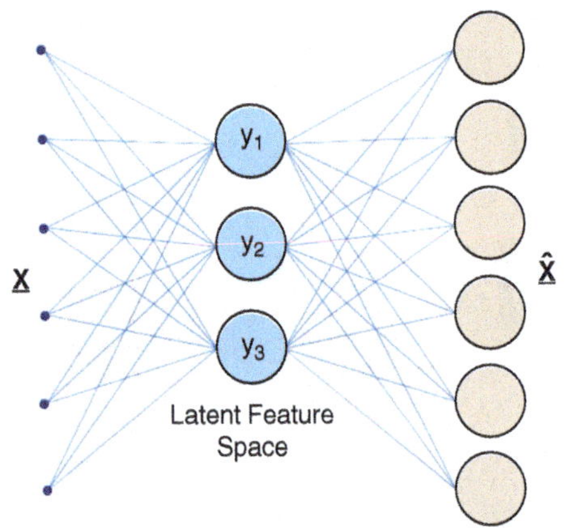

Figure 19.6 A nonlinear autoencoder for dimensionality reduction.

Thus, the autoencoder structure must have a bottleneck layer in order to generate an under-complete representation [10]. This bottleneck prevents the network from learning a trivial identity mapping and forces it to extract meaningful features from the data. Consequently, the latent representation may be viewed as a lossy encoding of the original input vector $\underline{x}$.

If we define a loss function $\mathcal{L}(\mathcal{W})$, where $\mathcal{W} = \{\mathbf{W}^{(1)}, \mathbf{W}^{(2)}\}$ denotes the set of unknown network parameters, our goal of training this basic nonlinear autoencoder is to find the optimal estimates of these parameters by minimizing (or maximizing) the loss function iteratively for any choice of the number of features or latent variables. Typical loss functions used for autoencoder training are the averaged squared error (ASE) and the cross-entropy between the original and reconstructed data. The latter loss function is used for binary input and output data, while the former is preferred for real-valued data. Given an unlabeled training dataset $\{\underline{x}(p)\}_{p=1}^{P}$, the ASE loss function is given by

$$\mathcal{L}(\mathcal{W}) = \frac{1}{P}\sum_{p=1}^{P}\sum_{n=1}^{N}(x_n(p) - \hat{x}_n(p))^2 = \frac{1}{P}\sum_{p=1}^{P}\left\|\underline{x}(p) - g(f(\underline{x}(p)))\right\|^2 \tag{19.20}$$

where x_n and $\hat{x}_n(p)$ are the nth element of vectors $\underline{x}(p)$ and $\hat{\underline{x}}(p)$, respectively. The SGD algorithm, or one of its variants discussed in the previous chapter together with the error back-propagation method, can be used to iteratively estimate the unknown parameters of the network. The gradient of (19.20) with respect to the weight vector of neuron n in the output layer is

$$\nabla_{\underline{w}_n^{(2)}}\mathcal{L}(\mathcal{W}) = -2(x_n(p) - \hat{x}_n(p))g'\left(\underline{w}_n^{(2)}\underline{y}(p)\right)\underline{y}(p) = -2r_n^{(2)}(p)\underline{y}(p) \tag{19.21}$$

where $r_n^{(2)}(p) = (x_n(p) - \hat{x}_n(p))g'\left(\underline{w}_n^{(2)}\underline{y}(p)\right)$ is the learning signal associated with output neuron n, and $\underline{w}_n^{(2)}$ is the nth row of matrix $\mathbf{W}^{(2)}$. For a sigmoidal activation function $g'\left(\underline{w}_n^{(2)}\underline{y}(p)\right) = (1 - \hat{x}_n(p))\hat{x}_n(p)$. Using error back-propagation, the gradient of (19.20) with respect to the weight vector of hidden neuron m becomes

$$\nabla_{\underline{w}_m^{(1)}}\mathcal{L}(\mathcal{W}) = -2\sum_{n=1}^{N} w_{n,m}^{(2)}r_n^{(2)}(p)f'\left(\underline{w}_m^{(1)}\underline{x}(p)\right)\underline{x}(p) = -2r_m^{(1)}(p)\underline{x}(p) \tag{19.22}$$

where $r_m^{(1)}(p) = \sum_{n=1}^{N} w_{n,m}^{(2)}r_n^{(2)}(p)f'(\underline{w}_m^{(1)}\underline{x})$ is the back-propagated learning signal for hidden neuron m, and $\underline{w}_m^{(1)}$ is the mth row of matrix $\mathbf{W}^{(1)}$. If the activation function of the hidden-layer neurons is also sigmoidal type, then $f'\left(\underline{w}_m^{(1)}\underline{x}(p)\right) = (1 - y_m(p))y_m(p)$, where $y_m(p)$ is the mth element of the latent feature vector $\underline{y}(p)$. Once these parameters are obtained, the encoder layer can be used to map any new data vector to its corresponding low-dimensional feature representation.

Autoencoders are typically used as a front-end feature extraction and dimensionality reduction system for classifiers such as CNNs. A study reported in [22] investigated the feature extraction capabilities of autoencoders with and without introducing inhibitory and excitatory interactions among hidden-layer neurons. The MNIST database, which consists of a collection of binary images of size 28×28 of handwritten digits, was used in the experiments. Figure 19.7 shows a small subset of MNIST images together with the learned encoder filter weights of a basic autoencoder without neuron interactions. These filter weights correspond to the rows of the encoder weight matrix $\mathbf{W}^{(1)}$. As can be seen, the learned filters capture a variety of local structures, including small pen strokes, edges, corners, and background regions. Consequently, the encoder is able to extract increasingly abstract and discriminative features that can subsequently be used for classification and other pattern-recognition tasks.

19.3.2 Sparse Autoencoders

So far, we have assumed that the number of neurons in the hidden layer is known a priori based on the properties of the data. In practice, however, it is often desirable to allow the network itself to determine the effective number of latent features required for accurate data representation. This can be achieved by posing the problem of minimizing the loss function as a constrained optimization problem, where the constraint is to enforce the smallest possible number of hidden-layer neurons. This can be achieved using a variant of the basic autoencoder by adding a *sparsity* constraint that leads to a slim hidden layer representation. One approach to enforce sparsity is based on keeping only those hidden-layer neurons that contribute toward learning and data reconstruction by monitoring their *activity* and automatically removing those that are not active during the training. The *activity* of a neuron can be determined by the running average of its output, i.e. $\bar{y}_m(p) = \frac{1}{p}\sum_{k=1}^{p} y_m(k)$ at iteration p. Neuron m is considered to be *inactive* when $\bar{y}_m(p) \approx 0$. Let us define an inactivity

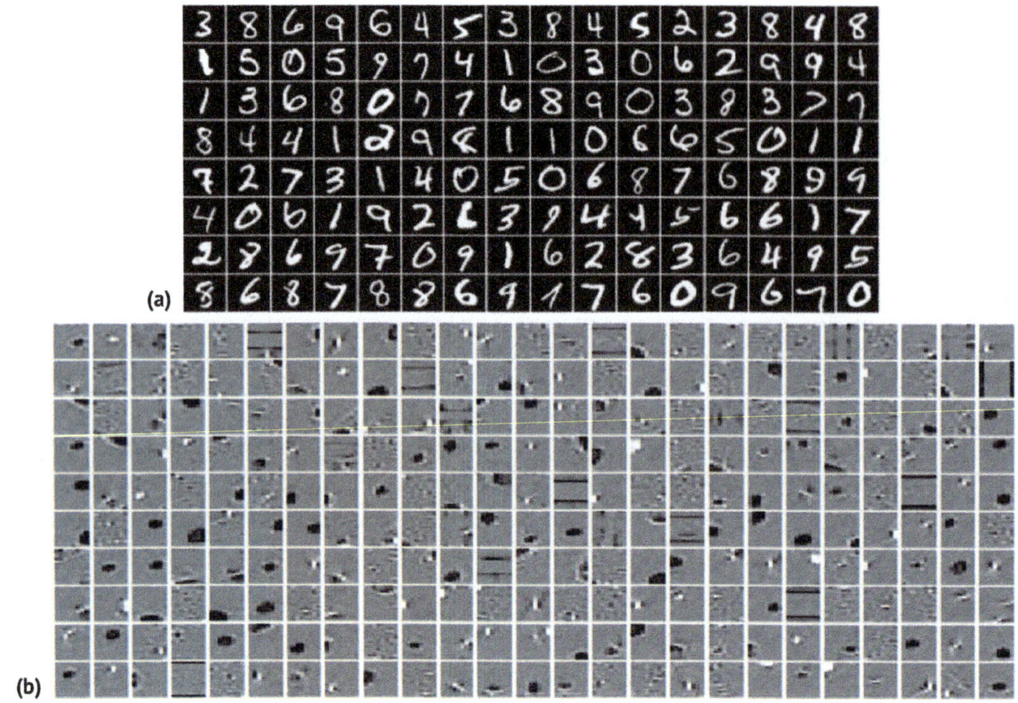

Figure 19.7 (a) A small subset of MNIST images. *Source:* [22]/Hugo Larochelle, Dumitru Erhan and Pascal Vincent. (b) Features extracted by the autoencoder. *Source:* [22]/Hugo Larochelle, Dumitru Erhan and Pascal Vincent.

threshold or the sparsity parameter τ. Now, the objective is to encourage as many neurons as possible to have $\bar{y}_m(p) \approx \tau$, i.e. inactive. Thus, we add a sparsity penalty to the loss function to enforce this constraint. One choice of sparsity constraint proposed in [23] is given by

$$\rho(\mathbf{W}^{(1)}) = \sum_{m=1}^{M} \left[\tau \log \frac{\tau}{\bar{y}_m(p)} + (1-\tau) \log \frac{(1-\tau)}{(1-\bar{y}_m(p))} \right] \tag{19.23}$$

where M is the number of neurons in the hidden layer. This penalty can be considered as the Kullback–Leibler (KL) divergence measure [23, 24] between the desired activity level τ and the observed average activity $\bar{y}_m(p)$. For inactive neurons $\bar{y}_m(p) \approx \tau$, the corresponding terms in the summation of (19.23) become zero. Because the sparsity constraint depends only on the hidden-layer activity, it affects the training of the first layer, while the training of the second layer is the same as that of the basic autoencoder. Adding this constraint to the loss function gives

$$\mathcal{L}_s(W) = \mathcal{L}(W) + \lambda \rho(\mathbf{W}^{(1)}) \tag{19.24}$$

where λ is the regularization parameter that controls the relative importance of the constraint in relation to the loss function. The sparse autoencoders provide slim hidden layers, which tend to capture the most salient features for data representation with better generalization performance [11].

To apply an SGD type learning algorithm to the constrained loss function (19.24), the gradient expression in (19.22) must be modified to account for the penalty term. The resulting gradient becomes:

$$\nabla_{\underline{w}_m^{(1)}} \mathcal{L}_s(W) = -2 r_m^{(1)}(p) \underline{x}(p) \tag{19.25}$$

where the modified hidden-layer learning signal $r_m^{(1)}(p)$ is

$$r_m^{(1)}(p) = \left(\sum_{n=1}^{N} w_{n,m}^{(2)} r_n^{(2)}(p) + \frac{\lambda}{2} \left(\frac{\tau}{\bar{y}_m(p)} - \frac{1-\tau}{1-\bar{y}_m(p)} \right) \right) f'\left(\underline{w}_m^{(1)} \underline{x}(p) \right). \tag{19.26}$$

Note that when using this sparsity-constrained loss function, it is recommended [23] to apply the entire training dataset to produce hidden-layer outputs in order to compute the average activity $\bar{y}_m(p)$. Once the estimate of this activity measures are generated, the forward pass is repeated followed by the error back-propagation using the gradient estimate in (19.26). One important consideration is to ensure that the sparsity-constrained loss function does not lead to a trivial solution in which all neurons become inactive, or alternatively, only a few neurons are active while the remaining neurons are eliminated. Therefore, the sparsity constraint should be chosen carefully to balance feature compactness and reconstruction accuracy.

19.3.3 Denoising Autoencoders

Consider a scenario where the training samples $\{\underline{z}(p)\}_{p=1}^{P}$ are noisy versions of clean data samples, i.e. $\underline{z}(p) = \underline{x}(p) + \underline{\eta}(p)$ where $\underline{\eta}(p)$, $p \in [1, P]$ are additive random noise vectors. The goal is to configure an autoencoder to remove additive noise effectively and generate an estimate of the original clean data vectors denoted by $\hat{\underline{x}}(p)$. In this framework, the autoencoder acts as a nonlinear low-pass filter for multivariate data $\underline{z}(p) \in \mathbb{R}^N$. The *denoising* autoencoder [13, 25] is trained using artificially corrupted versions of the clean training samples, while the desired output vectors are the corresponding clean samples. The network output $\hat{\underline{x}}(p) = g(f(\underline{z}(p)))$ is then compared with the clean data sample $\underline{x}(p)$ to generate the error vector that drives the parameter updating process.

There are several important differences between denoising autoencoders and those used for dimensionality reduction and feature extraction in the previous sections. First, the encoder is trained to extract latent features that represent the underlying clean data rather than the corruption process. In other words, the latent representation must ideally preserve the essential characteristic of the original data while discarding additive noise. Second, unlike autoencoders used for dimensionality reduction, where the decoder layer is not usually needed after training, in denoising autoencoders the decoder layer performs inference to restore signal or image from the latent features. Finally, denoising autoencoders are not typically under-complete, i.e. the latent variable space can be of the same dimension or larger than that of the input. Such models are known as *over-complete* models.

The cost function that is used to train a denoising autoencoder is typically based on minimizing the conditional entropy,

$$\mathcal{L}(W) = H(\underline{X}|\underline{Z}) = -E_{p(\underline{X},\underline{Z})}[\log p(\underline{X}|\underline{Z};\, W)] \tag{19.27}$$

where the expectation is taken with respect to the joint distribution $p(\underline{X}, \underline{Z})$. Note that minimizing this conditional entropy corresponds to maximizing the mutual information [24] between the clean data $\underline{X}$ and the noisy observations $\underline{Z}$ since we have $I(\underline{X}; \underline{Z}) = H(\underline{X}) - H(\underline{X}|\underline{Z})$ where the entropy of the data, $H(\underline{X})$, is fixed. Consequently, minimizing (19.27) encourages the network to preserve as much information as possible about the underlying clean data. Now, since the latent features extracted by the encoder are representative of the essential characteristics of the input data, we have $p(\underline{X}|\underline{Z}) = p(\underline{X}|\underline{Y} = f(\underline{z}; \mathbf{W}^{(1)}))$, where $f(.)$ is the encoder function.

Let $p(\underline{Z}|\underline{X})$ be the distribution of the process that corrupts the clean data $\underline{X}$ and generates the noisy observation $\underline{Z}$. The process of training the denoising autoencoder involves the following steps: (a) draw a sample $\underline{x} \sim p(\underline{X})$ from the training dataset; (b) generate a corrupted version of $\underline{z} \sim p(\underline{Z}|\underline{X} = \underline{x})$; and (c) use the pair of training examples $(\underline{x}, \underline{z})$ to train the denoising autoencoder to predict $\underline{x}$ from $\underline{z}$ by minimizing the loss function in (19.27) using SGD-based training, and repeat until convergence.

In [25], the authors derived an explicit expression for the *optimal reconstruction* function $r(\underline{z}) = g(f(\underline{z})) = \hat{\underline{x}}$ for the case where the additive noise is modeled as multivariate normal with zero mean and covariance matrix $\Sigma_\eta = \sigma_\eta^2 \mathbf{I}$ (see Problem 19.15).

Here, we use a widely adopted U-net architecture as a denoising AE. U-net is a deep CNN-based structure originally proposed for medical image segmentation [26, 27]. However, unlike the conventional CNNs that consist of multiple pairs of convolutional, pooling, and fully connected layers, U-net employs a U-shaped encoder–decoder architecture. The encoder consists of multiple convolutional and pooling layers that progressively extract multiresolution features corresponding to noise-free images. The decoder consists of upsampling followed by convolutional layers and concatenations of features with the same resolution from the encoder stages to reconstruct the filtered images at the output. The upsampling and subsequent local convolution operations are conceptually similar to the synthesis filter bank operations used in the DWT of Chapter 7.

Numerous variants of the U-net architectures have been proposed for image restoration, segmentation, and other image processing applications [26–28]. A good review of these networks can be found in [26]. In our experiment, we configured a U-net consisting of an encoder network with three pairs of convolutional and pooling layers with 3×3 kernels and 64, 128, and 256 feature maps, respectively. The decoder network consisted of three pairs of upsamplers and convolutional layers

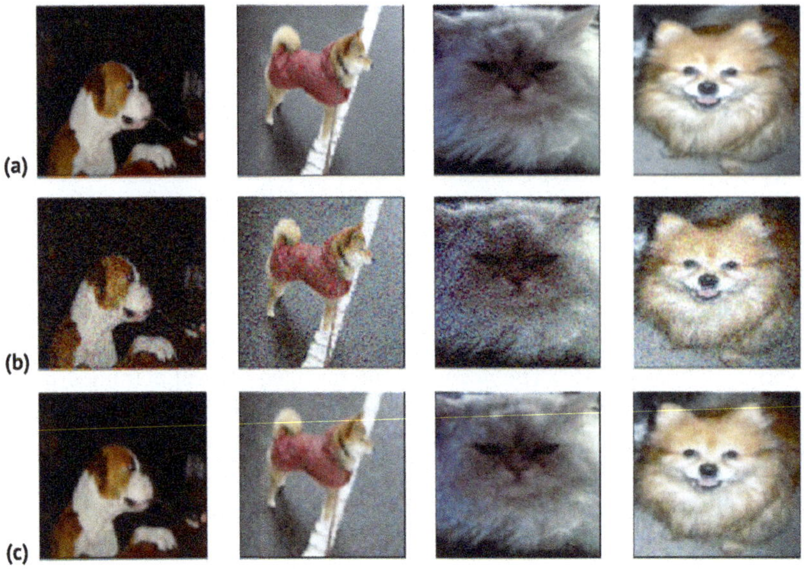

Figure 19.8 (a) Original noise-free color images of three dogs and one cat, (b) noisy versions of images in part (a), and (c) restored images using a U-net.

with 256, 128, and 64 feature maps. All neurons used ReLU activation functions, and max pooling is used in all encoder layers. Training is carried out using the Adam optimizer with the SSE as the cost function and an initial learning rate of 0.001.

The dataset used in this study was the Oxford-IIIT Pet Database, which contains images of different breeds of dogs and cats of about 200 images per class. The images were produced for different lighting condition, background, pose, and animal position. Figure 19.8a shows four original noise-free color images of one cat and three dogs. Figure 19.8b shows noisy versions of these images that were obtained by adding white Gaussian noise to produce average SNR of approximately 14.2 dB for all training and testing images. Figure 19.8c shows the restored images obtained using the trained U-net. The average output SNR value of the filtered test images was found to be approximately 23.4 dB, indicating an improvement of about 9 dB. Visual inspection of the filtered images also reveal excellent restoration quality with minimal blurring artifacts, which is common in traditional image filtering methods discussed in Chapters 8 and 10. The trained denoising AE can be applied to any other noisy image beyond this pet dataset.

19.3.4 Contractive Autoencoders

Contractive autoencoders are a class regularized autoencoders [10, 29] designed to improve the robustness of the learned representations, particularly to small perturbations of the input data. Consequently, they can also be viewed as a type of denoising autoencoder in which the *contractive penalty* term is added as a regularization term to the loss function to yield latent feature representation invariant, or robust, to small perturbations of the input data. In contrast, denoising autoencoders improve robustness of the reconstruction, that is, $r(\underline{z}) = g(f(\underline{z}))$, which may only indirectly encourage robustness of the representation. However, if the features extracted from the encoder are intended for classification purposes, the robustness of the extracted features is more important than that of the reconstruction [29].

The contractive penalty term corresponds to the Frobenius norm of the Jacobian matrix of the encoder outputs $\underline{y} = f(\underline{x})$ with respect to the input vector $\underline{x}$. Thus, the regularized loss function of this autoencoder is

$$\mathcal{L}_c(\mathcal{W}) = \mathcal{L}(\mathcal{W}) + \lambda\rho(\mathbf{W}^{(1)}), \tag{19.28}$$

where $\rho(\mathbf{W}^{(1)}) = ||\mathbf{J}_f(\underline{x})||_F^2$ with $||\mathbf{A}||_F^2 = tr(\mathbf{A}^T\mathbf{A})$ being the squared Frobenius norm and $\mathbf{J}_f(\underline{x}) = \frac{\partial f(\underline{x})}{\partial \underline{x}}$ is the Jacobian matrix. For sigmoidal hidden layer neurons, the closed-form expression for the penalty term in (19.28) is

$$||\mathbf{J}_f(\underline{x})||_F^2 = \sum_{i=1}^{M}[y_i(1-y_i)]^2\sum_{j=1}^{N} w_{i,j}^2 \tag{19.29}$$

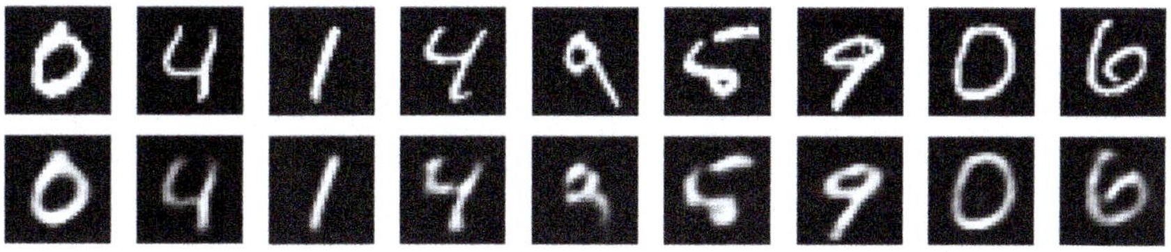

Figure 19.9 Some examples of MNIST images with minor perturbations and contractive autoencoder results. *Source:* Adapted from Suvanjanprasai.

where $w_{i,j}$ is the weight that connects the input x_j to the neuron i in the hidden layer. The first term $\mathcal{L}(\mathcal{W})$ in (19.28) corresponds to the reconstruction loss function defined in Section 19.3.1. The penalty term has the geometric effect of local contraction of the encoder mapping around each training sample, hence the name contractive. Enforcing small values of the first derivatives with respect to the input, provides invariance of the feature representation to small perturbations in the input. In the original paper on contractive autoencoders [29], the authors considered the tied-weight case, where $\mathbf{W}^{(2)} = \mathbf{W}^{(1)\mathrm{T}}$, i.e. the weight matrix of the decoder is the transpose of the weight matrix of the encoder. The computational complexity of generating this penalty and its gradient is comparable to that of computing the reconstruction error, namely $\mathcal{O}(MN)$ [29].

Contractive autoencoders are used extensively in image processing applications due to their ability to remain robust to small deformations and noise, which is a critical requirement for any practical pattern recognition system. In [29], the authors compared the performance of the contractive autoencoder with several other unsupervised feature learning methods, including Restricted Boltzmann Machine (RBM) trained using Contrastive Divergence, basic autoencoder, and denoising autoencoder. All the autoencoder variants used tied weights, sigmoid activation function for both the encoder and decoder, a cross-entropy reconstruction loss function, and SGD-based training. The trained encoder layer was then combined with a two-layer MLP classifier with one hidden layer of 1000 neurons. The parameters of the entire network were fine-tuned using some labeled data from both the MNIST and grayscale CIFAR-10 datasets. The results on both datasets indicated that the contractive autoencoder led to the lowest classification error. Compared with the denoising autoencoder aside from robustness of the latent features to input perturbations, the contractive autoencoder employs a deterministic gradient for the regularized loss function, which can be computed more efficiently using optimization algorithms such as the conjugate gradient. Nevertheless, the denoising autoencoder has a simpler architecture and does not require computation of the Jacobian of the encoder outputs.

Figure 19.9 shows some examples of MNIST images (top row) and the reconstructed versions (bottom row) obtained using a contractive autoencoder with 128 latent features. The network was trained using the contractive loss function in (19.28) on the MNIST dataset. The Adam optimizer was used to estimate the encoder and decoder parameters over 10 training epochs. As can be observed, some images in this dataset exhibited minor perturbations, which were removed by the contractive autoencoder in the reconstructed images.

There are many other autoencoders that are beyond the scope of this chapter. Examples include variational and adversarial autoencoders [10, 20], which are used as generative models; sequence-to-sequence autoencoders [30], which are applied to time-series or sequential data modeling; and stacked and convolutional autoencoders [11, 12]. In recent years, there has also been growing interest in developing autoencoder-based associative memories capable of recalling or retrieving stored patterns from distorted and incomplete versions. In [31], it was shown that over-parameterized deep autoencoders can store real-valued training patterns as attractor states, similar to the classical binary associative memories, and retrieve the original patterns through state propagation. In the next chapter, we will present a hybrid architecture for associative memories consisting of a deep CNN-based autoencoder together with a recurrent neural network that performs state propagation to equilibrium states associated with the stored or memorized patterns.

19.3.5 Relation of Autoencoders to Manifold Learning

There are many similarities, as well as important differences, between manifold learning methods and nonlinear autoencoders. As far as similarities are concerned, both methods perform a mapping from high-dimensional ambient space to a low-dimensional latent space in order to extract the underlying nonlinear structure of the data. Both methods leverage unsupervised learning to reduce the dimension of the data, although the learning strategies are vastly different.

Manifold learning methods mostly attempt to preserve certain local geometrical properties of the original data in the low-dimensional feature space via explicit non-iterative learning. They typically construct a nearest-neighbor graph where each node corresponds to a training sample, and directed edges represent nearest-neighbor relationships (see Figure 19.3). The ability of manifold learning to generalize to unseen data samples is directly dependent on the smoothness of the manifold and the density of the training samples, as demonstrated in the experiments of Section 19.2.1. In particular, most manifold learning methods [6, 7, 8, 9] rely on the locally flat nature of the low-dimensional manifold. Moreover, they generally require that the manifold be compact, sufficiently sampled, and free of holes, sharp valleys, or peaks. On the other hand, nonlinear autoencoders seek to capture the best set of low-dimensional features for data reconstruction, feature extraction, and classification purposes using iterative learning methods. Autoencoders consist of one or more encoder and decoder layers with nonlinear activation functions. In Section 19.2.1, we demonstrate that it is possible to embed new unseen data samples into the manifold without affecting the previously embedded points, or to synthesize a new data sample in the ambient space by interpolating among existing points on the manifold. This flexibility in learning new concepts is not an inherent property of conventional autoencoders, although continual-learning strategies may provide mechanisms for accomplishing such tasks.

19.4 Computer Projects

The computer projects in this chapter use Python libraries torch, sklearn, NumPy, datetime, tensorflow, and Matplotlib.

Computer Project 19.1: In this computer project, the goal is to use the multi-neuron extension of Oja's learning rule to train a linear autoencoder to extract principal components and the principal eigenvectors of an image database. The same Yalefaces image database used in Section 6.4.4 will also be used in this project. Please refer to Computer Project 6.4 for a description of this dataset and its properties. You can downsample all images in this dataset to size 64×64 pixels before any processing. The specific goals are essentially the same as those outlined in Computer Project 6.4, except the last part on classification, which is excluded. In this project, the LMS-based learning rule together with the deflation process in (19.7) is used, and the results are compared with those of the standard PCA extraction method in Computer Project 6.4. In particular, the convergence behavior of this learning algorithm for the principal eigenvectors and eigenvalues of the training dataset should be investigated. More specifically, the convergence of the principal eigenvectors can be monitored during training by plotting the norm of the error vector between the eigenvectors generated using the standard PCA method and those produced by this learning algorithm. Provide a detailed discussion on the strengths and weaknesses of the generalized Hebbian learning algorithm in terms of computational requirements, convergence, accuracy, and related factors.

Computer Project 19.2: The goals of this computer project are basically the same as those of Computer Project 19.1, but using the RLS learning rule presented in Section 19.1.1.

Computer Project 19.3: In this computer project, we investigate the application of the basic nonlinear autoencoder in Section 19.3.1 for dimensionality reduction using the same downsampled image dataset used in Computer Project 19.1. Please refer to Computer Project 6.4 for guidelines on how to choose the training and testing datasets. The loss function for the SGD-based algorithm is the SSE, as suggested in Section 19.3.1. The Adam optimizer can be used for more robust and efficient training. Once training is completed, reshape the encoder weight vectors into image format and visualize these *bases images*. Describe any interesting properties, similar to those in Figure 19.7 for the MNIST dataset. Next, map the images in the testing dataset to their corresponding low-dimensional latent feature space. Using the decoder layer, reconstruct the test images from their latent features. Compute the error image as the difference between the original and reconstructed images, and compute the SNR in dB over all images in the testing set. Repeat the same process for at least three choices of the number of hidden-layer neurons and compare the results in terms of the SNR values as well as their visual appearance. Provide a detailed discussion on your results, including the strengths and weaknesses of this nonlinear dimensionality reduction network.

Computer Project 19.4: This computer project aims to use a denoising autoencoder and compare its performance with traditional spatial filters discussed in Chapter 8. The datasets for training and testing remain the same as the previous three computer projects. However, before training and testing the system, corrupt the images using additive white Gaussian noise and salt-and-pepper noise. Choose the variance of the Gaussian noise so that the average SNR value is approximately 7 dB for all images. For salt-and-pepper noise, ensure that the percent of corrupted pixels is around 30–40% for all images. Use the corrupted versions of the training images as input to the encoder and the corresponding noise-free original images as targets at the output of the decoder layer. Again, use the SSE loss function together with the Adam optimizer for training of the network. Note that in this case, the number of neurons in the hidden layer should be the same as, or larger than, the input dimension, that is, the representation is *over-complete*. Once training is completed, evaluate the noise filtering capability of this network on both additive Gaussian and salt-and-pepper noise by applying the corrupted test images and computing the average SNR values after filtering. The variance of the remaining noise can be determined from the error images, defined as the differences between the original and filtered images. Compare the performance, in terms of SNR as well as visual inspection, of your results with those generated using spatial averaging and median filters in Section 8.2.1. For this comparison, use only 10 images from the testing dataset. Provide a comprehensive analysis of your results and discuss the advantages and limitations of the denoising autoencoder versus the traditional filtering methods.

19.5 Problems

Problem 19.1 Show that using Oja's learning rule the encoder weight vector converges to the first principal eigenvector of the data covariance matrix.

Problem 19.2 Develop the multi-neuron version of Oja's learning rule without any deflation process.

Problem 19.3 Consider a linear autoencoder with input vector $\underline{x} \in \mathbb{R}^N$ and reduced-dimensional feature vector $\underline{y} = \mathbf{W}\underline{x}$, where $\underline{y} \in \mathbb{R}^M$ and $\mathbf{W}$ is the weight matrix of the encoder layer. If the weight matrix of the decoder layer is $\mathbf{W}^T$, find the expression for the weight matrix $\mathbf{W}$ that minimizes the MSE given by $\varepsilon = E\left[\|\underline{x} - \hat{\underline{x}}\|^2\right]$.

Problem 19.4 For a linear autoencoder with two hidden-layer neurons, show that the deflation process achieves orthonormalization of the weight vectors or the principal eigenvectors.

Problem 19.5 Provide a justification for the choice of the loss function of the Laplacian eigenmaps in (19.12).

Problem 19.6 Show that the Laplacian eigenmaps loss function $\sum_i \sum_j w_{ij} \|\underline{y}_i - \underline{y}_j\|^2$ is equivalent to $tr(\mathbf{Y}^T\mathbf{L}\mathbf{Y})$ where $\mathbf{L} = \mathbf{D} - \mathbf{W}$ with $\mathbf{D}$ being the diagonal matrix with elements $D_{ii} = \sum_j w_{j,i}$ and illustrate the role of the constraint term $\mathbf{Y}^T\mathbf{D}\mathbf{Y} = \mathbf{I}$.

Problem 19.7 Show that minimizing (19.13) leads to eigenvalue problem in (19.14) and explain why the eigenvectors corresponding to the smallest eigenvalues are used to form the low-dimensional embedding matrix $\mathbf{Y}$.

Problem 19.8 Prove that minimizing (19.16) leads to (19.17) and the practical significance of this result.

Problem 19.9 Using the result in (19.17), provide all the steps in the derivation of (19.18).

Problem 19.10 Derive the synthesis equation (19.19).

Problem 19.11 Consider the basic autoencoder structure in Section 19.3.1. Assume that the cross-entropy loss function

$$\mathcal{L}(W) = -\frac{1}{P}\sum_{p=1}^{P}\sum_{n=1}^{N} x_n(p)\log \hat{x}_n(p) + (1 - x_n(p))\log(1 - \hat{x}_n(p))]$$

is used (binary case) instead of the SSE where x_n and $\hat{x}_n$ are the nth element of vectors $\underline{x}$ and $\hat{\underline{x}}$, respectively. Derive the expressions for the gradients of the loss function with respect to the weights of the output neuron n and hidden-layer neuron m, i.e. $\nabla_{\underline{w}_n^{(2)}}\mathcal{L}(W)$ and $\nabla_{\underline{w}_m^{(1)}}\mathcal{L}(W)$.

Problem 19.12 For the sparse autoencoder in Section 19.3.2, use the sparsity constraint in (19.23) and derive the expressions for the gradient of the loss functions in (19.24), where the learning signal is given by (19.25).

Problem 19.13 Repeat Problem 19.12 for the sparsity constraint $\rho(\mathbf{W}^{(1)}) = \sum_{n=1}^{N} \left(\|\underline{w}_n^{(1)}\|^2 - \tau \right)$. Explain your observations.

Problem 19.14 Repeat Problem 19.12 for the sparsity constraint $\rho(\mathbf{W}^{(1)}) = \|\mathbf{W}^{(1)}\|_F^2 - \tau$, where $\|\mathbf{W}^{(1)}\|_F^2 = tr[\mathbf{W}^{(1)^T}\mathbf{W}^{(1)}]$. Explain your observations.

Problem 19.15 Consider a denoising autoencoder where the additive noise is assumed to be zero mean multivariate normal with covariance matrix $\mathbf{\Sigma}_\eta = \sigma_\eta^2 \mathbf{I}$. If the MSE defined by $\varepsilon = E\left[||\underline{x} - r(\underline{z}; \mathcal{W})||^2\right]$ is used as the cost function for the training, derive the expression for the optimal reconstruction $r^*(\underline{z}; \mathcal{W}^*)$ that minimizes this cost function.

Problem 19.16 Show that if the reconstruction function of a denoising autoencoder is approximated around $\underline{x}$ using the truncated Taylor series expansion, $r(\underline{x} + \underline{\eta}) \approx r(\underline{x}) + \frac{\partial r(\underline{x})}{\partial \underline{x}} \underline{\eta}$, the autoencoder becomes a contractive autoencoder with the cost function

$$\mathcal{L} = E\left[\left\|\underline{x} - r(\underline{x})\right\|^2 + \sigma_\eta^2 \left\| \frac{\partial r(\underline{x})}{\partial \underline{x}} \right\|_F^2 \right].$$

Problem 19.17 Consider the regularized loss function for the contractive autoencoder given in (19.28). Assuming that the encoder neurons use sigmoidal activation function, derive the expression in (19.29) for the regularization term.

Problem 19.18 Repeat Problem 19.17 for $\tanh(x)$ activation function. Explain your results and the effects of using this activation function by comparing to (19.28) obtained using the sigmoid.

Problem 19.19 Repeat Problem 19.17 for the threshold logic activation function with slope a is the transition region. Explain your results and the effects of using this activation function by comparing to (19.28) obtained using the sigmoid.

Problem 19.20 Repeat Problem 19.17 for the ReLU activation function. Explain your results and the effects of using this activation function by comparing to (19.28) obtained using the sigmoid.

References

1 E. Oja, "A simplified neuron as a principal component analyzer," *Journal of Mathematical Biology*, vol. 15, pp. 267–273, 1982.

2 T. D. Sanger, "Optimal unsupervised learning in a single-layer linear feedforward neural network," *Neural Networks*, vol. 2, pp. 459–473, 1989.

3 P. Foldiak, "Adaptive network for optimal linear feature extraction," Proceedings of the International Joint Conference on Neural Networks, Washington, DC, pp. 1401–1406, 1989.

4 S. Y. Kung and K. I. Diamantaras, "A neural network learning algorithm for adaptive principal component extraction," Proceedings of the IEEE International Conference on Acoustics, Speech, and Signal Processing (ICASSP), Albuquerque, pp. 861–864, 1990

5 S. Bannour and M. R. Azimi-Sadjadi, "Principal component extraction using recursive least squares learning," *IEEE Transactions on Neural Networks*, vol. 6, pp. 457–469, 1995.

6 J. Tenenbaum, V. de Silva, and J. Langford, "A global geometric framework for nonlinear dimensionality reduction," *Science*, vol. 290, pp. 2319–2323, 2000.

7 S. Roweis and L. Saul, "Nonlinear dimensionality reduction by locally linear embedding," *Science*, vol. 290, pp. 2323–2326, 2000.

8 K. Weinberger and L. Saul, "Unsupervised learning of image manifolds by semidefinite programming," Proceedings of IEEE Conference on Computer Vision and Pattern Recognition, vol. 2, pp. 988–995, 2004.

9 M. Belkin and P. Niyogi, "Laplacian Eigenmaps and spectral techniques for embedding and clustering," *Advances in Neural Information Processing Systems*, vol. 14, pp. 585–591, 2001.

10 I. Goodfellow, Y. Bengio, and A. Courville, "*Deep Learning*," MIT Press, 2016.

11 I. D. Mienye and T. G. Swart, "Deep autoencoder neural networks: A comprehensive review and new perspectives," *Archives of Computational Methods in Engineering*, vol. 32, pp. 3981–4000, 2025.

12 K. Berahmand, F. Daneshfar, E. S. Salehi, Y. Li, and Y. Xu, "Autoencoders and their applications in machine learning: A survey," *Artificial Intelligence Review*, vol. 28, 2024.

13 P. Vincent, H. Larochelle, I. Lajoie, Y. Bengio and P. A. Manzagol, "Stacked denoising autoencoders: Learning useful representations in a deep network with a local denoising criterion", *Journal of Machine Learning Research*, vol. 11, pp. 3371–3408, 2010.

14 H. Bourland and Y. Kamp, "Auto-association by multilayer perceptrons and singular value decomposition," *Biological Cybernetics*, vol. 59, pp. 291–294, 1988.

15 S. Haykin, "*Adaptive Filter Theory*," Prentice-Hall, 3rd Edition, 1995.

16 E.K.P. Chong, W-S Lu, and S.H. Zak, "*An Introduction to Optimization with Applications to Machine Learning*," Wiley, 5th Edition, 2024.

17 G. H. Golub and C.F. Van Loan, "*Matrix Computation*," John Hopkins University Press, 1996.

18 M. A. Carreira-Perpinan, "Continuous latent variable models for dimensionality reduction and sequential data reconstruction," Ph.D. dissertation, Sheffield, Sheffield, U.K., 2001.

19 N. Klausner and M.R. Azimi-Sadjadi, "Manifold-based classification of unexploded ordnance in low-frequency sonar," *IEEE Journal on Oceanic Engineering*, vol. 45, pp. 1034–1044, 2020.

20 D.P. Kingma and M. Welling, "An introduction to variational autoencoders," *Foundations and Trends in Machine Learning*, pp. 1–18, 2019.

21 V. Rupapara, M. Narra, N. K. Gonda, K. Thipparthy, and S. Gandhi, "Auto-encoders for content-based image retrieval with its implementation using handwritten dataset," Proceedings of the IEEE International Conference on Communication and Electronics Systems (ICCES), 2020.

22 H. Larochelle, D. Erhan, and P. Vincent, "Deep learning using robust interdependent codes," JMLR, 2009.

23 A. Ng, "Sparse autoencoder," CS294A Lecture notes, 72, pp. 1–19, 2011.

24 T. M. Cover and J. A. Thomas, "*Elements of Information Theory*," Wiley, 2nd Edition, 2006.

25 G. Alain and Y. Bengio, "What regularized auto-encoders learn from the data-generating distribution," *Journal of Machine Learning Research*, vol. 15, pp. 3743–3773, 2014.

26 O. Ronneberger, P. Fischer, and T. Brox, "U-Net: Convolutional networks for biomedical image segmentation," In: N. Navab, J. Hornegger, W. Wells, and A. Frangi, *Medical Image Computing and Computer-Assisted Intervention*, Springer, vol 9351, pp. 234–241, 2015.

27 N. Siddique, S. Paheding, C.P. Elkin, and V. Devabhaktuni, "U-Net and its variants for medical image segmentation: A review of theory and applications," *IEEE Access*, vol. 9, pp. 82031–82057, 2021.

28 P. Dua, "*Image Denoising Using a U-net*," Stanford University, 2019.

29 S. Rifai, P. Vincent, X. Muller, X. Glorot, and Y. Bengio, "Contractive auto-encoders: Explicit invariance during feature extraction," Proceedings of the 28th International Conference on Machine Learning, Bellevue, WA, 2011.

30 S. Chen and W. Guo, "Auto-encoders in deep learning-A review with new perspectives," *MDPI Mathematics*, vol. 11, p. 1777, 2023.

31 A. Radhakrishnan, M. Belkin, and C. Uhler, "Overparameterized neural networks implement associative memory, "*Proceedings of the National Academy of Sciences*, vol. 117, pp. 27162–27170, 2020.

20

AI Applications in Digital Image Processing

This chapter delves into two additional applications of AI in digital image processing (DIP) that differ from the more conventional applications discussed in the previous chapters. While earlier chapters introduced traditional machine learning and AI applications in pattern classification, feature extraction, image encoding, and image restoration, this chapter broadens the scope to explore nontraditional and more advanced application areas. Our objective is to expand the reader's understanding of AI's full potential in DIP by demonstrating how these techniques can be applied to problems beyond the standard image processing tasks.

The two topics selected for this chapter are: disparity estimation in stereo images and associative memory for grayscale and color images. Disparity estimation, a fundamental problem in computer vision and robotics, involves calculating the pixel displacement between two stereo images to determine depth information. The incorporation of modern AI techniques offers more robust and efficient solutions than the traditional methods. Associative memory, on the other hand, attempts to emulate human-like memory mechanisms for storing and recalling visual information. By focusing on these two distinct and emerging applications, we aim to provide a comprehensive exploration of AI's expanding role in DIP, highlighting the advanced methodologies that will shape the future of this field.

20.1 Disparity Estimation from Stereo Images

Stereo imaging systems are used to provide high-resolution stereoscopic images for depth perception and data compression [1–6]. Some widely used applications include 3-D TV, video-conferencing, computer games, augmented reality, remote sensing, robotics, self-driving cars, and autonomous systems. In these applications, it is often necessary to utilize the correspondences between a pair of right and left images acquired by imaging systems with slightly displaced locations, usually along the horizontal direction. Disparity estimation aims at finding the position differences, or binocular parallaxes, between the corresponding points or blocks in the right and left images of a stereo pair. This problem is closely related to motion estimation in video sequences, where object displacements are measured across successive frames. However, disparity estimation typically requires greater accuracy than motion estimation. This is because human eyes recognize still objects more sharply than moving objects and are more sensitive to depth-related distortions in stereoscopic imagery than to distortions in video images. As a consequence, error in estimating disparity can greatly affect the quality of reconstructed 3-D images and depth perception. This problem is exacerbated by any mismatch between the left and right images due to such factors as reflectivity and illumination differences, blurring effects, partial or full occlusion, and noise.

There are two general categories of traditional methods that can be used for stereo disparity estimation: local and global methods [1]. A simple local method is stereo block matching [2], which is also used to estimate motion parameters in video sequences. In this method, each block in the right image is matched with the most similar block within a larger search region in the reference left image. The corresponding spatial displacement between the matched blocks is recorded as the disparity for the block in the right image. The process is repeated for all block in the right image, leading to a blockwise disparity map. However, the mismatching issues mentioned earlier severely impact the usefulness of this method. In [4, 5], the authors used 2-D block filtering and sequential orthogonal matching pursuit methods to alleviate some of these mismatching effects for stereo image compression. Global methods, on the other hand, provide better accuracy for estimating disparities at the expense of higher computational requirements. Among the global methods is the Markov random field (MRF) discussed in Section 12.1.4 for image segmentation. The energy function used by the MRF model [3] for the disparity estimation imposes a smoothness constraint among neighboring displacement vectors, hence encouraging a spatially smooth disparity map while using a measure of similarity between pixels in the left and right images. This method does not explicitly compensate for intensity variations between stereo images. Additionally, the computational cost associated with identifying depth discontinuities can be prohibitively high.

Humans have an innate ability to perceive depth in their binocular vision, even under the aforementioned difficult conditions. Binocular disparity between the images observed by the two eyes, along with monocular depth cues such as relative object size, illumination, occlusion, texture gradients, enables the brain to construct 3-D images of the scene. Many of these monocular cues are learned and continuously refined through experience and interaction with the environment. Inspired by the mechanisms of human visual perception, which rely on binocular vision and monocular cues, artificial neural networks can be developed to generate reliable disparity maps and subsequent depth perception in near real time.

Over the past decade, numerous stereo matching architectures based on deep convolutional neural networks (CNNs) have been developed to perform disparity estimation and depth perception in the hope of achieving human-like performance. The reader is referred to several survey papers in [6–8] for comprehensive reviews of modern neural network-based stereo matching systems. A major issue that plagues the applicability of some of these systems is the lack of generalization ability. This is primarily due to the fact that datasets with accurate ground-truth disparity maps are difficult to acquire. Therefore, these networks are typically trained on synthetic datasets, and they fail to generalize to unseen real data. Moreover, some systems fail to exploit contextual and semantic information that could further improve the quality and consistency of the estimated disparity maps.

Among the state-of-the-art frameworks developed for evaluating and benchmarking different deep learning models for stereo matching is OpenStereo [9], which was designed to address the lack of generalization of the earlier models. OpenStereo is a PyTorch-based toolbox that provides a platform for implementing a wide variety of stereo matching models and comparing their performance on several stereo databases. It offers many unique features, including modularity of configuration, compatibility with diverse stereo matching methods, versatility in using different structures and models, and support for a wide range of datasets for training and testing. These capabilities make OpenStereo a valuable research and development platform for advancing modern stereo vision systems.

20.1.1 Pyramid Stereo Matching Network

One of the models that can be implemented on OpenStereo platform is the Pyramid Stereo Matching Network (PSMNET) introduced in [10]. The architecture of the PSMNET is depicted in Figure 20.1. The left and right images are first applied to two separate deep CNNs for feature extraction at multiple scales and resolutions. The first three cascaded convolutional layers each contain 32 kernels of size 3×3. The output of the third layer consists of 32 feature maps, each of dimensions $\frac{N}{2} \times \frac{N}{2}$ pixels, where N is the dimension of the input image. Average pooling with a stride of 2 is performed in the second layer. The next four convolutional layers use residual learning [11] for better feature representation. These layers contain 32, 64, 128, and 128 filters, respectively. The third and fourth layers use dilated convolution to enlarge the receptive field of the filters and capture more global image features. Again, average pooling with a stride of 2 is performed in the second convolutional layer.

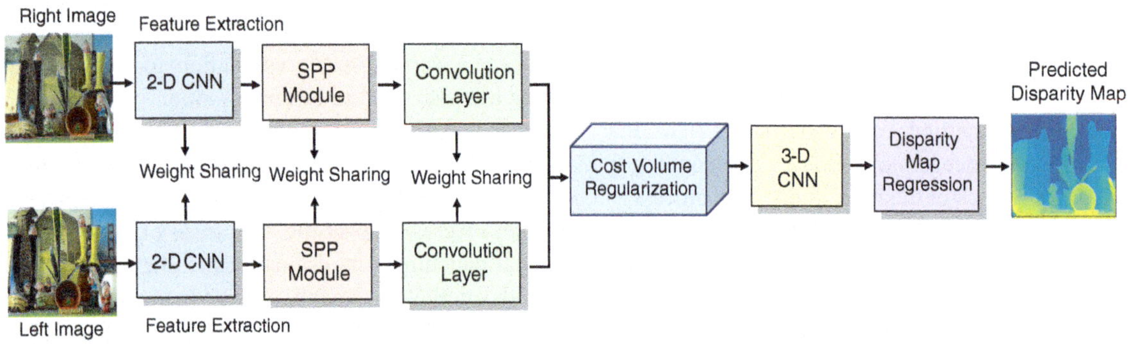

Figure 20.1 The architecture of the PSMNET and its components for stereo disparity estimation [10]. *Source:* (Photo) Anna Blasiak et al. 2005 / Middlebury Computer Vision Group.

The feature maps of the fourth layer of size $\frac{N}{4} \times \frac{N}{4}$ pixels are applied to the Spatial Pyramid Pooling (SPP) modules to exploit global contextual information in the hope of capturing the same type of monocular cues that human eyes use for depth perception. The contextual information extracted by the SPP module corresponds to region-based features at multiple scales and locations to build a more comprehensive feature representation and facilitate disparity estimation, especially under challenging conditions. For example, the SPP module can learn hierarchical contextual relationships associated with a car and its components, including windows, tires, and hoods. Such information can subsequently assist both object classification and stereo correspondence matching. The SPP module uses average pooling blocks of sizes 64×64, 32×32, 16×16, and 8×8 pixels, followed by 1×1 convolution and upsampling operations. The left and right feature maps of the SPP module are concatenated to form a 4-D *cost volume,* which represents the cost of matching features between corresponding pixels in the left and right images for a range of possible disparities. In other words, the cost volume quantifies how well a pixel in the left image corresponds to a pixel in the right image at a particular horizontal displacement.

The next component in Figure 20.1 is the 3-D CNN regularizer, which aggregates feature information along the disparity dimension and the two spatial coordinates. Consequently, it processes three variables of the cost volume data, namely height, width, and the disparity (depth). The structure of this 3-D CNN is typically a stacked encoder–decoder known as the *hourglass*, which resembles an actual hourglass with the narrowing in the middle. The encoder component is a deep 3-D CNN that captures high-level semantic information from the cost volume and helps resolve difficult matching problems arising from occluded objects, repetitive patterns, and illumination and reflection mismatches. The decoder network in this hourglass network is also a stacked 3-D CNN that consists of deconvolutional layers with upsampling operations. The output of the decoder has higher spatial resolutions, which facilitates more accurate and robust disparity estimation and subsequent depth perception. The hourglass network is trained using the Smooth l_1 loss function recommended in [10] because of its robustness and low sensitivity to outliers. This loss function is defined as

$$\mathcal{L}(d, \hat{d}) = \frac{1}{M} \sum_{i=1}^{M} s_{l_1}(d_i - \hat{d}_i), \tag{20.1}$$

where d_i and $\hat{d}_i$ are the actual (ground-truth) and estimated disparity for the labeled pixel i, respectively, M is the number of pixels with labels, and $s_{l_1}(x)$ is the smooth l_1 function defined as:

$$s_{l_1}(x) = \begin{cases} 0.5x^2, & \text{for } |x| < 1 \\ |x| - 0.5, & \text{otherwise.} \end{cases} \tag{20.2}$$

The estimated disparity $\hat{d}$ is generated by the disparity regression block using a fully differentiable function that promotes a smooth disparity map. However, the first step is to convert the predicted costs associated with each disparity in the cost volume block to a probability volume by negating each cost value and then normalizing using the softmax operation. The estimated disparity is then generated using the expected disparity, which is

$$\hat{d} = E[\Delta] = \sum_{\delta=0}^{\Delta_{\max}} \delta\, \sigma(-c_\delta), \tag{20.3}$$

where δ is the disparity with the associated cost c_δ of the cost volume, $\Delta_{\max}$ is the maximum possible disparity, and $\sigma(.)$ represents the softmax operation, which gives the probability $P(\Delta = \delta)$. The disparity regression produces a more consistent disparity map.

The PSMNET trained on the KITTI 2012 dataset [12] was used in our experiments to compute the disparity maps for a pair of left and right images of *dolls* from the Scenes 2005 database [13]. The images were taken under three different illuminations and three different exposure times of 250, 1000, and 4000 msec. Three representative cases were selected for evaluation. The first and third cases correspond to extreme acquisition conditions involving low illumination with short exposure time, and high illumination with long exposure time, respectively. The second case corresponds to nominal illumination and exposure time.

The original left and right images for these cases are shown in Figures 20.2a,b, 20.3a,b, and 20.4a,b, respectively. The ground-truth disparity map is shown in Figure 20.5. The estimated disparity maps generated using the trained model together with the error images (difference between the truth and estimated disparity maps) are also shown in Figures 20.2c,d, 20.3c,d, and 20.4c,d. The mean squared error (MSE) and End Point Error (EPE) values are also given in the

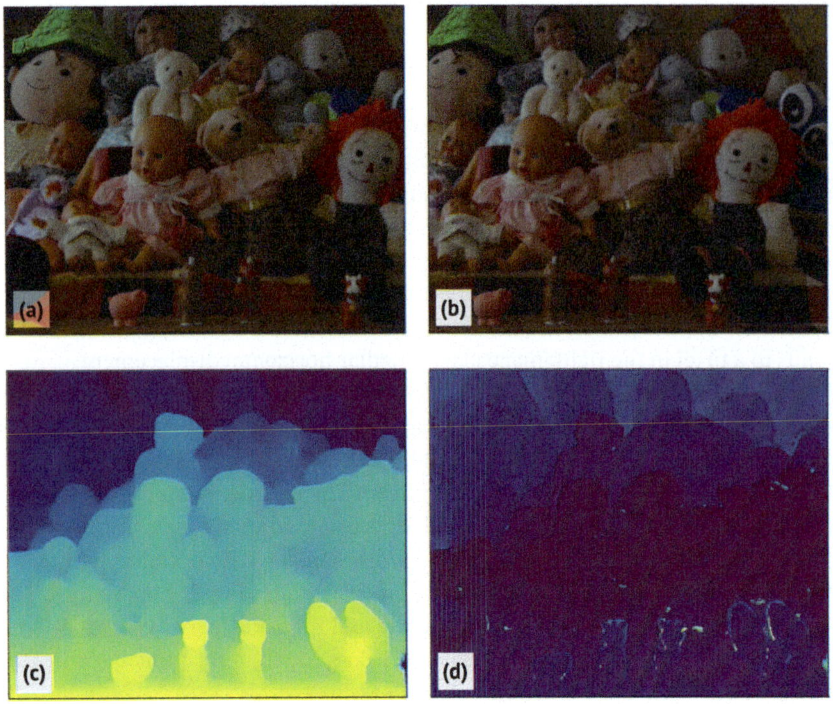

Figure 20.2 A pair of (a) left and (b) right stereo images taken under low illumination condition and short exposure time, (c) estimated disparity map, and (d) error image between truth and estimated disparity maps: MSE = 0.0185 and EPE = 0.106 [13]. *Source*: (A-B) D. Scharstein et al. (2005) / Middlebury Computer Vision Group.

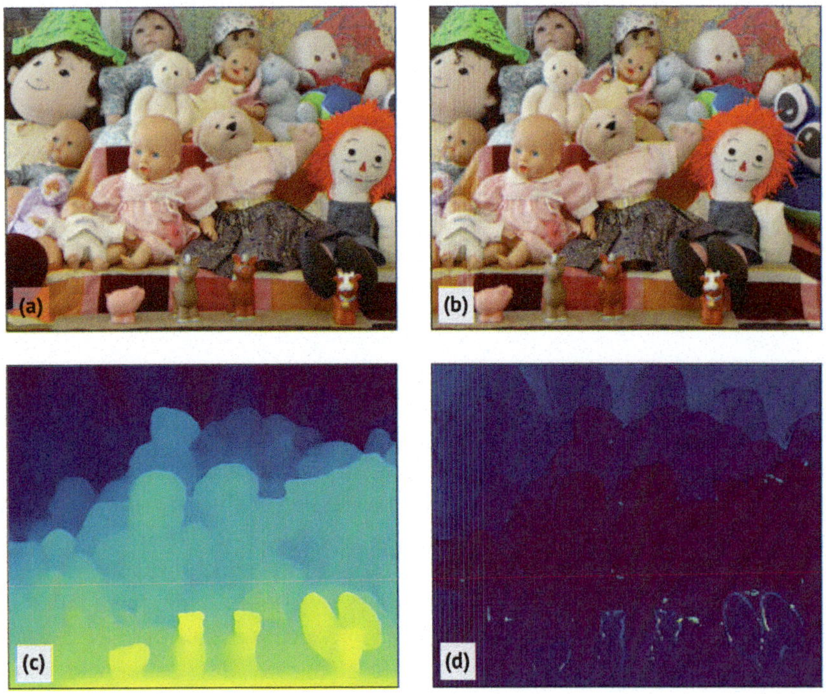

Figure 20.3 A pair of (a) left and (b) right stereo images taken under normal illumination conditions and exposure time, (c) estimated disparity map, and (d) error image between truth and estimated disparity maps: MSE = 0.0175 and EPE = 0.102 [13]. *Source*: (A-B) Anna Blasiak et al. 2005 / Middlebury Computer Vision Group.

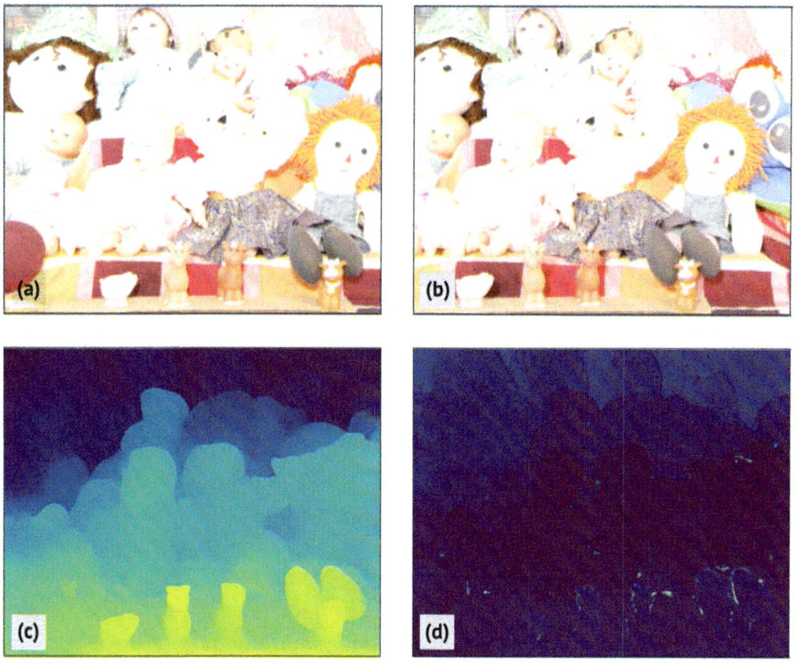

Figure 20.4 A pair of (a) left and (b) right stereo images taken under high illumination condition and long exposure time, (c) estimated disparity map, and (d) error image between truth and estimated disparity maps: MSE = 0.0158 and EPE = 0.095 [13]. *Source*: (A-B) Anna Blasiak et al. 2005 / Middlebury Computer Vision Group.

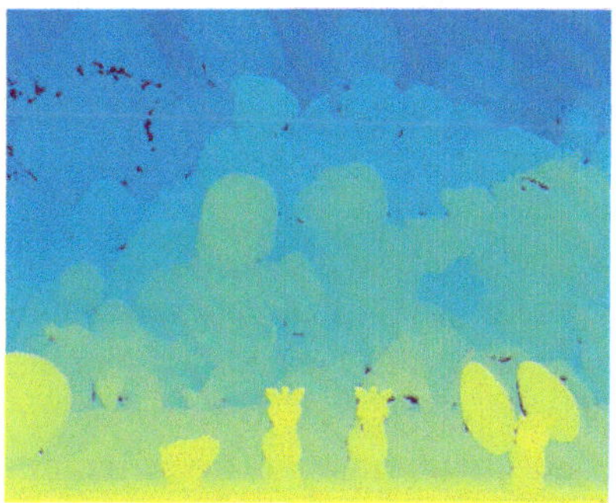

Figure 20.5 Ground-truth disparity map for the left image in Figure 20.3.

figure captions. The EPE is the average absolute disparity error across all image pixels and is a common metric used to evaluate the accuracy of disparity estimation. Comparisons of the ground-truth map with those generated using the PSMNET indicate consistently good performance of this system across a wide range of illumination and exposure time conditions. More specifically, as far as lighting and illumination mismatch scenarios are concerned, PSMNET provides robust and consistent disparity estimation. This observation is further supported by the relatively stable values of MSE and EPE obtained under the different imaging conditions. Finally the experimental results also attest to the generalization capability of the PSMNET model to challenging imaging scenarios.

20.2 Associative Memory for Pattern Retrieval

Associative memory is one of the fundamental functions of the human brain, enabling the recall of a complete sensory experience from a partial stimulus. For example, hearing a few bars of a familiar piece of music can recall associated scenes and sounds. Unlike computer memory, which uses a location-addressable memory (LAM) system, associative memory functions as a content-addressable memory (CAM). That is, rather than retrieving content by its storage location address, a stored pattern is retrieved based on its similarity to an input pattern. This capability enables a rapid, highly parallel search through stored memories to identify the closest match.

Associative memory systems possess several important functionalities and configurations. They can store multiple associated stimulus-response pattern pairs when used as hetero-association; or store multiple single patterns when employed for auto-association. Storage is accomplished by modifying the connectivity of neurons in a large, densely interconnected network, forming a robust and spatially distributed *long-term memory*. Recall, on the other hand, involves retrieving the correct stored pattern even when the input stimulus is distorted or incomplete, a capability known as error correction. This retrieval process occurs through state propagation over time, which constitutes the system's *short-term memory*. During recall, the system performs a parallel search over all stored patterns and converges toward the stored pattern that most closely resembles the input stimulus, retrieving it either fully or partially. Performance can deviate when memory patterns overlap, potentially causing less faithful regeneration of the output, or when too much distortion in the stimulus leads to an incorrect recall.

Pioneered by John Hopfield [14], Hopfield networks provide one of the earliest neural implementations of associative memory for binary patterns. The network's synaptic weights are determined during the storage phase using Hebbian learning, discussed in Section 17.3.3 of Chapter 17. These weights constitute the long-term memory of the system by establishing stable equilibrium states, or attractors, that represent the memorized patterns. The retrieval phase involves the network's state variables, which constitute the short-term memory, evolving through the state-space until they converge to one of the established attractors. The Hopfield network architecture is fully recurrent, connecting the output of each neuron to the input of all other neurons, with no self-feedback. The neurons outputs are generated using the McCulloch-Pitts activation function. When presented with a noisy or incomplete version of a stored pattern, the system's dynamics drive the state propagation toward one of the learned attractors, thereby completing or retrieving the associated memory. These attractors correspond to the local minima of an energy function constructed during the storage phase. The success of the recall process is contingent upon several factors, including the level of distortion in the input pattern, the storage capacity of the network, and the degree of correlation among the stored patterns.

The extension of associative memory to grayscale and color images can be accomplished using several different approaches. One possible approach is to decompose a grayscale image into its bit-plane representation and train a separate binary Hopfield network for each bit-plane image. Recall is then performed by reconstructing the grayscale image from all the recalled bit-plane images. Clearly, this approach becomes cumbersome, particularly for grayscale images with a large number of bits/pixel and for color images.

More recently, several new approaches have been developed that do not rely on the bit-plane representation of the original image. Among these methods is the over-parameterized auto-encoding network in [15] that does not require minimization of an energy function. They prove that training samples can be stored as attractors of a dynamical system, and iterating the learned mapping can achieve reliable pattern recall. In [16], a neuro-biologically inspired associative memory was introduced using a recurrent architecture that incorporates Hebbian plasticity and sparse, attention-driven retrieval mechanisms. This memory system, called the Engram Neural Network (ENN), utilizes memory-augmented recurrent network with a differentiable memory matrix, a Hebbian plasticity rule for memory updating, and a sparsity mechanism for content-based retrieval.

This section presents a hybrid associative memory architecture that combines a deep CNN-based autoencoder with a modified RNN that employs Gated Recurrent Units (GRUs). The use of GRUs is particularly important for iterative retrieval because they enable the network to retain information from previous states while mitigating the long-term dependency problems commonly encountered in conventional RNNs. The overall retrieval process works as follows. First, an encoder network maps the input pattern into a latent feature space. The GRU-based recurrent network then acts as a dynamical system that iteratively evolves the latent state toward an attractor. A decoder network reconstructs the pattern by mapping the converged latent state vector back to the image domain. Chapter 19 introduced different types of autoencoders and their applications. The next section provides a brief review of recurrent neural networks and the GRU mechanism that forms the basis of the proposed associative memory system.

20.2.1 Recurrent Neural Networks

Recurrent Neural Networks (RNNs) are a family of neural networks designed to process sequential data such as text, speech, or time series [17]. Consequently, they find numerous applications in machine translation, text summarization, question answering, dialogue modeling, and sequence classification such as sentiment analysis. RNNs have several differences with the feedforward networks discussed in Chapters 18 and 19. Unlike feedforward networks, RNNs incorporate feedback connections that enable them to maintain an internal *hidden state,* which acts as a memory, carrying information from previous elements of a sequence to influence the processing of the subsequent inputs. This recurrent architecture, combined with shared parameters across all steps, gives RNNs the flexibility to process variable-length sequences. At each time step, RNN takes both the current input and the past hidden states to produce updated hidden states and the corresponding output. In other words, it makes state prediction for every element of a sequence. In this manner, the hidden state serves as the network's memory and carries information learned from earlier elements of a sequence, enabling the network to use historical context when processing new information. RNNs use a distributed hidden state to store a compressed history of past elements in a sequence, although this capability is limited for long sequences. Parameter sharing in RNN also encourages the network to learn a common set of temporal relationships independent of sequence length and the order of occurrence of events.

RNNs can be configured in several different architectures adaptable to various applications. Common configurations include one-to-one, many-to-one, one-to-many, and many-to-many structures, distinguished by the number and dimensionality of their inputs and outputs. For example, image captioning exemplifies a one-to-many architecture, where a single image input is used to produce a sequential, word-by-word description of the scene.

A typical one-to-one RNN structure at a specific discrete time instance t, shown in Figure 20.6, consists of an input vector $\underline{x}_t \in \mathbb{R}^N$, an output vector $\underline{y}_t \in \mathbb{R}^L$, and a hidden state vector $\underline{s}_t \in \mathbb{R}^M$. The hidden state evolves over time due to feedback connections in the network. The general dynamical equation (nonlinear state equations) that governs this state evolution in time is given by

$$\underline{s}_t = f(\mathbf{W}\underline{s}_{t-1} + \mathbf{U}\underline{x}_t), \tag{20.4}$$

where $\mathbf{U}$ and $\mathbf{W}$ are trainable parameter matrices of the network, $\underline{s}_{t-1}$ is the previous hidden state vector, and $f(.)$ is an element-wise nonlinear activation function such as tanh(.). The hidden state in the recursive equation in (20.4) captures information about the past and present sequence of inputs, $\underline{x}_t, \underline{x}_{t-1}, \dots, \underline{x}_1$, hence forming the memory of the network. The output equation is

$$\underline{y}_t = g(\mathbf{V}\underline{s}_t), \tag{20.5}$$

where $\mathbf{V}$ is the weight matrix for the output layer and $g(.)$ is the element-wise output activation function, such as softmax. Unlike the feedforward neural networks, RNNs share the same parameters $\mathbf{U}$, $\mathbf{W}$, and $\mathbf{V}$ across all time steps. This parameter-sharing scheme allows the model to learn a time-invariant nonlinear dynamical system governed by the state equations (20.4) and (20.5).

To process an entire sequence, an RNN is unrolled or unfolded through time, which converts it into a directed acyclic graph (DAG) or a layered network. Figure 20.7a illustrates the RNN at a single time step t, with an input vector $\underline{x}_t$, state vector $\underline{s}_t$, and output vector $\underline{y}_t$. Figure 20.7b,c shows the unrolling process across three consecutive time steps: $t-1$, t, and $t+1$ and the resulting unrolled layered network. The shared parameters across all time are visible in the unrolled architecture. The unrolled architecture offers several key advantages:

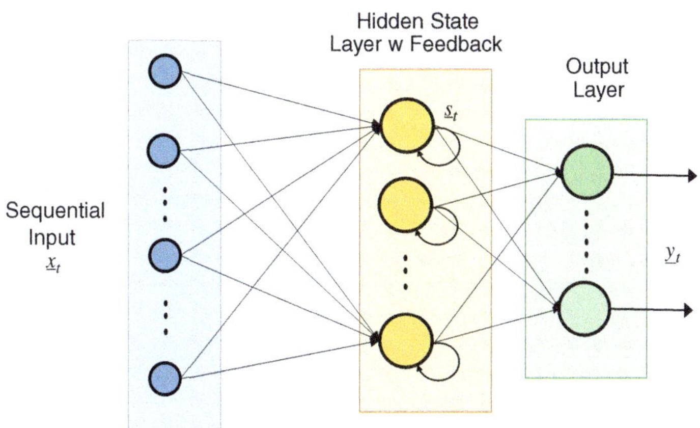

Figure 20.6 A one-to-one RNN structure at time instance t.

- The model can handle input sequences of any length because the same state transition function, $f(.)$, is applied at every time step.
- The number of trainable parameters remains constant regardless of the sequence length, reducing model complexity and improving scalability.
- To enable more abstract temporal feature extraction, multiple RNN layers can be stacked to create a deep RNN, as shown in Figure 20.8. In a stacked RNN, during the forward pass, state propagation moves through time (left-to-right) and across layers (bottom-up). During the backward pass, gradients (or learning signals) are propagated right-to-left through time and top-down through the layers, a process known as back-propagation through time (BPTT). A common choice for the cost function in these models is the categorical cross-entropy loss introduced in Section 18.2.1.

The BPTT originally developed in [18] extends the standard error back-propagation algorithm to RNNs by propagating gradients across all time steps of the unrolled architecture. Despite its effectiveness, the original RNN architecture trained with BPTT suffers from several well-known limitations [19]. A significant challenge arises from the behavior of learning signals (or gradients) propagating backward through many time steps, particularly over long sequences. They tend to either grow without bound, leading to unstable weight estimates and training divergence, or decay exponentially to zero causing vanishing gradients. The latter situation prevents effective learning of long-range dependencies because the learning signal becomes too weak to effectively update weights in lower layers, thereby hindering the network's ability to learn long-term dependencies. Furthermore, the original RNN's capacity for contextual information is restricted, typically to approximately 10 time steps between relevant input and target events. This limitation prevents standard RNNs from effectively predicting long-range relationships, such as those between distant words in a lengthy sentence. These inherent problems are addressed by more advanced RNN variants, such as GRUs, which incorporate gating mechanisms to mitigate the

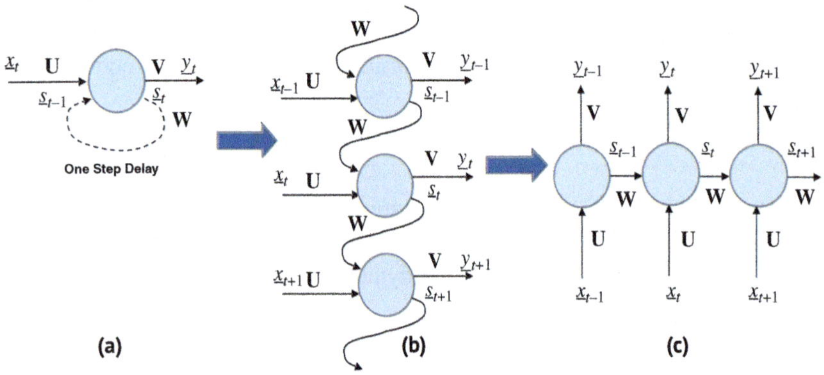

Figure 20.7 (a) RNN as a DAG at a time step t, (b) unrolling process in three consecutive time steps $t-1$, t and $t+1$, and (c) unrolled RNN. *Source*: [22]/Hugo Larochelle, Dumitru Erhan and Pascal Vincent.

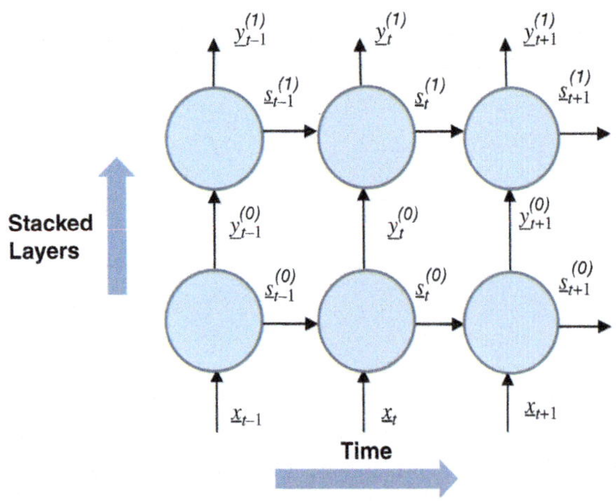

Figure 20.8 A stacked unrolled RNN extracts more abstract temporal representations.

issues of vanishing and exploding gradients, enabling more robust learning of long-term dependencies. The next section focuses on the GRU architecture, which forms the basis of the associative memory system developed in this chapter.

20.2.2 Gated Recurrent Units

Gated Recurrent Units (GRUs) were developed [20] to address the vanishing or exploding gradient problems that can occur during the training of conventional RNNs. Their effectiveness is due to two gating mechanisms, namely the *reset gate* and the *update gate*, which regulate the flow of information and manage dependencies over time. At each time step t, the GRU processes the current input vector $\underline{x}_t$ and the previous hidden state $\underline{s}_{t-1}$ through these two gates. The reset gate determines how much information from the previous hidden state should be retained when calculating the candidate hidden state. The update gate then controls how much of this candidate state is incorporated into updating the new hidden state, $\underline{s}_t$. This process effectively gives the GRU a mechanism for deciding which past information is relevant and should be carried forward, and which one is no longer relevant and should be discarded. The steps of a fully gated GRU are as follows:

1. Reset Gate: Using the previous hidden state vector $\underline{s}_{t-1}$ and current input vector $\underline{x}_t$ compute:

$$\begin{aligned} \underline{r}_t &= \sigma\left(\mathbf{W}_r \underline{s}_{t-1} + \mathbf{U}_r \underline{x}_t\right) \\ \underline{\hat{s}}_t &= \tanh\left(\mathbf{W}_h(\underline{r}_t \odot \underline{s}_{t-1}) + \mathbf{U}_h \underline{x}_t\right), \end{aligned} \tag{20.6}$$

where $\underline{r}_t$ is the reset vector, $\underline{\hat{s}}_t$ is the candidate new hidden state, $\sigma(.)$ represents a unipolar sigmoidal activation function, and $\odot$ denotes element-wise product operation. When the elements of the reset gate vector $\underline{r}_t$ are close to 0, the influence of the previous hidden state is effectively ignored, in which case candidate hidden state is determined primarily based on the current input vector. Conversely, when the elements of $\underline{r}_t$ are close to one, information from the previous hidden state is retained and contributes significantly to the candidate state.

2. Update Gate: Using the previous hidden state vector $\underline{s}_{t-1}$, the current input vector $\underline{x}_t$, and the candidate hidden state vector $\underline{\hat{s}}_t$, compute:

$$\begin{aligned} \underline{z}_t &= \sigma\left(\mathbf{W}_z \underline{s}_{t-1} + \mathbf{U}_z \underline{x}_t\right) \\ \underline{s}_t &= (\underline{1} - \underline{z}_t) \odot \underline{s}_{t-1} + \underline{z}_t \odot \underline{\hat{s}}_t, \end{aligned} \tag{20.7}$$

where $\underline{z}_t$ is the update gate vector and $\underline{s}_t$ is the new hidden state vector. This new hidden state is generated by a convex combination of the candidate hidden state vector and the previous hidden state vector weighted by the update gate vector $\underline{z}_t$. That is, the update gate determines the contribution of each component in this state update equation. If the elements of $\underline{z}_t$ are close to zero, the previous hidden state is largely preserved. Conversely, when they are close to one, the candidate hidden state replaces the previous state.

All the parameter matrices $\mathbf{W}_r$, $\mathbf{W}_h$, $\mathbf{W}_z$, $\mathbf{U}_r$, $\mathbf{U}_h$, and $\mathbf{U}_z$ are obtained during the training process using BPTT algorithm. Figure 20.9 shows the general structure of GRU cell with the reset and update gates at time instance t.

As in the standard RNN, a GRU network typically consists of stacked GRUs, where the hidden state vector from one GRU is passed to the next in the sequence. The initial hidden state is usually set to $\underline{s}_0 = \underline{0}$. The final hidden state vector $\underline{s}_T$ where T is the length of the input sequence is used as a compact representation of the entire sequence to generate the final output, such as a classification decision.

One question that may arise here is "what are the conditions for the asymptotic stability of GRU?" This important question is addressed as follows.

Stability Analysis of GRU: Let us first combine the state equations in (20.6) and (20.7)

$$\underline{s}_t = (\underline{1} - \underline{z}_t) \odot \underline{s}_{t-1} + \underline{z}_t \tanh\left(\mathbf{W}_h(\underline{r}_t \odot \underline{s}_{t-1}) + \mathbf{U}_h \underline{x}_t\right), \tag{20.8}$$

which represents a discrete nonlinear system when used in conjunction with other nonlinear equations for $\underline{z}_t$ and $\underline{r}_t$. The stability analysis of such systems has been studied using many different approaches, particularly for nonlinear control systems. In [21], the authors extended the Input-to-State stability (ISS) in [22] to single-and multi-layer GRU networks. The ISS condition is analogous to bounded-input bounded-output (BIBO) stability, though it applies to the state instead of the output. Here, we present a brief summary of this result in the following theorem. The reader is referred to [21] for proof.

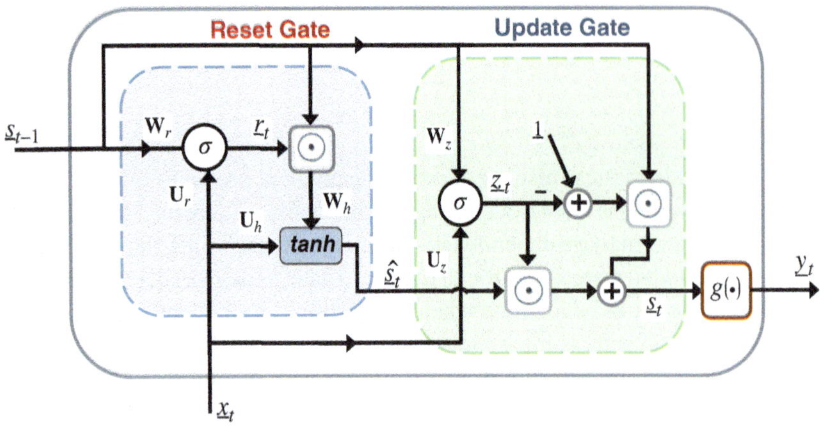

Figure 20.9 Structure of a GRU cell at time instance t with reset and update gates.

Theorem: ISS for Single-Layer GRU: A sufficient condition for the ISS of a nonlinear system in (20.8) is

$$||\mathbf{W}_h||_\infty \bar{\sigma}_r < 1, \quad \text{where} \quad \bar{\sigma}_r = \sigma(||\mathbf{U}_r\ \mathbf{W}_r||_\infty), \tag{20.9}$$

and $||\mathbf{A}||_\infty = \max\limits_{1\le i\le M} \sum\limits_{j=1}^{N} |a_{ij}|$.

This stability condition can be used to determine the stability of the trained network, or monitored during the training of a GRU.

20.2.3 A Hybrid Architecture for Auto- and Hetero-Association Learning

In this section, we present a hybrid deep learning model [23] that combines a CNN-based autoencoder with a GRU network to create an associative memory for grayscale or color images. The key idea is to perform iterative state propagation in the latent feature space generated by the encoder. The feedback mechanism of the GRU enables the iterative refinement of the latent representation, thereby improving the network's robustness to noise, distortion, occlusion, and incomplete input data.

The encoder consists of two convolutional layers with 32 and 64 kernels, respectively. Both layers use 4×4 kernel size with a stride of 2. For input RGB images of size 128×128, the output feature maps are flattened to a vector of dimension $64 \times 32 \times 32$. This vector is then passed through a fully connected layer to generate a 256-dimensional latent feature vector. This feature vector is processed by a GRU network, which has 256 hidden state variables, matching the dimensionality of the latent feature space. The decoder network is designed to perform the inverse operation of the encoder. More specifically, it inputs the converged latent representation generated by the GRU to reconstruct the original image. This decoder consists of a fully connected linear layer that maps the 256-dimensional latent vector to a tensor of dimensions $64 \times 32 \times 32$, followed by ReLU activation function, an unflattening operation to restore the original spatial dimensions, and two deconvolutional layers with 64 and 32 kernels, respectively. Neurons in the convolutional and deconvolutional layers use ReLU activation function. This hybrid architecture enables implementing both auto-associative and hetero-associative memory, which are briefly discussed next.

Given a dataset consisting of P pairs of exemplar patterns $\{\underline{x}(p), \underline{y}(p)\}_{p=1}^{P}$, a hetero-associative memory implements a mapping $\underline{y}(p) = \Psi(\underline{x}(p))$ such that if $\underline{x}$ is any unknown vector sufficiently close to a stored pattern $\underline{x}(k)$, then $\Psi(\underline{x}) = \underline{y}(k)$, i.e. it retrieves or recalls the pattern associated with $\underline{x}(k)$. An auto-associative memory is a special case of the hetero-associative memory where $\underline{y}(p) = \underline{x}(p)$. Therefore, if $\underline{x}$ is closer to the stored pattern $\underline{x}(k)$ than any other pattern $\underline{x}(l)$, then $\Psi(\underline{x}) = \underline{x}(k)$, i.e. the system recalls the original pattern.

The dataset generated for this study contained 20 semantically associated image pairs, each of size 128×128 pixels. Figure 20.10 shows these images with their semantically associated images right below them, e.g., seeds to plants. These image pairs were used to train the hetero-associative memory. For auto-association, all 40 images served as both the input patterns and desired outputs. The encoder, GRU, and decoder were trained jointly using the SSE loss function and the Adam optimizer with a learning rate of 0.0001. Each model was trained for 2000 epochs.

Figure 20.10 Original training images.

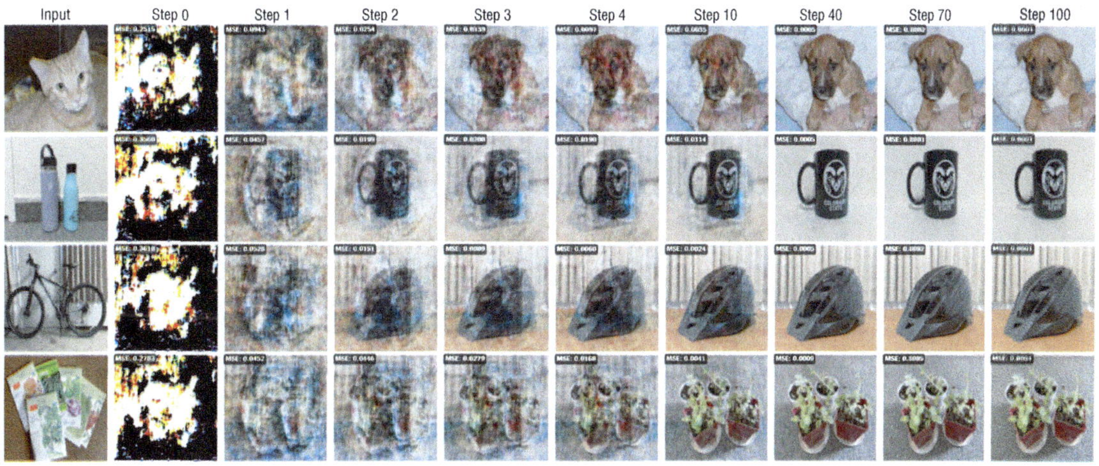

Figure 20.11 Progression of GRU's latent feature refinement and the retrieved associated images during various stages of state propagation and recall.

During retrieval, the GRU iteratively refined the latent features until convergence to a stable state or until a maximum of 100 iterations was reached. After convergence, the refined latent feature vector was then passed to the decoder for image reconstruction. Throughout the hetero-associative training, both the reconstructed images and the SSE loss were monitored. Figure 20.11 captures the progressive refinement of the retrieved images during the iterative state propagation within the GRU for specific original training images. These results indicate that iterative state propagation within the latent feature space led to high-fidelity retrieval of the stored associations in several iterations.

The trained model was then tested on the same image pairs in the presence of additive Gaussian noise, structured masking, and random white line perturbations. We first investigated the robustness of the hetero-associative memory by examining its ability to retrieve the correct associated images from noisy inputs. Additive Gaussian noise with varying signal-to-noise ratios (SNRs) was introduced into the input images. To further evaluate the robustness of the associative memory to structured degradations, experiments were also conducted using square black masks of varying sizes and random locations to simulate partial occlusion, together with random white-line patterns of different lengths and densities to emulate scratches commonly encountered in degraded photographs.

The corrupted images were then mapped to their corresponding retrieved outputs using the trained hetero-associative memory. Figure 20.12 shows several examples of corrupted images and their corresponding retrieved images. Despite the presence of moderate noise, masking, and structural perturbations, the iterative state propagation in the latent feature space converged to the correct attractors, resulting in faithful retrieval of the associated images.

To obtain the performance plots of the trained system, additive white Gaussian with varying noise variance was added to the images to produce SNR values ranging from 20 dB to −20 dB. Both the retrieved images and ground-truth images were

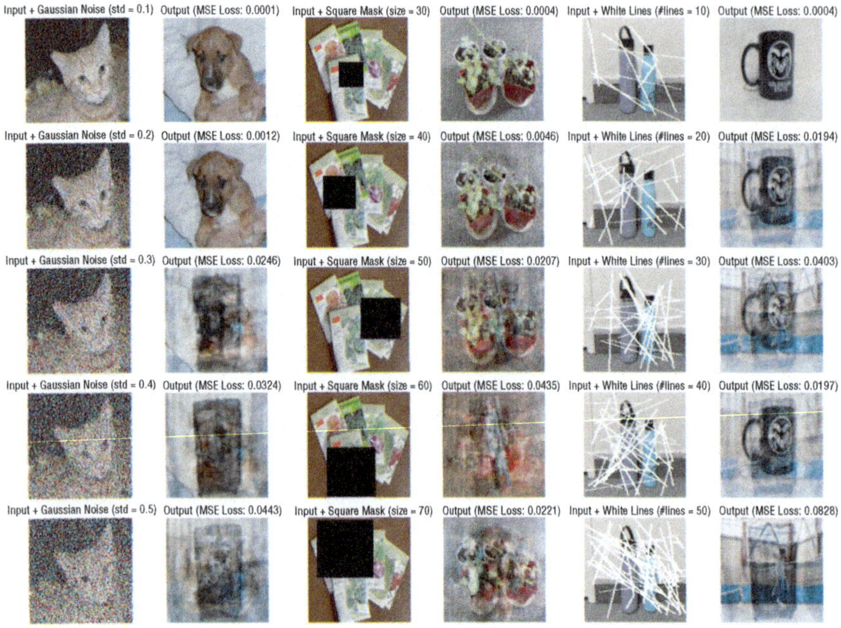

Figure 20.12 Hetero-association results of the trained system in presence of additive Gaussian noise and other structural perturbation.

normalized to the range [0, 1]. The quality of the hetero-associative recall was evaluated using the Structural Similarity Index Measure (SSIM) [24] measure. A successful associative recall was declared when the SSIM score exceeded 0.8 value determined experimentally. Recall accuracy was computed as the percentage of correctly recalled image pairs satisfying the SSIM criterion. Figure 20.13a gives the recall accuracy for different associative memory loads, corresponding to 4, 8, and 12 stored input–output pairs. For all three cases, the model achieves 100% correct recall for SNR values between 20 dB to 0 dB, demonstrating excellent robustness to moderate additive noise. However, as the SNR decreased below 0 dB, recall performance gradually deteriorated, with the deterioration becoming increasingly dependent on the memory load. For the 4-pair configuration, correct recall remained high under severe noisy conditions and declined only gradually as noise variance increased. In contrast, the 8-pair configuration exhibited a sharper drop in correct recall at negative SNR values, indicating increased sensitivity to noise. Although the 12-pair configuration maintained a relatively high recall at moderate noise levels, its recall accuracy dropped rapidly in the presence of more severe noise.

These results suggest that the learned hetero-associative mappings remain stable provided that the corrupted input remains within the basin of attraction of the corresponding stored memory. As the noise level increases, the latent representation eventually leaves the basin of attraction, leading to retrieval failures. Moreover, these results demonstrate an inherent trade-off between memory capacity and robustness to noise. While storing more input–output pairs enhances the model's representation capacity, it reduces recall reliability under high level of noise.

A similar study was conducted to evaluate robustness to partial occlusion. Square masks covering between 10% and 100% of the image support were applied in increments of 10%. The masked images were then presented to the trained hetero-associative memory, and recall accuracy was computed using the SSIM criterion with a threshold of 0.8, as in the previous study. The recall plot in Figure 20.13b illustrates that this hetero-associative memory remains highly robust under low to moderate levels of occlusion, particularly when fewer input–output pairs are stored. For the 4-pair configuration, correct recall remained at 100% up to 60% masking effects, after which performance degraded gradually. The 8-pair configuration exhibited a steady reduction in recall as the size of the masked area increased, with recall dropping significantly beyond 50% occlusion and reaching zero at extreme masking levels. The 12-pair configuration showed the most pronounced sensitivity to occlusion, with recall degrading rapidly beyond 30 – 40% masking and failing completely when more than 60% of the input support is masked.

Again, these results indicate that increasing the number of stored hetero-associative pairs reduces robustness to partial occlusion due to increased interference among stored associations, limiting the model's ability to reliably retrieve the outputs when large portions of the input are missing.

The sufficient condition for ISS stability, i.e. $||\mathbf{W}_h||_\infty \bar{\sigma}_r < 1$, in Theorem 1 was examined after GRU's convergence for different numbers of pairs of stored patterns. This measure was found to be $||\mathbf{W}_h||_\infty \bar{\sigma}_r = 0.108, 0.108$, and 0.132, for 4, 8, and 12 pairs of patterns, respectively indicating the stability of the trained GRUs.

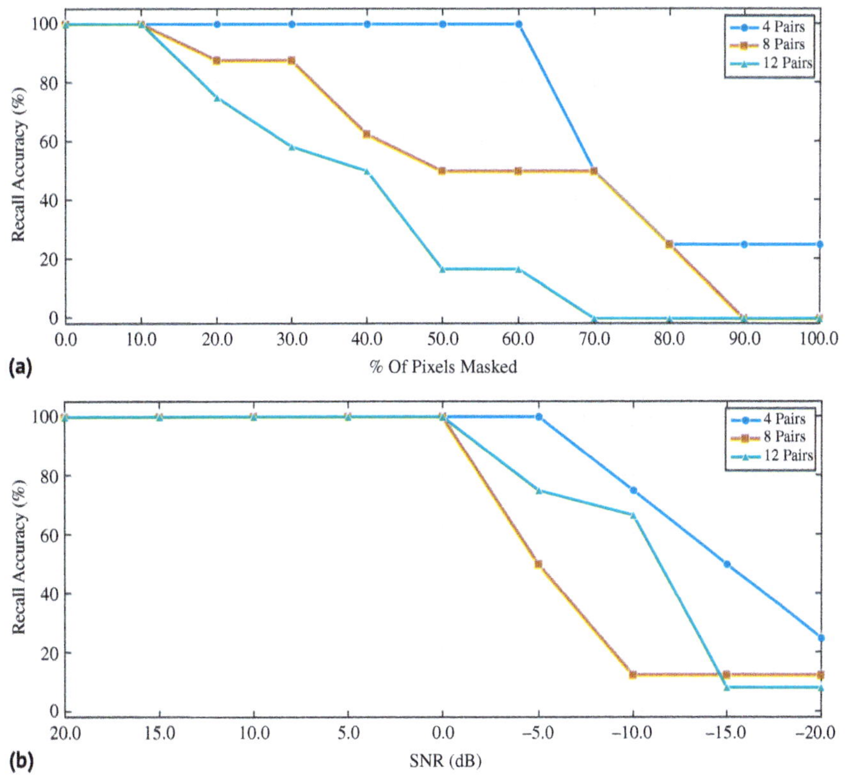

Figure 20.13 (a) Correct recall versus input SNR for different hetero-associative memory loads and (b) correct recall versus percentage of image masking.

Finally, a similar experimental protocol was repeated for the auto-associative memory, in which each image served as both input and desired output. The performance of the trained auto-associative model was subsequently evaluated on noisy and corrupted versions of the training images. Figure 20.14 presents the input images and their corresponding recalled images after the GRU's convergence. Notably, the iterative refinement of latent states in this configuration resulted in nearly perfect memory recall, even under conditions of significant noise and extensive masking effects. However, the model exhibited relatively poor recall when subjected to modest white line perturbations, producing less faithful reconstructions than those obtained using the hetero-associative memory. One potential contributing factor that merits further study is the effective memory capacity of the GRU as a function of the latent feature dimension in both hetero- and auto-associative configurations. More specifically, this dependency may influence the size and separation of the learned attractor basins. A detailed investigation of this relationship remains an important topic for future research.

20.3 Computer Projects

The computer projects in this chapter use Python libraries torch, cv2, argparse, random, os, math, PIL, sklearn, NumPy, datetime, and Matplotlib.

Computer Project 20.1: Using the OpenStereo platform [9] in PyTorch toolbox, implement two different stereo matching and disparity estimation models on the same dolls images in Figures 20.2–20.4. The networks to be implemented are the Adaptive Aggregation Network (AANET) in [25] and the Group-Wide Correlation Stereo Network (GwCSNET) in [26], both of which are available in the OpenStereo platform. Compare the performance of these networks with each other and also with the PSMNET results presented in Section 20.1. Provide a comprehensive benchmarking study of their disparity estimation accuracy under different mismatch conditions, including illumination differences, occlusions, and repetitive patterns. The study should include multiple stereo image pairs and should report appropriate quantitative performance measures, such as MSE and End Point Error (EPE). Discuss the strengths and weaknesses of each model in terms of accuracy, robustness, computational complexity, and suitability for practical stereo vision applications.

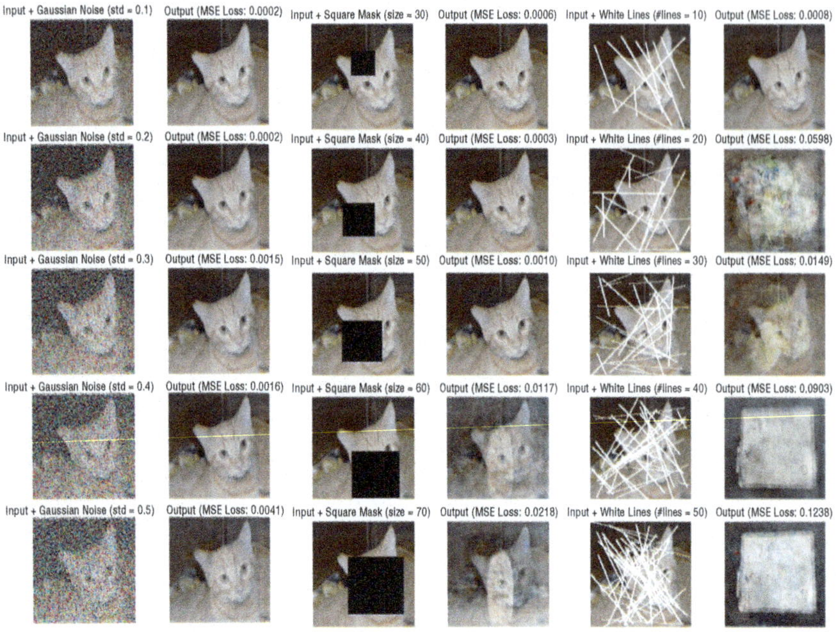

Figure 20.14 Auto-association results of the trained system in the presence of additive Gaussian noise and other corruptions.

Computer Project 20.2: The goal of this computer project is to repeat the study in Section 20.2.3, but using a larger and more diverse dataset, namely the CIFAR-10, which consists of 60,000 32×32 pixels color images across 10 classes. A subset of 1000 images (100 for each class) should be selected for this study. For hetero-association, the training dataset should be arranged into semantically related input–output pairs. For example, images from one animal class may be associated with those from another semantically related class. The same hybrid structure consisting of an autoencoder and GRU should be adopted. However, since CIFAR-10 images are much smaller in size than those used in Section 20.2.3, the encoder and decoder architectures may be modified by changing the number of layers and kernels. An appropriate dimension for the latent feature space (the input to the GRU) should be determined to provide a trained network with good recall performance. Various sources of degradation, including additive and multiplicative noise, LSI blur, and partial occlusion should be applied to the images at increasing density. For each degradation type and severity level, the trained network should be tested and the MSE between the retrieved image and the corresponding target image should be recorded. Performance plots should be generated for percent correct recall versus corruption rate. A correct recall is declared when the retrieved image is correctly associated with its intended target output. The study should also analyze the impact of varying the dimension of the latent feature space on the correct recall rate and examine how memory capacity of the network changes with the latent space dimension. A comprehensive performance analysis should be carried out to determine the ability of the hetero-associative memory to handle different image degradations and occlusion. Discuss the strengths and weaknesses of the system in terms of recall accuracy, robustness, memory capacity, and computational requirements.

Computer Project 20.3: Repeat Computer Project 20.2 for auto-associative memory. In this experiment, each image should serve as both the input and the desired output. The study should evaluate the ability of the system to reconstruct stored images from noisy, blurred, and partially occluded versions of the training images. As in Computer Project 20.2, the effects of degradation severity, latent feature dimension, recall accuracy, memory capacity, and computational requirements should be analyzed and discussed.

References

1 R. A. Hamzah and H. Ibrahim, "Literature survey on stereo vision disparity map algorithms," *Journal of Sensors*, vol. 2016, p. 8742920, 2016.

2 D. Tzovaras, M. G. Strintzis, and H. Sanhinoglou, "Evaluation of multiresolution block matching technique for motion and disparity estimation," *Signal Processing: Image Communications*, vol. 6, pp. 59–67, 1994.

3 L. Zhang and S. M. Seitz, "Estimating optimal parameters for MRF stereo from a single image pair," *IEEE Transactions on Pattern Analysis and Machine Intelligence*, vol. 29, pp. 331–342, 2007.

4 S. H. Seo, M. R. Azimi-Sadjadi, and B. Tian, "A least squared-based 2-D filtering for disparity estimation", *IEEE Transactions on Image Processing*, vol. 9, pp. 1967–1972, 2000.

5 S. H. Seo and M. R. Azimi-Sadjadi, "A 2-D filtering scheme for stereo image compression using sequential orthogonal subspace updating," *IEEE Transactions on Circuits and Systems for Video Technology*, vol. 11, pp. 52–64, 2001.

6 M. Poggi, F. Tosi, K. Batsos, P. Mordohai, and S. Mattoccia, "On the synergies between machine learning and binocular stereo for depth estimation from images: A survey," *IEEE Transactions on Pattern Analysis and Machine Intelligence*, vol. 44, pp. 5314–5334, 2022.

7 F. Tosi, L. Bartolomei, and Matteo Poggi, "A survey on deep stereo matching in the twenties," *International Journal of Computer Vision*, vol. 133, pp. 4245–4276, 2025.

8 K. Zhou, X. Meng, and B. Cheng, "Review of stereo matching algorithms based on deep learning," *Computational Intelligence and Neuroscience*, vol. 2020, p. 8562323, 2020.

9 X. Guo, C. Zhang, J. Lu, Y. Duan, Y. Wang, T. Yang, Z. Zhu, and L. Chen, "OpenStereo: A comprehensive benchmark for stereo matching and strong baseline," 2023.

10 J. R. Chang and Y. S. Chen, "Pyramid stereo matching network," Proceedings of the IEEE/CVF Conference on Computer Vision and Pattern Recognition, pp. 5410–5418, 2018.

11 K. He, X. Zhang, S. Ren, and J. Sun, "Deep residual learning for image recognition," Proceedings of the IEEE Conference on Computer Vision and Pattern Recognition, pp. 770–778, 2016.

12 A. Geiger, P. Lenz and R. Urtasun, "Are we ready for autonomous driving? The KITTI vision benchmark suite," Proceedings of the IEEE Conference on Computer Vision and Pattern Recognition, pp. 3354–3361, 2012.

13 D. Scharstein and C. Pal, "Learning conditional random fields for stereo,". Proc. of the IEEE Computer Society Conference on Computer Vision and Pattern Recognition (CVPR 2007), Minneapolis, MN, June 2007.

14 J. J. Hopfield, "Neural networks and physical systems with emergent collective computational abilities," *Proceedings of the National Academy of Sciences of the USA*, vol. 79, pp. 2554–2558, 1982.

15 A. Radhakrishnana, M. Belkin, and C. Uhler, "Overparameterized neural networks implement associative memory," *Proceedings of the National Academy of Sciences of the USA*, vol. 117, pp. 27162–27170, 2020.

16 D. J. Szelogowski, "Hebbian Memory-Augmented Recurrent Networks: Engram Neurons in Deep Learning," https://arxiv.org/abs/2507.21474.

17 R. M. Schmidt, "Recurrent Neural Networks (RNNs): A gentle Introduction and Overview," 2019. arXiv:1912.05911.

18 P. J. Werbos, "Generalization of back-propagation with application to a recurrent gas market model," *Neural Networks*, vol. 1, pp. 339–356, 1988.

19 Y. Bengio, P. Simard, and P. Frasconi, "Learning long-term dependencies with gradient descent is difficult," *IEEE Transactions on Neural Networks*, vol. 5, pp. 157–166, 1994.

20 K. Cho, B. Van Merrienboer, D. Bahdanau, and Y. Bengio, "On the properties of neural machine translation: Encoder-decoder approaches," *Proceedings of the Workshop on Syntax, Semantics and Structure in Statistical Translation*, pp. 103–111, 2014.

21 F. Bonassia, M.Farinaa, and R. Scattolinia, "On the stability properties of gated recurrent units neural networks," *Systems and Control Letters*, vol. 157, November 2021.

22 Z. Jiang, E. Sontag, and Y. Wang, "Input-to-state stability for discrete time nonlinear systems," *Automatica*, 37 (6), 857–869, 2001.

23 M. R. Azimi-Sadjadi and P. Ramicetty, "A Hybrid Autoencoder-Recurrent Network Structure for Associative Memory," *Proceedings of International Conference on Machine Learning Technologies (ICMLT)*. Berlin, 2026.

24 Z. Wang, A. C. Bovik, H. R. Sheikh, and E. P. Simoncelli, "Image quality assessment: From error visibility to structural similarity," *IEEE Transactions on Image Processing*, vol. 13, pp. 600–612, 2004.

25 H. Xu and J. Zhang, "Aanet: Adaptive aggregation network for efficient stereo matching," Proceedings of the IEEE Conference on Computer Vision and Pattern Recognition, 2020.

26 X. Guo, K. Yang, W. Yang, X. Wang, and H. Li, "Group-wise correlation stereo network," *Proceedings of the IEEE Conference on Computer Vision and Pattern Recognition*, pp. 3268–3277, 2019.

A

Review of 1-D z-Transform

This appendix briefly reviews the 1-D z-transform and summarizes its important properties and applications.
The role of the z-transform in discrete-time systems is analogous to that of the Laplace transform in continuous-time systems. The z-transform of a two-sided 1-D sequence $\{x(n)\}$ is defined by

$$X(z) = \sum_{n=-\infty}^{\infty} x(n)z^{-n}.$$

If $x(n)$ is a strictly right-sided sequence, i.e. $x(n) = 0, \ \ \forall n < 0$, which includes all causal sequences, then,

$$X(z) = \sum_{n=0}^{\infty} x(n)z^{-n}.$$

Example A.1 *Find $X(z)$ for: (a) $x(n) = a^n u(n)$; and (b) $x(n) = a^n u(-n)$ where $u(n)$ is the unit step function.*

Solution:

a. Using geometric-series identity $\sum_{n=0}^{\infty} \alpha^n = \frac{1}{1-\alpha}$ for $|\alpha| < 1$, we have

$$X(z) = \sum_{n=0}^{\infty} a^n z^{-n} = \sum_{n=0}^{\infty} \left(\frac{a}{z}\right)^n = \frac{1}{1-\frac{a}{z}} = \frac{z}{z-a} \quad \text{for } |\frac{a}{z}| < 1 \text{ or } |z| > a,$$

which converges for $|z| > a$. Thus, the region of convergence (RoC) for this sequence is outside of the circle with radius a and extends to ∞.

b. For $x(n) = a^n u(-n)$, we have

$$X(z) = \sum_{n=-\infty}^{0} a^n z^{-n} = \sum_{n=0}^{\infty} \left(\frac{z}{a}\right)^n = \frac{1}{1-\frac{z}{a}} = \frac{-a}{z-a} \quad \text{for } |\frac{z}{a}| < 1 \text{ or } |z| < a.$$

That is, the RoC for this sequence is inside the circle with radius a.

Example A.2 *Let $x(n) = A \sin(\Omega_0 nT)$, $\forall n \geq 0$. Find $X(z)$.*

Solution: Using the Euler formula, we can write

$$X(z) = \sum_{n=0}^{\infty} A \sin(\Omega_0 nT)z^{-n} = A \sum_{n=0}^{\infty} \frac{e^{j\Omega_0 nT} - e^{-j\Omega_0 nT}}{2j} z^{-n}$$

$$= \frac{A}{2j}\left[\sum_{n=0}^{\infty}\left(\frac{e^{j\Omega_0 T}}{z}\right)^n - \sum_{n=0}^{\infty}\left(\frac{e^{-j\Omega_0 T}}{z}\right)^n\right] = \frac{A}{2j}\left[\frac{1}{1-e^{j\Omega_0 T}z^{-1}} - \frac{1}{1-e^{-j\Omega_0 T}z^{-1}}\right]$$

$$= \frac{2Az^{-1}\sin(\Omega_0 T)}{1-2z^{-1}\cos(\Omega_0 T)+z^{-2}},$$

which exists when $|\frac{e^{\pm j\Omega_0 T}}{z}| < 1$ or $|z| > 1$ (since $|e^{\pm j\Omega_0 T}| = 1$). Thus, RoC is outside the unit circle and extends to ∞.

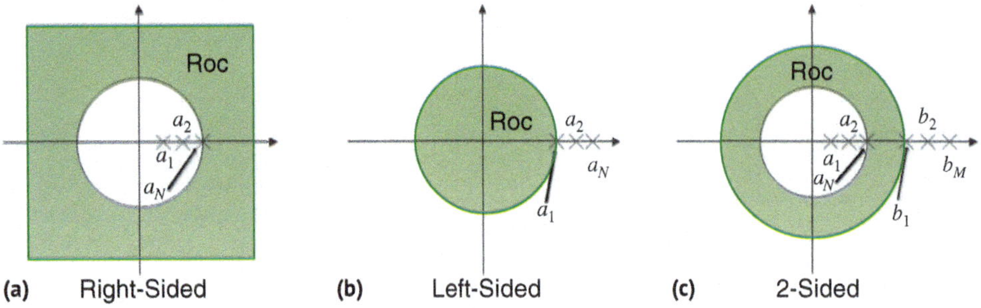

Figure A.1 RoC of right-sided (a), left-sided (b), and two-sided (c) sequences. In these figures $a_N > \cdots > a_2 > a_1$ and $b_1 < b_2 < \cdots < b_M$.

Example A.3 *Consider a two-sided sequence defined as:*

$$x(n) = \begin{cases} \left(\frac{1}{2}\right)^n, & n \geq 0 \\ 3^n, & n \leq -1 \end{cases}.$$

Find $X(z)$ and its RoC.

Solution: For this two-sided sequence, we can write

$$X(z) = \sum_{n=-\infty}^{-1} 3^n z^{-n} + \sum_{n=0}^{\infty} \left(\frac{1}{2}\right)^n z^{-n} = \sum_{n=-\infty}^{0} \left(\frac{3}{z}\right)^n - 1 + \sum_{n=0}^{\infty} \left(\frac{1}{2z}\right)^n$$

$$= \frac{1}{1-\frac{z}{3}} - 1 + \frac{1}{1-\frac{1}{2z}} = \frac{-\frac{5}{2}z}{(z-3)(z-1/2)},$$

which converges for $|z| > 1/2$ and $|z| < 3$, i.e. RoC is within a ring between two circles of radii 1/2 and 3.

Remarks:

The preceding examples illustrate the following general properties of the RoC.

1. For a right-sided sequence, RoC is outside a circle bounded on the inside by the largest magnitude pole of $X(z)$ and on the outside by ∞ (see Figure A.1a).
2. For a left-sided sequence, RoC is inside a circle bounded on the outside by the smallest magnitude pole of $X(z)$ and on the inside by 0 (see Figure A.1b).
3. For a two-sided sequence, RoC is within a ring between two circles bounded on the inside by the pole with the largest magnitude corresponding to the right-sided part and on the outside by the pole with the smallest magnitude corresponding to the left-sided part (Figure A.1c).
4. **Important Note:** RoC should NOT enclose a pole of the z-transform function.

A.1 z-Transform Properties

The usefulness of the z-transform for LSI system analysis lies in its properties. The key properties are summarized below.

1. **Linearity:** Let $x_1(n) \overset{z}{\leftrightarrow} X_1(z)$ with RoC $R_1 < |z| < R_2$ and $x_2(n) \overset{z}{\leftrightarrow} X_2(z)$ with RoC $R_3 < |z| < R_4$.

   Then, we have

   $$ax_1(n) + bx_2(n) \overset{z}{\leftrightarrow} aX_1(z) + bX_2(z) \text{ with RoC } R_5 < |z| < R_6,$$

   where $R_5 = \max(R_1, R_3)$ and $R_6 = \min(R_2, R_4)$.

Note: If the linear combination results in pole-zero cancellation, the RoC may be larger. For example, both $x_1(n) = a^n u(n)$ and $x_2(n) = a^n u(n-1)$ with z-transforms $X_1(z) = \frac{z}{z-a}$ and $X_2(z) = \frac{a}{z-a}$ have RoCs $|z| > a$ but $x_1(n) - x_2(n) = \delta(n)$ has a RoC, which is the entire z-plane. This is because the linear combination $X_1(z) - X_2(z) = \frac{z-a}{z-a} = 1$ led to pole-zero cancellation.

2. **Time Shift:** Let $x(n) \overset{z}{\leftrightarrow} X(z) \quad$ with RoC $R_1 < |z| < R_2$.

   Then, we have $x(n - n_0) \overset{z}{\leftrightarrow} z^{-n_0}X(z) \quad$ with RoC $R_1 < |z| < R_2$.

   Thus, RoCs are the same except possibly at $z = 0$ or $z = \infty$. To see this, consider the case where $x_1(n) = \delta(n) \overset{z}{\leftrightarrow} X_1(z) = 1$ where the RoC is everywhere on the z-plane, then $x_2(n) = \delta(n-1) \overset{z}{\leftrightarrow} X_2(z) = \frac{1}{z}$ has an RoC which is everywhere except at $z = 0$; while $x_3(n) = \delta(n+1) \overset{z}{\leftrightarrow} X_3(z) = z$ has an RoC which is everywhere except at $z = \infty$.

   Generalized Time Shift Property

   The preceding result assumes zero initial conditions. To impose the initial conditions when taking the z-transform, the following generalized versions can be used depending on whether the time shift is to the right or to the left.

   $$x(n - n_0) \overset{z}{\leftrightarrow} z^{-n_0}\left[X(z) + \sum_{k=-n_0}^{-1} x(k)z^{-k}\right] \quad \text{with initial conditions } x(-n_0), \dots, x(-1).$$

   $$x(n + n_0) \overset{z}{\leftrightarrow} z^{n_0}\left[X(z) - \sum_{k=0}^{n_0-1} x(k)z^{-k}\right] \quad \text{with initial conditions } x(0), \dots, x(n_0 - 1).$$

   These generalized versions are very useful, particularly when z-transform is used to find the responses of LSI systems described by difference equations.

3. **Multiplication by an Exponential Sequence:** Let $x(n) \overset{z}{\leftrightarrow} X(z) \quad$ with RoC $R_1 < |z| < R_2$. Then,

   $$a^n x(n) \overset{z}{\leftrightarrow} X(z/a) \quad \text{with RoC } |a|R_1 < |z| < |a|R_2.$$

4. **Differentiation:** Let $x(n) \overset{z}{\leftrightarrow} X(z) \quad$ with RoC $R_1 < |z| < R_2$. Then,

   $$nx(n) \overset{z}{\leftrightarrow} -z\frac{dX(z)}{dz} \quad \text{with RoC } R_1 < |z| < R_2.$$

5. **Conjugate (Complex Signals):** Let $x(n) \overset{z}{\leftrightarrow} X(z) \quad$ with RoC $R_1 < |z| < R_2$. Then,

   $$x^*(n) \overset{z}{\leftrightarrow} X^*(z^*) \quad \text{with RoC } R_1 < |z| < R_2.$$

6. **Initial Value Theorem:** If $x(n) = 0, \quad \forall n < 0$ (right-sided). Then, the initial value is $x(0) = \lim_{z\to\infty} X(z)$.

7. **Final Value Theorem (FVT):** If $x(n) = 0, \quad \forall n < 0$. Then, $\lim_{n\to\infty} x(n) = \lim_{z\to 1}(1 - z^{-1})X(z) = \lim_{z\to 1}(z-1)X(z)$.

   Condition: The theorem is valid only if $(1 - z^{-1})X(z)$ has no poles on or outside the unit circle (except possibly a simple pole at $z = 1$). This theorem is particularly useful for steady-state analysis of discrete-time control systems.

8. **Linear Convolution:** Consider the linear convolution

   $$y(n) = x(n) * h(n) = \sum_{k=-\infty}^{\infty} x(k)h(n-k) = \sum_{k=-\infty}^{\infty} x(n-k)h(k),$$

   of the input signal $x(n)$ and the impulse response $h(n)$ of an LSI system. If $x(n) \overset{z}{\leftrightarrow} X(z)$ with RoC $R_1 < |z| < R_2$ and $h(n) \overset{z}{\leftrightarrow} H(z)$ with RoC $R_3 < |z| < R_4$, where $H(z)$ is the transfer function of the LSI system, then we have

   $$Y(z) = X(z)H(z) \quad \text{with RoC} \quad \max[R_1, R_3] < |z| < \min[R_2, R_4].$$

   Thus, linear convolution can be carried out in the z-domain using algebraic product of the two z-transforms. Pole–zero cancellation may enlarge the resulting RoC.

9. **Complex Convolution Theorem:** This is the dual of the time-domain convolution property. Let $w(n) = y(n)x(n)$, where the RoC of $X(z)$ is $R_{x-} < |z| < R_{x+}$ and the RoC of $Y(z)$ is $R_{y-} < |z| < R_{y+}$. Then

$$W(z) = \frac{1}{2\pi j}\oint_{C_2} X(\nu)Y\left(\frac{z}{\nu}\right)\nu^{-1}d\nu,$$

where C_2 is a closed contour in the overlap region of the RoCs of $X(\nu)$ and $Y\left(\frac{z}{\nu}\right)$. Alternatively,

$$W(z) = \frac{1}{2\pi j}\oint_{C_1} X\left(\frac{z}{\nu}\right)Y(\nu)\nu^{-1}d\nu,$$

where C_1 is a closed contour in the overlap region of the RoCs of $X\left(\frac{z}{\nu}\right)$ and $Y(\nu)$. The RoC of $W(z)$ is $R_{x-}R_{y-} < |z| < R_{x+}R_{y+}$. Both contour integrations are along the counterclockwise direction.

10. **Parseval's Theorem:** The inner product of two signals in the time domain is equal to the inner product of their z-transforms,

$$\sum_{n=-\infty}^{\infty} x(n)y^*(n) = \frac{1}{2\pi j}\oint_C X(\nu)Y^*\left(\frac{1}{\nu^*}\right)\nu^{-1}d\nu,$$

where C lies in the overlap region of the RoCs of $X(\nu)$ and $Y^*\left(\frac{1}{\nu^*}\right)$. If $X(z)$ and $Y(z)$ converge on the unit circle, we can choose $\nu = e^{j\Omega}$, then

$$\sum_{n=-\infty}^{\infty} x(n)y^*(n) = \frac{1}{2\pi}\int_{-\pi}^{\pi} X(e^{j\Omega})Y^*(e^{j\Omega})d\Omega,$$

which is Parseval's theorem in the frequency domain. As a special case, if $x(n) = y(n)$, the inner product preservation becomes the energy preservation property:

$$\sum_n |x(n)|^2 = \frac{1}{2\pi}\int_{-\pi}^{\pi} |X(e^{j\Omega})|^2 d\Omega,$$

where $X(e^{j\Omega})$ is the discrete-time Fourier transform (DTFT) of $x(n)$.

A.2 Inverse z-Transform Using Partial Fraction Expansion

The partial fraction expansion (PFE) method for computing the inverse z-transform is analogous to the method used for the inverse Laplace transform. The key difference is that the expansion is performed on $\frac{X(z)}{z}$ rather than on $X(z)$. This arises because the z-transform of a discrete-time sequence typically contains an additional factor of z in the numerator. For example, Laplace transform of a continuous exponential function is $\mathcal{L}\{e^{-at}\} = \frac{1}{s+a}$ versus z-transform of discrete version, which is $\mathcal{Z}\{a^n u(n)\} = \frac{z}{z-a}$.

After expanding $\frac{X(z)}{z}$ into partial fractions, each term is multiplied by z, and the inverse z-transform is obtained directly from standard transform tables.

Example A.4 *Given* $X(z) = \frac{z^2+z}{(z-1)^2}$ *determine* $x(n)$ *using PFE.*

Solution: Expand $\frac{X(z)}{z} = \frac{z+1}{(z-1)^2} = \frac{A_1}{(z-1)} + \frac{A_2}{(z-1)^2}$, and find A_1 and A_2 using the standard PFE, i.e.

$$A_2 = (z-1)^2\frac{X(z)}{z}\Big|_{z=1} = (z+1)|_{z=1} = 2,$$

$$A_1 = \frac{d}{dz}\left[(z-1)^2\frac{X(z)}{z}\right]\Big|_{z=1} = 1.$$

Hence,

$$X(z) = \frac{z}{z-1} + \frac{2z}{(z-1)^2},$$

which using the z-transform tables gives

$$x(n) = \mathcal{Z}^{-1}\left\{\frac{z}{z-1}\right\} + \mathcal{Z}^{-1}\left\{\frac{2z}{(z-1)^2}\right\} = (1+2n)u(n).$$

The unit step term $u(n)$ sequence indicates that the result is right-sided.

A.3 Solution of Difference Equations Using z-Transform

The z-transform provides a systematic procedure for obtaining the complete response of a discrete-time LSI system, consisting of both the homogeneous (zero-input) and particular (zero-state) responses. The procedure is summarized as follows:

a. Take the z-transform of both sides of the difference equation and incorporate the initial conditions using the generalized time-shifting property (Property 2 in Section A.1).
b. Solve algebraically for $Y(z)$, and apply PFE to determine the inverse z-transform and complete response $y(n)$.

Example A.5 *Consider the second-order difference equation*

$$y(n+2) + 3y(n+1) + 2y(n) = x(n),$$

where $x(n) = \delta(n)$ *and* $y(0) = 1,\ \ y(1) = -1$. *Find the complete response of this system.*

Solution: Using the generalized time-shift property $\mathcal{Z}\{y(n+n_0)\} = z^{n_0}\left[Y(z) - \sum_{k=0}^{n_0-1} y(k)z^{-k}\right]$, and taking z-transform of both sides gives

$$z^2[Y(z) - y(0) - y(1)z^{-1}] + 3z[Y(z) - y(0)] + 2Y(z) = X(z).$$

Substituting $X(z) = 1,\ y(0) = 1$, and $y(1) = -1$, yields

$$(z^2 + 3z + 2)Y(z) = 1 + z^2 + 2z.$$

Thus,

$$Y(z) = \frac{1}{z^2+3z+2} + \frac{z^2+2z}{z^2+3z+2} = Y_p(z) + Y_h(z) = \frac{z^2+2z+1}{z^2+3z+2} = \frac{z+1}{z+2}.$$

Using PFE, we expand $\frac{Y(z)}{z}$ as:

$$\frac{Y(z)}{z} = \frac{z+1}{z(z+2)} = \frac{1/2}{z} + \frac{1/2}{z+2},$$

or in the time domain;

$$y(n) = \frac{1}{2}\delta(n) + \frac{1}{2}(-2)^n u(n).$$

The homogeneous and particular responses may also be obtained separately. The particular solution is $Y_p(z) = H(z)X(z)$, where $H(z) = \frac{1}{z^2+3z+2}$ is the transfer function, while the homogeneous solution $Y_h(z)$ depends only on the initial conditions. To find the homogeneous solution, we expand $\frac{Y_h(z)}{z} = \frac{z+2}{z^2+3z+2} = \frac{1}{(z+1)}$, which gives $y_h(n) = (-1)^n u(n)$. For the particular solution, we expand

$$\frac{Y_p(z)}{z} = \frac{1}{z(z^2+3z+2)} = \frac{1/2}{z} - \frac{1}{(z+1)} + \frac{1/2}{(z+2)},$$

which gives:

$$y_p(n) = \frac{1}{2}\delta(n) - (-1)^n u(n) + \frac{1}{2}(-2)^n u(n).$$

Example A.6 *Consider the difference equation*

$$y(n) - \frac{3}{2}y(n-1) + \frac{1}{2}y(n-2) = x(n),$$

where $x(n) = 1 + 3^{-n}, \forall n \geq 0$, *and the initial conditions are* $y(-2) = 0$ *and* $y(-1) = 2$. *Find the complete solution of this LSI system.*

Solution: Taking the z-transform and using the generalized time-shifting property, we obtain

$$Y(z) - \frac{3}{2}z^{-1}[Y(z) + y(-1)z] + \frac{1}{2}z^{-2}[Y(z) + y(-1)z + y(-2)z^2] = X(z).$$

But, $X(z)$ is

$$X(z) = \frac{z}{z-1} + \frac{z}{z-1/3} = \frac{2z(z-2/3)}{(z-1)(z-1/3)}.$$

Rearranging the above equation yields

$$(1 - 3/2z^{-1} + 1/2z^{-2})Y(z) = X(z) + 3 - z^{-1},$$

or

$$Y(z) = \frac{z^2X(z)}{z^2 - 3/2z + 1/2} + \frac{3z^2 - z}{z^2 - 3/2z + 1/2} = \frac{z[3(z-1)(z-1/3)^2 + 2z^2(z-2/3)]}{(z-1)^2(z-1/2)(z-1/3)}.$$

Note that the particular and homogeneous solutions are $Y_p(z) = \frac{z^2X(z)}{z^2-3/2z+1/2}$ and $Y_h(z) = \frac{3z^2-z}{z^2-3/2z+1/2}$, respectively. Using PFE, we expand $\frac{Y(z)}{z}$ as:

$$\frac{Y(z)}{z} = \frac{A_1}{z-1/2} + \frac{A_2}{z-1/3} + \frac{B_1}{z-1} + \frac{B_2}{(z-1)^2},$$

where

$$A_1 = (z-1/2)\frac{Y(z)}{z}\bigg|_{z=1/2} = -3 \qquad A_2 = (z-1/3)\frac{Y(z)}{z}\bigg|_{z=1/3} = 1$$

$$B_2 = (z-1)^2\frac{Y(z)}{z}\bigg|_{z=1} = 2 \qquad B_1 = \frac{d}{dz}\left[(z-1)^2\frac{Y(z)}{z}\right]\bigg|_{z=1} = 7$$

Thus, we get

$$Y(z) = \frac{-3z}{z-1/2} + \frac{z}{z-1/3} + \frac{7z}{z-1} + \frac{2z}{(z-1)^2}.$$

Using the inverse z-transform tables, we get the complete response as:

$$y(n) = \left(-3\,2^{-n} + 3^{-n} + 7 + 2n\right)u(n).$$

Reference

1 J. Proakis and D. Manolakis, "*Digital Signal Processing: Principles, Algorithms and Applications*," Pearson, 4th Edition, 2006.

B

2-D *z*-Transform

The 2-D z-transform of 2-D array $x(m, n)$ is defined as:

$$X(z_1, z_2) = \sum_{m=-\infty}^{\infty} \sum_{n=-\infty}^{\infty} x(m, n) z_1^{-m} z_2^{-n}.$$

The RoC is the set of points for which $|X(z_1, z_2)| < \infty$, i.e. bounded. The 2-D analog of 1-D annulus (ring) is known as the "Reinhardt domain" [1]. If a point (z_1, z_2) lies in the Reinhardt domain $\mathcal{R}$, then every point $(e^{j\mu_1} z_1, e^{j\mu_2} z_2)$, where $\mu_1, \mu_2 \in \mathbb{R}$, also lies in this domain since the magnitude remains unchanged. As in the 1-D case, the RoC of the 2-D z-transform can vary depending on the region of support (ROS) of the spatial domain function $x(m, n)$. In what follows, we consider several typical ROS scenarios [1].

1. Finite ROS: For a 2-D array with finite ROS, we have

$$X(z_1, z_2) = \sum_{m=-M_1}^{N_1} \sum_{n=-M_2}^{N_2} x(m, n) z_1^{-m} z_2^{-n},$$

which converges for all finite values of z_1, z_2 except possibly for $z_1 = 0$ or $z_2 = 0$.

2. First-Quadrant ROS: For a 2-D array with ROS in the first quadrant, we have

$$X(z_1, z_2) = \sum_{m=0}^{\infty} \sum_{n=0}^{\infty} x(m, n) z_1^{-m} z_2^{-n}.$$

In this case, if a point (z_{01}, z_{02}) lies in the RoC of $X(z_1, z_2)$, then every point (z_1, z_2) that satisfies $|z_1| \geq |z_{01}|$ and $|z_2| \geq |z_{02}|$ also lie in the RoC. As a consequence, the boundary of the RoC in the $(\ln |z_1|, \ln |z_2|)$ plane must have a non-positive slope. If the slope over some region is positive, some points may satisfy the inequalities, yet they lie outside the RoC (see Figure B.1a).

3. Half-Plane ROS: For a 2-D array with ROS in the upper half-plane, we have

$$X(z_1, z_2) = \sum_{m=-\infty}^{\infty} \sum_{n=0}^{\infty} x(m, n) z_1^{-m} z_2^{-n}.$$

In this case, if a point (z_{01}, z_{02}) lies in the RoC of $X(z_1, z_2)$, then every point (z_1, z_2) that satisfies $|z_1| = |z_{01}|$ and $|z_2| \geq |z_{02}|$ also lie in the RoC. Thus, the boundary of RoC is only a function of $\ln |z_1|$. An example is shown in Figure B.1b.

4. Full-Plane ROS: In this case, the RoC exhibits a variety of possible sizes and shapes. Such 2-D arrays can be expressed as the sum of four quadrant-supported arrays, and thus the overall RoC will be the intersection of the corresponding quadrant RoCs.

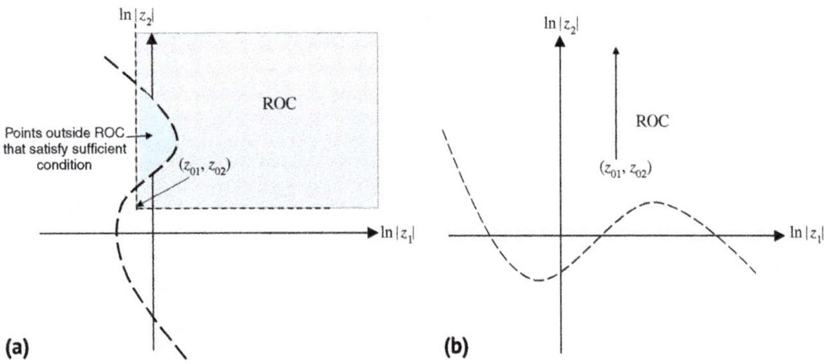

Figure B.1 The RoC of 2-D array with (a) first-quadrant support, and (b) half-plane support [1].

Example B.1 *Consider a 2-D array defined as:*

$$x(m,n)=\begin{cases}1, & m\geq 0, n\geq 0,\ \ n\leq m\\ 0, & \text{elsewhere}\end{cases}.$$

where the support lies below the line $m-n=0$ in the first quadrant. Find its 2-D z-transform and RoC.

Solution:

$$X(z_1,z_2)=\sum_{n=0}^{\infty}\sum_{m=n}^{\infty}z_1^{-m}z_2^{-n}=\sum_{l=0}^{\infty}\sum_{n=0}^{\infty}z_1^{-l}(z_1z_2)^{-n},\qquad \text{where } l=m-n$$

$$=\left(\sum_{l=0}^{\infty}z_1^{-l}\right)\left(\sum_{n=0}^{\infty}(z_1z_2)^{-n}\right)=\frac{1}{(1-z_1^{-1})(1-z_1^{-1}z_2^{-1})}.$$

which has a RoC for $|z_1|>1,\ \ |z_1z_2|>1$, or in the log domain, $\ln|z_1|>0,\ \ \ln|z_1|+\ln|z_2|>0$. Thus, the RoC is bounded by the vertical line $\ln|z_1|=0$ and the line $\ln|z_1|+\ln|z_2|=0$ with slope is -1.

Remarks:

1. If we let $z_1=e^{j\Omega_1}$ and $z_2=e^{j\Omega_2}$, the 2-D z-transform becomes 2-D DSFT discussed in Section 3.1.6 of Chapter 3 in which case the z_1,z_2-hyperplane is mapped to the *unit bicircle.*
2. If the 2-D array is separable, $x(m,n)=x_1(m)x_2(n)$, then its 2-D z-transform is also separable, i.e., $X(z_1,z_2)=X_1(z_1)X_2(z_2)$ (see Example B.2).

Example B.2 *Find the 2-D z-transform of the first-quadrant array $x(m,n)=\rho^{m+n}$, where $\rho<1$.*

Solution:

$$X(z_1,z_2)=\sum_{m=0}^{\infty}\sum_{n=0}^{\infty}\rho^{m+n}z_1^{-m}z_2^{-n}=\sum_{m=0}^{\infty}(\rho z_1^{-1})^m\sum_{n=0}^{\infty}(\rho z_2^{-1})^n$$

$$=\frac{1}{(1-\rho z_1^{-1})(1-\rho z_2^{-1})},$$

which has a RoC for $|z_1|>\rho$ and $|z_2|>\rho$.

Reference

1 D. E. Dudgeon and R. M. Mersereau, "*Multidimensional Digital Signal Processing,*" Prentice Hall, 1984.

Index

Note: Page numbers in *italics* and **bold** refers to figures and tables respectively.

a

acquisition. *see* image acquisition
activation functions
  bipolar sigmoidal 377–378, *378*, 397, 426
  hard-limiter 377, 378, 381, 386, 387
  multiquadrics 379
  radial basis function (RBF) 379
  ReLU 378, 407, 409, 416, 425, 427, 442, 446, 458
  softmax 378–379, 409–413, 416, 425, 427, 455
  unipolar sigmoidal 377, 378, 388, 404, 427
active contour (snakes)
  Euler-Lagrange equation 265, 267
  external energy 265, 267, 268, *268*, *269*
  force balance 265
  formulation 265–266
  internal energy 265
  iterative implementation 265
adaptive learning rate. *see* performance learning
Adaptive Linear Element (ADALINE) 381–386, 395, 397. *see also* performance learning
adaptive thresholding. *see* histogram thresholding
adaptive Wiener filter 204, 217
additive noise 6, 49, 82, 140, 144, 153, 154, 163, 167, 195, 198, 201, 203–206, *204*, 214, 217–220, *294*, 302–305, 352, 356, 441, 446, 460
adjacency 256
affine transformations 276
algebraic reconstruction technique (ART). *see* iterative reconstruction methods
aliasing
  antialiasing filter 71, *71*, 81
  effects in images 69–71
  error 70
  frequency domain 69–71
  spatial domain 70
all-pole models 172
all-zero models 172
alphabet of symbols 223, *224*
alternating flip. *see* filter banks (FB)
analytic 64, 175, 177, 179, 182, 184, 185, 187, 196
analytical image reconstruction. *see also* image reconstruction
  cubic interpolation 346
  discrete FBP 346
  filtered back-projection (FBP) 344–347
  linear interpolator 346
  nearest neighbor interpolator 346
  phantom head 346, *347*, 349, *350*
  Ram-Lak filter 355
  ramp filter 345, 346, 355
  Shepp-Logan filter 345
area conservation 34, 35, 51
arithmetic encoding 228, 230–231, *231*, 240, 243
artificial
  intelligence (AI) 1, 7–9, 53, 276, 333, 376, 402, 449–462
  neural networks (ANN). *see* neural networks
  neurons. *see* neuron models
associative memory
  auto-associative 438, 454, 458, 461, 462, *462*
  binary 443, 454
  grayscale or color 392, 449, 454, 458
  hetero-associative 391, 458–462
  Hopfield network 454
  hybrid structure 443, 458–462
  recall/retrieval accuracy 391–392, 443, 453–454
associativity 12, 34
auto-correlation function (ACF) 21–26, 29, 30, 45–50, *48*, 52, 172–175, 177, 178, 181–187, 191–193, 205, 206, 219, 220
autoencoders
  activity 439, 440
  basic 438–439
  contractive 442–443, *443*
  decoder layer 441, 444, 445
  denoising 441–442, *442*
  encoder layer 442–445
  inactive neuron 440
  linear 391, 431–434
  nonlinear 438–444
  optimal reconstruction function 441
  over-complete model 441
  sparse 439–441
  stacked and convolutional 443
  tied-weight 443
  under-complete model 439, 441

auto-regressive (AR)
  1-D models 171–178
  2-D models 178–187
  vector model 188–191
auto-regressive moving average (ARMA) models 171

b

back-propagation neural networks (BPNN)
  generalized delta rule 403–407, 425
  hidden layers 404–405, 425
  output layer 403–405, 413
  standardization 407
  topology 407
backward difference method 166
band-limited function 16, 40, 67, 202
basis functions
  CWT 110–114
  DCT 88–90, 95
  DFT 18–20, 42–45, 86, 87
  DSFT 40–42, 46–49, 51, 52, 141, 158, 168, 169, 219
  DTFT 15–18, 24–27, 29, 41, 42, 109
  DWS 115, 116–118
  DWT 118–120
  FT 15, 38–40, 42, 51, 108, 109, 112, 342
  Gabor 108, 287–290, 352
  ideal interpolation 69, 70
  KL transform 164, 236, 242
  Walsh transform (WT) 90–92
  WHT 92–95
  Zernike moments 302–305
basis images
  2-D DCT 87, 89–90, *91*, 95, 101, 103, 124, *125*, 159, *159*, 235, 236, 238, 239, *240*, 242
  2-D DFT 42–45, 47, 49, 52, 85, 87, 103, 154, 159, 164, 168, 169
  2-D KLT 85, 98
  2-D WHT 94–95, 101, 104
Basis vectors 86, 88, 101, 102, 406
Bayer color filter array 72
Bayes classifier. *see also* pattern classification
  conditional risk 363, 364
  decision threshold 364, *366*
  expected loss 363
  hyperquadrics 365
  likelihood ratio 364
  linear discriminant function 365
  Quadratic discriminant analysis (QDA) 365
Bayesian methods 255–256, 361, 363
Bayes rule 251, 261, 263, 363, 364
bending energy 265
Bernoulli distribution 367–369
Bessel function 350
between-class variance 251, 253
biological neurons
  axon 375, 376, 390
  dendrite 375, 376
  excitatory 376
  firing 54, 58, 376, 377, 403
  inhibitory 54, 376, 418
  neurotransmitter 375
  period of temporal/spatial summation 376
  refractory period 376
biorthogonal basis. *see* wavelet series
bit allocation
  Huang–Schultheiss 237–238
  Segall 237, 238, *239*
  Shannon 237
bit budget 227, 236, 237
bit extraction 163
bit-plane 163, 240, 454
bit-plane coding 163, 454
bit-plane encoding 240
bit rate 6, 124, 226, 240, 245
block matching. *see* disparity estimation algorithms
block matrices 180, 191, 213
block quantization 236
block Toeplitz 181
block transform coding 245
blur
  atmospheric turbulence 6, 195, **196**
  CCD local interactions **196**
  diffraction limited (coherent) **196**
  diffraction limited (incoherent) **196**
  motion 6, 195, **196**, 197, 203, *204*, 216
  rectangular scanning aperture **196**
boundary descriptors 57
boundary detection 330
bounded-input bounded-output (BIBO) stability 14, 15, 28, 35, 37, 50, 172, 457. *see also* stability
brightness 54–56, *57*, 58, 72
B-spline method
  basis functions 299
  control points 298–300, 307
  knots 299
Butterworth high-pass 157
Butterworth low-pass 157

c

Cartesian 61, 348
cascade 58, 282, 450
causality 11, 14, 31, 35, 171, 178, 190, 391
causal models
  1-D 172–176
  2-D 179–182
  anti-causal 12
  parameter estimation 173–174
center of mass 74
central limit theorem 214
cepstral domain 154, 160–162, *162*, 164
cepstrum 161, 162
charge-coupled devices (CCD) 72, 195
CIFAR-10 424, 425, 443, 462

circular convolution 19–20, 44–45, 88, 93, 104, 155, 159, 199, 201, 299, 346
circular shift 19, 44, 45, 296
classification/classifier
  binary hypothesis testing. *see* detection/detector
  confusion matrix 366
  decision boundary 359–361, *360*, 365, 377, 387, *388*, 402, 407
  decision rule 359, *360*, 361–364
  decision surface 386
  discriminant function 251, 360–362, *360*, 365, 367, 369, 387, 401
  hyperplane 361, 365, 377, 386–388, 402
  likelihood ratio 364
  linear 99, 360–362, 365, 369, 370, 372, 373, 377, 388
  linearly separable 361, 363, 401
  log-likelihood 368, 369
  M-ary hypothesis testing 255–256, 363, 366
  minimum distance 99, 102, 301, 362–364, 371
  nonlinear 332. *see also* neural networks
clustering
  agglomerative 258
  divisive 258
  hierarchical 258
  K-means 259
  Kohonen neural networks. *see* competitive learning
  partition-based (hard clustering) 259–260
  partition-based (soft clustering) 260
coarse-fine search 89, 241, 294
coarse quantization 241
codebook 80, 241–243, 370, 393, 394, *395*, *396*, 422–425
CODEC 227, 233, 241
coder 178, 225, *228*, 233–235, *235*
codeword 127, 223–225, 228–231, *228*, 233, 243
coding/encoding
  arithmetic 228, 230–231, *231*, 240
  Huffman 228–231
  run-length 228, 231–232
  subband 107, 238
coincidence learning
  Hebbian rule 431
  Oja's rule 432, 433
  postsynaptic cell 391, *391*
  presynaptic cell 391
color
  chromaticity diagram 60, *60*, 61
  chrominance 58, 60, 61, 63
  CIE 58, *59*, 60, *60*, 61
  CMYK space 62
  colorimetry 58
  correlogram 277–279, *278*, 305
  discrimination 58–61
  feature 277–279, 369
  gamut 61, 62
  HSI space 58, *59*, 63
  hue 58, 60, 62, 63, 72
  intensity 58, *59*, 62, 63, 72
  matching 59, *59*
  models 171, 185
  moments 277–279, 305
  palette 80, 278
  perception 58, 62
  primary colors 58–61
  RGB space 63
  saturation 58, *59*, 60
  secondary colors *59*, *61*, 61–63
  spaces 58, *59*, *61*, 62, 63, **63**, 72, 80, 240, 241, 255, 259, 277, 279
  triangle *59*, 61, 63
  trichromatic 58–61
  tristimulus values, space 59, 60
  vision 57
  YC_bC_r space 63
colored noise 177, 185, 187, 220
commutative 45, 313
compactness 441
compandor 78–79, *78*
competitive learning
  Kohonen neural networks 393, *393*
  LVQ. *see* learning vector quantization (LVQ)
  self-organized feature maps (SOFM) 418, 419, 422
  Winner-takes-all 392
completeness 256
component labeling 4. *see also* learning
compression
  lossless. *see* lossless compression
  lossy. *see* lossy compression
compressor 77–79, 82, *228*
computerized tomography (CT) 1, 6, 223, 268, 269, *270*, 306, 329, *330*, 339, 346, 354
computer vision 1, *2*, 7, 264, 311, 328, 449
concave 266
conditional expectation 441
conjugate gradient method
  Polak-Ribière parameter 353
  with preconditioning 353
  without preconditioning 353
connectivity 151, 256, 257, 291, 296, 297, *297*, 319, 326, 409, 410, 435, *435*, 436, 454
constrained least square 300
contour
  closed 291, *291*, 292, 294, 296, *297*, 298, 301, *309*, *468*
  features. *see* Fourier descriptors (FDs); wavelet descriptor (WD)
  linking 291
  simple 291
  tracing 291
contrast 4–7, 10, 55–58, 79, 132–134, *132*, *133*, 162–164, *162*, 216, 217, 267–269, 284, 354
contrast sensitivity function (CSF) 57–58, *57*
convex 77, 311, 318–319, *318*, 393, 402, 403, 413, 415, 457

convolutional neural networks (CNN). *see also* neural networks
- adaptive moment estimation (Adam) 415–417
- AlexNet 409
- average pooling 414, 450, 451
- convolutional layer 408–410, *409*, 413–417, 450
- feature maps 409, 413, 450, 458
- ImageNet 409
- LeNet-5 409, *409*, 410, *410*
- max pooling 409, 414, 416, 425
- pooling layer 408–411, *408*, 413, 414, 416, 441
- root-mean-square propagation (RMSProp) 415–416, 425
- weight sharing 409

convolution summation
- 1-D 12
- 2-D 32–34, 37, 48, 50

co-occurrence matrix. *see* gray level co-occurrence matrix (GLCM)
corners 144, 164, 314, 316, 320, 409, 439
covariance 96–98, 100, 103, 104, 164, 171, 178, 190, 191, 204, 206–208, 212, 213, 215, 217, 245, 261, 278–281, 292, 364, 365, 383, 431, 441
critical flicker frequency (CFF) 58
cross-correlation/cross-covariance 21–22, 25, 26, 39, 46–48, 50, 152, 159, 173, 200, 205, 382, 391
cross-power spectrum
- 1-D 26
- 2-D 47, 48

d

data compression 5, 9, 80, 85, 87, 88, 120, 122, 124, 126, 127, 176, 178, 223, 224, 226–228, 236, 242, 379, 380, 392, 393, 396, 418, 449
decorrelation 96, 276
detection/detector
- detection threshold 55
- dichotomizer 364, 386, 387, *388*
- false negative, missed detection 255, 365, 366
- false positive, false alarm 255, 365, 366
- receiver operating characteristic (ROC) curve 366, *366*
- true negative 365, 366
- true positive 365, 366

deterministic
- degradation 6, 195, 196
- image 197, 219, 351
- signal 31

De Vries-Rose law 55
diagonalization 98
diagonal matrix 96, 100, 280, 435
dictionary 230, 241, 242, 288
differential pulse coding modulation (DPCM) 233–235, 238
digital image processing (DIP)
- applications 1, 85, 171, 311, 418, 420, 449–462
- areas 5–7, 85, 449
- components 4–5

digital signal processing (DSP) 9, 10, 17, 20
dimensionality reduction. *see* autoencoders
directional smoothing 150
discrete cosine transform (DCT)
- 1-D 87–89
- 2-D 87, 89–90, *91*, 95, 101, 103, 124, *125*, 159, 235, 236, 238–240, 242
- basis functions 88, 89, 95
- basis images 89, 95, 101
- basis vectors 88
- energy packing 159, 236
- inverse 88–90, 124, 159, 239
- matrix representation 85, 89–90

discrete Fourier transform (DFT)
- 1-D 18, 42, 44, 85–87, 291, 346
- 2-D 42–45, 47, 49, 85, 87, 94, 154, 159, 164, 198, 199, 203, 282
- basis functions 15, 18, 42, 86, 88, 89
- basis images 87, 94
- basis vectors 86
- convolution. *see* circular convolution
- inverse 18, 87
- magnitude 43, *44*, 49
- matrix representation 85–87
- phase 296

Discrete Space Fourier transform (DSFT) 40–42, 46–49, 51, 141, 158, 169, 472
discrete space systems 31–45
discrete time Fourier transform (DTFT) 15–18, 24, 26, 27, 29, 41, 42, 109, 468
discrete time signals 9–11, 15, 20–28, 92, 107, 111, 115, 146
discrete wavelet series (DWS). *see* wavelet series
discrete wavelet transform (DWT). *see also* filter banks (FB)
- 1-D 107, *121*
- 2-D 6, 85, 123–127, *125*, *126*, 159, *160*, 232, 238, *240*, 282, 283, 306
- added details 120–124, *123*, 232, 238
- basis functions 119
- Daubechies' family 119
- denoising 124, 125, 127, 159
- Haar 118–119
- inverse 118
- lowest-order approximation 124, 126, 238, 295
- subband coding 107

disparity estimation algorithms. *see also* stereoscopic imaging
- binocular disparity 450
- block matching 449
- depth perception 449–451
- global methods 449
- local methods 449
- monocular depth cues 450
- Occlusion 449–450

distance
- Euclidean 259, 292, 303, 306, 307, 327, 362, 392
- geodesic 326–327
- Mahalanobis 362

distortion 54, 73–75, 77, 78, 80, 82–84, 122, 125, 144, 153, 225–228, *227*, 233, 234, 236, 237, 276, 394, 401, 408, 409, 449, 454, 458
distributive 12, 34, 45, 313
dynamic range 43, 77, 78, 134, 161, 235, 236, 240

e

edge detection
  exterior 329, *329*
  interior 329, *329*
edge extraction/detection
  Canny operator 150–152, *151*
  edge map 150, 267, 329
  Laplacian of Gaussian (LoG) operator 149–150, *149–151*, 250, 270
  Laplacian operator *147*, 148–149, *148*
  Prewitt operator *147*, 148, *148*
  Roberts operator 147–148, *147*
  Sobel operator *147*, 148, *148*
edge linking/edge tracing 151, 267, 291
edge preserving filter 144
eigen-
  decomposition 436
  faces 99–100, *100*, *101*, 102, 279
  function 39
  image 98, 99, 279, 280
  value 98, 99, 101, 102, 164, 279–282, 362, 385, 433, 435, 444
  vector 96–99, 104, 279, 280, 431–433, 435, 436, 444
emission tomography 339
energy
  compaction 88, 94, 95
  conservation 39, 86
  function 261, 263, 265, 266, 270, 450, 454
  preservation 97, 112, 468
ensemble
  auto-correlation 21
  auto-covariance 21
  cross-correlation 21
  cross-covariance 22
  mean 21
entropy 224–227, *226*, 229, 230, 232, 241, 251, *251*, 253–254, *255*, 269
entropy coding 225–226
ergodicity 24
estimation
  parameter 173–174, 176–177, 180–187, 190–191, 261, 298, 395, 396
  state 206, 208, 209, 212
Euler formula 19, 465
Eulerian 264
even signal 11
excitatory 54, 55, 376, 418, 419, 421, 439
expander 78, *78*, 79, 82

f

face recognition 99
Fast Fourier transform (FFT) 9, 18, 20, 45, 50, 346, 354
fast KL transform 6, 164, 236, 242
fast transform 45
feature extraction
  Bellman's curse of dimensionality 276
  chain codes 291, 296–297
  covariance, equivariance 276–277
  discriminatory 390
  feature space 164, 359–361, *360*, 363, 370, 386, 388, 403, 406, 417, 444, 454, 458, 459, 462
  invariance 276–277, 303
fidelity criterion 5
filter banks (FB). *see also* Discrete Wavelet transform (DWT)
  alternating flip 116–118
  analysis FB 109, 116–120, *121*, 123, *123*, 295, 296
  biorthogonal 118, 120, 126
  dilation equation 123
  downsampler 129
  half-band high-pass filter 115
  half-band low-pass filter 115
  iterated FB 120–122
  orthogonal FB. *see* Discrete Wavelet transform (DWT)
  quadrature mirror filters (QMF) 120
  regularity 121, 122
  synthesis FB *115*, 116–120, 124, 126, 127, 129, 295, 441
  time-reversed 119
  upsampler 116
  wavelet packets (WP) 122
filtering using image transforms 203, 442
finite impulse response (FIR) 15, 16, 20, 27, 37, 118–120, 129, 139, 155, 157, 200, 273, 409
first-order hold 69
Fisher discriminant function 252
fitting line segments 155, 314
fold-over frequencies 69
forward-difference method 265
Fourier descriptors (FDs)
  equal arc length resampling 292
  equal-point resampling 292
Fourier transform
  1-D continuous 38, *38*, 342, 344, 345
  2-D continuous 37–40, *38*, *40*
  magnitude 38
  phase 38
frames 114. *see also* wavelet series
  bounds 114–115
  tight 115
frequency
  cut-off 155
  domain 15–20, 25, 37–45, 49, 67, 68, 90, 107, 108, 110, 149, 150, 155, 167, 198–201, 205, 206, 216, 217, 287, 288, 346, 347, 468
  response 25, 26, 38–40, *40*, 41, 48, 50, 57, 58, 69, 70, 83, 108, 110, 118, 119, 126, 140, 141, 155, *155*, 157, *157*, 158, 167–169, *169*, **196**, 198, 200–204, 206, 217, 343, 345, 346, 354, 355
  sampling method 155
Frobenius matrix norm 300, 442

g

Gabor
  basis images 288, *289*
  filter 283, 287–290, 332, 390
  transform 108, 287
  wavelets 288, 289
Gamma function 283
Gated Recurrent Unit (GRU). *see also* recurrent neural networks (RNN)
  input-to-state stability (ISS) 457, 458, 460
  reset gate 457, *458*
  update gate 457, *458*
Gaussian distribution 262, 364, 365, 373
Gaussian kernel 435
generalized eigenvalue problem 435
generalized homomorphic filter 161–162, *162*. *see also* homomorphic filtering
generalized Hough transform (GHT) 298
generalized inverse 161, *162*
generalized linear filtering 389
generalized linear models (GLM)
  exponential family of distributions 389
  sufficient statistics 389
geodesic 326–328, *327*
geometric correction 1
geometric mean filter 203
geometry features 276, 277
Geostationary Satellite (GOES) 1, 3, *3*, 6, 125, *126*, 132, *133*, 134, *135*, 223, 280, 281, *281*. *see also* multispectral imaging
gradient operators 146, *147*, 151, 265, 329
gradient/steepest descent algorithm 351, 383–384. *see also* performance learning
Gradient vector flow (GVF) snake
  capture range 266, 267
  discrete implementation 267
  Euler-Lagrange equation 266
  regularization parameter 267
granular 195
graphical processing unit (GPU) 376
graphs
  directed acyclic graph (DAG) 455, *456*
  vertices 260
gray level co-occurrence matrix (GLCM)
  angular second moment 285
  contrast feature 284
  co-occurrence matrix 284
  correlation- feature 284–285
  dissimilarity feature 285
  entropy feature 285
  homogeneity feature 285

h

Haar transform 118–119
Hadamard transform 90–92
hexagonal sampling 350
high-pass filter 56, 115, 116, 120, 123, 146, 148, 164, 296
Hilbert transform 343, 344
histogram thresholding
  maximum entropy 253–255
  Otsu's method 251–253
homomorphic filtering 154, 161–162, *162*, 164
Hough transform 291, 297–298
human eye
  aberrations 53, 54
  accommodation 53
  aqueous humors 53
  complex cell 58
  cones 54
  cornea 53
  fovea 54
  ganglion cells 54
  Hubel-Weisel model 54
  iris 53
  mesoptic 54, *57*
  optical nerve 53
  photoptic 54
  photoreceptors 54
  pupil 53, 54, 56
  receptive field 54, *55*, 57
  retina 53–55
  rods 54
  scotopic 54
  simple cell 54
  visual cortex 53, 58
  vitreous humors 53
hybrid
  associative memory. *see* associative memory
  compression methods 228, 230, 233
  operations 153–154
hyperspectral 1, 259, 276, 280

i

ideal filter 69, 90, 155, *155*, 157
idempotent 320, 331
image acquisition 131, 347
image classification. *see* classification/classifier
image compression and encoding. *see* compression
image enhancement
  contrast 6, 131–133, *132*, *133*
  edge extraction. *see* edge detection
  histogram modeling, equalization, specification 135–139
  spatial filtering 152, 153, 408
  transformed-domain filtering 124, 154
image formation 38, 53–64, 195, *195*, 197, 275
image modeling
  causal 171–176, 179–182
  full-plane 37
  noncausal 176–178, 185–187
  nonsymmetric half plane (NSHP) 36, 37, 178, *179*, 189
  quarter plane (QP) 36
  semicausal 182–185
image perception 54, 223
image quantization 72–81

image reconstruction. *see also* analytical image reconstruction
- adjoint operator 343
- back-projection operator 6, 343–347
- convolution back-projection 344
- inverse Radon transform 341–343
- projection-slice theorem 341–344
- Radon transform 340–345, *347*, 354
- sinogram 339, 341, 342, 344, 346, *347*, 348, 349, 352, 354

image restoration. *see* Kalman filter; Wiener filter
image sampling 67–84
image segmentation. *see* segmentation
image transforms 85–104
image transmission 1, 3, 223
image understanding 375
impulse response 11, 12, 14, 15, 20, 25, 26, 33, 36, 56, 108, 115–121, 126, 155, 172, 173, 199–201, 295, 296, 341, 345, 467
independent identically distributed (IID) 49, 383
infinite impulse response (IIR) 15, 28, 36, 37
information measure 224
Infrared (IR) 1, *3*, 125–126, 280–281, *281*
inhibitory 54, 376, 418, 439
inner product 16, 39, 86–88, 116, 279, 344, 468
innovations process 208, 233
intensity level slicing 238
interpolation filter 69, 82, 344
inverse contrast ratio 216, 217
inverse filter
- direct 197–199, 202, 203, 216
- regularized 198–199, 203, 216. *see also* pseudo-inverse

iterative reconstruction methods. *see also* image reconstruction
- algebraic reconstruction technique (ART) 339, 347–351
- statistical reconstruction 351–354

j

Jacobian matrix 342, 344, 442
JPEG
- 2000 240
- XS 7, 232, 241

just noticeably different (JND) 55

k

Kalman filter. *see also* image restoration
- A posteriori error covariance matrix 208, 212, 215
- A priori error covariance matrix 208, 212, 215
- block 214
- gain 208, 212, 213
- innovation 208
- Reduced Update (RUKF) 209–217, *216*
- steady-state 208, 216
- strip 206–208, 212, 217
- updating 208, 215

Karhunen-Loeve (KL) transform. *see also* principal component analysis (PCA)
- 1-D 96–98
- 2-D 85, 98
- basis functions 96
- basis images 98
- energy packing 236
- inverse 96, 97, 242
- matrix representation 98

Kernel machines
- support vector machine (SVM) 379
- support vector regression (SVR) 379

Kohonen neural networks 393, *393*
Kronecker
- delta 10, 19, 32, 33, 47, 172
- products 93

Kullback-Leibler divergence (KLD) 440

l

Lagrangian multiplier 218
Laplace transform 465, 468
lateral inhibition 54, 56, 57
layered machines 376, 401–428. *see also* convolutional neural networks (CNN)
learning
- Amari's generalized rule 381, 384, 388, 391
- catastrophic forgetting 380
- continual 435, 444
- definition (Mitchell) 379
- discriminative 387
- epoch 386, 391, 393
- error back-propagation 403–406, 410–413
- generative 371
- incremental 380, 437
- learning rate or step size 381, 383, 392, 393, 405, 407, 415, 416, 419–422, 433, 442, 458
- local minima 259, 406, 407, 454
- momentum 407, 415, 416, 425
- overfitting 380, 407, 408
- overtraining 380, 407
- rule 380–381, 384–393, 395, 401, 403, 405, 410–413, 418, 420, 421, 431–434, 444
- signal 380, 381, 384, 386, 388, 391, 404, 405, 407, 414, 415, 439, 440, 456

learning vector quantization (LVQ). *see also* competitive learning
- LVQ1 422–424, *424*
- LVQ2 424, *424*, 425
- Reward & Punishment 422

Least Mean Squared (LMS) 382, 384–389
least square filters, constrained 300
Legendre Moments 302
light
- brightness 54–56
- dark 55, *56*
- luminance 55, 72
- pure white 60
- radiance 53
- reflection 55, 62
- wavelength 53, 58
- white 22, 25, 47, 58, 60, 63, 459–461

linearly separable 361, 363, 365, 387, 401
linear machine 386–389
linear prediction 5, 173, 188, 190, 384
linear systems 10, 11, 51, 171, 184, 200, 206
line detection 291, 297
Lloyd-Max 75–78, *76*, 80, *80*, 82, 228, 236, 237
loss functions. *see also* learning
  categorical cross-entropy 412, 413, 456
    see also convolutional neural networks (CNN)
  focal 412
  hinge 413. *see also* Kernel machines
  sum squared error (SSE) 413. *see also* back-propagation neural networks
lossless compression 230. *see also* coding/encoding; Pulse-Code Modulation (PCM)
lossy compression
  DPCM 234–235
  hybrid methods 242
  predictive encoding 233–235
  transform domain. *see* Discrete Cosine transform (DCT); Discrete Wavelet transform (DWT)

m

MacAdam ellipses *60*, 61
Mach bands 56, *57*, 58
magnetic resonance imaging (MRI) 1, 2, 6, 223, 249, 255, *255*, 257, *257*, 259, *259*, 260, 263, *264*
manifold learning
  ambient space 434–436, *435*
  edge weight 435
  graph Laplacian 435, 436
  helix 434, *434*
  Isometric Feature Mapping (IsoMap) 435
  k-nearest neighbor (k-NN) 435
  Laplacian eigenmaps 435–438
  latent feature space 434
  Locally Linear Embedding (LLE) 435
  local similarity 435
  manifold space 434, 436
  Maximum Variance Unfolding (MVU) 435
Markov random field (MRF)
  clique potential 262
  cliques 260, *261*, 262, 263
  Expectation Maximization (EM) 261, 262, 264
  Gaussian mixture models 261, 263
  Gibbs distribution 261
  Gibbs energy function 261, 270
  Iterated Conditional Modes (ICM) 261–264, *264*
  neighborhood system 260, *261*, 263
  partition function 261, 263
  Potts model 262
  Simulated Annealing (SA) 262
  sites 260–263
masking 146, 459–461, *461*
matched filtering 152–154, *152*, 159, 164, 298
matrix theory results 97, 365
maximum a posteriori (MAP) 261, 262, 364, 378
maximum entropy *251*, 253–255, *255*, 269
maximum likelihood (ML) 174, 219, 261, 283, 339, 364, 389
mean square error, estimate, extrapolation 74, 83, 175, 199, 381, 451
medial axis 320, 333
median 144, 267, 323
median filtering 144–146, *145*, *146*, 163, 164, 323, 445
medical imaging 1, 7, 8, 260, 262, 333, 345
memoryless 131, 224, 225, 241
Mexican hat *55*, 56, 149, 418, 419
minimum mean squared error (MMSE) estimator 199
minimum-phase 27
minimum variance representation (MVR) 177, 178, 180, 190, 191
modes 132, 134, 240, 250, 251, 253
modified Bessel function of first kind 350
Modified National Institute of Standards and Technology (MNIST) 371, 396, 410, 416, *417*, 439, *440*, 443, *443*, 444
modulation transfer function (MTF) 57
Moire effect 81, 347, 349
Moments
  central 278, 301–302
  features 301–305, *304*
  invariant 301–302
  matching 303
  Zernike. *see* Zernike moments
monochrome vision model 58
Moore-Penrose inverse, 348. *see also* pseudo-inverse, of matrix
morphological image processing-binary
  background pixels 316, 317
  closing 314–315, **316**
  convex deficiency 318
  convex hull 318–319, *319*
  dilation 312–314, *312*, **314**, 321, *322*
  distance transform 320, 321
  erosion 312–314, *312*, **314**, 316, 319, *320*, 321, *322*
  foreground pixels 317–319, 322
  geodesic dilation 326–327
  geodesic distance 326, 327
  geodesic erosion 326–327
  hit or miss 316
  idempotence 327
  maximal disk 319–321
  medial axis transform (MAT) 320–321
  minimal convex set 318, 319
  Minkowski addition 312, 313
  Minkowski subtraction 312, 313
  opening 314–315, **316**
  pruning 321–322, *321*
  reconstruction filters 319–321
  S-close 315
  skeleton 319–321, *320*
  S-open 315
  structuring element 312–322, *312*, *313*, *315*, *317–319*
  thickening 317–318
  thinning 316–317, *317*

Morphological image processing-grayscale
above K 328
background pixels 324
bottom-hat transform 324, 329, 333
closing 324, *324*
dilation 323, *323*
edge detection-exterior 329, *329*
edge detection-interior 329, *329*
erosion 323, *323*
flat structuring element 323, 328, 330, 331
foreground pixels 324, 325
geodesic reconstruction 327–328
geodesic transform 326–328
Hierarchical texture Decomposition (HTD) for image segmentation 330–332, **332**
idempotence 327, 328, 331
under K 328
marker 326–328, *327*
mask 323, 326–328, *327*
multi-scale decomposition 330, 331
non-flat structuring element 330
opening 324, *324*
primitives 330–331
skeleton & shape 325
structuring element 322–328, *323*
texture primitives 330, 331
threshold set 322
top-hat transform 324
top surface 325, *326*
umbra 325, *326*
moving average (MA) representations 171–174, 177, 179, 182, 184, 185, 207, 415, 416
multiresolution 107, 109, 110, 441
multispectral imaging 258
multivariate models 188–191. *see also* vector models
mutual information 227, 380, 441

n

Naive Bayes classification
Bernoulli features 368–369
Laplace smoothing 368–370
multi-class features, categorical 369–370
multinomial features 369
Natural Language Processing (NLP) 369
neural networks
autoencoders. *see* autoencoders
convolutional. *see* convolutional neural networks (CNN)
deep convolutional. *see* convolutional neural networks (CNN)
feedback 418
feedforward 401, 438, 455
multilayer Perceptron (MLP). *see* back-propagation neural networks
recurrent. *see* recurrent neural networks (RNN)
self-organized feature maps (SOFM). *see* Kohonen neural networks
neuron models
activation. *see* activation functions
bias 402–404, 411
McCulloch-Pitts 376, 454
Perceptron 376, 377, *377*
radial basis functions (RBF) 379
weights 376, 377
Newton–Raphson method 283
noiseless image 143
noise models 195, 201, 204–206
noncausal models
1-D 171, 176, 185
2-D 185, 186–187
parameter estimation 176–177, 185–187
nondestructive testing 339
nonlinear filters 144, 315
nonparametric techniques 359
non-recursive. *see* finite impulse response (FIR)
normal distribution 32, 256, 383, 389
normal equations 175, 176, 180, 183, 185, 190, 200
Nyquist
criterion 115
frequency 68–70
interval 68

o

objective function. *see* loss functions
OpenStereo platform. *see also* disparity estimation algorithms
Adaptive Aggregation Network (AANET) 461
Group-Wide Correlation Stereo Network (GwCSNET) 461
Pyramid Stereo Matching Network (PSMNET). *see* PSMNET
PyTorch module 450, 461
optical illusion 56
optical transfer function (OTF) 57
optimum mean square quantizer. *see* Lloyd-Max
orthogonal
complement 118, 382
decomposition 241
gradient 353
matrix 88, 118
moments 302
projection matrix 382
orthonormal 16, 18, 86, 88, 90, 92, 96, 99, 103, 114, 115, 118–124, 280, 391, 392, 432
Oxford-IIIT Pet Database 442

p

Parseval's theorem 39, 86, 112, 113, 199, 468
passband 69, 70, 155–157, 159, 202
pattern classification 107, 249, 276, 286, 305, 359–367, 375, 392, 401–403, 407–409, 431, 449. *see also* classification/classifier
peak signal-to-noise ratio (PSNR) 81, 154, 164, 293, 303

perfect reconstruction 97, 116
performance learning
  ADALINE 381–386
  delta rule 388
  Least Mean Squared (LMS) 382, 384–389. *see also* stochastic gradient descent
  logistic regression, binary 388–389
  logistic regression, multiclass 389–390
  misadjustment 385
  Perceptron rule 386, *387*
  performance measure. *see* loss functions
  regularized logistic regression 390
  sparse logistic regression 390
  stochastic gradient descent (SGD) 384, 387, 389, 390
periodic
  1-D sequence 10
  2-D doubly 32, *33*, 42
  2-D general 32–33, *33*
  replication 68, *68*, 69
point operations
  cumulative distribution function (CDF) 136–138
  digital negative 134, *135*
  grayscale mapping, linear and nonlinear 131–135, *132–135*
  histogram equalization 136, *137*
  histogram modeling 135–139
  histogram specification 137–139, *138*
point spread function (PSF) 33–41, 48, 49, 53, 54, 69, 149, 155, *155*, 157, *157*, 158, 168, 196–198, **196**, *197*, 203, 207, 212, 323, 343, 351, 409
Poisson distribution 195, 373
positive definite (PD) matrix 174, 176, 186, 218, 353, 354
power spectral density 27. *see also* spectral density function (SDF)
prediction, predictor 5, 171, 175–180, 188, 190, 208, 212, 214, 233–235, 245, 384, 385, 388, 395, 396, 404, 455
primary sources 65
primitives 291, 311, 330, 331, 401, 409
principal component analysis (PCA). *see also* autoencoders
  anti-Hebbian 432
  decorrelation 96, 276
  deflation 444
  generalized Hebbian learning 432, 444. *see also* coincidence learning
  invariance property 307
principal components 5, 96, 279, 431, 432, 444
principle of optimality 97
projection
  geometry 339, *340*
  operators 343–344
  theorems 344
pseudo-color 6, 131
pseudo-inverse
  filters 6, 197, 199, 203
  of matrix 300. *see also* Moore-Penrose inverse
PSMNET
  cost volume 451
  hourglass network 451
  KITTI 2012 dataset 451
  Scenes 2005 Dataset 451
  Spatial Pyramid Pooling (SPP) module 451
Pulse-Code Modulation (PCM) 7, 228, 230, 231, 233–235, 238

q

Quad-trees 375, 412
quantization/quantizer
  decision level 72, 74, 244
  distortion function 77, 236, 237, 246
  error 74, 76, 77, 83, 234
  high rate 74–77
  Lloyd-Max 75–78, *76*, 80, *80*, 82, 228, 236, 237
  optimal 236, 237
  reconstruction level 72–75, 77–79, 82, 238
  rule 72–73
  transition level 74, 75, 79, 82
  uniform 73–75, *76*, 77–80, *78*, *80*, 82, 228, 235, 238, 240, 241, 243

r

radar cross-section 5
radii 305, 334, *466*
random
  image field 45, 178, 193
  process 20–23, 26, 45–47, **46**, 96, 171, 181, 206, 383
  signal 20, 45, 96–98
rate-distortion function
  IID Gaussian sources 226–227, *227*
  multivariate Gaussian sources 227
rational SDF 171, 178, 185
readout noise 195
receptive field 54, *55*, 57, 288, 379, *408*, 409, 410, 450
reconstruction. *see* image reconstruction
reconstruction filter 70, 81, 116, 117, 326–328, 334. *see also* interpolation filter
recurrent neural networks (RNN). *see also* neural networks
  back-propagation through time (BPTT) 456, 457
  deep RNN 456
  directed acyclic graph (DAG) 455, *456*
  hidden state 455, 457
  many-to-many RNN 455
  many-to-one RNN 455
  nonlinear state equation 455
  one-to-one RNN 455, *455*
  parameter sharing 455
  sequence classification 455, 456
  stacked RNN 456, *456*, 457
  unfolded/unrolled RNN 455
  vanishing gradient 456
recursive filter. *see* infinite impulse response (IIR)
recursive least squares (RLS) algorithm 433–434

redundancy
  coding 223
  inter-pixel 223, 234
  psycho-visual 223, 233
  spatial Inter-frame 223
  temporal inter-frame 223
reflection coefficients 164
region-based approaches 249, 256–258, 260, 269, 451
region growing 256, 258, 269
region of support (ROS) 36, 37, 44, 45, 87, 152, 171, 178, *179*, 182, 185, 187–189, 190, 191, 206, 207, 209, 210, 212–214, 216, 233, 234, 235, 288, 322, 323, 343, 395, 396, 471
region splitting & merging 256–257, *257*, *258*
relaxation parameter 348, 349
resolution
  color 72
  grayscale 72
  spatial 5, 67, 223, 409, 422, 451
ripple 283, 329
risk level 256
root filter 160–162, *160*, *161*, 164
root mean squared (RMS) 347, 349, 415
rotation matrix 279, 280
row and column indexing 31

s

salt & pepper noise 144, 145, *145*, 163
sample
  auto-covariance 21
  covariance 98, 100, 191, 204, 279, 281, 407
  mean 98, 256, 262, 279, 281, 407
sampling/sampler 4, 9, *9*, 10, 15–18, 31, 40–43, 67–84, 114, 115, 150, 155, 262, 267, 291, 292, 307, 345–347, 350
saturation 55, *56*, 58, 60, 62, 63, 72, 132, 208, 391, 406
scaling 86, 110, 112, 119, 121–123, *122*, 126, 152, 164, 167, 276, 277, 292, 293, 297, 301–306, 331, 353, 373, 401, 408
segmentation. *see also* image segmentation
  completeness 256
  connectivity 256
  contour-based 264–269
  edge-based 249, 264–269
  feature-based 249, 258
  maximal homogeneity 256
  region-based 256–258, *257–258*
  similarity-based 250–256
self-organized feature maps (SOFM)
  boundary neurons 421, 422
  competition 418, 419
  cooperation 418, 419
  growing SOFM 421–422
  growth phase 421–422
  growth threshold 421
  Kohonen map 417
  smoothing phase 422
  spread factor 421
  topological ordering 417–420, 422
semantic 7, 249, 276, 375, 450, 451, 458, 462
semi-causal models
  1-D *36*
  2-D 36, 395
  parameter estimation 395
semi-minimum phase 184
separability 38, 40, 41, 85, 87, 89, 94, 107, 123, 251, 292, 304–307, 363
sequency 92, 93, 95, 101
Shannon's first theorem of source coding 225–226
shape features. *see also* Zernike moments
  moments invariant 301–302
  slenderness 302
  spread 302
shift invariant 11, 12, 299
short time Fourier transform (STFT). *see also* wavelet transform
  frequency localization 107
  frequency resolution 108–109
  Heisenberg inequality 108
  spectrogram *109*, *110*
  time-bandwidth product 108, 110
  time localization 107
  time resolution 108–109
similarity measure 285
singular value decomposition (SVD), thin SVD 100–101
sinusoidal grating 57, *57*
skewness 231, 278, 279, 305, 404
space invariant 33, 34, 37, 45
spatial
  averaging 50, 81, *140*, 141, 143, *143*, 144, 153, 164, 203, 216, 217, 409, 445
  correlation 223
  filtering 56, 144, 152, 153, 408
  frequency 5, 38, 57, 65, 81, 89, 161, 288, *289*
  high-pass filtering 56, 146
  invariance 276
  low-pass filtering 139–144
  varying 144
  vision 56–58
speckle noise 139, 164, 195, 196, 214–217, 322, 332, 333
spectral density function (SDF) 24, 25, 28, 46, 171, 172, 175–179, 181–187, 191
spectral estimation 171, 178
spectral factorization 27–28, 47, 49, 171, 172, 175–177, 181–182, 184, 185, 187
spectral features
  generalized Gaussian density (GGD) 282
  multi-channel 280–282
  single channel 282–283
spectral representation 107
spectral responses 58, 61
spectral sensitivity function (SSF) 58, 59
stability 14, 15, 31, 35–37, 171, 176, 321, 344, 381, 397, 415, 416, 457, 458, 460
state space models 454

stationary processes 192
stereoscopic imaging. *see also* disparity estimation algorithms
  3-D TV 449
  depth perception 449–451
Stevens' law 56
stochastic gradient descent (SGD) 384, 387, 389, 390. *see also* Least Mean Squared (LMS)
stochastic process. *see* random process
stopband 157
Structural Similarity Index Measure (SSIM) 460
Student's-*t* test 256
supervised learning. *see* performance learning
synthetic aperture radar (SAR) 6, 139, 164, 195, *216*, 217, 332, 333
synthetic aperture sonar (SAS) *3*, 139, 195

t
template matching 39, 139, 152, 153, 316
temporal contrast sensitivity function (TCSF) 58
temporal vision 56–58
test statistics 256
texture/textural features. *see* Gabor filter; gray level co-occurrence matrix (GLCM)
thresholding 124, *125*, 126, 127, 134, *134*, 151, 152, 159, 232, 238, 250–251, 253, 256–259, 269, 270
threshold intensity (TVI) curve 55, *56*
time average. *see* ergodicity
time-frequency 107–109, 113, 287
Toeplitz 96, 173, 174, 181, 183, 186, 200, 207, 266
training. *see* neural networks
transfer function 15, 16, 25–28, 37, 48, 57, 116, 119, 120, 171, 172, 175, 177, 181, 184, 186, 192, 193, 198, 199, 201, 203, 204, 216, 217, 296, 469
transform coding 245
transform coefficients, features 43, 85
transition band 56
translation invariant 301, 303, 313, 320
trapezoidal filter 157
two-dimensional systems 31, *33*, 33–37, 39, 41, 48, 152, 186, 312

u
unbiased 75, 425
uncorrelated 96, 102, 190, 197, 201, 205, 214, 217
undersampling. *see* aliasing
U-net 441, 442, *442*
uniqueness 38
unitary
  matrix 96, 97
  transforms 6, 85, 97, 98, 154, 155, 159, 161, 164, 235, 236, 241, 242
unit pulse. *see* impulse response
unit step 10, 42, 136, 414, 465, 469
unsharp masking 146
unsupervised learning
  coincidence. *see* coincidence learning
  competitive. *see* competitive learning

v
vector models
  AR 189–191, *189*
  MA 193
vector quantization (VQ) 80, 233, 241–242, 258, 259, 278, 371, 380, 392–394, 418, 422
  competitive learning. *see* Kohonen neural networks
  K-means 259
visibility function 333, 339
visible light. *see* light
visual perception 54–56, 249, 375, 408, 450

w
Walsh-Hadamard transform (WHT)
  1-D 92–94
  2-D 90–95, *95*
  basis functions 90, *92*, 92–94
  basis images 94, 95, *95*
  elementary matrix 93
  energy packing 94
  inverse 93–95
  matrix representation 94–95
  ordered Hadamard matrix 94
  sequency 93
Walsh transform
  basis functions 90, 92
  sequency 92
wavelet descriptor (WD) 291, 294–297. *see also* contour
wavelet packets 122. *see also* filter banks (FB)
wavelet series
  critical sampling 114, 115
  dyadic sampling 114
  frame bounds 114–115
  frames 114
  tight frames 115
wavelet transform. *see also* short time Fourier transform (STFT)
  admissibility 111, 113
  analysis equation 110, 114
  continuous 109–114
  Daubechies 119
  daughter wavelet 110, 112, 114
  dilation parameter 122
  father wavelet 121
  Haar 111, 118–119
  inverse 111, 112, 114
  Meyer 111
  Morlet 111
  mother wavelet 110–112, *111*, 121
  multiresolution 110
  scaling function 110, 112, 119, 121, 122, *122*
  Scalogram 112
  synthesis equation 111, 114, 115
  wavelet function 111, 112, 114, 121, 122, *122*
Weber-Fechner law 56
Weber's law 55, *56*

white light 22, 25, 47, 58, 60, 63
whitening 407
white noise 49, 127, 140, 141, 159, 163, 179, 182, 197, 199, 201–203, 208, 212, 214
white noise driven 182
wideband noise 71
wide-sense stationary (WSS) 45, 96, 172, 199, 283
Wiener–Doob method 172–173
Wiener filter
  1-D 199–203
  2-D 203–206
  general transfer function 199, 201, 204
  geometric mean 203
  global 206
  local 217
  power spectrum filter 201, 202, *202*, 203, 206
Wigner-Ville distribution (WVD). *see also* wavelet transform
  Moyal formula 113
windowing
  Hamming 81, 157, 345, 346
  Hanning 81, 345
  Kaiser-Bessel 349
World Model 4
wraparound error. *see* aliasing error

x

X-ray imaging 4, 5, 339

y

Yule-Walker method
  1-D models 173, 178
  2-D models 180, 233, 235, 395
  vector models 208

z

Zernike
  moments 301–305, *306*, 307
  polynomials 302, 303
zero crossing 92, 93, 147–150
zero-memory 73
zero-order hold (ZOH) 69, 71, 81, 82, 346
zonal filter 154, 155
zone plate 81, *81*
zooming 30
z-transform 28, 115, 120, 296
  1-D 14, 15, 172, 177, 465–470
  2-D 181, 182, 184, 186, 471–472
  one-sided 466, *466*
  Region of Convergence (RoC) 465–468, *466*
  two-sided 465, 466, *466*

Wiley The manufacturer's authorized representative according to the EU General Product Safety Regulation is Wiley-VCH GmbH, Boschstr. 12, 69469 Weinheim, Germany, e-mail: Product_Safety@wiley.com.

Printed and bound by CPI Group (UK) Ltd, Croydon, CR0 4YY
10/09/2026
02215036-0001